D1810922

7

Advances in
VIRUS RESEARCH

VOLUME 56

ADVISORY BOARD

Advances in
VIRUS RESEARCH

Edited by

KARL MARAMOROSCH
Department of Entomology
Rutgers University
New Brunswick, New Jersey

FREDERICK A. MURPHY
School of Veterinary Medicine
University of California, Davis
Davis, California

AARON J. SHATKIN
Center for Advanced Biotechnology and Medicine
Piscataway, New Jersey

VOLUME 56

NEUROVIROLOGY
Viruses and the Brain

Edited by
Michael J. Buchmeier and Iain L. Campbell
Department of Neuropharmacology
The Scripps Research Institute
La Jolla, California

ACADEMIC PRESS
San Diego   New York   Boston
London   Sydney   Tokyo   Toronto

Academic Press
A Harcourt Science and Technology Company
525 B Street, Suite 1900, San Diego, California 92101-4495, USA
http://www.academicpress.com

Academic Press
Harcourt Place, 32 Jamestown Road, London NW1 7BY, UK
http://www.academicpress.com

International Standard Serial Number: 0065-3527

International Standard Book Number: 0-12-039856-7

PRINTED IN THE UNITED STATES OF AMERICA
00 01 02 03 04 05 SB 9 8 7 6 5 4 3 2 1

CONTENTS

BASIC MODELS

Genetic Determinants of Neurovirulence of Murine Oncornaviruses

JOHN L. PORTIS

Pseudorabies Virus Neuroinvasiveness: A Window into the Functional Organization of the Brain

J. PATRICK CARD

Neurovirology and Developmental Neurobiology

JOHN K. FAZAKERLEY

VIRAL IMMUNE RESPONSES IN THE CENTRAL NERVOUS SYSTEM

Chemokines and Viral Diseases of the Central Nervous System

VALERIE C. ASENSIO AND IAIN L. CAMPBELL

Regulation of T Cell Responses During Central Nervous System Viral Infection

DAVID N. IRANI AND DIANE E. GRIFFIN

Virus-Induced Autoimmunity: Epitope Spreading to Myelin Autoepitopes in Theiler's Virus Infection of the Central Nervous System

STEPHEN D. MILLER, YAEL KATZ-LEVY, KATHERINE L.NEVILLE, AND CAROL L. VANDERLUGT

HIV: HUMAN IMMUNODEFICIENCY VIRUS

The Blood–Brain Barrier and AIDS

LISA I. STRELOW, DAMIR JANIGRO, AND JAY A. NELSON

Neuroimmune and Neurovirological Aspects of Human Immunodeficiency Virus Infection

CHRISTOPHER POWER AND RICHARD T. JOHNSON

Simian Immunodeficiency Virus Model of HIV-Induced Central Nervous System Dysfunction

E.M.E. BURUDI AND HOWARD S. FOX

Neuroendocrine-Immune Interactions during Viral Infections

BRAD D. PEARCE, CHRISTINE A. BIRON, AND ANDREW H. MILLER

PRECLINICAL AND CLINCAL MODELS

Role of Viruses in Etiology and Pathogenesis of Multiple Sclerosis

SAMANTHA S. SOLDAN AND STEVEN JACOBSON

Bornavirus Tropism and Targeted Pathogenesis: Virus–Host Interactions in a Neurodevelopmental Model

MADY HORNIG, THOMAS BRIESE, AND W. IAN LIPKIN

Paradigms for Behavioral Assessment of Viral Pathogenesis

MICHAEL R. WEED AND LISA H. GOLD

PREFACE

The idea for this volume arose during discussions between the editors in the spring of 1999 and was fueled by discussions with our colleagues here at The Scripps Research Institute and elsewhere. It was clear when this project began that long held beliefs and assumptions about the central nervous system, its accessibility to immune surveillance, and its response to infection required reassessment in light of advances in neurobiology and neuroimmunology. One goal of this book is to highlight the contemporary thinking in these areas. As is the case with any such endeavor we run the risk of omitting important viewpoints, or interpretations, but we are confident that the contributing authors have taken great care to present well-reasoned and balanced arguments on some of the most important issues existing in neurovirology.

Preparing a book such as this requires considerable help. The editors would like to express our appreciation first to all of the contributing authors and to our colleague Dr. Floyd Bloom. We would also like to acknowledge the efforts of Aaron Johnson and Traci John of Academic Press. Finally we would like to thank our spouses, Julia and Nancy, who have shown patience with our temporary insanity as this volume has evolved.

Iain L. Campbell
Michael J. Buchmeier

INTRODUCTION

The National Institutes of Health designated the 1990s as the decade of the brain. The intent of this notable action was to stimulate research in the neurosciences, thus it came to be that we witnessed a quantum increase in our knowledge of how the brain and the central nervous system (CNS) respond to viral infection, how viral infections trigger and sustain frank and subtle neuropathologies, and how these interactions manifest as degenerative and cognitive disorders. Viruses are among the most common environmental influences we encounter in the normal human life span, and these insidious agents are well adapted to infect and persist within the CNS. As a result of eons of evolution with viruses, the mammalian brain has in turn evolved unique mechanisms of defense to combat and control infection. Principal amongst these strategies are: First, a blood-brain barrier in which the endothelial cells form tight junctions and which serves as an efficient barrier to infectious agents and to cellular and macromolecular components of the immune system. Second, a relative deficiency of proteins needed for the recognition of antigen such as major histocompatibility complex molecules and activation of immune cells. Third, the absence of "professional" antigen presenting cells such as dendritic cells, and finally, the lack of a lymphatic drainage. However, in spite of these properties, it can no longer be presumed that the brain is an immunologically "privileged" site. Under the appropriate circumstances activated immune cells can and do enter and traffic through the brain. These unusual properties of the brain have many implications for the host and for any virus that may eventually gain entry. Viruses such as rabies and poliovirus that enter the brain by axonal transport and are sequestered inside immunologically incompetent neuronal cells, free to replicate and spread in an unfettered manner, eventually lysing their host cell and causing severe morbidity and mortality. At the other extreme, the unleashed fury of a targeted antiviral immune response in the brain, as occurs in, for example, HSV encephalitis, may also bring about the swift demise of the host. Between these extremes are viruses such as human immunodeficiency virus-1 that can enter the brain and persist for life, causing limited damage and provoking only mild to moderate host responses. In all, the recognition and elaboration of these mechanisms of defense and their consequences forms the crux of much of contemporary viral pathogenesis, and is the foundation for this volume.

At the beginning of this new millennium we find that the spectacular advances of the previous decade in the fields of genetics, immunology, and molecular biology and their application to virology have catapulted our understanding of the etiology and the pathogenesis of many viral diseases of the CNS. This volume brings together a distinguished group of authors to explore the many facets of contemporary neurovirology. As they make clear in this book, viral–host interactions in the brain are complex and dynamic and often on shifting ground. The environment of the CNS may not only mould the nature and function of antiviral T cells and other leucocytes that enter the brain, but it may also apply selective pressure on the infectious agent to rapidly evolve to escape the unwanted attentions of the host immune surveillance. A number of authors focus on more well established models of neurological infections to highlight the diverse mechanisms and consequences of persistent virus infection of the brain. Others provide fascinating insights into the discovery of new viral links to established human psychiatric and demyelinating diseases. The enigmatic spongiform encephalopathies, scrapie and BSE, are now thought to be due not to a classical viral agent but rather to an aberrant or "rogue" protein termed a prion. As outlined in this volume, elegant studies both *in vitro* and *in vivo* are removing the veil of mystery that has shrouded our understanding of the pathogenesis of these neurological diseases. Finally, we are beginning to see some benefits flowing out of the many advances in molecular virology with two examples highlighted in this volume illustrating the power of this technology. The unique specificities of viruses are already being harnessed to map circuitry within the brain and to deliver genetic material to specific populations of cells within the brain. With the adaptation of modern genomic technologies the future is certain to hold both new problems and answers to old questions about the nature of the interaction between viruses and the brain.

Iain L. Campbell
Michael J. Buchmeier
La Jolla, California, 2001

BASIC MODELS

ADVANCES IN VIRUS RESEARCH, VOL 56

GENETIC DETERMINANTS OF NEUROVIRULENCE OF MURINE ONCORNAVIRUSES

John L. Portis

Laboratory of Persistent Viral Diseases
Rocky Mountain Laboratories, NIAID
Hamilton, Montana 59840

I. INTRODUCTION

Much of our knowledge of mammalian retroviruses had its origins in the derivation of inbred strains of mice (Slye, 1914). It was through the use of the C3H strain, inbred for its high incidence of mammary adeno-carcinoma (Strong, 1935), but its low incidence of spontaneous leukemia, that Gross identified the first murine leukemia virus (MuLV)(Gross, 1951). However, the induction of cancers by these viruses in highly inbred mice was considered to be somewhat of a labo-ratory artifact of inbreeding, and there was great interest in determin-ing whether these viruses might also be the cause of cancers in outbred animals and perhaps humans. In the early 1970s, Murray Gardner and associates began a search for leukemia viruses in outbred animals (Gardner, 1994); and over a 10-year period they collected, and observed in the laboratory, over 10,000 mice from 15 different areas of southern

California (Gardner *et al.*, 1976; Gardner, 1978). Although mice from the majority of the collection areas were virus negative, they discovered several colonies that had an unusually high incidence of nonthymic lymphoma (Bryant *et al.*, 1981) and expressed lifelong high levels of infectious MuLV. Some of these mice also developed a neurologic disease manifested by hind limb paralysis (Gardner *et al.*, 1973). The field isolates were found to contain two types of murine retroviruses belonging to the ecotropic and amphotropic host range groups (Hartley and Rowe, 1976; Rasheed *et al.*, 1976). Ecotropic viruses infected only rodent cells *in vitro*, whereas the amphotropic viruses infected cells of a variety of species, including mice. The host range of MuLV is determined by the sequence of the envelope gene (Battini *et al.*, 1992). Based on viral interference analyses, these two viruses were shown to use different receptors to enter cells (Hartley and Rowe, 1976; Rein, 1982), and these receptors have now been cloned and characterized (Albritton *et al.*, 1989; Miller *et al.*, 1994). The ecotropic viruses were found to induce both the nonthymic lymphomas and the paralytic syndrome (Rasheed *et al.*, 1976; Rasheed *et al.*, 1983). The prototypic virus in this group was isolated from the brain of a mouse captured at a squab farm near Lake Casitas in southern California and was named CasBrM (Hartley *et al.*, 1976), the "M" designating it as mouse-tropic. When the nomenclature of mouse-tropic viruses was changed to the ecotropic designation, the name was changed to CasBrE.

Although the leukemia viruses first identified by Gross were integrated in the germ line as proviruses, this is not the case for the wild mouse viruses (Barbacid *et al.*, 1979; O'Neill *et al.*, 1987). Gardner and his colleagues found that these viruses are transmitted primarily from mother to offspring in the milk (Gardner *et al.*, 1979), although evidence of horizontal transmission among adults in external secretions is also seen (Gardner *et al.*, 1979; Portis *et al.*, 1987). Neonates become persistently infected and remain viremic for life. These mice are immunologically unresponsive to the virus. Thus, neither antiviral antibodies nor cytotoxic T lymphocytes are detectable at any time after neonatal inoculation, though more sensitive techniques have revealed evidence of weak antigen-specific T_H2 responses (Sarzotti *et al.*, 1996). Only in mice inoculated neonatally is the neurologic disease observed and in the wild the incidence is low (<20%) and the incubation period long (several months to more than a year) (Gardner *et al.*, 1973). Neurologic disease is virtually never seen in mice inoculated after they are 6–8 days of age (Hoffman *et al.*, 1981; Officer *et al.*, 1973). Although maturation of the immune response appears to be one component of this age-dependence (Hoffman *et al.*, 1981), this resistance has also been observed in athymic nude mice (Czub *et al.,* 1991)—an issue which will be discussed further in Section V.

The occurrence of the ecotropic virus in only some of the colonies of wild mice appears to be the consequence of the segregation of a host resistance gene, FV-4 (Gardner *et al.,* 1980; Odaka *et al.,* 1981). Interestingly, FV-4 is a truncated replication-defective ecotropic provirus (Kozak *et al.,* 1984) that encodes an endogenous ecotropic envelope protein (Ikeda and Odaka, 1983; Limjoco *et al.,* 1993) and that restricts virus spread by blocking the ecotropic receptor (viral interference). This resistance gene is undoubtedly the remnant of a remote infection, the critical component of which was conserved in the mouse genome because it endowed the host with a selective advantage.

The vast majority of diseases caused by the oncornaviruses are proliferative in nature. Infected cells are induced to proliferate by a number of mechanisms. Friend virus (SFFV) encodes a truncated envelope protein that binds to the erythropoietin receptor (Li *et al.,* 1990), inducing polyclonal proliferation of erythroid progenitors. Envelope proteins of other MuLVs of the polytropic host range are thought to cause polyclonal T cell proliferation by binding to cell surface receptors linked to signal transduction pathways (Li and Baltimore, 1991). Diseases characterized by clonal proliferation, such as leukemias and lymphomas, are a consequence of the capacity of these viruses to integrate a DNA copy of their genome into the host cell genome, thereby activating or inactivating cellular genes involved in growth control (Rosenberg *et al.,* 1997). The lymphomas and leukemias caused by CasBrE are of this type (Bergeron *et al.,* 1991; Bergeron *et al.,* 1993). Although much remains to be learned of the molecular details, at least conceptually, the role of virus in these proliferative diseases is understood. Even at the conceptual level, however, little is yet known of the mechanisms responsible for the neurologic diseases induced by these viruses.

Before we discuss these neurologic diseases, we should define a few terms, originally formulated by R. T. Johnson (1998), which will be used repeatedly in this chapter. "Neurovirulence" simply refers to the capacity of a virus to cause neurologic disease and does not imply any specific mechanisms. "Neuroinvasiveness" defines a virus's capacity to invade the nervous system, but does not imply any specific cellular tropism. "Neurotropism" means that a virus actually infects neurons. Thus, neurovirulence implies that the virus is neuroinvasive but does not always indicate that the virus is neurotropic. As will be discussed later (Section III), neuroinvasiveness does not always lead to neurovirulence.

II. Clinicopathological Manifestations of Neurovirulent
Murine Oncornavirus Infection

A. Ecotropic Viruses

The initial clinical sign of the disease caused by CasBrE is an abnormal abduction reflex of the hind limbs that occurs when the mice are lifted by the tail; it is generally associated with an action tremor (Andrews and Gardner, 1974). As the disease progresses, the mice exhibit hind limb weakness and, finally, frank spastic paralysis that may also involve the forelimbs (Table I). At this stage there is also wasting associated with neurogenic atrophy of skeletal muscles (Gardner et al., 1973). The predominant neuropathologic features are spongiosis (vacuolation) and gliosis involving preferentially the caudal regions of the nervous system. Demyelination has been observed (Andrews and Gardner, 1974; Oldstone et al., 1977) but appears to be secondary to the damage to the neuronal perikarya. Whether neurons actually die and drop out in this disease has been debated. It is clear, however, from ultrastructural studies that neurons and glial cells undergo degenerative changes (Brooks et al., 1980; Swarz et al., 1981), and in mice with long-standing chronic disease, there is a depopulation of neurons, at least in the anterior horns of the spinal cord (Andrews and Gardner, 1974; Oldstone et al., 1980).

Ultrastructurally, the vacuolar changed induced by CasBrE (Andrews et al., 1974; Swarz et al., 1981) bears a striking resemblance to scrapie and the other TSEs (Hoffman et al., 1982; Lampert et al., 1972). Vacuoles appear to represent swollen postsynaptic terminals (dendrites) and are often filled with membranous material. Interestingly, it was observed early on that virus particles appeared not to be associated with these vacuoles (Andrews and Gardner, 1974; Swarz et al., 1981). Instead, in a number of ultrastructural studies, virus budding was observed primarily in the central nervous system (CNS) microvasculature (Hoffman et al., 1992; Oldstone et al., 1977; Pitts et al., 1987), in both endothelial cells and pericytes (probably perivascular macrophages). In addition, however, budding virus was also noted in occasional large motor neurons of the ventral horns of the spinal cord (Andrews and Gardner, 1974; Gardner et al., 1973; Oldstone et al., 1977; Oldstone et al., 1980; Oldstone et al., 1983). Virus in these cells appeared to be budding primarily into intracellular vesicles, possibly endoplasmic reticulum. It was noted however, in the initial studies on the pathogenesis of CasBrE that the neurons exhibiting intracellular virus budding appeared not to be undergoing degeneration, and indeed, the neurons which did exhibit degenerative changes did not harbor such virus particles (Andrews and Gardner, 1974). Furthermore, the intra-neuronal virions were observed only in mice with chronic forms of disease and were never seen in mice developing the disease after short incubation periods (Swarz et al., 1981). These virus particles are likely the result of the activation of endogenous retroviruses and thus probably represent an epiphenomenon (see section X). These discrepancies led Andrews and Gardner (1974), in their original ultrastructural study of this disease, to suggest that the neuropathogenesis of CasBrE was a consequence of indirect effects of virus infection, a prediction which now seems to be correct. The nature of these effects is still a matter of speculation, but the capacity to manipulate both the viral genome and the host has brought us closer to understanding, at the molecular level, the virus–host interactions that lead to neuropathology.

Since the discovery of the wild-mouse ecotropic viruses, a number of other neurovirulent ecotropic viruses have been identified, including Moloney MuLV ts-1 (Wong et al., 1983); Friend MuLV PVC211 (Kai and Furuta, 1984; Masuda et al., 1992); and NT40 (Czub et al., 1995). All cause spongiform encephalomyelopathies indistinguishable clinically and pathologically from the disease caused by CasBrE. While ts-1 and CasBrE are pathogenic only for mice, PVC211 and NT40 cause disease in rats, having been selected by serial passage in that species.

B. Polytropic Viruses

Polytropic viruses have a broader host range than the ecotropic viruses but are distinct from the wild-mouse amphotropic viruses, which also have a wide host range (Cloyd et al., 1985). The polytropic receptor has recently been cloned by three independent groups (Battini et al., 1999; Tailor et al., 1999; Yang et al., 1999) and is also structurally and functionally distinct from the ecotropic and amphotropic receptors. These viruses are recombinants between ecotropic viruses and endogenous retroviral sequences resident in the mouse genome (Stoye and Coffin, 1987). The neurovirulent polytropic virus FMCF98 causes a disease which is different both clinically and pathologically from that caused by the ecotropic viruses (Table I). Mice manifest early hyperexcitability that characterizes them as "popcorn" mice (Portis et al., 1995). The mice jump out of the cage at the slightest auditory stimuli, such as snapping of the finger. This early stage is followed, within days or weeks, by evidence of imbalance, ataxia, and, finally, immobility associated with generalized seizures. Pathologically, there is an intense gliosis involving both astrocytes and microglia, but spongiosis is a rare and inconsistent finding (Portis et al., 1995).

C. Other Neurovirulent Viruses

All of these diseases are manifested by clinical signs that are easily recognized (e.g., paralysis, tremor, seizures). There is another murine retroviral disease that is manifested primarily by cognitive deficits. The virus is LP-BM5, a virus complex composed of a replication-defective virus containing extensive deletions (Aziz *et al.*, 1989) and a replication-competent helper virus. Mice inoculated with this virus complex as adults develop a lymphoproliferative disease associated with profound immunosuppression (Mosier *et al.*, 1985), and with opportunistic infections (Gazzinelli *et al.*, 1992). These effects are due to the presence of the defective virus, not the helper (Aziz *et al.*, 1989). The cognitive deficit is manifested by abnormalities in spatial learning (Sei *et al.*, 1992) and is associated with evidence of microglial and astrocytic activation (Kustova *et al.*, 1996); abnormalities in neurotransmitter levels (Espey *et al.*, 1998; Kustova *et al.*, 1997); signs of neuronal degeneration; and synaptic remodeling (Kustova *et al.*, 1998) (Table I).

In addition to the neurodegenerative diseases, murine oncornaviruses also cause acute fatal cerebrovascular hemorrhage. Intracerebral hemorrhage was found to be the cause of death of some mice infected with CasBrE (Portis *et al.*, 1990), and more recently a mutant of Friend MuLV (TR1.3) was found to induce a rapid widely disseminated cerebrovascular hemorrhage (Park *et al.*, 1993) (Table I). The hemorrhages caused by TR1.3 are the consequence of the fusigenicity of the envelope protein for endothelial cells in the brain, resulting in a disruption of vascular integrity (Park *et al.*, 1994b, 1994a).

III. Mapping of Viral Determinants of Neurologic Disease

Murine oncornaviruses are simple retroviruses with three structural genes (*gag, pol,* and *env*) that are bounded by long terminal repeats (LTRs) at the ends of the genome that carry the transcriptional promoter and termination sequences (Miller, 1997) (Fig. 1). The *gag* gene encodes the core proteins of the virus. The *pol* gene encodes the viral protease, polymerase, and integrase involved in replication of the viral genome. The *Env* proteins comprise the surface of the virion and are involved in receptor binding and fusion, both functions of which are critical for virus entry into the host cell. Between the *gag* gene and the 5′ LTR is the "5′ leader" sequence that contains the primer binding site, splice donor, and packaging sequences of the viral RNA (Fig. 1).

Table II is a compilation of the results of genetic studies in which viral determinants of neurovirulence have been mapped. One common

TABLE I
MANIFESTATIONS OF NEUROVIRULENCE

Virus	Host range	Clinical	Pathology	Reference
CasBrE	Ecotropic	Paralysis Tremor	Spongiosis Gliosis	Andrews and Gardner (1974)
CasBrE	Ecotropic	Early death	Cerebral hemorrhage	Portis et al. (1990)
Friend MuLV; PVC211	Ecotropic	Paralysis, tremor	Spongiosis, Gliosis	Hoffman et al. (1992)
Moloney MuLV; ts-1	Ecotropic	Paralysis, tremor	Spongiosis, Gliosis	Zachary et al. (1986)
Friend MuLV; NT-40	Ecotropic	Paralysis, tremor	Spongiosis, Gliosis	Czub et al. (1995)
Friend MuLV; TR1.3	Ecotropic	Early death	Cerebral hemorrhage	Park et al. (1993)
LPBM5[a]	Ecotropic	Cognitive deficit	Gliosis	Kustova et al. (1996); Sei et al. (1992)
MoAmpho V[b]	Amphotropic	Paralysis, tremor	Spongiosis, gliosis	Munk et al. (1997)
FMCF98	Polytropic	Hyper-excitability, ataxia	Gliosis	Portis et al. (1995)

[a] Virus complex consisting of replication-competent ecotropic helper virus and replication-defective virus (Aziz et al., 1989).
[b] Chimeric virus consisting of Moloney MuLV with envelope sequences derived from amphotropic virus 4070A.

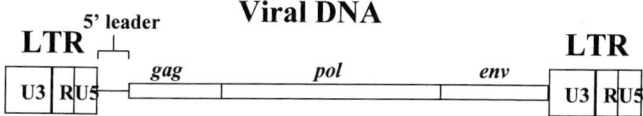

FIG 1. Schematic diagram of a typical murine oncornavirus genome. Represented is the organization of the viral DNA as it would exist as an integrated provirus in the genome of an infected cell. The long terminal repeat (LTR) at the 5′ end contains the transcriptional promoter (U3) and the LTR at the 3′ end contains the polyadenylation signal. The 5′ leader sequence contains the primer binding site for reverse transcription, and the packaging sequence required for proper incorporation of viral RNA into virus particles. The three structural genes are shown as blocks.

TABLE II
VIRAL DETERMINANTS OF NEUROVIRULENCE

Virus	Sequences	Reference
CasBrE	env	DesGroseillers et al. (1984); Paquette et al. (1989)
CasBrE	LTR	DesGroseillers et al. (1985); Portis et al. (1991)
CasBrE	5' leader	Portis et al. (1994)
MoMuLV ts-1	env	Szurek et al. (1988)
FMuLV PVC211	LTR, gag, env	Masuda et al. (1993)
FMuLV TR1.3	env	Park et al. (1994b)
MoAmphoV	env, non-env[a]	Munk et al. (1997)
FMCF98	env	Portis et al. (1995)

[a] The neuropathogenicity of MoAmpho V is a consequence of the combination of the envelope gene of amphotropic virus 4070A and the rest of the genome from Moloney MuLV. Neither of the parental viruses is neurovirulent (see the text).

feature for all of the viruses listed is the importance of the viral envelope gene in neurovirulence. It is also clear, however, that the envelope gene is not alone responsible. The sequence of the viral LTR, the 5'-leader sequence of the viral genome, as well as the *gag* gene, have also been shown to have an effect. This point is exemplified by the following observation: The virus MoAmpho V, listed in Tables I and II, is a chimeric virus consisting of the envelope gene from the wild-mouse amphotropic virus 4070A (Chattopadhyay *et al.*, 1981), and the rest of the genome from the Moloney MuLV clone MLV-K (Miller *et al.*, 1984). Neither of these parental viruses is neuropathogenic (DesGroseillers *et al.*, 1984; Munk *et al.*, 1997). Yet the chimeric virus caused a paralytic disease manifested by spongiosis and gliosis indistinguishable from that caused by CasBrE (Munk *et al.*, 1997; Munk *et al.*, 1998). This implies that the envelope gene of the parental amphotropic virus 4070A harbors sequences that can determine neurovirulence, given the appropriate context of other viral sequences—in this case donated by the Moloney MuLV parent.

In order to begin to simplify some of the complexities of the neurovirulent phenotype, we carried out a series of *in vivo* studies on CasBrE and derivatives thereof, using an inbred mouse strain (IRW) that had been shown to be highly susceptible to leukemogenesis induced by a variety of murine oncornaviruses. Like the disease seen in wild-mouse populations, our molecular clone of CasBrE induced neurologic disease in <10% of neonatally inoculated IRW mice, and in those mice that did develop the disease, incubation periods ranged from 3 to 6 months (Portis *et al.*,

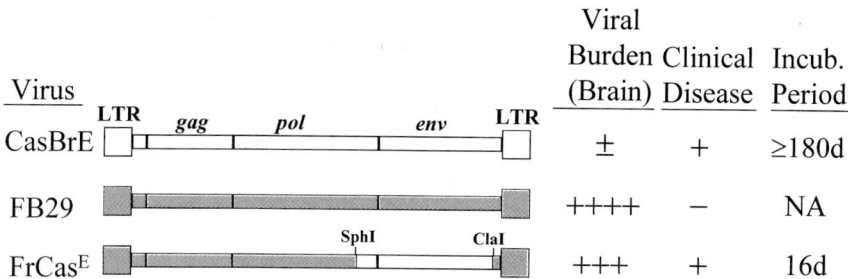

Virus	Viral Burden (Brain)	Clinical Disease	Incub. Period
CasBrE	±	+	≥180d
FB29	++++	−	NA
FrCas^E	+++	+	16d

FIG 2. Schematic diagram showing the parental viruses of the highly neuropathogenic chimeric virus FrCas^E. CasBrE is a molecular clone of the wild-mouse ecotropic virus isolated at Lake Casitas in California. FrCas^E consists of the genome of the non-neurovirulent, but highly neuroinvasive, virus FB29 containing an *Sph*I–*Cla*I fragment with the envelope gene and 3' *pol* sequences from CasBrE. The columns at the right represent the cumulative results of studies in which each virus was inoculated intraperitoneally into 24-to-48-hour-old neonates, and in which these mice were followed for signs of clinical disease (+ or −) or killed at fixed time points for measurement of brain viral burden. Viral burden was measured by a variety of techniques including Southern blot analysis for viral DNA, infectivity assays, and Western blot analysis for viral capsid and envelope proteins. Viral burden is graded on a relative scale. Incubation period (in days) is the time between virus inoculation and the first signs of clinical disease.

1990) The reason for the low incidence and long incubation period was evident when we found that viral burden achieved by CasBrE in the brain was relatively low (Portis *et al.*, 1990), indicating that the virus was not particularly neuroinvasive (Fig. 2).

We had been working with another murine oncornavirus, FB29 (Sitbon *et al.*, 1986), which is a strain of Friend MuLV and causes erythroleukemia 3–4 months after neonatal inoculation. Surprisingly, this virus was found to infect the brain at high levels, but did not induce a neurologic disease (Portis *et al.*, 1990). Since the neurovirulence of CasBrE had previously been mapped to the viral envelope gene (DesGroseillers *et al.*, 1984), we transferred the *env* gene of CasBrE to the genome of FB29 with the idea of perhaps generating a virus that might deliver the CasBrE *env* to the brain at high levels. This chimeric virus (FrCas^E), like FB29, infected the brain at high levels but, unlike FB29, all of the mice infected with FrCas^E succumbed to a severe tremulous paralysis associated with wasting, with an incubation period of only 14–16 days (Portis *et al.*, 1990) (Fig. 2). The neuropathology induced by FrCas^E consisted of widespread spongiosis involving not only the brainstem and subcortical gray matter, but also the deep layers of the cerebral cortex (Czub *et al.*, 1994) This result appeared to confirm the notion that the low incidence and long incubation period of CasBrE disease in the wild was due to its relatively low level of neuroinvasiveness. In addition, since FB29 was neuroinvasive but not neurovirulent, this implied that these two phenotypes were separable. It is, of course, possible that FB29 and FrCas^E infected different cells in the nervous system, but, as will be discussed (in Section VIII), this turned out not to be the case (Askovic et al., 2000).

IV. VIRAL SEQUENCES THAT DETERMINE NEUROINVASIVENESS

A. Envelope Gene

The influence of envelope sequences on neuroinvasiveness has been shown both for the ecotropic (Askovic et al., 2000; Masuda et al., 1996; Szurek et al., 1988) and the polytropic viruses (Hasenkrug et al., 1996; Poulsen et al., 1998). In the case of the ecotropic virus PVC211, mutations at two residues located in the receptor binding domain of the envelope protein ($Glu_{116} \rightarrow Gly$ and $Glu_{129} \rightarrow Lys$) have been shown to greatly increase the tropism for brain capillary endothelial cells (Masuda et al., 1996). These two residues are in close proximity to the residues that appear to form the contact points with the receptor (Davey et al., 1999; Zavorotinskaya et al., 1999) and thus it has been proposed that this might affect the avidity of binding to the receptor on brain endothelial cells.

In the comparisons shown in Fig. 2, the envelope gene of FB29 increases neuroinvasiveness (measured by viral burden in the brain) relative to FrCas^E. Here, there are demonstrable effects on replication in peripheral nonneuronal tissues as well. Peak viremia titers and splenic infectious centers in mice inoculated with FB29 are consistently higher by 0.5–1 log than peak titers seen in FrCas^E-inoculated mice (Czub et al., 1992). We suspect that the reason for this difference is the capacity of FB29 to induce hemolytic anemia (Sitbon et al., 1986), a phenotype which is not observed in FrCas^E-inoculated mice. The hemolytic anemia causes compensatory erythroproliferation (Sitbon et al., 1986), which in turn would increase the number of target cells available for infection. The nature of the effect of envelope sequences on neuroinvasiveness in the polytropic model has not been studied.

B. LTR and the 5' Leader

Nonenvelope viral sequences that control neuroinvasiveness were localized by constructing a series of chimeric viruses (Czub et al., 1992; Portis et al., 1991), summarized in Fig. 3. It is clear that sequences

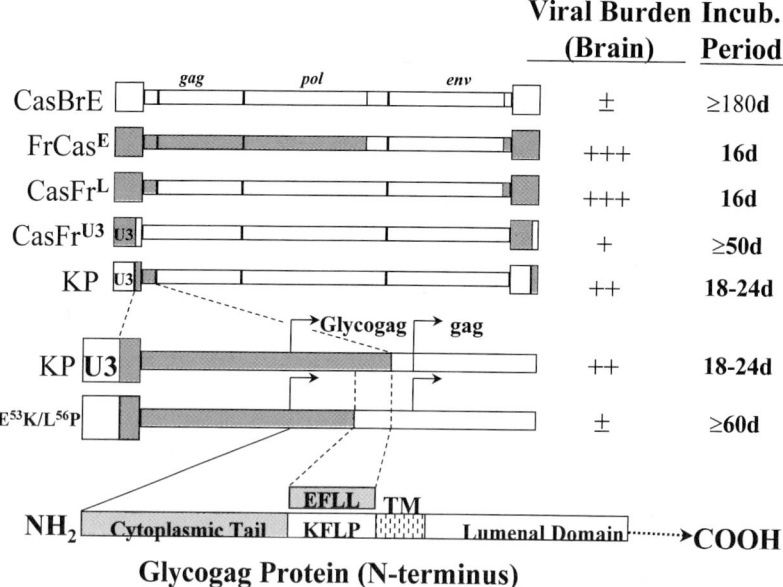

Glycogag Protein (N-terminus)

Fɪɢ 3. Mapping of the determinants of neuroinvasiveness. These schematic genomes all carry the envelope gene of CasBrE and thus are all potentially neurovirulent. Segments from FB29 are shown in gray. Viral burdens and incubation period are measured as described in the caption for Fig. 2. The high viral burden and short incubation period of FrCas^E are consequences of at least two sequences from FB29, one located in the U3 region of the LTR and the other in the 5′ leader sequence. The relevant sequence in the 5′ leader (shown in the two chimeric viruses at the bottom) encodes membrane-proximal amino acid residues EFLL in the cytoplasmic domain of Glycogag (shown schematically at the bottom). Mutation of this motif to KFLP (the sequence present in CasBrE) decreased viral loads in the brain and lengthened the incubation period. Note also that for viruses with the CasBrE envelope gene, higher viral burdens in the brain translate into shorter incubation periods.

from FB29 that so dramatically enhanced the neuroinvasiveness of FrCas^E relative to CasBrE (Fig. 2) are located within the U3 region of the viral LTR and within the 5′ leader sequence (see Fig. 1). Interestingly, the sequence of the 5′ leader had a stronger influence on neuroinvasiveness than did the U3 region (Portis et al., 1991).

These viruses all spread from the periphery to the brain by the hematogenous route. Thus any effect that increases virus titers in the blood will likely also increase the virus available for infection of the brain (although this principle has a caveat described in Section V). The viral LTR contains the transcriptional promoter sequences that control viral gene expression, and thus it is not surprising that LTR sequences might impact the level of brain infection simply through their effect on peripheral replication. In view of the tissue specificity of the transcriptional enhancers (Jaenisch *et al.*, 1976; Speck *et al.*, 1990), however, the possibility exists that some of the effect of LTR sequences on neuroinvasiveness may be brain-specific or perhaps even brain region–specific (DesGroseillers *et al.*, 1985).

The influence of the 5′ leader region of the viral genome on neuroinvasiveness also appears to act by affecting peripheral replication and/or viral spread. This region of the viral genome is involved in primer binding for reverse transcription, dimerization of viral RNA, viral RNA packaging, and mRNA splicing. In addition, both for the murine (Edwards and Fan, 1980) and feline oncornaviruses (Neil *et al.*, 1980), the 5′ leader region contains a coding sequence for the membrane protein called glycosylated *gag* (Glycogag). The mapping studies summarized in Fig. 3 identified a motif located in the N-terminal cytoplasmic tail of Glycogag that had a strong effect on neuroinvasiveness (Fujisawa *et al.*, 1998; Portis *et al.*, 1991; Portis *et al.*, 1994). The nature of this effect is not known, primarily because the function of Glycogag in the virus life cycle is not yet understood.

Glycogag is a Type II integral membrane protein ($N_{cyto}C_{exo}$) (Fujisawa *et al.*, 1997; Pillemer *et al.*, 1986), the initiation codon for which is an unusual CTG (*Prats et al.*, 1989) located 5′ of, but in-frame with, the ATG initiation codon of the *gag* proteins, which are constituents of the virion core. Thus, Glycogag contains the entire coding sequence of *gag*, but, in addition, has a unique N terminus of 88–100 residues that forms the cytoplasmic tail (Fig. 3). The protein is cleaved by a cellular protease near the middle of the molecule (Fujisawa *et al.*, 1997) (though cleavage patterns can vary between virus isolates). The N-terminal half of the protein, bearing the transmembrane domain, remains cell-associated and is expressed at the plasma membrane. The C-terminal half is secreted (Edwards and Fan, 1979; Fujisawa *et al.*, 1997) and associates avidly with extracellular matrix (Edwards *et al.*, 1982; Fujisawa *et al.*, 1997).

The sequence which influences neuroinvasiveness appears to represent a dileucine motif (EFLL) (Fujisawa *et al.*, 1998) that is located 11 residues N-terminal of the transmembrane domain of the protein. Dileucine motifs have been identified in a variety of membrane proteins, including CD4 (Aiken, C. *et al.*, 1994); CD3 (Letourneur and Klausner, 1992); and major histocompatibility complex (MHC) Class II β and invariant chains (Zhong, G. *et al.*, 1997), and appear to act as a sorting signal, interacting with adaptor proteins primarily located in the *trans*-Golgi (reviewed by Sandoval and Bakke, 1994). This motif in Glycogag downregulates the level of cell-surface protein (Fujisawa, R. *et al.*, 1998). We now think this occurs because of increased intracellular degradation (Fujisawa and Portis, unpublished), which leads to overall lower steady-state levels of the protein. Thus, viruses which encode this motif exhibit lower steady-state levels of Glycogag than those in which this sequence is mutated. Curiously, then, low steady-state levels of the protein correlate with high-level neuroinvasiveness.

How does this work? What we know from studies of null mutants of the virus KP (Fig. 3), in which Glycogag expression is eliminated, is that the protein, though dispensable for virus replication *in vitro*, has a strong influence on the kinetics of virus spread *in vivo* (Portis *et al.*, 1994). Mice inoculated with these mutant viruses exhibit slower spread of virus in the spleen and lower viremia titers, and the virus fails to infect the brain (Portis *et al.*, 1996). Likewise, mutations which disrupt the dileucine motif in the cytoplasmic tail of Glycogag⁺ viruses (Fig. 3) have a similar effect, slowing the spread of virus *in vivo* but not *in vitro* and restricting infection of the brain (Portis *et al.*, 1994). By extension, one can conclude that this dileucine motif is likely to play an important role in the function of this enigmatic protein in the virus life cycle, but so far that function has eluded us.

V. Host Factors and Neuroinvasiveness

Within a week after neonatal inoculation of Glycogag-null mutants of the virus KP, second-site revertants begin to appear in the spleen (Portis *et al.*, 1996). These viruses express a truncated, but apparently functional, form of the protein. These revertants continue to increase in frequency over several months of observation until at least half of the virus present in the blood encodes the truncated protein. However, this virus is never detected in the brain, and these mice never develop neurologic disease. Yet when these revertants were injected intraperitoneally back into neonatal mice, those mice developed a rapidly progressive spongiform neurodegenerative disease (Portis *et al.*, 1996). Clearly the adults harbored fully neurovirulent virus but were resistant to expression of disease.

The explanation of this observation comes from a consideration of the age-dependent resistance to neurologic disease that has been observed for most of the neurovirulent MuLVs that have been studied (see Section I). Mice inoculated as neonates are susceptible, while mice inoculated at 10 days of age or older are resistant. This resistance is a consequence of an intrinsic resistance of the brain to infection, which is noted first at approximately postnatal (P) day 6 and is complete by P10–12 (Czub *et al.*, 1991; Czub *et al.*, 1992). The resistance cannot be overcome by direct intracranial inoculation of virus, is observed in athymic nude mice, and appears to be due to a loss of susceptibility of microvascular endothelial cells in the brain to infection (Lynch *et al.*, 1995). Some cell populations, within the brain parenchyma, particularly microglial cells, remain susceptible to infection after P10, but this is demonstrable only after implantation of virus-producing cells (Lynch *et al.*, 1995). The mechanism of this resistance is not altogether clear. Since these viruses require cell division to complete their replication cycle, it is likely that the decreased rate of endothelial cell division that occurs during postnatal development of the nervous system explains the resistance. The resistance appears to be overcome by *in vitro* culture, since primary brain capillary endothelial cell cultures that are derived from fully resistant young adults regain the capacity to be infected (Masuda *et al.*, 1997a).

The intrinsic resistance of the brain to infection indicates that there is only a transient "window" of susceptibility during the first week to 10 days after birth, during which virus in the periphery has access to the brain. Since virus reaches the brain through the hematogenous route, this implies that the level of virus infection in the brain should be a function of the level of virus replication reached in the periphery before the "window" closes. Consistent with this hypothesis is the correlation we observe between the viremia titers reached in the first 5–6 days after neonatal inoculation and the level of virus infection of the brain achieved by the respective viruses (Czub *et al.*, 1992; Fujisawa *et al.*, 1998). Those viruses which reached peak titers earliest were found to infect the brain at the highest levels. As would be predicted, viremia titers measured at later time points (e.g., 15–20 days after inoculation) were found to be a poor correlate of neuroinvasiveness (Czub *et al.*, 1992).

This "window" of susceptibility would also explain the influence of the viral LTR, Glycogag, and viral envelope sequences on neuroinvasiveness, since each of these sequences affects the kinetics of virus spread in peripheral nonneuronal tissues, though through different mechanisms. It is also apparent that host genes or viral sequences that simply effect a delay in the spread of virus in peripheral nonneuronal tissues during this early postnatal period would be highly effective in preventing brain infection.

As mentioned above, the maintenance of CasBrE in wild-mouse populations occurs through transmission of virus in maternal milk to suckling newborns. CasBrE carries an envelope gene that, if delivered at high levels to the brain, would kill the mice before they reached sexual maturity. One might speculate that the coevolution of CasBrE and its feral mouse host has led to the selection of viral sequences that limit access of the virus to the brain. If one compares the sequence of the cytoplasmic domain of Glycogag, from a variety of murine oncornaviruses isolated from inbred laboratory mice, with that of CasBrE, it is apparent that CasBrE is the only virus in which the dileucine motif is mutated. This, then, could represent a sequence that was selected because it rendered the virus less fit, though perhaps facilitating its transmission to the next generation.

VI. Viral Envelope Sequences that Determine Neurotoxicity

It is clear, from the studies of FB29/CasBrE chimeric viruses discussed above, that neurovirulence (the capacity to induce neurologic disease) can be dissociated from neuroinvasiveness. Thus, virus infection of the central nervous system per se is not sufficient unless the virus carries a "neurovirulent" envelope gene. What is meant, however, by a "neurovirulent" envelope gene? As discussed in Section IVA, the envelope gene contains sequences that influence neuroinvasiveness, but there must be additional sequences that determine a phenotype that we shall call "neurotoxicity," for want of a better term. Fine mapping studies have been carried out in both the polytropic and ecotropic models and are summarized in Fig. 5.

The mapping studies on the neurovirulent polytropic virus FMCF98 (Table II) began with the somewhat perplexing result shown in Fig. 4. The envelope genes of FMCF98 and a nonneurovirulent polytropic virus FMCF54 were first cloned into the genome of FB29, generating the chimeric viruses Fr98 and Fr54 (Fig. 4). Neonates inoculated with Fr98 intraperitoneally develop ataxia and seizures with a short incubation period of 3 weeks. Fr54 is avirulent. Chimeric envelope genes were constructed using segments from these two viruses, while maintaining the same FB29 genetic background. There appeared to be two regions of *env* that conferred neurovirulence. One was located 5' and the other 3' of an *Eco*RI site near the middle of the gene (Fig. 4), and the effects of these sequences appeared to be additive (Hasenkrug *et al.*, 1996). Further studies revealed that the sequences 5' of the *Eco*RI site affected neuroinvasiveness (Robertson *et al.*, 1997). Viruses carrying these sequences from Fr54 were less neuroinvasive (achieved lower viral burdens in

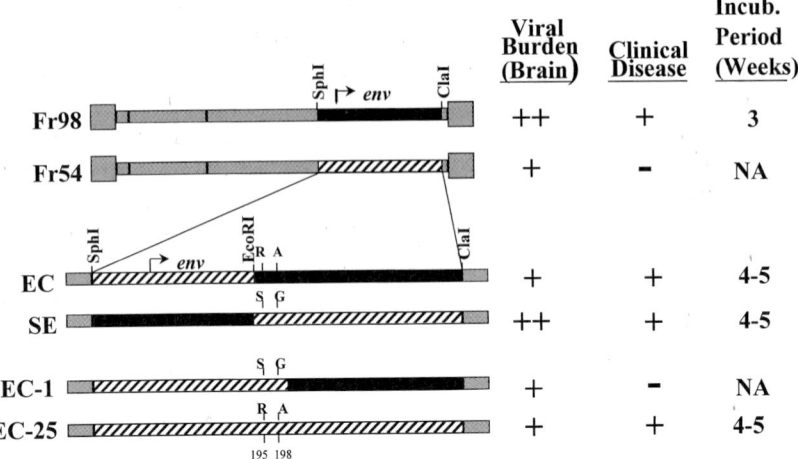

Fig 4. Mapping of the determinants of neurovirulence within the envelope gene of the polytropic virus FMCF98. The two genomes at the top Fr98 and Fr54 represent the parental viruses, each of which is itself a chimeric virus consisting of the envelope gene and 3′ *pol* sequences of FMCF98 (black) and FMCF54 (hatched), respectively, in the genetic background of FB29 (gray). FMCF54 is an avirulent polytropic virus. Viral burdens and incubation periods were determined as described in the caption for Fig. 2. Clinical disease is characterized by ataxia. The only differences between the genomes of EC and EC-1 are at positions 195 and 198. That these two residues from Fr98 (R_{195} and A_{198}) confer neurotoxicity is shown by the double mutant EC-25, which differs from Fr54 only at these two positions and causes ataxia.

the brain) than viruses carrying these sequences from Fr98. The nature of this effect is not yet known and the relevant sequences have not been further localized.

The sequences 3′ of the *Eco*RI site appeared to affect "neurotoxicity" specifically (Poulsen *et al.*, 1998). Thus, whether these sequences were derived from Fr54 or Fr98, neuroinvasiveness did not change. In the constructs shown in Fig. 4, this conclusion comes from a comparison of Fr98 and the construct SE. However, neurovirulence was conferred, even at relatively low viral burdens, when the sequences 3′ of the *Eco*RI site were derived from Fr98. This conclusion comes from a comparison of Fr54 and the construct EC (Fig. 4). Fine mapping studies have localized this effect to changes at amino acids 195 and 198 (compare constructs EC25 and EC in Fig. 4).

These observations, however, do not explain the neurovirulence of the virus SE (Fig. 4). This phenotype would suggest that sequence of Fr54 3′ of the *Eco*RI site might also contain a "neurotoxicity" determinant that perhaps was not as strong as that of Fr98, and that this toxicity was exhibited only when viral burden in the brain was increased by virtue of the presence of Fr98 sequences 5' of the EcoRI site. One way to test this hypothesis was to increase the viral burden of Fr54 through nongenetic means. This was accomplished by using a neural stem cell line called C17.2 (Ryder et al., 1990). These cells can be infected in vitro (Lynch et al., 1996) and have the capacity to engraft in the brains of neonatal mice after intraventricular injection. The cells are incorporated into the germinal subventricular zone of the developing brain and migrate along with the endogenous neural stem cells to the cerebral cortex and subcortex (Snyder and Wolfe, 1996). They differentiate into neurons and neuroglia and can persist for weeks to months in situ. Thus, this cell engraftment system can be used as a delivery system for virus (Lynch et al., 1996), bypassing the blood–brain barrier and any restrictions to viral entry into the brain from the periphery. When Fr54 was delivered to the brain at high levels, using this technique, the mice developed ataxia clinically identical to that induced by Fr98 (Poulsen et al., 1999). This study, thus, indicated that both Fr54 and Fr98 envelope genes contain sequences presumably located 3' of the EcoRI site in the envelope gene that confer the "neurotoxic" phenotype, but the effect of Fr98 is stronger, being expressed at a relatively low viral burden. Whether or not this interpretation turns out to be correct, it illustrates the complexities of the neurovirulent phenotype and the interplay of different sequences within the same gene in the expression of this phenotype.

Fine mapping studies have also been carried out for the neurovirulent ecotropic viruses Moloney ts-1, PVC211, and CasBrE (Table II). The "neurotoxicity" of Moloney ts-1 is determined by a $Val^-_{25} \rightarrow Ile$ change located near the N terminus of the envelope protein (Fig. 5) (Szurek et al., 1990a). This change was also found to interfere with transport of the protein from the endoplasmic reticulum (ER) to the Golgi complex leading to the intracellular accumulation of uncleaved envelope precursor polyprotein (Szurek et al., 1990a, 1990b). Defective proteolytic processing of the envelope protein of $FrCas^E$, accompanied by aberrant virus assembly, specifically in microglial cells in vitro, has also been reported (Lynch et al., 1994). To date, however, the evidence linking defective posttranslational processing of the envelope protein and the neurovirulence of these viruses remains anecdotal.

For PVC211, "neurotoxicity" determinants have been localized to a region 5' of a BamHI site (Fig. 5), inclusive of the signal peptide and 3' coding sequence of pol (Masuda et al., 1993). Within this region are located the two amino acid residues (Gly_{116} and Lys_{129}) that determine

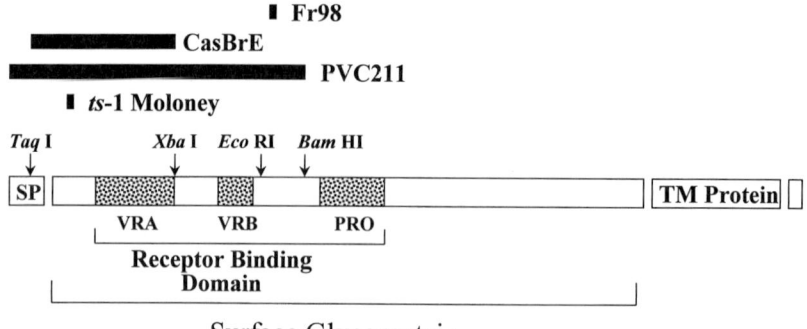

Surface Glycoprotein

Fɪɢ 5. A schematic diagram of the envelope protein of a generic murine oncornavirus showing the general organization of the receptor binding domain (Battini *et al.*, 1992). The black bars above show the location of the regions or specific sequences that have been found to determine the "neurotoxicity" of the neurovirulent ecotropic and polytropic viruses. SP, signal peptide; TM, transmembrane protein. The C terminus of the TM protein is cleaved off by the viral protease. The receptor binding domain is composed of three subdomains designated as variable region A (VRA), variable region B (VRB), and the proline-rich region (PRO).

tropism for brain capillary endothelial cells (Section IVA) and thus are important determinants of neuroinvasiveness. However, these two residues appear not to confer "neurotoxicity" (Masuda *et al.*, 1997b), implying that additional "neurotoxic" sequences within this region remain to be identified.

For CasBrE, the mapping studies just examined which regions of *env* were required to induce clinical disease (paralysis) (Paquette *et al.*, 1989). Thus, it is not known whether these sequences influenced "neurotoxicity" or neuroinvasiveness. Nevertheless, a region of *env* between a *Taq*I and *Xba*I site (Fig. 5), encoding the C-terminal half of the signal peptide through the VRA region, had the strongest influence on the incidence and tempo of the disease. It is worth noting, from Fig. 5, that these sequences that appear to determine "neurotoxicity" are concentrated in the N-terminal half of the surface glycoprotein, a region of the molecule that also harbors the receptor binding domain. Thus, it is possible that the neurovirulence of these viruses involves in some way the interaction between the envelope protein and viral receptor, either at the cell surface or intracellularly. Because the receptors for these viruses also perform normal physiologic functions, it is tempting to speculate that neurotoxicity may involve a subversion of those functions. The functional consequences of the binding of an ecotropic envelope protein to its receptor, CAT-1, have been explored *in* *vitro* in fibroblasts and *Xenopus* oocytes (Wang *et al.*, 1996). CAT-1 is a multiple-membrane-spanning cationic amino acid transporter (Albritton *et al.*, 1989; Wang *et al.*, 1991). Only weak inhibition of CAT-1-mediated arginine transport (on the order of 50%) was observed, but the effect was dramatically magnified when N-linked glycosylation of the receptor was prevented. This has suggested that receptor glycosylation could afford a level of protection from such potentially toxic effects of retrovirus infection (Wang *et al.*, 1996). Perhaps the "neurotoxic" envelope proteins bypass this protective mechanism. Clearly this issue needs to be examined by using neurovirulent viruses in cells more relevant to the nervous system.

VII. Cell Types Involved in Neurovirulence

Ultimately, genetic studies alone, even those that have identified specific amino acids that influence neuropathogenesis, will likely not uncover the mechanisms responsible for the neurotoxicity of these viruses. The advantage of animal models is the capacity to reduce the complexity of disease by identifying the minimum viral and host factors necessary for expression of neuropathology. Which cells are necessary and what viral proteins are required? What is the nature of the neurotoxicity induced by these viruses? Is the viral envelope protein directly neurotoxic *in vivo*? Developing approaches to answering these question has provided further clues to the pathogenesis of these diseases.

FrCas[E] virus, inoculated intraperitoneally into neonatal IRW mice, is first detectable in the brain at 6 days postinoculation (Czub *et al.*, 1994), the first cells infected being microvascular endothelial cells. There the virus amplifies and spreads to cells of the parenchyma including microglia (brain-resident macrophages); neurons (exclusively those that divide postnatally); and occasional oligodendrocytes (Lynch *et al.*, 1991). Infection of astrocytes has not been detected *in vivo* (Lynch *et al.*, 1991), although astrocytes grown *in vitro* are infectable (Lynch *et al.*, 1994).

In comparing the location of the infected cells relative to the location of the spongiform lesions, it became clear that the infected neurons were histologically and ultrastructurally normal, whereas neurons and astrocytes located near infected microglia exhibited vacuolar degenerative change (Lynch *et al.*, 1991). Similar observations have been made in the spongiform encephalomyelopathy induced by Moloney ts-1 as well (Baszler and Zachary, 1991). These, then, were the first clues that infection of microglia might be necessary for expression of pathology.

This hypothesis was tested directly by implanting infected microglia into the brains of mice in which virus spread within the brain was restricted by age. Microglia isolated from IRW neonates were infected with FrCas^E *in vitro* (Lynch *et al.*, 1994) and implanted into the brains of 10-day-old mice (Lynch *et al.*, 1995). When they were examined 2–3 weeks later, it was apparent that the implanted cells had survived and even migrated to locations distant from the implantation site. The cells were shown to continue to express viral antigens and, most importantly, foci of spongiosis were located in the immediate vicinity of the infected microglia (Lynch *et al.*, 1995). No other cell types were found to be infected in the brains of these mice, indicating that infection of microglia alone was responsible for the spongiosis.

This experiment, however, did not determine whether this phenomenon was specific for microglia. One interpretation of these results could be that the microglia were simply acting as virus factories, and that it was some viral product, such as the envelop protein, released by these cells that was neurotoxic. There is considerable support for this mechanism in HIV-mediated neurotoxicity in which the envelope protein gp 120 administered extracellularly has been shown to cause neuronal cell death *in vitro* (Lipton, 1992) and *in vivo* in mice (Toggas *et al.*, 1994) as well as in rats (Glowa *et al.*, 1992). To address this issue, FrCas^E-infected cells of neural lineages were implanted. These studies utilized the same C17.2 neural stem cell line (Ryder *et al.*, 1990), referred to above (Section VI), to deliver to the brain complete virus or viral envelope protein alone. Significantly, these cells are of neuroectodermal origin and differentiate into neurons and neuroglia (astrocytes and oligodendrocytes) (Snyder *et al.*, 1997; Taylor and Snyder, 1997) but not into microglia, which are derived from the mesenchyme. Because the C17.2 cells require implantation into neonates for successful engraftment, the age-dependent resistance used in the microglial implantation experiments could not be utilized. Instead, the host resistance gene Fv-1 (Jolicoeur and Baltimore, 1976) was used to restrict virus spread. Surprisingly, implantation of these cells resulted in no evidence of spongiosis, despite high-level virus expression in the brain tantamount to that observed after virus inoculation of mice in the absence of Fv-1 restriction (Lynch *et al.*, 1996). This result contrasted dramatically with the results of the microglial implants and provided compelling support for the notion that infection of microglial cells is necessary for induction of spongiosis in this system. These studies also demonstrated that expression of the envelope protein of

FrCas^E or, for that matter, any of the viral proteins in the brain per se, is not directly toxic to neurons *in vivo*.

The polytropic viruses, like the ecotropic viruses, gain access to the central nervous system by infecting microvascular endothelial cells (Robertson *et al.*, 1997), though the amplification step at this site is of a far lower magnitude (Robertson and Portis, unpublished). Microglial cells comprise the predominant parenchymal cell that is infected by these viruses (Robertson *et al.*, 1997), but, unlike the ecotropic viruses, infection of neurons has not been observed (Portis *et al.*, 1995). The kinetics and the severity of the ataxia induced by these viruses have been found to be functions of the frequency of infected microglia (Robertson *et al.*, 1997) and thus, microglial cell infection would appear to be an important determinant of the neurovirulence of these viruses as well. However, the implantation studies carried out in the ecotropic model have not been done with the polytropic viruses.

VIII. ENVELOPE EXPRESSION IN MICROGLIA

How can we reconcile the following observations?—(1) the induction of spongiosis by FrCas^E requires infection of microglial cells; (2) the induction of spongiosis is dependent on the sequence of the envelop protein; and (3) the envelope protein itself is not neurotoxic. In addition, studies on nonneurovirulent ecotropic viruses that are coisogenic with FrCas^E (differing only in *env* and 3′ *pol*) found that the frequency of microglial cells infected by these viruses was actually higher and more widely distributed in the brain than that of FrCas^E (Askovic *et al.*, 2000). Yet these viruses did not induce spongiosis. Thus, microglial infection per se is not sufficient for induction of lesions. The induction of spongiosis appears to require the expression in these cells of viral envelope protein containing specific "neurotoxic" sequences.

Whether expression of a "neurotoxic" envelope protein alone in microglial cells is sufficient was not addressed in the microglial implant experiments described above, because we have been unable to achieve high-level transduction of the FrCas^E envelope gene in primary microglial cell cultures. Microglial cell lines might be suitable for such a purpose. However, cell lines we have tested express endogenous oncornaviruses that would confound any attempt at interpreting the results.

Lynch *et al.*, (1999) have recently approached this question *in vivo* in a novel series of experiments in which C17.2 neural stem cells were converted into a retroviral packaging cell line and used to deliver the mRNA encoding the FrCas^E envelope protein to resident microglial cells. In these experiments the C17.2 packaging cells were transduced by using a retroviral vector carrying the *env* coding sequence, and produced infectious, but replication-defective, virus particles. These cells were then implanted intraventricularly into neonatal mice. Successful engraftment was demonstrated and viral envelope protein was detected both in the C17.2 cells and in resident microglia. Most importantly, these *env*-expressing microglial cells were found in abundance in regions of the brain known to be susceptible to the neurotoxicity of FrCas^E virus. Nevertheless, spongiosis was not observed. Thus, it would appear that neurovirulence requires not only the expression of a "neurotoxic" envelope gene in microglia, but also the expression of one or more other viral genes. How this might work is at the present time a matter of pure speculation. Other viral proteins may influence the trafficking of the envelope protein in microglia. Alternatively, elements of the virus assembly process may be required for expression of neurovirulence. Obviously, this lead should be followed and it promises to uncover more details of the molecular and cellular interactions responsible for the neurotoxicity induced by this virus.

The results of this C17.2 implant study, however, appear at odds with results of transgenic mouse studies in which the envelope genes of two neurovirulent oncornaviruses, CasBrE (Kay *et al.*, 1993) and Moloney ts-1 (Yu *et al.*, 1997), were expressed as transgenes in the brain. In both cases expression levels were very low, being detectable only by the RNase protection assay. This was our experience as well. A number of years ago, in collaboration with Drs. M. B. A. Oldstone and L. Mucke of the Scripps Clinic and Research Foundation, and with the Transgenic Facility of the NIAID, we generated transgenic mice carrying the CasBrE envelope gene that was driven by the retroviral promoter as well as by two brain-specific promoters (neuron-specific enolase and GFAP). Although transgenic mice were derived in each group, expression levels were either vanishingly low or not detectable, suggesting that the CasBrE envelope protein was lethal when expressed during gestation (unpublished data).

The transgenic studies cited above (Kay *et al.*, 1993; Yu *et al.*, 1997) utilized the respective viral promoters (CasBrE and Moloney ts-1) and indeed, clinical signs of neurologic dysfunction and pathologic changes were observed, and resembled those induced by inoculation of the respective viruses. Although the sites of *env* expression were not determined, in each case the promoter is known to function in microglial cells. The one caveat in these experiments was the long incubation periods observed, ranging from 8 to 12 months, in the case of the Moloney ts-1 *env* transgenics, to 15 to 24 months in the CasBrE *env* mice. Endogenous viral gene expression under these circumstances, even in the absence of infectious virus production, could supply the necessary *trans*-acting effect predicted by the implant experiments (Lynch *et al.*, 1999). Resolution of the difference in results obtained in these experiments might come from the generation of transgenic mice by using an inducible promoter, the technology for which is now available (Sauer, 1998).

IX. ROLE OF INFLAMMATION

In none of the neurologic diseases induced by the murine oncornaviruses is there evidence of inflammatory cell infiltrates in the central nervous system. However, the definition of inflammation in the brain has broadened in recent years due in large part to the realization that proinflammatory cytokines and chemokines are expressed by cells intrinsic to the brain. In fact, some consider the induction of gliosis to be a telltale histopathological sign of CNS inflammation, because astroglial activation is induced by a number of proinflammatory cytokines including interleukin (IL)-1 (Giulian *et al.*, 1994), IL-6 (Chiang *et al.*, 1994) and tumor necrosis factor, or TNFα (Kahn *et al.*, 1995).

By extension, then, inflammation must represent a component of the neuropathogenesis of all of the murine oncornaviruses, since gliosis is always observed in these diseases. Whether inflammation is the cause or the result of the neuronal and glial pathology induced by these viruses, however, is a matter of debate. TNFα mRNA (Choe *et al.*, 1998) and protein (Nagra *et al.*, 1994), as well as IL-6 protein (Nagra *et al.*, 1994), have been shown to be upregulated in the spongiform diseases caused by Moloney ts-1 and CasBrE. In both cases, however, the brains were assayed at a time when extensive spongiosis had been manifested for an indeterminate period of time. In order to gain further insight into this question, we have begun studies on the two chimeric viruses Fr98 and FrCas^E, which induce rapid and highly predictable disease courses. We are using the RNAase protection assay to quantify the expression of genes encoding a variety of inflammatory mediators and their receptors at early time points in the disease process. Furthermore, the coisogenic viruses F43 and FB29, which are highly neuroinvasive but not neurovirulent (Askovic *et al.*, 2000), serve as controls to distinguish changes which are disease-specific from those which are simply a response to virus infection. While studies are ongoing, there are a number of preliminary findings which are of interest. In mice infected with FrCas^E and in the terminal stage of disease (19–21 days postinoculation), TNFα

and β mRNA's were consistently upregulated >3-fold (Askovic *et al.*, 2001), in agreement with the prior studies cited above. Neither of these genes were found to be upregulated in the brains of age-matched mice infected with F43, and thus they would appear to be disease-specific. However, at early time points (14 days postinoculation), there was no evidence of upregulation of either of these genes, despite the presence of extensive spongiosis. While it is certainly possible that the TNFs contribute to some aspects of disease progression, it would appear that these cytokines are not responsible for the spongiosis induced by this virus. The only other inflammatory mediators that appear to be upregulated (at the level of mRNA) at early time points, and in a lesion-specific fashion, are the β-chemokines MIP-1α and β (Askovic and Portis, manuscript in preparation). Interestingly, at no times during the course of the disease have we observed upregulation of proinflammatory cytokines such as IL-1, IL-6, IL-12, or IL-18 (Askovic and Portis, manuscript in preparation), and there is also no evidence of upregulation of F4/80 mRNA (Askovic *et al.*, 2000) or of protein (Lynch, W. P. *et al.*, 1991). F4/80 is a member of the EGF-TM7 family of membrane proteins (McKnight and Gordon, 1998) which is restricted in the mouse to cells of the macrophage lineage, and its upregulation is a sign of microglial activation (Andersson *et al.*, 1991). This suggests that microglia have not undergone the activation responses that are generally one of the earliest signs of either chemical or traumatic injury to the nervous system (Kreutzberg, 1996). Although these studies are not complete, at this point it is fair to say that the apparent lack of response of microglia is unexpected, in view of their high level of infection by FrCas[E] and their location within areas of extensive neuronal and astroglial damage. These results raise several questions. First, does FrCas[E] actively suppress microglial activation? Second, does this phenomenon play a role in the pathogenesis of the disease? Third, in view of the upregulation of β-chemokines, why are there no cellular inflammatory infiltrates in the brain? Fourth, recent demonstration of functional CC chemokine receptors on both neurons and astrocytes (Klein *et al.*, 1999; Tanabe *et al.*, 1997) raises the question of whether the β-chemokines may have direct effects on these cells.

The story appears to be different for the rapid disease induced by the polytropic virus Fr98. Here, microglial activation manifested by upregulation of CD11b is observed at a time coincident with the onset of clinical signs (Robertson *et al.*, 1997a). The glial response is associated with an upregulation of TNFα, IL-1 α and β, IL-6 and IL-12p40 mRNA's (Roberston, 1997) and, in addition, there is a positive correlation between the level of TNFα mRNA and the severity of clinical disease (Robertson, 1997). Thus, while microglial cell infection plays a central role in the pathogenesis of both the ecotropic and polytropic viral diseases, the pathophysiology of the two diseases may be different. Furthermore, since the genomes of FrCas^E and Fr98 differ only in *env* and 3' *pol*, it would appear that the qualitative difference in the inflammatory responses induced by these viruses also maps to this region of the viral genome.

At this time one can only speculate on the nature of the inflammatory stimulus induced by these viruses. Mice are inoculated neonatally; there is little in the way of acquired immunity to these viruses (see Section I); and infiltration of inflammatory cells from the periphery is consistently absent. There is evidence that viruses such as Borna (Sauder and de la Torre, 1999), LCMV (Asensio *et al.*, 1999) and MHV (Lane *et al.*, 1998) may directly stimulate cytokine and chemokine expression from cells intrinsic to the nervous system, without apparent participation of infiltrating leukocytes. It is of interest, in this regard, that the envelope proteins of Polytropic retroviruses are known to interact with a variety of receptors involved in signal transduction in lymphocytes and erythroid cells (Li *et al.*, 1991). Indeed, the polytropic receptor used by these virus to enter cells (Battini *et al.*, 1999; Tailor *et al.*, 1999; Yang *et al.*, 1999) appears to be related to the protein Sig 1p, which is involved in G-protein-coupled mitogenic signaling in *Saccharomyces cerevisiae* (Spain *et al.*, 1995). In addition, T lymphocytes infected with polytropic, but not ecotropic, retroviruses overexpress IL-9 (Flubacher *et al.*, 1994), a TH$_2$ lymphokine that is thought to play an important role in the pathogenesis of inflammatory respiratory diseases such as asthma (Temann *et al.*, 1998). There is, thus, at least a theoretical basis for the idea that these viruses might directly stimulate expression of inflammatory mediators.

X. RETROVIRUSES AND MULTIPLE SCLEROSIS

Two neurologic diseases of humans are known to be caused by retroviruses: human immunodeficiency virus (HIV)-associated dementia, and tropical spastic paraparesis caused by HTLV-1. Evidence of the involvement of retroviruses in other diseases of the nervous system is loose at best. The most recent example is the suggested link between the expression of an endogenous virus belonging to the HERV-W family (Komurian-Pradel *et al.*, 1999) and multiple sclerosis (Perron *et al.*, 1997). Viral RNA was identified by RT-PCR in culture supernatants of

leptomeningeal cells and B cells of MS patients but not in patients
with a variety of other brain diseases. Using techniques of comparable
sensitivity, others have failed to find any evidence that expression of a
retrovirus is linked to MS (Hackett et al., 1996). One possible scenario
is that an endogenous retrovirus may be activated as a secondary consequence of the inflammatory brain disease. As discussed in Section
IIA, this appears to happen in the more chronic forms of neurodegenerative disease caused by CasBrE.

The recent finding, however, that the expression of an endogenous
retrovirus (HERV-K) may be causally linked to the occurrence of
insulin-dependent diabetes mellitus (Conrad et al., 1997) has reinforced the notion that the expression of endogenous retroviruses in
humans may have untoward consequences. Two examples of the role of
endogenous viruses in mouse diseases are worth considering. Neurovirulent murine polytropic viruses are recombinants between exogenous and endogenous retroviral sequences (Stoye and Coffin, 1987).
Furthermore, the regions of the envelope gene that determine neurovirulence (Hasenkrug et al., 1996; Poulsen et al., 1998) are of
endogenous origin. This implies that there are endogenous retroviral
envelope genes present in the mouse genome that are capable of inducing neurologic disease so long as they are mobilized—in this case, by
recombination with a replication-competent and neuroinvasive exogenous retrovirus. Furthermore, the disease induced by these viruses
appears to have an inflammatory component.

The other example is the paralytic disease caused by lactate dehydrogenase elevating virus (LDV). LDV is a positive-strand RNA
virus that is a member of the Arteriviridae (Plagemann and Moennig, 1992) and infection of mice results in lifelong viremia. The virus
is nonpathogenic except in C58 and AKR mice, in which it causes a
poliomyelitis (Stroop and Brinton, 1983). Both of these mouse
strains have a high incidence of spontaneous leukemia and express
replication-competent endogenous ecotropic oncornaviruses. Genetic
studies indicate that susceptibility to the paralytic disease segre-
gates with endogenous virus expression in the brain (Anderson et
al., 1995). The nature of this effect is not known, but it indicates
that endogenous retroviruses expressed in the brain might interact
with other viruses in unpredictable ways. It is certainly prudent to
exercise due caution in drawing any conclusions as to a causal relationship between expression of endogenous retroviruses and brain
diseases of humans (Stoye, 1999). Nevertheless, there is some prece-
dent from the world of mouse oncornaviruses that suggests that
such a relationship could exist.

XI. CONCLUDING REMARKS

Although it is difficult to make generalizations that would fit each of the mouse models described in this chapter, there are several features that appear to characterize the neurovirulence of many of the murine oncornaviruses: (1) neurovirulence is determined by multiple viral sequences that are located both within and outside the viral envelope gene; (2) there is a clear cut dissociation between neuroinvasiveness and neurovirulence, and viral sequences can be identified that determine one, independent of the other phenotype; (3) neuropathogenicity depends on the infection of microglial cells by viruses carrying specific "neurotoxic" envelope sequences; (4) the viral envelope protein itself appears not to be directly neurotoxic; and (5) the tempo and severity of the disease caused by viruses carrying such "neurotoxic" sequences is enhanced by viral sequences that increase viral burden in the brain.

The central role played by microglial cells in these diseases suggests that these viruses may, in some way, subvert the normal functioning of these cells. This begs the question, What do these cells do? It has been estimated that parenchymal microglial cells comprise 5–10% of all cells in the brain (Lawson *et al.*, 1990) and exhibit relatively long turnover rates (Lawson *et al.*, 1992). Their delicate, highly branched cellular processes form a rather uniform network extending around and between neurons in the gray matter and axons in the white matter.

Our thinking about the function of microglial cells is currently in a state of flux. While it is clear that these cells are the scavengers of the brain and also can act, under certain special situations, as antigen-presenting cells (reviewed in Kreutzberg [1996]), it now appears that they may play a critical role in the maintenance of immune privilege in the CNS (Bauer *et al.*, 1999; Ford *et al.*, 1996). These newly discovered functions are beginning to impact our understanding of both neuroinflammatory diseases such as multiple sclerosis and HIV-associated dementia, and neurodegenerative diseases such as Alzheimer's and the prion diseases. Defining the nature of the perturbations of microglial function induced by the murine oncornaviruses promises to provide further clues to the roles of these cells in maintaining normal homeostasis within the CNS.

REFERENCES

Aiken, C., Konner, J., Landau, N. R., Lenburg, M. E., and Trono, D. (1994). Nef induces CD4 endocytosis: requirement for a critical dileucine motif in the membrane-proximal CD4 cytoplasmic domain. *Cell* **76:**853–864.

Albritton, L. M., Tseng, L., Scadden, D., and Cunningham, J. M. (1989). A putative murine ecotropic retrovirus receptor gene encodes a multiple membrane-spanning protein and confers susceptibility to virus infection. *Cell* **57:**659–666.

Anderson, G. W., Palmer, G. A., Rowland, R. R., Even, C., and Plagemann, P. G. (1995). Infection of central nervous system cells by ecotropic murine leukemia virus in C58 and AKR mice and in in utero-infected CE/J mice predisposes mice to paralytic infection by lactate dehydrogenase-elevating virus. *J. Virol.* **69:**308–319.

Andersson, P. B., Perry, V. H., and Gordon, S. (1991). The kinetics and morphological characteristics of the macrophage–microglial response to kainic acid-induced neuronal degeneration. *Neuroscience* **42:**201–214.

Andrews, J. M., and Gardner, M. B. (1974). Lower motor neuron degeneration associated with type C RNA virus infection in mice: neuropathological features. *J. Neuropathol. Exp. Neurol.* **33:**285–307.

Asensio, V. C., Kincaid, C., and Campbell, I. L. (1999). Chemokines and the inflammatory response to viral infection in the central nervous system with a focus on lymphocytic choriomeningitis virus. *J. Neurovirol.* **5:**65–75.

Askovic, S., McAtee, F. J., Favara, C., and Portis, J. L. (2000). Brain infection by neuroinvasive but avirulent murine oncornaviruses. *J. Virol.* **74:**465–473.

Askovic, S., Favara, C, McAtee, F. J. and Portis, J. L. (2001). Increased expression of MIP-1α and MIP-1β in the brain correlates spatially and temporally with the spongiform neurodegeneration induced by a murine oncornavirus. *J. Viro.* **75:** in press.

Aziz, D. C., Hanna, Z., and Jolicoeur, P. (1989). Severe immunodeficiency disease induced by a defective murine leukaemia virus. *Nature* **338:**505–508.

Barbacid, M., Robbins, K. C., and Aaronson, S. A. (1979). Wild mouse RNA tumor viruses. A nongenetically transmitted virus group closely related to exogenous leukemia viruses of laboratory mouse strains. *J. Exp. Med.* **149:**254–266.

Baszler, T. V., and Zachary, J. F. (1991). Murine retroviral neurovirulence correlates with an enhanced ability of virus to infect selectively, replicate in, and activate resident microglial cells. *Am. J. Pathol.* **138:**655–671.

Battini, J.-L., Heard, J. M., and Danos, O. (1992). Receptor choice determinants in the envelope glycoproteins of amphotropic, xenotropic, and polytropic murine leukemia viruses. *J. Virol.* **66:**1468–1475.

Battini, J. L., Rasko, J. E., and Miller, A. D. (1999). A human cell-surface receptor for xenotropic and polytropic murine leukemia viruses: possible role in G protein-coupled signal transduction. *Proc. Natl. Acad. Sci. U.S.A.* **96:**1385–1390.

Bauer, J., Stadelmann, C., Bancher, C., Jellinger, K., and Lassmann, H. (1999). Apoptosis of T lymphocytes in acute disseminated encephalomyelitis. *Acta Neuropathol. (Berlin)* **97:**543–546.

Bergeron, D., Houde, J., Poliquin, L., Barbeau, B., and Rassart, E. (1993). Expression and DNA rearrangement of proto-oncogenes in Cas-Br-E-induced non-T-, non-B-cell leukemias. *Leukemia* **7:**954–962.

Bergeron, D., Poliquin, L., Kozak, C. A., and Rassart, E. (1991). Identification of a common viral integration region in Cas-Br-E murine leukemia virus-induced non-T-, non-B-cell lymphomas. *J. Virol.* **65:**7–15.

Brooks, B. R., Swarz, J. R., and Johnson, R. T. (1980). Spongiform polioencephalomyelopathy caused by a murine retrovirus. I. Pathogenesis of infection in newborn mice. *Lab. Invest.* **43:**480–486.

Bryant, M. L., Scott, J. L., Pal, B. K., Estes, J. D., and Gardner, M. B. (1981). Immunopathology of natural and experimental lymphomas induced by wild mouse leukemia virus. *Am. J. Pathol.* **104:**272–282.

Chattopadhyay, S. K., Oliff, A. I., Linemeyer, D. L., Lander, M. R., and Lowy, D. R. (1981). Genomes of murine leukemia viruses isolated from wild mice. *J. Virol.* **39:**777–791.

Chiang, C. S., Stalder, A., Samimi, A., and Campbell, I. L. (1994). Reactive gliosis as a consequence of interleukin-6 expression in the brain: studies in transgenic mice. *Dev. Neurosci.* **16:**212–221.

Choe, W., Stoica, G., Lynn, W., and Wong, P. K. (1998). Neurodegeneration induced by MoMuLV-ts1 and increased expression of Fas and TNF-alpha in the central nervous system. *Brain Res.* **779:**1–8.

Cloyd, M. W., Thompson, M. M., and Hartley, J. W. (1985). Host range of mink cell focus-inducing viruses. *Virology* **140:**239–248.

Conrad, B., Weissmahr, R. N., Boni, J., Arcari, R., Schupbach, J., and Mach, B. (1997). A human endogenous retroviral superantigen as candidate autoimmune gene in type I diabetes. *Cell* **90:**303–313.

Czub, M., Czub, S., McAtee, F. J., and Portis, J. L. (1991). Age-dependent resistance to murine retrovirus-induced spongiform neurodegeneration results from central nervous system–specific restriction of virus replication. *J. Virol.* **65:**2539–2544.

Czub, M., Czub, S., Rappold, M., Mazgareanu, S., Schwender, S., Demuth, M., Hein, A., and Dorries, R. (1995). Murine leukemia virus-induced neurodegeneration of rats: enhancement of neuropathogenicity correlates with enhanced viral tropism for macrophages, microglia, and brain vascular cells. *Virology* **214:**239–244.

Czub, M., McAtee, F. J., and Portis, J. L. (1992). Murine retrovirus-induced spongiform encephalomyelopathy: host and viral factors which determine the length of the incubation period. *J. Virol.* **66:**3298–3305.

Czub, S., Lynch, W. P., Czub, M., and Portis, J. L. (1994). Kinetic analysis of spongiform neurodegenerative disease induced by a highly virulent murine retrovirus. *Lab. Invest.* **70:**711–723.

Davey, R. A., Zuo, Y., and Cunningham, J. M. (1999). Identification of a receptor-binding pocket on the envelope protein of friend murine leukemia virus. *J. Virol.* **73:**3758–3763.

DesGroseillers, L., Barrette, M., and Jolicoeur, P. (1984). Physical mapping of the paralysis-inducing determinant of a wild mouse ecotropic neurotropic virus. *J. Virol.* **52:**356–363.

DesGroseillers, L., Rassart, E., Robitaille, Y., and Jolicoeur, P. (1985). Retrovirus-induced spongiform encephalopathy: The 3′-end long terminal repeat-containing viral sequences influence the incidence of disease and the specificity of the neurological syndrome. *Proc. Natl. Acad. Sci. U.S.A.* **82:**8818–8822.

Edwards, S. A., and Fan, H. (1979). gag-Related polyproteins of Moloney murine leukemia virus: evidence for independent synthesis of glycosylated and unglycosylated forms. *J. Virol.* **30:**551–563.

Edwards, S. A., and Fan, H. (1980). Sequence relationship of glycosylated and unglycosylated gag polyproteins of Moloney murine leukemia virus. *J. Virol.* **35:**41–51.

Edwards, S. A., Lin, Y. C., and Fan, H. (1982). Association of murine leukemia virus gag antigen with extracellular matrices in productively infected mouse cells. *Virology* **116:**306–317.

Espey, M. G., Kustova, Y., Sei, Y., and Basile, A. S. (1998). Extracellular glutamate levels are chronically elevated in the brains of LP-BM5-infected mice: a mechanism of retrovirus-induced encephalopathy. *J. Neurochem.* **71:**2079–2087.

Flubacher, M. M., Bear, S. E., and Tsichlis, P. N. (1994). Replacement of interleukin-2 (IL-2)-9-dependent autocrine loop: implications for MCF virus-induced leukemogenesis. *J. Virol.* **68:**7709–7716.

Ford, A. L., Foulcher, E., Lemckert, F. A., and Sedgwick, J. D. (1996). Microglia induce CD4 T lymphocyte final effector function and death. *J. Exp. Med.* **184:**1737–1745.

Fujisawa, R., McAtee, F., and Portis, J. L. (1998). Neuroinvasiveness of a murine retrovirus is influenced by a dileucine-containing sequence in the cytoplasmic tail of Glycosylated Gag. *J. Virol.* **72:**5619–5625.

Fujisawa, R., McAtee, F. J., Zirbel, J. H., and Portis, J. L. (1997). Characterization of glycosylated Gag expressed by a neurovirulent murine leukemia virus: identification of differences in processing in vitro and in vivo. *J. Virol.* **71:**5355–5360.

Gardner, M. B. (1978). Type C viruses of wild mice: characterization and natural history of amphotropic, ecotropic, and xenotropic MuLv. *Curr. Top. Microbiol. Immunol.* **79:**215–259.

Gardner, M. B. (1994). Retroviruses and wild mice: an historical and personal perspective. *Adv. Cancer Res.* **65:**169–201.

Gardner, M. B., Chiri, A., Dougherty, M. F., Casagrande, J., and Estes, J. D. (1979). Congenital transmission of murine leukemia virus from wild mice prone to the development of lymphoma and paralysis. *J. Natl. Cancer Inst.* **62:**63–70.

Gardner, M. B., Henderson, B. E., Estes, J. D., Rongey, R. W., Casagrande, J., Pike, M., and Huebner, R. J. (1976). The epidemiology and virology of C-type virus-associated hematological cancers and related diseases in wild mice. *Cancer Res.* **36:**574–581.

Gardner, M. B., Henderson, B. E., Officer, J. E., Rongey, R. W., Parker, J. C., Oliver, C., Estes, J. D., and Huebner, R. J. (1973). A spontaneous lower motor neuron disease apparently caused by indigenous type-C RNA virus in wild mice. *J. Natl. Cancer Inst.* **51:**1243–1254.

Gardner, M. B., Rasheed, S., Pal, B. K., Estes, J. D., and O'Brien, S. J. (1980). *Akvr-1,* a dominant murine leukemia virus restriction gene, is polymorphic in leukemia-prone wild mice. *Proc. Natl. Acad. Sci. U.S.A.* **77:**531–535.

Gazzinelli, R. T., Hartley, J. W., Fredrickson, T. N., Chattopadhyay, S. K., Sher, A., and Morse, H. C., III (1992). Opportunistic infections and retrovirus-induced immunodeficiency: studies of acute and chronic infections with Toxoplasma gondii in mice infected with LP-BM5 murine leukemia viruses. *Infect. Immun.* **60:**4394–4401.

Giulian, D., Li, J., Xia, L., George, J., and Rutecki, P. A. (1994). The impact of microglial-derived cytokines upon gliosis in the CNS. *Dev. Neurosci.* **16:**128–136.

Glowa, J. R., Panlilio, L. V., Brenneman, D. E., Gozes, I., Fridkin, M., and Hill, J. M. (1992). Learning impairment following intracerebral administration of the HIV envelope protein gp120 or a VIP antagonist. *Brain Res.* **570:**49–53.

Gross, L. (1951). Pathogenic properties and "vertical" transmission of the mouse leukemia agent. *Proc. Soc. Exp. Biol. Med.* **78:**342–348.

Hackett, J., Jr., Swanson, P., Leahy, D., Anderson, E. L., Sato, S., Roos, R. P., Decker, R., and Devare, S. G. (1996). Search for retrovirus in patients with multiple sclerosis. *Ann. Neurol.* **40:**805–809.

Hartley, J. W., and Rowe, W. P. (1976). Naturally occurring murine leukemia viruses in wild mice: characterization of a new "amphotropic" class. *J. Virol.* **19:**19–25.

Hasenkrug, K., Robertson, S. J., Portis, J., McAtee, F., Nishio, J., and Chesebro, B. (1996). Two separate envelope regions influence induction of brain disease by a polytropic murine retrovirus (FMCF98). *J. Virol* **70:**4825–4828.

Hoffman, P. M., Cimino, E. F., Robbins, D. S., Broadwell, R. D., Powers, J. M., and Ruscetti, S. K. (1992). Cellular tropism and localization in the rodent nervous system of a neuropathogenic variant of Friend murine leukemia virus. *Lab. Invest.* **67:**314–321.

Hoffman, P. M., Pitts, O. M., Rohwer, R. G., Gajdusek, D. C., and Ruscetti, S. K. (1982). Retrovirus antigens in brains of mice with scrapie- and murine leukemia virus-induced spongiform encephalopathy. *Infection & Immunity.* **38:**396–398.

Hoffman, P. M., Ruscetti, S. K., and Morse, H. C., III. (1981). Pathogenesis of paralysis and lymphoma associated with a wild mouse retrovirus infection. Part I. Age- and dose-related effects in susceptible laboratory mice. *J. Neuroimmunol.* **1:**275–285.

Ikeda, H., and Odaka, T. (1983). Cellular expression of murine leukemia virus gp70-related antigen on thymocytes of uninfected mice correlates with Fv-4 gene-controlled resistance to Friend leukemia virus infection. *Virology* **128:**127–139.

Jaenisch, R., Dausman, J., Cox, V., and Fan, H. (1976). Infection of developing mouse embryos with murine leukemia virus: tissue specificity and genetic transmission of the virus. *Hamatol. Bluttransfus.* **19:**341–356.

Johnson, R. T. (1998). Pathogenesis of central nervous system infections. In "Viral Infections of the Nervous System" 35–82. Lippincott-Raven, Philadelphia.

Jolicoeur, P., and Baltimore, D. (1976). Effect of Fv-1 gene product on proviral DNA formation and integration in cells infected with murine leukemia viruses. *Proc. Natl. Acad. Sci. U.S.A.* **73:**2236–2240.

Kahn, M. A., Ellison, J. A., Speight, G. J., and De Vellis, J. (1995). CNTF regulation of astrogliosis and the activation of microglia in the developing rat central nervous system. *Brain Res.* **685:**55–67.

Kai, K., and Furuta, T. (1984). Isolation of paralysis-inducing murine leukemia viruses from Friend virus passaged in rats. *J. Virol.* **50:**970–973.

Kay, D. G., Gravel, C., Pothier, F., Laperriere, A., Robitaille, Y., and Jolicoeur, P. (1993). Neurological disease induced in transgenic mice expressing the *env* gene of the Cas-Br-E murine retrovirus. *Proc. Natl. Acad. Sci. U.S.A.* **90:**4538–4542.

Klein, R. S., Williams, K. C., Alvarez-Hernandez, X., Westmoreland, S., Force, T., Lackner, A. A., and Luster, A. D. (1999). Chemokine receptor expression and signaling in macaque and human fetal neurons and astrocytes: implications for the neuropathogenesis of AIDS. *J. Immunol.* **163:**1636–1646.

Komurian-Pradel, F., Paranhos-Baccala, G., Bedin, F., Ounanian-Paraz, A., Sodoyer, M., Ott, C., Rajoharison, A., Garcia, E., Mallet, F., Mandrand, B., and Perron, H. (1999). Molecular cloning and characterization of MSRV-related sequences associated with retrovirus-like particles. *Virology* **260:**1–9.

Kozak, C. A., Gromet, N. J., Ikeda, H., and Buckler, C. E. (1984). A unique sequence related to the ecotropic murine leukemia virus is associated with the Fv-4 resistance gene. *Proc. Natl. Acad. Sci. U.S.A.* **81:**834–837.

Kreutzberg, G. W. (1996). Microglia: a sensor for pathological events in the CNS. *Trends. Neurosci.* **19:**312–318.

Kustova, Y., Espey, M. G., Sei, Y., and Basile, A. S. (1997). Regional decreases [corrected] in AMPA receptor density in mice infected with the LP-BM5 murine leukemia virus. *Neuroreport.* **8:**1243–1247.

Kustova, Y., Espey, M. G., Sung, E. G., Morse, D., Sei, Y., and Basile, A. S. (1998). Evidence of neuronal degeneration in C57B1/6 mice infected with the LP-BM5 leukemia retrovirus mixture. *Mol. Chem. Neuropathol.* **35:**39–59.

Kustova, Y., Sei, Y., Goping, G., and Basile, A. S. (1996). Gliosis in the LP-BM5 murine leukemia virus-infected mouse: an animal model of retrovirus-induced dementia. *Brain Res.* **742:**271–282.

Lampert, P. W., Gajdusek, D. C., and Gibbs, C. J., Jr. (1972). Subacute spongiform virus encephalopathies. Scrapie, Kuru and Creutzfeldt-Jakob disease: a review. *Am. J. Pathol.* **68:**626–652.

Lane, T. E., Asensio, V. C., Yu, N., Paoletti, A. D., Campbell, I. L., and Buchmeier, M. J. (1998). Dynamic regulation of α- and β-chemokine expression in the central nervous system during mouse hepatitis virus-induced demyelinating disease. *J. Immunol.* **160:**970–978.

Lawson, L. J., Perry, V. H., Dri, P., and Gordon, S. (1990). Heterogeneity in the distribution and morphology of microglia in the normal adult mouse brain. *Neuroscience* **39:**151–170.

Lawson, L. J., Perry, V. H., and Gordon, S. (1992). Turnover of resident microglia in the normal adult mouse brain. *Neuroscience* **48:**405–415.

Letourneur, F., and Klausner, R. D. (1992). A novel dileucine motif and a tyrosine-based motif independently mediate lysosomal targeting and endocytosis of CD3 chains. *Cell* **69:**1143–1157.

Li, J.-P., D'Andrea, A. D., Lodish, H. F., and Baltimore, D. (1990). Activation of cell growth by binding of Friend spleen focus-forming virus gp55 glycoprotein to the erythropoietin receptor. *Nature* **343:**762–764.

Li, J. P., and Baltimore, D. (1991). Mechanism of leukemogenesis induced by mink cell focus-forming murine leukemia viruses. *J. Virol.* **65:**2408–2414.

Limjoco, T. I., Dickie, P., Ikeda, H., and Silver, J. (1993). Transgenic Fv-4 mice resistant to Friend virus. *J. Virol.* **67:**4163–4168.

Lipton, S. A. (1992). Requirement for macrophages in neuronal injury induced by HIV envelope protein gp120. *Neuroreport.* **3:**913–915.

Lynch, W. P., Brown, W. J., Spangrude, G. J., and Portis, J. L. (1994). Microglial infection by a neurovirulent murine retrovirus results in defective processing of envelope protein and intracellular budding of virus particles. *J. Virol.* **68:**3401–3409.

Lynch, W. P., Czub, S., McAtee, F. J., Hayes, S. F., and Portis, J. L. (1991). Murine retrovirus-induced spongiform encephalopathy: productive infection of microglia and cerebellar neurons in accelerated CNS disease. *Neuron* **7:**365–379.

Lynch, W. P., Robertson, S. J., and Portis, J. L. (1995). Induction of focal spongiform neurodegeneration in developmentally restricted mice by implantation of murine retrovirus-infected microglia. *J. Virol.* **69:**1408–1419.

Lynch, W. P., Sharpe, A. H., and Snyder, E. Y. (1999). Neural stem cells as engraftable packaging lines can mediate gene delivery to microglia: evidence from studying retroviral env-related neurodegeneration. *J. Virol.* **73:**6841–6851.

Lynch, W. P., Snyder, E. Y., Qualtiere, L., Portis, J. L., and Sharpe, A. H. (1996). Late virus replication events in microglia are required for neurovirulent retrovirus-induced spongiform neurodegeneration: Evidence from neural progenitor-derived chimeric mouse brains. *J. Virol.* **70:**8896–8907.

Masuda, M., Hanson, C. A., Alvord, W. G., Hoffman, P. M., and Ruscetti, S. K. (1996). Effects of subtle changes in the SU protein of ecotropic murine leukemia virus on its brain capillary endothelial cell tropism and interference properties. *Virology* **215:**142–151.

Masuda, M., Hanson, C. A., Dugger, N. V., Robbins, D. S., Wilt, S. G., Ruscetti, S. K., and Hoffman, P. M. (1997a). Capillary endothelial cell tropism of PVC-211 murine leukemia virus and its application for gene transduction. *J. Virol.* **71:**6168–6173.

Masuda, M., Hoffman, P. M., and Ruscetti, S. K. (1993). Viral determinants that control the neuropathogenicity of PVC-211 murine leukemia virus in vivo determine brain capillary endothelial cell tropism of the virus in vitro. *J. Virol.* **67:**4580–4587.

Masuda, M., Remington, P. M., Hoffman, P. M., and Ruscetti, S. K. (1992). Molecular characterization of a neuropathogenic and nonerythroleukemogenic variant of Friend murine leukemia virus PVC-211. *J. Virol.* **66:**2798–2806.

Masuda, M., Ruscetti, S. K., and Hoffman, P. M. (1997b). Molecular mechanism for retroviral neuropathogenesis: possible involvement of capillary endothelial cells. *Leukemia* **11 Suppl 3:**233–5:233–235.

McKnight, A. J., and Gordon, S. (1998). The EGF-TM7 family: unusual structures at the leukocyte surface. *J. Leukoc. Biol.* **63:**271–280.

Miller, A. D. and Verma, I. M. (1984). Two base changes restore infectivity to a noninfectious molecular clone of Moloney murine leukemia virus (pMLV-1). *J. Virol.* **49:**214–222.

Miller, D. G., Edwards, R. H., and Miller, A. D. (1994). Cloning of the cellular receptor for amphotropic murine retroviruses reveals homology to that for gibbon ape leukemia virus. *Proc. Natl. Acad. Sci. U.S.A.* **91:**78–82.

Mosier, D. E., Yetter, R. A., and Morse, H. C., III. (1985). Retroviral induction of acute lymphoproliferative disease and profound immunosuppression in adult C57BL/6 mice. *J. Exp. Med.* **161**:766–784.

Munk, C., Lohler, J., Prassolov, V., Just, U., Stockschlader, M., and Stocking, C. (1997). Amphotropic murine leukemia viruses induce spongiform encephalomyelopathy. *Proc. Natl. Acad. Sci. U.S.A.* **94**:5837–5842.

Munk, C., Thomsen, S., Stocking, C., and Lohler, J. (1998). Murine leukemia virus recombinants that use phosphate transporters for cell entry induce similar spongiform encephalomyelopathies in newborn mice. *Virology* **252**:318–323.

Nagra, R. M., Heyes, M. P., and Wiley, C. A. (1994). Viral load and its relationship to quinolinic acid, TNF alpha, and IL-6 levels in the CNS of retroviral infected mice. *Mol. Chem. Neuropathol.* **22**:143–160.

Neil, J. C., Smart, J. E., Hayman, M. J., and Jarrett, O. (1980). Polypeptides of feline leukemia virus: a glycosylated *gag*-related protein is released into culture fluids. *Virology* **105**:250–253.

Odaka, T., Ikeda, H., Yoshikura, H., Moriwaki, K., and Suzuki, S. (1981). Fv-4: gene controlling resistance to NB-tropic Friend murine leukemia virus. Distribution in wild mice, introduction into genetic background of BALB/c mice, and mapping of chromosomes. *J. Natl. Cancer Inst.* **67**:1123–1127.

Officer, J. E., Tecson, N., Estes, J. D., Fontanilla, E., Rongey, R. W., and Gardner, M. B. (1973). Isolation of a neurotropic type C virus. *Science* **181**:945–947.

Oldstone, M. B. A., Jensen, F., Dixon, F. J., and Lampert, P. W. (1980). Pathogenesis of the slow disease of the central nervous system associated with wild mouse virus. II. Role of virus and host gene products. *Virology* **107**:180–193.

Oldstone, M. B. A., Jensen, F., Elder, J., Dixon, F. J., and Lampert, P. W. (1983). Pathogenesis of the slow disease of the central nervous system associated with wild mouse virus. III. Role of input virus and MCF recombinants in disease. *Virology* **128**:154–165.

Oldstone, M. B. A., Lampert, P. W., Lee, S., and Dixon, F. J. (1977). Pathogenesis of the slow disease of the central nervous system associated with WM 1504 E virus. I. Relationship of strain susceptibility and replication to disease. *Am. J. Pathol.* **88**:193–212.

O'Neill, R. R., Hartley, J. W., Repaske, R., and Kozak, C. A. (1987). Amphotropic proviral envelope sequences are absent from the Mus germ line. *J. Virol.* **61**:2225–2231.

Paquette, Y., Hanna, Z., Savard, P., Brousseau, R., Robitaille, Y., and Jolicoeur, P. (1989). Retrovirus-induced murine motor neuron disease: mapping the determinant of spongiform degeneration within the envelope gene. *Proc. Natl. Acad. Sci. U.S.A.* **86**:3896–3900.

Park, B. H., Lavi, E., Stieber, A., and Gaulton, G. N. (1994a). Pathogenesis of cerebral infarction and hemorrhage induced by a murine leukemia virus. *Lab. Invest.* **70**:78–85.

Park, B. H., Matuschke, B., Lavi, E., and Gaulton, G. N. (1994b). A point mutation in the *env* gene of a murine leukemia virus induces syncytium formation and neurologic disease. *J. Virol.* **68**:7516–7524.

Park, B. H., Lavi, E., Blank, K. J., and Gaulton, G. N. (1993). Intracerebral hemorrhages and syncytium formation induced by endothelial cell infection with a murine leukemia virus. *J. Virol.* **67**:6015–6024.

Perron, H., Garson, J. A., Bedin, F., Beseme, F., Paranhos-Baccala, G., Komurian-Pradel, F., Mallet, F., Tuke, P. W., Voisset, C., Blond, J. L., Lalande, B., Seigneurin, J. M., and Mandrand, B. (1997). Molecular identification of a novel retrovirus repeatedly isolated from patients with multiple sclerosis. The Collaborative Research Group on Multiple Sclerosis. *Proc. Natl. Acad. Sci. U.S.A.* **94**:7583–7588.

Pillemer, E. A., Kooistra, D. A., Witte, O. N., and Weissman, I. L. (1986). Monoclonal antibody to the amino-terminal L sequence of murine leukemia virus glycosylated *gag* polyproteins demonstrates their unusual orientation in the cell membrane. *J. Virol.* **57:**413–421.

Pitts, O. M., Powers, J. M., Bilello, J. A., and Hoffman, P. M. (1987). Ultrastructural changes associated with retroviral replication in central nervous system capillary endothelial cells. *Lab. Invest.* **56:**401–409.

Plagemann, P. G., and Moennig, V. (1992). Lactate dehydrogenase-elevating virus, equine arteritis virus, and simian hemorrhagic fever virus: a new group of positive-strand RNA viruses. *Adv. Virus Res.* **41:**99–192.

Portis, J. L., Czub, S., Garon, C. F., and McAtee, F. J. (1990). Neurodegenerative disease induced by the wild mouse ecotropic retrovirus is markedly accelerated by long terminal repeat and *gag-pol* sequences from nondefective Friend murine leukemia virus. *J. Virol.* **64:**1648–1656.

Portis, J. L., Czub, S., Robertson, S. J., McAtee, F. J., and Chesebro, B. (1995). Characterization of a Neurologic Disease Induced by a Polytropic Murine Retrovirus: Evidence for Differential Targeting of Ecotropic and Polytropic Viruses in the Brain. *J. Virol.* **69:**8070–8075.

Portis, J. L., Fujisawa, R., and McAtee, F. J. (1996). The glycosylated gag protein of MuLV is a determinant of neuroinvasiveness: Analysis of second site revertants of a mutant MuLV lacking expression of this protein. *Virology* **226:**384–392.

Portis, J. L., McAtee, F. J., and Hayes, S. F. (1987). Horizontal transmission of murine retroviruses. *J. Virol.* **61:**1037–1044.

Portis, J. L., Perryman, S., and McAtee, F. J. (1991). The R-U5-5' leader sequence of neurovirulent wild mouse retrovirus contains an element controlling the incubation period of neurodegenerative disease. *J. Virol.* **65:**1877–1883.

Portis, J. L., Spangrude, G. J., and McAtee, F. J. (1994). Identification of a sequence in the unique 5' open reading frame of the gene encoding glycosylated gag which influences the incubation period of neurodegenerative disease induced by a murine retrovirus. *J. Virol.* **68:**3879–3887.

Poulsen, D. J., Favara, C., Snyder, E. Y., Portis, J., and Chesebro, B. (1999). Increased neurovirulence of polytropic mouse retroviruses delivered by inoculation of brain with infected neural stem cells. *Virology* **263:**23–29.

Poulsen, D. J., Robertson, S. J., Favara, C. A., Portis, J. L., and Chesebro, B. W. (1998). Mapping of a neurovirulence determinant within the envelope protein of a polytropic murine retrovirus: induction of central nervous system disease by low levels of virus. *Virology* **248:**199–207.

Prats, A. C., De Billy, G., Wang, P., and Darlix, J. L. (1989). CUG initiation codon used for the synthesis of a cell surface antigen coded by the murine leukemia virus. *J. Mol. Biol.* **205:**363–372.

Rasheed, S., Gardner, M. B., and Chan, E. (1976). Amphotropic host range of naturally occuring wild mouse leukemia viruses. *J. Virol.* **19:**13–18.

Rasheed, S., Gardner, M. B., and Lai, M. M. (1983). Isolation and characterization of new ecotropic murine leukemia viruses after passage of an amphotropic virus in NIH Swiss mice. *Virology* **130:**439–451.

Rein, A. (1982). Interference grouping of murine leukemia viruses: A distinct receptor for the MCF-recombinant viruses in mouse cells. *Virology* **120:**251–257.

Robertson, S. J. Murine retrovirus-induced neurologic disease: Characterization of host and viral factors involved in neurovirulence, (Ph.D Thesis). University of Montana, Missoula, Montana, 1997.

Robertson, S. J., Hasenkrug, K. J., Chesebro, B., and Portis, J. L. (1997). Neurologic disease induced by polytropic murine retroviruses: neurovirulence determined by efficiency of spread to microglial cells. *J. Virol.* **71:**5287–5294.

Rosenberg, N. and Jolicoeur, P. (1997). Retroviral pathogenesis. In " Retroviruses" (Coffin, J., Hughes, S. H., and Varmus, H.), 475–585. Cold Spring Harbor Lavnoratories Press.

Ryder, E. F., Snyder, E. Y., and Cepko, C. L. (1990). Establishment and characterization of multipotent neural cell lines using retrovirus vector-mediated oncogene transfer. *J. Neurobiol.* **21:**356–375.

Sandoval, I. V., and Bakke, O. (1994). Targeting of membrane proteins to endosomes and lysosomes. *Trends in Cell Biology* **4:**292–297.

Sarzotti, M., Robbins, D. S., and Hoffman, P. M. (1996). Induction of protective CTL responses in newborn mice by a murine retrovirus. *Science* **271:**1726–1728.

Sauder, C., and de la Torre, J. C. (1999). Cytokine expression in the rat central nervous system following perinatal Borna disease virus infection. *J. Neuroimmunol.* **96:**29–45.

Sauer, B. (1998). Inducible gene targeting in mice using the Cre/lox system. *Methods* **14:**381–392.

Sei, Y., Arora, P. K., Skolnick, P., and Paul, I. A. (1992). Spatial learning impairment in a murine model of AIDS. *FASEB Journal.* **6:**3008–3013.

Sitbon, M., Sola, B., Evans, L., Nishio, J., Hayes, S. F., Nathanson, K., Garon, C. F., and Chesebro, B. (1986). Hemolytic anemia and erythroleukemia, two distinct pathogenic effects of Friend MuLV: mapping of the effects to different regions of the viral genome. *Cell* **47:**851–859.

Slye, M. (1914). The incidence of inheritability of spontaneous tumors in mice. *J. Med. Res.* **30:**281–298.

Snyder, E. Y., and Wolfe, J. H. (1996). Central nervous system cell transplantation: a novel therapy for storage diseases? *Curr. Opin. Neurol.* **9:**126–136.

Snyder, E. Y., Yoon, C., Flax, J. D., and Macklis, J. D. (1997). Multipotent neural precursors can differentiate toward replacement of neurons undergoing targeted apoptotic degeneration in adult mouse neocortex. *Proc. Natl. Acad. Sci. U.S.A.* **94:**11663–11668.

Spain, B. H., Koo, D., Ramakrishnan, M., Dzudzor, B., and Colicelli, J. (1995). Truncated forms of a novel yeast protein suppress the lethality of a G protein alpha subunit deficiency by interacting with the beta subunit. *J. Biol. Chem.* **270:**25435–25444.

Speck, N. A., Renjifo, B., and Hopkins, N. (1990). Point mutations in the Moloney murine leukemia virus enhancer identify a lymphoid-specific viral core motif and 1,3-phorbol myristate acetate-inducible element. *J. Virol.* **64:**543–550.

Stoye, J. P. (1999). The pathogenic potential of endogenous retroviruses: a sceptical view. *Trends Microbiol.* **7:**430.

Stoye, J. P., and Coffin, J. M. (1987). The four classes of endogenous murine leukemia virus: structural relationships and potential for recombination. *J. Virol.* **61:**2659–2669.

Strong, L. C. (1935). The establishment of the C3H inbred strain of mice for the study of spontaneous carcinoma of the mammary gland. *Genetics* **20:**586–591.

Stroop, W. G., and Brinton, M. A. (1983). Mouse strain-specific central nervous system lesions associated with lactate dehydrogenase-elevating virus infection. *Lab. Invest.* **49:**334–345.

Swarz, J. R., Brooks, B. R., and Johnson, R. T. (1981). Spongiform polioencephalomyelopathy caused by a murine retrovirus. II. Ultrastructural localization of virus replication and spongiform changes in the central nervous system. *Neuropathol. Appl. Neurobiol.* **7:**365–380.

Szurek, P. F., Floyd, E., Yuen, P. H., and Wong, P. K. (1990a). Site-directed mutagenesis of the codon for Ile-25 in gPr80env alters the neurovirulence of ts1, a mutant of Moloney murine leukemia virus TB. *J. Virol.* **64:**5241–5249.

Szurek, P. F., Yuen, P. H., Ball, J. K., and Wong, P. K. (1990b). A Val-25-to-Ile substitution in the envelope precursor polyprotein, gPr80env, is responsible for the temperature sensitivity, inefficient processing of gPr80env, and neurovirulence of ts1, a mutant of Moloney murine leukemia virus TB. *J. Virol.* **64:**467–475.

Szurek, P. F., Yuen, P. H., Jerzy, R., and Wong, P. K. Y. (1988). Identification of point mutations in the envelope gene of Moloney murine leukemia virus TB temperature-sensitive paralytogenic mutant ts1: molecular determinants for neurovirulence. *J. Virol.* **62:**357–360.

Tailor, C. S., Nouri, A., Lee, C. G., Kozak, C., and Kabat, D. (1999). Cloning and characterization of a cell surface receptor for xenotropic and polytropic murine leukemia viruses [see comments]. *Proc. Natl. Acad. Sci. U.S.A.* **96:**927–932.

Tanabe, S., Heesen, M., Berman, M. A., Fischer, M. B., Yoshizawa, I., Luo, Y., and Dorf, M. E. (1997). Murine astrocytes express a functional chemokine receptor. *J. Neurosci.* **17:**6522–6528.

Taylor, R. M., and Snyder, E. Y. (1997). Widespread engraftment of neural progenitor and stemlike cells throughout the mouse brain. *Transplant. Proc.* **29:**845–847.

Temann, U. A., Geba, G. P., Rankin, J. A., and Flavell, R. A. (1998). Expression of interleukin 9 in the lungs of transgenic mice causes airway inflammation, mast cell hyperplasia, and bronchial hyperresponsiveness. *J. Exp. Med.* **188:**1307–1320.

Toggas, S. M., Masliah, E., Rockenstein, E. M., Rall, G. F., Abraham, C. R., and Mucke, L. (1994). Central nervous system damage produced by expression of the HIV-1 coat protein gp120 in transgenic mice. *Nature* **367:**188–193.

Wang, H., Kavanaugh, M. P., North, R. A., and Kabat, D. (1991). Cell-surface receptor for ecotropic murine retroviruses is a basic amino-acid transporter. *Nature* **352:**729–731.

Wang, H., Klamo, E., Kuhmann, S. E., Kozak, S. L., Kavanaugh, M. P., and Kabat, D. (1996). Modulation of ecotropic murine retroviruses by N-linked glycosylation of the cell surface receptor/amino acid transporter. *J. Virol.* **70:**6884–6891.

Wong, P. K. Y., Soong, M. M., MacLeod, R., Gallick, G. E., and Yuen, P. H. (1983). A group of temperature-sensitive mutants of Moloney leukemia virus which is defective in cleavage of *env* precursor polypeptide in infected cells also induces hind-limb paralysis in newborn CFW/D mice. *Virology* **125:**513–518.

Yang, Y. L., Guo, L., Xu, S., Holland, C. A., Kitamura, T., Hunter, K., and Cunningham, J. M. (1999). Receptors for polytropic and xenotropic mouse leukaemia viruses encoded by a single gene at Rmc1. *Nat. Genet.* **21:**216–219.

Yu, Y. E., Choe, W., Zhang, W., Stoica, G., and Wong, P. K. (1997). Development of pathological lesions in the central nervous system of transgenic mice expressing the *env* gene of ts1 Moloney murine leukemia virus in the absence of the viral *gag* and *pol* genes and viral replication. *J. Neurovirol.* **3:**274–282.

Zachary, J. F., Knupp, C. J., and Wong, P. K. Y. (1986). Noninflammatory spongiform polioencephalomyelopathy caused by a neurotropic temperature-sensitive mutant of Moloney murine leukemia virus TB. *Am. J. Pathol.* **124:**457–468.

Zavorotinskaya, T., and Albritton, L. M. (1999). A hydrophobic patch in ecotropic murine leukemia virus envelope protein is the putative binding site for a critical tyrosine residue on the cellular receptor. *J. Virol.* **73:**10164–10172.

Zhong, G., Romagnoli, P., and Germain, R. N. (1997). Related leucine-based cytoplasmic targeting signals in invariant chain and major histocompatibility complex class II molecules control endocytic presentation of distinct determinants in a single protein. *J. Exp. Med.* **185:**429–438.

PSEUDORABIES VIRUS NEUROINVASIVENESS: A WINDOW INTO THE FUNCTIONAL ORGANIZATION OF THE BRAIN

J. Patrick Card

Department of Neuroscience
University of Pittsburgh
Pittsburgh, Pennsylvania 15260

I. Viruses and the Nervous System

The tropism and invasiveness of neurotropic viruses have been increasingly exploited in studies of the nervous system in order to define connections among ensembles of neurons. The allure of this experimental approach is directly related to the ability of these pathogens to invade neurons and to produce infectious progeny that subsequently pass through neuronal connections to infect other neurons within a circuit. In effect, the method takes advantage of the natural biology of the virus to produce a signal that is amplified, through viral replication, by each neuron in the circuit. Because no other neuronal tracer possesses these attributes, the viral transneuronal tracing method has become a popular tool for defining the functional architecture of the brain. Advances in our understanding of the mechanisms that contribute to the replication and dissemination of virus in the brain have also been an important component of this work. In fact, acceptance of this method by the neurobiology community can be traced directly to mechanistic studies that have established the specificity of viral transport through circuits and the nonneuronal

39

responses to viral infection that contribute to this specificity. These mechanistic studies of viral invasiveness of the brain, in turn, have revealed unique insights into aspects of the life cycle of these viruses that could not be defined by using *in vitro* approaches. Thus, analysis of the invasive properties of neurotropic viruses has fostered experimental approaches that have been of tremendous value to both neuroscientists and virologists.

Recognition that neurotropic viruses might be useful for mapping neuronal circuitry emerged from early studies that demonstrated that virus-induced neuropathology occurred in system-specific patterns. For example, Goodpasture and Teague (Goodpasture and Teague, 1923) reported that intravitreal injection of herpes simplex virus (HSV) produced cytopathology among cell groups along the course of the centrally projecting axons of the optic nerve. This and related literature, summarized by Howe and Bodian (1942), was instrumental in initiating investigations that established the axon as the conduit through which neurotropic viruses invade the nervous system, a finding that has obvious import for viral tracing studies (see Kristensson [1996] for a review). However, the virus-induced pathology demonstrated in this and subsequent work also revealed the potential confounds that could undermine the usefulness of neurotropic viruses for circuit definition and emphasized the importance of characterizing the mechanisms underlying viral invasion and the brain response to infection. Indeed, the monograph published by Howe and Bodian (1942) made the important point that "an understanding of the virus–host reaction must depend as much upon a knowledge of the physiological mechanisms of the host as upon the determination of the nature and behavior of the virus." Recognition of this important principle has been instrumental in establishing the viral transneuronal tracing method as an effective tool for analysis of brain organization. Additionally, it defined the importance of multidisciplinary analysis that has increasingly become the cornerstone of neurovirology and has established this field at the intersection of neurobiology, virology, and immunology.

The goal of this chapter is to demonstrate how the use of neurotropic viruses for analysis of brain organization has emerged from an integrated, multidisciplinary examination of the mechanisms that determine viral invasiveness. The focus will be on studies that have employed neurotropic alpha herpesviruses, since these studies form the bulk of the literature that has established the method. Particular emphasis will be placed on the experimental use of pseudorabies virus (PRV), for two reasons. First, attenuated strains of this virus have been widely used to study the organization of a variety of systems. Second, and most important for the purposes of this chapter, there is an ever-expanding literature in which well-documented models of PRV invasiveness have been used to probe the role of virally encoded proteins in viral invasiveness and virulence. Nevertheless, it is important to emphasize that work with HSV continues to be instrumental in establishing the utility of viral transneuronal tracing (Kuypers and Ugolini, 1990; Strick and Card, 1992), and that other neurotropic viruses, particularly rabies virus (Dolivo, Kucera, and Bommeli, 1982; Kucera *et al.*, 1985), have also proven to be quite useful for this purpose.

II. Neuronal Architecture and Neurotropic Viruses

Neurons exhibit an enormous functional diversity and this is reflected in the tremendous morphological diversity of different classes of neurons. Nevertheless, fundamental aspects of cellular structure are common to all neurons and some of these morphological features offer distinct advantages to a neurotropic pathogen. First and foremost, the polarized morphology and intercellular connections that provide the most fundamental means of information exchange in the nervous system are ideal for dissemination of infectious progeny. This polarity is reflected in the two distinct classes of processes, axons and dendrites, which emanate from the perikarya of most neurons, and which provide the principal substrate for cellular communication. A single axon, or axis cylinder, emerges from each cell and contacts other neurons and their dendrites through specialized appositions known as synapses. Different classes of neurons will exhibit varying degrees of axonal branching and it is not uncommon for a single neuron to contact literally thousands of other neurons. In fact, the complex branching patterns of axons characteristically include the formation of "collaterals" that allow a single axon to innervate multiple spatially separate regions of the brain. Finally, the intracellular machinery of neurons is specialized to transport proteins, vesicles, and organelles bidirectionally between the perikarya of neurons and the numerous varicose endings that constitute the projection field of an axon. Thus, the polarized morphology and cellular machinery that support synaptic communication between groups of neurons provide an ideal means for a pathogen to disseminate infectious progeny to other permissive hosts that are often in distant areas of the brain. This spatial separation of synaptically linked neurons has obvious advantages for a pathogen in that it allows progeny virus to elude the defense responses marshaled by the nervous system against infection.

An illustration of the diversity of neuronal morphology is shown in Fig. 1 (see also color section). Although Fig. 1 does not capture the

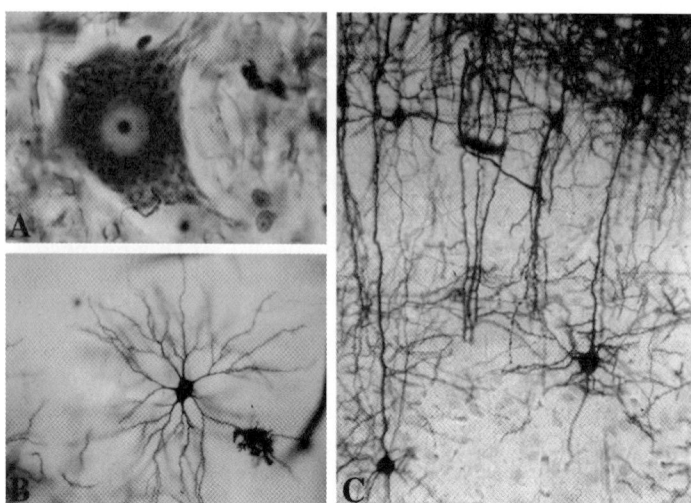

FIG 1. [For color reproduction, see color section.] The characteristic morphological diversity of neurons in the central nervous system as revealed by Nissl stains (A) and by silver impregnation methods (B, C) is illustrated. The differential staining of ribosomes, rough endoplasmic reticulum, and RNA by the basophilic Nissl stain provides a beautiful presentation of neuronal perikarya but does not reveal the extensive process networks characteristically associated with the neuron. Nevertheless, this is the routine stain used to demonstrate cytoarchitecture in histological preparations of sections through the brain (see Fig. 3). Silver impregnation, or Golgi preparations, provide an exceptional illustration of the polarized nature of neurons. Note the differences in the organization of the dendritic arbors of the local circuit neuron (B) and the cortical pyramidal neurons (C). Whereas the dendrites of the local circuit neuron (B) extend radially from the cell body, the dendrites arising from the pyramidal neurons are polarized and far more extensive.

tremendous variability of cell size and process branching of neurons (see Parent [1996] for a detailed consideration of this topic), it does provide an illustration of aspects of neuronal morphology that are exploited by alpha herpesviruses to gain access to the brain. Additionally, it provides a framework for defining the nomenclature that is commonly used to describe the direction of viral transport within a neuron or through a polysynaptic circuit.

Movement in the nervous system, whether it is the propagation of action potentials, the trafficking of proteins, or the movement of virions, is defined in relation to the perikarya (cell bodies) of neurons. In this nomenclature, movement away from the cell body along the axon is described as "anterograde" and movement from axon terminals to the cell body is referred to as "retrograde." In most instances, the direction of viral transport is readily defined by these terms since the majority of neurons exhibit only minor variations from the polarized morphology illustrated in Fig. 1. However, some confusion can arise when one is dealing with sensory neurons of the peripheral nervous system. These cells are a natural conduit for alpha herpesvirus invasion in the wild and are characterized by a unique morphology that differs substantially from the majority of neurons. Known as "pseudounipolar" neurons, these cells lack dendrites and give rise to a single axon that divides into two branches that travel to different targets. The cell bodies also reside in ganglia outside the confines of the central nervous system (CNS). One of the axonal branches travels to the peripheral target (e.g., mucosa of the nasal cavity or a specific domain of the skin) and is responsible for sensory transduction at that site. The other branch travels to the central nervous system, where it terminates within a specific region of the spinal cord or brainstem. Understanding the direction implied by anterograde or retrograde viral transport in these sensory neurons depends on the stage of the viral life cycle. For example, virion invasion of a peripheral axon innervating the cornea will lead to retrograde transport of the capsid to the parent cell body in the trigeminal ganglion. Replication and spread of virus in the same neuron will lead to anterograde transport of progeny virus to either the cornea or brainstem. Thus, movement in the portion of the axon that innervates the corneas can be described as either anterograde or retrograde, depending on the context of the discussion.

Studies using alpha herpesviruses for circuit analysis exploit the natural tropism of these viruses for neurons but employ inoculation procedures that introduce the virus to the brain through routes that are generally not encountered in the wild. In the wild, the primary route of viral invasion is through the pseudounipolar neurons that provide sensory innervation to mucous membranes or the dermis (Gustafson, 1975; Whitley, 1990). Animal models employing this route of invasion, which include the well-known corneal scarification method for infection of sensory neurons in the trigeminal ganglion, have been particularly valuable for the study of viral latency and reactivation. However, they differ, in a very important respect, from the procedures commonly employed for neuronal tract tracing. Specifically, the inoculation procedures used in tracing studies are more invasive and introduce virus to the central nervous system through the peripherally projecting processes of neurons whose cell bodies reside in the brain, or by direct injection of virus into brain parenchyma. Importantly, these routes of invasion generally produce productive replication and spread of infectious progeny through neuronal circuits (Enquist *et al.*, 1999). Nevertheless, the direction of viral spread and the magnitude of infection produced by these inoculation procedures are heavily influenced by both the architecture of the circuitry in the area of injection and the strain of virus employed in the experiment.

III. IMPORTANCE OF USING WELL-CHARACTERIZED
STRAINS OF VIRUS

Alpha herpesviruses are neurotropic pathogens that belong to the family *Herpesviridae*. These enveloped DNA viruses have the capacity to infect and replicate within a variety of cells but have a particular tropism for neurons (Enquist *et al.*, 1999). PRV virions exhibit a structure characteristic of all members of this family of viruses (Fig. 2). The viral genome is sequestered within a capsid that is composed of virally encoded proteins and that in turn, is contained within an envelope acquired from the host cell. It is not the purpose of this chapter to provide a detailed consideration of the organization of the PRV genome, as this is available from a number of other excellent sources (Enquist *et al.*, 1999; Mettenleiter, 1991; Mettenleiter, 1995). However, it is useful to provide a basic overview of the genomic organization of PRV as a prelude to considering how mutations and deletions of the PRV genome can alter invasiveness.

Alpha herpesvirus envelope protein genes have been divided into "essential" and "nonessential" categories that are based on their influence on viral replication and invasiveness in cultured cells (Mettenleiter, 1991). This large body of literature is beyond the scope of this chapter, but it is important to note that genes designated as nonessential for viral replication and spread *in vitro* have been shown to subserve important roles *in vivo*. Fig. 2 illustrates the basic organization of the PRV genome along with that of several PRV strains commonly used in neuronal circuit analysis. Both classes of viral envelope genes are distributed within unique long and unique short regions of the viral genome, along with other genes that are necessary for viral replication and assembly. Most viral tracing studies utilize an attenuated strain of virus that was developed as a vaccine in the early 1960s. This strain, known as PRV-Bartha (Bartha, 1961), harbors a number of mutations and deletions that distinguish it from wild-type virus. In particular, a large deletion in the unique short region of the genome eliminates genes implicated in PRV tropism and virulence (Lomniczi *et al.*, 1987). Although available evidence supports the conclusion that

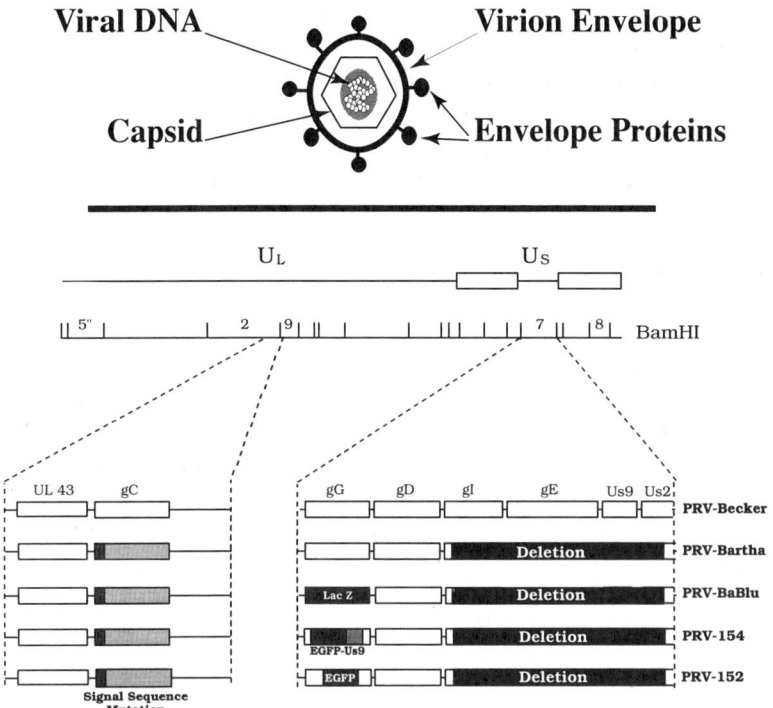

FIG 2. The basic structure of alpha herpesvirus virions, along with the genomic organization of strains of pseudorabies virus commonly used for neuronal circuit analysis, are illustrated. The attenuated vaccine strain of PRV known as Bartha has been used most extensively for circuit analysis. This is principally due to the low virulence and reduction of cytopathogenicity of this strain as compared to that of wild-type viruses such as PRV-Becker. PRV-Bartha has also been used as the parental virus to construct recombinant viruses that are useful for analysis of complex circuits in dual infection paradigms. See the text for further details and Fig. 7 for an illustration of a dual-infection paradigm.

PRV-Bartha has the capability to replicate in all CNS neurons, the deletions and mutations present in the PRV-Bartha genome do restrict the direction of virion transport through neuronal circuits. Whereas wild-type strains of PRV move bidirectionally (anterogradely and retrogradely) through CNS circuitry, PRV-Bartha is restricted to retrograde transsynaptic transport through the majority of CNS neurons (Card, Levitt, and Enquist, 1998). Anterograde transport of PRV-Bartha through a subpopulation of retinal ganglion cells in the visual system is an important exception to this rule and will be discussed in

greater detail later in this chapter. For the purpose of the present discussion it is useful to simply make two points. First, *in vivo* models
provide powerful means of assessing the functions of specific virally
encoded gene products. This is apparent in several studies over the
past decade that have utilized well-characterized *in vivo* models to
demonstrate specific roles for viral genes in specific aspects of viral
invasion (e.g., first-order invasiveness versus subsequent transsynaptic spread). Second, because of the complex interconnectivity of the
nervous system, the restricted retrograde movement of PRV-Bartha
and related strains through neuronal circuitry is an essential aspect of
the usefulness of this virus for neuronal circuit analysis. From the perspective of circuit analysis, this point cannot be overemphasized since
the complexity of connections in the brain would be essentially indeci-
pherable if virus were transported bidirectionally through a circuit.
This was eloquently demonstrated in the pioneering study of Strick
and colleagues that used directionally selective strains of HSV to
probe the organization of motor circuitry in brains of nonhuman primates (Zemanick, Strick, and Dix, 1991). That study revealed differences in the direction of transport of the McIntyre B and H129 strains
of HSV type 1 that have been confirmed and extended in a number of
subsequent studies (Barnett, Cassell, and Perlman, 1993; Barnett,
Jacobsen, and Evans, 1994; Hoover and Strick, 1993; Hoover and
Strick, 1999; Middleton and Strick, 1994; Middleton and Strick, 1996;
Sun, Cassell, and Perlman, 1996).

Collectively, these observations emphasize the importance of using
well-characterized strains of virus for circuit analysis. They also identify issues relating to the natural biology of alpha herpesvirus replication and invasion that can exert profound influences on the outcome of
infection within a circuit.

IV. REPLICATION AND INTRACELLULAR SPREAD OF PSEUDORABIES
VIRUS IN THE BRAIN

Intracellular assembly and trafficking of infectious progeny are of
fundamental importance to the spread of virus through a neuronal circuit. Over the past decade, directed analysis of this issue within the
context of viral invasion of the nervous system has identified three
aspects of viral invasiveness that have direct bearing on the interpretation of patterns of infection produced in viral tracing studies. First,
it is important to recognize that virion affinities for proteins in the
extracellular matrix, as well as differential affinity for cellular profiles

(e.g., axons versus perikarya) at the injection site, determine the extent of viral invasion of neural circuitry. Thus, the architecture of the region of injection will exert a profound influence on the outcome of infection. Second, the intracellular pathways of virion replication and assembly have an important influence on the subsequent spread of virus through a polysynaptic circuit. This is particularly important from the standpoint of trafficking of virions through the somatodendritic compartment since axonal contacts from different populations of neurons are often differentially distributed on neuronal somata and dendrites of infected cells. Third, the sites of virion egress from infected neurons are fundamental to determining the specificity of transneuronal passage of virus through a circuit. These issues form the basis for the following discussion of the viral life cycle in the nervous system.

A. Virion Affinities and Viral Invasiveness

In viral tracing studies the pattern of infection produced in any experiment is directly dependent on two factors: virion affinities and zone of diffusion from the injection site. In effect, virion affinities will determine the cellular profiles that become infected and, by extension, will determine the zone of virus diffusion. However, since tracing of neuronal circuits is dependent on eliciting a productive infection in permissive neuronal profiles, one is immediately presented with the problem of defining this zone. This is a difficult proposition under the best of circumstances and has become even more challenging in the light of data suggesting that not all neurons that sequester virus will enter into a productive state of viral replication. In this context, it has been demonstrated that neurons with sparse projections to an area of infection may not accumulate sufficient virus to induce a productive infection, particularly if they contribute to a dense neuropil with which they have to compete for virus. This was directly demonstrated in studies that examined the effect of viral concentration on the uptake and spread of PRV following injection of virus into rat striatum or thalamus (Card, Enquist, and Moore, 1999; O'Donnell et al., 1997). In the studies of viral invasion of striatal circuitry, neurons that densely innervate the area of injection (e.g., dopaminergic neurons of the midbrain) were shown to be resistant to reduction of the concentration of injected virus. That is, a two log unit reduction in viral concentration exerted only minimal effects upon the ability of PRV to replicate in these neurons. In contrast, neurons that contributed sparse axonal projections to the same area of injection were extremely susceptible to reductions in viral concentration, in some cases failing to replicate virus. Since the same neurons are capable of being infected when virus is injected into other areas in which the neurons maintain denser projection fields, this defect cannot be attributed to the cells' refractoriness to infection. Thus, the data are consistent with the interpretation that neurons must accumulate a minimum concentration of virus to elicit a productive infection and that, at least in this circuitry, the ability to achieve that threshold depends upon the density of the terminal field. This suggests that determining the "effective zone of viral uptake" depends upon the amount of virus that is injected as well as the cellular constituents at the injection site.

The aforementioned data are also consistent with work by Vahlne and colleagues with HSV (Vahlne *et al.*, 1978; Vahlne *et al.*, 1980), and with work by Marchand and Schwab (1987), using PRV—all the data demonstrate that these alpha herpesviruses have highest affinity for axon terminals, and lesser affinities for other neuronal compartments and glia. Subsequent *in vivo* analyses have provided further support for this conclusion. For example, at short postinoculation intervals after injection of PRV into the striatum, only scattered infected neurons are present at the area of injection, and these cells exhibit the morphological features of local circuit neurons rather than the more numerous population of neurons that project to distant sites in the brain (Card, Enquist, and Moore, 1999). In the same animals, numerous infected neurons are present in areas that project to the striatum. This occurs in spite of the fact that the projection neurons are interspersed among the aggregations of axon terminals and have equal access to virus. It is also noteworthy that the striatal projection neurons remain free of infection through the longest postinoculation survival intervals. The results achieved in the striatum contrast with those produced by injection of PRV into a hypothalamic nucleus (the suprachiasmatic nucleus, or SCN) whose neurons project to distant sites but also give rise to recurrent collaterals that terminate among SCN neurons. Injection of PRV into the SCN leads to an immediate infection of all SCN neurons as well as an infection of distant neurons that project to this cell group. Since both areas contain a prominent neuropil (i.e., a dense accumulation of axon terminals) and all infected neurons at short survival had axons that contributed to this neuropil, these data provide a dramatic illustration of the preferential affinity of PRV for axon terminals.

In spite of the demonstrable affinity of PRV and HSV for axon terminals, early *in vivo* studies of viral invasiveness of the visual system demonstrated that PRV and HSV can infect neurons by invading their

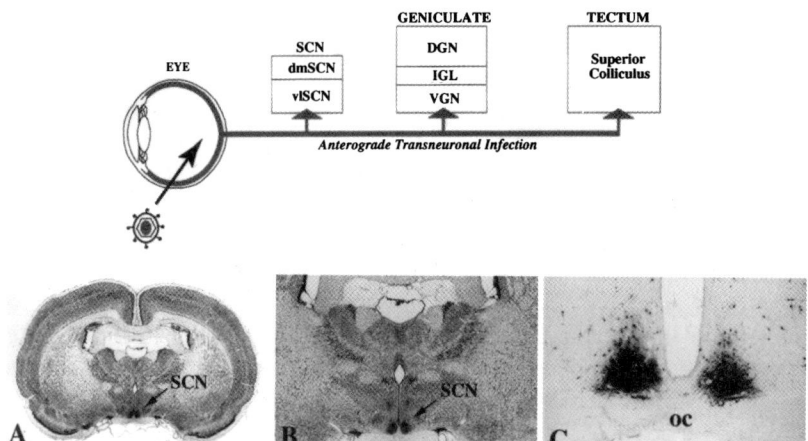

FIG 3. The schematic diagram illustrates an animal model that has been widely applied to investigate the synaptic organization of visual projection systems and to define the role of virally encoded envelope proteins in viral invasiveness. Whereas PRV-Becker infects all areas of brain that receive retinal input, PRV-Bartha only infects a functionally distinct subset of these projections. These include a region of the hypothalamus (the suprachiasmatic nuclei) that is important for the temporal organization of behavior and hormonal secretion. The SCN, shown in the Nissl stains (A and B), are intimately associated with the retinal axons in the optic chiasm (oc) and become infected by PRV-Bartha within the same postinoculation survival interval produced by wild-type virus (C). In contrast, areas involved in visual perception (DGN) and reflex movement of the eyes (tectum) do not become infected by PRV-Bartha, even at postinoculation intervals that are longer than those required for wild-type infection of these regions. See Card and Enquist (1995) and Enquist et al. (1999) for a detailed review of the literature that has established this model.

perikarya (Card et al., 1991; Norgren et al., 1992). These experiments expanded on the classic studies of Goodpasture and Teague (1923), in which central visual pathways were infected by injecting virus into the vitreous body of the eye. The goal of the PRV studies was to establish an animal model in which the molecular basis of anterograde transport of virus could be analyzed in a well-characterized system (Card et al., 1991). Thus, the paradigm incorporated an intravitreal injection (Fig. 3) in an attempt to produce a differential infection of the retinal ganglion cells that project to areas of the central nervous system that are involved in visual perception, reflex eye movements, and the temporal organization of behavior. These retinal ganglion cells reside at the interface of the vitreous body and retina and are separated from the deeper, axon terminal–rich layers of retina by a substantial barrier of glial cells. Thus, intravitreal injection exposed retinal ganglion cells to high concentrations of virus under circumstances where there was no competing pool of axon terminals. The resulting studies demonstrated first-order infection of retinal ganglion cells and subsequent anterograde transneuronal infection of neurons in the CNS that receive direct retinal innervation. Furthermore, temporal analysis of viral replication demonstrated that the infection of retinal ganglion cells was established prior to the appearance of viral immunoreactivity in any other population of cells in the retina (Card et al., 1991). These data strongly support the conclusion that neurons can become infected through their perikarya in circumstances where the cellular architecture of the injected region permits the cells to be exposed to high concentrations of virus.

Further evidence supporting the ability to infect neurons through their perikarya after intracerebral injection is apparent in a recent study in which PRV was injected into the locus ceruleus (LC), a noradrenergic cell group characterized by dense cellular packing and a sparse axonal innervation (Chen et al., 1999). This study demonstrated that localized injections of small amounts of PRV (60 nl) infected LC neurons at the injection site at short postinoculation intervals. Given the paucity of axon terminals within this nucleus it is logical to conclude that the LC neurons became infected by direct viral invasion of the neuronal perikarya, although one cannot exclude the possibility that some of the neurons may have been infected through recurrent collaterals of LC axons.

Collectively, these findings strongly support the conclusion that axons are the preferred site of virion entry and that the concentration of injected virus and the architecture at the site of injection will be an important determinant of the outcome of infection. In this context, it is important to point out that diffusion of virus through the extracellular space will also be restricted by the affinity of viral envelope proteins for extracellular matrix proteins such as heparin sulfate proteoglycan (Flynn and Ryan, 1996; Karger and Mettenleiter, 1993; Mettenleiter et al., 1990; Sawitzky, Voigt, and Habermehl, 1993). This affinity certainly contributes to the ability to make localized injections of virus in experimental paradigms involving direct injection of virus into the brain, and will also influence the amount of virus available for invasion of permissive neuronal profiles.

Although these data are of preeminent importance for deciphering the zone of viral uptake and the route of viral transport in circuit studies, they also have important implications for experimental applications involving the use of replication-deficient alpha herpesviruses. A considerable literature continues to explore the utility of using these strains as vectors for delivering transgenes to the nervous system (cf. Glorioso et al. [1995], for a recent review). Although the vectors are genetically modified to eliminate their capacity to replicate in vivo, they are generally grown on complementing cell lines that endow the viral envelopes with a full complement of envelope proteins. As a result, they exhibit the same affinities of wild-type virus and will be subject to the same influences of neuronal architecture on the invasion of permissive profiles. Thus, it is important to consider this information in designing a route of inoculation that will lead to expression of the transgene in the desired population of neurons rather than in neurons that are simply projecting to the area within which those cells reside.

B. Intracellular Replication and Trafficking of Pseudorabies Virus

Because of the polarized architecture of neurons and the fact that neuronal inputs often terminate differentially on the somata and dendrites of neurons, an accurate map of a neuronal circuit is dependent on the transport of infectious progeny through the full extent of the somatodendritic compartment. Certainly, the extensive intracellular distribution of viral immunoreactivity that has been demonstrated in many transneuronal analyses indicates that this is generally the case (cf., for reviews, Card [1995]; Enquist et al. [1999]; Loewy [1995]; Mettenleiter [1995]; Toth and Palkovits [1997]; Ugolini [1995]). Nevertheless, comparative analyses have shown that the intracellular distribution of virally encoded proteins can vary considerably, depending on the infecting strain of PRV (Card, Levitt, and Enquist, 1998; Standish et al., 1995). Recognition of these factors has led to two important lines of analysis. First, several investigators have conducted systematic analyses of the replication and intracellular spread of virus in infected neurons. Their data have provided interesting insights into the life cycle of the virus that have contributed to the considerable in vitro literature that has addressed this issue. Second, directed analysis of viral spread through neurons has established model systems in which null mutants can be used to explore the role of virally encoded proteins in viral replication and spread. The following discussion will address each of these issues.

Light and electron microscopic analyses demonstrate a progressive appearance of viral immunoreactivity in different compartments of the neuron that would be predicted on the basis of what is known about the sequential transcription of viral genes and the assembly of progeny virus. These studies reveal an initial accumulation of viral immunore-

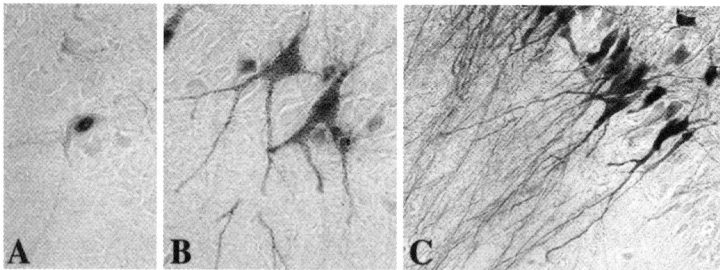

FIG 4. These immunoperoxidase preparations illustrate the distribution of viral antigens in hippocampal neurons at early (A), intermediate (B), and late (C) stages of PRV-Bartha replication. Note that viral immunoreactivity is initially confined to the nucleus of the infected neuron (A) and then progressively moves through the somata and the proximal dendrites (B). The punctate staining on the somatodendritic tree at these early stages of infection marks sites where transsynaptic passage of virus has left patches of highly antigenic envelope proteins in pre- and postsynaptic membranes. With advancing replication and vesicular trafficking of virally encoded proteins, the full dendritic tree of infected neurons becomes homogeneously stained (C).

activity in the nuclei of infected neurons that is quickly followed by the appearance of viral antigens in the cytoplasm and proximal dendrites (Fig. 4). This pattern of cellular staining is common to all infected neurons and reflects the encoding of the proteins necessary for replication and assembly of infectious progeny. It is the subsequent appearance of viral immunoreactivity in the dendritic tree that differs among strains (Card, Levitt, and Enquist, 1998). Neurons infected with the wild-type Becker strain of PRV maintain this limited distribution of viral immunoreactivity throughout the longest postinoculation intervals. In startling contrast, PRV-Bartha produces extensive staining of the dendritic arbor at postinoculation intervals that are shorter than those for the longest surviving animals infected with PRV-Becker. The appearance of viral immunoreactivity is progressive and ultimately involves the full extent of the dendritic arbor, including the most distal processes and appendages. This has been demonstrated directly in two ultrastructural analyses that took advantage of well-characterized neuronal systems (Card et al., 1993). Fig. 5 illustrates the results of this investigation, in which the distribution of progeny virus was examined in the somatodendritic compartment of brainstem neurons that project to the stomach. These neurons exhibit a polarized dendritic tree that ramifies within a circumscribed region that is amenable to serial ultrastructural analysis. In addition, the dendrites of these neurons can be simultaneously filled with a retrograde tracer that permits localization of distal processes at the ultrastructural

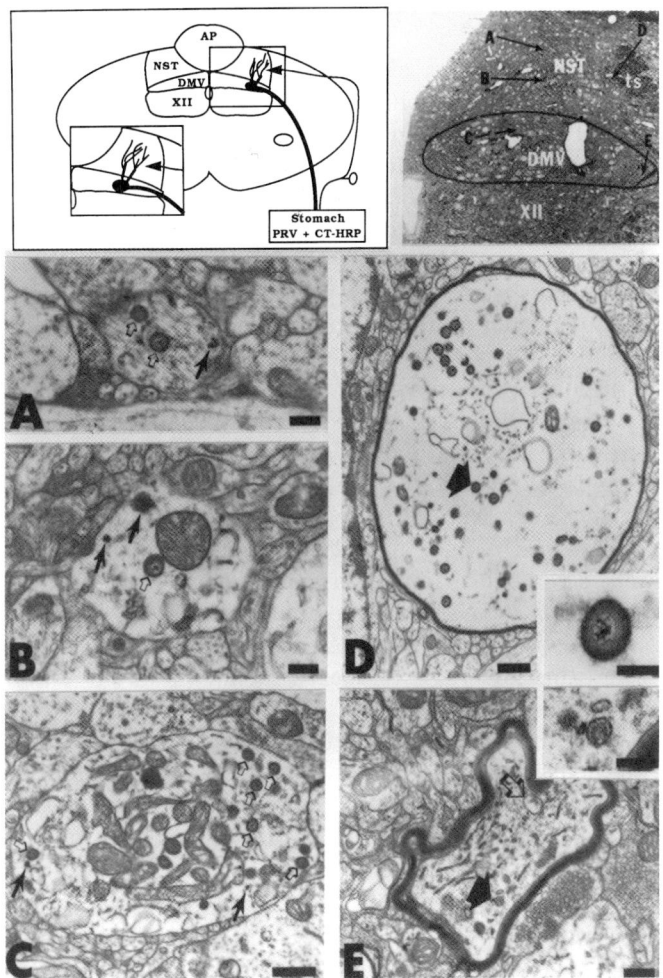

FIG 5. Ultrastructural data demonstrating that infectious progeny traffic into the most distal appendages of the dendritic tree are illustrated. These data, taken, with permission, from a study published in 1993 (Card *et al.*, 1993), employed a model of retrograde transsynaptic infection in which a "covert" infection of the rat brainstem is produced by peripheral inoculation of the stomach wall. The organization of the circuit and the experimental paradigm are illustrated in the schematic diagram. Dual injections of PRV and a classical neuronal tracer (cholera toxin conjugated to horseradish peroxidase [CT-HRP]) produced a retrograde infection of brainstem neurons that innervated the stomach and also endowed the somatodendritic compartment of these neurons with an electron dense tracer that marked all branches of the dendritic tree for ultrastructural analysis. Transmission electron microscopic analysis revealed envelope virions throughout the full extent of the dendritic tree, including the distalmost branches (A–C). Enveloped virions were also observed in the central process of the pseudounipolar sensory neurons (D). In contrast, only capsids were observed in the axonal process of the gastric motor neurons (E). The thick section in the upper right-hand corner illustrates the location of the electron micrographs shown in (A)–(E).

level. Thus, administration of the retrograde tracer along with the simultaneous viral inoculation of the stomach allowed for the systematic analysis of the distribution of progeny virus with identified dendrites of brainstem neurons innervating the stomach. By the use of this approach and analysis of serial sections cut through the dendritic field, it was possible to demonstrate that progeny virus invades the most distal processes of these neurons and therefore has the capacity to infect all neurons that form synaptic contacts with the infected cell. The fact that this is not unique to stomach motor neurons was recently demonstrated by Carr and colleagues (1999). These investigators used ultrastructural analysis and dual labeling methods to demonstrate two important points. First, they demonstrated that viral immunoreactivity extended into the distal dendrites and associated spines of pyramidal neurons of the prefrontal cortex that were retrogradely infected by injecting virus into one of their projection targets. This is important since much of the synaptic input to cortical pyramidal neurons terminates on these spinous appendages, and failure of virus to invade this portion of the somatodendritic compartment would eliminate the ability of virus to infect the neurons that give rise to this input. Second, this analysis provided evidence that virus was, in fact, passing transsynaptically through spines to infect neurons that synapsed on them. This was accomplished through the use of dual labeling methods that demonstrated synaptic associations between dopaminergic terminals and spines containing PRV immunoreactivity. Considered along with data demonstrating that midbrain dopaminergic neurons that project to the prefrontal cortex are infected in this and other paradigms, these data support the conclusion that progeny virus trafficks through all portions of the dendritic compartment, including the smallest appendages arising from the most distal processes. This is important from the standpoint of viral circuit analysis but also reveals an aspect of the viral life cycle that is essential for evaluating viral invasiveness.

As noted earlier, characterization of viral invasiveness *in vivo* has provided powerful models that complement studies that have historically used *in vitro* systems and null mutants to probe the functions of virally encoded gene products. Whereas *in vitro* analyses provide important insights into the role of viral genes in invasion and spread, their ability to predict the function of genes in viral invasiveness of neuronal circuitry is limited. This is apparent from the early characterization of viral genes as "essential" or "nonessential" for replication and spread *in vitro,* and from the subsequent demonstration that "nonessential" genes contribute to these functions *in vivo.* Additionally, an emerging literature has established the utility of *in vivo* models in defining the role of virally encoded genes in intracellular trafficking and virulence (cf. Enquist *et al.* [1999]), for a recent review).

Anterograde models of viral invasiveness of visual circuitry have proven to be particularly useful for establishing invasive and virulence phenotypes for nonessential gene products of the unique short region of the viral genome. An early comparison of the invasiveness of PRV-Becker and PRV-Bartha revealed a remarkable difference in invasiveness that had not been expected on the basis of *in vitro* studies (Card *et al.*, 1991). Following intravitreal injection, PRV-Becker was shown to infect all areas of the CNS that receive retinal innervation, but it did so in a curious fashion that discriminated among functionally distinct components of the visual projection systems. Areas of the brain involved in visual perception (the dorsal geniculate nucleus, or DGN) and reflex movements of the eye (the tectum) were infected within 48 hours of intravitreal injection, but infection of regions of the brain that orchestrate the temporal order of behavior and hormone secretion (the SCN and the intergeniculate leaflet, or IGL) did not occur until approximately 24 hours later. Analysis of the invasiveness of PRV-Bartha in the same circuitry revealed that this virus maintained the ability to infect the SCN (Fig. 3) and IGL, and did so in the same temporal framework that PRV-Becker did, but lacked the capacity to infect the DGN and the tectum. Over the years this has become an extremely informative model for characterizing the molecular mechanisms that contribute to anterograde transsynaptic infection by PRV and has also provided a vivid illustration of the important influence of the host cell on the outcome of infection (Brideau, Card, and Enquist, 2000; Card *et al.*, 1992; Enquist *et al.*, 1994; Tirabassi *et al.*, 1997; Whealy *et al.*, 1993). The restricted invasive phenotype characteristic of PRV-Bartha has also provided a tool for dissecting the organization of ensembles of neurons synoptically associated with the SCN and IGL (Leak, Card, and Moore, 1998; Moore, Speh, and Card, 1995), thereby providing functional insights into this system that could not be defined with conventional tracing methods.

Analysis of retrograde infection of the nervous system has been similarly informative in defining the role of viral gene products in replication and spread as well as in the definition of the functional organization of brain circuitry. A revealing series of studies early in the 1990s utilized retrograde models of viral replication and spread to demonstrate divergent roles for essential PRV genes that had not been suspected from *in vitro* analyses (Babic *et al.*, 1996; Heffner *et al.*, 1993; Peeters *et al.*, 1992a; Peeters *et al.*, 1992b; Peeters *et al.*, 1993; Rauh and Mettenleiter,

1991). Tissue culture studies have demonstrated that mutants lacking gB, gD, and gH are not infectious due to an inability of the null mutants lacking these gene products to gain access to permissive cells. However, *in vivo* analysis of the invasiveness of these null mutants has demonstrated that all three gene products are essential for invasion of permissive cells, but that only gB and gH are required for subsequent retrograde spread through CNS circuitry. These conclusions are based on the previously demonstrated ability to produce a predictable temporal progression of viral transport through CNS circuitry, following peripheral projection targets such as the tongue. These types of retrograde analysis of viral transport through CNS circuitry are becoming increasingly utilized in mechanistic studies of viral invasiveness and are fundamental to the literature that has established viral transneuronal tracing as an effective means of studying brain organization. Analysis of CNS circuitry that modulates the activity of peripheral autonomic circuits has benefited enormously from this experimental approach, and the many studies that have focused on this system have provided a solid foundation for mechanistic analysis.

C. Viral Egress from Infected Neurons

Certainly, sites of viral egress are among the most important considerations in evaluating the specificity of viral spread through neuronal circuitry. If virions were released indiscriminately from parent neurons, this would seriously undermine the probability that viral transport through the brain would be restricted to synaptically linked neurons. Indeed, the appearance of viral antigen in glial cells adjacent to infected neurons in advanced stages of viral replication has been offered as evidence that viral egress is indiscriminate and fostered the longest-standing debate regarding the specificity of the viral transneuronal tracing method. Nevertheless, available evidence supports the conclusion that release of virions from infected neurons occurs selectively in the vicinity of synaptic contacts. For example, light microscopic localization of viral antigens in infected neurons reveals that viral immunoreactivity on the somatodendritic compartment is initially present as patches that are first observed proximally but move progressively into the dendritic tree with advancing viral replication. These patches of staining are similar in both size and distribution to the expected pattern of axon terminals on the cell surface. Although there are differing opinions on the intracellular pathways of virion assembly and envelopment, there can be little doubt that the process of release and transsynaptic passage of virions involves sequential membrane fusion events that would leave focal concentrations of the viral envelope in the plasma membrane of the parent cell as well as in the membrane of the target axon terminal. Since these membranes are highly enriched in envelope proteins, it seems reasonable to assume that differential release of virions at axon terminals would lead to patches of immunoreactivity on the neuron surface. Although inferential, these light microscopic observations are supported by the findings of a systematic serial ultrastructural analysis of viral egress from brainstem neurons that innervate the stomach (Card et al., 1993). In that analysis all instances of virion egress occurred adjacent to axon terminals.

Using the logic advanced to explain the patchy staining, one would interpret the fact that neurons ultimately exhibit uniform staining, after prolonged viral replication, as being indicative of virion release throughout the plasma membrane of the parent cell. However, this was not observed in the ultrastructural analysis of viral egress (Card et al., 1993), and other studies of the trafficking of virally encoded envelope proteins indicate that some of these proteins are transported to the plasma membrane of infected neurons (cf. Enquist et al. [1999] for a review). Collectively, these data suggest that virion trafficking and egress in infected neurons is a highly organized process that leads to differential release of infectious progeny at regions of the cell surface contacted by axon terminals. The mechanisms that could produce this type of selected release remain to be established. Nevertheless, it is useful to note that the cytoskeleton at sites of synaptic contact differs from that in other areas of the neuron, and that intracellular pathways exist for differential transport of receptors and other proteins to the postsynaptic membrane. As a result, targeted transport of virions to the postsynaptic membrane could be achieved if the virus subverted existing intracellular transport mechanisms, just as it does other cellular processes during viral replication and assembly. This is yet another example of how efforts to demonstrate the specificity of viral transport through neuronal circuitry have revealed targets that can further our understanding of the mechanisms that contribute to viral neuroinvasiveness.

V. Brain Defenses and Neuroinvasiveness

The brain is very effective in mounting a local response to viral infection. This response is multicellular, temporally organized, and in most instances is initiated prior to the appearance of cytopathic changes in infected neurons (cf. Card and Enquist [1995] for a review). It also effectively restricts the local dissemination of virus through the extracellular space at long postinoculation intervals where infected neurons begin to exhibit pathological changes. Fundamentally, however, it is the covert entry of the virus into the nervous system that allows virions to initially elude the local antiviral response and spread through neuronal connections prior to the appearance of cellular pathology. In this fashion, PRV effectively parasitizes the polarized architecture and synaptic connections of the nervous system in a manner that is extremely valuable for defining the organization of neuronal circuitry.

Although it is now commonly accepted that the brain response to neuronal infection is fundamental to biasing spread of virus through neuronal circuits, that was not always the case. In fact, for quite some time the appearance of viral antigen in glial cells responding to infection was used as the principal criticism against the usefulness of neurotropic viruses for transneuronal analysis. Simply stated, the demonstration of viral immunoreactivity in glial cells surrounding infected neurons was cited as clear evidence that virions escape the confines of neuronal circuitry and spread through the extracellular space to infect neighboring neurons not involved in the circuit. In retrospect, this argument overlooked important evidence that supported the conclusion that the principal route of viral spread was through synaptically linked neurons rather than via the extracellular space. For example, a careful consideration of the distribution of infected neurons at increasingly longer postinoculation intervals largely supported the conclusion that virus spread through synaptic connections rather than by neuronal lysis. In the latter circumstance, one would predict that neuronal lysis and release of virus into the extracellular space would lead to an ever-expanding spread of infection in the zone surrounding the initially infected neurons. This has repeatedly been shown *not* to be the case in a very large body of literature extending back to Dolivo's early studies with PRV (Dolivo, 1980; Dolivo *et al.*, 1979; Martin and Dolivo, 1983) and the studies of Stevens and colleagues with HSV (Bak *et al.*, 1977). In essentially every case, virus has been shown to move sequentially through the nervous system in a pattern consistent with the known connections established with classical neuronal tracers. A particularly vivid example of this was revealed in an analysis of the spread of virulent and attenuated strains of PRV through caudal brainstem circuitry (Card *et al.*, 1990). In these experiments, virus was injected into the targets of caudal brainstem neurons that control the functional activity of the viscera. This route of infection has a number of attributes that continue to

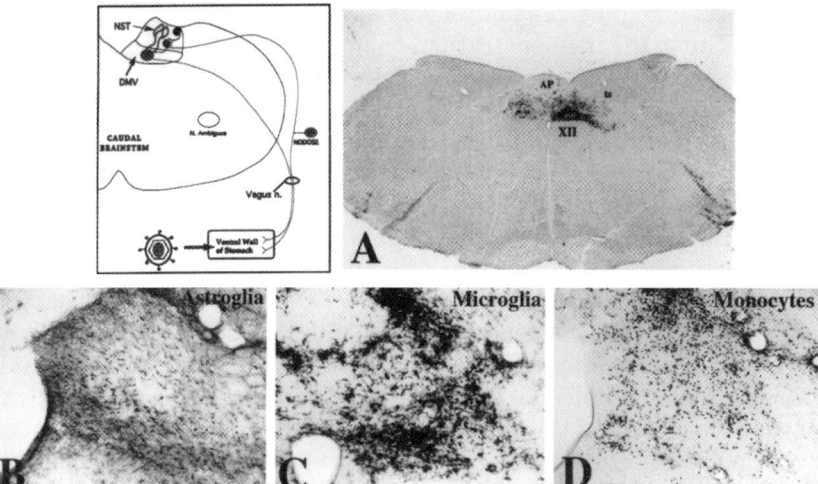

Fig 6. The model for retrograde infection of rodent brainstem following stomach inoculation (illustrated in the schematic) has also proven to be instrumental in demonstrating the specificity of transsynaptic passage of PRV in the central nervous system. Note the selective infection of gastric motor neurons in the dorsal motor vagal (DMV) nucleus that results from peripheral inoculation of the stomach (A). With advancing survival, progeny virus passes transsynaptically to infect neurons in the immediately adjacent nucleus of the solitary tract (NST) but does not spread to infect tongue motor neurons in the immediately adjacent nucleus of the 12th cranial nerve (XII) that have no synaptic relations with the gastric motor neurons. Immunohistochemical localization of astroglia (B), microglia (C), and monocytes (D) demonstrates that these cells differentially accumulate in the region of infection. This "reactive" nonneuronal response contributes to the specificity of viral transsynaptic passage. See the text and Rinaman et al. (1993) for further details.

make it a very effective model for defining brain responses to infection. First and foremost is the ability to reproducibly infect neurons in the brain in a noninvasive manner (Fig. 6). This is possible because the virus invades the nervous system through the peripherally projecting processes of the neurons that project out of the brain to the abdomen. Another powerful attribute of this model is that the neurons infected by injection of virus into the viscera are situated immediately adjacent to other neurons that are responsible for controlling the movement of the tongue. Thus, tongue motor neurons can only become infected by spread of virus through the extracellular space. Because these tongue motor neurons are permissive to infection by PRV and HSV (Fay and Norgren, 1997; Ugolini, Kuypers, and Simmons, 1987), this system provides a rigorous means of determining if

PRV infects neighboring neurons by spread through the extracellular space. Thus, the demonstration that tongue motor neurons were never infected in this model, even at the longest postinoculation intervals or after inoculation with virulent strains of PRV, provided a dramatic illustration of the effectiveness of the nonneuronal response in restricting extracellular spread of PRV (Card et al., 1993; Rinaman, Card, and Enquist, 1993; Yang et al., 1999).

The cellular basis of the brain's ability to restrict the extracellular spread of virus has been revealed in a number of investigations. Early studies by Weinstein and colleagues (1990) demonstrated that HSV infection of the brain elicited local reactive responses of glial cells in the vicinity of the infection and also led to recruitment of immune cells into the region of infection from the vasculature. The localized nature of that response and the absence of viral antigens outside the immediate area of infection suggested that it was very effective in limiting necrotizing spread of virus. Support for this conclusion has been provided by numerous studies of viral replication and invasion using PRV in rodent brain (Rassnick et al., 1998; Rinaman and Levitt, 1993; Rinaman, Levine et al., Roesch, and Card, 1999). In addition to the aforementioned evidence for transneuronal spread of virus, these studies have provided important insights into the cellular mechanisms that influence viral spread in the brain. It is now well established that infection of CNS neurons with PRV elicits a predictable and temporally organized response (Fig. 6). This consists of an early reactive astrogliosis in the vicinity of infection that precedes any overt neuropathological changes in the infected neurons. The astroglia that participate in this early response, by virtue of their extensive process networks and close association with neurons and the vasculature, are well situated to isolate the infected neurons and also participate in the recruitment of other cells that participate in the nonneuronal response to infection. The latter include resident microglial cells, which are well known for their reactive responses to neuronal injury and disease and immune cells that are focally recruited to the site of infection from the vasculature. These populations of cells are sequentially recruited to the site of infection, where they form dense aggregations around infected cells. Furthermore, the magnitude of the reactive response, and the number of cells recruited to the site of infection, correlate with the virulence of the infecting strain of virus.

Each class of nonneuronal cells appears to subserve a distinct component of the antiviral response and the cells are differentially susceptible to infection by PRV (cf. (Card and Enquist [1995] for a detailed review). In this context, astroglia appear to play a preeminent role in isolating infected neurons and preventing spread of virus through the extracellular space. However, it is also the appearance of viral antigens within these cells that has been the focus of the debate regarding dissemination of virus through the extracellular space. Quite reasonably, these data have been interpreted as being indicative of indiscriminate release and spread of virus through the extracellular space. Additionally, this observation has been offered as evidence that infected astroglia are a secondary source of progeny virus that further undermines the specificity of circuit-related viral transport. A directed analysis of these issues has demonstrated otherwise. As noted above, astroglia exhibit reactive response to viral infection that precedes the appearance of pathological changes in infected neurons as well as the recruitment of microglia and peripheral immune cells to the site of infection. A hallmark of this reactive astrogliosis is a prominent increase in the size and density of process networks emanating from individual cells. These filament-rich processes of reactive astroglia ensconce infected cells, providing a formidable barrier between the infected neurons and neighboring neurons. Evidence that this barrier of reactive astroglia contributes to the specificity of transsynaptic spread of virus emerges from two important observations. First, in studies with HSV, Vahlne and colleagues (Vahlne *et al.*, 1978; Vahlne *et al.*, 1980) have shown that virions have a high affinity for astrocytic membrane that is only exceeded by the affinity of the virus for synaptosomes. Thus, axon terminals and astroglia are the most probable sites of virion uptake following release from parent neurons. Second, strong evidence supports the conclusion that astrocytes *in vivo* are incapable of producing infectious progeny. This was initially suggested in studies with HSV that were conducted by Stevens and colleagues (Bak *et al.*, 1977), and was subsequently demonstrated in a systematic ultrastructural examination of PRV replication in rodent brain (Card *et al.*, 1993). The latter analysis demonstrated that astroglia initiate viral replication but harbor a defect that prevents them from applying an envelope to capsids. Thus, three important aspects of virus-induced reactive gliosis— the early initiation of the isolating reactive astroglial response; the high affinity of virions for astrocytic membrane; and the replication defect—argue convincingly that these cells act to limit nonspecific dissemination of virions and bias transport of progeny virus through synaptically linked neurons. These observations are also consistent with the many studies that have shown that virus spreads through the nervous system in a pattern consistent with known connections rather than in a radially expanding, necrotizing spread.

VI. RECOMBINANT VIRUSES AND THE NERVOUS SYSTEM

Recombinant alpha herpesvirus stains that are engineered to express unique reporter proteins represent a recently introduced variation of the viral transneuronal method. This approach utilizes dual injection of recombinant viruses to investigate complex synaptic relationships in which one population of neurons exerts a common influence on functionally distinct populations of neurons (Fig. 7, see also color section). It has been applied by a limited number of laboratories utilizing either HSV or PRV and is based on the ability of the two recombinants to produce unique reporters that can be distinguished with monospecific antibodies (Jansen et al., 1995; Kim et al., 1999; Levatte et al., 1998; Mabon, Weaver, and Dekaban, 1999). To date, studies with PRV have used PRV-Bartha as the parental strain for the recombinants, but have inserted the transgenes at different sites. For example, Jansen and colleagues (1995) used two PRV-Bartha derivatives that could be distinguished by the unique expression of either wild-type gC or β-galactosidase by transgenes inserted into the unique long and unique short regions of the genome, respectively. Following injection of these strains into different peripheral targets of the autonomic nervous system, Jansen and co-workers were able to coinfect neurons in the CNS, and thereby reveal the tremendous potential of this experimental approach. Their pioneering analysis also stimulated subsequent efforts to address fundamental issues that impact on the design of such studies as well as the interpretation of data derived from application of this method.

Successful application of the dual-inoculation method is dependent on the ability to infect neurons with more than one strain of virus, and on the efficiency of expression of the transgenes that provide the unique reporters of each virus. Interference of one strain of virus with the replication of the second strain, or inefficient expression of one or both of the transgenes, could lead to false negatives that would produce misleading results. Demonstrations that these are viable concerns are evident in two recent studies. First, in an investigation employing two recombinants constructed from PRV-Bartha, Kim and colleagues (1999) demonstrated that prior infection of neurons with one recombinant virus could interfere with the ability of a second recombinant to replicate in the same neurons. Analysis of the invasiveness of the two recombinants in single-injection paradigms also revealed that they differed in virulence and in the rate at which they were transported through the same circuitry, even though they were constructed from the same parental virus. Thus, even though coinfec-

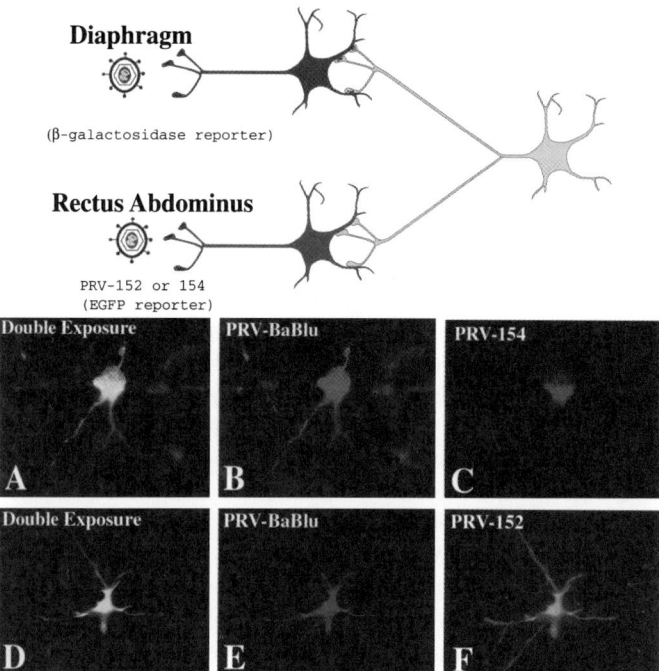

FIG 7. [For a color reproduction, see the color section.] The utility of using antigenically distinct recombinant viruses for transneuronal analysis is illustrated. In this experiment, PRV recombinants constructed by L. W. Enquist and colleagues were used to identify neurons in the CNS that branch to synapse on motor neurons that control the activity of the diaphragm and rectus abdominus muscles of the ferret (see Billig *et al.* [1999a] and Yates *et al.* [1999] for a definition of the circuits in single injection viral studies). The genomic organization of each recombinant is illustrated in Fig. 2. Simultaneous inoculation of the two recombinants into separate muscles as shown in the schematic diagram led to retrograde infection of the motor neurons that innervate them and to transsynaptic infection of neurons whose axons branched to synapse on both populations of motor neurons. Examples of dual-infected neurons using immunofluorescence localization of the unique reporters (β-galactosidase and enhanced green fluorescent protein) are shown in (A)–(F). PRV-BaBlu and PRV-152 produce extensive staining of the somatodendritic compartment of infected neurons (B, E, F), whereas the EGFP-Us9 fusion protein reporter of PRV-154 is differentially transported to the trans-Golgi network (C). See the text for further details and also Billig *et al.*, 2000.

tion of neurons did occur in this investigation, it was apparent that the number of cells replicating both strains underrepresented the population of neurons that should have been infected with both viruses. Second, Mabon Weaver, and Dekaban (1999) further demonstrated that the immune response that is marshaled in response to viral infection can compromise transgene expression and thereby produce misleading results.

In spite of the limitations noted above, the dual-infection paradigm remains an informative means of addressing issues of synaptic connectivity that cannot be examined with any other method. When neurons are shown to replicate both strains of virus, as they were in the study of Jansen and colleagues (Jansen *et al.*, 1995) and in other studies that have successfully applied this approach, one can have assurance that they became infected by virtue of shared synaptic connections with a common path of transport of the two recombinants. However, no credence can be given to a negative result. Recent data also suggest that it is possible to improve the efficiency of coinfection. A notable feature of the dual-injection studies reported to date is that they employ recombinants in which the transgenes are inserted into different sites of the viral genome. Furthermore, some of these investigations inserted viral genes that are lacking in the parental virus. For example, Jansen and colleagues (1995) inserted a wild-type gC gene into the PRV-Bartha genome, and Kim and colleagues repaired a portion of the deletion in the unique short region of PRV-Bartha and used an antisera monospecific for one of the reconstituted genes (gI) as the unique reporter of this recombinant. As Kim and colleagues (1999) and others (cf. Enquist *et al.* [1999] for a review) have demonstrated, these types of manipulations alter the virulence of the recombinant and the rate of transport through circuitry, thereby increasing the probability that one recombinant will interfere with the replication of the second. A recent study indicates that excluding viral genes as reporters and using the same site of transgene insertion reduce interference and produce recombinants that are more closely matched in their rate of transport through neuronal circuitry. In that study, recombinants in which *lacZ* (for production of β-galactosidase) or the enhanced green fluorescent protein (EGFP) genes were inserted into the gG locus of the PRV-Bartha genome and were used to explore the organization of CNS circuitry that controls outflow to abdominal musculature of the ferret (Billig *et al.*, 2000). Prior studies using PRV-Bartha had defined the CNS circuitry controlling motor neurons that innervate either the diaphragm (Yates *et al.*, 1998) or the rectus abdominis muscle (Billig *et al.*, 1999), and analysis of the invasiveness of the recombinants in single-injection studies demonstrated the same circuitry. Slight differences in the rate of transport of the two recombinants were noted in these single-injection studies. Nevertheless, when the two recombinants were simultaneously injected into either the diaphragm or the rectus abdominis, efficient retrograde transsynaptic coinfection of brainstem neurons was achieved (Fig. 7). These data suggest that the use of isogenic viruses that only differ in their transgenes provides the optimal approach for dual-injection studies. However, this does not preclude the possibility that some interference may occur and that negative data must be interpreted conservatively.

VII. Viral Circuit Analysis: Future Applications

The establishment of viral transsynaptic tracing as a method of defining polysynaptic circuits in the brain opens many doors for providing unique insights into brain organization and function. To date, the method has been largely applied to define the "normal" organization and synaptology of functionally distinct systems. However, an obvious extension of this methodology is to examine the reorganization of circuitry in response to injury or disease. Use of the procedure to examine the development of neuronal circuitry in fetuses and young animals offers yet another important application of the method. Two recent observations support the conclusion that viral transneuronal analysis will be particularly valuable for defining the developmental assembly of neuronal circuit. The first derives from a study published recently by Rinaman, Roesch, and Card (1999), demonstrating that the nonneuronal response that contributes to the specificity of viral transport through neuronal circuits in adults is also operative in neonatal rats. This is an important observation, given the fact that this response plays such a central role in maintaining the specificity of viral transport through a circuit. A second series of studies from the same laboratory (Rinaman, Levitt, and Card, 2000) has established that synaptic connections are necessary for virus to pass transneuronally in the developing brain. It is well known, from the developmental literature, that axons will grow into target areas of the brain well before they establish synaptic connections. This developmental process has made it difficult to study the functional maturation of the brain because it has been difficult to establish when the ingrowing axons actually form synapses. Using the brainstem circuitry previously used to establish the specificity of PRV transport in the adult, Rinaman and co-workers have shown that there is temporal separation in the transneuronal passage of virus into different forebrain nuclei known to innervate the brainstem neurons that project to the stomach. This occurs in spite of the fact that axons from all of these systems have previously grown into the brainstem cell group that contains infected neurons. This suggests that it will be possible to use viral transneuronal analysis as a means of defining the synaptic assembly of neuronal circuits, thereby gaining essential information on the functional assembly and maturation of the brain.

VIII. Conclusions

Use of neurotropic viruses to improve our understanding of the organization of the nervous system has undergone an explosive growth over the past decade and the prospects for further advances in our ability to exploit this technology continue to improve. Importantly, the use of viruses for circuit analysis has gone hand in hand with the investigation of the mechanisms that underlie viral neurotropism and invasiveness. The result has been the creation of a field of analysis that continues to provide fundamental insights in the fields of neurobiology, virology, and immunology. The firm foundation established by these investigations promises to foster continued interdisciplinary growth that will advance our understanding in each of these fields.

Acknowledgments

I gratefully acknowledge the many collaborative associations that have made possible our studies of the neuroinvasiveness of PRV. I am particularly indebted to Dr. Lynn Enquist and Dr. Linda Rinaman for the many contributions they made to the published work referred to in this chapter, and to Dr. Bill Yates for the productive association that has led to our recent use of recombinant viruses in dual-injection paradigms. Preparation of this chapter was supported by a grant from the John D. and Catherine T. MacArthur Foundation Research Network on Early Experience and Brain Development.

References

Babic, N., Klupp, B., Brack, A., Mettenleiter, T. C., Ugolini, G., and Flamand, A. (1996). Deletion of glycoprotein gE reduces the propagation of pseudorabies virus in the nervous system of mice after intranasal inoculation. *Virology* **219:**279–284.

Bak, I. J., Markham, C. H., Cook, M. L., and Stevens, J. G. (1977). Intra-axonal transport of herpes simplex virus in the rat central nervous system. *Brain Research* **136:**361–368.

Barnett, E. M., Cassell, M. D., and Perlman, S. (1993). Two neurotropic viruses, herpes simplex virus type 1 and mouse hepatitis virus, spread along different neural pathways from the main olfactory bulb. *Neuroscience* **57:**1007–1025.

Barnett, E. M., Jacobsen, G., and Evans, G. (1994). Herpes simplex encephalitis in the temporal cortex and limbic system after trigeminal nerve inoculation. *Journal of Infectious Diseases* **169:**782–786.

Bartha, A. (1961). Experimental reduction of virulence of Aujeszky's disease virus. *Magy. Allatorv. Lapja* **16:**42–45.

Billig, I., Foris, J., Card, J. P., and Yates, B. J. (1999). Transneuronal tracing of neural pathways controlling the rectus abdominis, in the ferret. *Brain Research* **820**:31–44.

Billig, I., Foris, J. M., Enquist, L. W., Card, J. P., and Yates, B. J. (2000). Definition of neuronal circuitry controlling the activity of phrenic and abdominal motoneurons in ferret using recombinant strains of pseudorabies virus. *Journal of Neuroscience* **20**:7446–7454.

Brideau, A. D., Card, J. P., and Enquist, L. W. (2000). Role of pseudorabies virus Us9, a type II membrane protein, in infection of tissue culture cells and the rat nervous system. *Journal of Virology* **74**:834–845.

Card, J. P. (1995). Pseudorabies virus replication and assembly in the rodent central nervous system. *In* "Viral Vectors: Tools For Study And Genetic Manipulation Of The Nervous System" (M. G. Kaplitt and A. D. Loewy, eds.), pp. 319–347. Academic Press, Orlando.

Card, J. P., and Enquist, L. W. (1995). Neurovirulence of pseudorabies virus. *Critical Reviews in Neurobiology* **9**(2 & 3): 137–162.

Card, J. P., Enquist, L. W., and Moore, R. Y. (1999). The neuroinvasiveness of pseudorabies virus injected intracerebrally is dependent upon viral concentration and terminal field density. *Journal of Comparative Neurology* **407**:438–452.

Card, J. P., Levitt, P., and Enquist, L. W. (1998). Different patterns of neuronal infection after intracerebral injection of two strains of pseudorabies virus. *Journal of Virology* **72**(5): 4434–4441.

Card, J. P., Rinaman, L., Lynn, R. B., Lee, B.-H., Meade, R. P., Miselis, R. R., and Enquist, L. W. (1993). Pseudorabies virus infection of the rat central nervous system: Ultrastructural characterization of viral replication, transport, and pathogenesis. *Journal of Neuroscience* **13**(6):2515–2539.

Card, J. P., Rinaman, L., Schwaber, J. S., Miselis, R. R., Whealy, M. E., Robbins, A. K., and Enquist, L. W. (1990). Neurotropic properties of pseudorabies virus: Uptake and transneuronal passage in the rat central nervous system. *Journal of Neuroscience* **10**(6):1974–1994.

Card, J. P., Whealy, M. E., Robbins, A. K., and Enquist, L. W. (1992). Pseudorabies virus envelope glycoprotein gI influences both neurotropism and virulence during infection of the rat visual system. *Journal of Virology* **66**(5):3032–3041.

Card, J. P., Whealy, M. E., Robbins, A. K., Moore, R. Y., and Enquist, L. W. (1991). Two alpha-herpesvirus strains are transported differentially in the rodent visual system. *Neuron* **6**:957–969.

Carr, D. B., O'Donnell, P., Card, J. P., and Sesack, S. R. (1999). Dopamine terminals in the rat prefrontal cortex synapse on pyramidal cells that project to the nucleus accumbens. *Journal of Neuroscience* **19**(24):11049–11060.

Chen, S., Yang, M., Miselis, R. R., and Aston-Jones, G. (1999). Characterization of transsynaptic tracing with central application of pseudorabies virus. *Brain Research* **838**:171–183.

Dolivo, M. (1980). A neurobiological approach to neurotropic viruses. *Trends in Neuroscience* **3**:149–152.

Dolivo, M., Honegger, P., George, C., Kiraly, M., and Bommeli, W. (1979). Enzymatic activity, ultrastructure and chemical specificity of neurons. *In* "Progress in Brain Research" (M. Cuenod, G. W. Kreutzberg, and F. E. Bloom, eds.), vol. 51, pp. 51–57. Elsevier/North-Holland Biomedical Press, Amsterdam.

Dolivo, M., Kucera, P., and Bommeli, W. (1982). Progression of the rabies virus in the visual system of the rat. *Comp. Immunol. Microbiol. Infect. Dis.* **5**:67–69.

Enquist, L. W., Dubin, J., Whealy, M. E., and Card, J. P. (1994). Complementation analysis of pseudorabies virus gE and gI mutants in retinal ganglion cell neurotropism. *Journal of Virology* **68**(8):5275–5279.

Enquist, L. W., Husak, P. J., Banfield, B. W., and Smith, G. A. (1999). Infection and spread of alphaherpesviruses in the nervous system. *Advances in Virus Research* **51**:237–347.

Fay, R. A., and Norgren, R. (1997). Identification of rat brainstem multisynaptic connections to the oral motor nuclei using pseudorabies virus 1. Lingual muscle motor systems. *Brain Research Review* **25**:291–311.

Flynn, S. J., and Ryan, P. (1996). The receptor-binding domain of pseudorabies virus glycoprotein C is composed of multiple discrete units that are functionally redundant. *Journal of Virology* **70**:1355–1364.

Glorioso, J. C., Bender, M. A., Goins, W. F., Fink, D. J., and DeLuca, N. (1995). Herpes simplex virus as a gene-delivery vector for the central nervous system. *In* "Viral Vectors" (M. G. Kaplitt and A. D. Loewy, eds.), pp. 1–24. Academic Press, San Diego.

Goodpasture, E. W., and Teague, O. (1923). Transmission of the virus of herpes along nerves in experimentally infected rabbits. *J. Med. Res.* **44**:139–184.

Gustafson, D. P. (1975). Pseudorabies. *In* "Diseases of Swine" (H. W. Dunne and A. D. Leman, eds.), pp. 391–410. Iowa State University Press, Ames.

Heffner, S., Kovacs, F., Klupp, B. G., and Mettenleiter, T. C. (1993). Glycoprotein gp50-negative pseudorabies virus: a novel approach toward a nonspreading live herpesvirus vaccine. *Journal of Virology* **67**:1529–1537.

Hoover, J. E., and Strick, P. L. (1993). Multiple output channels in the basal ganglia. *Science* **259**:819–821.

Hoover, J. E., and Strick, P. L. (1999). The organization of cerebellar and basal ganglia outputs to primary motor cortex as revealed by retrograde transneuronal transport of herpes simplex virus type 1. *Journal of Neuroscience* **15**(4):1446–1463.

Howe, H. A., and Bodian, D. (1942). "Neural Mechanisms in Poliomyelitis." The Commonwealth Fund. London.

Jansen, A. S. P., Van Nguyen, X., Karpitskiy, V., Mettenleiter, T. C., and Loewy, A. D. (1995). Central command neurons of the sympathetic nervous system: Basis of the fight-or-flight response. *Science* **270**:253–260.

Karger, A., and Mettenleiter, T. C. (1993). Glycoproteins gIII and gp50 play dominant roles in the biphasic attachment of pseudorabies virus. *Virology* **194**:654–664.

Kim, J.-S., Moore, R. Y., Enquist, L. W., and Card, J. P. (1999). Circuit-specific co-infection of neurons in the rat central nervous system with two pseudorabies virus recombinants. *Journal of Virology* **73**:9521–9531.

Kristensson, K. (1996). Sorting signals and targeting of infectious agents through axons: an annotation to the 100 years' birth of the name "axon." *Brain Research Bulletin* **41**:327–333.

Kucera, P., Dolivo, M., Coulon, P., and Flamand, A. (1985). Pathways of the early propagation of virulent and avirulent rabies strains from the eye to the brain. *Journal of Virology* **55**:158–162.

Kuypers, H. G. J. M., and Ugolini, G. (1990). Viruses as transneuronal tracers. *TINS* **13**:71–75.

Leak, R. K., Card, J. P., and Moore, R. Y. (1998). Suprachiasmatic pacemaker organization analyzed by viral transsynaptic transport. *Brain Research* **819**:23–32.

Levatte, M. A., Mabon, P. J., Weaver, L. C., and Dekaban, G. A. (1998). Simultaneous identification of two populations of sympathetic preganglionic neurons using recombinant herpes simplex virus type 1 expressing different reporter genes. *Neuroscience* **82**(4):1253–1267.

Levine, J. M., Enquist, L. W., Card, J. P. (1998). Reactions of oligodendrocyte precursor cells to alpha herpesvirus infection of the central nervous system. *Glia* **23**:316–328.

Loewy, A. D. (1995). Pseudorabies virus: A transneuronal tracer for neuroanatomical studies. *In* "Viral Vectors. Gene Therapy and Neuroscience Applications" (M. G. Kaplitt and A. D. Loewy, eds.), pp. 349–366. Academic Press, San Diego.

Lomniczi, B., Watanabe, S., Ben-Porat, T., and Kaplan, A. S. (1987). Genome location and identification of functions defective in the Bartha vaccine strain of pseudorabies virus. *Journal of Virology* **61**:796–801.

Mabon, P. J., Weaver, L. C., and Dekaban, G. A. (1999). Cyclosporin A reduces the inflammatory response to a multi-mutant herpes simplex virus type-1 leading to improved transgene expression in sympathetic neurons in hamsters. *Journal of Neurovirology* **5**:268–279.

Marchand, C. F., and Schwab, M. E. (1987). Binding, uptake and retrograde axonal transport of herpes virus suis in sympathetic neurons. *Brain Research* **383**:262–270.

Martin, X., and Dolivo, M. (1983). Neuronal and transneuronal tracing in the trigeminal system of the rat using the herpes virus suis. *Brain Research* **273**:253–276.

Mettenleiter, T. C. (1991). Molecular biology of pseudorabies virus (Aujeszky's Disease). *Comp. Immun. Microbiol. Inf. Dis.* **14**:151–163.

Mettenleiter, T. C. (1995). Molecular properties of alpha herpesviruses used in transneuronal pathway tracing. *In* "Viral Vectors. Gene Therapy and Neuroscience Applications" (M. G. Kaplitt and A. D. Loewy, eds.), pp. 367–393. Academic Press, San Diego.

Mettenleiter, T. C., Zsak, L., Zuckermann, F., Sugg, N., Kern, H., and Ben-Porat, T. (1990). Interaction of gIII with a cellular heparinlike substance mediates adsorption of pseudorabies virus. *Journal of Virology* **64**:278–286.

Middleton, F. A., and Strick, P. L. (1994). Anatomical evidence for cerebellar and basal ganglia involvement in higher cognitive function. *Science* **266**(5184):458–461.

Middleton, F. A., and Strick, P. L. (1996). The temporal lobe is a target of output from the basal ganglia. *PNAS USA* **93**(16):8683–8687.

Moore, R. Y., Speh, J. C., and Card, J. P. (1995). The retinohypothalamic tract originates from a distinct subset of retinal ganglion cells. *Journal of Comparative Neurology* **352**:351–366.

Norgren, R. B., McLean, J. H., Bubel, H. C., Wander, A., Bernstein, D. I., and Lehman, M. N. (1992). Anterograde transport of HSV-1 and HSV-2 in the visual system. *Brain Research Bulletin* **28**:393–399.

O'Donnell, P., Lavin, A., Enquist, L. W., Grace, A. A., and Card, J. P. (1997). Interconnected parallel circuits between rat nucleus accumbens and thalamus revealed by retrograde transsynaptic transport of pseudorabies virus. *Journal of Neuroscience* **17**:2143–2167.

Parent, A. (1996). Neurons. *In* "Carpenter's Human Neuroanatomy," 9th ed., pp. 131–198. Williams & Wilkins, Baltimore.

Peeters, B., de Wind, N., Broer, R., Gielkens, A., and Moormann, R. (1992a). Glycoprotein H of pseudorabies virus is essential for entry and cell-to-cell spread of virus. *Journal of Virology* **66**(3888).

Peeters, B., de Wind, N., Hooisma, M., Wagenaar, F., Gielkens, A., and Moormann, R. (1992b). Pseudorabies virus envelope glycoproteins gp50 and gII are essential for virus penetration, but only gII is involved in membrane fusion. *Journal of Virology* **66**:894–905.

Peeters, B., Pol, J., Gielkens, A., and Moorman, R. (1993). Envelope glycoprotein gp50 of pseudorabies virus is essential for virus entry but is not required for spread in mice. *Journal of Virology* **67**:170–177.

Rassnick, S., Enquist, L. W., Sved, A. F., and Card, J. P. (1998). Pseudorabies virus induced leukocyte trafficking into rat CNS. *Journal of Virology* **72**(11):9181–9191.

Rauh, I., and Mettenleiter, T. C. (1991). Pseudorabies virus glycoprotein gII and gp50 are essential for virus penetration. *Journal of Virology* **65**:5348–5356.

Rinaman, L., Card, J. P., and Enquist, L. W. (1993). Spatiotemporal responses of astrocytes, ramified microglia, and brain macrophages to central neuronal infection with pseudorabies virus. *Journal of Neuroscience* **13**(2):685–702.

Rinaman, L., and Levitt, P. (1993). Establishment of vagal sensorimotor circuits during fetal development. *Journal of Neurobiology* **24**:641–659.

Rinaman, L., Levitt, P., and Card, J. P. (2000). Progressive postnatal assembly of limbic–autonomic circuits revealed by central transneuronal transport of pseudorabies virus. *Journal of Neuroscience* **20**:2731–2741.

Rinaman, L., Roesch, M. R., and Card, J. P. (1999). Retrograde transsynaptic pseudorabies virus infection of central autonomic circuits in neonatal rats. *Developmental Brain Research* **114**:207–216.

Sawitzky, D., Voigt, A., and Habermehl, K. O. (1993). A peptide-model for the heparinbinding property of pseudorabies virus glycoprotein gIII. *Med. Microbiol. Immunol.* **182**:285–292.

Standish, A., Enquist, L. W., Miselis, R. R., and Schwaber, J. S. (1995). Dendritic morphology of cardiac related medullary neurons defined by circuit-specific infection by a recombinant pseudorabies virus expressing β-galactosidase. *Journal of Neuro Virology* **1**:359–368.

Strick, P. L., and Card, J. P. (1992). Transneuronal mapping of neural circuits with alpha herpesviruses. *In* "Experimental Neuroanatomy. A Practical Approach" (J. P. Bolam, ed.), pp. 81–101. IRL Press at Oxford University Press, Oxford.

Sun, N., Cassell, M. D., and Perlman, S. (1996). Anterograde, transneuronal transport of herpes simplex virus type 1 strain H129 in the murine visual system. *Journal of Virology* **70**:5405–5413.

Tirabassi, R. S., Townley, R. A., Eldridge, M. G., and Enquist, L. W. (1997). Characterization of pseudorabies virus mutants expressing carboxy-terminal truncations of gE: Evidence for envelope incorporation, virulence, and neurotropism domains. *Journal of Virology* **71**:6455–6464.

Toth, I. E., and Palkovits, M. (1997). Viral labelling of synaptically connected neurons. *Neurobiology* **5**(1):17–41.

Ugolini, G. (1995). Transneuronal tracing with Alpha-herpesviruses: A review of the methodology. *In* "Viral Vectors. Gene Therapy and Neuroscience Applications" (M. G. Kaplitt and A. D. Loewy, eds.), pp. 293–318. Academic Press, San Diego.

Ugolini, G., Kuypers, H. G. J. M., and Simmons, A. (1987). Retrograde transneuronal transfer of herpes simplex virus type 1 (HSV1) from motoneurons. *Brain Research* **422**:242–256.

Vahlne, A., Nystrom, B., Sandberg, M., Hamberger, A., and Lycke, E. (1978). Attachment of herpes simplex virus to neurons and glial cells. *Journal of General Virology* **40**:359–371.

Vahlne, A., Svennerholm, B., Sandberg, M., Hamberger, A., and Lycke, E. (1980). Differences in attachment between herpes simplex type 1 and type 2 viruses to neurons and glial cells. *Infection and Immunity* **28**:675–680.

Weinstein, D. L., Walker, D. G., Akiyama, H., and McGeer, P. L. (1990). Herpes simplex virus type 1 infection of the CNS induces major histocompatibility complex antigen expression on rat microglia. *Journal of Neuroscience Research* **26**:55–65.

Whealy, M. E., Card, J. P., Robbins, A. K., Dubin, J. R., Rziha, H.-J., and Enquist, L. W. (1993). Specific pseudorabies virus infection of the rat visual system requires both gI and gp63 glycoproteins. *Journal of Virology* **67**(7):3786–3797.

Whitley, R. J. (1990). Herpes Simplex Virus. *In* "Virology" (B. Fields, D. M. Knipe, *et al.*, eds.), pp. 1843–1887. Raven Press, New York.

Yang, M., Card, J. P., Tirabassi, R. S., Miselis, R. R., and Enquist, L. W. (1999). Retrograde, transneuronal spread of pseudorabies virus in defined neuronal circuitry of the rat brain is facilitated by gE mutations that reduce virulence. *Journal of Virology* **73:**4350–4359.

Yates, B. J., Smail, J. A., Stocker, S. D., and Card, J. P. (1998). Transneuronal tracing of neural pathways controlling activity of diaphragm motoneurons in the ferret. *Neuroscience* **90:**1501–1513.

Zemanick, M. C., Strick, P. L., and Dix, R. D. (1991). Direction of transneuronal transport of herpes simplex virus 1 in the primate motor system is strain-dependent. *PNAS* **88:**8048–8051.

ADVANCES IN VIRUS RESEARCH, VOL 56

NEUROVIROLOGY AND DEVELOPMENTAL NEUROBIOLOGY

John K. Fazakerley

Laboratory for Clinical and Molecular Virology
University of Edinburgh
United Kingdom

I. INTRODUCTION

Many viruses can replicate in the developing central nervous system (CNS) and many virus infections of this system are fatal or result in devastating disease with major sequelae. Indeed, the developing rodent brain is so effective in replicating a wide variety of viruses that before tissue culture systems were available, and for some time afterward, many viruses were amplified by passage in suckling mouse

73

brain and were titrated by determining an intracerebral LD_{50}. The situation in the mature CNS is different: whereas some viruses remain capable of spread and destruction in this tissue, many that replicate and produce disease in the developing CNS are not a problem in the mature CNS; some appear not to be able to replicate there, while others do so but only to a limited extent. Furthermore, there are several viruses that are able to persist in the mature adult CNS, often with minimal or no apparent tissue damage. These age-related changes in viral infection, replication, and virulence in the CNS have been the subject of study for over 50 years. For an early view of this field, see Lennette and Koprowski (1944) and Sigel (1952). For an update on neurobiology, see Zigmond et al. (1999). For a general review on virus infections of the fetus and neonate, see Arvin (1997). For a detailed account of virus infections of the nervous system, see Johnson (1998).

This chapter will first present an overview of CNS developmental events for the virologists (Section II); and then, for the neurobiologists, an overview of how virus infections depend on, and can alter, cellular differentiation states (Section III). This is followed (in Section IV) by a consideration of alphavirus infection of the developing CNS that provides a detailed example of the complexity of events. Section V reviews a few other selective examples of infections of the developing CNS that are chosen to illustrate that this is highly dependent on the time of infection and the area infected. Section VI briefly considers important infections of the developing human CNS.

II. Central Nervous System Development

In vertebrates the brain is by far the most complex organ in the body. In the adult human the CNS contains in excess of 10^{10} cells, with several different principal cell types, including hundreds of different neurons, connected into fantastically complex networks by some 10^{14} synaptic contacts. In the human genome, estimates suggest there are somewhere between 30,000 and 60,000 genes, around 80% of which may be expressed in the CNS at some time in life. As would be expected from these figures, development of this system is extremely complex. In comparison, our understanding remains relatively basic. The mature CNS is comprised of several principal cell types including neurons, astrocytes, oligodendrocytes, microglia, and supporting cells such as meningeal, ependymal, choroid plexus, and cerebral endothelial cells.

Further complexity is readily apparent in that hundreds of different subtypes of neurons are recognized; these are based, for example, on morphology, location, or neurotransmitter system. There are subtypes of glial cells too. Each of these major cell types and their subpopulations are generated, differentiated, and activated at different and sometimes multiple times in development, with major variations found between different brain regions. Development of the vertebrate nervous system starts early in embryogenesis with formation of the neural plate early after gastrulation, and in mammals it continues until well after birth. As an aid to the discussion and understanding of this complex developmental process, events can be divided up in a number of ways. A useful one is to consider three major series of events: first, cytogenesis and histogenesis; second, growth and differentiation; and third, sculpting of neural circuits.

A. Cytogenesis and Histogenesis

In mammals, cytogenesis and histogenesis are accomplished relatively quickly. Most cells are formed prior to birth. In the human forebrain for example, cytogenesis and histogenesis are largely completed by midgestation. However, this is a broad simplification of events, and certain regions of the mammalian brain undergo cytogenesis well after this time—in some cases, into postnatal life. For example, the dentate gyrus, the cerebellum, and the olfactory bulb have significant postnatal neurogenesis (Das and Altman, 1971; Altman, 1972a,b,c; Bayer, 1980a,b; Brunjes and Frazier, 1986).

The majority of CNS cells, both neurons and glia, are generated from progenitor cells in the pseudostratified neuroepithelium of the neural tube. These specific sites of neurogenesis require that cells migrate, sometimes considerable distances, to reach their final locations. These histogenic migrations have been particularly well studied in the cerebral cortex, where cells generated in the germinal ventricular zone migrate outward along radial glia to their specifically determined final positions (Rakic, 1972, 1988). Successively generated cells push through cells already present and come to lie in more superficial layers (Angevine and Sidman, 1961; Berry et al., 1964; Rakic, 1972, 1974; Tan and Breen, 1993). This radial inside-out formation results in least mature neurons being in layer II and most mature neurons, in layers I and VI in a cortex which is largely composed of its ontogenic columns (Rakic, 1988). This early period of primary neurogenesis and migration generates the principal neurons of the forebrain, midbrain, and cerebellar cortex. At later time points, secondary germinal zones form in the subventricular zone that overlies the ventricular zone. These give rise to late developing neuronal and glial subpopulations such as granule cells of the cerebellum, dentate gyrus, and the olfactory bulb, and astrocytes and oligodendrocytes in the cerebral cortex (Altman, 1969; Sidman and Rakic, 1973; Levison and Goldman, 1993; Lois and Alvarez-Buylla, 1994). In humans, this secondary neurogenesis continues into the second year of life. In the mouse it continues into the second postnatal week. In the mouse, low levels of neurogenesis continue into adulthood in the subventricular zone of the forebrain. Some of these neurons migrate long distances to the olfactory bulb (Lois and Alvarez-Buylla, 1994). In the adult mouse, this migration pathway, known as the rostral migratory stream, forms a continuous chain of neuronal precursor cells, from the subventricular zone of the forebrain to the olfactory bulb, that continues throughout life (Lois et al., 1996).

B. Growth and Differentiation

After completion of mitosis and generally after migration and expansion of the brain tissue, neurons undergo a phase of growth and differentiation. This is largely determined by a combination of lineage-dependent and often fixed gene-expression patterns directed by specific complements of transcription factors and inherited cytoplasmic factors that, to some extent, are modified in response to environmental signals. These signals originate from the substratum, adjacent cells, or soluble factors and clearly vary with location and with time as the system develops. Formation of neurites, both axons and dendrites, is directed by growth cones (Speidel, 1933; Landis, 1983). These are terminally expanded specialized organelles that guide the extending neurites in response to environmental signals such as the extracellular matrix molecules laminin and tenascin; cell surface molecules such as cadherins and anchored semaphorins; or soluble molecules such as netrins and soluble semaphorins (Hammarback et al., 1985; Kolodkin et al., 1993; Serafini et al., 1994; Takeichi, 1995). Growth cones are powered by actin polymerization (Condeelis, 1993). Behind the leading edge of the growth cone there are large numbers of mitochondria for energy generation and stacks of multilaminate membrane vesicles that fuse with the plasma membrane to facilitate neurite extension (Dai and Sheetz, 1995).

C. Sculpting of Neuronal Circuits, Axonal Pruning, and Developmental Cell Death

Neurons interact via synapses, the formation and function of which involve an elaborate exchange of signals between two cells. A single

CNS neuron may receive input from thousands of synapses activated by a variety of transmitter systems that may be excitatory or inhibitory. During development a great excess of synaptic connections is formed and, in some brain structures, twice as many connections as are required in the mature system (Huttenlocher, 1984). Late in development many of these synapses and indeed many neurons are eliminated. In the rodent, a relatively trivial proportion of the full synaptic complement is formed during histogenesis. It is only later in development, as the neurons complete growth and differentiation, particularly their late axonogenesis, that the majority of synapses are formed. As neuronal circuits become established the axons become myelinated. In the mouse these processes occur in the first two weeks after birth and maximal connectivity is reached at this time (Aghajainian and Bloom, 1967, Hinds and Hinds, 1976; Larramendi et al., 1969). Peak synaptic complement in the human is reached in the first 6 to 12 months after birth. This is followed by a period of synapse elimination resulting from axonal pruning and cell death. Axonal pruning can be extensive—for example, in the neonatal mouse up to 50% of the somatosensory neocortical spinal projections are lost by this process (Crandall et al., 1985).

During CNS development many more neurons are produced than are required in the mature adult system. Following axonogenesis and synaptogenesis, those neurons which are not required are eliminated by programmed cell death. This process is a normal developmental feature of most tissue systems. Programmed cell death is generally equated with apoptosis, a well-characterized form of cell death that is defined by a series of morphological and biochemical events. However, apart from studies on often transformed neural-derived cells in culture, there has been relatively little direct characterization of the details of the cell death process of CNS cells in tissues during developmental processes. Clearly this is not an easy issue to address, but based on morphological changes, there is good reason to challenge whether all of the developmental programmed cell death seen in the CNS is consistent with the by now classical descriptions of apoptosis derived from studies of cell death in continuously cultured cells and other tissues (Clarke, 1998; Oppenheim, 1999). It is nevertheless clear that there is developmentally regulated death of cells at all stages in CNS development, from mitotic progenitor cells in the ventricular zone to differentiating neurons and glial cells in the postnatal brain (Cowan and O'Leary, 1984; Oppenheim, 1991; Barres et al., 1992; Blaschke et al., 1996). This developmental programmed cell death functions to eliminate excess cells and it has been observed in many vertebrate brain structures including the cerebral cortex, dentate gyrus, cerebellum, thalamus, striatum, globus pallidus, and substantia nigra (Ferrer et al., 1990; Ashwell, 1990; Waite et al., 1992; Gould and McEwan, 1993; Janec and Burke, 1993; Waters et al., 1994).

The programmed cell death of postmitotic neurons that occurs late in development, for example in the postnatal rodent brain, can eliminate up to 50% of all neurons. The process is regulated by competition between neurons for neurotrophic survival factors, of which the earliest described and best-known example is the competition between peripheral neurons for limited supplies of nerve growth factor (NGF) produced by innervated target tissues (Levi-Montalcini, 1987; Crowley et al., 1994). Many such neurotrophic factors are now recognized and the same survival mechanisms are at work in the CNS as in the peripheral nervous system. In addition to NGF, CNS survival factors include brain-derived growth factor (BDNF) and neurotrophins (NT) 3, 4/5, and 6. Specific neuronal populations require neurotrophin support at different times in their development and may require different neurotrophins at different times or more than one neurotrophin at any one time (Barde, 1989; Davies, 1994a,b; Klein, 1994). For example, NT-3 mRNA is most abundant in embryonic CNS development, whereas BDNF mRNA is maximal in the postnatal brain. Neurotrophins and their receptors are distributed throughout the mature brain but are particularly abundant in the hippocampus, where they may be required for the continuing plasticity of this system. Neurotrophins such as NT-3 also have a role in oligodendrocyte differentiation. Neurotrophins bind to members of the trk receptor family. An accessory binding molecule, p75[LNTR], has also been described. Many other soluble molecules have been identified that have been shown to affect survival, growth, or differentiation of at least some CNS cells either in vivo or in vitro; these include ciliary neurotrophic factor, glial-derived neurotrophic factor; transforming growth factor-β; fibroblast growth factors; insulin-like growth factors; leukemia inhibitory factor; platelet-derived growth factor. For a review, see Johnson (1999).

D. Changes in Blood–Brain Barrier and Other Developmental Events

In the mouse brain, endothelial cell precursors are identifiable as early as E10.5 (Qin and Sato, 1995), and it is clear that a fully formed impermeable blood–brain barrier is in place in the adult; surprisingly, though, it remains controversial as to the time in brain development at which this barrier becomes fully functional (Saunders *et al.*, 1991; Rubin and Staddon, 1999). In the rodent, serum proteins are excluded from the brain at relatively early times in brain development (Saunders *et al.*, 1991); for example, horseradish peroxidase is excluded from the rat brain by E16 (Risau *et al.*, 1986); however, ionic permeability continues into early postnatal life (Butt *et al.*, 1990). Developmental events in the brain may be affected by events in other tissues. For example, failure to develop, or abnormal development of a particular organ, can result in reduced supplies of neurotrophins, absence of support of innervating neurons, and subsequent death of CNS neurons. Such an effect was demonstrated experimentally in the early classic experiments leading to the discovery of NGF, where removal of the limb bud of the chick embryo resulted in reduced numbers of sensory and motor neurons in the region of the spinal cord, giving rise to limb innervation (Hamburger, 1934). Changes in immune responses are also important when considering age-related changes in the outcome of CNS virus infections. Immune-response changes have, for example, been suggested to be responsible for age-related changes in responses to mumps in hamsters, mouse hepatitis virus in mice, and herpes simplex virus in mice (Hirsch *et al.*, 1970; Morgenson, 1979).

III. Virus Replication

A. Cell Growth and Differentiation

Viruses are confined to using the metabolic and biosynthetic pathways of the cells they infect. These pathways vary not only between cell types but also within any given cell lineage, with stage of differentiation and between activated and resting cells. There are many well-documented examples of viruses that replicate in specific cell types and at specific stages in cell growth, differentiation, or activation. Examples would be reactivation of cytomegalovirus from latency by host cell differentiation (Dutko and Oldstone, 1981); activation of Maedi-Visna virus replication in monocytes during their differentiation into macrophages (Gendelman *et al.*, 1986); initiation of papillomavirus replication on differentiation of keratinocytes (Bedell *et al.*, 1991; Dollard *et al.*, 1993); and replication of minute virus in testicular cells on differentiation (Guetta *et al.*, 1986). One key mechanism mediating these effects in the above examples is regulation of viral gene transcription. This has been well documented for some DNA viruses such as papovaviruses and herpesviruses and for the reverse transcribing retroviruses. In these cases, specific enhancers and transcriptional activators have been shown to be important and to vary between cell types and stages of cellular differentiation. For example, replication of polyomavirus varies both with tissue type and with time in a specific tissue and is dependent on host factors that affect viral transcription (Wirth et al., 1992).

Events are clearer for DNA viruses but RNA virus replication may also depend on host cell differentiation or activation state. For example, vesicular stomatitis virus replicates only in activated but not resting T lymphocytes (Bloom et al., 1970), and measles virus, Semliki Forest virus and lymphocytic choriomeningitis virus infections are restricted in mature neurons (Miller and Carrigan, 1982; Fazakerley et al., 1993; De La Torre et al., 1993). Like DNA viruses, this host effect on RNA virus replication is also likely to depend on specific factors for RNA replication, transcription, or translation (Lai, 1998). Cellular proteins may be required as integral components of the RNA-dependent RNA polymerase or they may bind to the viral RNA to initiate replication. For example, the RNA-dependent RNA polymerase of vesicular stomatitis virus requires the EF-1 eukaryotic translation factor for activity (Das et al., 1998); hn-RNP-E binds to poliovirus RNA and then forms a complex with the viral polymerase (Gamarnik and Andino, 1997); and the cellular La antigen binds to many viral RNA molecules and may be required for RNA synthesis (Kurilla and Keene, 1983; Wilusz et al., 1983; Kurilla et al., 1984; Meerovitch et al., 1993; Pardigon and Strauss, 1996). Other cellular factors such as tubulin, actin, and membranous vesicles are also required by many viruses (Moyer et al., 1986; De et al., 1991; Egger et al., 1996). The importance of membranes is discussed further in Section IV,C,2.

B. Viral Gene Modulation of Host Cells

In many cases, viruses do not just rely on host cell mechanisms to support their replication; they can be proactive, and many viral genomes contain genes whose products function to change various physiological parameters of the cell in order to initiate or optimize virus replication. These genes are targeted at a number of cellular functions. A prime example would be viral genes such as the adenovirus E1A or the polyomavirus large T, which serve to push the host cell into replication, thus generating the key metabolites and enzymes required for the viruses' own replication (reviewed by Nevins and Vogt [1996]). Viral replication can also be enhanced by genes that inhibit processing of cellular mRNAs or that inhibit host cell protein synthesis (Crone and Keene, 1989; Fortes et al., 1994; Leibig et al., 1993). Another example would be genes such as adenovirus E1B-19K or the gammaherpesvirus BHRF1 genes that function to prevent the host cell from undergoing apoptosis (Rao et al., 1992; Tarodi et al., 1994). There are also viral genes aimed at disrupting host immune response, such as the UL18 gene of cytomegalovirus or the ICP47 gene of Herpes simplex virus, which inhibit MHC-I presentation of peptides; or the various genes of poxviruses designed to counter the action of cellular cytokines and interferons (Spriggs et al., 1992; Hill et al., 1994, 1995; Fruh et al., 1995).

C. Viruses and Host Cell Death

As discussed in Section II,C, programmed cell death is a normal developmental event in many tissues including the CNS. Over recent years it has become clear that on viral infection, many cells die by apoptosis. In a complex multicellular organism, rapid cell suicide on virus infection can be seen as an altruistic response (Allsopp and Fazakerley, 2000; Fazakerley and Allsopp, 2000). If it occurs before complete virus replication and assembly, it will be highly effective in limiting virus spread. Suicide of virally infected cells can therefore be considered to be an early, perhaps the earliest, defense against viral infection. Many, and possibly most, large viruses that, due to their size and complexity, have relatively long replication times, encode antiapoptotic genes. It can be argued that these were acquired and maintained in the viral genome as a result of their provision of a strong selective advantage against the highly effective strategy of the host cell to initiate cell suicide.

Theoretically, "altruistic cell suicide" would seem to be a highly effective antiviral strategy for cell types that are replaceable, but this argument is less convincing for cell types that are irreplaceable and even less so for cells that are both irreplaceable and vital. For example, rapid apoptosis on infection of a group of skin epithelial cells might prevent spread of infection and these cells will be rapidly replaced, whereas apoptotic death on infection of postmitotic mature neurons in the respiratory center could be life-threatening. Theoretically, it would seem advantageous to the survival of the organism if the postmitotic, long-lived neurons of the mature CNS were exceptions to the concept of altruistic suicide upon infection (Allsopp and Fazakerley, 2000). Indeed, it seems that mature, terminally differentiated neurons are in fact relatively resistant to undergoing cell suicide in response to some virus infections (Fazakerley and Allsopp, 2000). In keeping with this, neither do mature neurons participate in fratricidal T-cell immune surveillance; CNS neurons fail to

display MHC-I molecules at the cell surface, although dysfunctional neurons that no-longer fire action potentials do so in response to interferon-γ
(Neumann et al., 1995). Thus it can be argued that for a mature postmitotic neuron, infection and survival have been selected over death, so
long as the neuron remains functional. A developmental change from
apoptosis susceptible to relative resistance could be a key determinant in
age-related changes for several neurotropic viruses and this is issue is
discussed further in Section IV,C,3.

IV. ALPHAVIRUS INFECTIONS AS EXAMPLES OF AGE-RELATED NEUROVIRULENCE

A. Alphavirus Encephalitis in Humans

In humans, eastern equine encephalitis (EEE) and western equine
encephalitis (WEE) have a propensity to attack the very young. With
EEE, about two-thirds of cases are seen in children under 2 years of
age. The mortality rate is very high and most survivors have severe
sequelae. In general, the younger the child, the more severe the sequelae (Farber et al., 1940; Fothergill et al., 1938). WEE also shows a
predilection for young children. Of 375 confirmed cases in California in
1952, 104 (28%) were seen in children under 1 year of age at onset
(Aguilar et al., 1968). Around 50% of children who survived this epidemic developed sequelae with an inverse relationship between age at
infection and incidence of these sequelae (Table I) (Aguilar et al., 1968).
The sequelae in WEE and EEE are similar and include recurrent
seizures, motor disturbances, behavioral and emotional problems, mental retardation, and cerebellar disorders (Noran and Baker, 1943; Mulder et al., 1951; Palmer and Finley, 1956). Sensory deficits are rare. In
many cases the sequelae become increasingly apparent with age. As in
humans, young horses are more susceptible than old ones to these
viruses (Eklund and Blumstein, 1938).

The higher incidence of disease in children could stem either from
decreased susceptibility to CNS disease processes with age or, alternatively, from decreased frequency of CNS infection with age. The latter
could be due to changes at the blood–brain barrier or to extraneural
factors—for example, an age-related change in the ability of virus to
replicate in other tissues due to developmental events, an increased
protective immunity in adults, or a greater frequency of infection in
children. Although infections from these mosquito-borne viruses may
be more common in children, WEE and EEE are not endemic in
human populations; increased specific immunity with age is therefore

TABLE I

WESTERN EQUINE ENCEPHALITIS INFECTION[a]

Age at onset (months)	With sequelae	Without sequelae
Under 1	23 (79%)	6 (21%)
1	34 (51%)	33 (49%)
2 to 5	22 (40%)	33 (60%)
6 to 11	5 (36%)	9 (64%)
All ages	84 (51%)	81 (49%)

[a] Number of children under 1 year of age with sequelae following an outbreak of western equine encephalitis infection in California in 1952. Data taken from Aguilar et al. (1986).

unlikely to be a factor. As described in Section II, in the first 2 years after birth, there is much ongoing development in the human brain. This includes axonogenesis, synaptogenesis, myelination, and regressive events such as axonal pruning and apoptotic death of excess cells. It seems likely that the developing brain may be better able to replicate virus than the mature brain and may be more prone to disease processes. This has been observed in experimental animal systems and is described in detail in the next section. The decreased severity of disease and sequelae with increasing age of infection in infants is entirely consistent with progressive completion of CNS developmental changes during the first two postnatal years.

B. Semliki Forest Virus Infection and Developmental Changes in Mouse Brain

Semliki Forest virus (SFV), Sindbis virus (SV), and Venezuelan equine encephalitis (VEE) virus are all able to infect mice and have each been extensively studied as experimental infections of the laboratory mouse. All three viruses demonstrate marked age-related changes in neurovirulence (Reinarz et al., 1971; Johnson et al., 1972; Fleming, 1977; Woodward et al., 1978; Oliver et al., 1997). In each case this is strain dependent. Some strains are virulent for mice of all ages from the neonate to the adult while others are virulent in neonates but not adults. The transition in susceptibility occurs around the age of weaning. From the point of view of developmental changes in the CNS, events have been studied in the greatest detail for SFV and these are presented in the following discussion.

SFV infection of the CNS has been studied in fetal, neonatal, suckling, and adult mice. In pregnant mice the A7 strain infects the placenta and then the developing embryos; infection is lethal (Atkins *et al.*, 1982; Milner and Marshall, 1984). In contrast, infection of pregnant mice with the ts22 mutant of A7 can result in congenital defects including neural tube abnormalities (Mabruk *et al.*, 1988, 1989). Interestingly, these do not seem to result from direct effects of virus infection on developing neural cells, but from secondary effects of infection of surrounding mesenchymal cells (Mabruk *et al.*, 1989). A similar situation is observed with influenza virus infection where infection of chick embryo chorionic and amniotic membranes results in abnormal early CNS development including neural tube defects without apparent infection of the neuroectoderm (Hamburger and Habel, 1947; Johnson *et al.*, 1971).

Neonatal mice inoculated intraperitoneally with the A7 or A7(74) strain of SFV rapidly die from fulminant encephalitis, whereas intraperitoneal infection of 3- to 4-week-old mice results in a subclinical encephalitis. In each case a plasma viremia is established, from which virus crosses cerebral capillaries to initiate small perivascular foci of CNS infection (Pusztai *et al.*, 1971; Fleming, 1977; Pathak and Webb, 1974; Fazakerley *et al.*, 1993). In mice inoculated at postnatal day 12 (P12) or before, these foci rapidly enlarge, resulting in a panencephalitis (Oliver *et al.*, 1997). In contrast, in 3- to 4-week-old mice the perivascular foci do not enlarge with time, and after 10 days postinfection no virus can be detected in the brain by infectivity assay or *in situ* hybridization (Fazakerley *et al.*, 1993). These weaned and adult mice develop a demyelinating encephalomyelitis, the lesions of which are apparent by 14 days postinfection. In weaned mice, clearance of infectious virus from the CNS and the lesions of demyelination that develop are dependent on specific immune responses (Jagelman *et al.*, 1978; Fazakerley and Webb, 1987a,b; Subak-Sharpe *et al.*, 1993). However, the age-related difference in virus spread and virulence cannot be attributed to maturity of specific immunity since, in 3- to 4-week-old athymic *nu/nu* mice (with no T lymphocytes), or in 3- to 4-week-old mice with severe combined immunodeficiency (SCID, with no T or B lymphocytes), CNS infection does not become widespread and virus persists for many months in small foci scattered throughout the CNS (Fazakerley and Webb, 1987a; Fazakerley *et al.*, 1993; Amor *et al.*, 1996).

Detailed studies of the spread of SFV infection in the brains of mice of different ages, following intraperitoneal or intranasal inoculation, demonstrates a correlation between infectivity and maturation of neural pathways (Oliver *et al.*, 1997; Oliver and Fazakerley, 1998). This correlation is striking and particularly apparent in the cerebral cortex, hippocampus, cerebellum, and olfactory bulb. These studies provide one of the best examples of age-related changes in the course of a neurotropic virus infection and will therefore be considered here in some detail. Fig. 1 shows an overview of the spread of SFV in the brains of mice of different ages.

Age-related changes in the distribution of SFV infection in the cerebral cortex can be seen in Figs. 1 and 2 (Oliver *et al.*, 1997). The early columnar distribution of infection correlates well with the known columnar interconnectivity between neurons in layers II, III, and V and with the functional organization of the cortex (Purves *et al.*, 1992; Malach, 1994), and is consistent with virus spread along developing neuronal processes. The columns of infected cells are broad and overlapping in the prefrontal, frontal, forelimb, and hindlimb cortical areas, and narrow in the occipital cortex (Fig. 2). This reflects the respective broad and narrow banding characteristic of cortical motor and sensory neuronal groupings (Purves *et al.*, 1992). During the first week after birth in the mouse, widespread axonogenesis and synaptogenesis take place to establish connectivity between the functional columns of cortical neurons (Eayrs and Goodhead, 1959; Rees *et al.*, 1976; Kristt, 1978; Landis, 1983; Agmon *et al.*, 1993). The reduction in the number of infected columns between P4 and P8 correlates with the completion of columnar connectivity at this time.

The change in infectivity of the hippocampal system also correlates with the time of its maturation. Most of the afferent inputs and synapses in the hippocampal cortex have been described in the rodent (Bayer, 1980a,b, 1985; Brown and Zador, 1990). Direct and indirect pathways connect the hippocampal formation and the subiculum with the septal nuclei, hypothalamus, thalamus, and widespread regions of the cerebral cortex and midbrain reticular formation. In mice inoculated at P2, P4, and to a lesser extent P6, viral RNA-positive cells are present in each of these regions and infection extends into the processes of these cells (Fig. 2). As the animals age and the system matures, spread of virus along hippocampal pathways is progressively reduced such that by P8, infected cells are often seen only in initial afferent entry points. This reduction in infectivity corresponds to the time at which hippocampal connectivity is completed (Bayer, 1985; Brown and Zador, 1990).

The cerebellum is one of the brain regions that develops late and to a large extent postnatally. As with granule cells of the dentate gyrus and olfactory bulb, cerebellar granule cells are still actively being produced at birth (Altman, 1972a,c). In the first two postnatal weeks, these cells migrate through the Purkinje cell layer from the external to the internal granule cell layer. During this period of migration in which the cells are also undergoing differentiation and synaptogenesis, these cells remain infectable by SFV (Fig. 2); after completion of migration and establishment of connections (P10–P12), granule cell

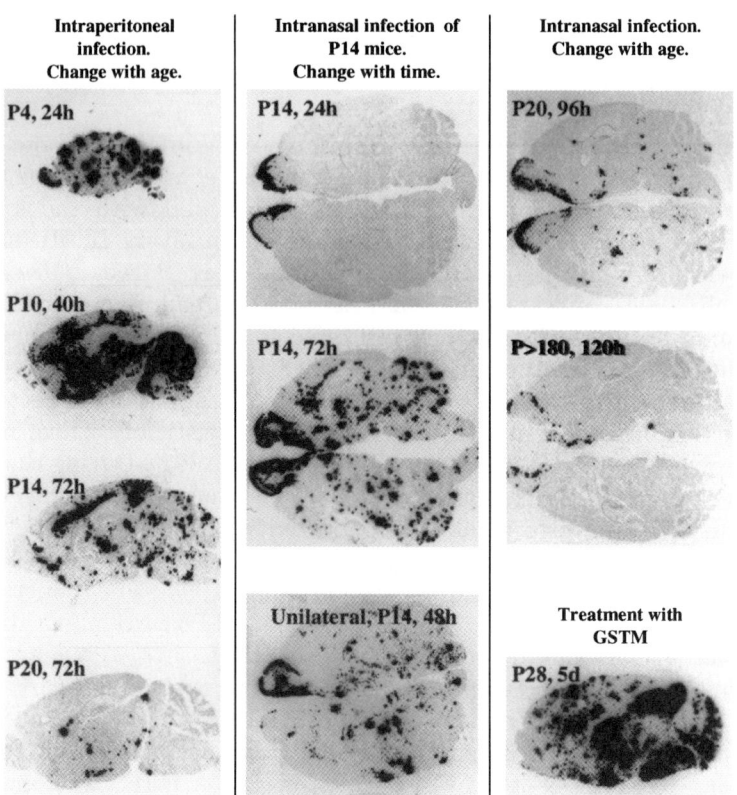

Intraperitoneal infection. Change with age.	Intranasal infection of P14 mice. Change with time.	Intranasal infection. Change with age.
P4, 24h	P14, 24h	P20, 96h
P10, 40h	P14, 72h	P>180, 120h
P14, 72h		
P20, 72h	Unilateral, P14, 48h	Treatment with GSTM P28, 5d

FIG 1. Autoradiographic images demonstrating the distribution of Semliki Forest virus–infected cells in mouse brains. In all cases the black signal demonstrates areas of hybridization of a ^{35}S-labeled riboprobe to viral genomic sense RNA. All sections are sagittal and are taken close to the midline. The olfactory bulb is to the left and the cerebellum and brain stem are to the right.

The four panels of single sections in the left-hand column show age-related changes in virus distribution following intraperitoneal infection. In the mouse inoculated at 4 days of age (P4) and studied at the earliest time point, 24h postinfection, columns of infected cells are clearly visible in the cortex; the hippocampus and cerebellum are also infected. As the mice age (P10, P14, P20), even with increased time after infection (40h and 72h are shown), fewer columns are infected in the cerebral cortex. In the mouse inoculated at P10, the hippocampus is no longer infected but the cerebellum remains heavily infected. In the mouse inoculated at P14, only the deep layers of the cerebral cortex are infected and there are now only foci of infection in the cerebellum and the hippocampus is again uninfected. Inoculation at P20 or later results in small, scattered foci of infection that never enlarge with time, even in mice with severe combined immunodeficiency (SCID) (Fazakerley *et al.*, 1993; Amor *et al.*, 1996). The changes in cortical and hippocampal distribution are shown in more detail in Fig. 2 and these studies are described in full in Oliver *et al.* (1997).

The first two panels in the middle column show changes in virus distribution with time, following intranasal inoculation of P14 mice. Each panel shows the left and right sides of the brain. At 24h only the main olfactory bulb is infected. By 72h the infection

infection is markedly reduced (Oliver *et al.*, 1997). Purkinje cells receive input from mossy fibers (via granule cells) and climbing fibers (directly). In the mature cerebellum, Purkinje cells receive input from only one climbing fiber originating in the contralateral inferior olive. In the neonate there are inputs from several climbing fibers that are progressively lost in the first few weeks of postnatal life. Around P12 there is a reduction in Purkinje cell apical cone and dendrite growth, and a reduction in synaptogenic rate between mossy fibers and granule cells (Larramendi, 1969; Altman, 1972b; Crepel *et al.*, 1976; Mariani and Changeux, 1981). In parallel with the maturation of this system the extent of SFV infection of Purkinje cells is progressively reduced from infection of the majority of cells in neonates and in mice infected at P10, to infection of only a few Purkinje cells in mice infected at P20 mice (Fig. 2) (Oliver *et al.*, 1997).

◄───

has spread into the anterior olfactory nucleus, the olfactory tubercle, and connected structures throughout the brain. In the main olfactory bulb, the banding represents infection of the neuronal cell bodies in the periglomerular and mitral cell layers that are separated by the external plexiform layer. The third panel in this column shows the distribution of infection following unilateral inoculation of virus into the animal's right nostril. At P14 the connections between the main olfactory bulb on one side of the brain and the anterior olfactory nucleus on the contralateral side are maturing and virus passes along these and spreads into connections on the contralateral side of the brain. After maturation of this connection at P16 (not shown here), unilateral inoculation results in infection of secondary and tertiary connections on the inoculated side only. The course of infection, following intranasal inoculation into mice of different ages, is described in detail in Oliver and Fazakerley (1998).

The first two panels in the right-hand column demonstrate the correlation between reduction in the spread of the infection and maturation of olfactory pathways with age. In the mice inoculated at P20 and studied at 96h postinfection, there is spread of infection from the main olfactory bulb into the anterior olfactory nucleus, the olfactory tubercle, and connected structures. The extent of the spread, both number and size of foci, is, however, less in the mice inoculated at P20 than in the mice inoculated at P14 (see the panels in the previous column). In mice inoculated at >P180, the olfactory connections are mature and in the majority of mice there is no spread beyond the main olfactory bulb. In the main olfactory bulb there is occasionally spread from the olfactory nerve layer into the periglomerular, mitral, and granule cell layers. Only rarely, as shown in panel 2 (>P180, 120h), is there spread beyond the main olfactory bulb into the olfactory tubercle. The final panel shows the extent of infection in the brain of an adult (P28) mouse infected intraperitoneally with SFV and given 10 mg gold sodium aurothiomalate (GSTM) 4 h prior to infection. Gold compounds result in spread of infection in the adult mouse brain (Scallan and Fazakerley, 1999). Compare the extent of infection in this brain with that shown in the bottom panel of the left-hand column, which shows the restricted focal distribution normally observed in immunocompetent or SCID mice inoculated at P20 or later. From Oliver *et al.* (1997) and Oliver and Fazakerley (1998). Reproduced with permission.

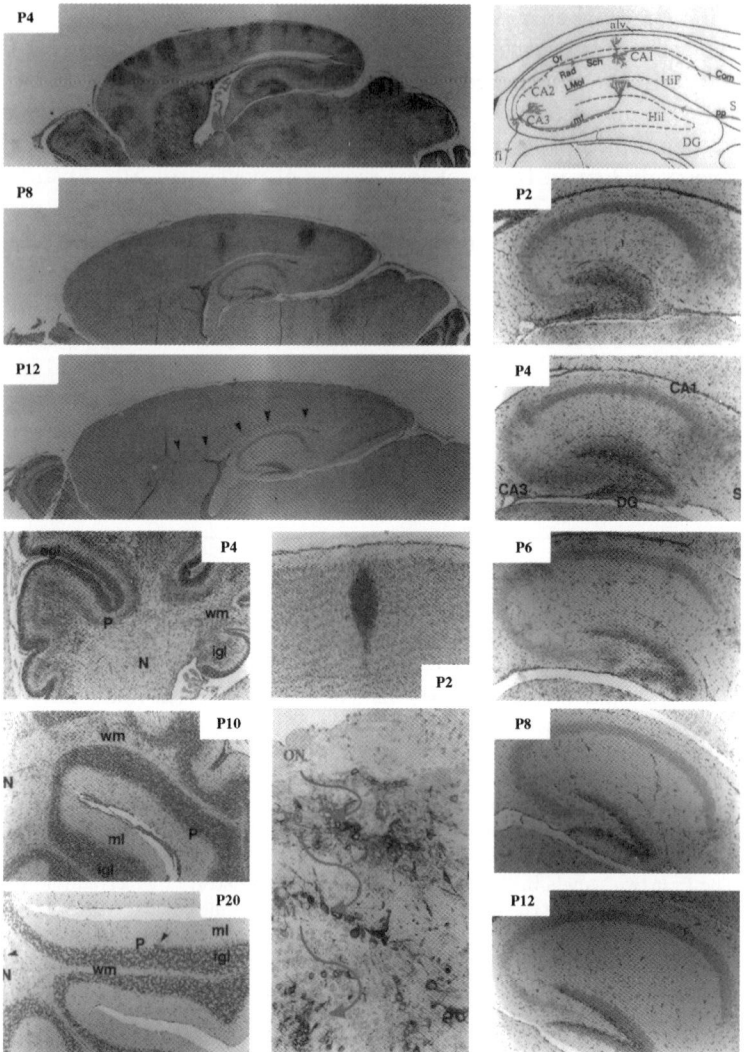

FIG 2. First column (top left): representative images showing sagittal sections close to the midline of the cerebral cortices of animals inoculated at P4, P8, and P12 and studied at 36, 40, and 48 hours postinfection, respectively. Viral RNA was detected with a digoxigenin-labeled riboprobe developed to give a brown signal, and sections were counterstained with hematoxylin and eosin. Mice were inoculated intraperitoneally. Virus reaches CNS cells via passage across cerebral endothelial cells. In the mouse inoculated at P4, columns of infected cells are apparent in the cortex. The columns are wider in the frontal and narrower in the occipital cortex and increase in number and extent with time after infection. Infected columns are fewer following inoculation at P8 and are not observed following inoculation at

P12. At P12 only cells in the deepest regions of layer VI of the cortex, and in the underlying cingulate gyrus and corpus callosum, are infected (arrowheads). These may be immature populations of cells. Note that the whole hippocampus is infected in the mouse inoculated at P4, but that only the subiculum is infected in the mouse inoculated at P8 and that there is no infection of the hippocampus in the mouse inoculated at P12. These studies are described in detail in Oliver *et al.* (1997).

Below the series of panels showing changes in the cortex are 3 panels showing age-related changes in the distribution of virus in the cerebellum. The different regions of the cerebellum are marked as follows: egl, external granule cell layer; igl, internal granule cell layer; ml, molecular layer; N, deep cerebellar nucleus; P, Purkinje cell layer; wm, white matter. In the mouse inoculated at P4 (studied at P6), the external granule cell layer remains extensive and widespread infection is apparent in the cells of the forming internal granule cell layer. These cells have recently migrated from the external granule cell layer and are undergoing axonogenesis and synaptogenesis at this time. In the mouse inoculated at P10 (studied at P12), formation of the internal granule cell layer is virtually complete and the external granule cell layer has almost disappeared. Foci of infection are still observed in the internal granule cell layer and the Purkinje cell remain infectable at this time. At late time points (e.g., P20) there, occasional foci of granule cell infection or infection of Purkinje cells are observed.

The upper of the two panels in the central portion of the lower half of the figure shows a column of infected cortical neurons at higher power (mouse is infected intraperitoneally at P2 and sampled at 24h). The panel beneath this shows widespread infection of the interconnected layers of the olfactory bulb and their connections (mouse is infected intranasally at P8 and sampled at 48h).

The column of panels on the right-hand side of this composite show age-related changes in the distribution of SFV infection in the hippocampus. The panel illustrates the structure and connectivity of the hippocampus in sagittal section (modified from Paxinos and Watson [1986], with permission). The structural regions of the hippocampus are marked: alv, alvius hippocampus; CA1, CA2, and CA3, pyramidal neuron layers; DG, dentate gyrus; fi, fimbria; HiF, hippocampal fissure; Hil, hilus dentate gyrus; Lmol, lacunosum moleculare layer; Or, oriens layer; S, subiculum; Rad, stratum radiatum. The principal input and output connections of the hippocampus are shown in red. The main inputs are the perforant pathway (pp), which connects with mossy fibers (mf), and the commissural input (com), which connects with Schaffer collaterals (Sch). Mice inoculated at P2 and P4 showed almost identical patterns of hippocampal infection and evidence of perforant pathway transmission. Neurons in the subiculum, granule cells in the external (upper) limb of the dentate gyrus, their dendrites in the molecular layer and hilus (Hil), mossy fibers and CA3 pyramidal cells and their tufts reaching into the radiatum (Rad), lacunosum moleculare, and stratum oriens are all positive. At this time, axons from the entorhinal cortex are entering the dentate gyrus and forming connections with granule cell dendrites in the molecular layer. The distribution of infection in the brain of mice inoculated at P6 is consistent with this spread, although the extent of infection is less. Mice inoculated at P8 had fewer SFV RNA-positive cells and those that were observed were usually confined to one defined area with little or no spread—for example, as shown here, infection of neurons in the subiculum. With the exception of rare, single neurons, cells positive for viral RNA were never observed in the hippocampus in animals inoculated at P12 and older. From Oliver *et al.* (1997) and Oliver and Fazakerley (1998), with permission.

Following intranasal infection, SFV can spread to the CNS along the olfactory nerve (Kaluza *et al.*, 1987). The course of SFV infection in the developing olfactory system has been studied in detail (Oliver and Fazakerley, 1998). Following intranasal inoculation of mice at P4 to P10, virus was observed to travel along olfactory pathways from the main olfactory bulb to secondary olfactory structures, including the anterior olfactory nucleus and the olfactory tubercle, and then to connections scattered throughout the brain (Fig. 1). In the main olfactory bulb, infection was first observed in the olfactory nerve layer; this was followed successively by infection in the periglomerular, mitral, and granule cell layers (Fig. 2). With increasing age at infection, the rate of spread through the system, the amount of viral RNA in infected cells, and the extent of infection of each of these layers were progressively reduced (Fig. 1). Between P10 and P20 there was a marked reduction in the extent of infection in the granule cell layer and by 6 weeks of age, infection of the main olfactory bulb was confined to a few cells that were increasingly rare with connective distance from the site of virus input. These age-related changes in the extent of infection of the main olfactory bulb correlated with changes in the expression of GAP-43, a marker of axonogenesis. In all layers of the main olfactory bulb except the olfactory nerve layer, GAP-43 expression decreased dramatically between P4 and P10 and was undetectable by P20. In the olfactory nerve layer, GAP-43 expression was observed even at late time points (Oliver and Fazakerley, 1998). The correlation between GAP-43 staining and ability of the virus to spread indicates that spread is curtailed as connectivity is completed. The GAP-43 results are consistent with previous studies on the development of the olfactory bulb. The main olfactory bulb of the mouse increases in size throughout the first two postnatal weeks, during which time there is extensive dendrogenesis, axonogenesis, and synaptogenesis (Hinds and Hinds, 1976; Brunjes and Frazier, 1986).

Ability of the infection to spread beyond the main olfactory bulb also correlated with maturation of neural pathways. Olfactory information is sent out of the main olfactory bulb to fore-, mid-, and hind-brain regions via axons of tufted and mitral cells. The regions which receive input from these cells include the anterior olfactory nucleus, olfactory tubercle, piriform cortex, amygdala, and entorhinal cortex. The anterior olfactory nucleus and piriform cortex acquire adultlike characteristics at the beginning and the end of the second postnatal week, respectively (Brunjes and Frazier, 1986; Schwob and Price, 1984). Cells in these regions were extensively SFV RNA-positive following intranasal inoculation at

P13 or before but showed a progressively decreasing infection between P14 and P20. The main olfactory bulb receives input from the anterior olfactory nucleus, tenia tecta, piriform cortex, and lateral olfactory tract and extensive infection of all of these areas was observed in mice inoculated before P7; thereafter, infection was reduced and was absent after P14. This correlates with the maturation of these connections, which occurs between P7 and P14 (Brunjes and Frazier, 1986; Schwob and Price, 1984). A particularly striking example of the correlation between virus spread and maturation of connectivity was seen following unilateral intranasal inoculation of virus. The cross-connections between the main olfactory bulb on one side of the brain and the anterior olfactory nucleus on the other side (contralateral) form at P14 and are mature by P16. Following unilateral inoculation of virus into one nostril, in P14 mice, extensive infection of the ipsilateral main olfactory bulb and the contralateral anterior olfactory nucleus were observed in the absence of widespread infection of the contralateral main olfactory bulb or of evidence of infection of the contralateral olfactory nerve layer (Fig. 1). In contrast, in P16 mice, widespread infection of the ipsilateral main olfactory bulb resulted in no infected cells in the contralateral anterior olfactory nucleus (Oliver and Fazakerley, 1998).

C. Understanding Events in Developing Brain Following Infection with Semliki Forest and Other Viruses

1. Do Viruses Require Specific Receptors to Infect Developing Neurons?

The above studies demonstrate that the distribution of SFV in the developing brain correlates with maturation of neuronal circuits, and suggests virus is only transmitted by or between neurons undergoing axonogenesis and synaptogenesis. One explanation of this would be a developmentally regulated change in the expression of the virus receptor molecule on neurons. The receptor for SFV remains unknown. The virus has been reported to bind to major histocompatibility molecules (Helenius et al., 1978), though this finding remains controversial. The neuronal receptor for SV has been suggested to be developmentally regulated (Ubol and Griffin, 1991). A second possibility is that virus may be able to enter immature developing neurons via a nonspecific mechanism. Extensive endocytosis occurs at developing postsynaptic membranes and this process ceases on maturation (Altman, 1971; Rees et al., 1976; Vaughn and Sims, 1978). This endocytosis could be sufficiently extensive to allow virus entry in the absence of a specific receptor. On cessation of this endocytosis at neuronal maturation, entry could thereafter depend on specific receptor-mediated endocytosis, which has been well studied for the prototype strain of SFV (Helenius *et al.*, 1980). In this regard, it is worth noting that the A7(74) strain of SFV was derived by serial passage through suckling mouse brain. In the absence of appropriate selective pressures, the virus could have lost its ability to bind, or to bind efficiently, to its receptor on mature neurons. Indeed, it is interesting to speculate that nonspecific uptake of viruses by developing neurons could explain, as noted in the introductory paragraph, the ability of the suckling mouse brain to replicate many, indeed nearly all, viruses tested. Given the active differentiation and, in some populations, the ongoing mitosis, and given the large number of different cell types, it seems likely that most viruses, if they have no block to entry, would be able to find, in the developing CNS, cells with a suitable transcriptional environment for their replication.

2. The Importance of Membranes

Another explanation for the age-related changes in SFV infection would be a developmentally related change affecting the ability of the virus to replicate and produce infectious virions in neurons. As discussed in Section II,A, this could result from changes in the complement of transcription factors that presumably occurs in many neuronal types with maturation. Ultrastructural studies and biochemical observations suggest that changes in membrane synthesis may be one such factor. Electron microscopic studies indicate that although SFV A7(74) can infect neurons in the adult mouse brain, this infection is rarely productive (Pathak and Webb, 1978, 1988a,b; Fazakerley *et al.*, 1993). For SFV, as with many other viruses, the processes of RNA replication, virion maturation, and budding each demonstrate an absolute requirement for continual smooth membrane production and phospholipid synthesis. Electron microscopic and biochemical studies have shown that SFV replicates on, and requires intracellular membranes derived from, endosomes and lysosomes (Acheson and Tamm, 1967; Grimley *et al.*, 1968, 1972; Grimley and Friedman, 1970; Friedman *et al.*, 1972; Pathak *et al.*, 1976; Froshauer *et al.*, 1988; Pathak and Webb, 1978; 1988a; Kuge *et al.*, 1989; Peranen and Kaarianen, 1991; Perez *et al.*, 1991; Peranen *et al.*, 1995). A similar requirement for smooth membranes is observed with several other viruses, even nonenveloped viruses, and in many cases these membranes must be recently synthesized and inhibition of this can prevent virus replication (Dales and Siminovitch, 1961; Amako and Dales, 1967; Caliguri and Tamm, 1970; Schlesinger and Malfer, 1982; Katoh *et al.*, 1986; Kuge *et al.*, 1989;

Guinea and Carrasco, 1990). Membrane synthesis is active in immature neurons undergoing axonogenesis and synaptogenesis. Neurite extension involves the fusion of membrane vesicles in the proximal growth cone with the plasma membrane, and these neurons undergo active endocytosis (Dailey and Bridgman, 1993; Dai and Sheetz, 1995; Altman, 1971; Rees *et al.*, 1976; Vaughn and Sims, 1978). As discussed above, in the mouse these events are particularly extensive in the first two postnatal weeks (Section II,C). The virus may utilize these newly synthesized smooth membranes for replication, maturation, or budding. With neuronal maturation the membranogenesis and endocytosis cease (Altman, 1971; Rees *et al.*, 1976). The absence of suitable membranes or components of their associated biosynthetic pathways could restrict virus replication or maturation, resulting in the observed switch from widespread infection to restricted foci of infection.

An interesting observation in support of a role for membrane synthesis comes from a series of experiments spanning several decades, in which it has been observed that administration of gold-containing compounds to adult mice infected with SFV A7(74) results in conversion of this nominally avirulent infection to a virulent infection (Allner *et al.*, 1974; Mehta and Webb, 1987; Scallan and Fazakerley, 1999). Gold, in common with other heavy metals, appears to affect neuronal differentiation state and the neurons of adult mice treated with gold compounds show a dramatic increase in smooth membrane production and a dramatic spread of virus around the brain (Pathak and Webb, 1983; Mehta *et al.*, 1990; Scallan and Fazakerley, 1999). SFV A7(74)–infected adult mice treated with gold compounds die within a few days but interestingly, at least by the time of death, at 4 to 5 days postinfection, this widespread infection in the adult mouse brain was not associated with neuronal apoptosis (Scallan and Fazakerley, 1999).

3. Changes in Susceptibility to Neuronal Apoptosis on Infection

SFV and SV infection of dividing cells in continuous culture results in death by apoptosis (Levine *et al.*, 1993; Scallan *et al.*, 1997). SFV infection of primary cultures of trigeminal sensory neurons in neonatal mice also results in death by apoptosis, as does SFV infection of neurons in the neonatal mouse brain (Allsopp *et al.*, 1998). Explanted dorsal root ganglia in neonatal mice undergo apoptosis if infected with SV at an early point after explantation, but not if infected after *in vitro* maturation (Levine *et al.*, 1993). Infection of neonatal, but not adult, mouse brains with the avirulent 633 strain of SV results in apoptosis (Lewis *et al.*, 1996). Studies with the A7(74) strain of SFV indicate that with age, brain neuronal populations differ in their susceptibility to virus-induced apoptosis. As discussed above, olfactory bulb neurons and migrating cerebellar granule cells undergo widespread SFV infection in mice inoculated between P1 and P10 (Oliver *et al.*, 1997; Oliver and Fazakerley, 1998). This infection is associated with widespread apoptosis of these developing neuronal populations, whereas more mature neurons that become infected in other parts of these same brains—for example, in areas of the cerebral cortex—do not readily undergo apoptosis (Fazakerley *et al.*, 2000). As the brain matures, the cells infected are more resistant to apoptotic death and in the adult, SFV A7(74) readily persists for life in the brains of immunocompromised mice without any apparent cell or tissue destruction (Fazakerley and Webb, 1987a; Amor *et al.*, 1996).

As introduced in Section II,C, apoptosis is a normal event in the developing nervous system and is particularly pronounced in the first two weeks of postnatal development, when many neurons that do not form correct connections are eliminated. In addition to the alphaviruses, a number of other CNS virus infections result in the induction of apoptosis in the developing CNS. These include La Crosse virus, the T3D (Dearing) strain of reovirus type 3, and human isolates of Dengue virus (Pekosz *et al.*, 1996; Oberhaus *et al.*, 1998; Despres *et al.*, 1998). In the mature CNS, induction of apoptosis varies between viruses and, importantly, between closely related strains of the same virus (Fazakerley and Allsopp, 2000). For example, the TE strain of Sindbis virus induces apoptosis in both the neonatal and the adult mouse brain, whereas the 633 strain of this virus, which differs from TE at only one amino acid, induces widespread apoptosis in the developing, but not the adult, mouse brain (Lewis *et al.*, 1996). In some cases, a difference in the rate or extent of apoptosis on infection has been observed with development. In the developing brains of suckling mice infected with the CVS strain of rabies virus, there is rapid (within 25h) and extensive apoptosis in some infected neuronal populations—for example, infected dentate gyrus cells undergo apoptosis while infected Purkinje cells do not (Jackson and Park 1998; Theerasurakarn and Ubol, 1998). In adult mice, intracerebral inoculation of rabies virus also results in apoptosis of some neurons but this occurs more slowly than in the developing brain (Jackson and Rossiter 1997). Infection of the mouse CNS with vesicular stomatitis virus results in apoptosis that is particularly striking in the olfactory bulb (Bi *et al.*, 1995).

4. Synthesis of Events

Fig. 3, which is based on the foregoing discussion in this section, provides a synthesis of events that seem at present the most likely explanation for the changes in the distribution and outcome of SFV infection in the developing mouse brain. This series of events may also occur with other CNS infections.

5. Stress, Shock, and Cytokines

Alphavirus infection of neonatal mice results in widespread infection of many tissues in addition to the developing brain. Whereas the apoptotic loss of crucial developing neuronal populations could be fatal, this has not been shown to be the cause of death. Death could equally result from pathology in other tissues. SV and SFV infections trigger cytokine gene transcription both in the CNS and in other tissues (Wesselingh et al., 1994; Trgovcich et al., 1997). Infection with the TRSB strain of SV is instructive in these matters; this strain is lethal for neonatal mice but produces only minimal encephalitis. Infection is associated with lesions in the thymus and hematopoietic tissues and with high systemic levels of tumor necrosis factor-α (TNFα), adrenocorticotropin-releasing hormone, and corticosterone, which together constitute a severe and potentially fatal systemic inflammatory response syndrome (Trgovcich et al., 1996, 1997). It has been suggested that this response and the associated hypovolemic shock may be more important than CNS events for the virulence of this virus in neonatal mice (Klimstra et al., 1999). Ascertaining the cause of death in a situation such as this is not easy since events are occurring in many tissues. Whatever the cause of death, though, these studies serve to remind us, first, that events in the developing brain do not necessarily occur in isolation, and second, that soluble factors may be important mediators. Infection alone may be sufficient to trigger neuronal apoptosis in susceptible cells of the developing mouse brain, but a soluble factor such as TNFα could also be important for inducing susceptibility. The TNFα-induced cell death pathway is well characterized (as reviewed by Ashkenazi and Dixit [1998]). In the olfactory bulb of P4 and P6 mice infected with SFV A7(74), there is, as described previously, widespread infection and apoptosis; however, many of the apoptotic cells do not stain positive for viral proteins (Fazakerley et al., 2000; Allsopp and Fazakerley, 2000). This could result from increased sensitivity of cells to apoptosis following exposure to cytokines, resulting in rapid apoptosis on infection before viral transcripts or viral proteins reach detectable levels. Apoptosis could also be triggered by

Developing brain (neonatal mice)

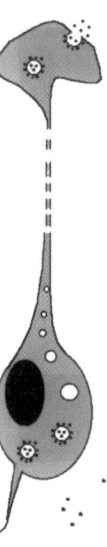

Axonogenesis, synaptogenesis and associated endocytosis and membrane synthesis are very active. Cells are relatively susceptible to apoptosis and many naturally die by this process. Virus uptake by endocytosis. Receptor mediated ?

Virus replicates on smooth membrane vesicles. Mature virus particles bud into these vesicles and at the plasma membrane.

Viral proteins are readily observed in neurites.

Infection spreads between connected cells.

Late developing regions such as the olfactory bulb rapidly undergo apoptosis (blue) following infection

maturation

Mature brain (adult mice)

Axonogenesis and synaptogenesis are complete. Endocytosis and membrane synthesis are down-regulated. Increased resistance to apoptosis ?

Membranous virus replication structures, mature virus particles and budding virions are not observed but large accumulations of viral RNA and protein are observed in the cell body.

Neurites do not stain for viral proteins.

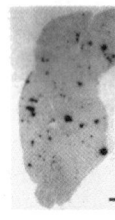

Infection remains focal even in SCID mice and can persist for life without apparent damage.

Treatment of adult mice with gold compounds results in widespread infection but this is not associated with apoptosis (absence of blue nuclei) in infected cells.

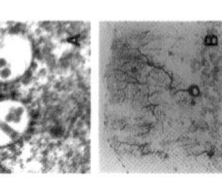
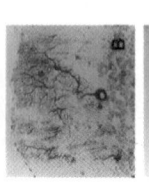
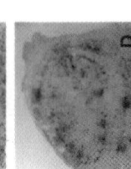
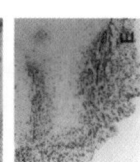
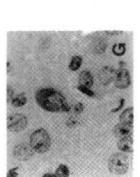
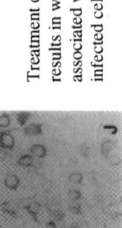

sufficiently high local levels of cytokines without any direct viral involvement. Alternatively, apoptosis of uninfected cells could be triggered by a fall in neurotrophic survival factors previously supplied by connected cells but lacking as infected connected cells die. Indeed, it would not be beyond the complexity of the systems for all of these events to be occurring sequentially or simultaneously.

V. Virus Infections of Developing Nervous System

The previous section reviewed age-related changes that occur in the well-studied alphavirus experimental model systems, posed some general questions, and presented some possible explanations. There are many virus infections that have been associated with CNS malformations or neurologic deficits that follow fetal or neonatal infection in both man and animals. The understanding of these varies enormously. For some there are only descriptions of the pathology and the clinical disease that range from brief and sporadic reports to extensive studies. For others there are, in addition, experimental studies on the pathogenesis. As a rule, the malformations observed reflect the region of the brain infected and the stage of gestation at which infection occurs. Infections early in gestation often destroy mitotic cells in the ventricular and subventricular germinal zones, resulting in cavitary lesions of the cortex. Infections later in gestation or in early postnatal life often destroy cells in the proliferating and differentiating secondary germinal zones; the most apparent clinical manifestation of this is often ataxia following disruption of cerebellar development. The following sections selectively review a few

$\longleftarrow$ ——

FIG 3. Schematic representation of age-related changes in mouse brains and their effect on SFV replication and the outcome of infection. A, Replication of SFV on smooth membrane vesicles known as cytopathic vacuoles. Nucleocapsids can be seen around the outside and mature virus particles within the vesicle; neonatal mouse brain. B, SFV infected Purkinje cell in the cerebellum of a mouse inoculated at P4. C, autoradiographic image showing the distribution of SFV infected cells, 24th after intraperitoneal infection of a P4 mouse. D, SFV-infected cells in the olfactory bulb, 48h postinfection, P8 mouse. E, widespread apoptosis in the olfactory bulb at 60h following intranasal inoculation at P8. F, large aggregates of viral proteins and RNA accumulate in the cell body of infected mature neurons. G, SFV-infected neuron in the brain of an adult mouse at 6 days postinfection. H, single SFV positive neuron in the brain of an adult mouse with severe combined immunodeficiency (SCID), 28 days postinfection. I, autoradiographic image showing the distribution of SFV-infected cells, 4 days after intraperitoneal infection of an adult mouse. A and F are kindly provided by Dr. S. Pathak. G and J are taken from Scallan et al. (1999), with permission. H is taken from Oliver et al. (1997), with permission.

additional examples of infections, both natural and experimental, in the developing CNS. The examples successively illustrate infections that give rise to malformations of multiorgan systems including the CNS; infections early in gestation that give rise to cavitary CNS lesions; and infections that affect the late-generating or late-maturing brain structures such as the cerebellum, the dentate gyrus, and the cortex.

A. Virus Infections Associated with Congenital Malformations of Multiorgan Systems Including CNS

Rubella is perhaps the best known of the viruses that infect the developing CNS. This was first recognized in 1941, after an epidemic of rubella in Australia that resulted in a number of children with cataracts and other congenital abnormalities (Gregg, 1941). Rubella virus produces a viremia in infected adults, and in pregnant women the virus can establish an infection first in the placenta and then in the developing embryo. Many embryonic tissues can be infected (Bellanti *et al.*, 1965). Maternal infection in the first trimester generally (>70%) results in congenital damage with sequelae apparent in the newborn (Munro *et al.*, 1987). The frequency of congenital defects declines progressively throughout the second trimester and these are rare in the third (Miller *et al.*, 1982). Early infection can result in resorption of the embryo, but premature delivery and stillbirth also occur. Congenital neurologic problems include cataracts, glaucoma, retinopathy, microphthalmia, microcephaly, hearing loss, bulging anterior fontanelle, lethargy, irritability, mental retardation, motor disabilities, abnormal posture and movements, and abnormal electroencephalograms (Miller *et al.*, 1982; Desmond *et al.*, 1967, 1969). Effects in other tissue systems include heart disease such as ductus arteriosus or valvular stenosis, thrombocytopenia, hepatomegaly, and splenomegaly (Cooper *et al.*, 1969). At term and in survivors for many months after, rubella virus can frequently be isolated from a number of organs and may be present in cerebrospinal fluid; virus is often excreted in nasopharyngeal secretions, feces, and urine.

The pathogenesis of these congenital malformations is not clear and a number of mechanisms may be operating. Infection can lead to destruction of infected cells, and studies of both aborted fetal and neonatal material demonstrate areas of focal necrosis in several tissues including the CNS and the vascular system (Desmond *et al.*, 1969; Rorke and Spiro, 1967). However, tissues often show only small numbers of scattered and morphologically normal rubella-positive cells. In one study of tissues from a first trimester abortus, it was estimated that only in the order of 1:1,000 to 1:100,000 cells was virus infected, but cultures derived from these tissues had reduced growth rates (Rawls and Melnick, 1966). *In vitro*, rubella virus is generally noncytopathic, but infection of human cells with rubella virus has been observed to reduce cell growth and perturb normal cell structure (Yoneda *et al.*, 1986). Both of these processes, cell death and disrupted cell growth, could be operating to produce the congenital problems associated with rubella virus infection. It is readily apparent that infections at different periods of organogenesis or histogenesis will result in varying outcomes. No experimental animal model system of congenital rubella virus infection has been studied in any great detail. Experimentally, rubella virus can infect baboons, chimpanzees, marmosets, ferrets, rabbits, and rats. Infection of rabbits produces congenital ocular malformations and infection of ferrets produces congenital CNS vascular lesions (Kono *et al.*, 1969; Rorke *et al.*, 1968).

The pestiviruses, classical swine fever virus, bovine viral diarrhea virus, and border disease virus have all been associated with congenital malformations. As with rubella virus, the fetus is infected via the placenta and the outcome of infection depends on the time of gestation (vanOirschot and Terpstra, 1977; Casaro *et al.*, 1971; Done *et al.*, 1980). For classical swine fever virus, infection in early gestation (before day 41) usually results in abortion or stillbirth; infection in midgestation (days 41 to 85), in live births with congenital malformations and a persistent viremia; and infection in late gestation (after day 85), in normal nonviremic animals. The CNS congenital malformations include microcephaly and cerebellar hypoplasia with destruction of granule cells and hypomyelination (Emerson and Delez, 1967). CNS malformations following bovine viral diarrhea virus infection of the fetus include hydranencephaly, hydrocephalus, hypomyelination, retinal degeneration, and cerebellar hypoplasia (Brown *et al.*, 1973; Trautwein *et al.*, 1985). The cerebellar defects increase in severity with time of infection in gestation. Similar lesions result in lambs being infected with border disease virus (Barlow *et al.*, 1980; Potts *et al.*, 1985).

B. Cavitary Cerebral Abnormalities

A number of virus infections can cause cavitary malformations of the forebrain in cattle, sheep, and goats. Examples include Akabane, Cache Valley, Wesselsbron, Main Drain, San Angelo, La Crosse, Aino, Venezuelan equine encephalitis viruses, vaccine strains of Rift Valley fever, and bluetongue viruses (Kurogi *et al.*, 1975; Coetzer and Barnard, 1977; London *et al.*, 1977; Konno *et al.*, 1982, 1988; Edwards *et al.*, 1989; Calisher and Sever, 1995; Kitano *et al.*, 1996; Edwards *et al.*, 1997). Immunization of sheep with live attenuated bluetongue virus vaccine resulted in the birth of lambs with cavitary defects of the brain (Schultz and DeLay, 1955). Again, the nature and extent of the defect depend on the period of gestation at which infection occurs (Osburn *et al.*, 1971b). The virus infects the subventricular zone germinal neuroepithelium and disrupts histogenesis of the cerebral cortex (Section II,A). Infection in the first trimester gives rise to massive necrosis, inflammation, and cavitation of the cerebral hemispheres. By birth the inflammation has resolved, necrotic material has been cleared, and the animal has a large noninflammatory cavity in the forebrain. The progenitor cells in the subventricular zone also give rise to glial cells and infection in the second trimester results in cavitation in the cortical white matter (Osburn *et al.*, 1971a). Experimental inoculation of this virus into mice also results in cavitary forebrain lesions. Interestingly, in the mouse there is also infection and destruction of the olfactory bulb, although there is no cerebellar pathology (Narayan and Johnson, 1972).

C. Hydrocephalus

A number of natural virus infections have been associated with hydrocephalus and several have been shown to produce this malformation experimentally. Experimentally, hydrocephalus has been reported in fetal or neonatal rodents infected with mumps, reovirus 1, influenza, respiratory syncytial, Ross River, St. Louis encephalitis, vaccinia, and herpes simplex viruses (Johnson and Johnson, 1968; Margolis and Kilham, 1969; Anderson and Hanson, 1975; Lagace-Simard *et al.*, 1982; Mims *et al.*, 1973; Davis, 1981; Hayashi *et al.*, 1986). In many of these cases, virus infects ependymal cells, although meningeal cells may also be infected. Destruction of these cells and the associated inflammation often result in hydrocephalus following aqueductal stenosis (Johnson *et al.*, 1967). Hydrocephalus following aqueductal stenosis and mumps infection is found in children, and ependymal cells containing viral nucleocapsids have been observed in samples of cerebrospinal fluid (Herndon *et al.*, 1974). It is not clear why hydrocephalus is observed primarily in neonatal rodents and young children. It could reflect postnatal developmental changes in susceptible cell types—in this case, ependymal cells; alternatively, or in addition, it could reflect age-related changes in the ability of viruses to enter the CNS. In the case of natural infections, immunity in adults may be sufficient to prevent virus from reaching the CNS or may more rapidly control the CNS infection, resulting in a less severe disease. In experimental systems, at least in mice, direct intracerebral inoculation of influenza virus can also result in hydrocephalus (Johnson and Johnson, 1972).

D. Infections of Late-Generating Brain Regions Including Cerebellar Hypoplasia

As discussed in Sections IIA,C and IV,B, secondary neurogenesis of some cell populations in the cerebellum, the olfactory bulb, and the dentate gyrus continues into the second year of life in the human and into the second postnatal week in the mouse. Parvoviruses only replicate in cells undergoing DNA synthesis (Tattersall, 1972). In the postnatal brain this includes cells of the external granule cell layer that give rise to the internal granule cells. Intracerebral infection of neonatal rats, hamsters, cats, and ferrets with rat parvovirus results in cerebellar hypoplasia with partial or complete destruction of the granule cell layer and abnormal connectivity (Margolis and Kilham, 1968a,b; Herndon et al., 1971). A naturally occurring example of this is feline panleukopenia virus that can cause cerebellar hypoplasia and resultant ataxia in kittens (Kilham and Margolis, 1966).

Cerebellar hypoplasia and ataxia are also seen in the rat following neonatal (P1 to P7) intracerebral infection with the arenavirus, lymphocytic choriomeningitis virus (LCMV) (Monjan et al., 1971), and the related Tamiami virus (Friedman et al., 1975). In LCMV infection of the neonatal rat, cerebellar lesions are maximal following inoculation at P4; inoculation at P1 or P7 results in less severe disease; and inoculation at P14, P21, or P90 produces no disease (Monjan et al., 1971). Following infection of P4 rats, LCMV infects the external granule cells and can persist in cerebellar neurons for months after infection (Monjan et al., 1971; Del Cerro et al., 1975; Baldridge et al., 1993). Death of neural cells is apparent from 1 to 3 weeks postinfection and the internal granule cell layer is noticeably depleted by 2 weeks (Del Cerro et al., 1975; Baldridge et al., 1993). A mononuclear cell infiltrate is apparent within 5 days of infection, and the cerebellar disease appears to require lymphocytes since it can be suppressed by treatment of infected rats with antilymphocyte serum (Monjan et al., 1974). The disease can also be prevented by treating infected rat pups with neutralizing monoclonal antibodies directed to the LCMV glycoprotein, which presumably acts by limiting the extent of the infection (Baldridge et al., 1993). Interestingly, in LCMV-infected P4 rats in both of these studies (Monjan, 1973; Baldridge, 1993), there was also clear and extensive infection of neurons in the dentate gyrus and the olfactory bulb, which, as discussed above (Sections IIA and IVB), are also active areas of neurogenesis during the first two postnatal weeks in the rodent. There are a few reports linking LCMV exposure to cases of human hydrocephalus (Ackermann et al., 1974; Barton et al., 1995; Sheinberg, 1975).

Borna virus infection of neonatal rats results in apoptotic loss of cerebellar granule and of Purkinje cells and dentate gyrus neurons (Bautista *et al.*, 1995; Rubin *et al.*, 1999a; Eisenman *et al.*, 1999). The progressive loss of dentate gyrus neurons correlates with a progressive deficit in spatial learning and memory, as assessed by a water maze test (Rubin *et al.*, 1999b). In addition to these learning and memory deficits, these rats also display abnormal righting reflexes, abnormal developmental motor milestones, hyperactivity, abnormal play behavior, inhibition of exploratory behavior and stereotypic behavior (Dittrich *et al.*, 1989; Pletnikov *et al.*, 1999a,b; Hornig *et al.*, 1999). It has been suggested that several of these features, both pathological and clinical, parallel those in human autism and that this experimental infection provides a model of this human disease (Hornig *et al.*, 1999; Pletnikov 1999b). Schizophrenia and autism are both associated with cerebellar hypoplasia and both have been suggested, at least for some cases, to have a viral etiology (Mason-Brothers *et al.*, 1990; Waltrip *et al.*, 1990; Lyon *et al.*, 1989; Bailey *et al.*, 1998). Cerebellar hypoplasia of unknown etiology occurs in humans and may have several different causes (Sarnat and Alcala, 1980).

E. Maturation of Cerebral Cortex

There is a well-documented, age-related virulence of Japanese encephalitis virus (JEV) in both rats and mice (Grossberg and Scherer, 1966; Ogata *et al.*, 1991). Intracerebral inoculation is lethal in Fisher rats aged P13 or less and avirulent in older animals (Ogata *et al.*, 1991). Following a detailed study of the distribution of viral proteins in the cerebral cortex in rats inoculated at various ages from P1 to P17, it was clearly shown that age-related changes paralleled neuronal maturation patterns. As described in sections IIA and IVB, the cerebral cortex matures in an inside-out pattern with the most mature cells in the inner layers and the least mature cells in the outer layers. With increasing age at infection between P1 and P9, JEV-infected cells become progressively confined to the outer, less mature layers. Furthermore, transplantation of developing cerebral cortex E19 neurons into the cerebral cortex of a P15 rat brain, followed by JEV infection 3 days later, resulted in infection of only the immature developing neurons in the graft and not of the surrounding, more mature neurons (Ogata *et al.*, 1991). Perhaps surprisingly, neurons in the basal ganglia and deep cerebral nuclei remained susceptible to infection at slightly later times (P12) than did the majority of cortical neurons, and infection of olfactory bulb neurons was not observed after P4.

JEV is widely distributed throughout Asia, where it is annually responsible for approximately 35,000 reported cases and 10,000 deaths. In endemic areas nearly all persons become infected at some point in childhood and most cases requiring hospitalization occur in children. During epidemics in nonendemic areas, the elderly are also at increased risk. The related Murray Valley encephalitis virus (MEV) from Australia also produces disease more commonly in children than in adults; rates again increase in the elderly (Doherty et al., 1976). JEV and MEV infect a variety of animal species. Pigs and birds are particularly important for the JEV zoonotic cycle and have thus been the main naturally infected animals to be studied, and CNS malformations have been recorded in fetal and newborn pigs (Innes and Saunders, 1962). As discussed in Section IV,A, for the alphavirus encephalitides, the relative contributions of age-related changes in the CNS, in other tissues, in immune responses, and in the frequency of infection, to the more frequent and more severe disease observed in children are not clear. In the case of JEV, it is likely that increased specific immunity with age is an important factor in endemic areas. However, children are still at increased risk in nonendemic areas and, as detailed above, experimental studies with JEV demonstrate that developmental changes in the susceptibility of CNS cells to infection are a likely determinant of this age-related susceptibility to disease.

In contrast to the age-related decrease in the susceptibility to Japanese encephalitis and to the alphavirus equine encephalitides (Section IV,A), there is St. Louis encephalitis (SLE). St. Louis encephalitis virus has been recognized to produce human epidemics in the United States since 1932, the largest of which was in 1975, with 1,815 reported cases. Disease is rare in children and more common in the elderly. The ratio of inapparent to apparent infection varies from 806:1 in children to 85:1 in the elderly and the case fatality rate increases with age from almost zero in children to 27% in the elderly (Henderson et al., 1970). A number of strains of SLE virus are recognized with varying virulence in mice (Monath, 1980). The neuropathogenesis of these has not been studied experimentally in any detail. As with SLE, the recent outbreak of West Nile virus in the northeastern United States mostly affected the elderly.

F. Murine Retroviruses and Spongiform Degeneration

Neonatal mice inoculated with the Cas-Br-E and some other related murine retroviruses develop a spongiform degeneration and motor neuron disease in the brain and the spinal cord that results in progressive paralysis and death (Gardner et al., 1973; Gardner, 1985). Histologically, the disease is characterized by spongiform degeneration, neuronal loss, and gliosis. These are observed in a number of motor areas including the anterior horns of the spinal cord, the brainstem, the deep cerebellar nuclei, and motor areas of the cortex. The appearance and magnitude of the lesions and the resultant clinical disease correlate with the extent of infection in the affected areas (Kay et al., 1991; Czub et al., 1992, 1994). However, infected cells are not confined to diseased areas but are also found to be scattered throughout other areas of the CNS and are particularly notable in the cerebellar granule cell layer, the dentate gyrus, the olfactory bulb, and the corpus callosum, although these areas rarely show pathological changes even during advanced stages of disease (Kay et al., 1991; Lynch et al., 1991; Lynch and Portis, 1993). In all brain areas the main cell types infected are microglial, endothelial, meningeal, and ependymal cells (Kay et al., 1991; Gravel et al., 1993). The extent of neuronal infection has been a controversial issue but most studies conclude that although neuronal populations distant from the degenerative changes are infected, in the areas of degeneration, neurons are not infected (Kay et al., 1991; Lynch et al., 1991; Pitts et al., 1987; Swarz et al., 1981).

An age-related change in susceptibility to this disease occurs between P6 and P10; mice inoculated at P6 develop disease while mice inoculated at P10 do not (Hoffman et al., 1981; Czub et al., 1991). Studies on genetically engineered viruses, immunocompromised mice, and intracerebral inoculations demonstrate that the age-related change cannot be attributed to immune responses (Czub et al., 1991). A fascinating finding is that the onset of neuropathology and the onset of clinical disease appear to be fixed since the earliest time of onset is independent of the time of inoculation and the extent of infection. Inoculation of midgestation embryos with a genetically modified virus (Fr-Cas^E) results in an earlier and more extensive CNS infection than does inoculation of neonates, but in both cases neuropathology is first observed around P10 and clinical disease around P15 (Portis et al., 1990; Lynch and Portis, 1993). These results suggest that the neuropathology and subsequent clinical disease are dependent on some postnatal neural developmental event, which occurs before or around P10. In rodents, changes in motor neurons have been documented around this time and these include changes in the expression of surface proteoglycans and N-methyl-D-aspartate receptors (Kalb and Hockfield, 1988, 1990, 1992). Coincident with the onset of neuropathology is a change in the glycosylation pattern of the viral envelope glycoprotein (Czub et al., 1994). Different regions of the infected brain express different glycosylation isoforms. The areas of neuronal infection without degeneration (cerebellum, olfactory bulb) have a gp65 isoform of the envelope glycoprotein, the same form that is found in virions, whereas areas of microglial infection and spongiform change have a gp70 isoform (Lynch and Sharpe, 2000).

The onset of spongiform degeneration in motor areas occurs in all affected regions at the same time. As noted by Czub and colleagues (1994), this could result from distant events that influence each of these motor areas. They suggested a prime candidate would be the Purkinje cells and cortex of the cerebellum. Consistent with this, expression of viral envelope proteins by cerebellar granule neurons (around P8 and P9) immediately precedes the onset of the distant motor region degeneration (P10); granule cells have multiple and complex connections with Purkinje cells and at around this time, Purkinje cell axons are establishing connections with affected motor regions (Czub et al., 1994). That virus replication in the developing early postnatal cerebellar granule cells affects the intimately connected Purkinje cells, which in turn affects the motor connections of these cells, is an entirely plausible and satisfying explanation consistent with developmental neurobiology (Sections II,C and IV,B). However, spongiform degeneration can also be induced in the absence of cerebellar infection by transplantation of virally infected microglial cells into degeneration-susceptible regions of the brain of P10 or older mice (Lynch et al., 1995).

G. Extraneural Developmental Changes

West Nile virus infection of mice shows an age-related change in the outcome of infection that is determined in large part by extraneural events (Weiner et al., 1970). Fig. 4 illustrates the age-related virulence and virus titers in the blood and the CNS following intracerebral and intraperitoneal inoculations. In this case age-related changes in virulence are related to the ability of the infection to establish a viremia. This situation is a good contrast to that described in section IV,B for the A7(74) strain of SFV, where a high titer plasma viremia is generated on infection in all ages of mice, the virus is neuroinvasive at all ages, and age-related changes in neurovirulence are determined principally by developmental events within the CNS (Oliver et al., 1997; Oliver and Fazakerley, 1998).

VI. OTHER IMPORTANT INFECTIONS OF THE DEVELOPING HUMAN CENTRAL NERVOUS SYSTEM

It is probably worth adding a few brief words about additional and important infections of the developing human nervous system that are not covered in the examples above. The most common fetal infection in humans is cytomegalovirus (CMV). One percent of newborns have evidence of fetal infection with this virus. Viremia at any time during

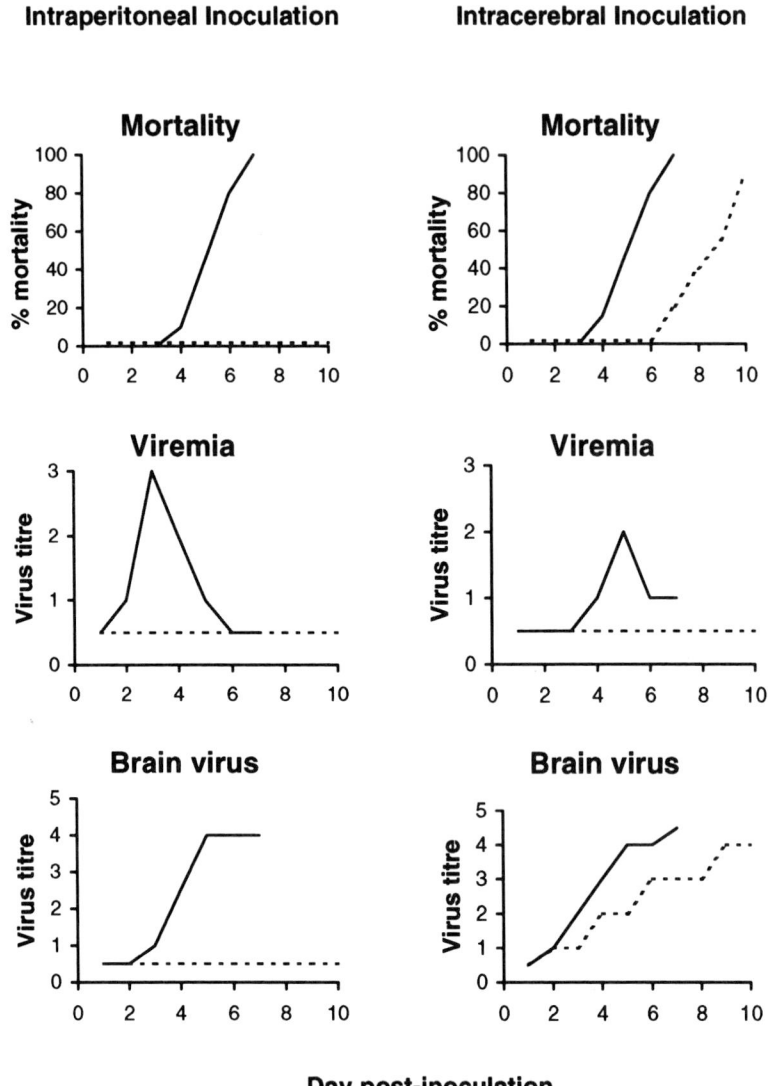

Intraperitoneal Inoculation   **Intracerebral Inoculation**

Day post-inoculation

Fig 4. Age-related changes in the outcome and kinetics of West Nile virus infection of mice. Data are taken from Weiner *et al.* (1970), with permission. In this study newborn (P1 to P3) and 12-week-old Swiss mice were inoculated either intraperitoneally or intracerebrally with the Egypt 101 strain of West Nile virus. Mortality, blood virus titers (viremia), and brain virus titers were determined at various days postinfection. The solid lines show the data for newborn mice and the dashed lines, the data for adult mice. The first column demonstrates that in newborn mice, following intraperitoneal inoculation of virus, an early viremia is established, virus replicates in the brain, and the mice

pregnancy can spread virus to the fetus (Davis *et al.*, 1971). Most (>95%) newborns with congenital CMV infection appear normal, but severe and subtle neurologic deficits become apparent with age in 5% to 15% of cases; these include mental retardation, microcephaly, and progressive encephalopathy (Alford *et al.*, 1990; Koeda *et al.*, 1993). The most severe consequence of CMV infection is cytomegalic inclusion disease of the newborn, which affects several organs including the CNS and has a high (30%) mortality rate. Only a small percentage (15%) of survivors develop normally (MacDonald and Tobin, 1978). CNS involvement leads to microcephaly, hydrocephalus, periventricular calcification, seizures, and severe psychomotor retardation (Alford *et al.*, 1990). Deafness is common and chorioretinitis can result in severe visual impairment. Experimentally, murine cytomegalovirus infection of the developing mouse brain results in apoptosis, although this is likely to result from an indirect mechanism since the apoptotic cells appear not to be virally infected (Kosugi *et al.*, 1998).

If CMV is the most common fetal infection in humans, herpes simplex virus (HSV) is the most important perinatal infection. HSV can infect the CNS of newborns, children, and adults. In the neonate, HSV encephalitis is more severe than in children or in adults and brain lesions are diffuse, whereas in children and adults lesions are predominantly found in the orbital–frontal and temporal lobes and are usually unilateral. Varicella zoster virus can infect the fetus by crossing the placenta and infection in early gestation can lead to severe brain damage (Srabstein *et al.*, 1974; Nicola and Hanshaw, 1979).

There is a higher frequency of antibodies to echovirus 7 in mothers giving birth to newborns with gross malformations of the CNS than in controls (Lapinleimu *et al.*, 1972). Interestingly, echovirus sequences have also been observed with higher frequency in spinal cord material from cases of amyotrophic lateral sclerosis than in spinal cord material from other neurologic diseases (Woodall *et al.*, 1994; Berger *et al.*,

---

die. In contrast, intraperitoneal infection of adult mice is not lethal; there is no viremia and no replication of virus in the brain. The second column shows that the most important age-related change is not in the brain but in the ability of the virus to replicate outside the brain. The adult brain retains the ability to replicate this virus, as shown by direct intracerebral inoculation of virus, which results in both newborn and adult mice in virus replication and in death of the animals. It should be noted, though, that although extraneural changes are keys to the age-related change in the outcome of the infection, there also appear to be age-related changes in virus replication in the brain; virus replication in the adult brain is slower than in the newborn brain.

2000). Perhaps, over long periods of time infection with these viruses can affect neuronal survival.

Human immunodeficiency virus (HIV) is growing in importance as a fetal infection. Transmission from mother to fetus or to newborn occurs *in utero* or probably more commonly at birth, with the risk of infection being related to virus plasma levels in the mother (Calvelli and Rubinstein, 1990; Sperling *et al.,* 1996). Some HIV-infected infants demonstrate a rapid and progressive encephalopathy and microcephaly with weakness, ataxia, blindness, and seizures. Other children develop, with increasing years, subtle cognitive deficits and delays in motor and mental functions (Gay *et al.,* 1995). Neuropathologic studies demonstrate diffuse cortical gliosis and perivascular inflammation, microglial nodules, vasculitis, and vascular and juxtavascular mineralization in the frontal lobe white matter (Budka, 1989; Sharer, 1992). Transmission from mother to fetus has been observed in the second trimester (Lyman *et al.,* 1990). That HIV may be able to enter and replicate in the proliferating and differentiating neurons present at this time of brain development is suggested by *in vitro* studies that demonstrate that the virus can replicate in at least some neuronal and glial cell lines in culture (Ensoli *et al.,* 1994, 1995, 1997; Sharpless *et al.,* 1992; Truckenmiller *et al.,* 1993). Differential ability to replicate in neuronal cell lines is determined by genetic elements within the viral LTR (Corboy *et al.,* 1992; Ensoli *et al.,* 1997). As discussed in section III,A, this probably reflects the presence or absence of cellular transcription factors that can activate this viral promoter. Most studies on HIV infection of the adult brain conclude that the virus does not infect neurons and that the main cell type infected is the microglial cell (Peudenier *et al.,* 1991). In most studies, the degree of neuronal loss appears to be more extensive than the infection and there is no spatial correlation between the two. In one study of pediatric HIV-1 encephalitis and progressive encephalopathy, a spatial association was observed between HIV-infected inflammatory cells and apoptotic neurons (Gelbard *et al.,* 1995). It has been suggested that, as in adults, the neuropathology observed in pediatric cases is also dependent on HIV infection of microglial cells and production of toxic factors that, in the early postnatal period are more likely to induce, for example, excitotoxic neuronal injury in the apoptosis-susceptible differentiating neuronal populations present at this time (Epstein and Gelbard, 1999). DNA fragmentation, a characteristic of apoptosis, occurs in the neurons, macrophages, and microglia of these brains (Krajewski *et al.,* 1997). Soluble viral products such as the surface envelope glycoprotein gp120 or the viral Tat protein, or soluble cellular products such as cytokines, particularly TNFα or NO, have been postulated to mediate this death (Lannuzel *et al.*, 1997; New *et al.*, 1997; Shi *et al.*, 1998; Hesselgesser *et al.*, 1998).

REFERENCES

Acheson, N. H., and Tamm, I. (1967). Replication of Semliki Forest virus: An electron microscopic study. *Virology* **32:**128–143.

Ackermann, P., Korver, G., Turss, R., Wonne, R., and Hochgesand, P. (1974). Pranatale infektion mit dem virus der lymphocytaren choriomenigitis. *Dtsch. Med. Wochenschr.* **99:**529–632.

Aghajainian, G. K., and Bloom, F. E. (1967). The formation of synaptic junctions in developing rat brain: a quantitative electron microscopic study. *Brain Research* **6:**716–727.

Agmon, A., Yang, L. T., O'Dowd, D. K., and Jones, E. G. (1993). Organized growth of thalamocortical axons from the deep tier of terminations into layer IV of developing mouse barrel cortex. *J. Neurosci.* **13:**5365–5382.

Aguilar, M. J., Calanchini, P. R., and Finley, K. H. (1968). Perinatal arbovirus encephalitis and its sequelae. *In "Infections of the nervous system"* (H. M. Zimmerman, ed.), pp. 216–235. New York.

Alford, C. A., Stagno, S., Pass, R. F., and Britt, W. J. (1990). Congenital and perinatal cytomegalovirus infections. *Rev. Infect. Dis.* **12(Suppl):**S745–S753.

Allner, K., Bradish, C. J., Fitzgeorge, R., and Nathanson, N. (1974). Modifications by sodium aurothiomalate of the expression of virulence in mice by defined strains of Semliki Forest virus. *J. Gen. Virol.* **24:**221–228.

Allsopp, T. E. and Fazakerley, J. K. (2000). Altruistic cell suicide and the specialised case of the virus-infected central nervous system. *Trends in Neurosci.* **23:**280–286.

Allsopp, T. E., Scallan, M. F., Williams, A., and Fazakerley, J. K. (1998). Virus infection induces neuronal apoptosis: A comparison with trophic factor withdrawal. *Cell Death and Differentiation* **5:**50–59.

Altman, J. (1969). Autoradiographic and histological studies of postnatal neurogenesis. IV. Cell proliferation and migration in the anterior forebrain with special reference to persisting neurogenesis in the olfactory bulb. *J. Comp. Neurol.* **136:**433–458.

Altman, J. (1971). Coated vesicles and synaptogenesis. A developmental study in the cerebellar cortex of the rat. *Br. Res.* **30:**311–322.

Altman, J. (1972a). Postnatal development of the cerebellar cortex in the rat. I. The external germinal layer and the transitional molecular layer. *J. Comp. Neurol.* **145:**353–398.

Altman, J. (1972b). Postnatal development of the cerebellar cortex in the rat. II. Phases in the maturation of Purkinje cells and of the molecular layer. *J. Comp. Neurol.* **145:**399–464.

Altman, J. (1972c). Postnatal development of the cerebellar cortex in the rat. III. Maturation of the components of the granule layer. *J. Comp. Neurol.* **145:**465–514.

Amako, K. and Dales, S. (1967). Cytopathology of Mengovirus infection. II. Proliferation of membranous cisternae. *Virology* **32:**201–215.

Amor, S., Scallan, M. F., Morris, M. M., Dyson, H., and Fazakerley, J. K. (1996). Role of immune responses in protection and pathogenesis during Semliki Forest virus encephalitis. *J. Gen. Virol.* **77:**281–291.

Anderson, A. A. and Hanson, R. P. (1975). Intrauterine infection of mice with St. Louis encephalitis virus: immunological, physiological, neurological and behavioral effects on progeny. *Infect. Immun.* **12:**1173–1183.

Angevine, J. B. and Sidman, R. L. (1961). Autoradiographical study of cell migration during histogenesis in cerebral cortex of the mouse. *Nature (London)* **192**:766–768.

Arvin, A. (1997). Viral infections of the fetus and neonate. In: "Viral Pathogenesis" (N. Nathanson, ed.), pp. 801–814. Lippincott-Raven, Philadelphia.

Ashkenazi, A. and Dixit, V. M. (1998). Death receptors: signaling and modulation. *Science* **281**:1305–1308.

Ashwell, K. (1990). Microglia and cell death in the developing mouse cerebellum. *Dev. Brain Res.* **55**:219–230.

Atkins, G. J., Carter, J., and Sheahan, B. J. (1982). Effect of alphavirus infection on mouse embryos. *Infect. Immun.* **38**:1285–1290.

Bailey, A., Luthert, P., Dean, A., Harding, B., Janota, I., Montgomery, M., Rutter, M., and Lantos, P. (1998). A clinicopathological study of autism. *Brain* **121**:889–905.

Baldridge, J. R., Pearce, B. D., Parekh, B. S., and Buchmeier, M. (1993). Teratogenic effects of neonatal arenavirus infection on the developing rat cerebellum are abrogated by passive immunotherapy. *Virology* **197**:669–677.

Barde, Y. A. (1989). Trophic factors and neuronal survival. *Neuron* **2**:1525–1534.

Barlow, R. M., Rennie, J. C., Gardiner, A. C., and Vantsis, J. T. (1980). Infection of pregnant sheep with NADL strain of bovine virus diarrhoea virus and their subsequent challenge with border disease. *J. Comp. Pathol.* **90**:67–72.

Barres, B. A., Hurt, I. K., Coles, H. S. R., Burne, J., Voyvodic, J. T., Richardson, W. D., and Raff, M. C. (1992). Cell death and control of cell survival in the oligodendrocyte lineage. *Cell* **70**:31–46.

Barton, L. L., Peters, C. J., and Ksiazek, T. G. (1995). Lymphocytic choriomeningitis virus: an unrecognized teratogenic pathogen. *Emer. Infect. Dis.* **1**:152–153.

Bautista, J. R., Rubin, S. A., Moran, T. H., Schwartz, G. J., and Carbone, K. M. (1995). Developmental injury to the cerebellum following perinatal Borna disease virus infection. *Dev. Br. Res.* **90**:45–53.

Bayer, S. A. (1980a). Development of the hippocampal region in the rat. I. Neurogenesis examined with 3H Thymidine autoradiography. *J. Comp. Neurol.* **190**:87–114.

Bayer, S. A. (1980b). Development of the hippocampal region in the rat. II. Morphogenesis during embryonic and early postnatal life. *J. Comp. Neurol.* **190**:115–134.

Bayer, S. A. (1985): Hippocampal region. In "The Rat Nervous System" (G. Paxinos, ed.), pp. 335–348. Academic Press, Sydney Australia.

Bedell, M. A., Hudson, J. B., and Golub, T. R. (1991). Amplification of human papillomavirus genomes in vitro is dependent on epithelial differentiation. *J. Virol.* **65**:2254–2260.

Bellanti, J. A., Artenstein, M. S., Olsen, L. C., Buescher, E. L., Luhrs, C. E., and Milstead, K. L. (1965). Congenital rubella: clinicopathologic, virologic and immunologic studies. *Am. J. Dis. Child.* **110**:464–472.

Berger, M. M., Kopp, N., Vital, C., Redl, B., Aymard, M., and Lina, B. (2000). Detection and cellular localization of enterovirus RNA sequences in spinal cord of patients with ALS. *Neurology* **54**:20–25.

Berry, M., Rogers, A. W., and Eayrs, J. T. (1964). Pattern of cell migration during cortical histogenesis. *Nature* **203**:591–593.

Bi, Z. B., Barna, M., Komatsu, T., and Reiss, C. S. (1995). Vesicular stomatitis-virus infection of the central nervous system activates both innate and acquired immunity. *J. Virol.* **69**:6466–6472.

Blaschke, A. J., Staley, K., and Chun, J. (1996). Widespread programmed cell death in proliferative and postmitotic regions of the fetal cerebral cortex. *Development (Cambridge, Eng.)* **122**:1165–1174.

Bloom, B., Jimenez, L., and Marcus, P. I. (1970). A plaque assay for enumerating antigen positive cells in delayed type hypersensitivity. *J. Exp. Med.* **132**:16.

Brown, T. H. and Zador, A. M. (1990). Hippocampus. *In* "The Synaptic Organisation of the Brain" (G. M. Shepherd, ed.), pp. 346–388. Oxford University Press, Oxford.

Brown, T. T., de Lahunta, A., Scott, F. W., Kahrs, R. F., McEntee, K., and Gillespie, J. H. (1973). Virus induced congenital anomalies of the bovine fetus. II. Histopathology of cerebellar degeneration (hypoplasia) induced by the virus of bovine viral diarrhea-mucosal disease. *Cornell Vet* **63:**561–578.

Brunjes, P. C. and Frazier, L. L. (1986). Maturation and plasticity in the olfactory system of vertebrates. *Br. Res. Rev.* **11:**1–45.

Budka, H. (1989). Human immunodeficiency virus (HIV)-induced disease of the central nervous system: pathology and implications for pathogenesis. *Acta Neuropathol.* **77:**225–236.

Butt, A. M., Jones, H. C., and Abbott, N. J. (1990). Electrical resistance across the blood-brain barrier in anaesthetized rats: a developmental study. *J. Physiol. (London)* **429:**47–62.

Caliguri, L. A. and Tamm, I. (1970). The role of cytoplasmic membranes in poliovirus biosynthesis. *Virology* **4:**100–111.

Calisher, C. H. and Sever, J. L. (1995). Are North American Bunyamwera serogroup viruses etiologic agents of human congenital defects of the central nervous system? *Emer. Infect. Dis.* **1:**147–151.

Calvelli, T. A. and Rubinstein, A. (1990). Pediatric HIV infection: a review. *Immunodefic.* **2:**83–127.

Casaro, A. P. E., Kendrick, J. W., and Kennedy, P. C. (1971). Response of the bovine fetus to bovine viral diarrhea-mucosal disease virus. *Am. J. Vet. Res.* **32:**1543–1561.

Clarke, P. G. H. (1998): Apoptosis versus necrosis: How valid a dichotomy for neurons. *In* "Cell Death in Diseases of the Nervous System" (V. Koliatsos, ed.), pp. 3–28. Humana Press, Totowa, N.J.

Coetzer, J. A. W. and Barnard, B. J. H. (1977). Hydrops amnii in sheep associated with hydranencephaly and arthrogryposis with Wesselsbron disease and Rift Valley fever viruses as aetiological agents. *Onderstepoort. J. Vet. Res.* **44:**119–126.

Condeelis, J. (1993). Life at the leading edge. *Annu. Rev. Cell Biol.* **9:**411–444.

Cooper, L. Z., Ziring, P. R., Ockerse, A. B., Kiely, B., and Krugman, S. (1969). Rubella: clinical manifestations and management. *Am. J. Dis. Child.* **118:**18–29.

Corboy, J. R., Buzy, J. M., and Zink, M. C. (1992). Expression directed from HIV long terminal repeats in the central nervous system of transgenic mice. *Science* **258:**1804–1808.

Cowan, W. M. and O'Leary, D. (1984): Cell death and process elimination: The role of regressive phenomena in the development of the vertebrate nervous system. *In:* "Medicine, Science and Society. Celebrating the Harvard Medical School Bicentennial." New York.

Crandall, J. E., Whitcomb, J. M., and Caviness, V. S. (1985). Development of the spinal-medullary projection from the mouse barrel field. *J. Comp. Neurol.* **239:**205–215.

Crepel, F., Mariani, J., and DelhayeBouchaud, N. (1976). Evidence for a multiple innervation of Purkinje cells by climbing fibres in the rat cerebellum. *J. Neurobiol.* **7:**567–578.

Crone, D. E. and Keene, J. D. (1989). Viral transcription is necessary and sufficient for vesicular stomatitis virus to inhibit maturation of small nuclear ribonucleoproteins. *J. Virol.* **63:**4172–4180.

Crowley, C., Spencer, S. D., Nishimura, M., Chen, K. S., PittsMeek, S., Armanini, M. P., Ling, L. H., McMahon, S. B., Shelton, D. L., Levinson, A. D., and Phillips, H. S. (1994). Mice lacking nerve growth factor display perinatal loss of sensory and sympathetic neurons yet develop basal forebrain cholinergic neurons. *Cell* **76:**1–20.

Czub, M., Czub, S., McAtee, F., and Portis, J. L. (1991). Age-dependent resistance to murine retrovirus-induced spongiform neurodegeneration results from central nervous system-specific restriction of virus replication. *J. Virol.* **65:**2539–2544.

Czub, M., McAtee, F. J., and Portis, J. L. (1992). Murine retrovirus-induced spongiform encephalomyelopathy: host and viral factors which determine the length of the incubation period. *J. Virol.* **66:**3298–3305.

Czub, S., Lynch, W. P., Czub, M., and Portis, J. L. (1994). Kinetic analysis of spongiform neurodegenerative disease induced by a highly virulent murine retrovirus. *Lab. Invest.* **70:**711–723.

Dai, J. and Sheetz, M. P. (1995). Axon membrane flows from the growth cone to the cell body. *Cell* **83:**693–701.

Dailey, M. E. and Bridgman, P. C. (1993). Vacuole dynamics in growth cones: correlated EM and video observations. *J. Neurosci.* **13:**3375–3393.

Dales, S. and Siminovitch, L. (1961). The development of vaccinia virus in Earl's strain cells as examined by electron microscopy. *J. Biophys. Biochem. Cytology.* **10:**475–500.

Das, G. D. and Altman, J. (1971). Postnatal neurogenesis in the cerebellum of the cat and tritiated thymidine autoradiography. *Br. Res.* **30:**323–330.

Das, T., Mathur, M., Gupta, A. K., Janssen, G. M. C., and Banerjee, A. K. (1998). RNA polymerase of vesicular stomatitis virus specifically associates with translation elongation factor-1 or its activity. *Proc. Natl. Acad. Sci. U.S.A.* **95:**1460–1465.

Davies, A. M. (1994a). Switching neurotrophin dependence. *Curr. Biol.* **4:**273–276.

Davies, A. M. (1994b). The role of neurotrophins in the developing nervous system. *J. Neurobiol.* **25:**1334–1348.

Davis, L. E. (1981). Communicating hydrocephalus in newborn hamsters and cats following vaccinia virus infection. *J. Neurosurg.* **54:**767–772.

Davis, L. E., Harms, A. C., and Chin, T. D. Y. (1971). Transient cortical blindness and cerebellar ataxia associated with mumps. *Arch. Ophthalmol.* **85:**366–368.

De, B. P., Lesson, A., and Banerjee, A. K. (1991). Human parainfluenza virus type 3 transcription in vitro: Role of cellular actin in mRNA synthesis. *J. Virol.* **65:**3268–3275.

De La Torre, J. C., Rall, G., and Oldstone, C. (1993). Replication of lymphocytic choriomeningitis virus is restricted in terminally differentiated neurons. *J. Virol.* **67:**7350–7359.

Del Cerro, M., Nathanson, N., and Monjan, A. A. (1975). Pathogenesis of cerebellar hypoplasia produced by lymphocytic choriomeningitis virus infection of neonatal rats. *Lab. Invest.* **33:**608–617.

Desmond, M. M., Montgomery, J. R., and Melnick, J. L. (1969). Congenital rubella encephalitis. *Am. J. Dis. Child.* **118:**30–31.

Desmond, M. M., Wilson, G. S., and Melnick, J. L. (1967). Congenital rubella encephalitis. *J. Pediatr.* **71:**311–331.

Despres, P., Frenkiel, M. P., Ceccaldi, P. E., DosSantos, C. D., and Deubel, V. (1998). Apoptosis in the mouse central nervous system in response to infection with mouse-neurovirulent dengue viruses. *J. Virol.* **72:**823–829.

Dittrich, W., Bode, L., Ludwig, H., Kao, M., and Schneider, K. (1989). Learning deficiencies in Borna disease virus-infected but clinically healthy rats. *Biol. Psychiatry* **26:**818–828.

Doherty, R. L., Carley, J. G., and Fillippich, C. (1976). Murray Valley encephalitis in Australia, 1974: antibody response in cases and community. *Aust. NZ. J. Med.* **6:**446–453.

Dollard, S. C., Broker, T. R., and Chow, L. T. (1993). Regulation of the human papillomavirus type 11 E6 promoter by viral and host transcription factors in primary human keratinocytes. *J. Virol.* **67:**1721–1726.

Done, J. T., Terlecki, S., Richardson, C., Harkness, J. W., Sands, J. J., Patterson, D. S., Sweasey, D., Shaw, I. G., Winkler, C. E., and Duffell, S. J. (1980). Bovine virus diarrhoea-mucosal disease virus: pathogenicity for the fetal calf following maternal infection. *Vet. Rec.* **106:**473–479.

Dutko, F. J., and Oldstone, M. B. A. (1981). Cytomegalovirus causes a latent infection in undifferentiated cells and is activated by induction of a cell differentiation. *J. Exp. Med.* **154**:1636–1651.

Eayrs, J. T., and Goodhead, B. (1959). Postnatal development of the cerebral cortex in the rat. *J. Anat.* **93**:385–402.

Edwards, J. F., Karabatsos, N., Collisson, E. W., and De La Concha Bermejillo, A. (1997). Ovine fetal malformations induced by in utero inoculation with Main Drain, San Angelo and La Crosse viruses. *Am. J. Trop. Med. Hyg.* **56**:171–176.

Edwards, J. F., Livingston, C. W., Chung, S. I., and Collison, E. C. (1989). Ovine arthrogryposis and central nervous system malformations associated with in utero Cache Valley virus infection: spontaneous disease. *Vet. Pathol.* **26**:33–39.

Egger, D., Pasamontes, L., Bolten, R., Boyko, V., and Bienz, K. (1996). Reversible dissociation of the poliovirus replication complex: Functions and interactions of its components in viral RNA synthesis. *J. Virol.* **70**:8675–8683.

Eisenman, L. M., Brothers, R., Tran, M. H., Kean, R. B., Dickson, G. M., Dietzschold, B., and Hooper, D. C. (1999). Neonatal borna disease virus infection in the rat causes a loss of Purkinje cells in the cerebellum. *J. NeuroVirol.* **5**:181–189.

Eklund, C. M. and Blumstein, A. (1938). Relation of human encephalitis to encephalomyelitis in horses. *J.A.M.A.* **111**:1734.

Emerson, J. L., and Delez, A. L. (1967). Cerebellar hypoplasia, hypomyelinogenesis and congenital tremors in pigs, associated with prenatal hog cholera vaccination of sows. *J. Am. Vet. Med. Assoc.* **147**:47–54.

Ensoli, F., Cafaro, A., Fiorelli, V., Vannelli, B., Ensoli, B., and Thiele, C. J. (1995). HIV-1 infection of primary human neuroblasts. *Virology* **210**:221–225.

Ensoli, F., Ensoli, B., and Thiele, C. J. (1994). HIV-1 gene expression and replication in neuronal and replication in neuronal and glial cell lines with immature phenotype: effects of nerve growth factor. *Virology* **200**:668–676.

Ensoli, F., Wang, H., Fiorelli, V., Zeichner, S. L., De Cristofaro, M. R., Luzi, G., and Thiele, C. J. (1997). HIV-1 infection and the developing nervous system: lineage-specific regulation of viral gene expression and replication in distinct neuronal precursors. *J. NeuroVirol.* **3**:290–298.

Epstein, L. G. and Gelbard, H. A. (1999). HIV-1-induced neuronal injury in the developing brain. *Journal Of Leukocyte Biology* **65**:453–457.

Farber, S., Hill, A., Connerly, M. L., and Dingle, J. H. (1940). Encephalitis in infants and children caused by virus of eastern variety of equine encephalitis. *J.A.M.A.* **114**:1725.

Fazakerley, J. K. and Allsopp, T. E. (2000). Programmed cell death in virus infections of the nervous system. *Curr. Top. Microbiol. Immunol.* (in press).

Fazakerley, J. K., Pathak, S., Scallan, M., Amor, S., and Dyson, H. (1993). Replication of the A7(74) Strain of Semliki Forest virus is restricted in neurons. *Virology* **195**:627–637.

Fazakerley, J. K. and Webb, H. E. (1987a). Semliki Forest virus-induced, immune-mediated demyelination – adoptive transfer studies and viral persistence in nude-mice. *J. Gen. Virol.* **68**:377–385.

Fazakerley, J. K. and Webb, H. E. (1987b). Semliki Forest virus induced, immune mediated demyelination: the effect of irradiation. *Br. J. Exp. Pathol.* **68**:101–113.

Fazakerley, J. K., Allsopp, T., Williams, A., and Oliver, K. (2000) (unpublished observations).

Ferrer, I., Bernet, E., Soriano, E., Del Rio, T., and Fonseca, M. (1990). Naturally occuring cell death in the cerebral cortex of the rat and removal of the dead cells by transitory phagocytes. *Neurosci.* **39**:451–458.

Fleming, P. (1977). Age-dependent and strain-related differences of virulence of Semliki Forest virus in mice. *J. Gen. Virol.* **37**:93–105.

Fortes, P., Beloso, A., and Ortin, J. (1994). Influenza virus NS1 protein inhibits prem-RNA splicing and blocks mRNA nucleocytoplasmic transport. *EMBO J.* **13**:704–712.

Fothergill, L. D., Dingle, J. H., Farber, S., and Connerly, M. L. (1938). Human encephalitis caused by the eastern variety of equine encephalitis. *New England J. Med.* **219**:411

Friedman, H. M., Gilden, D. H., Roosa, R. A., and Nathanson, N. (1975). The effect of neonatal thymectomy on Tamiami virus-induced central nervous system disease. *J. Neuropathol. Exp. Neurol.* **34**:159–166.

Friedman, R. M., Levin, J. G., Grimley, P. M., and Berezesky, I. K. (1972). Membrane-associated replication complex in arbovirus infection. *J. Virol.* **10**:504–515.

Froshauer, S., Kartenbeck, J., and Helenius, A. (1988). Alphavirus RNA replicase is located on the cytoplasmic surface of endosomes and lysosomes. *J. Cell Biol.* **107**:2075–2086.

Fruh, K., Ahn, K., and Djaballah, H. (1995). A viral inhibitor of peptide transporters for antigen presentation. *Nature* **375**:415–418.

Gamarnik, A. V. and Andino, R. (1997). Two functional complexes formed by KH domain containing proteins with the 5′ noncoding region of poliovirus RNA. *RNA* **3**:882–892.

Gardner, M. B. (1985). Retroviral spongiform polioencephalomyelopathy. *Rev. Infect. Dis.* **7**:99–110.

Gardner, M. B., Henderson, B. E., Officer, J. E., Rongey, R. W., Parker, J. C., Oliver, C., Estes, J. D., and Huebner, R. J. (1973). A spontaneous lower motor neuron disease apparently caused by indigenous type-C RNA virus in wild mice. *J. Natl. Cancer Inst.* **51**:1243–1254.

Gay, C. L., Armstrong, F. D., and Cohen, D. (1995). The effects of HIV on cognitive and motor development in children born to HIV-seropositive women with no reported drug use: birth to 24 months. *Pediatrics* **96**:1078–1082.

Gelbard, H. A., James, H. J., Sharer, L. R., Perry, S. W., Saito, Y., Kazee, A. M., Blumberg, B. M., Epstein, L. G., Shi, B., Raina, J., Lorenzo, A. E., Busciglio, J., and Gabuzda, D. (1995). Apoptotic neurons in brains from pediatric-patients with HIV-1 encephalitis and progressive encephalopathy: Neuronal apoptosis induced by HIV-1 Tat protein and TNF-alpha: potentiation of neurotoxicity mediated by oxidative stress and implications for HIV-1 dementia. *Neuropathology And Applied Neurobiology* **21**:208–217.

Gendelman, H. E., Narayan, O., and Kennedy-Stoskopf, S. (1986). Tropism of sheep lentiviruses for monocytes: susceptibility to infection and virus gene expression increase during maturation of monocytes to macrophages. *J. Virol.* **58**:67–74.

Gould, E. and McEwen, B. S. (1993). Neuronal birth and death. *Curr. Opin. Neurobiol.* **3**:676–682.

Gravel, C., Kay, D. G., and Jolicoeur, P. (1993). Identification of the target cell type in spongiform myeloencephalopathy induced by the neurotropic Cas-Br-E murine leukemia virus. *J. Virol.* **67**:6648–6658.

Gregg, N. M. (1941). Congenital cataract following German measles in the mother. *Trans. Ophthalmol. Soc. Aust.* **3**:35–46.

Grimley, P. M., Berezesky, I. K., and Friedman, R. M. (1968). Cytoplasmic structures associated with an Arbovirus infection: loci of viral ribonucleic acid synthesis. *J. Virol.* **2**:1326–1338.

Grimley, P. M. and Friedman, R. M. (1970). Development of Semliki Forest virus in mouse brain—An electron microscopic study. *Exp. Mol. Pathol.* **12**:1–13.

Grimley, P. M., Levin, J. G., Berezesky, I. K., and Friedman, R. M. (1972). Specific membranous structures associated with the replication of group A Arboviruses. *J. Virol.* **10**:492–503.

NEUROVIROLOGY AND DEVELOPMENTAL NEUROBIOLOGY   115

Grossberg, S. E. and Scherer, W. F. (1966). The effect of host age, virus dose, and route of inoculation on inapparent infection in mice with Japanese encephalitis virus. *Proc. Soc. Exp. Biol. Med.* **123:**118–124.

Guetta, E., Ron, D., and Tal, J. (1986). Developmental-dependent replication of minute virus of mice in differentiated mouse testicular lines. *J. Gen. Virol.* **67:**2549–2554.

Guinea, R. and Carrasco, L. (1990). Phospholipid biosynthesis and poliovirus genome replication, two coupled phenomena. *EMBO J.* **9:**2001–2016.

Hamburger, V. (1934). The effects of wing bud extirpation on the development of the central nervous system in chick embryos. *J. Exp. Zool.* **68:**449–494.

Hamburger, V. and Habel, K. (1947). Teratogenetic and lethal effects of influenza-A and mumps viruses on early chick embryos. *Proc. Soc. Exp. Biol. Med.* **66:**608–617.

Hammarback, J. A., Palm, S. L., Furcht, L. T., and Letourneau, P. C. (1985). Guidance of neurite outgrowth by pathways of substratum-adsorbed laminin. *J. Neurosci. Res.* **13:**213–220.

Hayashi, K., Iwasaki, Y., and Yanagi, K. (1986). Herpes simplex virus type 1-induced hydrocephalus in mice. *J. Virol.* **57:**942–951.

Helenius, A., Marsh, M., and White, J. (1980). The entry of viruses into animal cells. *Trends. Biochem.* **5:**104–106.

Helenius, A., Morein, B., Fries, E., Simons, K., Robinson, P., and Schirmaer, V. (1978). Human (HLA-A and HLA-B) and murine (H-2k) and H-2d) histocompatibility antigens are cell surface receptors for Semliki Forest virus. *Proc. Natl. Acad. Sci. U.S.A.* **75:**3846–3850.

Henderson, B. E., Metselaar, D., Kirya, G. B., and Timms, G. L. (1970). Investigations into yellow fever virus and other arboviruses in the northern regions of Kenya. *Bull. World Hlth. Org.* **42:**787–795.

Herndon, R. M., Johnson, R. T., Davis, L. E., and Descalzi, L. R. (1974). Ependymitis in mumps virus meningitis: electron microscopical studies of cerebrospinal fluid. *Arch. Neurol.* **30:**475–479.

Herndon, R. M., Margolis, G., and Kilham, L. (1971). The synaptic organization of the malformed cerebellum induced by perinatal infection with feline panleukopenia virus (PLV). Elements forming the cerebella glomeruli. *J. Neuropathol. Exp. Neurol.* **30:**196–205.

Hesselgesser, J., Taub, D., Baskar, P., Greenberg, M., Hoxie, J., Kolson, D. L., and Horuk, R. (1998). Neuronal apoptosis induced by HIV-1 gp120 and the chemokine SDF-1 alpha is mediated by the chemokine receptor CXCR4. *Current Biology* **8:**595–598.

Hill, A., Jugovic, P., and York, I. (1995). Herpes simplex virus turns off the TAP to evade host immunity. *Nature* **375:**411–415.

Hill, A. B., Barnett, B. C., McMichael, A. J., and McGeoch, D. J. (1994). HLA class I molecules are not transported to the cell surface in cells infected with herpes simplex virus types 1 and 2. *J. Immunol.* **152:**2736–2741.

Hinds, J. W. and Hinds, P. L. (1976). Synapse formation in the mouse olfactory bulb. I. Quantitative studies. *J. Comp. Neurol.* **169:**15–40.

Hirsch, M. S., Zisman, B., and Allison, A. (1970). Macrophages and age-dependent resistance to herpes simplex virus in mice. *J. Immunol.* **104:**1160–1165.

Hoffman, P. M., Ruscetti, S. K., and Morse, H. C. (1981). Pathogenesis of paralysis and lymphoma associated with a wild mouse retrovirus infection. Age and dose-related effects in susceptible laboratory mice. *J. Neuroimmunol.* **1:**275–285.

Hornig, M., Weissenbock, H., Horscroft, N., and Lipkin, W. I. (1999). An infection-based model of neurodevelopmental damage. *Proc-Natl-Acad-Sci-U-S-A.* **96:**12102–12107.

Huttenlocher, P. R. (1984). Synapse elimination and plasticity in developing human cerebral cortex. *Am. J. Ment. Defic.* **88:**488–496.

Innes, J. R. and Saunders, I. Z. (1962): "Comparative Neuropatholgy." Academic Press, New York.

Jackson, A. C. and Rossiter, J. P. (1997). Rabies virus infection produces apoptosis. *Neurology* **48**:3032

Jackson, A. C. and Park, H. (1998). Apoptotic cell death in experimental rabies in suckling mice. *Acta Neuropath.* **95**:159–164.

Jagelman, S., Suckling, A. J., Webb, H. E., and Bowen, E. T. W. (1978). The pathogenesis of avirulent Semliki Forest virus infections in athymic nude mice. *J. Gen. Virol.* **41**:599–607.

Janec, E. and Burke, R. E. (1993). Naturally occurring cell death during postnatal development of the substantia nigra pars compacta of rats. *Molec. Cell. Neurosci.* **4**:30–35.

Johnson, J. E. (1999): Neurotrophic factors. In "Fundamental Neuroscience" (M. J. Zigmond, F. E. Bloom, S. C. Landis, J. L. Roberts, L. R. Squire, eds.), pp. 611–635. Academic Press, San Diego.

Johnson, K. P. and Johnson, R. T. (1972). Granular ependymitis: occurrence in myxovirus infected rodents and prevalence in man. *Am. J. Pathol.* **67**:511–526.

Johnson, R. T. (1998): "Viral Infections of the Nervous System," 2nd ed. Lippincott-Raven; Philadelphia.

Johnson, R. T. and Johnson, K. P. (1968). Hydrocephalus following viral infection: the pathology of aqueductal stenosis developing after experimental mumps virus infection. *J. Neuropathol. Exp. Neurol.* **27**:591–606.

Johnson, R. T., Johnson, K. P., and Edmonds, C. J. (1967). Virus-induced hydrocephalus: development of aqueductal stenosis in hamsters after mumps infection. *Science* **157**:1066–1067.

Johnson, K. P., Klasnja, R., and Johnson, R. T. (1971). Neural tube defects of chick embryos: an indirect result of influenza A virus infection. *J. Neuropathol. Exp. Neurol.* **30**:68–74.

Johnson, R. T., McFarland, H. F., and Levy, S. E. (1972). Age-dependent resistance to viral encephalitis: Studies of infections due to Sindbis virus in mice. *J. Infect. Dis.* **125**:257–262.

Kalb, R. G. and Hockfield, S. (1988). Molecular evidence for early activity-dependent development of hamster motor neurons. *J. Neurosci.* **8**:2350–2360.

Kalb, R. G. and Hockfield, S. (1990). Large diameter primary afferent input is required for expression of the Cat-301 proteoglycan on the surface of motor neurons. *Neurosci.* **34**:391–401.

Kalb, R. G. and Hockfield, S. (1992). N-Methyl-D-aspartate receptors are transiently expressed in the developing spinal cord ventral horn. *Proc-Natl-Acad-Sci-U-S-A.* **88**:8502–8506.

Kaluza, G., Lell, G., Reinacher, M., Stitz, L., and Willems, W. R. (1987). Neurogenic spread of Semliki Forest virus in mice. *Arch. Virol.* **93**:97–110.

Katoh, I., Yoshinaka, Y., and Luftig, R. B. (1986). The effect of cerulenin on Moloney murine leukemia virus morphogenesis. *Virus Res.* **5**:265–276.

Kay, D. G., Gravel, C., Robitaille, Y., and Jolicoeur, P. (1991). Retrovirus-induced spongiform myeloecephalopathy in mice: Regional distribution of infected target cells and neuronal loss occurring in the absence of viral expression in neurons. *Proc-Natl-Acad-Sci-U-S-A.* **88**:1281–1285.

Kilham, L. and Margolis, G. (1966). Viral etiology of spontaneous ataxia of cats. *Am. J. Pathol.* **48**:991–1011.

Kitano, Y., Ohzono, H., Yasuda, N., and Shimizu, T. (1996). Hydrencephaly, cerebellar hypoplasia, and myopathy in chick embryos infected with Aino virus. *Vet. Pathol.* **33**:672–681.

Klein, R. (1994). Role of neurotrophins in mouse neuronal development. *FASEB J.* **8:**738–744.

Klimstra, W. B., Ryman, K. D., Bernard, A., Nguyen, K. B., Biron, C. A., and Johnston, R. E. (1999). Infection of neonatal mice with Sindbis virus results in a systemic inflammatory response syndrome. *J. Virol.* **73:**10387–10398.

Koeda, T., Inagaki, M., and Kawahara, H. (1993). Progressive encephalopathy associated with cytomegalovirus infection without immune deficiency. *J. Child. Neurol.* **8:**373–377.

Kolodkin, A. L., Matthes, D. J., and Goodman, C. S. (1993). The semaphorin genes encode a family of transmembrane and secreted growth cone guidance molecules. *Cell* **75:**1389–1399.

Konno, S., Koeda, T., and Madarame, H. (1988). Myopathy and encephalopathy in chick embryos experimentally infected with Akabane virus. *Vet. Pathol.* **25:**1–8.

Konno, S., Moriwaki, M., and Nakagawa, M. (1982). Akabane disease in cattle: congenital abnormalities caused by viral infection: spontaneous disease. *Vet. Pathol.* **19:**246–266.

Kono, R., Hayakawa, Y., Hibi, M., and Ishii, K. (1969). Experimental vertical transmission of rubella virus in rabbits. *Lancet* **1:**343–347.

Kosugi, I., Shinmura, Y., Li, R. Y., AibaMasago, S., Baba, S., Miura, K., and Tsutsui, Y. (1998). Murine cytomegalovirus induces apoptosis in non-infected cells of the developing mouse brain and blocks apoptosis in primary neuronal culture. *Acta. Neuropath.* **96:**239–247.

Krajewski, S., James, H. J., Ross, J., Blumberg, B. M., Epstein, L. G., Gendelman, H. E., Gummuluru, S., Dewhurst, S., Sharer, L. R., Reed, J. C., and Gelbard, H. A. (1997). Expression of pro- and anti-apoptosis gene products in brains from paediatric patients with HIV-1 encephalitis. *Neuropathol. Appl. Neurobiol.* **23:**242–253.

Kristt, D. A. (1978). Neuronal differentiation in somatosensory cortex of the rat. I. Relationship to synaptogenesis in the first postnatal week. *Brain Res.* **150:**467–486.

Kuge, O., Akamatsu, Y., and Nishijima, M. (1989). Abortive infection with Sindbis virus of a Chinese hamster ovary cell mutant defective in phosphatidylserine and phosphatidylethanolamine biosynthesis. *Biochem. Biophys. Acta.* **986:**61–69.

Kurilla, M. G., Cabradilla, C. D., Holloway, B. P., and Keene, J. D. (1984). Nucleotide sequence and host La protein interactions of rabies virus leader RNA. *J. Virol.* **50:**773–778.

Kurilla, M. G. and Keene, J. D. (1983). The leader RNA of vesicular stomatitis virus is bound by a cellular protein reactive with anti-La lupus antibodies. *Cell* **34:**837–845.

Kurogi, H., Inaba, Y., and Goto, Y. (1975). Serologic evidence for etiologic role of Akabane virus in epizootic abortion-arthrogryposis-hydranencephaly in cattle in Japan, 1972–1974. *Arch. Virol.* **47:**71–83.

Lagace-Simard, J., Descoteaux J. P., and Lussier, G. (1982). Experimental pneumovirus infections. 2. Hydrocephalus of hamsters and mice due to infection with human respiratory syncytial virus (RS). *Am. J. Pathol.* **107:**36–40.

Lai, M. C. (1998). Cellular factors in the transcription and replication of viral RNA genomes: A parallel to DNA-Dependent RNA transcription. *Virology.* **244:**1–12.

Landis, S. C. (1983). Neuronal growth cones. *Ann. Rev. Physiol.* **45:**567–580.

Lannuzel, A., Barnier, J. V., Hery, C., VanTan, H., Guibert, B., Gray, F., Vincent, J. D., and Tardieu, M. (1997). Human immunodeficiency virus type 1 and its coat protein gp120 induce apoptosis and activate JNK and ERK mitogen-activated protein kinases in human neurons. *Ann. Neurol.* **42:**847–856.

Lapinleimu, K., Koskimies, O., Cantell, K., and Saxen, L. (1972). Viral antibodies in mothers of defective children. *Teratology* **5:**345–352.

Larramendi, L. M. H. (1969): Analysis of synaptogenesis in the cerebellum of the mouse. *In* "Neurobiology of Cerebellar Evolution and Development" (R. Llinas, ed.), pp. 803–843. American Medical Association Education and Research Foundation, Chicago.

Leibig, H. D., Ziegler, E., and Yan, R. (1993). Purification of two picornaviral 2A proteinases: Interaction with eIF-4 gamma influence on in vitro translation. *Biochemistry* **32:**7581–7588.

Lennette, E. H. and Koprowski, H. (1944). Influence of age on the susceptibility of mice to infection with certain neurotropic viruses. *J. Immunol.* **49:**175.

Levi-Montalcini, R. (1987). The nerve growth factor 35 years later. *Science* **237:**1154–1162.

Levine, B., Huang, Q., Isaacs, J. T., Reed, J. C., Griffin, D. E., and Hardwick, J. M. (1993). Conversion of lytic to persistent alphavirus infection by the bcl-2 cellular oncogene. *Nature* **361:**739–742.

Levison, S. W. and Goldman, J. E. (1993). Both oligodendrocytes and astrocytes develop from progenitors in the subventricular zone of postnatal rat brain. *Neuron* **10:**201–212.

Lewis, J., Wesselingh, S. L., Griffin, D. E., and Hardwick, J. M. (1996). Alphavirus induced apoptosis in mouse brains correlates with neurovirulence. *J. Virol.* **70:**1828–1835.

Lois, C. and Alvarez-Buylla, A. (1994). Long-distance neuronal migration in the adult mammalian brain. *Science* **264:**1145–1148.

Lois, C., Garcia Verdugo, J. M., and Alvarez-Buylla, A. (1996). Chain migration of neuronal precursors. *Science* **271:**978–981.

London, W. T., Levitt, N. H., Kent, S. G., Wong, V. G., and Sever, J. L. (1977). Congenital cerebral and ocular malformations induced in rhesus monkeys by Venezuelan equine encephalitis virus. *Teratology* **16:**285–290.

Lyman, W. D., Kress, Y., Kure, K., Rashbaum, W. K., Rubinstein, A., and Soeiro, R. (1990). Detection of HIV in fetal central nervous system tissue. *AIDS* **4:**917–920.

Lynch, W. P., Czub, M., McAtee, F., Hayes, S. F., and Portis, J. L. (1991). Murine retrovirus-induced spongiform encephalopathy: productive infection of microglia and cerebellar neurons in accelerated disease. *Neuron* **7:**365–379.

Lynch, W. P. and Portis, J. L. (1993). Murine retrovirus-induced spongiform encephalopathy: Disease expression is dependent on postnatal development of the central nervous system. *J. Virol.* **67:**2601–2610.

Lynch, W. P., Robertson, S., and Portis, J. L. (1995). Induction of focal spongiform neurodegeneration in developmentally resistant mice by implantation of murine retrovirus-infected microglia. *J. Virol.* **69:**1408–1419.

Lynch, W. P. and Sharpe, A. H. (2000). Differential glycosylation of the Cas-Br-E env protein is associated with retrovirus induced spongiform neurodegeneration. *J. Virol.* **74:**1558–1565.

Lyon, M., Barr, C., Cannon, T., Mednick, S., and Shore, D. (1989). Fetal neural development and schizophrenia. *Schizophrenia Bull* **15:**149–160.

Mabruk, M. J., Flack, A. M., Glasgow, G. M., Smyth, J. M., Folan, J. C., O Sullivan, M. A., Sheahan, B. J., and Atkins, G. J. (1988). Teratogenicity of the Semliki Forest virus mutant ts22 for the foetal mouse: induction of skeletal and skin defects. *J. Gen. Virol.* **69:**2755–2762.

Mabruk, M. J., Glasglow, G. M., Flack, A. M., Folan, J. C., Bannigan, J. G., O Sullivan, M. A., Sheahan, B. J., and Atkins, G. J. (1989). Effect of infection with the ts22 mutant of Semliki Forest virus on development of the central nervous system in the fetal mouse. *J. Virol.* **63:**4027–4033.

MacDonald, H. and Tobin, J. O. H. (1978). Congenital cytomegalovirus infection: a collaborative study on epidemiological, clinical and laboratory findings. *Dev. Med. Child. Neurol.* **20:**471–482.

Malach, R. (1994). Cortical columns as devices for maximizing neuronal diversity. *Trends In Neurosciences* **17:**101–105.

Margolis, G. and Kilham, L. (1968a). In pursuit of an ataxic hamster, or virus induced cerebellar hypoplasia. *International Academy of Pathology Monograph 9: Central Nervous system* 157–183.

Margolis, G. and Kilham, L. (1968b): Virus-induced cerebellar hypoplasia. *In* "Infections of the nervous system" (H. M. Zimmerman, ed.), pp. 113–146. Williams and Wilkins, Baltimore.

Margolis, G. and Kilham, L. (1969). Hydrocephalus in hamsters, ferrets, rats and mice following inoculations with reovirus type I. II. Pathologic studies. *Lab. Invest.* **21:**189–198.

Mariani, J. and Changeux, J. -P. (1981). Ontogenesis of olivocerebellar relationships. I. Studies by intracellular recordings of the multiple innervation of Purkinje cells by climbing fibres in the developing rat cerebellum. *J. Neurosci.* **1:**696–702.

Mason-Brothers, A., Ritvo, E., Pingree, C., Peterson, P., Jenson, W., McMahon, W., Freeman, B., Jorde, L., Spencer, M., Mo, A., and Ritvo, A. (1990). The UCLA-University of Utah epidemiologic survey of autism: Prenatal, perinatal and postnatal factors. *Pediatrics* **86:**514–519.

Meerovitch, K., Svitkin, Y. V., Lee, H. S., Lejbkowics, F., Kenan, D. J., Chan, E.-K., Agol, V. I., Keene, J. D., and Sonenberg, N. (1993). La autoantigen enhances and corrects aberrant translation of poliovirus RNA in reticulocyte lysate. *J. Virol.* **67:**3798–3807.

Mehta, S. and Webb, H. E. (1987). The effect of gold sodium thiomalate in adult Swiss/A2G mice infected with Togaviruses and Flaviviruses. *J. Gen. Virol.* **68:**2665–2668.

Mehta, S., Pathak, S., and Webb, H. E. (1990). Induction of membrane proliferation in mouse CNS by gold sodium thiomalate with reference to increased virulence of the avirulent Semliki Forest virus. *Biosci. Rep.* **10:**271–279.

Miller, C. A. and Carrigan, D. R. (1982). Reversible repression and activation of measles virus infection in neural cells. *Proc. Natl. Acad. Sci. U.S.A.* **79:**1629–1633.

Miller, E., Cradock-Watson, J. E., and Pollock, T. M. (1982). Consequences of confirmed maternal rubella at successive stages of pregnancy. *Lancet* **2:**781–784.

Milner, A. R. and Marshall, I. D. (1984). Pathogenesis of in utero infections with arbortigenic and nonarbortigenic alphaviruses in mice. *J. Virol.* **50:**66–72.

Mims, C. A., Murphy, F. A., Taylor, W. P., and Marshall, I. D. (1973). The pathogenesis of Ross River virus infections in mice. I. Ependymal infection, cortical thinning and hydrocephalus. *J. Infect. Dis.* **127:**121–128.

Monath, T. P. (1980): Epidemiology. In: *St. Louis encephalitis,* edited by T. P. Monath, pp. 239–312. APHA, Washington DC.

Monjan, A. A., Cole, G. A., Gilden, D. H., and Nathanson, N. (1973). Pathogenesis of cerebellar hypoplasia produced by lymphocytic choriomeningitis virus infection of neonatal rats. I. Evolution of disease following infection of 4 days of age. *J. Neuropathol. Exp. Neurol.* **32:**110.

Monjan, A. A., Cole, G. A., and Nathanson, N. (1974). Pathogenesis of cerebellar hypoplasia produced by lymphocytic choriomenigitis virus infection of neonatal rats: protective effect of immunosuppression with antilymphoid serum. *Infect. Immun.* **10:**499–502.

Morgenson, S. (1979). Role of macrophages in natural resistance to virus infection. *Microbiol. Rev.* **43:**1–26.

Moyer, S. A., Baker, S. C., and Lessard, J. L. (1986). A factor necessary for the synthesis of both Sendai virus and vesicular stomatitis virus RNA. *Proc. Natl. Acad. Sci. U.S.A.* **83:**5405–5409.

Mulder, D. W., Parrott, M., and Thaler, M. (1951). Sequelae of western equine encephalitis. *Neurology* **1**:318

Munro, N. D., Sheppard, S., Smithells, R. W., Holzel, H., and Jones, G. (1987). Temporal relations between maternal rubella and congenital defects. *Lancet* **2**:201–204.

Narayan, O. and Johnson, R. T. (1972). Effects of viral infection on nervous system development. I. Pathogenesis of bluetongue virus infection in mice. *Am. J. Pathol.* **68**:1–14.

Neumann, H., Cavalie, A., Jenne, D. E., and Wekerle, H. (1995). Induction of MHC class I genes in neurons. *Science* **269**:549–552.

Nevins, J. R. and Vogt, P. K. (1996): Cell transformation by viruses. *In* "Fields virology" (B. N. Fields ed.), pp. 301–344. Lippincott-Raven, Philadelphia.

New, D. R., Ma, M. H., Epstein, L. G., Nath, A., and Gelbard, H. A. (1997). Human immunodeficiency virus type 1 Tat protein induces death by apoptosis in primary human neuron cultures. *J. Neurovirol.* **3**:168–173.

Nicola, L. and Hanshaw, S. (1979). Congenital and neonatal varicella. *J. Pediatr.* **94**:175–176.

Noran, N. H. and Baker, A. B. (1943). Sequels of equine encephalomyelitis. *Arch. Neurol. Psychiat.* **49**:398.

Oberhaus, S. M., Dermody, T. S., and Tyler, K. L. (1998). Apoptosis and the cytopathic effects of reovirus. *Curr. Top. Microbiol. Immunol.* **233**:23–49.

Ogata, A., Nagashima, K., Hall, W. W., Ichikawa, M., Kimura-Kuroda, J., and Yasui, K. (1991). Japanese encephalitis virus neurotropism is dependent on the degree of neuronal maturity. *J. Virol.* **65**:880–886.

Oliver, K. R. and Fazakerley, J. K. (1998). Transneuronal spread of Semliki Forest virus in the developing mouse olfactory system is determined by neuronal maturity. *Neurosci.* **82**:867–877.

Oliver, K. R., Scallan, M. F., Dyson, H., and Fazakerley, J. K. (1997). Susceptibility to a neurotropic virus and its changing distribution in the developing brain is a function of CNS maturity. *J. Neurovirol.* **3**:38–48.

Oppenheim, R. W. (1991). Cell death during development of the nervous system. *Ann. Rev. Neurosci.* **14**:453–501.

Oppenheim, R. W. (1999): Programmed cell death. *In* "Fundamental neuroscience", (M. J. Zigmond, F. E. Bloom, S. C. Landis, J. L. Roberts, L. R. Squire Eds.), pp. 581–609. Academic Press, San Diego.

Osburn, B. I., Silverstein, A. M., Prendergast, R. A., Johnson, R. T., and Parshall, C. J. (1971a). Experimental viral-induced congenital encephalopathies. I. Pathology of hydranencephaly and porencephaly caused by bluetongue vaccine virus. *Lab. Invest.* **25**:197–205.

Osburn, B. I., Johnson, R. T., Silverstein, A. M., Prendergast, R. A., Jochim, M. M., and Levy, S. E. (1971b). Experimental viral induced congenital encephalopathies. II. The pathogenesis of bluetongue vaccine virus infection in fetal lambs. *Lab. Invest.* **25**:206–210.

Palmer, R. J. and Finley, K. H. (1956). Sequelae of encephalitis: report of a study after the California epidemic. *California Medicine* **84**:98

Pardigon, N. and Strauss, J. H. (1996). Mosquito homolog of the La autoantigen binds to Sindbis virus RNA. *J. Virol.* **70**:1173–1187.

Pathak, S. and Webb, H. E. (1974). Possible mechanisms for the transport of Semliki Forest virus into and within mouse brain: An electron microscopic study. *J. Neurol. Sci.* **23**:175–184.

Pathak, S. and Webb, H. E. (1978). An electron-microscopic study of avirulent and virulent Semliki forest virus in the brains of different ages of mice. *J. Neurol. Sci.* **39**:199–211.

Pathak, S. and Webb, H. E. (1983). Effect of myocrisin (sodium aurothiomalate) on the morphogenesis of avirulent Semliki Forest virus in mouse brain: An electron microscopical study. *Neuropathol. Appl. Neurobiol.* **9**:313–327.

Pathak, S. and Webb, H. E. (1988a). An electron microscopical study of the replication of avirulent Semliki Forest virus in the retina of mice. *J. Neurol. Sci.* **85**:87–96.

Pathak, S. and Webb, H. E. (1988b). Cytoplasmic viral core aggregates and budding of mature virus and spherules in mouse brain following Semliki Forest virus infections. *Inst. Phys. Conf. Ser.* **3**:213–214.

Pathak, S., Webb, H. E., Oaten, S. W., and Bateman, S. (1976). An electron-microscopic study of the development of virulent and avirulent strains of Semliki Forest virus in mouse brain. *J. Neurol. Sci.* **28**:289–300.

Paxinos, G. and Watson, C. (1986). "The rat brain in stereotaxic coordinates," 2nd ed. Academic Press, San Diego.

Pekosz, A., Phillips, J., Pleasure, D., Merry, D., and Gonzalez-Scarano, F. J. (1996). Induction of apoptosis by La Crosse virus infection and role of neuronal differentiation and human bcl-2 expression in its prevention. *J. Virol.* **70**:5329–5335.

Peranen, J. and Kaariainen, L. (1991). Biogenesis of type I cytopathic vacuoles in Semliki Forest virus-infected BHK cells. *J. Virol.* **65**:1623–1627.

Peranen, J., Laakkonen, P., Hyvonen, M., and Kaariainen, L. (1995). The alphavirus replicase protein nsp1 is membrane-associated and has affinity to endocytic organelles. *Virology* **208**:610–620.

Perez, L., Guinea, R., and Carrasco, L. (1991). Synthesis of Semliki Forest virus RNA requires continuous lipid synthesis. *Virology.* **183**:74–82.

Peudenier, S., Hery, C., Montagnier, J., and Tardieu, M. (1991). Human microglial cells: characterization in cerebral tissue and in primary culture, and study of their susceptibility to HIV-1 infection. *Ann. Neurol.* **29**:152–161.

Pitts, O. M., Powers, J. M., Bilello, J. A., and Hoffman, P. M. (1987). Ultrastructural changes associated with retroviral replication in the central nervous system capillary endothelial cells. *Lab. Invest.* **56**:401–409.

Pletnikov, M. V., Rubin, S. A., Schwartz, G. J., Moran, T. H., Sobotka, T. J., and Carbone, K. M. (1999a). Persistent neonatal Borna disease virus (BDV) infection of the brain causes chronic emotional abnormalities in adult rats. *Physiology & Behavior* **66**:823–831.

Pletnikov, M. V., Rubin, S. A., Vasudevan, K., Moran, T. H., and Carbone, K. M. (1999b). Developmental brain injury associated with abnormal play behavior in neonatally Borna disease virus-infected Lewis rats: a model of autism. *Behavioural Brain Research* **100**:43–50.

Portis, J. L., Czub, S., Garon, C. F., and McAtee, F. (1990). Neurodegenerative disease induced by the wild mouse ecotropic retrovirus is markedly accelerated by long term terminal repeat and gag-pol sequences from nondefective friend murine leukemia virus. *J. Virol.* **64**:1648–1656.

Potts, B. J., Berry, L. J., Osburn, B. I., and Johnson, K. P. (1985). Viral persistence and abnormalities of the central nervous system after congenital infection of sheep with border disease virus. *J. Infect. Dis.* **151**:337–341.

Purves, D., Riddle, D. R., and LaMantia, A.-S. (1992). Iterated patterns of brain circuitry (or how the cortex gets its spots). *Trends Neurol. Sci.* **15**:332–368.

Pusztai, R., Gould, E., and Smith, H. (1971). Infection pattern in mice of an avirulent and virulent strain of Semliki Forest virus. *Br. J. Exp. Pathol.* **52**:669–677.

Qin, Y. and Sato, T. N. (1995). Mouse multidrug resistance la/3 gene is the earliest known endothelial cell differentiation marker during blood-brain barrier development. *Dev. Dyn.* **202**:172–180.

Rakic, P. (1972). Mode of cell migration to the superficial layers of fetal monkey neocortex. *J. Comp. Neurol.* **145**:61–84.

Rakic, P. (1974). Neuron in rhesus monkey visual cortex: Systematic relation between time of origin and eventual disposition. *Science* **183**:425–427.

Rakic, P. (1988). Specification of cerebral cortical areas. *Science* **241**:170–176.

Rao, L., Debbas, M., Sabbatini, P., Hockenbery, D., Korsmeyer, S., and White, E. (1992): The adenovirus E1A proteins induce apoptosis, which is inhibited by the E1B 19-kDa and Bcl-2 proteins. *Proc. Natl. Acad. Sci. U.S.A.*, pp. 7742–7746.

Rawls, W. E. and Melnick, J. L. (1966). Rubella virus carrier cultures derived from congenitally infected infants. *J. Exp. Med.* **123**:795–816.

Rees, R. P., Bunge, M. B., and Bunge, R. P. (1976). Morphological changes in the neuritic growth cone and target neuron during synaptic junction development in culture. *J. Cell Biol.* **68**:240–263.

Reinarz, A. B., Broome, M. G., and Sagik, B. P. (1971). Age resistance of mice to Sindbis virus infection: viral replication as a function of host age. *Infect. Immun.* **3**:268–273.

Risau, W., Hallmann, R., and Albrecht, U. (1986). Differentiation-dependent expression of proteins in brain endothelium during development of the blood-brain barrier. *Developmental Biology* **117**:537–545.

Rorke, L. B., Fabiyi, A., Elizan, T. S., and Sever, J. L. (1968). Experimental cerebrovascular lesions in congenital and neonatal rubella-virus infections of ferrets. *Lancet* **2**:153–154.

Rorke, L. B. and Spiro, A. J. (1967). Cerebral lesions in congenital rubella syndrome. *J. Pediatr.* **70**:243–255.

Rubin, L. L. and Staddon, J. M. (1999). The cell biology of the blood brain barrier. *Neurosci.* **22**:11–28.

Rubin, S. A., Bautista, J. R., Moran, T. H., Schwartz, G. J., and Carbone, K. M. (1999a). Viral teratogenesis: brain developmental damage associated with maturation state at time of infection. *Brain Res. Dev. Br. Res.* **112**:237–244.

Rubin, S. A., Sylves, P., Vogel, M., Pletnikov, M., Moran, T. H., Schwartz, G. J., and Carbone, K. M. (1999b). Borna disease virus-induced hippocampal dentate gyrus damage is associated with spatial learning and memory deficits. *Brain. Res. Bull.* **48**:23–30.

Sarnat, H. B. and Alcala, H. (1980). Human cerebellar hypoplasia: a syndrome of diverse causes. *Arch. Neurol.* **37**:300–305.

Saunders, N. R., Dziegielewska, K. M., and Mollgard, K. (1991). The importance of the blood-brain barrier in fetuses and embryos. *Trends In Neurosciences* **14**:14–15.

Scallan, M. F., Allsopp, T. E., and Fazakerley, J. K. (1997). Bcl-2 acts early to restrict Semliki Forest virus replication and delays virus-induced programmed cell death. *Journal of Virology* **71**:1583–1590.

Scallan, M. F. and Fazakerley, J. K. (1999). Aurothiolates enhance the replication of Semliki Forest virus in the CNS and the exocrine pancreas. *Journal Of Neurovirology* **5**:392–400.

Schlesinger, M. J. and Malfer, C. (1982). Cerulenin blocks fatty acid acylation of glycoproteins and inhibits Vesticular Stomatitis and Sindbis virus particle formation. *J. Biol. Chem.* **257**:9887–9890.

Schultz, G. and DeLay, P. D. (1955). Losses in newborn lambs associated with bluetongue vaccination of pregnant ewes. *J. Am. Vet. Assoc.* **127**:224–226.

Schwob, J. E. and Price, J. L. (1984). The development of axonal connections in the central olfactory system of rats. *J. Comp. Neurol.* **223**:177–202.

Serafini, T., Kennedy, T. E., Galko, M. J., Mirzayan, C., Jessell, T. M., and TessierLavigne, M. (1994). The netrins define a family of axon outgrowth-promoting proteins homologous to C. elegans UNC-6. *Cell* **78**:409–424.

Sharer, L. R. (1992). Pathology of HIV-1 infection of the central nervous system: a review. *J. Neuropathol. Exp. Neurol.* **51**:3–11.

Sharpless, N., Gilbert, D., Vandercam, B., Zhou, J. M., Verdin, E., Ronnett, G., Friedman, E., and Dubois-Dalcq, M. (1992). The restricted nature of HIV-1 tropism for cultured neural cells. *Virology* **191**:813–825.

Sheinberg, M. M. (1975). Antibody to lymphocytic choriomeningitis virus in children with congenital hydrocephalus. *Acta Virol.* **19**:165–166.

Shi, B., Raina, J., Lorenzo, A. E., Busciglio, J., and Gabuda, D. (1998). Neuronal apoptosis induced by HIV-1 Tat protein and TNF-alpha: potentiation of neurotoxicity mediated by oxidative stress and implications for HIV-1 dementia. *J. Neurovirol.* **4**:281–290.

Sidman, R. L. and Rakic, P. (1973). Neuronal migration with special reference to developing human brain: A review. *Brain. Res.* **62**:1–35.

Sigel, M. M. (1952). Influence of age on susceptibility to virus infections with particular reference to laboratory animals. *Ann. Rev. Microbiol.* **6**:247–280.

Speidel, C. C. (1933). Studies of living nerves: II. Activities of ameboid growth cones, sheath cells, and myelin segments, as revealed by prolonged observation of individual nerve fibers in frog tadpoles. *Am. J. Anat.* **52**:1–79.

Sperling, R. S., Shapiro, D. E., and Coombs, R. W. (1996). Maternal viral load, zidovudine treatment, and the risk of transmission of human immunodeficiency virus type 1 from mother to infant. *N. Engl. J. Med.* **335**:1621–1629.

Spriggs, M. K., Hruby, D. E., and Maliszewski, C. R. (1992). Vaccinia and cowpox viruses encode a novel secreted interleukin-1-binding protein. *Cell* **71**:145–152.

Srabstein, J. C., Morris, N., Larke, R., deSa, D. J., Castelino, B. B., and Sum, E. (1974). Is there a congenital varicella syndrome? *J. Pediatr.* **84**:239–294.

Subak-Sharpe, I., Dyson, H., and Fazakerley, J. K. (1993). In vivo depletion of CD8+ T cells prevents lesions of demyelination in Semliki Forest virus infection. *J. Virol.* **67**:7629–7633.

Swarz, J. R., Brooks, B. R., and Johnson, R. T. (1981). Spongiform polioencephalomyelopathy caused by a murine retrovirus. II. Ultrastructural localization of virus replication and spongiform changes in the central nervous system. *Neuropathol. Appl. Neurobiol.* **7**:365–380.

Takeichi, M. (1995). Morphogenetic roles of classic cadherins. *Curr. Opin. Cell. Biol.* **7**:619–627.

Tan, S.-S. and Breen, S. (1993). Radial mosaicism and tangential cell dispersion both contribute to mouse neocortical development. *Nature* **362**:638–640.

Tarodi, B., Subramanian, T., and Chinnadurai, G. (1994). Epstein-Barr virus BHRF1 protein protects against cell death induced by DNA-damaging agents and heterologous virus infection. *Virology* **201**:404–407.

Tattersall, P. (1972). Replication of the parvovirus MVM. I. Dependence of virus multiplication and plaque formation on cell growth. *J. Virol.* **10**:586–590.

Theerasurakarn, S. and Ubol, S. (1998). Apoptosis induction in brain during the fixed strain of rabies virus infection correlates with onset and severity of illness. *J. Neurovirol.* **4**:407–414.

Trautwein, G., Hewicker, M., Liess, B., Orban, S., and Grunert, E. (1985). Studies on transplacental transmissibility of a bovine virus diarrhoea (BVD) vaccine virus in cattle. III. Occurrence of central nervous system malformations in calves born from vaccinated cows. *J. Vet. Med.* **33**:260–268.

Trgovcich, J., Aronson, J. F., and Johnston, R. E. (1996). Fatal Sindbis infection of neonatal mice in the absence of encephalitis. *Virology* **224**:73–83.

Trgovcich, J., Ryman, K., Extrom, P., Eldridge, C., Aronson, J. F., and Johnston, R. E. (1997). Sindbis virus infection of neonatal mice results in a severe stress response. *Virology* **227**:234–238.

Truckenmiller, M. E., Kulaga, H., Coggiano, M., Wyatt, R., Snyder, S. H., and Sweetnam, P. M. (1993). Human cortical neuronal cell line: a model for HIV-1 infection in an immature nervous system. *AIDS Research and Human Retroviruses* **9:**445–453.

Ubol, S. and Griffin, D. (1991). Identification of a putative alphavirus receptor on mouse neural cells. *J. Virol.* **65:**6913–6921.

vanOirschot, J. T. and Terpstra, C. (1977). A congenital persistent swine fever infection. I. Clinical and virological observations. II. Immune response to swine fever virus and unrelated antigens. *Vet. Microbiol.* **2:**121–142.

Vaughn, J. E. and Sims, T. J. (1978). Axonal growth cones and developing axonal collaterals form synaptic junctions in embryonic mouse spinal cord. *J. Neurocytol.* **7:**337–363.

Waite, P. M. E., Lixin, L., and Ashwell, K. W. S. (1992). Development and lesion induced cell death in the rat ventrobasal complex. *Neuroreport* **3:**485–488.

Waltrip, R. W., Carrigan, D. R., and Carpenter, W. (1990). Immunopathology and viral reactivation: A general theory schizophreni. *J. Nerv. Ment. Dis.* **178:**729–738.

Waters, C. M., Moser, H. W., Walkinshaw, G., and Mitchell, I. G. (1994). Death of neurons in the neonatal rodent and primate globus pallidus occurs by a mechanism of apoptosis. *Neurosci.* **63:**881–894.

Weiner, L. P., Cole, G. A., and Nathanson, N. (1970). Experimental encephalitis following peripheral inoculation of West Nile virus in mice of different ages. *J. Hyg. (London)* **68:**435–446.

Wesselingh, S. L., Levine, B., Fox, R. J., Choi, S., and Griffin, D. E. (1994). Intracerebral cytokine messenger-RNA expression during fatal and nonfatal alphavirus encephalitis suggests a predominant type-2 T-cell response. *J. Immunol.* **152:**1289–1297.

Wilusz, J., Kurilla, M. G., and Keene, J. D. (1983). A host protein (La) binds to a unique species of minus-sense leader RNA during replication of vesticular stomatitis virus. *Proc. Natl. Acad. Sci. U.S.A.* **80:**5827–5831.

Wirth, J. J., Amalfitano, A., and Gross, R. (1992). Organ- and age-specific replication of polyomavirus in mice. *J. Virol.* **66:**3278–3286.

Woodall, C. J., Riding, M. H., Graham, D. I., and Clements, G. B. (1994). Sequences specific for enterovirus detected in spinal cord from patients with motor neurone disease. *Br. Med. J.* **308:**1541–1543.

Woodward, C. G., Marshall, I. D., and Smith, H. (1978). Investigations of reasons for the avirulence of the A7 strain of Semliki Forest virus in adult mice. *Br. J. Exp. Pathol.* **58:**616–624.

Yoneda, T., Urade, M., Sakuda, M., and Miyazaki, T. (1986). Altered growth, differentiation, and responsiveness to epidermal growth factor of human embryonic mesenchymal cells of palate by persistent rubella virus infection. *J. Clin. Invest.* **77:**1613–1621.

Zigmond, M. J., Bloom, F. E., Landis, S. C., Roberts, J. L., and Squire, L. R. (1999): "Fundamental Neuroscience". Academic Press, San Diego.

VIRAL IMMUNE RESPONSES
IN THE CENTRAL NERVOUS SYSTEM

ADVANCES IN VIRUS RESEARCH, VOL 56

CHEMOKINES AND VIRAL DISEASES OF THE CENTRAL NERVOUS SYSTEM

Valérie C. Asensio and Iain L. Campbell

Department of Neuropharmacology, SP-315
The Scripps Research Institute
La Jolla, California 92037

I. INTRODUCTION

The central nervous system (CNS) has been viewed as a relatively immune privileged organ. In physiologic states, the CNS contains few, if any, leukocytes, and an absence of professional antigen-presenting cells (APC); it is barren for the expression of key immune accessory molecules such as major histocompatibility molecules (MHCs), and is fortified by an effective blood–brain barrier. Nevertheless, the past decade has witnessed a revision of this concept, as evidenced by the fact that activated T lymphocytes, regardless of their antigen specificity, migrate effectively through the blood–brain barrier (BBB), but also by the fact that the CNS contains numerous resident cells, such as microglia, cerebral endothelial cells, and astrocytes, that can, under the appropriate conditions, acquire the expression of immune accessory molecules and may function as APC (Hickey, 1999; Owens et al., 1994).

Significantly, in numerous pathological conditions, immune cells are readily recruited to and accumulate in the CNS. Infection of the human

127

CNS with many different viruses (e.g., human T cell leukemia virus type I [HTLV-I]; measles virus, human immunodeficiency virus [HIV] type 1; and herpes simplex virus [HSV] type 2), or of the rodent CNS (e.g., Theiler's murine encephalomyelitis virus [TMEV], mouse hepatitis virus [MHV], lymphocytic choriomeningitis virus [LCMV] and vesicular stomatitis virus) induces vigorous host inflammatory responses with recruitment of large numbers of leukocytes, particularly T lymphocytes and macrophages. This host response can be a two-edged sword that, on the one hand, is needed to control and extinguish the invading virus but, on the other, can produce immune pathology and tissue injury resulting in mental and physical debilitation, and often death, of the host organism. Therefore, increasing our knowledge of the mechanisms that control the trafficking of immune cells into the CNS, and knowledge of the subsequent interactions between these cells that contribute to CNS damage, is an important objective.

Our understanding of the factors that govern leukocyte trafficking has been advanced considerably with the discovery of a large superfamily of mostly small proteins (termed chemokines) that function as leukocyte chemoattractants in immunoinflammatory states (for reviews, see Bacon and Oppenheim, 1998; Rollins, 1997; Taub and Oppenheim, 1994). Chemokines coordinate trafficking of peripheral blood leukocytes by stimulating their chemotaxis, adhesion, extravasation, and other effector functions. In view of these properties, research efforts have turned increasingly to a focus on the possible involvement of chemokines in regulating both peripheral tissue and CNS leukocyte migration during viral infection. However, it is clear that the scope of chemokine involvement in the pathogenesis of host/viral interactions extends well beyond leukocyte migration. With studies on HIV-1 in the vanguard, new roles for chemokines and their receptors in viral infection became apparent with the demonstration that certain classes of chemokine receptors functioned as essential coreceptors for HIV-1 binding and entry into the cell. Yet more recent work has indicated that the genomes of some large viruses, belonging to the herpesvirus and poxvirus families, contain homologs of the mammalian chemokines and their receptors. These homologs may function to modulate local inflammatory responses and favor viral survival and spread. From the host perspective, accumulating evidence indicates that chemokines are plurifunctional molecules that may have a significant impact on the CNS, regulating cellular communication in the developing and the normal adult CNS (Asensio and Campbell, 1999).

Here, we review the subject of chemokines and their receptors in the evolution of viral infectious diseases of the CNS. As a prelude to the

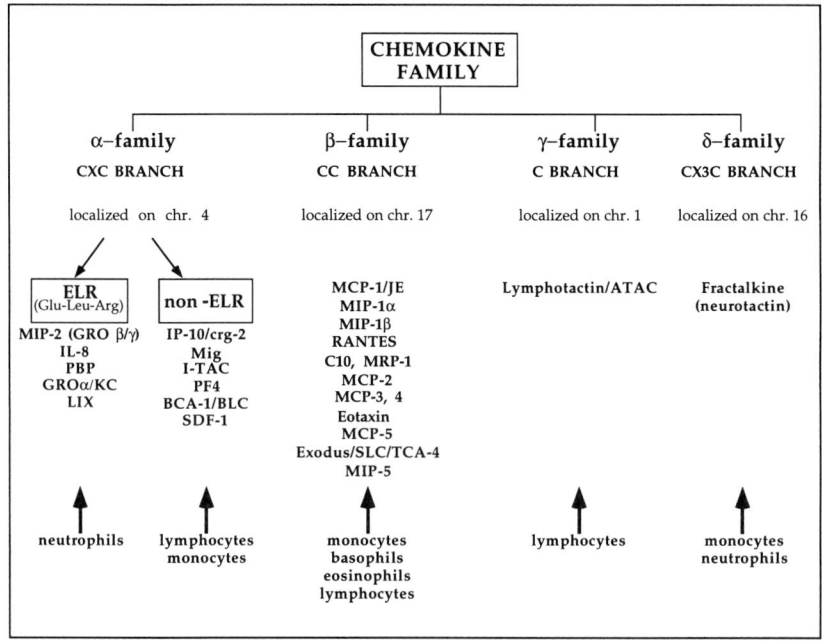

FIG 1. Classification and broad cellular specificities of the CXC, CC, C, and CX₃C chemokines. ATAC, activation-induced chemokine-related molecule exclusively expressed in CD8⁺ T lymphocytes; BCA-1/BLC, B cell–attracting chemokine/B-lymphocyte chemoattractant; GRO, growth-regulated oncogene; IL-8, interleukin-8; IP-10, inflammatory protein 10 kD; I-TAC, interferon-inducible T cell alpha-chemoattractant; LIX, LPS-induced CXC chemokine; MCP, monocyte chemotactic protein; Mig, monokine induced by interferon-gamma; MIP, macrophage inflammatory protein; MRP-1, MIP-related protein 1; PBP, platelet basic protein; PF4, platelet factor 4; RANTES, regulated on activation-normal T cell expressed and secreted; SDF-1, stromal cell–derived factor-1; SLC, secondary lymphoid tissue chemokine; chr., chromosome.

main subject of this chapter, our discussion initially will provide the reader with a general background to the rapidly expanding area of chemokines and their receptors and the expression of these molecules in the CNS.

II. CHEMOKINES AND THEIR RECEPTORS: AN OVERVIEW

The chemokines are currently divided into 4 families (Fig. 1): the CXC or α family; the CC or β family; the C or γ family; and the CX₃C or δ family (Zlotnik *et al.*, 1999). In general, within each chemokine subfamily, the individual members show considerable homology in their amino acid sequence and often possess overlapping chemoattractant specificity. Chemokines share a common three-dimensional structure including an NH_2-terminal loop and three antiparallel β sheets that are followed by a C-terminal α helix. The amino acid backbone of chemokines belonging to the CC, CXC, and CX_3C families contains four conserved cysteines, while the C family chemokine, lymphotactin, contains only two cysteines, which correspond to the first and third cysteines in the other groups.

Chemokines can be produced by a variety of immune cells such as T lymphocytes, macrophages, and natural killer (NK) cells, but also by organ-specific cell types such as resident glial cells in the CNS (see the discussion below). Depending on the chemokine and its cellular source, the expression of chemokines may be constitutive or inducible, and a new classification of chemokines, defined according to their expression pattern, has been proposed (Mantovani, 1999a; Sallusto et al., 1999a). The regulation of chemokine expression can be mediated by soluble factors such as inflammatory cytokines, e.g., interleukin (IL)-1, tumor necrosis factor (TNF)-α, interferon (IFN)-γ, as well as by microbial products, e.g., bacterial lipopolysaccharide (LPS) or HIV coat protein gp120 (Baggiolini, 1998; Baggiolini et al., 1997; Premack and Schall, 1996; Strieter et al., 1996). In contrast, counterregulatory cytokines can downregulate the production of chemokines. For example, IL-10 and transforming growth factor (TGF)-β have been shown to inhibit LPS-induced microglial cell RANTES mRNA expression and protein release (Hu et al., 1999). The glucocorticoid dexamethasone is also a particularly potent inhibitor for CC and CXC chemokine production (Tobler et al., 1992; Villiger et al., 1992).

The biological effects of chemokines are mediated via their interaction with receptors belonging to the family of seven transmembrane (7TM)-spanning, G-protein-coupled receptors (GPCRs). All chemokine receptors contain two conserved cysteines, one in the NH_2-terminal domain and the other in the third extracellular loop; these are assumed to form disulfide bonds critical for the formation of the ligand binding pocket. In accordance with the chemokine classification, the receptors are divided into four major groups, CXCR, CCR, XCR, and CX_3CR (Table I). Individual chemokine receptors may often be promiscuous, meaning that they are capable of binding several different chemokines, and conversely, individual chemokines can often bind to several different receptors. Interaction between a chemokine and its receptor leads to generation of a coordinated series of signaling events such as mobilization and influx of Ca^{2+} and also activation of various signal transduction pathways (Premack and Schall, 1996).

TABLE I
CHEMOKINE RECEPTOR FAMILIES AND THEIR LIGANDS

Receptor family	Ligand[a]
CC family	
CCR1	MIP-1α, β, RANTES, MCP-2, MCP-3, MCP-4, MIP-5
CCR2	MCP-1, MCP-2, MCP-3, MCP-4
CCR3	RANTES, eotaxin, MCP-3, MCP-4, MIP-5
CCR4	MDC, RANTES, TARC
CCR5	RANTES, eotaxin, MIP-1α, MIP-1β, MCP-2, MCP-4
CCR6	LARC (MIP-3α)
CCR7	ELC, SLC
CCR8	I-309, TARC, MIP-1β
CCR9/10	TECK, MCP-1, MCP-2, MCP-4, MIP-1 α, eotaxin
CXC family	
CXCR1 (IL-8RA)	IL-8, GRO-α
CXCR2 (IL-8RB)	GRO-α, β, γ, IL-8, MIP-2, PBP, LIX, ENA-78
CXCR3	IP-10, Mig, I-TAC, **[MCP-4, SLC, eotaxin]**
CXCR4	SDF-1
CXCR5	BCA-1/BLC
CX$_3$C family	
CX$_3$CR1	Fractalkine-neurotactin
C family	
XCR1	Lymphotactin

[a] Chemokines in boldface type are cross-subfamily ligand-binding receptors. BCA-1/BLC, B cell-attracting chemokine/B-lymphocyte chemoattractant; ELC, Epstein-Barr virus–induced receptor ligand chemokine; ENA-78, epithelial cell–derived neutrophil-activating factor-78; GCP-2, granulocyte chemoattractant protein-2; GRO, growth-regulated oncogene; IL-8, interleukin-8; IP-10, inflammatory protein 10 kD; I-TAC, interferon-inducible T cell alpha-chemoattractant; LARC, liver- and activation-related chemokine; LIX, LPS-induced CXC chemokine; MCP, monocyte chemotactic protein; Mig, monokine induced by interferon-gamma; MIP, macrophage inflammatory protein; MRP-1, MIP-related protein 1; PBP, platelet basic protein; PF4, platelet factor 4; RANTES, regulated on activation-normal T cell expressed and secreted; SDF-1, stromal cell–derived factor-1; SLC, secondary lymphoid tissue chemokine; TECK, thymus-expressed chemokine.

A. CXC (or α-) Chemokines and Receptors

Chemokines of the CXC or α family have one amino acid (X) separating the first and second N-terminal cysteine residues. Well-known members of this family include IL-8 (interleukin-8); GRO-α,-β, and -γ (growth-regulated oncogene); MIP-2 (macrophage inflammatory protein 2); SDF-1 (stromal-derived factor-1); IP-10 (interferon-inducible protein 10 kDa); and Mig (monokine induced by interferon-γ), (see Fig. 1). The presence or absence of an ELR motif (glutamic acid-leucine-arginine sequence) preceding the first cysteine further subdivides the CXC chemokines into ELR or non-ELR groups. Examples of some common ELR chemokines include IL-8, GRO-α, and MIP-2, while non-ELR chemokines include IP-10, Mig and SDF-1. This grouping not only reflects a structural compartmentalization, but also a functional one in that ELR CXC chemokines are predominantly chemoattractant for polymorphonuclear leukocytes and bind to the CXCR1 and CXCR2 receptors, while non-ELR chemokines are chemoattractant for T and B cells and bind to the CXCR3 or CXCR4 receptors. In addition, members of the CXC chemokine family can either promote or inhibit angiogenesis, depending on whether they are ELR or non-ELR CXC chemokines, respectively. The ELR-chemokines are generally produced during inflammatory responses provoked by bacterial infections. On the other hand, production of the non-ELR chemokines, IP-10/crg-2 and Mig, is commonly associated with viral infections where they likely function to recruit T cells and macrophages that predominate at sites of infection (Amichay et al., 1996; Asensio and Campbell, 1997; Asensio et al., 1999a; Carr et al., 1998; Charles et al., 1999; Fisher et al., 1995; Lahrtz et al., 1997; Nazar et al., 1997; Vanguri and Farber, 1994). This scenario is borne out by the recent demonstration of pronounced antiviral activity of Mig and IP-10 when it is expressed ectopically by recombinant vaccinia viruses (Mahalingam et al., 1999). The antiviral actions of Mig and IP-10 in this setting are mediated indirectly and require NK cells and IFN-α, IFN-β, and IFN-γ. SDF-1 was originally isolated from bone marrow stromal cells as a pre-B cell growth-stimulating factor. In contrast to IP-10 and Mig, SDF-1 is produced constitutively in many organs (e.g., heart, lung, and CNS) and binds to the CXCR4 receptor. Animals lacking SDF-1 or its receptor invariably die at or soon after birth and have major abnormalities affecting the immune system (an impairment in B lymphocyte development) as well as various organs including the heart and CNS (Ma et al., 1998; Nagasawa et al., 1996; Tachibana et al., 1998; Zou et al., 1998). These studies revealed that SDF-1 is a crucial mediator of cellular migration and patterning during organogenesis.

To date, 5 CXC chemokine receptors have been described and their ligands identified (see Table I) (Baggiolini et al., 1997; Mackay, 1997; Moser et al., 1998). Most of the early characterization of the CXC chemokine receptors was done with the IL-8 receptors CXCR1 (also known as IL-8RA) and CXCR2 (also known as IL-8RB). CXCR2 is expressed on a variety of cells including T cells, monocytes, melanoma cells, synovial cells, neutrophils, and myeloid precursor cells. Expression of CXCR1 is restricted largely to myeloid cells and neutrophils. While CXCR2 binds all the ELR-chemokines, CXCR1 binds only IL-8. Mice lacking CXCR2 show normal development of neutrophils but have decreased numbers of myeloid progenitors (Broxmeyer *et al.*, 1996; Cacalano *et al.*, 1994). CXCR2-deficient mice do not exhibit any other overt pathologic abnormalities. CXCR3 is the receptor for several of the non-ELR CXC chemokines, IP-10, Mig and I-TAC. Surprisingly, some CC chemokines such as eotaxin, MCP-4 (Weng *et al.*, 1998), and 6Ckine (also called exodus 2 or SLC) (Soto *et al.*, 1998) also bind to CXCR3. Through this interaction, eotaxin does not activate the CXCR3 receptor but effectively blocks IP-10 binding and prevents receptor activation. Thus, these CC chemokines may serve as natural CXCR3 antagonists highlighting complexity in the regulation of chemokine actions. Human CXCR3 has been shown to be produced specifically by IL-2-activated CD4+ T cells, B cells, and NK cells. CD8+ T cell and NK cell clones have also been shown to express the CXCR3 receptor (Loetscher *et al.*, 1996). This suggests non-ELR chemokines such as IP-10 may exert a selective influence on Th1 and cytotoxic lymphocyte and NK cell recruitment, which may be important in the development of innate and adaptive immune responses to viral infections. CXCR4 and CXCR5 bind the other non-ELR chemokines SDF-1 and BCA, respectively. CXCR4, also known as LESTR or fusin, serves as a major coreceptor for lymphotropic strains of HIV-1 (Deng *et al.*, 1996; Feng *et al.*, 1996) and is expressed in a variety of tissues including brain, heart, liver, and colon and in nonhematopoietic cells (Lavi *et al.*, 1997; Ohtani *et al.*, 1998). CXCR4 mRNA expression is also found in murine lymphocytes, macrophages, neutrophils, and glial cells (Nagasawa *et al.*, 1998). As indicated above, disruption of the CXCR4 gene causes defects in vascular development, hematopoiesis, cardiogenesis, and neurogenesis of the cerebellum. CXCR5, originally described as BLR-1 (Burkitt's lymphoma receptor-1), is highly related to the IL-8 receptors (Dobner *et al.*, 1992). CXCR5 mRNA is expressed by B cell lymphomas and by granule and Purkinje cells of the cerebellum. Therefore, in addition to regulation of recirculating mature B lymphocytes (Forster *et al.*, 1996; Forster *et al.*, 1994; Kaiser *et al.*, 1993), this receptor may be involved in some CNS functions. However, mice lacking CXCR5 show significant phenotypic changes in lymphoid organs but not in the brain, suggesting that this receptor, unlike CXCR4, does not play an obligatory role in CNS development (Forster *et al.*, 1996).

B. CC (or β-) Chemokines and Receptors

The CC or β-chemokines contain the largest number of members and are so named because the NH_2-terminal cysteines are adjacent to each other. Individual CC chemokines may exert their actions on multiple leukocyte subtypes, including monocytes, basophils, eosinophils, T cells, dendritic cells, and natural killer cells. Well-known members of this family include MCP-1, MIP-1α, C10, eotaxin, and RANTES. A more comprehensive list of the members of this family can be found in Fig. 1. One of the most studied chemokines of this family is MCP-1, which attracts T cells and monocytes but not neutrophils (Carr et al., 1994; Matsushima et al., 1989). Mice with targeted disruption of the MCP-1 gene (Lu et al., 1998) have defects in monocyte/macrophage recruitment in a variety of inflammatory and immunological models. Conversely, transgenic mice with expression of MCP-1 targeted to pancreatic islets (Grewal et al., 1997) or oligodendrocytes in the brain (Fuentes et al., 1995) show an accumulation of monocytes at these sites. Interestingly, these accumulating monocytes showed little evidence of activation, indicating that MCP-1 is an effective monocyte chemoattractant but lacks the ability to further activate these cells. Another well-studied CC chemokine is MIP-1α, which induces migration of monocytes, T lymphocytes, and eosinophils (Baggiolini et al., 1994). Mice with a targeted disruption of MIP-1α show no obvious hematopoietic abnormalities; however, they are resistant to coxsackievirus-induced myocarditis and clear influenza at a delayed rate (Cook et al., 1995).

To date, 9 CC receptors (CCR1-CCR9) have been identified (Table I). CCR1 was originally designated as an MIP-1α/RANTES receptor (Gao et al., 1993; Gao and Murphy, 1995; Neote et al., 1993) and was later shown to also bind MCP-2 and MCP-3 (Ben-Baruch et al., 1995). The CCR1 receptor is mainly found on circulating mononuclear cells. Mice lacking CCR1 (receptor for MIP-1α, MIP-5, MCP-2, MCP-3, and RANTES) have a phenotype similar to that of MIP-1α deficient mice with a defect in inflammatory responses to microbial challenge such as coxsackie B and influenza viruses (Gao et al., 1997). In addition, these animals have impaired trafficking of subsets of myeloid progenitor cells. The primary receptor for MCP-1, CCR2, is expressed on monocytes, myeloid precursor cells, activated T lymphocytes, and B lymphocytes but not on neutrophils (Bonecchi et al., 1999; Frade et al., 1997; Myers et al., 1995). Mice with targeted disruption of the CCR2 receptor have defects in monocyte/macrophage recruitment in a variety of inflammatory and immunological models (Boring et al., 1998; Kurihara et al., 1997; Kuziel et al., 1997). CCR3, which is the eotaxin receptor, is prominently expressed by eosinophils and is involved in the recruitment of these cells during allergic reactions. CCR4 is the main receptor for RANTES and MIP-1α, while CCR5, expressed on monocytes/macrophages, T cells, and granulocyte precursors (Alkhatib et al., 1996; Deng et al., 1996), binds MIP-1α, MIP-1β, and RANTES. The receptors CCR6 and CCR7, originally identified as orphan receptors, bind MIP-3α, and MIP-3β and 6Ckine respectively (Varona et al., 1998; Yoshida et al., 1997). CCR8 binds I-309 (Goya et al., 1998), TARC, and MIP-1β (Bernardini et al., 1998) and is expressed in the thymus (Zingoni et al., 1998). CCR9, also designated as GPR-9-6, was shown to specifically bind TECK (Zaballos et al., 1999).

Compartmentalization of CC chemokine receptor expression is emerging as an important regulatory component during the adaptive immune response. CCR5, the receptor for RANTES, MIP-1β, and MIP-1α is expressed by memory and activated T cells but not by naive T cells (Bleul et al., 1997). Moreover, CCR5 is preferentially expressed by human Th1 cells, in contrast to CCR4 and CCR3, which are found predominantly on Th2 cells (Bonecchi et al., 1998; Qin et al., 1998). Concomitant with their differential chemokine receptor expression, Th1 and Th2 cells selectively migrate in response to the corresponding chemokine ligand. More recently, Sallusto et al. showed that memory T cells can be distinguished with respect to their effector function in peripheral organs by their expression of the chemokine receptor CCR7 (Sallusto et al., 1999b). CCR7+ memory T cells retain lymph node homing receptors and lack immediate effector function, whereas CCR7− memory T cells express receptors for migration to inflamed tissues and possess effector function. CCR8, like CCR4 (Bonecchi et al., 1998), is preferentially expressed by activated Th2 cells. CCR8 may be important in lymphocyte development as well as in Th2-immune responses. Polarized Th2 cells have been shown to differentially express the chemokine receptors, CCR8 (Zingoni et al., 1998), CCR4 (Bonecchi et al., 1998; Sallusto et al., 1999a) and CCR3 (Sallusto et al., 1997). In the murine system, a recent study showed that Th1 and Th2 cells expressed comparable levels of CCR1, CCR2, and CCR4 mRNA. Similar to human T helper cells, murine Th1 cells preferentially expressed CCR7 and CCR5, whereas Th2 cells expressed more CCR3 (Randolph et al., 1999). In conclusion, depending on their antigenic stimulation and/or polarization toward Th1 versus Th2 subsets, chemokine receptors are differentially expressed, which, in addition to their characteristic cytokine profiles, may thus further distinguish these cells (Mantovani, 1999b).

C. C (or γ-) and CX₃C (or δ-) Chemokines and Receptors

C. C (or γ-) and CX_3C (or δ-) Chemokines and Receptors

Lymphotactin, the lone member of the C-chemokine family, is produced by activated mouse T cells (Kelner et al., 1994). Lymphotactin is unique because (1) it has only two cysteines, the second and the fourth, of the four cysteines conserved in the CXC, CC, and CX_3C chemokines; and (2) the C-terminal sequence is much longer than those of other chemokines. Lymphotactin, and the subsequently reported single motif cysteine (SMC) chemokine-1 and the activation-induced, T-cell-derived, and chemokine-related molecule (ATAC), are all identical (Muller et al., 1995; Yoshida et al., 1995). Lymphotactin expression on activation is extremely rapid, which suggests that it has a role in the early phases of an inflammatory response. When lymphotactin is injected into peritoneum, it causes an influx of T lymphocytes and NK cells by 24h (Hedrick et al., 1997). Yoshida et al. (1998) identified an orphan receptor, GPR5 (Heiber et al., 1995), as being a high-affinity functional receptor for lymphotactin—this receptor was renamed XCR1.

The single member of the CX_3C or δ family, named fractalkine, is also unique with a novel arrangement of three amino acids separating the first two conserved cysteines. Furthermore, fractalkine exists in both soluble and membrane-bound forms (Bazan et al., 1997; Pan et al., 1997). This chemokine is found on activated endothelial cells and is tethered to the membrane by a mucinlike stalk that suggests a novel role for this chemokine in juxtacrine signaling. The interaction of endothelial-cell-expressed fractalkine with its receptor, CX_3CR1, on leukocytes mediates the initial capture, firm adhesion, and activation of circulating leukocytes independently of integrin or adhesion molecule involvement (Fong et al., 1998). Thus, fractalkine not only induces the migration of leukocytes but also facilitates their adhesion directly to the endothelium at sites of inflammation. Fractalkine has been shown to foster the migration of monocytes, T cells, and NK cells (Hedrick et al., 1997; Kelner et al., 1994). The receptor for fractalkine was originally cloned from a rat brainstem cDNA library and named RBS11 (Harrison et al., 1994). Subsequently, the human receptor was identified as V28 (Raport et al., 1995), or CMKBLR1, which is highly expressed in the brain and in neutrophils, monocytes/macrophages, and T lymphocytes (Combadiere et al., 1995). Imai et al. (1997) identified this molecule as being the fractalkine receptor (CX_3CR1) that is expressed mainly by NK cells and monocytes and to a lesser extent in CD8+ T cells.

D. Viral Chemokine and Chemokine Receptor Homologs

Many of the large DNA viruses are infamous for their ability to undermine host immunity. Two genres of viruses, the herpesviruses and poxviruses, were recently shown to contain open reading frames (ORFs) that encode homologs of mammalian chemokines and chemokine receptors (for detailed reviews, see Ahuja *et al.*, 1994; Dairaghi *et al.*, 1998; Lalani *et al.*, 2000; Murphy, 1994; Smith *et al.*, 1997).

Kaposi's sarcoma–associated herpes virus (KSHV) encodes three CC-like chemokines, vMIP-I, vMIP-II, and vMIP-III. The vMIP-I and vMIP-II show 60% sequence identity to each other and 43% and 52% identity to MIP-1α, while vMIP-III is more distantly related to its viral and mammalian counterparts (Stine *et al.*, 1999). Of the three viral chemokines, the function of vMIP-II is the best characterized. Receptor binding studies suggest vMIP-II is a ligand for CCR3 (Boshoff *et al.*, 1997). In addition to being a potent inhibitor of HIV infection mediated principally through this receptor (Boshoff *et al.*, 1997), vMIP-II is able to mobilize calcium and stimulate the chemotaxis of eosinophils where CCR3 is predominantly expressed (Boshoff *et al.*, 1997). In contrast to this agonist action on CCR3, vMIP-II also binds with a high degree of affinity to a number of other CC (CCR1, 2, 5) and CXC chemokine receptors (CXCR4), where it acts as an antagonist (Kledal *et al.*, 1997), as well as to CX$_3$CR1 (Chen *et al.*, 1998). Human monocyte chemotaxis induced by RANTES, MIP-1α, and MIP-1β is inhibited by vMIP-II, suggesting that this viral mediator may be a broad-spectrum chemokine antagonist. Recently, vMIP-I was shown to selectively bind to CCR8 (a CC chemokine receptor associated with Th2 lymphocytes), acting as an agonist and stimulating Ca^{2+} mobilization on human T cells (Dairaghi *et al.*, 1999). Interestingly, vMIP-II also binds to CCR8; however, this is associated with antagonistic actions (Sozzani *et al.*, 1998). Kaposi's sarcoma is an angioproliferative disorder, so it is interesting that in addition to deviation of the host chemokine response, vMIP-I and vMIP-II both induce angiogenesis in a chick chorioallantoic assay (Boshoff *et al.*, 1997). Thus these viral chemokines might contribute directly to the pathogenesis of the proliferative angiopathy in Kaposi's sarcoma.

In addition to these chemokine homologs, herpesviruses also contain open reading frames (ORFs) that share significant sequence conservation with the chemokine receptors. KSHV and *Herpesvirus saimiri* (HVS) contain ORF 74 in HSV, whose sequence is most similar to the human IL-8 receptors CXCR1 and CXCR2 (Ahuja and Murphy, 1993). Like CXCR1 and CXCR2, ORF 74 binds IL-8 as well as the other ELR CXC chemokines, GRO-α and PF-4 (Ahuja and Murphy, 1993). However unlike CXCR1 and CXCR2, ORF 74 can signal in a constitutive agonist–independent manner and stimulate the proliferation of kidney fibroblasts (Arvanitakis *et al.*, 1997). In addition to the stimulation of proliferation of nontransformed cells, ORF 74 signaling may also lead to cellular transformation and tumorigenicity (Bais *et al.*, 1998). Thus, it would appear that ORF 74 may contribute to the development of the transformed cell phenotype found with KSHV-infected cells in Kaposi's sarcoma lesions.

Cytomegalovirus (CMV) is a common opportunistic pathogen of immunocompromised hosts in whom the virus exhibits tropism for leukocytes. Human and murine CMV possess genes that encode a variety of chemokine homologs (MacDonald *et al.*, 1997). The murine CMV chemokine 1 and 2 (MCK-1 and MCK-2) peptides induce calcium signaling and adherence in peritoneal macrophages (Saederup *et al.*, 1999). The human CMV gene UL146 encodes a protein designated as vCXC-1 and provides the first example of a virus-encoded CXC chemokine (Penfold *et al.*, 1999). The vCXC-1 glycoprotein shows high-affinity binding to the CXCR1 and CXCR2 chemokine receptors, inducing calcium mobilization and chemotaxis of neutrophils. Thus, the common property of the CMV chemokine homologs is to function as chemokine agonists. This in turn may promote leukocyte migration to sites of infection and may be responsible for more efficient dissemination of CMV in the host. In support of this notion, MCK-1/MCK-2 mutant CMV displays markedly reduced peak levels of monocyte-associated viremia in experimentally infected mice (Saederup *et al.*, 1999).

A gene encoding a viral homolog of CCR1 was found within the genome of the human CMV. This gene, ORF US28, of human cytomegalovirus (HCMV) was shown to bind MIP-1α,-β, MCP-1, RANTES (Gao and Murphy, 1994), and MCP-3 (Bodaghi *et al.*, 1998) with higher affinity than their cognate receptors. Thus, it is attractive to speculate that HCMV-infected cells expressing US28 may be used for sequestration of chemokines from the extracellular milieu. US28 can also serve as a coreceptor for HIV infection of CD4 T cells and may facilitate an interplay between HIV and HCMV (Pleskoff *et al.*, 1997).

The poxvirus molluscum contagiosum virus (MCV) infects humans, causing papules of the skin associated with virtually no immune response to the virus in the skin lesion. The genome of MCV contains an ORF that encodes a protein (MC148R) that is a CC chemokine homolog (Senkevich *et al.*, 1996). MC148R has homology to MIP-1α/β but does not induce chemotaxis and antagonizes MIP-1α-induced chemotaxis (Krathwohl *et al.*, 1997). Furthermore, MC148R inhibits the leukocyte response to several CC and CXC chemokines (Damon *et al.*, 1998), indicating that it can act as a broad-spectrum chemokine antagonist. Therefore, a primary function of this virus-encoded chemokine homolog might be to provide an immune evasion strategy that acts to limit the recruitment of effector leukocytes to MCV-infected epidermal cells.

In conclusion, unquestionably these large DNA viruses encode functional homologs of chemokines and chemokine receptors. Although the precise role of the viral chemokine and chemokine receptor homologs (discussed above) in infectious processes remains to be determined, they clearly could contribute to the strategies employed by these viruses to subvert host immunity. The examples discussed here indicate that these strategies are diverse and, depending on the virus, range from wholesale suppression of leukocyte trafficking—providing immune evasion—to promoting selective leukocyte chemotaxis fostering the more efficient infection and dissemination of virus via its preferred target host cell.

III. CHEMOKINES AND THEIR RECEPTORS
IN THE CENTRAL NERVOUS SYSTEM

A concerted research effort has been mounted recently to ascertain the role of chemokines in immunoinflammatory disorders of the CNS, such as multiple sclerosis (MS). As a result, it has become clear that most cells resident in the CNS, including neurons, astrocytes, and microglia, can synthesize and secrete various classes of chemokines. Importantly, these same cells are endowed with a variety of different chemokine receptors and therefore have the potential to respond to chemokines in their local environment. As recently speculated by us (Asensio and Campbell, 1999) and by others (Mennicken et al., 1999), it is likely the CNS has its own chemokine ligand/receptor network, the function of which could extend well beyond the regulation of leukocyte trafficking in antiviral and other neuroimmune responses (see Fig. 3).

A. Constitutive and Inducible Expression of Chemokines in the CNS

At present, the genes for only three chemokines, MCP-1, SDF-1, and fractalkine, are known to be expressed constitutively in the CNS. Evidence suggests MCP-1 is expressed in the cortex and hippocampus of the rat brain during development as well as in the adult animal (Pousset, 1994). More recently, MCP-1 expression at the protein level was shown in the human cerebellum, medulla oblongata, and pons (Meng et al., 1999). However, the function of constitutively expressed MCP-1 in the brain is unknown.

The SDF-1 gene is found under physiological conditions in many organs including murine (Tashiro et al., 1993) and human (Shirozu et al., 1995) brain. SDF-1 has two isoforms, SDF-1α and SDF-1β, that are produced by alternative splicing from a single gene (Tashiro et al., 1993). SDF-1α RNA is present at high levels in cultured rat astrocytes, at low levels in neurons, and is not detectable in microglia, while SDF-1β RNA is found at low levels in all these cell types (Ohtani et al., 1998). Tanabe et al. (1997b) showed that SDF-1α was able to induce migration of microglial cells but not astrocytes, even though the SDF-1 receptor, CXCR4, was expressed on both cell types. Furthermore, human SDF-1α was able to induce calcium flux in cultured astrocytes (Bajetto et al., 1999). It was hypothesized that astrocyte-secreted SDF-1 could facilitate interneural cell communication (Ohtani et al., 1998). This is certainly the case during CNS development, since, as noted above, knocking out CXCR4 results in a major defect in granule cell migration and cerebellar development (Ma et al., 1998; Zou et al., 1998).

The CX_3C chemokine fractalkine is also found constitutively in the brain (Bazan et al., 1997; Pan et al., 1997). Fractalkine mRNA is expressed at high levels by neurons in the olfactory bulb, cerebral cortex hippocampus, caudate putamen, and nucleus accumbens (Harrison et al., 1998; Nishiyori et al., 1998; Schwaeble et al., 1998). In vitro studies have confirmed that fractalkine is constitutively expressed by neurons and, further, can be induced in astrocytes by TNF-α and IL-1β (Maciejewski-Lenoir et al., 1999). In a model of peripheral nerve injury, Harrison et al. have shown that fractalkine and its receptor were highly upregulated in the brain, with motor neurons being the primary source of the chemokine. Fractalkine treatment of primary cultured rat microglial cells induces vigorous increases in intracellular Ca^{2+} levels and chemotaxis, which are inhibited by an antibody against the fractalkine (CX_3CR1) receptor (Harrison et al., 1998). Although the precise function of fractalkine in the CNS awaits clarification, the localization of the fractalkine receptor on microglia and the expression of its ligand by neurons clearly establish a spatial framework that could facilitate juxtacrine communication and migration between these cells (see Fig. 3) (Harrison et al., 1998; Nishiyori et al., 1998).

A majority of chemokines are not detectable in the CNS under physiological conditions, but are present during diverse pathologic states including, as discussed below, viral diseases. It is beyond the scope of this chapter to detail all of the findings in these different pathologic states and the reader is encouraged to consult many excellent recent reviews dealing with this subject (Asensio and Campbell, 1999; Glabinski and Ransohoff, 1999; Karpus and Ransohoff, 1998; Mennicken et al., 1999; Ransohoff, 1997; Ransohoff and Tani, 1998). It is clear from most, if not all, cases examined that glial cells represent an important source for the localized production of specific chemokines. For example, in MS, expression of IP-10 (Balashov *et al.*, 1999; Sorensen *et al.*, 1999) and MCP-1 (Simpson *et al.*, 1998; Van Der Voorn *et al.*, 1999) is prominent in astrocytes while MIP-1α expression is strongly associated with macrophage/microglia (Balashov *et al.*, 1999). In support of these clinicopathologic findings, studies *in vitro* demonstrate that expression of various CC and CXC chemokines can be induced in astrocytes and microglia by proinflammatory cytokines and LPS. Although differential expression of MCP-1 by astrocytes, and of MIP-1α by microglia, was reported for murine glial cells (Hayashi *et al.*, 1995), this was not the case with human microglia, with expression of both these chemokines being induced in these cells by LPS (McManus *et al.*, 1998). Similarly, expression of IP-10 by murine astrocytes and microglia can be induced by IFN-γ or LPS (Luo *et al.*, 1998; Majumder *et al.*, 1998; Ren *et al.*, 1998; Vanguri, 1995). In contrast to the glial cells, few inducible chemokines have been reported to be expressed by neuronal cells either *in vitro* or *in vivo*.

How does CNS chemokine expression contribute to the pathogenesis of neuroinflammatory disease states? Clearly, one key role involves recruitment of leukocytes to the brain. Differences in the chemokine gene expression patterns can be found in different neuroinflammatory diseases and correlate with a bias toward the involvement of particular leukocytes. In experimental and clinical bacterial meningoencephalitis, where CNS lesions consist predominantly of neutrophils and monocytes, there is dominant cerebral production of the neutrophil and monocyte attractant chemokines IL-8, MIP-2, and GROα, and MCP-1, MIP-1α, and MIP-1β, respectively (Lahrtz *et al.*, 1998; Spanaus *et al.*, 1997; Sprenger *et al.*, 1996). In contrast, as discussed below, in viral meningoencephalitis, mostly lymphocytes and monocytes are found in the CNS, in parallel with the cerebral expression of the chemokine genes encoding IP-10, MCP-1, and RANTES, which are effective lymphocyte and monocyte chemoattractants (Asensio and Campbell, 1997; Lahrtz *et al.*, 1997). Transgenic studies show that individual chemokines can have a specific chemoattractant "signature." Thus, CNS expression of either the GROα genes (Tani *et al.*, 1996) or the MCP-1 genes (Fuentes *et al.*, 1995) under the control of the myelin basic protein (MBP) promoter induces robust leukocyte infiltration of the CNS that is composed predominantly of either neutrophils or monocytes, respectively. Studies with intracerebral microinjection also demonstrate that the acute presence of chemokines in the CNS leads to robust and cell-specific leukocyte recruitment. For example, MIP-2 is more potent than IL-8, IP-10, or MCP-1 in stimulating polymorphonuclear leukocyte recruitment to the brain (Bell *et al.*, 1996), while the CC chemokine C10, when injected into the lateral ventricle, is able to induce a recruitment of macrophages, but not T cells (Asensio *et al.*, 1999b). Finally, neutralization of the chemokine MIP-1α, which is expressed in EAE, is associated with a reduction in disease severity and in CNS inflammation (Karpus *et al.*, 1995). In all, these studies indicate that the qualitative makeup of chemokine production in the CNS during disease can dictate the recruitment of selected leukocytes to the CNS, and that targeting these molecules may provide an effective therapeutic approach to suppress CNS inflammation.

B. Chemokine Receptor Expression in the Brain

The CC and CXC receptors are expressed on a number of different cell types and their expression is often regulated by exogenous and endogenous stimuli. Numerous *in vitro* and *in vivo* studies show that different neural cells express a variety of chemokine receptors under normal conditions and could therefore facilitate a direct interaction of these cells with chemokines. Neurons have prominent expression of many chemokine receptors including CXCR2, CXCR4, CCR1, CCR4, CCR5, CCR9/10, CX_3CR1, and Duffy antigen-related chemokine (DARC) (Horuk *et al.*, 1997; Klein *et al.*, 1999; Lavi *et al.*, 1998; Meucci *et al.*, 1998; Sanders *et al.*, 1998; Vallat *et al.*, 1998). Immunohistochemical staining of the human brain revealed CXCR1 is not detectable, whereas CXCR2 is expressed at high levels by subsets of projection neurons in different regions of the brain including the hippocampus, dentate nucleus, pontine nuclei, locus coeruleus, and paraventricular nucleus, and also in the spinal cord (Horuk *et al.*, 1996; Horuk *et al.*, 1997). DARC is expressed by subsets of endothelial cells and Purkinje cells in the cerebellum, suggesting that this enigmatic receptor that lacks signaling function may have multiple roles in normal and pathological physiology, perhaps as a chemokine clearance receptor (Horuk *et al.*, 1996; Horuk *et al.*, 1997). Functional CCR3, CCR5, and CXCR4 receptors, with different patterns of distribution, were demonstrated on cultured fetal human and macaque neurons (Klein *et al.*, 1999). Expression of CXCR4 has also been detected by immunohistochemical staining on subpopulations of neurons in the normal adult brain (Lavi *et al.*, 1997; Sanders *et al.*, 1998; Vallat *et al.*, 1998), while in the rat, levels of the CXCR4 receptor RNA are higher on specific neurons, which include cerebellar Purkinje cells, hippocampal hilar neurons, and cerebral cortical neurons (Wong *et al.*, 1996). A clear theme emerging from all these studies is that neurons express a variety of chemokine receptors and that this expression in the CNS exhibits marked regional and cellular diversity, suggesting that chemokines could have differential effects on different neurons.

Other than neurons, glial cells also express a variety of chemokine receptors. The fractalkine receptor CX_3CR1 is found at high levels in rodent and human brain and shows strong localization to microglia (Combadiere et al., 1998; Harrison et al., 1998; Imai et al., 1997). The fact that fractalkine is expressed by neurons (as noted above), and its receptor by microglia, suggests a possible important role for this chemokine and its receptor in communication between neurons and microglia. A similar scenario may follow with SDF-1, whose receptor, CXCR4, in addition to expression by neurons, is found on astrocytes (Heesen et al., 1996) and on microglial cells (Tanabe et al., 1997b). Finally, the CC chemokine receptors CCR1, CCR5, and CCR3 have been shown to be expressed by astrocytes and/or microglial cells both in vitro and in vivo (Boddeke et al., 1999; He et al., 1997; Lavi et al., 1998; Tanabe et al., 1997a; Vallat et al., 1998; Westmoreland et al., 1998; Xia et al., 1998).

The expression of many of these chemokine receptors in the CNS is not static and can be dynamically regulated by various stimuli. For example, in a model of excitotoxic brain injury induced by intracerebral injection of NMDA in the rat, marked upregulation of CCR5 expression was observed by microglia and neurons (Galasso et al., 1998). In brain from Alzheimer's cases, plaque-associated reactive microglia also show increased CCR5 expression (Xia et al., 1998). Expression of the fractalkine receptor CX_3CR1 is increased on perineural microglia following facial nerve axotomy in the rat (Harrison et al., 1998). Boddeke et al., have shown, by RT-PCR (reverse transcription-polymerase chain rear) analysis, that CCR1, CCR2, and CCR5 expression in cultured rat microglial cells is significantly upregulated after LPS treatment (Boddeke et al., 1999; Spleiss et al., 1998). The pathophysiological significance of altered CNS chemokine receptor expression in disease states remains to be determined. However, it might be speculated that this provides yet another level of regulation in the already complex cellular communication processes mediated by the chemokines.

IV. CHEMOKINES AND THEIR RECEPTORS IN VIRAL DISEASES OF THE CENTRAL NERVOUS SYSTEM

CNS infection, with a number of different classes of viruses, can provoke vigorous inflammatory responses with subsequent recruitment of large numbers of leukocytes (for a detailed discussion of this topic, see Chapter 6 in this volume). In view of their functional properties, increasing interest has focused on the possible involvement of chemokines in regulating CNS leukocyte migration during viral infection. However, as discussed below, studies of HIV-1 have revealed that chemokine involvement in the pathogenesis of host–virus interactions extends well beyond leukocyte migration.

A. HIV- and SIV-Associated Neurological Disorders

The lentivirus HIV-1 and its simian counterpart, simian immunodeficiency virus (SIV), both infect the CNS and are responsible for causing neurologic disease. In the case of HIV, a vast amount of research effort focused on this virus led to the discovery of new chemokine receptors that proved to be critical for understanding the process of HIV-1 entry into cells (Bleul et al., 1996; Feng et al., 1996; Oberlin et al., 1996). The interactions between this virus and the host CNS have also been studied extensively, and more recently the focus has turned to the chemokines and their receptors. As discussed below, the findings point to a significant role for the chemokines and the chemokine receptors in HIV-1 neuropathogenesis.

1. HIV-Associated Cognitive and Motor Disorder

Although CD4 was identified initially as a primary receptor for HIV-1 binding and entry, it was clear this could not account for many of the characteristics of HIV-1 infectivity, such as the existence of different strains of HIV-1 with selective tropism for either T lymphocytes (T-tropic) or monocytes (M-tropic). A clue to the identity of other possible cofactors for HIV-1 entry into cells came with the discovery that the chemokines MIP-1α, MIP-1β, and RANTES produced by CD8$^+$ T cells could suppress HIV-1 infection of cultured cells (Cocchi et al., 1995). A seminal advance then came when it was shown that in addition to CD4, CXCR4 was required for T-tropic HIV-1 strains to infect host cells (Bleul et al., 1996; Feng et al., 1996; Oberlin et al., 1996). The identification of CCR5 as a receptor for MIP-1α, MIP-1β, and RANTES soon led to the demonstration that CCR5 served as a prominent coreceptor for cell entry by M-tropic strains of HIV-1 (Alkhatib et al., 1996; Choe et al., 1996; Deng et al., 1996; Doranz et al., 1996; Dragic et al., 1996). The factors that govern HIV-1 entry into cells are clearly very complex, and while CXCR4 and CCR5 are the major coreceptors for T-tropic and M-tropic HIV-1 strains, a large number of other chemokine receptors including CCR2b, CCR3, and CCR8, the US28 chemokine receptor homolog encoded by cytomegalovirus, as well as the orphan chemokine receptors gpr1, gpr15/BOB, and STRL33/BONZO, can function as coreceptors for HIV-1 infection (Deng *et al.*, 1997; Farzan *et al.*, 1997; Heiber *et al.*, 1996; Liao *et al.*, 1997; Littman, 1998; Marchese *et al.*, 1994; Pleskoff *et al.*, 1997). The importance of the chemokine receptors for HIV-1 infection *in vivo* is illustrated in the case of individuals that are homozygous for a 32 base pair mutation in the CCR5 gene and who have marked resistance toward HIV infection (Biti *et al.*, 1997; O'Brien *et al.*, 1997; Theodorou *et al.*, 1997). In addition, it has been reported that individuals with mutations in the CXCR4 and CCR2 genes also exhibit a retarded progression of HIV-1-associated disease (Michael *et al.*, 1997).

A significant number of patients with HIV-1 infection suffer from a variety of neurological problems known collectively as HIV-1-associated cognitive and motor disorder, or "neuroAIDS" (see Chapter 18 in this volume). In its severest form, neuroAIDS can be responsible for subcortical dementia, memory impairment, and motor disorder. Although the neuropathology of neuroAIDS is often variable, typical features include formation of multinucleated giant cells, diffuse gliosis, myelin pallor, and increased blood–brain barrier permeability. Neurodegeneration, with disrupted synaptodendritic organization and loss of neurons, is observed in many but not all cases (Everall *et al.*, 1993). The cause of neuroAIDS and the mechanisms that lead to brain injury are not well understood, although a combination of both host- and HIV-derived toxic factors is likely to be involved (Gendelman *et al.*, 1997). It is clear, however, that productive HIV-1 infection of the CNS is largely restricted to macrophage/microglia and infrequently to neurons or astrocytes. Infection of the CNS is thought to result from the entry of HIV-1-infected lymphocytes or macrophages—the subsequent presence of HIV-1 proteins in microglia and multinucleated giant cells suggests spread of the virus from the infiltrating infected leukocytes.

As discussed above, resident cells of the CNS are well endowed with a variety of different chemokine receptors. The levels and type of coreceptor expression are likely to be pivotal determinants of HIV-1 tropism in the brain. The coreceptors for M-tropic variants of HIV-1, CCR5, CCR3, and CXCR4 are expressed by microglia (He *et al.*, 1997; Lavi *et al.*, 1997; Vallat *et al.*, 1998). Studies with primary cultures of human microglia have shown that CCR5 and CCR3 can serve as coreceptors with CD4 for HIV-1 infection, while CXCR4 is relatively ineffective in this capacity. Microglia have very low surface-expressed CD4 (Buttini *et al.*, 1998) and this may thus preclude the efficient use of CXCR4 as a coreceptor. HIV-1 infection of microglial cultures can be inhibited by the cognate ligands MIP-1β (CCR5) and eotaxin (CCR3), further verifying the key role of the microglial-expressed CC-chemokine receptors for HIV-1 entry into cells.

Interestingly, the use of both CCR3 and CCR5 for HIV-1 infection of microglia contrasts with blood-derived macrophages in which CCR5, but not CCR3, functions as a coreceptor (Alkhatib et al., 1996; Deng et al., 1996; Wu et al., 1997). This suggests that M-tropic HIV-1 variants may arise in the brain that have distinct coreceptor requirements. In support of this, HIV-1 in the brain is typically M-tropic (Korber et al., 1994; Power et al., 1994). Moreover, He and colleagues (1997) demonstrated that YU2 and JRL Env proteins that are cloned directly from HIV-1-infected brain use CCR5 or CCR3, together with CD4, for virus entry into transfected cells.

A further determinant of HIV-1 entry and infection in the CNS may be the chemokines themselves. Increased levels of MIP-1α, MIP-1β, RANTES, MCP-1, IP-10, and SDF-1 have been reported in the brain or the CSF of individuals with neuroAIDS (Conant et al., 1998; Kelder et al., 1998; Kolb et al., 1999; Letendre et al., 1999; Sanders et al., 1998; Schmidtmayerova et al., 1996; Zheng et al., 1999b). HIV-1 may directly regulate the expression of these chemokines. Thus, infection of cultured microglia and/or astrocytes markedly increases the expression of MIP-1α and MIP-1β (Schmidtmayerova et al., 1996), while expression of SDF-1 RNA decreases in HIV-1-infected microglia but increases in similarly infected astrocytes (Zheng et al., 1999b). In recent unpublished studies, we have observed induction of IP-10 gene expression in astrocytes in the brain of transgenic mice with astrocyte-targeted expression of HIVgp120. These transgenic mice exhibit similar neurodegenerative and other pathologic changes (e.g., astrocytosis) to neuroAIDS (Toggas et al., 1994)—our findings suggest that a further action of HIVgp120 is the induction of IP-10. The CNS expression of IP-10 and other chemokines, such as MCP-1, correlates with viral load and increases with the progression of brain injury, indicating a possible role for these chemokines in the pathogenesis of neuroAIDS (Kelder et al., 1998; Kolb et al., 1999). Other than their potential contribution to brain injury (discussed in more detail below), the increased presence of these chemokines in the CNS during HIV-1 infection may promote, via their chemoattractant activity, the further recruitment and accumulation of T lymphocytes and macrophages in the brain. In support of this idea, HIV-Tat, protein-induced MCP-1 expression by astrocytes can stimulate the transmigration of monocytes across an in vitro model of the human BBB (Weiss et al., 1999). Finally, since MIP-1α and RANTES can effectively inhibit CCR5 coreceptor–mediated cell entry by HIV-1 (Cocchi et al., 1996; Dragic et al., 1996), expression of these and other chemokines in the CNS in neuroAIDS might limit the spread of HIV-1 by antagonizing infection of target microglia.

The wide distribution of chemokine receptors in the normal and HIV-infected human brain, as well as the increased expression of their cognate ligands, not only provides an avenue for HIV entry into the brain via the microglia, but might also contribute to the neuronal injury and loss that ensue. As indicated above, current evidence supports the idea that neuronal injury and death in neuroAIDS are consequences of indirect mechanisms involving a variety of host-derived molecules—for example, cytokines and excitotoxic amino acids or HIV-encoded factors such as gp120; it is probable that the host and viral products work in combination. Loss of neurons, astrocytes, and microglia through apoptosis appears to be prominent in neuroAIDS (Adle-Biassette *et al.*, 1995; Gelbard *et al.*, 1995; Shi *et al.*, 1996) and can be induced by HIV-1 infection of these cells *in vitro* (Shi *et al.*, 1996). Analysis of a panel of diverse HIV-1 primary isolates or strains revealed that, compared with blood-derived or T-tropic HIV-1 isolates, brain-derived M-tropic HIV-1 isolates are ineffective at inducing neuronal and astrocyte apoptosis (Ohagen *et al.*, 1999; Zheng *et al.*, 1999b). Thus, the primary determinant of neurodegeneration may not be the M-tropic HIV forms constituting the major CNS reservoir of HIV-1, but rather, it resides with blood-derived viruses that are prominent during later stages of disease. Furthermore, the Env V3 region is the major determinant of neuronal apoptosis and also determines chemokine coreceptor usage for infection (Ohagen *et al.*, 1999), implicating these viral coreceptors in the effector pathways that mediate neuronal damage during neuroAIDS.

As mentioned above, the blood-derived, T-tropic HIV-1 isolates rely primarily on CXCR4 as a coreceptor, whereas the brain-derived, M-tropic isolates use CCR5 and CCR3. In view of the congruity between the induction of neuronal apoptosis and chemokine coreceptor usage, and the fact that neurons express CXCR4 (as noted above), many recent studies have focused on the possible role of CXCR4 in mediating brain injury in neuroAIDS. Hesselgesser and colleagues showed that signaling by CXCR4, following the binding of T-tropic HIVgp120, activates apoptotic cell death in human hNT neuronal cells (Hesselgesser *et al.*, 1998). This does not appear to reflect an unusual property of hNT cells, which have a transformed phenotype, as induction of apoptotic cell death is also induced in rat hippocampal neurons exposed to HIVgp120 (Meucci *et al.*, 1998). In these latter experiments, pretreatment of the neurons with the cognate ligand for CXCR4, SDF-1, partially protected against gp120-induced neuronal apoptosis, suggesting that the effects of the viral product are mediated, in part, by CXCR4 binding. More recently, purified virions from T-tropic, but not M-tropic, HIV isolates were shown to induce CD4 independent neuronal signaling and apoptosis that could be blocked by antibodies to CXCR4 (Zheng et al., 1999a). In these studies, the purified T-tropic HIVgp120 protein proved to be far less effective in affecting neuronal signaling and cell death than the whole virions. Since the effects of the virions could be blocked by antibodies to gp120 and gp41, the findings indicate that the envelope proteins in their native state are more potent mediators than their recombinant or purified derivatives. In all, the implication from these studies, which utilized either human neuronal cell lines or highly purified rodent or human primary neuronal cultures, is that direct binding of HIV gp120 to the CXCR4 found on neurons activates apoptotic cell death in these cells. Contrary to this notion, in mixed brain-cell cultures derived from embryonic rat cerebrocortex, gp120-induced neuronal apoptosis was completely blocked by the tripeptide TKP, which blocks macrophages microglial activation (Kaul and Lipton, 1999). This blockade by TKP was specific for gp120 as SDF-1-induced neuronal apoptosis (as noted below) was unaffected by the peptide. Interestingly, gp120-induced neuronal apoptosis in this model system was also blocked by the CC chemokines RANTES and MIP-1β, which bind to CCR5, suggesting there may be regulatory cross talk between the CCR5 and CXCR4 receptors. However, the conclusion from this study is that gp120-induced neuronal apoptosis depends predominantly on an indirect pathway via activation of chemokine receptors on macrophages microglia. Many possible explanations exist as to why there are differences between these studies with respect to the observed effects of gp120's being conveyed by direct or indirect pathways. These include differences in the species and types of neuronal or brain-cell cultures used and the types, form, and levels of gp120 added to the cultures.

While the question of whether the induction of neuronal apoptosis by gp120 is consequent to interaction directly with the CXCR4 receptor on neurons or occurs indirectly, following activation of the macrophage/microglia, is unresolved, there is general agreement that HIVgp120 binding to CXCR4 can regulate a number of cellular signaling pathways (Kaul and Lipton, 1999; Meucci et al., 1998; Zheng et al., 1999a; Zheng et al., 1999b). These include inhibition of cyclic AMP; activation of phosphoinositol (PI) hydrolysis, which is partially prevented by antibody to CXCR4 (Zheng et al., 1999a; Zheng et al., 1999b); activation of p38 mitogen–activated kinase (p38 MAPK) (Kaul and Lipton, 1999); and increased Ca^{2+} mobilization (Meucci et al., 1998). HIV gp120 neuronal apoptosis can be inhibited by inhibitors of p38 MAPK (Kaul and Lipton, 1999) and by calcium/calmodulin-dependent protein kinase II, protein kinase A, and protein kinase C (Zheng *et al.,* 1999a), linking neuronal apoptosis, in part, to these signal transduction pathways.

The preceding discussion indicates that HIV gp120 neurotoxicity likely involves, in part, the chemokine receptor CXCR4, raising the possibility that HIV-1 subverts a key physiological cell death pathway. As we noted above, SDF-1/CXCR4 signaling is crucial for normal cerebellar neuronal migration and patterning during CNS development. SDF-1 is expressed constitutively in the adult CNS predominantly by astrocytes and to a lesser degree by neurons (Ohtani *et al.,* 1998), raising the likelihood that this chemokine may regulate physiological processes in the adult brain. In support of this, SDF-1α regulates a number of signal transduction pathways in neurons, causing decreased cyclic AMP, increased PI hydrolysis, activation of p38MAPK, and increased Ca^{2+} mobilization (Kaul and Lipton, 1999; Meucci *et al.,* 1998; Zheng *et al.,* 1999a). Functionally, exposure of hippocampal neurons to SDF-1 results in increased synaptic transmission, which is blocked by antibodies to CXCR4 (Zheng *et al.,* 1999a). The similar effects of SDF-1 and HIV gp120 in regulating CXCR4 signaling raise the question as to whether SDF-1 might also induce neuronal apoptosis. Here, the findings have been conflicting. Four studies revealed that SDF-1 promoted increased neuronal apoptosis (Hesselgesser *et al.,* 1998; Kaul and Lipton, 1999; Zheng *et al.,* 1999b) and activation of a key cell death gene, caspase 3 (Zheng *et al.,* 1999a). In contrast to these findings, Meucci and colleagues reported that, following treatment with SDF-1, there was marked protection of neurons from culture- and HIV gp120-induced apoptosis (Meucci *et al.,* 1998). The reason for these discrepancies is not known but again likely reflect differences in cell preparations and culture conditions.

2. SIV Encephalitis

Similar to HIV, SIV infection of macaques can lead to the development of an AIDS-associated encephalitis that is virtually indistinguishable from neuroAIDS (Raghavan *et al.,* 1999; Sasseville and Lackner, 1997). Both T-tropic and M-tropic SIV variants have been isolated and, like neuroAIDS, macrophage/microglia constitute the primary infected target cell population in the brain of SIV-infected monkeys. The host immune system plays a major role in the infection of the CNS with SIV, where perivascular macrophages are continuously replaced via recruitment from the circulating monocyte pool through the blood–brain barrier (Lane *et al.,* 1996). Thus, infected T cells and monocytes likely enter the CNS parenchyma during the asymptomatic phase of infection.

In addition to leukocyte and endothelial adhesion molecules, chemokines could be critical components involved in SIV infection of the brain. Using immunohistochemical staining to investigate chemokine expression in the encephalitic brain of SIV-infected macaques, Sasseville and coworkers reported upregulation in the expression of a number of chemokines, including MIP-1α, MIP-1β, RANTES, MCP-3, and IP-10, by vascular endothelium and/or perivascular mononuclear cells (Sasseville et al., 1996). In contrast to neuro AIDS, MCP-1 was not elevated in the SIV-encephalitic monkey brain. With the exception of MCP-1, expression of these proinflammatory chemokines in the SIV-infected brain is similar to that found in the human brain in neuroAIDS (as noted above). However, like neuroAIDS, the in vivo function of these chemokines in SIV-encephalitis remains unknown; it is reasonable speculation however, to suggest that leukocyte recruitment to the SIV-infected CNS may be directed by these chemokines. In a follow-up study by this group, examining chemokine receptor expression in SIV-AIDS encephalitis, the chemokine receptors CCR3, CCR5, CXCR3, and CXCR4 were found to be expressed by inflammatory cells within perivascular lesions (Westmoreland et al., 1998). In addition, expression of CCR3, CCR5, and CXCR4 was detected on subpopulations of large hippocampal and neocortical pyramidal neurons and on glial cells in both the normal and the SIV-encephalitic brain. Again, these findings show an overlap with those reported for the normal and the HIV-infected brain in humans (as noted above). M-tropic strains of SIV utilize CCR5 as a coreceptor with CD4 for infection of macrophages (Chen et al., 1997; Deng et al., 1997). Expression of CCR5 is enriched on microglial cells in the macaque brain and may therefore provide the molecular framework for infection by SIV in the monkey CNS.

Overall, the preceding discussion highlights the overlapping expression patterns and potentially similar roles for the chemokines and their receptors in the pathogenesis of SIV-encephalitis and neuroAIDS. This idea is further underscored by a recent study in which chemokine receptor expression and signaling were examined in cultured neurons derived from either macaque or human brains (Klein et al., 1999). Thus, both the qualitative pattern of expression and the functional behavior of the chemokine receptors, CCR3, CCR5, and CXCR4, in the two species are remarkably alike.

B. Virus-Induced CNS Demyelinating Diseases

In humans, multiple sclerosis is the most common and the best-known CNS demyelinating disorder. MS is thought to be an autoimmune T-cell-mediated disease that is targeted at the myelin sheath. The etiology of MS is unknown (Hafler, 1999), although at one time or another, different viruses have been implicated in its pathogenesis, including Epstein-Barr virus (EBV), measles virus, and recently, HHV-6. The human T lymphotropic virus type I (HTLV-1) infection of the CNS can cause a progressive inflammatory demyelinating disease and shares many similarities with MS. In view of a possible viral role in MS, research has focused on several experimental viral models associated with the development of neuroimmune responses and primary demyelination that show many clinicopathologic similarities with MS. Among the best-known and most widely studied of these experimental models are Theiler's murine encephalomyelitis virus (TMEV)- and murine hepatitis virus (MHV)-induced demyelinating disease.

1. HTLV-1-Associated Myelopathy / Tropical Spastic Paraperesis (HAM / TSP)

HTLV-1 is the etiologic agent for adult T cell leukemia (ATL) and for HAM/TSP, which shows similar features to primary progressive MS (Calabresi *et al.,* 1999). HAM-TSP is a chronic progressive neurological disorder characterized by a perivascular demyelination and axonal degeneration, along with the presence of inflammatory infiltrating cells such as macrophages and CD8$^+$ T cells in the damaged areas (Moore *et al.,* 1989; Umehara *et al.,* 1993; Wu *et al.,* 1993). These infiltrating cells produce several inflammatory mediators including the chemokine MCP-1 (Umehara *et al.,* 1993). In addition, HTLV-1-infected T cell lines produce a number of chemokines including SDF-1, RANTES, MIP-1α, and MIP-1β (Arai *et al.,* 1998; Baba *et al.,* 1996; Mendez *et al.,* 1997). HAM/TSP patients have a high frequency of CD8$^+$ CTL in their peripheral blood and cerebrospinal fluid (CSF). Biddison and coworkers have shown that HTLV-I-specific CD8$^+$ CTL clones secrete MIP-1α, MIP-1β, and IL-16 (Biddison *et al.,* 1997). Thus, these HTLV-1 CTLs may be an important source of chemokines in HAM/TSP, where they might contribute to the pathogenesis of this disorder. Consistent with this, elevated MIP-1α protein was found in the CSF in two of three patients with HAM/TSP (Miyagishi *et al.,* 1995). A further clinical study showed that MCP-1 was detectable on perivascular inflammatory cells and the vascular endothelium in active–chronic lesions of spinal cords of HAM/TSP patients (Umehara *et al.,* 1996).

2. MHV Encephalomyelitis

MHV belongs to the coronaviruses, a ubiquitous group of positive-stranded RNA viral pathogens of man and animals associated with a variety of respiratory, gastrointestinal, and neurological disorders.

CNS infection of susceptible murine hosts with neuro-adapted strains of MHV leads to a robust initial recruitment of leukocytes to the brain that follows an early peak of viral replication. This acute encephalomyelitis phase, with its resultant tissue injury, often leads to death of the host. However, in surviving animals, further viral replication is controlled by the host response and a chronic mononuclear cell inflammation, which occurs in the brain and spinal cord, is linked to a progressive demyelinating disease (Buchmeier and Lane, 1999; Lane and Buchmeier, 1997; Weiner, 1973). The mechanisms of demyelination in this model are not clear, although there is a requirement for a competent immune response.

In a study of the kinetics and histological localization of chemokine gene expression in the brain and spinal cord of MHV-infected mice, Lane and colleagues showed that a number of α- and β-chemokines were expressed during acute or chronic stages of MHV-infection (Lane et al., 1998). Expression of the transcripts for IP-10, MIP-2, MCP-1, MCP-3, MIP-1β, and RANTES overlapped with the occurrence of acute viral encephalomyelitis, being present by day 3 postinfection and peaking at day 7 postinfection. During the chronic demyelinating phase of the disease, both IP-10 and RANTES transcripts remained elevated. IP-10 RNA colocalized with viral RNA and was present in astrocytes and microglia associated with demyelinating lesions (Lane et al., 1998). With the exception of RANTES, a similar pattern of chemokine expression was observed with cultured astrocytes infected with active, but not UV-inactive, MHV, indicating that viral replication in these cells can directly stimulate chemokine gene expression. Therefore, in MHV infection of the brain, early viral-induced chemokine gene expression by resident CNS cells such as astrocytes might promote the recruitment of T lymphocytes and macrophages and contribute to the maintenance of the chronic inflammatory response leading to demyelination.

Further studies in the MHV model using CD4 KO and CD8 KO mice show an important role for CD4+ T-cells in the pathogenesis of inflammatory disease and demyelination (Lane et al., 2000). Thus, MHV-infected CD4 KO mice have a marked reduction in the number of activated macrophage/microglia within their brains and spinal cords and significantly less demyelination. Concomitant with the reduction in CNS inflammatory disease in the MHV-infected CD4 KO mice, lower levels of RANTES, but not other chemokine transcripts and protein were found indicating that CD4+ T-cells represent one major source of RANTES in the CNS during MHV encephalomyelitis. Administration of RANTES neutralizing antisera to MHV-infected mice is associated with a significant reduction in macrophage infiltration and demyelination compared to control mice (Lane *et al.*, 2000). These data clearly indicate that CD4[+] T-cells have a pivotal role in accelerating CNS inflammation and demyelination within infected mice possibly by regulating RANTES expression.

3. Theiler's Murine Encephalomyelitis Virus-Induced Demyelinating Disease

TMEV-induced demyelinating disease is characterized by CNS mononuclear cell infiltration resulting in a chronic CD4[+] T-cell-mediated demyelinating disease (for more discussion, see Chapter 7 in this volume). A study of the expression of CC and CXC chemokines in the CNS throughout the disease course showed that expression of C10, IP-10, MCP-1, MIP-1α, MIP-1β, and RANTES mRNA transcripts overlapped with the development of disease (Hoffman *et al.*, 1999). Expression of MIP-1α and MIP-1β protein was also detectable, being present in the spinal cord before the onset of disease and persisting throughout disease progression. These findings are somewhat reminiscent of those reported for MHV encephalomyelitis (discussed above) and reiterate the theme that chemokine expression in these viral-induced immune-mediated demyelinating diseases is a complex process that is clearly associated with disease progression. Identifying the key chemokine protagonists in TMEV-induced demyelinating disease awaits clarification.

C. Viral Meningoencephalitides

Meningoencephalitis represents one of the most devastating consequences of viral infection of the CNS, in which a concerted immunoinflammatory response by the host produces severe injury to the brain, leading to debilitating neurological disease and often death. In the viral meningoencephalitides, the specific mechanisms underlying the localization, extravasation, and activation of immune cells in the CNS and the subsequent interactions between these cells are not well understood.

1. Lymphocytic Choriomeningitis

Studies from our own laboratory examined chemokine gene expression in lymphocytic choriomeningitis (LCM) that was induced by intracranial inoculation of mice with LCMV (Asensio and Campbell, 1997; Asensio *et al.*, 1999a). Following inoculation, LCMV infects and rapidly replicates in the meninges, the choroid plexus, and the ependymal membranes lining the ventricles. In this very well characterized model, immunocompetent adult mice infected with LCMV develop an acute monophasic disease characterized by the presence of infiltrating mononuclear cells in these same regions of the brain, and this leads to convulsive seizures and death by 6–8 days later (Buchmeier et al., 1980; Doherty et al., 1990). Infiltrating cells are predominantly T lymphocytes as well as macrophages. MHC class-I-restricted anti-LCMV CD8+ cytotoxic lymphocytes (CTLs), in addition to clearance of the virus, are also the primary effectors of LCM (Buchmeier et al., 1980; Doherty et al., 1990). Immune-compromised mice that are unable, or fail, to mount anti-LCMV CD8+ CTL responses do not develop immune pathology in the brain and survive.

We hypothesized that chemokines may be important regulatory signals for the cerebral recruitment and extravasation of leukocytes in LCM (Asensio and Campbell, 1997; Asensio et al., 1999a). In examining this, we observed that the pattern of chemokine gene expression in LCM is dynamic and complex, with often overlapping expression of a number of different subclasses of chemokine genes. Thus, by day 3 postinfection the expression of a number of chemokine genes is evident, including C10, MCP-3, MIP-1β, MCP-1, CRG-2/IP-10, and RANTES. By day 6 postinfection the expression of all these chemokine genes increases markedly and the expression of the lymphotactin gene is also evident in the brain. A qualitatively similar but markedly decreased level of chemokine gene expression is observed in the brain of LCMV-infected athymic mice. A similar pattern of cerebral chemokine gene expression is also found in LCMV-infected IFN-γ KO mice, although the levels of expression in the absence of IFN-γ are about 50% of those in LCMV-infected wild-type controls. In both euthymic and athymic mice, expression of IP-10 was predominant at both early and late time points after infection and preceded detectable increases in proinflammatory and interferon cytokine gene expression and CNS leukocyte recruitment in euthymic mice. In all,

⟶

FIG 2. The CXC chemokine IP-10 is expressed at high levels in the brain, following infection by many viruses. In this example mice were infected intracranially with LCMV, and the viral nucleoprotein (NP) and IP-10 RNA expression in the brain were analyzed by in situ hybridization. High expression of IP-10 RNA was apparent by day 3 (A) postinfection and increased further by day 6 (B) postinfection. Regional expression of IP-10 RNA, on the whole, overlapped with sites of viral infection, as indicated by the expression of the LCMV-NP gene. However, parenchymal expression of the IP-10 gene was also found in the absence of local LCMV-NP expression. By using dual-label analysis to identify the cellular localization of IP-10 RNA expression, these parenchymal cells were identified as GFAP-positive astrocytes (C; arrows). From Asensio, V. C., Kincaid, C., and Campbell, I. L. Chemokines and the inflammatory response to viral infection in the central nervous system with a focus on lymphocytic choriomeningitis virus. J. Neurovirol. 5:65–75, 1999. Copyright 1999, American Society for Investigative Pathology; with permission.

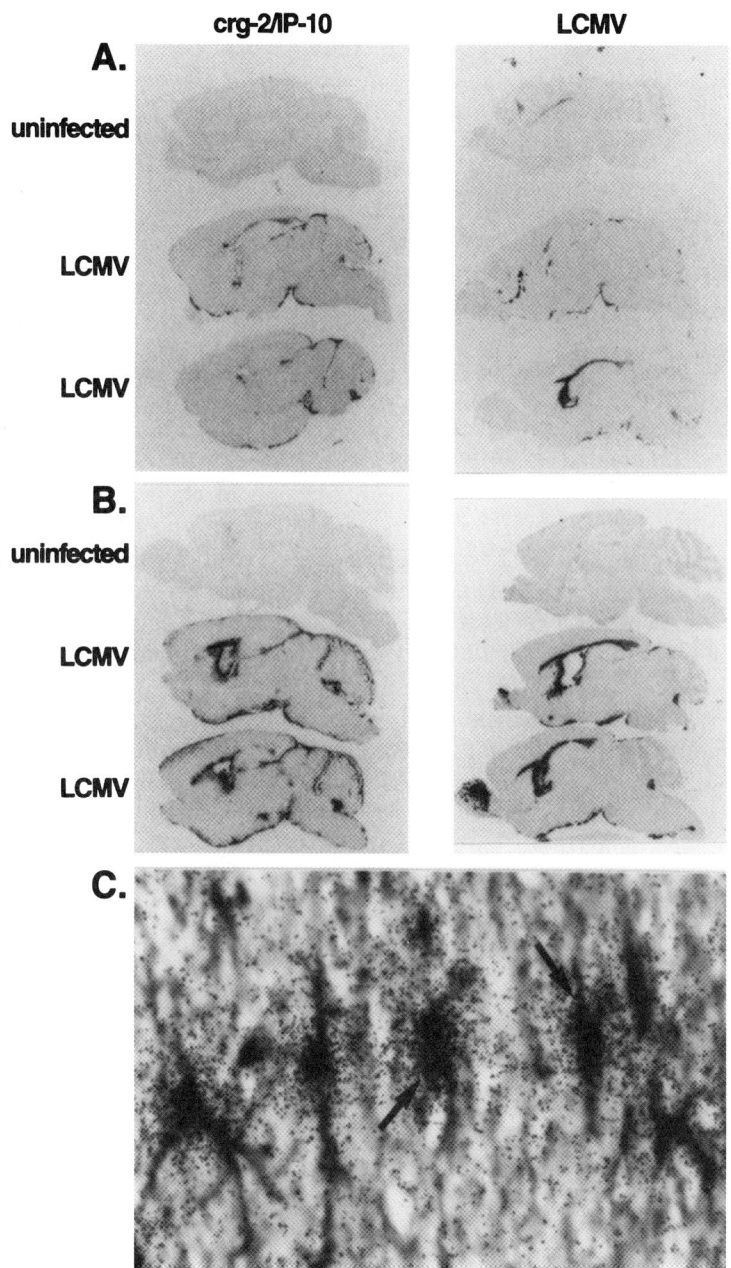

these observations, together with the finding that CRG-2/IP-10, a prominently expressed chemokine gene in many different CNS viral infections, is expressed by resident CNS cells including astrocytes (see Fig. 2), suggest that activation of chemokine gene expression may be a direct, early, and localized host response to LCMV infection. These findings are consistent with the proposed involvement of chemokines as key signaling molecules for the subsequent migration of leukocytes to the CNS, following LCMV infection. However, formal demonstration of the precise chemotactic, and possibly of other functions of chemokines in LCM, presently awaits verification.

2. HSV Encephalitis

Herpes simplex virus (HSV) is a common etiologic agent of acute focal encephalitis. Pathologically, in HSV encephalitis, leukocytes infiltrate the subarachnoid space as a hallmark of meningitis and encephalitis, while CSF analysis typically reveals a predominantly lymphocytic pleocytosis. To evaluate the possible contribution of chemokines to HSV encephalitis, Rosler and coworkers examined the spectrum, quantity, and time course of CSF chemokines in three patients with proven HSV type 1 encephalitis (HSE-1) (Rosler et al., 1998). High chemokine levels were present in the CSF of all HSE-1 patients. Peak levels occurred at the time of admission, with MCP-1 being greater than either MIP-1α or RANTES. IL-8 levels were elevated at 4 to 8 hours after admission. By contrast, plasma chemokine levels were considerably lower than CSF levels, pointing to the localized nature of the CNS chemokine response in these HSE-1 patients. A comparison of MCP-1 levels with clinical status revealed a high reciprocal correlation. This single clinical study implicates chemokines, particularly MCP-1, in the pathogenesis of HSE-1 and suggests these mediators may be useful indicators to determine the stage and severity of HSE-1.

Experimental studies of HSV-1 in mice, following ocular infection, highlight the rapid and sustained upregulation of a number of chemokine genes including GRO-α, MIP-1β, MIP-2, MCP-1, IP-10, and RANTES (Carr et al., 1998; Thomas et al., 1998). HSV stromal keratitis, associated with productive infection in the eye, results in significant accumulation of PMN, which may be driven by the presence in particular of the CXC chemokines GRO-α and MIP-2 (Thomas et al., 1998).

3. Paramyxovirus and Enterovirus Meningitis

A clinical survey of chemokine expression in CSF in patients with viral meningitis, due to infection with paramyxoviruses or enteroviruses, implicated IP-10 and MCP-1 as important contributory factors in the accumulation of activated T cells and monocytes in the

CNS (Lahrtz *et al.*, 1997). Importantly, this study formally demonstrated that the IP-10 and MCP-1 levels in CSF correlated with leukocyte chemotactic activity. MCP-1 was identified as an important chemoattractant for PBMCs while the combination of MCP-1 and IP-10 was required for migration of activated T cells. This study clearly establishes a direct link between the CNS expression of the chemokines IP-10 and MCP-1 and CNS leukocytosis in viral meningitis.

In experimental studies, the paramyxovirus Newcastle disease virus (NDV) induces IP-10 gene expression in infected astrocytes and microglia detectable by 3 hours postinfection (Vanguri and Farber, 1994). UV-inactivated NDV proved to be as effective as live virus in inducing IP-10 gene expression and was not blocked by the protein synthesis inhibitor cycloheximide, indicating that the IP-10 gene functions as an immediate early response gene, following NDV infection. Similar to NDV, measles virus (MV) is a paramyxovirus that induces IP-10 gene expression in human glioblastoma cells (Nazar *et al.*, 1997). The promoter requirements for the induction of IP-10 gene expression are similar for IFN-γ and MV; however, these two stimuli use different combinations of DNA binding factors, with STAT 1 or NFκB being more important in the direct induction of IP-10 by IFN-γ or MV, respectively. While it is presently unknown whether these paramyxoviruses induce the expression of other chemokine genes, their potent ability to directly induce IP-10 expression by glial cells may constitute a significant CNS source of production of this chemokine during viral meningitis.

D. Other Viral Diseases

Mouse adenovirus-type 1 (MAV-1) is a double-stranded DNA virus that causes a fatal hemorrhagic encephalopathy in C57B1/6 mice within 4–6 days (Guida *et al.*, 1995), along with infection of cerebrovascular endothelial cells. Chemokine gene expression was investigated and prominent induction of IP-10 was observed in the spleen and CNS of susceptible mice, whereas MCP-1 and MIP-2, respectively, were found in the spleen and brain of resistant mice (Charles *et al.*, 1999). Vascular endothelium and CNS glia were identified as sites of IP-10 mRNA expression in susceptible animals.

V. CONCLUDING REMARKS

In the normal mammalian CNS, the number of leukocytes present in the brain is scant. However, these cells are attracted to, and accumulate in, a variety of pathologic states, many involving viral infection. Although leukocyte migration into local tissue compartments such as the CNS is a multifactorial process, it has become clear that chemokines are pivotal components of this process, providing a necessary chemotactic signal for leukocyte recruitment. Generation of this signal in CNS viral infection may involve localized production of proinflammatory chemokines by cells intrinsic to the brain, including the astrocytes and microglia. The activated glia have the potential to produce a spectrum of CC and CXC chemokines that may vary, depending on the nature of the invading pathogen. This in turn will determine the qualitative makeup of the leukocytes recruited to the brain. It is important to note, however, that how a chemokine that is secreted by a parenchymally located glial cell is "sensed" by circulating peripheral leukocytes is unclear at this time.

Roles beyond leukocyte chemoattraction are implied for the chemokines and their receptors. Studies of HIV highlight the dynamic relationship between chemokine receptor usage and tropism for different isolates of HIV. On the other hand, the fact that chemokine receptor ligands are effective inhibitors of HIV entry into target cells suggests their local production may serve as a host response to limit further spread of the infectious agent. Some chemokines such as IP-10 are documented as having direct antiviral functions. Whether this is important in CNS defense against viral infection where IP-10 is highly expressed, and whether any other chemokines have an antiviral function, remain open questions.

A further consequence of the localized production of chemokines in the brain is likely the direct modulation of CNS cell function. Various classes of chemokine receptors are expressed by CNS cells including neurons and the glia. In many respects the CNS possesses its own chemokine network permitting autocrine and paracrine regulation of cellular activity. Chemokines are implicated in many physiological functions including development, cell growth, angiogenesis, and cellular migration. Perturbation of the CNS chemokine network, resulting from viral infection, could therefore either be beneficial by promoting reparative processes or detrimental by contributing to tissue injury and loss.

In conclusion, it is now apparent that, like their proinflammatory cytokine counterparts, the chemokines are important plurifunctional mediators in the host response to viral infection of the CNS (see Fig. 3 and color section). As such, a greater understanding of the neurocellular and viral targets and functions of the chemokines holds the promise of revealing new molecular focal points for therapeutic inter-

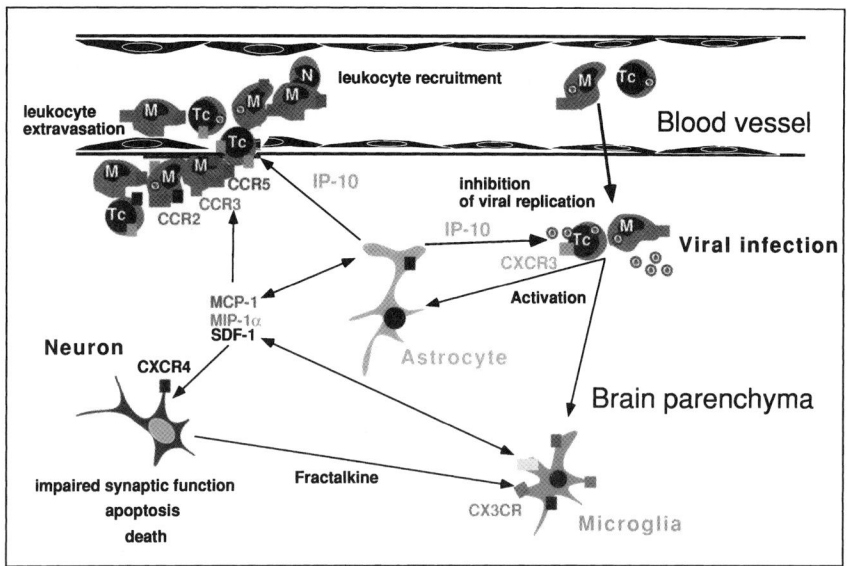

FIG 3. [For color reproduction, see color section.] Chemokines are potentially multifunctional effectors in CNS viral infections. In this schema, initial infection of the CNS with a virus results in the localized production of chemokines by reactive astrocytes and microglia. Some chemokines such as SDF-1 and fractalkine are produced constitutively in the CNS and their production may also be altered by viral infection. Increased chemokine production may then promote the further recruitment and extravasation of antiviral immune effector cells from the periphery. Chemokine receptors are widely expressed by neuronal and glial cells and additional functions of the chemokines are likely. These may include direct antiviral actions (for example, by IP-10), and intercellular communication (for example, neuronally derived fractalkine can stimulate microglia). In addition, chemokines may exert detrimental actions; for example, SDF-1 can impair neuronal activity and stimulate apoptosis of these cells. M, macrophage; Tc, T lymphocyte; N, neutrophil.

vention that could more effectively control CNS viral infections and prevent tissue injury.

ACKNOWLEDGMENTS

Valérie Asensio is a postdoctoral fellow of the National Multiple Sclerosis Society. The authors' studies were supported by NIH grants MH 50426; MH 47680 and NS 36979 to ILC. This chapter is publication No. 13255-NP, from the Scripps Research Institute.

REFERENCES

Adle-Biassette, H., Levy, Y., Colombel, M., Poron, F., Natchev, S., Keohane, C., and Gray, F. (1995). Neuronal apoptosis in HIV infection in adults. *Neuropathol. Appl. Neurobiol.* **21:**218–227.

Ahuja, S. K., Gao, J. L., and Murphy, P. M. (1994). Chemokine receptors and molecular mimicry. *Immunol. Today* **15**:281–287.

Ahuja, S. K. and Murphy, P. M. (1993). Molecular piracy of mammalian interleukin-8 receptor type B by herpesvirus saimiri. *J. Biol. Chem.* **268**:20691–20694.

Alkhatib, G., Combadiere, C., Broder, C. C., Feng, Y., Kennedy, P. E., Murphy, P. M., and Berger, E. A. (1996). CC CKR5: a RANTES, MIP-1α, MIP-1β receptor as a fusion cofactor for macrophage-Tropic HIV-1. *Science* **272**:1955–1958.

Amichay, D., Gazzinelli, R. T., Karupiah, G., Moench, T. R., Sher, A., and Farber, J. M. (1996). Genes for chemokines MuMig and Crg-2 induced in protozoan and viral infections in response to IFN-γ with patterns of tissue expression that suggest nonredundant roles in vivo. *J. Immunol.* **157**:4511–4520.

Arai, M., Ohashi, T., Tsukahara, T., Murakami, T., Hori, T., Uchiyama, T., Yamamoto, N., Kannagi, M., and Fujii, M. (1998). Human T-cell leukemia virus type 1 Tax protein induces the expression of lymphocyte chemoattractant SDF-1/PBSF. *Virology* **241**:298–303.

Arvanitakis, L., Geras-Raaka, E., Varma, A., Gershengorn, M. C., and Cesarman, E. (1997). Human herpesvirus KSHV encodes a constitutively active G-protein coupled receptor linked to cell proliferation. *Nature* **385**:347–350.

Asensio, V. C. and Campbell, I. L. (1997). Chemokine gene expression in the brains of mice with lymphocytic choriomeningitis. *J. Virol.* **71**:7832–7840.

Asensio, V. C. and Campbell, I. L. (1999). Chemokines and their receptors in the CNS: directing cellular communication. *Trends Neurosci.* **22**:504–512.

Asensio, V. C., Kincaid, C., and Campbell, I. L. (1999a). Chemokines and the inflammatory response to viral infection in the central nervous system with a focus on lymphocytic choriomeningitis virus. *J. Neurovirol.* **5**:65–75.

Asensio, V. C., Lassmann, S., Pagenstecher, A., Steffensen, S. C., Henriksen, S. J., and Campbell, I. L. (1999b). C10 is a novel chemokine expressed in experimental inflammatory demyelinating disorders that promotes the recruitment of macrophages to the central nervous system. *Am. J. Pathol.* **154**:1181–1191.

Baba, M., Imai, T., Yoshida, T., and Yoshie, O. (1996). Constitutive expression of various chemokine genes in human T-cell lines infected with human T-cell leukemia virus type 1: role of the viral transactivator Tax. *Int. J. Cancer* **66**:124–129.

Bacon, K. B., and Oppenheim, J. J. (1998). Chemokines in disease models and pathogenesis. *Cytokine Growth Factor Rev.* **9**:167–173.

Baggiolini, M. (1998). Chemokines and leukocyte traffic. *Nature* **392**:565–568.

Baggiolini, M., Dewald, B., and Moser, B. (1994). Interleukin-8 and related chemotactic cytokines–CXC and CC chemokines. *Adv. Immunol.* **55**:97–179.

Baggiolini, M., Dewald, B., and Moser, B. (1997). Human chemokines: an update. *Annu. Rev. Immunol.* **15**:675–705.

Bais, C., Santomasso, B., Coso, O., Arvanitikas, L., Geras-Raaka, E., Gutkind, J. S., Asch, A. S., Cesarman, E., Gershengorn, M. C., and Mesri, E. A. (1998). G-protein-coupled receptor of Kaposi's sarcoma-associated herpesvirus is a viral oncogene and angiogenesis activator. *Nature* **391**:86–89.

Bajetto, A., Bonavia, R., Barbero, S., Florio, T., Costa, A., and Schettini, G. (1999). Expression of chemokine receptors in the rat brain. *Ann. N. Y. Acad. Sci.* **876**:201–209.

Balashov, K. E., Rottman, J. B., Weiner, H. L., and Hancock, W. W. (1999). CCR5(+) and CXCR3(+) T cells are increased in multiple sclerosis and their ligands MIP-1alpha and IP-10 are expressed in demyelinating brain lesions. *Proc. Natl. Acad. Sci. U.S.A.* **96**:6873–6878.

Bazan, J. F., Bacon, K. B., Hardiman, G., Wang, W., Soo, K., Rossi, D., Greaves, D. R., Zlotnik, A., and Schall, T. J. (1997). A new class of membrane-bound chemokine with a CX3C motif. *Nature* **385**:640–644.

Bell, M. D., Taub, D. D., and Perry, V. H. (1996). Overriding the brain's intrinsic resistance to leukocyte recruitment with intraparenchymal injections of recombinant chemokines. *Neuroscience* **74:**283–292.

Ben-Baruch, A., Xu, L., Young, P. R., Bengali, K., Oppenheim, J. J., and Wang, J. M. (1995). Monocyte chemotactic protein-3 (MCP3) interacts with multiple leukocyte receptors. C-C CKR1, a receptor for macrophage inflammatory protein-1 alpha/Rantes, is also a functional receptor for MCP3. *J. Biol. Chem.* **270:**22123–22128.

Bernardini, G., Hedrick, J., Sozzani, S., Luini, W., Spinetti, G., Weiss, M., Menon, S., Zlotnik, A., Mantovani, A., Santoni, A., and Napolitano, M. (1998). Identification of the CC chemokines TARC and macrophage inflammatory protein-1 beta as novel functional ligands for the CCR8 receptor. *Eur. J. Immunol.* **28:**582–588.

Biddison, W. E., Kubota, R., Kawanishi, T., Taub, D. D., Cruikshank, W. W., Center, D. M., Connor, E. W., Utz, U., and Jacobson, S. (1997). Human T cell leukemia virus type I (HTLV-I)-specific CD8+ CTL clones from patients with HTLV-I-associated neurologic disease secrete proinflammatory cytokines, chemokines, and matrix metalloproteinase. *J. Immunol.* **159:**2018–2025.

Biti, R., Ffrench, R., Young, J., Bennetts, B., Stewart, G., and Liang, T. (1997). HIV-1 infection in an individual homozygous for the CCR5 deletion allele. *Nat. Med.* **3:**252–253.

Bleul, C. C., Farzan, M., Choe, H., Parolin, C., Clark-Lewis, I., Sodroski, J., and Springer, T. A. (1996). The lymphocyte chemoattractant SDF-1 is a ligand for LESTR/fusin and blocks HIV-1 entry. *Nature* **382:**829–833.

Bleul, C. C., Wu, L., Hoxie, J. A., Springer, T. A., and Mackay, C. R. (1997). The HIV coreceptors CXCR4 and CCR5 are differentially expressed and regulated on human T lymphocytes. *Proc. Natl. Acad. Sci. USA* **94:**1925–1930.

Bodaghi, B., Jones, T. R., Zipeto, D., Vita, C., Sun, L., Laurent, L., Arenzana-Seisdedos, F., Virelizier, J. L., and Michelson, S. (1998). Chemokine sequestration by viral chemoreceptors as a novel viral escape strategy: withdrawal of chemokines from the environment of cytomegalovirus-infected cells. *J. Exp. Med.* **188:**855–866.

Boddeke, E. W., Meigel, I., Frentzel, S., Gourmala, N. G., Harrison, J. K., Buttini, M., Spleiss, O., and Gebicke-Harter, P. (1999). Cultured rat microglia express functional beta-chemokine receptors. *J. Neuroimmunol.* **98:**176–184.

Bonecchi, R., Bianchi, G., Bordignon, P. P., D'Ambrosio, D., Lang, R., Borsatti, A., Sozzani, S., Allavena, P., Gray, P. A., Mantovani, A., and Sinigaglia, F. (1998). Differential expression of chemokine receptors and chemotactic responsiveness of type 1 T helper cells (Th1s) and Th2s. *J. Exp. Med.* **187:**129–134.

Bonecchi, R., Polentarutti, N., Luini, W., Borsatti, A., Bernasconi, S., Locati, M., Power, C., Proudfoot, A., Wells, T. N., Mackay, C., Mantovani, A., and Sozzani, S. (1999). Upregulation of CCR1 and CCR3 and induction of chemotaxis to CC chemokines by IFN-gamma in human neutrophils. *J. Immunol.* **162:**474–479.

Boring, L., Gosling, J., Cleary, M., and Charo, I. F. (1998). Decreased lesion formation in CCR2-/-mice reveals a role for chemokines in the initiation of atherosclerosis. *Nature* **394:**894–897.

Boshoff, C., Endo, Y., Collins, P. D., Takeuchi, Y., Reeves, J. D., Schweickart, V. L., Siani, M. A., Sasaki, T., Williams, T. J., Gray, P. W., Moore, P. S., Chang, Y., and Weiss, R. A. (1997). Angiogenic and HIV-inhibitory functions of KSHV-encoded chemokines. *Science* **278:**290–294.

Broxmeyer, H. E., Cooper, S., Cacalano, G., Hague, N. L., Bailish, and Moore, M. W. (1996). Involvement of interleukin (IL) 8 receptor in negative regulation of myeloid progenitor cells in vivo: evidence from mice lacking the murine IL-8 receptor homologue. *J. Exp. Med.* **184:**1825–1832.

Buchmeier, M. J., and Lane, T. E. (1999). Viral-induced neurodegenerative disease. *Curr. Opin. Microbiol.* **2:**398–402.

Buchmeier, M. J., Welsh, R. M., Dutko, F. J., and Oldstone, M. B. (1980). The virology and immunobiology of lymphocytic choriomeningitis virus infection. *Adv Immunol* **30:**275–331.

Buttini, M., Westland, C. E., Masliah, E., Yafeh, A. M., Wyss-Coray, T., and Mucke, L. (1998). Novel role of human CD4 molecule identified in neurodegeneration. *Nature Med.* **4:**441–446.

Cacalano, G., Lee, J., Kikly, K., Ryan, A. M., Pitts-Meek, S., Hultgren, B., Wood, W. I., and Moore, M. W. (1994). Neutrophil and B cell expansion in mice that lack the murine IL-8 receptor homolog. *Science* **265:**682–684.

Calabresi, P. A., Martin, R., and Jacobson, S. (1999). Chemokines in chronic progressive neurological diseases: HTLV-1 associated myelopathy and multiple sclerosis. *J. Neurovirol.* **5:**102–108.

Carr, D. J., Noisakran, S., Halford, W. P., Lukacs, N., Asensio, V., and Campbell, I. L. (1998). Cytokine and chemokine production in HSV-1 latently infected trigeminal ganglion cell cultures: effects of hyperthermic stress. *J. Neuroimmunol.* **85:**111–121.

Carr, M. W., Roth, S. J., Luther, E., Rose, S. S., and Springer, T. A. (1994). Monocyte chemoattractant protein 1 acts as a T-lymphocyte chemoattractant. *Proc. Natl. Acad. Sci. U.S.A.* **91:**3652–3656.

Charles, P. C., Chen, X., Horwitz, M. S., and Brosnan, C. F. (1999). Differential chemokine induction by the mouse adenovirus type-1 in the central nervous system susceptible and resistant strains of mice. *J. Neurovirol.* **5:**55–64.

Chen, J. D., Bai, X., Yang, A. G., Cong, Y., and Chen, S. Y. (1997). Inactivation of HIV-1 chemokine co-receptor CXCR-4 by a novel intrakine stategy. *Nature Med.* **3:**1110–1116.

Chen, S., Bacon, K. B., Li, L., Garcia, G. E., Xia, Y., Lo, D., Thompson, D. A., Siani, M. A., Yamamoto, T., Harrison, J. K., and Feng, L. (1998). In vivo inhibition of CC and CX3C chemokine-induced leukocyte infiltration and attenuation of glomerulonephritis in Wistar-Kyoto (WKY) rats by vMIP-II. *J. Exp. Med.* **188:**193–198.

Choe, H., Farzan, M., Sun, Y., Sullivan, N., Rollins, B., Ponath, P. D., Wu, L., Mackay, C. R., LaRosa, G., Newman, W., Gerard, N., Gerard, C., and Sodroski, J. (1996). The beta-chemokine receptors CCR3 and CCR5 facilitate infection by primary HIV-1 isolates. *Cell* **85:**1135–1148.

Cocchi, F., DeVico, A. L., Garzino-Demo, A., Arya, S. K., Gallo, R. C., and Lusso, P. (1995). Identification of RANTES, MIP-1 alpha, and MIP-1 beta as the major HIV- suppressive factors produced by CD8+ T cells. *Science* **270:**1811–1815.

Cocchi, F., DeVico, A. L., Garzino-Demo, A., Cara, A., Gallo, R. C., and Lusso, P. (1996). The V3 domain of the HIV-1 gp120 envelope glycoprotein is critical for chemokine-mediated blockade of infection. *Nature Med.* **2:**1244–1247.

Combadiere, C., Ahuja, S. K., and Murphy, P. M. (1995). Cloning, chromosomal localization, and RNA expression of a human β chemokine receptor-like gene. *DNA Cell Biol.* **14:**673–680.

Combadiere, C., Gao, J., Tiffany, H. L., and Murphy, P. M. (1998). Gene cloning, RNA distribution, and functional expression of mCX3CR1, a mouse chemotactic receptor for the CX3C chemokine fractalkine. *Biochem. Biophys. Res. Commun.* **253:**728–732.

Conant, K., Garzino-Demo, A., Nath, A., McArthur, J. C., Halliday, W., Power, C., Gallo, R. C., and Major, E. O. (1998). Induction of monocyte chemoattractant protein-1 in HIV-1 Tat-stimulated astrocytes and elevation in AIDS dementia. *Proc. Natl. Acad. Sci. USA* **95:**3117–3121.

Cook, D. N., Beck, M. A., Coffman, T. M., Kirby, S. L., Sheridan, J. F., Pragnell, I. B., and Smithies, O. (1995). Requirement of MIP-1α for an inflammatory response to viral infection. *Science* **269:**1583–1585.

Dairaghi, D. J., Fan, R. A., McMaster, B. E., Hanley, M. R., and Schall, T. J. (1999). HHV8-encoded vMIP-I selectively engages chemokine receptor CCR8. Agonist and antagonist profiles of viral chemokines. *J. Biol. Chem.* **274:**21569–21574.

Dairaghi, D. J., Greaves, D. R., and Schall, T. J. (1998). Abduction of chemokine elements by Herpesviruses. *Semin. Virol.* **8:**377–385.

Damon, I., Murphy, P. M., and Moss, B. (1998). Broad spectrum chemokine antagonistic activity of a human poxvirus chemokine homolog. *Proc. Natl. Acad. Sci. USA* **95:**6403–6407.

Deng, H., Liu, R., Ellmeier, W., Choe, S., Unutmaz, D., Burkhart, M., Di Marzio, P., Marmon, S., Sutton, R. E., Hill, C. M., Davis, C. B., Peiper, S. C., Schall, T. J., Littman, D. R., and Landau, N. R. (1996). Identification of a major co-receptor for primary isolates of HIV-1. *Nature* **381:**661–666.

Deng, H. K., Unutmaz, D., KewalRamani, V. N., and Littman, D. R. (1997). Expression cloning of new receptors used by simian and human immunodeficiency viruses. *Nature* **388:**296–300.

Dobner, T., Wolf, I., Emrich, T., and Lipp, M. (1992). Differentiation-specific expression of a novel G protein-coupled receptor from Burkitt's lymphoma. *Eur. J. Immunol.* **22:**2795–2799.

Doherty, P. C., Allan, J. E., Lynch, F., and Ceredig, R. (1990). Dissection of an inflammatory process induced by CD8+ T-cells. *Immunol. Today* **11:**55–59.

Doranz, B. J., Rucker, J., Yi, Y., Smyth, R. J., Samson, M., Peiper, S. C., Parmentier, M., Collman, R. G., and Doms, R. W. (1996). A dual-tropic primary HIV-1 isolate that uses fusin and the beta-chemokine receptors CKR-5, CKR-3, and CKR-2b as fusion cofactors. *Cell* **85:**1149–1158.

Dragic, T., Litwin, V., Allaway, G. P., Martin, S. R., Huang, Y., Nagashima, K. A., Cayanan, C., Maddon, P. J., Koup, R. A., Moore, J. P., and Paxton, W. A. (1996). HIV-1 entry into CD4+ cells is mediated by the chemokine receptor CC-CKR-5. *Nature* **381:**667–673.

Everall, I., Luthert, P., and Lantos, P. (1993). A review of neuronal damage in human immunodeficiency virus infection: its assessment, possible mechanism and relationship to dementia. *J. Neuropathol. Exp. Neurol.* **52:**561–566.

Farzan, M., Choe, H., Martin, K., Marcon, L., Hofmann, W., Karlsson, G., Sun, Y., Barrett, P., Marchand, N., Sullivan, N., Gerard, N., Gerard, C., and Sodroski, J. (1997). Two orphan seven-transmembrane segment receptors which are expressed in CD4-positive cells support simian immunodeficiency virus infection. *J. Exp. Med.* **186:**405–411.

Feng, Y., Broder, C. C., Kennedy, P. E., and Berger, E. A. (1996). HIV-1 entry cofactor: functional cDNA cloning of a seven-transmembrane, G protein-coupled receptor. *Science* **272:**872–877.

Fisher, S. N., Vanguri, P., Shin, H. S., and Shin, M. L. (1995). Regulatory mechanisms of MuRantes and CRG-2 chemokine gene induction in central nervous system glial cells by virus. *Brain Behav. Immun.* **9:**331–344.

Fong, A. M., Robinson, L. A., Steeber, D. A., Tedder, T. F., Yoshie, O., Imai, T., and Patel, D. D. (1998). Fractalkine and CX3CR1 mediate a novel mechanism of leukocyte capture, firm adhesion, and activation under physiologic flow. *J. Exp. Med.* **188:**1413–1419.

Forster, R., Mattis, A. E., Kremmer, E., Wolf, E., Brem, G., and Lipp, M. (1996). A putative chemokine receptor, BLR1, directs B cell migration to defined lymphoid organs and specific anatomic compartments of the spleen, *Cell* **87:**1037–1047.

Forster, R., Wolf, I., Kaiser, E., and Lipp, M. (1994). Selective expression of the murine homologue of the G-protein-coupled receptor BLR1 in B cell differentiation, B cell neoplasia and defined areas of the cerebellum. *Cell. Mol. Biol.* **40:**381–387.

Frade, J. M., Mellado, M., del Real, G., Gutierrez-Ramos, J. C., Lind, P., and Martinez, A. C. (1997). Characterization of the CCR2 chemokine receptor: functional CCR2 receptor expression in B cells. *J. Immunol.* **159:**5576–5584.

Fuentes, M. E., Durham, S. K., Swerdel, M. R., Lewin, A. C., Barton, D. S., Megill, J. R., Bravo, R., and Lira, S. A. (1995). Controlled recruitment of monocytes and macrophages to specific organs through transgenic expression of monocyte chemoattractant protein-1. *J. Immunol.* **155:**5769–5776.

Galasso, J. M., Harrison, J. K., and Silverstein, F. S. (1998). Excitotoxic brain injury stimulates expression of the chemokine receptor CCR5 in neonatal rats. *Am. J. Pathol.* **153:**1631–1640.

Gao, J.-L., Wynn, T. A., Chang, Y., Lee, E. J., Broxmeyer, H. E., Cooper, S., Tiffany, H. L., Westphal, H., Kwon-Chung, J., and Murphy, P. M. (1997). Impaired host defense, hematopoiesis, granulomatous inflammation and type 1–type 2 cytokine balance in mice lacking CC chemokine receptor 1. *J. Exp. Med.* **185:**1959–1968.

Gao, J. L., Kuhns, D. B., Tiffany, H. L., McDermott, D., Li, X., Francke, U., and Murphy, P. M. (1993). Structure and functional expression of the human macrophage inflammatory protein 1 alpha/RANTES receptor. *J. Exp. Med.* **177:**1421–1427.

Gao, J. L., and Murphy, P. M. (1994). Human cytomegalovirus open reading frame US28 encodes a functional β chemokine receptor. *J. Biol. Chem.* **269:**28539–28542.

Gao, J. L., and Murphy, P. M. (1995). Cloning and differential tissue-specific expression of three mouse β chemokine receptor-like genes, including the gene for a functional macrophage inflammatory protein-1α receptor. *J. Biol. Chem.* **270:**17494–17501.

Gelbard, H. A., James, H. J., Sharer, L. R., Perry, S. W., Saito, Y., Kazee, A. M., Blumberg, B. M., and Epstein, L. G. (1995). Apoptotic neurons in brains from paediatric patients with HIV-1 encephalitis and progressive encephalopathy. *Neuropathol. Appl. Neurobiol.* **21:**208–217.

Gendelman, H. E., Persidsky, Y., Ghorpade, A., Limoges, J., Stins, M., Fiala, M., and Morrisett, R. (1997). The neuropathogenesis of the AIDS dementia complex. *AIDS* **11:**S35–45.

Glabinski, A. R., and Ransohoff, R. M. (1999). Chemokines and chemokine receptors in CNS pathology. *J. Neurovirol.* **5:**3–12.

Goya, I., Gutierrez, J., Varona, R., Kremer, L., Zaballos, A., and Marquez, G. (1998). Identification of CCR8 as the specific receptor for the human β- chemokine I-309: cloning and molecular characterization of murine CCR8 as the receptor for TCA-3. *J. Immunol.* **160:**1975–1981.

Grewal, I. S., Rutledge, B. J., Fiorillo, J. A., Gu, L., Gladue, R. P., Flavell, R. A., and Rollins, B. J. (1997). Transgenic monocyte chemoattractant protein-1 (MCP-1) in pancreatic islets produces monocyte-rich insulitis without diabetes: abrogation by a second transgene expressing systemic MCP-1. *J. Immunol.* **159:**401–408.

Guida, J. D., Fejer, G., Pirofski, L. A., Brosnan, C. F., and Horwitz, M. S. (1995). Mouse adenovirus type 1 causes a fatal hemorrhagic encephalomyelitis in adult C57BL/6 but not BALB/c mice. *J. Virol.* **69:**7674–7681.

Hafler, D. A. (1999). The distinction blurs between an autoimmune versus microbial hypothesis in multiple sclerosis. *J. Clin. Invest.* **104:**527–529.

Harrison, J. K., Barber, C. M., and Lynch, K. R. (1994). cDNA cloning of a G-protein-coupled receptor expressed in rat spinal cord and brain related to chemokine receptors. *Neurosci. Lett.* **169:**85–89.

Harrison, J. K., Jiang, Y., Chen, S., Xia, Y., Maciejewski, D., McNamara, R. K., Streit, W. J., Salafranca, M. N., Adhikari, S., Thompson, D. A., Botti, P., Bacon, K. B., and Feng, L. (1998). Role for neuronally derived fractalkine in mediating interactions between neurons and CX3CR1-expressing microglia. *Proc. Natl. Acad. Sci. U.S.A.* **95:**10896–10901.

Hayashi, M., Luo, Y., Laning, J., Strieter, R. M., and Dorf, M. E. (1995). Production and function of monocyte chemoattractant protein-1 and other β-chemokines in murine glial cells. *J. Neuroimmunol.* **60:**143–150.

He, J., Chen, Y., Farzan, M., Choe, H., Ohagen, A., Gartner, S., Busciglio, J., Yang, X., Hofmann, W., Newman, W., Mackay, C. R., Sodroski, J., and Gabuzda, D. (1997). CCR3 and CCR5 are co-receptors for HIV-1 infection of microglia. *Nature* **385:**645–649.

Hedrick, J. A., Saylor, V., Figueroa, D., Mizoue, L., Xu, Y., Menon, S., Abrams, J., Handel, T., and Zlotnik, A. (1997). Lymphotactin is produced by NK cells and attracts both NK cells and T cells in vivo. *J. Immunol.* **158:**1533–1540.

Heesen, M., Berman, M. A., Benson, J. D., Gerard, C., and Dorf, M. E. (1996). Cloning of the mouse fusin gene, homologue to a human HIV-1 co-factor. *J. Immunol.* **157:**5455–5460.

Heiber, M., Docherty, J. M., Shah, G., Nguyen, T., Cheng, R., Heng, H. H., Marchese, A., Tsui, L. C., Shi, X., George, S. R., and *et al.* (1995). Isolation of three novel human genes encoding G protein-coupled receptors. *DNA Cell. Biol.* **14:**25–35.

Heiber, M., Marchese, A., Nguyen, T., Heng, H. H., George, S. R., and O'Dowd, B. F. (1996). A novel human gene encoding a G-protein-coupled receptor (GPR15) is located on chromosome 3. *Genomics* **32:**462–465.

Hesselgesser, J., Taub, D., Baskar, P., Greenberg, M., Hoxie, J., Kolson, D. L., and Horuk, R. (1998). Neuronal apoptosis induced by HIV-1 gp120 and the chemokine SDF-1 alpha is mediated by the chemokine receptor CXCR4. *Curr. Biol.* **8:**595–598.

Hickey, W. F. (1999). Leukocyte traffic in the central nervous system: the participants and their roles. *Semin. Immunol.* **11:**125–137.

Hoffman, L. M., Fife, B. T., Begolka, W. S., Miller, S. D., and Karpus, W. J. (1999). Central nervous system chemokine expression during Theiler's virus-induced demyelinating disease. *J. Neurovirol.* **5:**635–642.

Horuk, R., Martin, A., Hesselgesser, J., Hadley, T., Lu, Z. H., Wang, Z. X., and Peiper, S. C. (1996). The Duffy antigen receptor for chemokines: structural analysis and expression in the brain. *J. Leukoc. Biol.* **59:**29–38.

Horuk, R., Martin, A. W., Wang, Z., Schweitzer, L., Gerassimides, A., Guo, H., Lu, Z., Hesselgesser, J., Perez, H. D., Kim, J., Parker, J., Hadley, T. J., and Peiper, S. C. (1997). Expression of chemokine receptors by subsets of neurons in the central nervous system. *J. Immunol.* **158:**2882–2890.

Hu, S., Chao, C. C., Ehrlich, L. C., Sheng, W. S., Sutton, R. L., Rockswold, G. L., and Peterson, P. K. (1999). Inhibition of microglial cell RANTES production by IL-10 and TGF-beta. *J. Leukoc. Biol.* **65:**815–821.

Imai, T., Hieshima, K., Haskell, C., Baba, M., Nagira, M., Nishimura, M., Kakizaki M, Takagi, S., Nomiyama, H., Schall, T. J., and Yoshie, O. (1997). Identification and molecular characterization of fractalkine receptor CX3CR1, which mediates both leukocyte migration and adhesion. *Cell* **91:**521–530.

Kaiser, E., Forster, R., Wolf, I., Ebensperger, C., Kuehl, W. M., and Lipp, M. (1993). The G protein-coupled receptor BLR1 is involved in murine B cell differentiation and is also expressed in neuronal tissues. *Eur. J. Immunol.* **23:**2532–2539.

Karpus, W. J., Lukacs, N. W., McRae, B. L., Strieter, R. M., Kunkel, S. L., and Miller, S. D. (1995). An important role for the chemokine macrophage inflammatory protein-1-alpha in the pathogenesis of the T cell-mediated autoimmune disease, experimental autoimmune encephalomyelitis. *J. Immunol.* **155:**5003–5010.

Karpus, W. J., and Ransohoff, R. M. (1998). Chemokine regulation of experimental autoimmune encephalomyelitis: temporal and spatial expression patterns govern disease pathogenesis. *J. Immunol.* **161:**2667–2671.

Kaul, M., and Lipton, S. A. (1999). Chemokines and activated macrophages in HIV gp120-induced neuronal apoptosis. *Proc. Natl. Acad. Sci. U.S.A.* **96:**8212–8216.

Kelder, W., McArthur, J. C., Nance-Sproson, T., McClernon, D., and Griffin, D. E. (1998). Beta-chemokines MCP-1 and RANTES are selectively increased in cerebrospinal fluid of patients with human immunodeficiency virus-associated dementia. *Ann. Neurol.* **44:**831–835.

Kelner, G. S., Kennedy, J., Bacon, K. B., Kleyensteuber, S., Largaespada, D. A., Jenkins, N. A., Copeland, N. G., Bazan, J. F., Moore, K. W., Schall, T. J., and Zlotnik, A. (1994). Lymphotactin: a cytokine that represents a new class of chemokine. *Science* **266:**1395–1399.

Kledal, T. N., Rosenkilde, M. M., Coulin, F., Simmons, G., Johnsen, A. H., Alouani, S., Power, C. A., Luttichau, H. R., Gerstoft, J., Clapham, P. R., Clark-Lewis, I., Wells, T. N. C., and Schwartz, T. W. (1997). A broad-spectrum chemokine antagonist encoded by Kaposi's sarcoma-associated herpesvirus. *Science* **277:**1656–1659.

Klein, R. S., Williams, K. C., Alvarez-Hernandez, X., Westmoreland, S., Force, T., Lackner, A. A., and Luster, A. D. (1999). Chemokine receptor expression and signaling in macaque and human fetal neurons and astrocytes: implications for the neuropathogenesis of AIDS. *J. Immunol.* **163:**1636–1646.

Kolb, S. A., Sporer, B., Lahrtz, F., Koedel, U., Pfister, H.-W., and Fontana, A. (1999). Identification of a T cell chemotactic factor in the cerebrospinal fluid of HIV-1-infected individuals as interferon-γ inducible protein 10. *J. Neuroimmunol.* **93:**172–181.

Korber, B. T. M., Kunstman, K. J., Patterson, B. K., Furtaldo, M., McEvilly, M. M., Levy, R., and Wolinsky, S. M. (1994). Genetic differences between blood- and brain-derived viral sequences from human immunodeficiency virus type 1-infected patients: evidence of conserved elements in the V3 region of the envelope protein of brain-derived sequences. *J. Virol.* **68:**7467–7481.

Krathwohl, M. D., Hromas, R., Brown, D. R., Broxmeyer, H. E., and Fife, K. H. (1997). Functional characterization of the C-C chemokine-like molecules encoded by molluscum contagiosum virus types 1 and 2. *Proc. Natl. Acad. Sci. U.S.A.* **94:**9875–9880.

Kurihara, T., Warr, G., Loy, J., and Bravo, R. (1997). Defects in macrophage recruitment and host defense in mice lacking the CCR2 chemokine receptor. *J. Exp. Med.* **186:**1757–1762.

Kuziel, W. A., Morgan, S. J., Dawson, T. C., Griffin, S., Smithies, O., Ley, K., and Maeda, N. (1997). Severe reduction in leukocyte adhesion and monocyte extravasation in mice deficient in CC chemokine receptor 2. *Proc. Natl. Acad. Sci. USA* **94:**12053–12058.

Lahrtz, F., Piali, L., Nadal, D., Pfister, H. W., Spanaus, K. S., Baggiolini, M., and Fontana, A. (1997). Chemotactic activity on mononuclear cells in the cerebrospinal fluid of patients with viral meningitis is mediated by interferon-gamma inducible protein-10 and monocyte chemotactic protein-1. *Eur. J. Immunol.* **27:**2484–2489.

Lahrtz, F., Piali, L., Spanaus, K. S., Seebach, J., and Fontana, A. (1998). Chemokines and chemotaxis of leukocytes in infectious meningitis. *J. Neuroimmunol.* **85:**33–43.

Lalani, A. S., Barrett, J. W., and McFadden, I. (2000). Modulating chemokines: more lessons from viruses. *Immunol. Today* **21:**100–106.

Lane, J. H., Sasseville, V. G., Smith, M. O., Vogel, P., Pauley, D. R., Heyes, M. P., and Lackner, A. A. (1996). Neuroinvasion of simian immunodeficiency virus coincides with increased numbers of perivascular macrophages/microglia and intrathecal immune activation. *J. Neurovirol.* **2:**423–432.

Lane, T. E., Asensio, V. C., Yu, N., Paoletti, A., Campbell, I. L., and Buchmeier, M. J. (1998). Dynamic regulation of α- and β-chemokine expression in the central nervous system during mouse hepatitis virus-induced demyelinating disease. *J. Immunol.* **160:**970–978.

Lane, T. E., and Buchmeier, M. J. (1997). Murine coronavirus infection: a paradigm for virus-induced demyelinating disease. *Trends Microbiol.* **5:**9–14.

Lane, T. E., Liu, M. T. C., Chen, B. P. C., Asensio, V. C., Samawi, R. M., Paoletti, A. D., Campbell, I. L., Kunkel, S. L., Fox, H. S., and Buchmeier, M. J. (2000). A central role for CD4+ T-cells and RANTES in virus-induced central nervous system inflammation and demyelination. *J. Virol.* **74:**1415–1424.

Lavi, E., Kolson, D. L., Ulrich, A. M., Fu, L., and Gonzalez-Scarano, F. (1998). Chemokine receptors in the human brain and their relationship to HIV infection. *J. Neurovirol.* **4:**301–311.

Lavi, E., Strizki, J. M., Ulrich, A. M., Zhang, W., Fu, L., Wang, Q., O'Connor, M., Hoxie, J. A., and Gonzalez-Scarano, F. (1997). CXCR-4 (Fusin), a co-receptor for the type 1 human immunodeficiency virus (HIV-1), is expressed in the human brain in a variety of cell types, including microglia and neurons. *Am. J. Pathol.* **151:**1035–1042.

Letendre, S. L., Lanier, E. R., and McCutchan, J. A. (1999). Cerebrospinal fluid beta chemokine concentrations in neurocognitively impaired individuals infected with human immunodeficiency virus type 1. *J. Infect. Dis.* **180:**310–319.

Liao, F., Alkhatib, G., Peden, K. W., Sharma, G., Berger, E. A., and Farber, J. M. (1997). STRL33, A novel chemokine receptor-like protein, functions as a fusion cofactor for both macrophage-tropic and T cell line-tropic HIV-1. *J. Exp. Med.* **185:**2015–2023.

Littman, D. R. (1998). Chemokine receptors: keys to AIDS pathogenesis? *Cell* **93:**677–680.

Loetscher, M., Gerber, B., Loetscher, P., Jones, S. A., Piali, L., Clark-Lewis, I., Baggiolini, M., and Moser, B. (1996). Chemokine receptor specific for IP10 and Mig: structure, function, and expression in activated T-lymphocytes. *J. Exp. Med.* **184:**963–969.

Lu, B., Rutledge, B. J., Gu, L., Fiorillo, J., Lukacs, N. W., Kunkel, S. L., North, R., Gerard, C., and Rollins, B. J. (1998). Abnormalities in monocyte recruitment and cytokine expression in monocyte chemoattractant protein 1-deficient mice. *J. Exp. Med.* **187:**601–608.

Luo, Y., Kim, R., Gabuzda, D., Mi, S., Collins-Racie, L. A., Lu, Z., Jacobs, K. A., and Dorf, M. E. (1998). The CXC-chemokine, H174: expression in the central nervous system. *J. Neurovirol.* **4:**575–585.

Ma, Q., Jones, D., Borghesani, P. R., Segal, R. A., Nagasawa, T., Kishimoto, T., Bronson, R. T., and Springer, T. A. (1998). Impaired B-lymphopoiesis, myelopoiesis, and derailed cerebellar neuron migration in CXCR4- and SDF-1-deficient mice. *Proc. Natl. Acad. Sci. U.S.A.* **95:**9448–9453.

MacDonald, M. R., Li, X. Y., and Virgin, H. W. (1997). Late expression of a beta chemokine homolog by murine cytomegalovirus. *J. Virol.* **71:**1671–1678.

Maciejewski-Lenoir, D., Chen, S., Feng, L., Maki, R., and Bacon, K. B. (1999). Characterization of fractalkine in rat brain cells: migratory and activation signals for CX3CR-1-expressing microglia. *J. Immunol.* **163:**1628–1635.

Mackay, C. R. (1997). Chemokines: what chemokine is that? *Curr. Biol.* **7:**R384–R386.

Mahalingam, S., Farber, J. M., and Karupiah, G. (1999). The interferon-inducible chemokines MuMig and Crg-2 exhibit antiviral activity in vivo. *J. Virol.* **73:**1479–1491.

Majumder, S., Zhou, L. Z. H., Chaturvedi, P., Babcock, G., Aras, S., and Ransohoff, R. M. (1998). Regulation of human IP-10 gene expression in astrocytoma cells by inflammatory cytokines. *J. Neurosci. Res.* **54:**169–180.

Mantovani, A. (1999a). The chemokine system: redundancy for robust outputs. *Immunol. Today* **20:**254–257.

Mantovani, A. (1999b). Chemokines. Introduction and overview. *Chem. Immunol.* **72:**1–6.

Marchese, A., Docherty, J. M., Nguyen, T., Heiber, M., Cheng, R., Heng, H. H., Tsui, L. C., Shi, X., George, S. R., and O'Dowd, B. F. (1994). Cloning of human genes encoding novel G protein-coupled receptors. *Genomics* **23:**609–618.

Matsushima, K., Larsen, C. G., DuBois, G. C., and Oppenheim, J. J. (1989). Purification and characterization of a novel monocyte chemotactic and activating factor produced by a human myelomonocytic cell line. *J. Exp. Med.* **169:**1485–1490.

McManus, C., Berman, J. W., Brett, F. M., Staunton, H., Farrell, M., and Brosnan, C. F. (1998). MCP-1, MCP-2 and MCP-3 expression in multiple sclerosis lesions: an immunohistochemical and in situ hybridization study. *J. Neuroimmunol.* **86:**20–29.

Mendez, E., Kawanishi, T., Clemens, K., Siomi, H., Soldan, S. S., Calabresi, P., Brady, J., and Jacobson, S. (1997). Astrocyte-specific expression of human T-cell lymphotropic virus type 1 (HTLV-1) Tax: induction of tumor necrosis factor alpha and susceptibility to lysis by CD8+ HTLV-1-specific cytotoxic T cells. *J. Virol.* **71:**9143–9149.

Meng, S. Z., Oka, A., and Takashima, S. (1999). Developmental expression of monocyte chemoattractant protein-1 in the human cerebellum and brainstem. *Brain Dev.* **21:**30–35.

Mennicken, F., Maki, R., de Souza, E. B., and Quirion, R. (1999). Chemokines and chemokine receptors in the CNS: a possible role in neuroinflammation and patterning. *Trends Pharmacol. Sci.* **20:**73–78.

Meucci, O., Fatatis, A., Simen, A. A., Bushell, T. J., Gray, P. W., and Miller, R. J. (1998). Chemokines regulate hippocampal neuronal signaling and gp120 neurotoxicity. *Proc. Natl. Acad. Sci. U.S.A.* **95:**14500–14505.

Michael, N. L., Louie, L. G., Rohrbaugh, A. L., Schultz, K. A., Dayhoff, D. E., Wang, C. E., and Sheppard, H. W. (1997). The role of CCR5 and CCR2 polymorphisms in HIV-1 transmission and disease progression. *Nature Med.* **3:**1160–1162.

Miyagishi, R., Kikuchi, S., Fukazawa, T., and Tashiro, K. (1995). Macrophage inflammatory protein-1α in the cerebrospinal fluid of patients with multiple sclerosis and other inflammatory neurological diseases. *J. Neurol. Sci.* **129:**223–227.

Moore, G. R., Traugott, U., Scheinberg, L. C., and Raine, C. S. (1989). Tropical spastic paraparesis: a model of virus-induced, cytotoxic T-cell-mediated demyelination? *Ann. Neurol.* **26:**523–530.

Moser, B., Loetscher, M., Piali, M., and Loetscher, P. (1998). Lymphocyte responses to chemokines. *Int. Rev. Immunol.* **16:**323–344.

Muller, S., Dorner, B., Korthauer, U., Mages, H. W., D'Apuzzo, M., Senger, G., and Kroczek, R. A. (1995). Cloning of ATAC, an activation-induced, chemokine-related molecule exclusively expressed in CD8+ T lymphocytes. *Eur. J. Immunol.* **25:**1744–1748.

Murphy, P. M. (1994). Molecular piracy of chemokine receptors by herpesviruses. *Infectious Agents Dis.* **3:**137–154.

Myers, S. J., Wong, L. M., and Charo, I. F. (1995). Signal transduction and ligand specificity of the human monocyte chemoattractant protein-1 receptor in transfected embryonic kidney cells. *J. Biol. Chem.* **270:**5786–5792.

Nagasawa, T., Hirota, S., Tachibana, K., Takakura, N., Nishikawa, S., Kitamura, Y., Yoshida, N., Kikutani, H., and Kishimoto, T. (1996). Defects of B-cell lymphopoiesis and bone-marrow myelopoiesis in mice lacking the CXC chemokine PBSF/SDF-1. *Nature* **382:**635–638.

Nagasawa, T., Tachibana, K., and Kishimoto, T. (1998). A novel CXC chemokine PBSF/SDF-1 and its receptor CXCR4: their functions in development, hematopoiesis and HIV infection. *Semin. Immunol.* **10:**179–185.

Nazar, A. S., Cheng, G., Shin, H. S., Brothers, P. N., Dhib-Jalbut, S., Shin, M. L., and Vanguri, P. (1997). Induction of IP-10 chemokine promoter by measles virus: comparison with interferon-gamma shows the use of the same response element but with differential DNA-protein binding profiles. *J. Neuroimmunol.* **77:**116–127.

Neote, K., DiGregorio, D., Mak, J. Y., Horuk, R., and Schall, T. J. (1993). Molecular cloning, functional expression, and signaling characteristics of a C-C chemokine receptor. *Cell* **72:**415–425.

Nishiyori, A., Minami, M., Ohtani, Y., Takami, S., Yamamoto, J., Kawaguchi, N., Kume, T., Akaike, A., and Satoh, M. (1998). Localization of fractalkine and CX3CR1 mRNAs in rat brain: does fractalkine play a role in signaling from neuron to microglia? *FEBS Lett.* **429:**167–172.

O'Brien, T. R., Winkler, C., Dean, M., Nelson, J. A., Carrington, M., Michael, N. L., and White, G. C. (1997). HIV-1 infection in a man homozygous for CCR5 delta 32. *Lancet* **349:**1219.

Oberlin, E., Amara, A., Bachelerie, F., Bessia, C., Virelizier, J. L., Arenzana-Seisdedos, F., Schwartz, O., Heard, J. M., Clark-Lewis, I., Legler, D. F., Loetscher, M., Baggiolini, M., and Moser, B. (1996). The CXC chemokine SDF-1 is the ligand for LESTR/fusin and prevents infection by T-cell-line-adapted HIV-1. *Nature* **382:**833–835.

Ohagen, A., Ghosh, S., He, J., Huang, K., Chen, Y., Yuan, M., Osathanondh, R., Gartner, S., Shi, B., Shaw, G., and Gabuzda, D. (1999). Apoptosis induced by infection of primary brain cultures with diverse human immunodeficiency virus type 1 isolates: evidence for a role of the envelope. *J. Virol.* **73:**897–906.

Ohtani, Y., Minami, M., Kawaguchi, N., Nishiyori, A., Yamamoto, J., Takami, S., and Satoh, M. (1998). Expression of stromal cell-derived factor-1 and CXCR4 chemokine receptor mRNAs in cultured rat glial and neuronal cells. *Neurosci. Lett.* **249:**163–166.

Owens, T., Renno, T., Taupin, V., and Krakowski, M. (1994). Inflammatory cytokines in the brain: does the CNS shape immune responses? *Immunol. Today* **15:**566–571.

Pan, Y., Lloyd, C., Zhou, H., Dolich, S., Deeds, J., Gonzalo, J. A., Vath, J., Gosselin, M., Ma, J., Dussault, B., Woolf, E., Alperin, G., Culpepper, J., Gutierrez-Ramos, J. C., and Gearing, D. (1997). Neurotactin, a membrane-anchored chemokine upregulated in brain inflammation. *Nature* **387:**611–617.

Penfold, M. E., Dairaghi, D. J., Duke, G. M., Saederup, N., Mocarski, E. S., Kemble, G. W., and Schall, T. J. (1999). Cytomegalovirus encodes a potent alpha chemokine. *Proc. Natl. Acad. Sci. U.S.A.* **96:**9839–9844.

Pleskoff, O., Treboute, C., Brelot, A., Heveker, N., Seman, M., and Alizon, M. (1997). Identification of a chemokine receptor encoded by human cytomegalovirus as a cofactor for HIV-1 entry. *Science* **276:**1874–1878.

Pousset, F. (1994). Developmental expression of cytokine genes in the cortex and hippocampus of the rat central nervous system. *Brain Res. Dev. Brain Res.* **81:**143–146.

Power, C., McArthur, J. C., Johnson, R. T., Griffin, D. E., Glass, J. D., Perryman, S., and Chesebro, B. (1994). Demented and non-demented patients with AIDS differ in brain-derived human immunodeficiency virus type 1 envelope sequences. *J. Virol.* **68:**4643–4649.

Premack, B. A., and Schall, T. J. (1996). Chemokine receptors: gateways to inflammation and infection. *Nat. Med.* **2:**1174–1178.

Qin, S., Rottman, J. B., Myers, P., Kassam, N., Weinblatt, M., Loetscher, M., Koch, A. E., Moser, B., and Mackay, C. R. (1998). The chemokine receptors CXCR3 and CCR5 mark subsets of T cells associated with certain inflammatory reactions. *J. Clin. Invest.* **101:**746–754.

Raghavan, R., Cheney, P. D., Raymond, L. A., Joag, S. V., Stephens, E. B., Adany, I., Pinson, D. M., Li, Z., Marcario, J. K., Jia, F., Wang, C., Foresman, L., Berman, N. E., and Narayan, O. (1999). Morphological correlates of neurological dysfunction in macaques infected with neurovirulent simian immunodeficiency virus. *Neuropathol. Appl. Neurobiol.* **25:**285–294.

Randolph, D. A., Huang, G., Carruthers, C. J., Bromley, L. E., and Chaplin, D. D. (1999). The role of CCR7 in T(H)1 and T(H)2 cell localization and delivery of B cell help in vivo. *Science* **286:**2159–2162.

Ransohoff, R. M. (1997). Chemokines in neurological disease models: correlation between chemokine expression patterns and inflammatory pathology. *J. Leukoc. Biol.* **62:**645–652.

Ransohoff, R. M., and Tani, M. (1998). Do chemokines mediate leukocyte recruitment in post-traumatic CNS inflammation? *Trends Neurosci.* **21:**154–159.

Raport, C. J., Schweickart, V. L., Eddy, R. L., Shows, T. B., and Gray, P. W. (1995). The orphan G-protein-coupled receptor-encoding gene *V28* is closely related to genes for chemokine receptors and is expressed in lymphoid and neural tissues. *Gene* **163:**295–299.

Ren, L. Q., Gourmala, N., Boddeke, H. W. G. M., and Gebicke-Haerter, P. J. (1998). Lipopolysaccharide-induced expression of IP-10 mRNA in rat brain and in cultured rat astrocytes and microglia. *Molec. Brain Res.* **59:**256–263.

Rollins, B. J. (1997). Chemokines. *Blood* **90:**909–928.

Rosler, A., Pohl, M., Braune, H.-J., Oertel, W. H., Gemsa, D., and Sprenger, H. (1998). Time course of chemokines in the cerebrospinal fluid and serum during herpes simplex type 1 encephalitis. *J. Neurol. Sci.* **157:**82–89.

Saederup, N., Lin, Y. C., Dairaghi, D. J., Schall, T. J., and Mocarski, E. S. (1999). Cytomegalovirus-encoded beta chemokine promotes monocyte-associated viremia in the host. *Proc. Natl. Acad. Sci. U.S.A.* **96:**10881–10886.

Sallusto, F., Lanzavecchia, A., and Mackay, C. R. (1999a). Chemokines and chemokine receptors in T-cell priming and Th1/Th2-mediated responses. *Immunol. Today* **19:**568–574.

Sallusto, F., Lenig, D., Forster, R., Lipp, M., and Lanzavecchia, A. (1999b). Two subsets of memory T lymphocytes with distinct homing potentials and effector functions. *Nature* **401:**708–712.

Sallusto, F., Mackay, C. R., and Lanzavecchia, A. (1997). Selective expression of the eotaxin receptor CCR3 by human T helper 2 cells. *Science* **277:**2005–2007.

Sanders, V. J., Pittman, C. A., White, M. G., Wang, G., Wiley, C. A., and Achim, C. L. (1998). Chemokines and receptors in HIV encephalitis. *AIDS* **12:**1021–1026.

Sasseville, V. G., and Lackner, A. A. (1997). Neuropathogenesis of simian immunodeficiency virus infection in macaque monkeys. *J. Neurovirol.* **3:**1–9.

Sasseville, V. G., Smith, M. M., Mackay, C. R., Pauley, D. R., Mansfield, K. G., Ringler, D. J., and Lackner, A. A. (1996). Chemokine expression in simian immunodeficiency virus-induced AIDS encephalitis. *Am. J. Pathol.* **149:**1459–1467.

Schmidtmayerova, H., Nottet, H., Nuovo, G., Raabe, T., Flanagan, C. R., Dubrovsky, L., Gendelman, H. E., Cerami, A., Bukrinsky, M., and Sherry, B. (1996). Human immunodeficiency virus type 1 infection alters chemokine beta peptide expression in human monocytes—implications for recruitment of leukocytes into brain and lymph nodes. *Proc. Natl. Acad. Sci. USA* **93:**700–704.

Schwaeble, W. J., Stover, C. M., Schall, T. J., Dairaghi, D. J., Trinder, P. K., Linington, C., Iglesias, A., Schubart, A., Lynch, N. J., Weihe, E., and Schafer, M. K. (1998). Neuronal expression of fractalkine in the presence and absence of inflammation. *FEBS Lett.* **439:**203–207.

Senkevich, T. G., Bugert, J. J., Sisler, J. R., Koonin, E. V., Darai, G., and Moss, B. (1996). Genome sequence of a human tumorigenic poxvirus: prediction of specific host response-evasion genes. *Science* **273:**813–816.

Shi, B., De Girolami, U., He, J., Wang, S., Lorenzo, A., Busciglio, J., and Gabuzda, D. (1996). Apoptosis induced by HIV-1 infection of the central nervous system. *J. Clin. Invest.* **98:**1979–1990.

Shirozu, M., Nakano, T., Inazawa, J., Tashiro, K., Tada, H., Shinohara, T., and Honjo, T. (1995). Structure and chromosomal localization of the human stromal cell-derived factor 1 (SDF1) gene. *Genomics* **28**:495–500.

Simpson, J. E., Newcombe, J., Cuzner, M. L., and Woodroofe, M. N. (1998). Expression of monocyte chemoattractant protein-1 and other β-chemokines by resident glia and inflammatory cells in multiple sclerosis lesions. *J. Neuroimmunol.* **84**:238–249.

Smith, G. L., Symons, J. A., Khanna, A., Vanderplasschen, A., and Alcami, A. (1997). Vaccinia virus immune evasion. *Immunol. Rev.* **159**:137–154.

Sorensen, T. L., Tani, M., Jensen, J., Pierce, V., Lucchinetti, C., Folcik, V. A., Qin, S., Rottman, J., Sellebjerg, F., Strieter, R. M., Frederiksen, J. L., and Ransohoff, R. M. (1999). Expression of specific chemokines and chemokine receptors in the central nervous system of multiple sclerosis patients. *J. Clin. Invest.* **103**:807–815.

Soto, H., Wang, W., Strieter, R. M., Copeland, N. G., Gilbert, D. J., Jenkins, N. A., Hedrick, J., and Zlotnik, A. (1998). The C-C chemokine 6Ckine binds the CXC chemokine receptor CXCR3. *Proc. Natl. Acad. Sci. U.S.A.* **95**:8205–8210.

Sozzani, S., Luini, W., Bianchi, G., Allavena, P., Wells, T. N., Napolitano, M., Bernardini, G., Vecchi, A., D'Ambrosio, D., Mazzeo, D., Sinigaglia, F., Santoni, A., Maggi, E., Romagnani, S., and Mantovani, A. (1998). The viral chemokine macrophage inflammatory protein-II is a selective Th2 chemoattractant. *Blood* **92**:4036–4039.

Spanaus, K. S., Nadal, D., Pfister, H. W., Seebach, J., Widmer, U., Frei, K., Gloor, S., and Fontana, A. (1997). C-X-C and C-C chemokines are expressed in the cerebrospinal fluid in bacterial meningitis and mediate chemotactic activity on peripheral blood-derived polymorphonuclear and mononuclear cells in vitro. *J. Immunol.* **158**:1956–1964.

Spleiss, O., Gourmala, N., Boddeke, H. W., Sauter, A., Fiebich, B. L., Berger, M., and Gebicke-Haerter, P. J. (1998). Cloning of rat HIV-1-chemokine coreceptor CKR5 from microglia and upregulation of its mRNA in ischemic and endotoxinemic rat brain. *J. Neurosci. Res.* **53**:16–28.

Sprenger, H., Rosler, A., Tonn, P., Braune, H. J., Huffmann, G., and Gemsa, D. (1996). Chemokines in the cerebrospinal fluid patients with meningitis. *Clin. Immunol. Immunopathol.* **80**:155–161.

Stine, J. T., Chantry, D., and Gray, P. (1999). Virally encoded chemokines and chemokine receptors: genetic embezzlement of host DNA. *Chem. Immunol.* **72**:161–180.

Strieter, R. M., Srandiford, T. J., Huffnagle, G. B., Colletti, L. M., Lukacs, N. W., and Kunkel, S. L. (1996). "The good, the bad, and the ugly." The role of chemokines in models of human disease. *J. Immunol.* **156**:3583–3586.

Tachibana, K., Hirota, S., Iizasa, H., Yoshida, H., Kawabata, K., Kataoka, Y., Kitamura, Y., Matsushima, K., Yoshida, N., Nishikawa, S., Kishimoto, T., and Nagasawa, T. (1998). The chemokine receptor CXCR4 is essential for vascularization of the gastrointestinal tract. *Nature* **393**:591–594.

Tanabe, S., Heesen, M., Berman, M. A., Fischer, M. B., Luo, Y., and Dorf, M. E. (1997a). Murine astrocytes express a functional chemokine receptor. *J. Neurosci.* **17**:6522–6528.

Tanabe, S., Heesen, M., Yoshizawa, I., Berman, M. A., Luo, Y., Bleul, C. C., Springer, T. A., Okuda, K., Gerard, N., and Dorf, M. E. (1997b). Functional expression of the CXC-chemokine receptor-4/fusin on mouse microglial cells and astrocytes. *J. Immunol.* **159**:905–911.

Tani, M., Fuentes, M. E., Peterson, J. W., Trapp, B. D., Durham, S. K., Loy, J. K., Bravo, R., Ransohoff, R. M., and Lira, S. A. (1996). Neutrophil infiltration, glial reaction, and neurological disease in transgenic mice expressing the chemokine N51/KC in oligodendrocytes. *J. Clin. Invest.* **98**:529–539.

Tashiro, K., Tada, H., Heilker, R., Shirozu, M., Nakano, T., and Honjo, T. (1993). Signal sequence trap: a cloning strategy for secreted proteins and type I membrane proteins. *Science* **261**:600–603.

Taub, D. D., and Oppenheim, J. J. (1994). Chemokines, inflammation and the immune system. *Ther. Immunol.* **1:**229–246.

Theodorou, I., Meyer, L., Magierowska, M., Katlama, C., and Rouzioux, C. (1997). HIV-1 infection in an individual homozygous for CCR5 delta 32. Seroco Study Group. *Lancet* **349:**1219–1220.

Thomas, J., Kanangat, S., and Rouse, B. T. (1998). Herpes simplex virus replication-induced expression of chemokines and proinflammatory cytokines in the eye: implications in herpetic stromal keratitis. *J. Interferon Cytokine Res.* **18:**681–690.

Tobler, A., Meier, R., Seitz, M., Dewald, B., Baggiolini, M., and Fey, M. F. (1992). Glucocorticoids downregulate gene expression of GM-CSF, NAP-1/IL-8, and IL-6, but not of M-CSF in human fibroblasts. *Blood* **79:**45–51.

Toggas, S. M., Masliah, E., Rockenstein, E. M., Rall, G. F., Abraham, C. R., and Mucke, L. (1994). Central nervous system damage produced by expression of the HIV-1 coat protein gp120 in transgenic mice. *Nature* **367:**188–193.

Umehara, F., Izumo, S., Nakagawa, M., Ronquillo, A. T., Takahashi, K., Matsumuro, K., Sato, E., and Osame, M. (1993). Immunocytochemical analysis of the cellular infiltrate in the spinal cord lesions in HTLV-I-associated myelopathy. *J. Neuropathol. Exp. Neurol.* **52:**424–430.

Umehara, F., Izumo, S., Takeya, M., Takahashi, K., Sato, E., and Osame, M. (1996). Expression of adhesion molecules and monocyte chemoattractant protein-1 (MCP-1) in the spinal cord lesions in HTLV-I-associated myelopathy. *Acta Neuropathol.* **91:**343–350.

Vallat, A. V., De Girolami, U., He, J., Mhashilkar, A., Marasco, W., Shi, B., Gray, F., Bell, J., Keohane, C., Smith, T. W., and Gabuzda, D. (1998). Localization of HIV-1 co-receptors CCR5 and CXCR4 in the brain of children with AIDS. *Am. J. Pathol.* **152:**167–178.

Van Der Voorn, P., Tekstra, J., Beelen, R. H., Tensen, C. P., Van Der Valk, P., and De Groot, C. J. (1999). Expression of MCP-1 by reactive astrocytes in demyelinating multiple sclerosis lesions. *Am. J. Pathol.* **154:**45–51.

Vanguri, P. (1995). Interferon-—inducible genes in primary glial cells of the central nervous system: comparisons of astrocytes with microglia and Lewis with brown Norway rats. *J. Neuroimmunol.* **56:**35–43.

Vanguri, P., and Farber, J. M. (1994). IFN and virus-inducible expression of an immediate early gene, *crg-2/IP-10,* and a delayed gene, *I-Aα,* in astrocytes and microglia. *J. Immunol.* **152:**1411–1418.

Varona, R., Zaballos, A., Gutierrez, J., Martin, P., Roncal, F., Albar, J. P., Ardavin, C., and Marquez, G. (1998). Molecular cloning, functional characterization and mRNA expression analysis of the murine chemokine receptor CCR6 and its specific ligand MIP-3α. *FEBS Lett.* **440:**188–194.

Villiger, P. M., Terkeltaub, R., and Lotz, M. (1992). Monocyte chemoattractant protein-1 (MCP-1) expression in human articular cartilage. Induction by peptide regulatory factors and differential effects of dexamethasone and retinoic acid. *J. Clin. Invest.* **90:**488–496.

Weiner, L. P. (1973). Pathogenesis of demyelination induced by a mouse hepatitis virus (JHM virus). *Arch. Neurol.* **28:**298–303.

Weiss, J. M., Nath, A., Major, E. O., and Berman, J. W. (1999). HIV-1 Tat induces monocyte chemoattractant protein-1-mediated monocyte transmigration across a model of the human blood-brain barrier and up-regulates CCR5 expression on human monocytes. *J. Immunol.* **163:**2953–2959.

Weng, Y., Siciliano, S. J., Waldburger, K. E., Sirotina-Meisher, A., Staruch, M. J., Daugherty, B. L., Gould, S. L., Springer, M. S., and DeMartino, J. A. (1998). Binding and functional properties of recombinant and endogenous CXCR3 chemokine receptors. *J. Biol. Chem.* **273:**18288–18291.

Westmoreland, S. V., Rottman, J. B., Williams, K. C., Lackner, A. A., and Sasseville, V. G. (1998). Chemokine receptor expression on resident and inflammatory cells in the brain of macaques with simian immunodeficiency virus encephalitis. *Am. J. Pathol.* **152:**659–665.

Wong, M. L., Xin, W. W., and Duman, R. S. (1996). Rat LCR1: cloning and cellular distribution of a putative chemokine receptor in brain. *Mol. Psychiatry* **1:**133–140.

Wu, E., Dickson, D. W., Jacobson, S., and Raine, C. S. (1993). Neuroaxonal dystrophy in HTLV-1-associated myelopathy/tropical spastic paraparesis: neuropathologic and neuroimmunologic correlations. *Acta Neuropathol.* **86:**224–235.

Wu, L., Paxton, W. A., Kassam, N., Ruffing, N., Rottman, J. B., Sullivan, N., Choe, H., Sodroski, J., Newman, W., Koup, R. A., and Mackay, C. R. (1997). CCR5 levels and expression pattern correlate with infectability by macrophage-tropic HIV-1, in vitro. *J. Exp. Med.* **185:**1681–1691.

Xia, M. Q., Qin, S. X., Wu, L. J., Mackay, C. R., and Hyman, B. T. (1998). Immunohistochemical study of the beta-chemokine receptors CCR3 and CCR5 and their ligands in normal and Alzheimer's disease brains. *Am. J. Pathol.* **153:**31–37.

Yoshida, R., Imai, T., Hieshima, K., Kusuda, J., Baba, M., Kitaura, M., Nishimura, M., Kakizaki, M., Nomiyama, H., and Yoshie, O. (1997). Molecular cloning of a novel human CC chemokine EBI1-ligand chemokine that is a specific functional ligand for EBI1, CCR7. *J. Biol. Chem.* **272:**13803–13809.

Yoshida, T., Imai, T., Kakizaki, M., Nishimura, M., Takagi, S., and Yoshie, O. (1998). Identification of single C motif-1/lymphotactin receptor XCR1. *J. Biol. Chem.* **273:**16551–16554.

Yoshida, T., Imai, T., Kakizaki, M., Nishimura, M., and Yoshie, O. (1995). Molecular cloning of a novel C-λ or γ-type chemokine, SCM-1. *FEBS Lett.* **360:**155–159.

Zaballos, A., Gutierrez, J., Varona, R., Ardavin, C., and Marquez, G. (1999). Cutting edge: identification of the orphan chemokine receptor GPR-9-6 as CCR9, the receptor for the chemokine TECK. *J. Immunol.* **162:**5671–5675.

Zheng, J., Ghorpade, A., Niemann, D., Cotter, R. L., Thylin, M. R., Epstein, L., Swartz, J. M., Shepard, R. B., Liu, X., Nukuna, A., and Gendelman, H. E. (1999a). Lymphotropic virions affect chemokine receptor-mediated neural signaling and apoptosis: implications for human immunodeficiency virus type 1-associated dementia. *J. Virol.* **73:**8256–8267.

Zheng, J., Thylin, M. R., Ghorpade, A., Xiong, H., Persidsky, Y., Cotter, R., Niemann, D., Che, M., Zeng, Y. C., Gelbard, H. A., Shepard, R. B., Swartz, J. M., and Gendelman, H. E. (1999b). Intracellular CXCR4 signaling, neuronal apoptosis and neuropathogenic mechanisms of HIV-1-associated dementia. *J. Neuroimmunol.* **98:**185–200.

Zingoni, A., Soto, H., Hedrick, J. A., Stoppacciaro, A., Storlazzi, C. T., Sinigaglia, F., D'Ambrosio, D., O'Garra, A., Robinson, D., Rocchi, M., Santoni, A., Zlotnik, A., and Napolitano, M. (1998). The chemokine receptor CCR8 is preferentially expressed in Th2 but not Th1 cells. *J. Immunol.* **161:**547–551.

Zlotnik, A., Morales, J., and Hedrick, J. A. (1999). Recent advances in chemokines and chemokine receptors. *Crit. Rev. Immunol.* **19:**1–47.

Zou, Y. R., Kottmann, A. H., Kuroda, M., Taniuchi, I., and Littman, D. R. (1998). Function of the chemokine receptor CXCR4 in haematopoiesis and in cerebellar development. *Nature* **393:**595–599.

REGULATION OF T CELL RESPONSES DURING CENTRAL NERVOUS SYSTEM VIRAL INFECTION

David N. Irani*,† and Diane E. Griffin†

*Department of Neurology
The Johns Hopkins University School of Medicine
Baltimore, Maryland 21287
†Department of Molecular Microbiology and Immunology
The Johns Hopkins University School of Hygiene and Public Health
Baltimore, Maryland 21205

I. Introduction
II. Effector Functions of T Cells During CNS Viral Infection
   A. T Cells Facilitate Recruitment of Other Inflammatory Cells into CNS
   B. T Cells Promote Clearance of Viruses from Brain
   C. T Cells Can Protect Hosts from Lethal CNS Viral Infections and Promote Neuronal Survival
   D. T Cells Can Induce Immune-Mediated Damage
III. Regulation of T Cell Responses in the Brain during CNS Viral Infection
   A. T Cell Responses in Brain Versus the Periphery During CNS Viral Infection: Selective Recruitment, Preferential Retention, or Local Regulation
   B. CTL Activity in Brain
   C. T Cell Proliferation in CNS
   D. Local T Cell Cytokine Production in Brain
   E. Control of T Cell Survival Within CNS
IV. Concluding Remarks
   References

I. Introduction

T cells can serve a variety of functions as part of the host immune response during central nervous system (CNS) viral infection. They can participate directly in viral clearance from the brain or they can promote the survival of the host without exerting any direct effect on virus replication. T cell responses elicited during CNS viral infection may sometimes actually cause neuropathological injury above and beyond what is induced by the pathogen itself. Such variability in the effects on outcome may in large part be dictated by the nature of the specific pathogen, but host factors also exert quantitative and qualitative effects on the types of T cell responses elicited during disease. When considering the various roles that T cells may play during CNS viral infections, one important and interesting issue is whether and

175

how different T cell responses are regulated within the brain. After briefly reviewing how T cells may contribute to the pathogenesis of neurologic disease, we will discuss specific examples of how individual T cell effector functions can be regulated during CNS viral infections. Although our current understanding of the mechanism(s) that underlie such regulatory events is limited, further characterization of these processes will be relevant in neurovirology as well as in the broader fields of basic and clinical neuroimmunology.

II. Effector Functions of T Cells during CNS Viral Infection

A. T Cells Facilitate Recruitment of Other Inflammatory Cells into CNS

Only a small number of T cells infiltrate the brain under normal circumstances; this paucity of immune surveillance of baseline is one of several reasons why the CNS has often been characterized as an "immunologically privileged" site. Yet during the course of CNS viral infection (whether restricted to the meninges, leading to an influx of inflammatory cells into the cerebrospinal fluid (CSF), or following viral spread to the brain and spinal cord, leading to actual parenchymal inflammation), the processes that normally exclude immune elements from the CNS are overcome and inflammatory cells rapidly enter these tissue compartments (brain, CSF, meninges). Parenchymal inflammation during viral encephalitis is most common around blood vessels in the form of perivascular infiltrates. Such an extravasation of cells from the bloodstream into the perivascular space, and then into the parenchyma itself, is likely to depend on specific molecular interactions between circulating immune cells and the cerebrovascular endothelium (Griffin et al., 1992; Irani and Griffin, 1996). In several well-studied experimental paradigms, CD4+ and CD8+ T cells, NK cells, and γδ+ T cells constitute many of the earliest inflammatory cells that infiltrate the brain during CNS viral infection (Moench and Griffin, 1984; Williamson et al., 1991; Irani and Griffin, 1991; Griffin et al., 1992). The proportion of these earliest-appearing T cells that are specific for viral antigens is not known, but it is likely that these virus-specific cells release inflammatory mediators, thereby contributing to subsequent inflammatory cell recruitment into the brain. For example, when T cell-deficient athymic nude mice are infected with the encephalitic alphavirus, Sindbis, animals develop markedly less perivascular inflammation, and recruitment of virus-specific B cells into the CNS decreases as compared to immunologically normal hosts (Hirsch and Griffin, 1979; Tyor et al., 1989).

Likewise, during murine lymphocytic choriomeningitis virus (LCMV) infection, activated, CD8[+], virus-immune T cells are the principal regulators of how much inflammation subsequently develops in the CSF of infected animals (Doherty *et al.*, 1988). Thus, T cells play an important role in facilitating the recruitment of other inflammatory cell types into the CNS during viral infection, thereby determining the overall level of inflammation.

B. T Cells Promote Clearance of Viruses from Brain

T cell-mediated lysis of infected cells has been demonstrated to be an important mechanism of viral clearance from tissues other than the CNS (Yap *et al.*, 1978; Dharakul *et al.*, 1990; Young *et al.*, 1990; Munoy *et al.*, 1991, Ando *et al.*, 1994). Both CD4[+] and CD8[+] virus-specific cytotoxic T lymphocytes (CTLs) have been described. The cytolytic actions of these T cells require that compatible major histocompatibility complex (MHC) class II and class I molecules, respectively, are expressed on virally infected target cells. Yet within the CNS, overall MHC expression is restricted, with glial cells expressing more of these molecules than neurons (Mauerhoff *et al.*, 1980; Wong *et al.*, 1984; Joly *et al.*, 1991). The paucity of MHC expression has been proposed as one means by which viruses may escape immune recognition and persist in the brains of infected hosts. Nevertheless, these limitations can be overcome and T cell-mediated clearance of viruses from the CNS has been demonstrated in a number of experimental systems. Following the inoculation of the JHM strain of mouse hepatitis virus (JHMV) into susceptible mice, for example, an acute encephalomyelitis is elicited with robust infection of astrocytes, oligodendrocytes, and, to a lesser degree, neurons. Using an adoptive transfer strategy, viral clearance from the brains of these hosts has been shown to be mediated by an *H2-D*-restricted, virus-specific CD8[+] T cell population (Sussman *et al.*, 1989). Interestingly, these CD8[+] CTLs have an absolute requirement for CD4+ T cells that appear to support their local survival within the brain (Stohlman *et al.*, 1998).

Likewise, during persistent LCMV infection of mice, the adoptive transfer of H-2-compatible, virus-immune T cells results in clearance of infectious virus from many tissues including the brain (Oldstone *et al.*, 1986). Most notable in this particular model system were tissue-specific differences in the apparent mechanism of viral clearance; this process took much longer in the CNS as compared to other organs but occurred without any histological evidence of neuronal destruction, while clearance from other organs involved the actual cytolysis of infected cells (Oldstone *et al.*, 1986; Tishon *et al.*, 1993). Whether this is the result of organ-specific differences in levels of MHC antigen expression is unclear, but viral clearance without the actual destruction of infected neurons, a nonrenewable cell population, would certainly be advantageous to the host. There now is convincing evidence in other systems that the antiviral actions of CD8+ T cells within the CNS depend on their local production of soluble mediators such as interferon-γ (IFN-γ), and not on direct perforin-mediated cytolysis of target cells (Kündig et al., 1993; Lin et al., 1997). In sum, while more than one mechanism may underlie their effects, it is clear that T cells can promote viral clearance from the CNS in many situations.

C. T Cells Can Protect Hosts from Lethal CNS Viral Infections and May Promote Neuronal Survival

The adoptive transfer of primed T cells can protect naive mice from a variety of otherwise lethal viral infections (Zinkernagel and Welsh, 1976; Yap et al., 1978; Jacoby et al., 1980; Sethi et al., 1983; Stohlman et al., 1986; Stohlman et al., 1995). In some situations, transferred T cells play an active role in suppressing virus replication, serving as the primary means through which they protect their hosts. In other cases, however, no effect on virus replication is seen and protection has to be explained through another mechanism, such as causing a change in the cell tropism of the virus or inhibiting virus-induced cell death. There are several notable examples where virus-immune T cells serve in this capacity during CNS viral infection. In the JHMV model, for example, a virus-specific, CD4+ T cell clone was found to protect mice from an otherwise lethal viral challenge without actually decreasing the amount of infectious virus in the brain (Stohlman et al., 1986). Protective T cells had to be administered intracerebrally in this system; their beneficial effect also depended on the recipient's own immune system as it could be ablated by pretreating the host with cyclophosphamide (Stohlman et al., 1986).

In more recent experiments using an otherwise lethal strain of Sindbis virus, animals preimmunized against nonstructural viral proteins were highly protected against permanent paralysis and death (Gorrell et al., 1997). This protection was not associated with either reduced virus replication in the brain or altered virus tropism (neurons of the brain and spinal cord remained the principal target of infection), nor was it due to an augmented humoral response against structural viral proteins that could also protect lethally infected animals (Stanley et al., 1986; Gorrell et al., 1997). Instead, this nonstructural protein-mediated protection proved to be a function of immune T cells and occurred by promoting the survival and functional recovery of virus-infected neurons (Gorrell *et al.*, 1997). Because virus-induced neuronal apoptosis is believed to be the cellular substrate underlying the neurovirulence of Sindbis virus *in vivo* (Lewis *et al.*, 1996), it was suggested that this particular form of protection occurred through local T cell cytokine release within the CNS early during infection that somehow interrupted apoptotic signaling in infected neurons. Although the mechanism of protection in this model remains incompletely understood, such a finding highlights the fact that T cells may also serve to promote the survival of target cells within the brain during CNS viral infection, until other immune elements can control virus replication and achieve viral clearance.

D. T Cells Can Induce Immune-Mediated Damage

In several well-characterized animal models of CNS viral infection, part of the elicited T cell response actually contributes to the pathology and adverse outcome of disease. Neurotropic LCMV infection of adult mice is perhaps the premier example of this phenomenon; the resulting fatal choriomeningitis that develops is mediated by the same virus-specific CD8+ T cells that participate in viral clearance (Cole *et al.*, 1972; Mims and Blanden, 1972; Buchmeier *et al.*, 1980; Oldstone *et al.*, 1986). In another model system, virus-specific CD4+ T cells clearly mediate the immunopathology seen in adult rats infected with Borna disease virus (BDV) and can actually transfer disease to another host in an MHC class II–restricted manner (Richt *et al.*, 1989). Finally, during chronic Theiler's murine encephalomyelitis virus (TMEV) infection that occurs when virus is not fully cleared from the CNS following acute encephalitis, CD4+ T cells, initially directed against viral antigens but later including T cells specific for various myelin epitopes, cause destruction of oligodendrocytes and produce demyelination (Lipton and Dal Canto, 1976; Pope *et al.*, 1996, Miller *et al.*, 1997). Other experimental examples also exist; there is no lack of evidence that T cells can sometimes induce immune-mediated damage to the CNS during viral infection.

III. REGULATION OF T CELL RESPONSES DURING CNS VIRAL INFECTION

A. T Cell Responses in Brain versus the Periphery During CNS Viral Infection: Selective Recruitment, Preferential Retention, or Local Regulation

Many arguments have been used to characterize the CNS as an "immunologically privileged" tissue: the restrictive nature of the blood–brain barrier that normally excludes circulating immune elements; the paucity of MHC antigen expression that minimizes local T cell reactivity; the lack of a formal lymphatic drainage system that limits immune exposure to antigens present in the CNS; and even the constitutive expression of inhibitory cytokines, such as transforming growth factor-β (TGF-β), in the brain that may serve local immunoregulatory functions. Yet despite all these apparent obstacles, vigorous immune responses are rapidly mounted inside the CNS to a variety of stimuli including local viral infection. The various roles that T cells may serve during these infections have been briefly reviewed, but whether and how particular T cell responses are regulated during CNS viral infection are less clear. One way to examine whether such regulation takes place is to characterize local T cell responses in the brain and compare them to responses that are elicited in peripheral immune compartments. Differences found between T cell responses inside and outside the CNS may then be explained in several ways including: (a) the preferential recruitment of specific subpopulations of effector T cells from the circulation into the brain; (b) a selective retention of T cell subpopulations within the CNS, once the blood–brain barrier has been traversed; and/or (c) the local alteration of T cell effector function within the CNS microenvironment. Since the molecular mechanisms that underlie the processes of T cell recruitment into, and retention within, the CNS are only now beginning to emerge, and since it is unclear to what degree the brain may actively influence local T cell effector function, details regarding how T cell responses may be regulated during CNS viral infection are correspondingly incomplete. Nevertheless, specific examples showing that individual T cell responses (CTL activity, proliferation, cytokine production, and apoptosis) differ between CNS and peripheral immune compartments provide preliminary evidence that regulation of T cell responses during viral infections of the nervous system can and does occur.

B. CTL Activity in Brain

Extensive characterizations of virus-specific CTL responses that develop on either side of the blood–brain barrier of mice with viral encephalitis have been undertaken in several experimental systems. Such studies serve as important illustrations of how this particular T cell function may be regulated during CNS viral infection. During acute JHMV infection, for example, several groups have found that the virus-specific CD8+ CTLs present in the brain are of restricted antigenic specificity (they respond mainly to the viral nucleocapsid protein but not to the spike, membrane, or hemagglutinin–esterase proteins);

CTLs with this same nucleocapsid specificity are not found, or are present at much lower levels, in lymphoid organs outside the CNS, including cervical lymph nodes (Stohlman *et al.*, 1993; Castro *et al.*, 1994). Such differences in specificity suggest that either these particular T cells are present at very low levels in the periphery and get rapidly and selectively recruited into the CNS or that they are actually generated locally within the brain. Some preliminary studies that address this question now suggest that JHMV-specific cytotoxic T lymphocyte precursor (CTLp) are, in fact, primed within cervical lymph nodes before migrating into the CNS (C. Bergmann and S. Stohlman, unpublished observations). This implies that these T cells possess some property that makes them particularly capable of entering or being retained within the brain. Whether priming in cervical lymph nodes (rather than in the spleen or other lymph node chains) causes their preferential recruitment into the CNS is not clear. Existing evidence, however, points to cervical lymph nodes as a site where antigen accumulates following its inoculation into the CNS and where the virus-specific CTLp are generated following the intracerebral inoculation of mice with neurotropic LCMV (Harling-Berg *et al.*, 1989; Lynch *et al.*, 1989).

In the chronic phase of JHMV infection that occurs as a result of incomplete viral clearance from glial cells such as oligodendrocytes, the fine antigen specificity of brain-derived CTLs is even more restricted than during the acute phase (Marten *et al.*, 1999). Thus, while CTLs from chronically infected animals recognize the same viral nucleocapsid antigens as T cells found in the CNS during the acute illness, they are much less tolerant of amino acid mutations in the specific epitope peptides (Marten *et al.*, 1999). This "focusing" of the antigenic specificity of local CTL over time during JHMV infection has been proposed to occur through a preferential selection (survival or even selective expansion) within the brain of CTLs with higher T cell receptor affinities. The forces behind such a selection process are not well understood, but they may in part relate to the fact that the supply of MHC-bound viral peptides becomes extremely limited as viral load diminishes and cellular tropism narrows in the brain over time. Whether such a focusing event is more likely to occur in the sequestered environment of the CNS (where complete viral clearance is more difficult to achieve relative to other tissues) is not known.

In a murine model of neurovirulent influenza virus infection, activated, virus-specific CTLs were found to persist in the brain long after viral clearance had been achieved (Hawke *et al.*, 1998). In this paradigm, it was proposed that these T cells were not actively recirculating into and out of the brain since the systemic spread of infection was minimal. Instead, it seemed more likely that CTLs were being activated and retained within the CNS by the ongoing expression of MHC class I–viral peptide complexes that were simply below the threshold of detection (Hawke et al., 1998). Importantly, the selective retention of influenza-specific CTLs in the lungs of mice inoculated intranasally with the same virus was not found (CD8+ T cells persisted within this tissue well after viral clearance had been achieved but had lost their capacity to lyse virus-infected targets), leading to the conclusion that factors which promote long-term CTL activation are present in the brain but not in other tissues (Hawke et al., 1998).

It is provocative to consider that persistent viral RNA itself, in the absence of detectable virus replication, may provide enough of a stimulus to activate selected CD8+ T cell effector functions in the CNS. While such a local T cell response may in fact be the reason why virus replication remains at undetectable levels, by definition it contributes to a chronic inflammation in the brain. Such an experimental finding then makes CNS viral persistence a much more plausible basis for explaining how immune-mediated neurologic diseases such as multiple sclerosis may be triggered. Indeed, the process of local antigenic focusing noted among T cells during the transition from acute to chronic JHMV infection may in part serve to reduce the risk that cross-reactive or even autoreactive CTLs evolve among the cells that persist within the brain. Such an evolution can occur (although, in this case, mediated by CD4+ T cells); CNS autoimmunity has now been shown to develop in mice persistently infected with TMEV through an immunologic process referred to as "epitope spreading" (Miller et al., 1997). Nevertheless, the long-term retention of activated, but nonproliferating, CTLs in the brain may provide a necessary component of immunity that is important for this "immunologically privileged" tissue (Hawke et al., 1998).

C. T Cell Proliferation in CNS During Viral Infection

The ex vivo proliferative capacity of brain-infiltrating T cells has been investigated in various experimental models of CNS inflammation. In many of these studies, the goal has been to characterize the antigenic specificity of T cells within the CNS as compared to those obtained from various extracerebral sites. In experimental autoimmune encephalomyelitis (EAE), for example, the reactivity of brain-derived T cells to various myelin antigens has been examined in efforts to try to understand how demyelination that may accompany this disorder can occur. During CNS viral infection, questions regarding the frequency of virus-reactive T cells in the brain are common, and in cases where virus-induced inflammation leads to immunopathology within the CNS, the issue of whether these T cells cross-react with neural antigens has also been examined. In the present context, however, examining such studies may serve to help understand how another T cell response (i.e., clonal expansion) may potentially be regulated during CNS viral infection.

T cells appear to be recruited nonspecifically into the CNS during chronic TMEV infection in susceptible SJL mice, at least as measured by a polymerase chain reaction (PCR)-based assay that determines T cell receptor (TcR)β-chain diversity among T cells from the spleens and brains of infected hosts (Musette *et al.*, 1995). By extending this molecular strategy, however, it was possible to demonstrate that populations of T cells in the CNS expressing certain TcR β-chains were markedly expanded while others were not; such an expansion did not occur among splenic T cells, suggesting that it was due to local T cell proliferation in the brain (Musette *et al.*, 1995). While this local CNS expansion was assumed to be antigen-driven, it was not possible to determine the specificity of this event because brain-derived T cells were not actually being cultured *in vitro*. Other studies, however, have shown that CD4+ T cells isolated from the CNS of TMEV-infected animals proliferate *in vitro* to nearly the same degree as splenic T cells when stimulated with viral antigens (Pope *et al.*, 1996). By themselves, these data would suggest that local regulation of T cell proliferation does not occur to any degree during CNS viral infection.

Other experimental results, however, suggest that this may not always be the case. In the murine model of neurovirulent influenza virus infection discussed above, proliferation among the activated, CD8+ T cells that persisted in the brains of infected animals, following viral clearance, was examined. Using an *in vivo* 5-bromo-2'-deoxyuridine (BrdU) incorporation technique, these investigators showed that despite their activated phenotype and their preserved ability to lyse virus-infected targets *in vitro,* the vast majority of these CD8+ T cells were not proliferating locally within the brain (Hawke *et al.*, 1998). Although an explanation for this finding was not offered in this report, the effect appeared specific to the CNS because many splenic T cells from these animals were actively dividing throughout disease (Hawke *et al.*, 1998). The effect also seemed to be related to how long the T cells had been retained within the brain (Hawke *et al.*, 1998).

In a murine alphavirus encephalitis model using an avirulent strain of Sindbis virus, T cells isolated from the brains of infected BALB/c animals did not proliferate *ex vivo* in response to either viral antigens or mitogenic lectins (Irani *et al.*, 1997). Notably, virus-induced immunopathology or neurologic sequelae are not induced following infection in these hosts. In fact, many of these brain-derived T cells appeared to have arrested in the cell cycle despite the fact that peripheral lymphocytes from infected mice showed no defect in proliferation (Irani et al., 1997). The same virus inoculated into SJL mice, however, elicited more intense and prolonged CNS inflammation as well as immune-mediated paralysis, despite equally effective viral clearance from the brain (Mokhtarian et al., 1989; Irani, 1998; Rowell and Griffin, 1999). One notable difference between the inflammatory responses in the brains of these two hosts was an increased proliferative capacity of brain-infiltrating T cells in the SJL mice (Fig. 1). Such a difference, although not proven to be related to the immune-mediated CNS damage observed in SJL mice, was attributed to intrinsic properties of the T cells and not something different about the local environment of the brain that was more permissive for local T cell proliferation in this mouse strain (Irani, 1998). Nevertheless, if T cells in SJL mice are more capable of proliferating within the brain than T cells within BALB/c mice following the same CNS viral infection, it is possible that a defect in local T cell regulation during this infection may predispose these hosts to immune-mediated neurologic injury. In this sense, deficient local regulation of certain T cell effector functions may actually contribute to the pathogenesis of disease in CNS viral infection.

D. Local T Cell Cytokine Production in Brain

Another important T cell effector function that occurs during CNS viral infection is the local production of cytokines within the brain. Like the T cell responses already discussed (CTL activity, proliferation), this process may also be controlled within the CNS in certain disease situations. However, regulation of T cell cytokine production during CNS viral infection must be considered in light of the fact that these mediators exert pleiotropic effects in different model systems; some cytokines (notably IFN-γ) may contribute directly to the control of virus replication in the brain (Kündig et al., 1993), while others may serve immunoregulatory functions such as promoting B cell survival and differentiation in the CNS (Tyor and Griffin, 1993; Wesselingh et al., 1994), and still others may act on infected neural cells in such a way that alters the cellular response to infection (Doherty et al., 1989; Gorrell et al., 1997). Despite these differences, examining some of these studies may reveal how this T cell effector response can be regulated during CNS viral infection.

During acute Sindbis virus encephalitis in mice, mRNAs encoding a number of cytokines are produced within the brain (Wesselingh et al.,

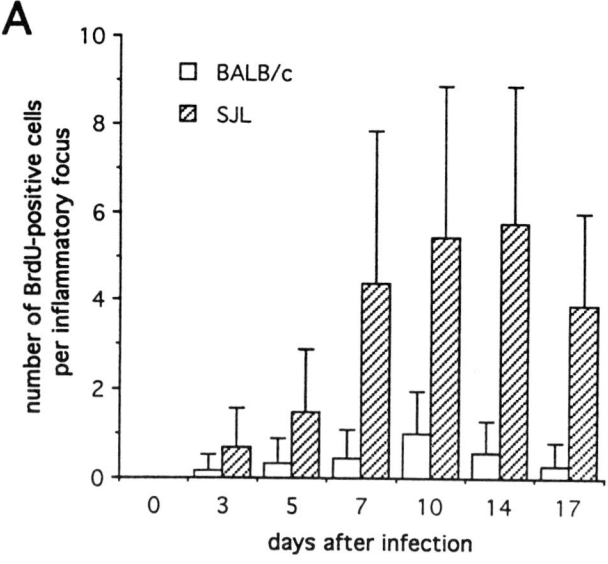

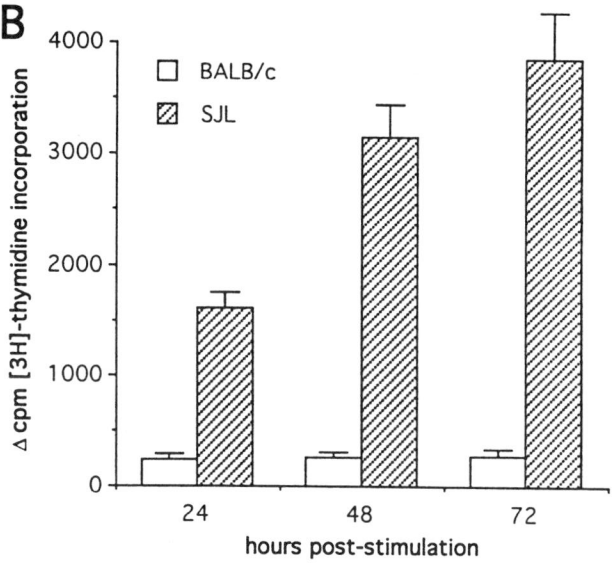

FIG 1. Proliferation of CNS inflammatory cells during acute Sindbis virus encephalitis in BALB/c and SJL mice. Using an *in vivo* BrdU labeling technique, more proliferating cells were found within the brains of SJL mice than those of BALB/c mice (A). Similarly, T cells isolated from the brains of infected SJL mice incorporated more tritiated thymidine per 10^5 viable cells than did T cells derived from the brains of BALB/c mice (B). Adapted with permission from Irani, 1998.

1994). While some (e.g., TNF-α, interleukin-1β [IL-1β] IL-6) are upregulated in advance of any histologic evidence of inflammation, suggesting production by intrinsic neural cells, most (e.g., IFN-γ) parallel the degree of brain parenchymal inflammation and are markedly reduced in the brains of severe combined immunodeficient (SCID) mice as compared to immunocompetent control animals (Wesselingh *et al.*, 1994). Quantitation of transcripts encoding T cell–derived cytokines in this disease suggests a predominant Th2-type response (high IL-4 and IL-10, lower IFN-γ, minimal IL-2) (Fig. 2). Since antibodies specific for the viral glycoproteins are the primary effectors of viral clearance from infected neurons (Levine *et al.*, 1991), and since virus-specific antibody-secreting B cells are retained in the CNS of animals for months after infectious virus has been cleared (Tyor *et al.*, 1992), it seems likely that factors which help to enrich the environment of the brain with Th2-type cytokines facilitate these virus-specific antibody responses.

How the local T cell population becomes skewed toward a Th2-type response is not known. Since activated T cells traffic into the brains of Sindbis-infected animals nonspecifically (Irani and Griffin, 1996), regulation of local cytokine production may be exerted within the CNS after T cells have accumulated at that site (Irani *et al.*, 1996; Irani *et al.*, 1997). Evidence to support this hypothesis comes from the observation that once T cells have crossed the blood–brain barrier and entered the perivascular and parenchymal compartments of the CNS, it is the virus-immune cells that are selectively retained within the brains of infected animals (Fig. 3). When T cells are adoptively transferred into the circulation of infected mice, those that traffic into, and are retained within, the brains selectively downregulate their expression of Th1-type cytokines (principally IL-2) over time (Irani *et al.*, 1997). This supports the hypothesis that, at least during acute Sindbis virus encephalitis, the brain can exert an active regulatory influence over the cytokine production by infiltrating T cells. How this regulation occurs is not known, but certain glycolipids (gangliosides) that are abundant on the external cell membranes of neural cells selectively inhibit Th1- but not Th2-type cytokine production by T cells activated *in vitro* (Fig. 4). Thus, during some viral infections of the CNS, T cell cytokine production can be regulated locally in the brain by lipid molecules that selectively inhibit the production of Th1-type cytokines. The net effect of this regulation is to enhance the relative levels of Th2-type cytokines. Others have also suggested that the CNS is intrinsically a "Th2-type" environment that naturally supports humoral immune responses and suppresses cellular responses (Cserr and Knopf, 1992).

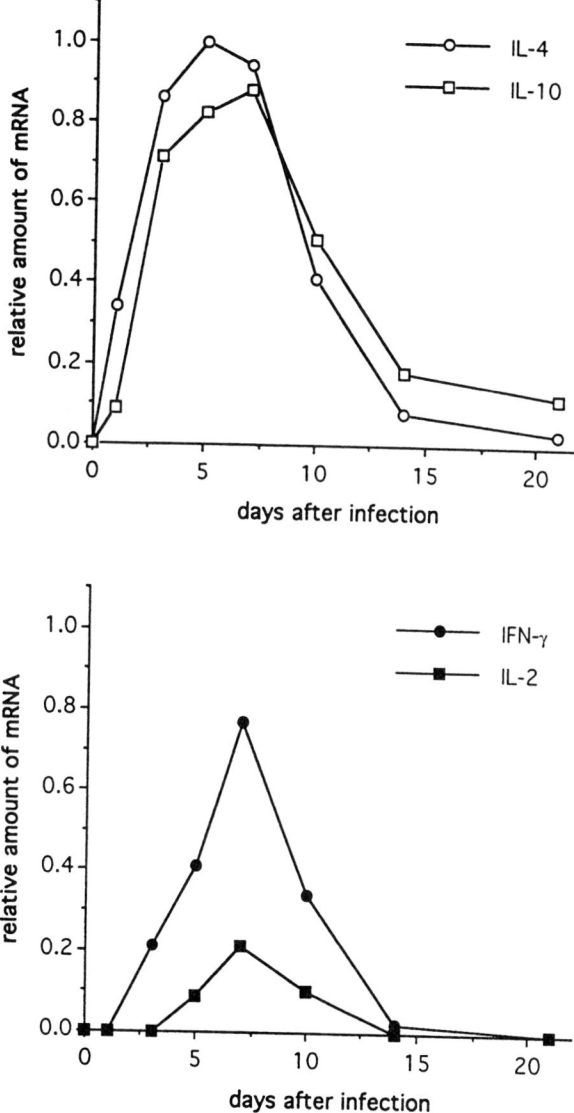

Fig 2. Quantitation of T cell cytokine mRNA levels present in the brain parenchyma of mice with acute Sindbis virus encephalitis. Individual cytokine transcripts were quantitated by RT-PCR relative to the constitutively expressed glyceraldehyde-3-phosphate dehydrogenase gene. Th2-type cytokines were present at higher levels. Adapted with permission from Wesselingh et al., 1994.

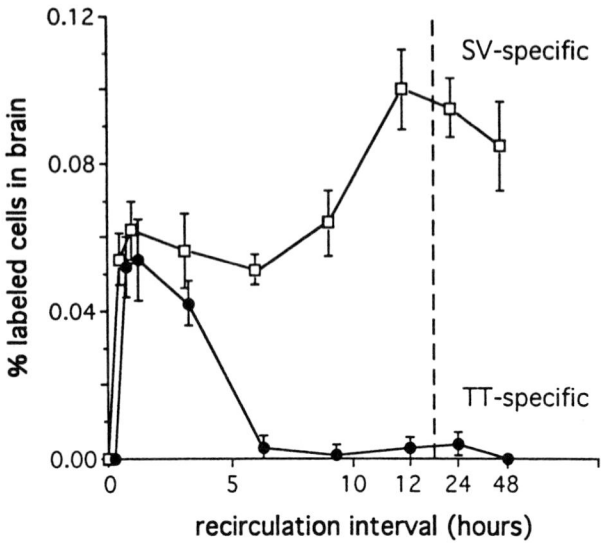

FIG 3. Kinetics of lymphocyte entry into the CNS of mice with Sindbis virus encephalitis. Fluorescently labeled T cells specific either for Sindbis virus (SV-specific) or tetanus toxoid (TT-specific) were inoculated intravenously into infected recipients. Inflammatory cells were then isolated from the brains of mice at various intervals and the percent of labeled cells present in each isolate was measured by flow cytometry. While T cells of either specificity rapidly entered the brain to equivalent degrees, virus-specific T cells were preferentially retained within this tissue while nonspecific T cells disappeared. Adapted with permission from Irani and Griffin, 1996.

E. Control of T Cell Survival Within CNS

In EAE, lymphocytes that infiltrate the brain may undergo apoptosis during the remission phase of disease (Schmeid et al., 1993; Bauer et al., 1995; Bonetti et al., 1997). Lymphocytic inflammation may also be cleared via an apoptotic process from the CNS of animals recovering from both acute coronavirus- and alphavirus-induced encephalitis (Barac-Latas et al., 1995; Irani, 1998). Based on these observations, it has been proposed that the induction of T cell apoptosis locally within the brain may be a generalized mechanism through which the brain terminates many inflammatory responses. Conversely, the persistence of CNS inflammation during viral encephalitis may in part result from a relative lack of apoptosis among brain-infiltrating T cells (Irani, 1998). In this situation, were myelin-reactive T cells to gain entry into the CNS and not be subjected to a process that normally downregulates inflammation, autoimmune injury could ensue. Examining how the survival of brain-infiltrating T cells is controlled during CNS viral

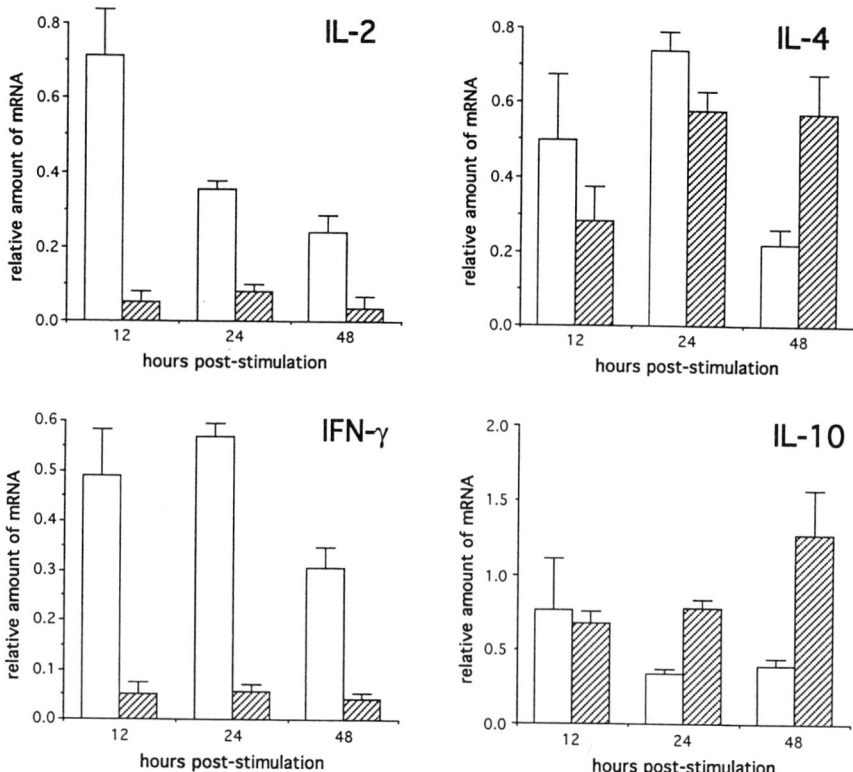

FIG 4. Quantitation of cytokine mRNA production by murine T cells stimulated in the absence (open bars) or presence (hatched bars) of normal brain-derived gangliosides. Individual cytokine transcripts were quantitated by RT-PCR relative to the constitutively expressed glyceraldehyde-3-phosphate dehydrogenase gene. Gangliosides selectively inhibited the accumulation of Th1-type cytokine mRNAs. Adapted with permission from Irani *et al.*, 1996.

infection may serve as yet another example of how T cell responses can be regulated during these diseases.

   Irrespective of the experimental system being examined, there is a paucity of information regarding the control of T cell apoptosis in the brain. In animals with EAE, it appears that only T cells that have actually infiltrated the brain parenchyma undergo apoptosis; those present in the meninges or in the perivascular space seem to escape apoptotic destruction (Schmeid *et al.*, 1993; Bauer *et al.*, 1995). This finding implies either that brain-infiltrating T cells are exposed to something in the CNS environment that induces apoptosis or that they no longer have access to survival factors that inhibit their apoptotic cell death.

Any number of potential mechanisms to explain this finding may be invoked: aberrant antigen presentation without costimulatory signals that are necessary for optimal T cell activation; downregulation of T cell survival factors such as IL-2; and/or local exposure to proapoptotic mediators such as TGF-β, Fas ligand (FasL), and possibly others.

In the context of CNS viral infection, T cell apoptosis in the brain has been investigated during acute Sindbis virus encephalitis. BALB/c and SJL mice initiate comparable CNS mononuclear cell inflammatory responses following this infection, which lead to viral clearance from the brain with identical kinetics (Mokhtarian *et al.*, 1989; Irani, 1998; Rowell and Griffin, 1999). Yet despite these equivalent antiviral host immune responses, CNS inflammation is more intense, persists much longer, and may actually contribute to paralysis in SJL mice, while cellular infiltrates disappear in a predictable manner from the brains of BALB/c mice that remain asymptomatic throughout disease (Mokhtarian *et al.*, 1989; Irani, 1998; Rowell and Griffin, 1999). When T cells isolated from the brain parenchyma of these two hosts at late stages of infection are examined *ex vivo*, those from BALB/c mice are more frequently apoptotic, suggesting that this process contributes to the rate at which local immune responses are terminated (Fig. 5). Brain-derived T cells from BALB/c mice also express significantly higher intracellular levels of the proapoptotic mediator, Bax (Fig. 6), and peripheral T cells from these hosts are much more susceptible to apoptosis induced *in vitro*, following incubation with brain tissue extracts (Fig. 7). These data imply that substances present in the brain may directly induce T cell apoptosis, perhaps through the induction of proapoptotic mediators such as Bax, and that T cells from different strains of mice show variable susceptibilities to this apoptotic stimulus. Indeed, T cells from SJL mice, a host that is uniquely susceptible to immune-mediated CNS disease, show a particular resistance to local apoptosis in the brain. While it remains to be proven whether and how the brain directly controls the survival of infiltrating lymphocytes in such a tissue-specific manner, it is attractive to speculate that such a mechanism exists and that a defect in this process may contribute to CNS autoimmunity.

Other host factors also contribute to the survival of brain-infiltrating T cells during CNS viral infection. For example, following the inoculation of JHMV into susceptible mice, virus-specific CD8+ CTLs within the brain have an absolute requirement for CD4+ T cells that primarily serve to support their survival in this tissue compartment (Stohlman *et al.*, 1998). Thus, lymphocyte apoptosis in the brains of JHMV-infected mice that have been depleted of CD4+ T cells is significantly increased (Stohlman *et al.*, 1998). While specific survival factors

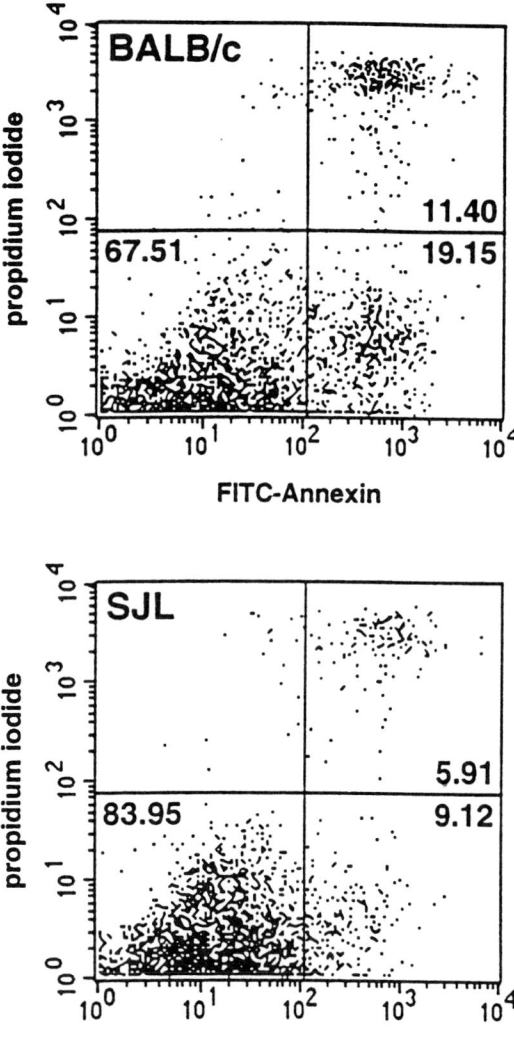

FIG 5. Flow cytometric analysis shows that a greater percentage of cells isolated from the brains of Sindbis virus-infected BALB/c mice are preapoptotic (annexin V-positive, propidium iodide-negative) or nonviable (annexin V-positive, propidium iodide-positive) as compared to cells recovered directly from the brains of infected SJL mice. Adapted with permission from Irani, 1998.

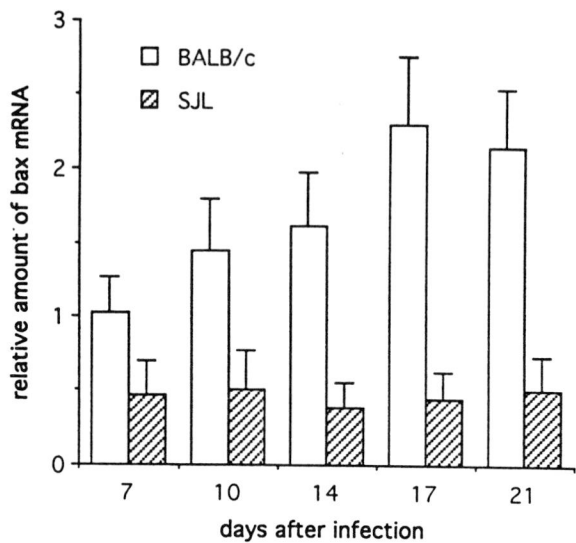

FIG 6. By RT-PCR, Bax-specific mRNAs are more abundant in T cells isolated from the brains of Sindbis virus-infected BALB/c mice as compared to T cells derived from the brains of infected SJL mice. Bax-specific transcripts were quantitated by RT-PCR relative to the constitutively expressed glyceraldehyde-3-phosphate dehydrogenase gene. Adapted with permission from Irani, 1998.

generated directly or indirectly through the actions of CD4+ T cells that prevent local CD8+ T cell apoptosis in the brain were not identified, their existence is strongly suggested. Such factors would be assumed to work in direct opposition to any proapoptotic influence exerted by the CNS microenvironment; it seems likely that the overall process that controls T cell survival within the CNS involves weighing the relative contributions of multiple pro- and anti-death influences. Such influences may in fact be integrally entwined as the brain may regulate the survival of infiltrating T cells, not by directly inducing their death, but by controlling how T cells produce their own survival factors (see Section III, D). Understanding these events will be crucial to improving our ability to gain pharmacologic control over pathologic, T cell–driven inflammation in the brain.

IV. CONCLUDING REMARKS

Studies in a number of experimental systems have shown that T cells can serve many different functions as part of the host immune

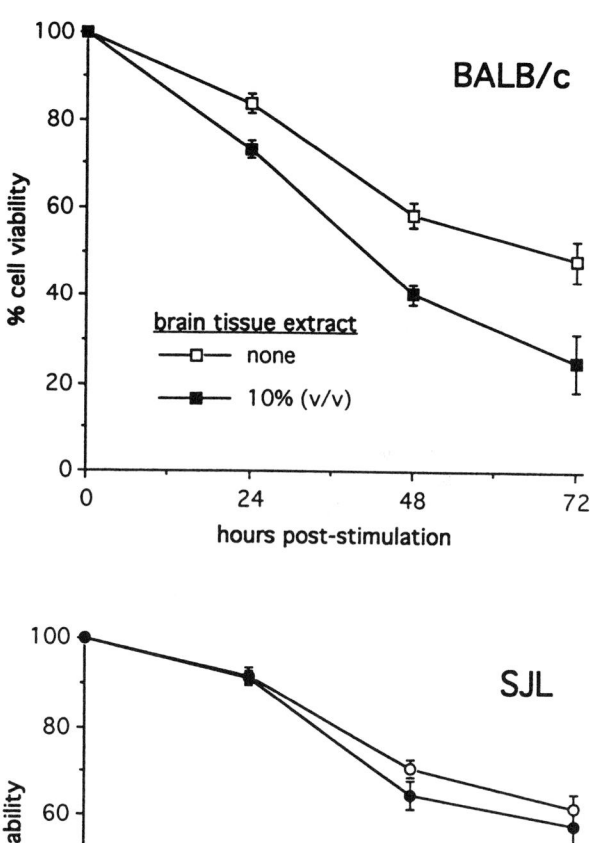

FIG 7. Tissue extracts prepared from the brains of uninfected BALB/c or SJL mice accelerate the activation-induced cell death of mitogen-stimulated peripheral T cells from BALB/c mice to a greater degree than T cells from SJL mice. Cell viability was measured in these experiments by using trypan blue exclusion. Adapted with permission from Irani, 1998.

response to CNS viral infection. Of particular interest here is how individual T cell responses may be actively regulated within the CNS during these diseases. We propose that such immunoregulatory events occur in CNS viral infection and we have focused on the control of local T cell responses within the brain, to support the idea that such regulation does occur. Our current understanding of the mechanisms that underlie such regulatory events is limited, but further study of these processes may provide broader insight into how pathologic T cell responses within the CNS, following a variety of stimuli, may or may not occur. An improved understanding of these regulatory events should be of interest to the entire neuroimmunology community.

REFERENCES

Ando, K., Guidotti, L. G., Wirth, S., Ishikawa, T., Missale, G., Moriyama, T., Schreiber, R. D., Schlicht, H.-J., Huang, S., and Chisari, F. V. (1994). Class I-restricted cytotoxic T lymphocytes are directly cytopathic for their targets in vivo. *J. Immunol.* **152:**3245–3252.

Barac-Latas, V., Wege H., and Lassmann H. (1995). Apoptosis of T lymphocytes in coronavirus-induced encephalomyelitis. *Reg. Immunol.* **6:**355–357.

Bauer, J., Wekerle, H., and Lassmann, H. (1995). Apoptosis in brain-specific autoimmune disease. *Curr. Opinion Immunol.* **7:**839–843.

Bonetti, B., Pohl, J., Gao, Y.-L., and Raine, C. S. (1997). Cell death during autoimmune demyelination: effector but not target cells are eliminated by apoptosis. *J. Immunol.* **159:**5733–5741.

Buchmeier, M. J., Welsh, R. M., Dutko, F. J., and Oldstone, M. B. A. (1980). The virology and immunobiology of lymphocytic choriomeningitis virus infection. *Adv. Immunol.* **30:**275–331.

Castro, R. F., Evans, G. D., Jaszewski, A., and Perlman, S. (1994). Coronavirus-induced demyelination occurs in the presence of virus-specific cytotoxic T cells. *Virology* **200:**733–743.

Cole, G. A., Nathanson, N., and Prendergast, R. A. (1972). Requirements for θ-bearing cells: lymphocytic choriomeningitis virus induced central nervous system disease. *Nature (London)* **238:**335–337.

Cserr, H. F., and Knopf, P. M. (1992). Cervical lymphatics, the blood–brain barrier, and the immunoreactivity of the brain: a new view. *Immunol. Today* **13:**507–512.

Dharakul, T., Rott, L., and Greenberg, H. B. (1990). Recovery from chronic rotavirus infection in mice with severe combined immunodeficiency: virus clearance mediated by adoptive transfer of immune CD8+ T lymphocytes. *J. Virol.* **64:**4375–4382.

Doherty, P. C., Allan, J. E., and Ceredig, R. (1988). Contributions of host and donor T cells to the inflammatory process in murine lymphocytic choriomeningitis. *Cell. Immunol.* **116:**475–481.

Doherty, P. C., Allan, J. E., and Clark, I. A. (1989). Tumor necrosis factor inhibits the development of viral meningitis or induces rapid death depending on the severity of inflammation at the time of administration. *J. Immunol.* **142:**3576–3582.

Gorrell, M. D., Lemm, J. A., Rice, C. M., and Griffin, D. E. (1997). Immunization with nonstructural proteins promotes functional recovery of alphavirus-infected neurons. *J. Virol.* **71:**3415–3419.

Griffin, D. E., Levine, B., Tyor, W. R., and Irani, D. N. (1992). The immune response in viral encephalitis. *Sem. Immunol.* **4:**111–119.

Harling-Berg, C., Knopf, P. M., Merriam, J., and Cserr, H. F. (1989). Role of cervical lymph nodes in the systemic humoral immune response to human serum albumin microinfused into rat cerebrospinal fluid. *J. Neuroimmunol.* **25:**185–193.

Hawke, S., Stevenson, P. G., Freeman, S., and Bangham, C. R. M. (1998). Long-term persistence of activated cytotoxic T lymphocytes after viral infection of the central nervous system. *J. Exp. Med.* **187:**1575–1582.

Hirsch, R. L., and Griffin, D. E. (1979). The pathogenesis of Sindbis virus infection in athymic nude mice. *J. Immunol.* **123:**1215–1219.

Irani, D. N., and Griffin, D. E. (1991). Isolation of brain parenchymal lymphocytes for flow cytometric analysis: application to acute viral encephalitis. *J. Immunol. Meth.* **39:**223–232.

Irani, D. N., and Griffin, D. E. (1996). Regulation of lymphocyte homing into the brain during viral encephalitis at various stages of infection. *J. Immunol.* **156:**3850–3857.

Irani, D. N., Lin, K.-I., and Griffin, D. E. (1996). Brain-derived gangliosides regulate the cytokine production and proliferation of activated T cells. *J. Immunol.* **157:**4333–4340.

Irani, D. N., Lin, K.-I., and Griffin, D. E. (1997). Regulation of brain-derived T cells during acute central nervous system inflammation. *J. Immunol.* **158:**2318–2326.

Irani, D. N. (1998). The susceptibility of mice to immune-mediated neurologic disease correlates with the degree to which their lymphocytes resist the effects of brain-derived gangliosides. *J. Immunol.* **161:**2746–2752.

Jacoby, R. O., Bhatt, P. N., and Schwartz, A. (1980). Protection of mice from lethal flavivirus encephalitis by adoptive transfer of splenic cells from donors infected with live virus. *J. Infect. Dis.* **141:**617–627.

Joly, E., Mucke, L., and Oldstone, M. B. A. (1991). Viral persistence in neurons explained by lack of major histocompatibility class I expression. *Science* **253:**1283–1285.

Kündig, T. M., Hengartner H., and Zinkernagel, R. M. (1993). T cell-dependent IFN-γ exerts an antiviral effect in the central nervous system but not in peripheral solid organs. *J. Immunol.* **150:**2316–2321.

Levine, B., Hardwick, J. M., Trapp, B. D., Crawford, T. O., Bollinger, R. C., and Griffin, D. E. (1991). Antibody-mediated clearance of alphavirus infection from neurons. *Science* **254:**856–860.

Lewis, J. Wesselingh, S. L., Griffin, D. E., and Hardwick, J. M. (1996). Alphavirus-induced apoptosis in mouse brains correlates with neurovirulence. *J. Virol.* **70:**1828–1835.

Lin, M. T., Stohlman, S. A., and Hinton, D. R. (1997). Mouse hepatitis virus is cleared from the central nervous system of mice lacking perforin-mediated cytolysis. *J. Virol.* **71:**383–389.

Lipton, H. L., and Dal Canto, M. C. (1976). Theiler's virus-induced demyelination: prevention by immunosuppression. *Science* **192:**62–64.

Lynch, F., Doherty, P. C., and Ceredig, R. (1989). Phenotypic and functional analysis of the cellular response in regional lymphoid tissue during an acute virus infection. *J. Immunol.* **142:**3592–3598.

Marten, N. W., Stohlman, S. A., Smith-Begolka, W., Miller, S. D., Dimacali, E., Yao, Q., Stohl, S., Goverman, J., and Bergmann, C. C. (1999). Selection of CD8+ T cells with highly focused specificity during viral persistence in the central nervous system. *J. Immunol.* **162:**3905–3914.

Mauerhoff, R., Pujol-Bonell, R., Mirakian, R., and Bottazzo, G. F. (1980). Differential expression and regulation of major histocompatibility complex (MHC) products in neural and glial cells of human fetal brain. *J. Neuroimmunol.* **18:**271–289.

Miller, S. D., VanderLugt, C. L., Begolka, W. S., Pao, W., Yauch, R. L., Neville, K. L., Katz-Levy, Y., Carrizosa, A., and Kim, B. S. (1997). Persistent infection with Theiler's virus leads to CNS autoimmunity via epitope spreading. *Nature Med.* **3:**1133–1136.

Mims, C. A., and Blanden, R. V. (1972). Antiviral action of immune lymphocytes in mice infected with lymphocytic choriomeningitis virus. *Infect. Immunity* **6:**695–698.

Moench, T. R., and Griffin, D. E. (1984). Immunocytochemical identification and quantitation of mononuclear cells in cerebrospinal fluid, meninges and brain during acute viral encephalitis. *J. Exp. Med.* **159:**77–88.

Mokhtarian, F., Grob, D., and Griffin, D. E. (1989). Role of the immune response in Sindbis virus-induced paralysis of SJL/J mice. *J. Immunol.* **143:**633–637.

Munoy, J. L., McCarthy, C. A., Clark, M. E., and Hall, C. B. (1991). Respiratory syncytial virus infection in C57B1.6 mice: clearance of virus from the lungs with virus-specific cytotoxic cells. *J. Virol.* **65:**4494–4497.

Musette, P., Bureau, J.-F., Gachelin, G., Kourilsky, P., and Brahic, M. (1995). T lymphocyte repertoire in Theiler's virus encephalomyelitis: the nonspecific infiltration of the central nervous system of infected SJL/J mice is associated with a selective local T cell expansion. *Eur. J. Immunol.* **25:**1589–1593.

Oldstone, M. B. A., Blount, P., Southern, P. J., and Lampert, P. W. (1986). Cytoimmunotherapy for persistent virus infection reveals a unique clearance pattern from the central nervous system. *Nature (London)* **321:**239–243.

Pope, J. G., Karpus, W. J., VanderLugt, C., and Miller, S. D. (1996). Flow cytometric and functional analysis of central nervous system-infiltrating cells in SJL/J mice with Theiler's virus-induced demyelinating disease. Evidence for a CD4+ T cell-mediated pathology. *J. Immunol.* **156:**4050–4058.

Richt, J. A., Stitz, L., Wekerle, H., and Rott, R. (1989). Borna disease, a progressive meningoencephalomyelitis as a model for CD4+ T cell-mediated immunopathology in the brain. *J. Exp. Med.* **170:**1045–1050.

Rowell, J. F., and Griffin, D. E. (1999). The inflammatory response to nonfatal Sindbis virus infection of the nervous system is more severe in SJL than in BALB/c mice and is associated with low levels of IL-4 mRNA and high levels of IL-10-producing CD4+ T cells. *J. Immunol.* **162:**1624–1632.

Schmeid, M., Breitschopf, H., Gold, R., Zischler, H., Rothe, G., Wekerle, H., and Lassmann, H. (1993). Apoptosis of T lymphocytes in experimental autoimmune encephalomyelitis: evidence for programmed cell death as a mechanism to control inflammation in the brain. *Am. J. Path.* **143:**446–452.

Sethi, K. K., Omata, Y., and Schneweiss, K. E. (1983). Protection of mice from fatal herpes simplex virus type I infection by adoptive transfer of cloned virus-specific and H-2-restricted cytotoxic T lymphocytes. *J. Gen. Virol.* **64:**443–452.

Stanley, J., Cooper, S. J., and Griffin, D. E. (1986). Monoclonal antibody cure and prophylaxis of lethal Sindbis virus encephalitis in mice. *J. Virol.* **58:**107–115.

Stohlman, S. A., Matsushima, G. K., Casteel, N., and Weiner, L. P. (1986). In vivo effects of coronavirus-specific T cell clones: DTH induced cells prevent a lethal infection but do not inhibit virus replication. *J. Immunol.* **136:**3052–3056.

Stohlman, S. A., Kyuwa, S., Polo, J. M., Brady, D., Lai, M. M., and Bergmann, C. C. (1993). Characterization of mouse hepatitis virus-specific cytotoxic T cells derived from the central nervous system of mice infected with the JHM strain. *J. Virol.* **67:**7050–7059.

Stohlman, S. A., Bergmann, C. C., van der Veen, R. C., and Hinton, D. R. (1995). Mouse hepatitis virus-specific cytotoxic T lymphocytes protect from lethal infection without eliminating virus from the central nervous system. *J. Virol.* **69:**684–694.

Stohlman, S. A., Bergmann, C. C., Lin, M. T., Cua, D. J., and Hinton, D. R. (1998). CTL effector function within the central nervous system requires CD4+ T cells. *J. Immunol.* **160:**2896–2904.

Sussman, M. A., Shubin, R. A., Kyuwa, S., and Stohlman, S. A. (1989). T cell-mediated clearance of mouse hepatitis virus strain JHM from the central nervous system. *J. Virol.* **63:**3051–3056.

Tishon, A., Eddleston, J., de la Torre, J. C., and Oldstone, M. B. A. (1993). Cytotoxic T lymphocytes cleanse viral gene products from individually infected neurons and lymphocytes in mice persistently infected with lymphocytic choriomeningitis virus. *Virology* **197:**463–467.

Tyor, W. R., Moench, T. R., and Griffin, D. E. (1989). Characterization of the local and systemic B cell response of normal and athymic nude mice with Sindbis virus encephalitis. *J. Neuroimmunol.* **24:**207–215.

Tyor, W. R., Wesselingh, S. L., Levine, B., and Griffin, D. E. (1992). Long-term intraparenchymal Ig secretion after acute viral encephalitis in mice. *J. Immunol.* **149:**4016–4023.

Tyor, W. R., and Griffin, D. E. (1993). Virus specificity and isotype expression of intraparenchymal antibody-secreting cells during Sindbis virus encephalitis in mice. *J. Neuroimmunol.* **177:**475–484.

Wesselingh, S. L., Levine, B., Fox, R. J., Choi, S., and Griffin, D. E. (1994). Intracerebral cytokine mRNA expression during fatal and nonfatal alphavirus encephalitis suggests a predominant type 2 T cell response. *J. Immunol.* **152:**1289–1297.

Williamson, J. S. P., and Stohlman, S. A. (1990). Effective clearance of mouse hepatitis virus from the central nervous system requires both CD4+ and CD8+ T cells. *J. Virol.* **64:**4589–4592.

Williamson, J. S. P., Sykes, K. C., and Stohlman, S. A. (1991). Characterization of brain-infiltrating mononuclear cells during infection with mouse hepatitis virus strain, JHM. *J. Neuroimmunol.* **32:**199–207.

Wong, G. H. W., Bartlett, P. F., Clark-Lewis, I., Battye, F., and Schrader, J. W. (1984). Inducible expression of H-2 and Ia antigens on brain cells. *Nature (London)* **310:**688–691.

Yap, K. L., Ada, G. L., and McKenzie, I. F. C. (1978). Transfer of specific cytotoxic T lymphocytes protects mice inoculated with influenza virus. *Nature (London)* **273:**238–242.

Young, D. F., Randall, R. E., Hoyle, J. A., and Souberbielle, B. E. (1990). Clearance of persistent paramyxovirus infection is mediated by cellular immune responses but not by serum neutralizing antibody. *J. Virol.* **64:**5403–5411.

Zinkernagel, R. M., and Welsh, R. M. (1976). H-2 compatibility requirement for virus-specific T cell-mediated effector functions in vivo. I. Specificity of T cells conferring antiviral protection against lymphocytic choriomeningitis virus is associated with H-2K and H-2D. *J. Immunol.* **117:**1495–1502.

ADVANCES IN VIRUS RESEARCH, VOL 56

VIRUS-INDUCED AUTOIMMUNITY: EPITOPE SPREADING TO MYELIN AUTOEPITOPES IN THEILER'S VIRUS INFECTION OF THE CENTRAL NERVOUS SYSTEM

Stephen D. Miller, Yael Katz-Levy, Katherine L. Neville, and Carol L. Vanderlugt

Department of Microbiology-Immunology and Interdepartmental Immunobiology Center
Northwestern University Medical School
Chicago, Illinois 60611

I. Introduction

The mechanisms underlying the initiation and progression of autoimmune diseases are not well understood. Clinical and epidemiological evidence indicate that viral infections may be important in induction of autoimmunity. Postulated mechanisms by which virus infections may lead to autoimmune disease include *molecular mimicry*—activation of autoreactive T cells secondary to an encounter with a pathogen that is shared or cross-reactive with self-antigens (Fujinami and Oldstone, 1985); *epitope spreading*—de novo activation of autoreactive T cells by release of sequestered antigens secondary to tissue destruction that is mediated by virus-specific T cells (Vanderlugt *et al.*, 1998; McRae *et al.*, 1995; Miller and Karpus, 1994); and *viral superantigens*—nonspecific activation of autoreactive T cells via stimulation of T cells bearing particular Vβ receptors (Scherer *et al.*, 1993).

199

Evidence for a viral etiology is particularly extensive for multiple sclerosis (MS) (Challoner *et al.*, 1995; Kurtzke, 1993), a human CD4[+] T cell-mediated demyelinating disease associated with antimyelin responses (Bernard and de Rosbo, 1991; Ota *et al.*, 1990). MS is an immune-mediated disease of the central nervous system (CNS) characterized by perivascular CD4[+] T cell and mononuclear cell infiltration, with subsequent primary demyelination of axonal tracks, leading to progressive paralysis (Wekerle, 1991). Despite decades of intensive research, the inducing antigen(s) and precise immunologic mechanisms involved in the induction and chronic course of MS are still poorly understood and there are limited therapeutic options for managing this disease. MS is generally considered to involve an autoimmune pathology. Regardless of many reports demonstrating elevated humoral and/or T cell-mediated responses to the major myelin proteins—myelin basic protein (MBP) and/or proteolipid protein (PLP)— in MS patients (Bernard and de Rosbo, 1991; Sun *et al.*, 1991; Allegretta *et al.*, 1990; Link *et al.*, 1990; Ota *et al.*, 1990), a clear-cut cause–effect relationship between myelin reactivity and disease pathology has yet to be demonstrated. In addition to a genetic component (Ebers *et al.*, 1995), epidemiological studies provide strong circumstantial evidence for an environmental trigger, most likely viral, in the induction of MS (Waksman, 1995; Kurtzke, 1993), with the caveat that no virus has been consistently isolated from MS lesions. One of the latest candidate agents is human herpesvirus type 6 (HHV6), a common childhood infection. HHV6 antigens have been demonstrated in MS plaques, but not in tissues from patients with other neurological diseases (Challoner *et al.*, 1995), but definitive association of HHV6 infection with MS awaits independent confirmation. It is thus highly probable that CNS pathology in certain forms of MS may be triggered by activation of myelin-reactive T cells, secondary to a virus infection, by molecular mimicry, or by epitope spreading.

Our laboratory has utilized both the relapsing experimental autoimmune encephalomyelitis (R-EAE) model of MS and the Theiler's murine encephalomyelitis virus (TMEV)-induced demyelinating disease (IDD) model of MS in SJL mice to examine mechanisms of autoimmune disease initiation and progression. Our recent studies have examined the role of epitope spreading as a major pathologic contributor to chronic disease progression in both R-EAE and TMEV-IDD (as summarized below). The TMEV-IDD model has been particularly useful in dissecting the relative contributions of antivirus and antiself responses to the chronic demyelinating process. In addition, this model provides an excellent system for understanding the immunologic events that regulate the induction of proinflammatory, autoreactive CD4+ T cells following infection with the wild-type BeAn strain of TMEV. This chapter will concentrate on a discussion of the mechanisms leading to autoimmunity in TMEV-IDD, which appears to be induced by epitope spreading.

II. RELEVANCE OF MURINE TMEV-INDUCED DEMYELINATING DISEASE TO HUMAN MULTIPLE SCLEROSIS

Although no animal model is an exact replica of MS, which itself is a highly variable disease, both virus-induced and autoimmune models exist. Murine TMEV infection appears to be the most promising of the virus-induced experimental animal models for the following reasons: (a) chronic pathological involvement, accompanied by inflammation, is limited to the white matter of the CNS; (b) myelin breakdown clearly leads to clinical disease, including gait spasticity, inducible extensor spasms of the limbs, and urinary incontinence; (c) TMEV persists only in the CNS at barely detectable levels for the life of the mouse; (d) demyelination, as is the case in human MS (Paterson and Day, 1981), appears to be mediated primarily by an inflammatory CD4+ Th1 cell-mediated pathologic process; and (e) as the mouse is the natural host for TMEV, and as CNS involvement occurs spontaneously in colony-reared mice, conclusions derived from experimental findings are more likely to be directly relevant to human disease.

III. TMEV INFECTION AS MODEL OF PERSISTENT VIRUS-INDUCED, CD4+ T CELL-MEDIATED DEMYELINATION

TMEV is a picornavirus and an enteric pathogen of mice and rats (Lipton and Friedmann, 1980). Certain strains of TMEV—e.g., DA—produce a biphasic disease following intracerebral (IC) inoculation (Lehrich et al., 1976; Dal Canto and Lipton, 1975). Early disease (1–30 days postinoculation) is characterized by virus replication in the CNS gray matter, particularly in motor neurons (2, 11, 12), which results in flaccid paralysis (poliomyelitis). Microglial proliferation and neuronal necrosis are predominant. Although a limited viremia occurs, virus replication in extraneural organs is limited (Lipton, 1975; Theiler, 1937) and a high percentage of animals survive and recover from paralysis. Mice surviving the early phase of infection with the DA strain of TMEV, or those infected with the tissue culture-adapted BeAn strain used in our laboratory (which causes only limited early gray matter disease and thus a better model of demyelination), develop a persistent CNS infection accompanied by a chronic, progressive, inflammatory demyelinating disease clinically recognized by problems in gait spasticity arising about 30–35 days postinfection (Lehrich *et al.*, 1976; Dal Canto and Lipton, 1975; Lipton, 1975). TMEV persists for the lifetime of the host via continuous, low-level virus replication (Lipton, 1980). Histopathologically, chronically infected animals exhibit mononuclear cell infiltrates in the meninges and white matter as well as areas of primary focal demyelination (Lehrich *et al.*, 1976; Dal Canto and Lipton, 1975). Myelin destruction is accomplished by the stripping of myelin lamellae by mononuclear processes and vesiculation of myelin independent of stripping (Dal Canto and Lipton, 1975). Pathogenesis during the early and late phases of the disease differs. Detection of DA virus protein and mRNA in degenerating neurons suggests a cytocidal infection in the early acute gray matter lesions (Brahic *et al.*, 1981; Dal Canto and Lipton, 1975; Lipton, 1975). In contrast, the demyelinating phase does not appear to result from cytocidal infection of oligodendrocytes in immunocompetent hosts (Clatch *et al.*, 1985; Dal Canto and Lipton, 1975). Rather, demyelination is immune-mediated and appears to be initiated by CD4+ T cell–mediated immune responses against TMEV epitopes, causing *bystander* damage to myelinated axons via the action of activated microglia/macrophages (Clatch *et al.*, 1985).

Multiple lines of evidence support a major pathologic role for virus-specific CD4+ T cell responses in the TMEV-induced demyelinating disease process. Nonspecific immunosuppression with cyclophosphamide or antithymocyte serum (Roos *et al.*, 1982; Lipton and Dal Canto, 1977; Lipton and Dal Canto, 1976) and CD4+, but not CD8+, T cell depletion (Gerety *et al.*, 1994a), after resolution of the initial viremia, results in a dramatic reduction in mononuclear cell infiltration and prevention of demyelination in surviving mice. Virus antigens are found primarily in macrophages within demyelinating lesions and to a lesser extent in astrocytes and neurons, but are not consistently found in oligodendrocytes (Clatch *et al.*, 1990; Dal Canto and Lipton, 1982), although it has been reported that DA virus antigen can be found in oligodendrocytes in late chronic disease (Rodriguez *et al.*, 1983). Thus, myelin is destroyed even though myelin-producing cells are not damaged to any significant extent, at least early in the disease process (Dal Canto and Lipton, 1980; Dal Canto and Lipton, 1979). TMEV-induced demyelination is under multigenic control (Lipton and Melvold, 1984). Identified susceptibility loci include: the H-2D region (Clatch *et al.*, 1987b; Clatch *et al.*, 1985; Rodriguez and David, 1985), an MHC class I gene, the effect of which may relate to a CD8$^+$ regulatory cell that prevents demyelination in resistant strains (Nicholson *et al.*, 1996; Nicholson *et al.*, 1994; Olsberg *et al.*, 1993; Pelka *et al.*, 1993); Tmevd-1 on chromosome 6 near the genes encoding the β chain of the T cell receptor (Melvold *et al.*, 1987); and Tmevd-2 on chromosome 3 near the *Car-2* locus (Melvold *et al.*, 1990).

Susceptibility to TMEV-IDD correlates temporally with the development of chronic high levels of class II-restricted, TMEV-specific DTH (Clatch *et al.*, 1987a; Clatch *et al.*, 1986; Clatch *et al.*, 1985) and is characterized by the predominant production of Th1-like cytokines (Begolka and Miller, 1998; Karpus *et al.*, 1994; Peterson *et al.*, 1993) and antivirus antibody of the IgG$_{2a}$ subclass (Peterson *et al.*, 1992). Transfer of TMEV-specific CD4$^+$ T cell blasts (either polyclonal or long-term T cell lines) results in an increase in disease incidence and severity in syngeneic recipients infected with a suboptimal dose of TMEV (Gerety *et al.*, 1994a). *In vivo* depletion of CD4$^+$ T cells, but not CD8$^+$ T cells, in SJL/J mice infected with the BeAn strain of TMEV, results in a decreased incidence of demyelinating disease as well as a slower onset of disease in those mice that eventually become clinically affected (Gerety *et al.*, 1994a; Borrow *et al.*, 1992; Welsh *et al.*, 1987). In addition, we have shown that essentially all of the activated (i.e., high-affinity IL-2R-bearing) CNS-infiltrating T cells are CD4$^+$ (Pope *et al.*, 1996; Clatch *et al.*, 1990), and that clinical and pathological disease are as severe in β$_2$-microglobulin-deficient SJL mice lacking CD8$^+$ T cells, or even more severe than in their +/− littermates (data are from a manuscript in preparation).

Polyclonal T cell responses in TMEV-infected and immunized mice cross-react to a significant extent with the closely related encephalomyocarditis virus (EMCV), but only marginally with poliovirus, a more distantly related picornavirus (Miller *et al.*, 1987). In susceptible SJL/J mice, VP2 contains the immunodominant T cell determinant(s), as 80–90% of the DTH response is directed against this virion protein (Gerety *et al.*, 1991). More recent studies have mapped the SJL/J T cell epitope to a 13 amino acid peptide on VP2 (VP2$_{70-86}$) (Gerety *et al.*, 1994b). Systemic transfer of an SJL/J-derived, VP2$_{70-86}$-specific Th1 cell line enhances the incidence and severity of disease in suboptimally infected mice demonstrating the immunopathologic potential of VP2-specific CD4$^+$ T cells (Gerety *et al.*, 1994a). The Vβ gene repertoire used in recognition of this epitope appears to be diverse (these are unpublished results). Less dominant CD4$^+$ T cell epitopes have been identified on VP1 (VP1$_{233-250}$) (Yauch and Kim, 1994) and VP3 (VP3$_{24-337}$) (Yauch *et al.*, 1995).

IV. VIRUS-SPECIFIC CD4$^+$ T CELL RESPONSES INITIATE DISEASE

As mentioned previously, the Theiler's virus-induced demyelinating disease model is ideal for addressing questions relating to the potential role of myelin-specific autoreactive T cell responses to the chronic demyelinating process. We therefore have employed several approaches for determining the role of autoreactivity in disease initiation and progression. These studies have shown that neuroantigen-specific T cells do not appear to play a role in disease initiation. Time-course studies comparing the development of T cell responses to both virus and myelin epitopes have shown that autoreactivity to myelin epitopes is not detected prior to disease onset (30–35 days postinfection) (Miller et al., 1987; Barbano and Dal Canto, 1984), while responses to TMEV epitopes are clearly demonstrable by 5–7 days postinfection (Clatch et al., 1986). In support of lack of myelin-specific autoreactivity at disease initiation, induction of peripheral tolerance to mouse spinal cord homogenate (MSCH, a heterogeneous mixture of multiple neuroantigens) does not affect the clinical onset of demyelination and the accompanying virus-specific T cell and antibody responses in TMEV-infected SJL/J mice (Miller et al., 1990). However, this tolerogenic regimen is extremely effective in preventing clinical and histologic signs of MSCH-induced relapsing EAE (R-EAE) and the accompanying neuroantigen-specific DTH responses (Miller et al., 1990; Kennedy et al., 1988). In contrast, tolerance to intact TMEV virions coupled to syngeneic splenocytes, which specifically anergizes virus-specific Th1 responses (Karpus et al., 1994; Peterson et al., 1993), results in a dramatic reduction in the incidence and severity of demyelinating lesions and clinical disease in SJL/J mice subsequently infected with TMEV (Karpus et al., 1995). Collectively, these results strongly support the hypothesis that TMEV-IDD is initiated by virus-specific T cells that target viral epitopes presented by CNS-resident APCs that harbor persistent virus for many months following infection (Pope et al., 1996; Clatch et al., 1990; Miller and Gerety, 1990). Chemokines and proinflammatory cytokines produced by these virus-specific T cells (Hoffman et al., 1999; Begolka et al., 1998) then lead to the attraction and activation of microglia and mononuclear inflammatory cells that mediate the initial myelin destruction.

V. MYELIN-SPECIFIC CD4$^+$ T CELL RESPONSES: PATHOLOGIC ROLE IN CHRONIC THEILER'S VIRUS-INDUCED DEMYELINATING DISEASE

Our laboratory and others have demonstrated that changes occur in class II restriction and antigen specificity of neuroantigen-specific T cell responses during the course of chronic EAE in the SJL mouse (Perry *et al.*, 1991; Tan *et al.*, 1991; McCarron *et al.*, 1990; Perry and Barzaga, 1987). We have employed the R-EAE model of CD4+ T cell-mediated demyelination to study the functional significance of epitope spreading, i.e., induction of T cell responses to endogenous myelin epitopes, to the relapsing pathogenesis of chronic R-EAE. R-EAE in the SJL mouse is a Th1-mediated autoimmune demyelinating disease, useful in dissecting the role of epitope spreading to chronic tissue damage because disease can be induced in a peptide-specific manner, and multiple encephalitogenic epitopes and their relative dominance on a variety of myelin proteins, including MBP, PLP, and myelin oligodendrocyte glycoprotein (MOG), have been well defined. Epitope spreading resulting from autoimmune tissue destruction plays an important role in the progression of ongoing disease in R-EAE. We have shown a role for both intermolecular and intramolecular epitope spreading in two different models of myelin peptide-induced R-EAE (McRae *et al.*, 1995; Miller *et al.*, 1995a) (see Fig. 1). Intramolecular epitope spreading was demonstrated by showing that T cells reactive to a secondary encephalitogenic PLP epitope (PLP178–191) are found in the spleens and CNS of mice in which acute EAE was induced by T cells reactive with the dominant, non-cross-reactive   PLP139–151   epitope.   PLP178–191-specific responses are activated as a result of, and correlate with, the degree of acute CNS tissue damage because they do not arise in mice tolerized with the inducing PLP139–151 epitope prior to acute disease. Most importantly, these T cells possess encephalitogenic potential as PLP178–191-specific T cells isolated from mice with PLP139–151-induced R-EAE can initiate disease on secondary serial adoptive transfer to naive recipients, and mice tolerized with PLP178–191 during remission from the acute stage of PLP139–151-induced R-EAE are

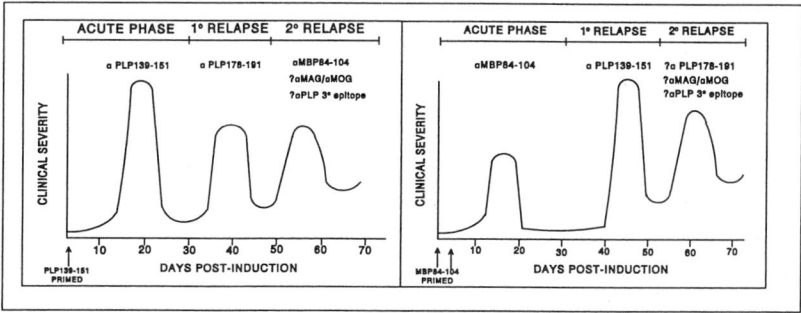

FIG 1. Pattern of intramolecular and intermolecular epitope spreading in PLP139–151-and MBP84–104-induced R-EAE in SJL mice.

protected from disease relapses (Vanderlugt *et al.*, 1999; McRae *et al.*, 1995). More recently, we have shown that responses to the immunodominant epitopes on MOG (MOG92–106) and MBP (MBP84–104) develop following the primary relapse, indicating intermolecular epitope spreading. Intermolecular epitope spreading was also demonstrated by showing the development of T cell responses to PLP139–151 in both the periphery and in the CNS following acute disease in mice in which R-EAE was initiated by the transfer of T cells specific for the non-cross-reactive MBP84–104 determinant (Vanderlugt *et al.*, 1999; McRae *et al.*, 1995). Thus, the spread of immune responses to the noninducing epitopes, presented endogenously during ongoing tissue destruction, appears to be predominantly responsible for chronic disease progression in this relevant MS model. Epitope spreading has also been demonstrated to play an important pathologic role in the pathogenesis of spontaneous autoimmune diabetes in the nonobese diabetic (NOD) mouse (Kaufman *et al.*, 1993; Tisch *et al.*, 1993). More relevant to the potential role of epitope spreading in human autoimmune disease, a recent report suggests that a similar pattern of epitope focusing and spreading may occur during the transition from isolated monosymptomatic demyelinating syndromes (a group of distinct clinical disorders with variable rates of progression to MS) to clinically defined MS (Tuohy *et al.*, 1997).

Considering the importance of epitope spreading in disease progression and relapses in R-EAE, we examined TMEV-infected SJL mice for the possible appearance of myelin-specific autoreactive T cells during the later stages of TMEV-IDD. Clinical disease begins approximately 30 days following infection and displays a chronic-progressive course, with 100% of the animals being affected by 40–50 days postinfection. T cell proliferative responses to UV-inactivated TMEV were demonstrable in the spleens of infected mice at both day 33 postinfection, concomitant with the onset of clinical signs, and at day 87. In contrast, T cell proliferation to the major encephalitogenic epitopes on PLP (PLP139–151 and PLP178–191) (Kennedy *et al.*, 1990) and on MBP (MBP84–104) were not demonstrable in spleen, cervical, or pooled peripheral lymph nodes at day 33, but a response to PLP139–151 was found in all lymphoid compartments at day 87 postinfection. A temporal analysis of DTH reactivity showed that approximately 50–60 days postinfection (i.e., 3–4 weeks after the onset of clinical disease), when myelin damage reaches a critical threshold, $CD4^+$ T cell–mediated DTH responses to the immunodominant PLP139–151 epitope were demonstrable. As disease progresses further, DTH responses to multiple myelin epitopes arise (Miller *et al.*, 1997). As we have recently reported in the SJL R-EAE model (Vanderlugt *et al.*, 1999), temporal

TABLE I

TEMPORAL APPEARANCE OF T CELL RESPONSES TO VIRUS AND MYELIN EPITOPES
IN THEILER'S VIRUS-INFECTED MICE

	T Cell Responses to TMEV and Myelin Epitopes							
	TMEV Epitopes		Myelin Protein Epitopes					
Days Post Infection	$VP2_{70-86}$	$VP3_{24-37}$	PLP 56–70	PLP 104–117	PLP 139–151	PLP 178–191	MOG 92–106	MBP 84–104
5–10	+[a]	—	—	—	—	—	—	—
10–14	+	+	—	—	—	—	—	—
21–35	+	+	—	—	—	—	—	—
48–64	+	+	—	—	+	—	—	—
65–95	+	+	+	—	+	—	+	—
100–150	+	+	+	—	+	+	+	—
150–200	+	+	+	—	+	+	+	—
200–300	+	+	+	—	+	+	+	+

[a] + indicates T cell response to the indicated TMEV or myelin protein epitope, as determined by delayed-type hypersensitivity and/or T cell proliferative assay.

analyses of the appearance of myelin-specific T cell responses in TMEV-infected SJL mice indicate that they arise in an ordered progression comparable to their relative encephalitogenic dominance (Table I) (Katz-Levy *et al.*, 1999b). Approximately one month following the appearance of PLP139–151-specific T cells, responses to PLP56–71 and MOG92–106 are demonstrable. This is followed by the development of T cells reactive to PLP178–191 within 5 months postinfection, and the responses to MBP84–104 arise another month later. Responses to the cryptic PLP104–117 epitopes were not seen in TMEV-infected mice. Thus, chronic inflammation initiated by virus-specific T cell responses leads to the induction and persistence of autoreactive T cells during the chronic stage of TMEV-IDD. Critically, more recent studies (a manuscript is in preparation) have indicated that the autoreactive T cells arising during chronic TMEV-IDD play an important pathologic role in chronic disease progression. The induction of peripheral tolerance to MP4 (a recombinant fusion protein comprising the 21.5 kDa isoform of human MBP and a recombinant variant of human PLP) at 45 days postinfection (just prior to the initial appearance of PLP139–151-specific responses) ameliorates further disease progression. In addition, PLP139–151-specific IFN-γ-producing Th1 cells can be recovered from the CNS of TMEV-infected mice beginning at approximately 45–50 days postinfection. Both of these results support an important role for these autoreactive T cells in pathogenesis of

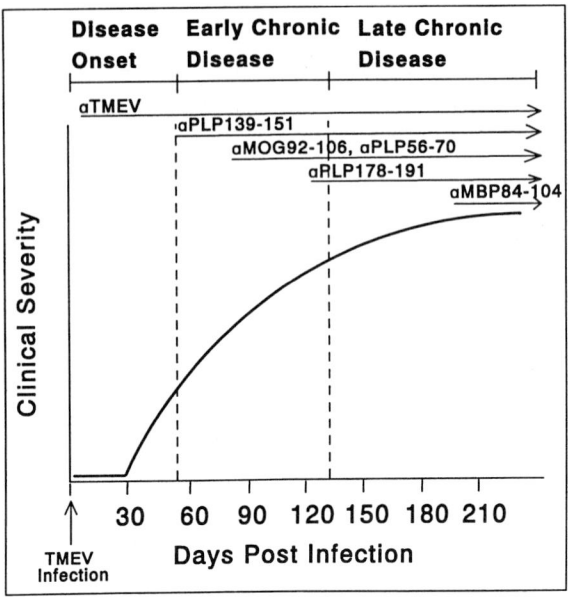

FIG 2. Pattern of epitope spreading during the course of chronic-progressive TMEV-induced demyelinating disease in SJL mice.

the chronic stages of disease. Collectively, these results suggest a model wherein virus-specific T cells drive early myelin destruction, leading to the activation and recruitment of myelin-specific T cells that contribute to chronic disease pathology (Fig. 2).

VI. MYELIN EPITOPE-SPECIFIC CD4[+] T CELL RESPONSES IN TMEV-INFECTED MICE ARISING VIA EPITOPE SPREADING

The difference in the temporal appearance of T cell reactivity to TMEV (within 7 days postinfection) (Clatch *et al.*, 1986) (Katz-Levy *et al.*, 1999b) and the involved myelin epitopes (≥50 days postinfection) (Table I) argues against molecular mimicry as inducing antimyelin responses in infected mice because both responses should arise concomitantly if mimicry were operative. In addition, as discussed previously, pretolerization with TMEV epitopes (Karpus *et al.*, 1995), but not with myelin epitopes (Miller *et al.*, 1990), inhibits disease induction, suggesting no cross-tolerance between myelin and TMEV epitopes. We also employed an algorithm previously used to detect regions of sequence homology between the immunodominant MBP85–99 epitope in HLA-DR2[+] MS patients and epitopes on several infectious organisms (Wucherpfennig and Stro-minger, 1995). Using the native PLP139–151 sequence (HSLGKALGH-PDKF), no homologies were found to TMEV viral sequences (Miller *et al.*, 1997). However, using a degenerate motif of PLP139–151, which allowed for tolerated amino acid substitutions within the PLP139–151 sequence, two sequences within the GDVII strain of TMEV were identified as dis-playing some homology (SPQSAPTGYRYD and GEQAAYAGRARA). However, these peptides did not immunize SJL/J mice for a T cell prolif-erative response, nor did the peptides induce EAE. In addition, only the latter peptide showed some detectable binding to the I-A^s, but the bind-ing affinity was 50-fold less than the native PLP139–151 peptide.

To directly assess potential cross-reactivity, the specificity of a panel of T cell hybridoma clones specific for the immunodominant TMEV epitopes (VP1, $_{233-250}$; VP2, $_{70-86}$, and VP3, $_{24-37}$) and for PLP139–151 was examined (Miller *et al.*, 1997). IL-2 production from representa-tive TMEV VP-specific T cell hybrids was elicited by UV-inactivated virions and the relevant capsid epitopes, but not by MBP, PLP, a panel of encephalitogenic PLP epitopes (PLP56–70, PLP104–117, PLP139–151, and PLP178–191), MBP84–104, MOG92–106, or by a mouse spinal cord homogenate. Similarly, representative PLP139–151-specific T cell hybrid clones were not activated in a cross-reactive fashion by any of the other encephalitogenic myelin epitopes or by intact TMEV, GST-fusion proteins encompassing the VP1, VP2, or VP3 capsid proteins, or by the immunodominant VP capsid epi-topes. Lack of cross-reactivity with TMEV epitopes was also observed when employing T cell hybrids specific for PLP178–191 and PLP104–117. Lastly, no cross-reactive responses were observed among polyclonal lymph node T cells from separate groups of SJL mice primed with intact TMEV, TMEV capsid proteins, TMEV immun-odominant epitopes, myelin proteins, or individual encephalitogenic myelin epitopes on MBP, PLP, and MOG. Collectively, these results indicate no apparent T cell cross-reactivity between epitopes on TMEV and on MBP, PLP, MOG, or MSCH and are consistent with the hypoth-esis that myelin responses in mice with chronic TMEV-IDD arise via epitope spreading and not as the result of molecular mimicry.

VII. Endogenous Presentation of Virus and Myelin Epitopes by CNS-Resident Antigen-Presenting C in TMEV-Infected Mice

It is of obvious interest to determine the anatomic location (peripheral lymphoid organs and/or the CNS) where T cells specific for endogenous myelin epitopes are primed, and this question is currently under investigation. In conjunction with this analysis, we have also put considerable effort into identifying the ability of antigen-presenting cells (APCs) isolated from the CNS target organ to endogenously present virus and myelin epitopes at various stages of the TMEV-IDD disease (Katz-Levy *et al.*, 1999a; Pope *et al.*, 1998). Our analyses have shown that macrophages/microglia isolated from the CNS of TMEV-infected SJL mice have the ability to endogenously process and present several virus epitopes, including VP2, $_{70-86}$ and VP3, $_{24-37}$, at both acute and chronic stages of the disease. This supports the conclusion that CNS-infiltrating macrophages/resident microglia are persistently infected with virus and can readily present viral epitopes either directly and/or by phagocytosing debris containing virus antigens. Relevant to the initiation of virus-induced autoimmune disease, only CNS APCs isolated from TMEV-infected mice with preexisting myelin damage ($\geq$60–80 days postinfection), but not those isolated from naive mice or mice with early-stage disease, were able to endogenously present a variety of PLP epitopes, including PLP56–70, PLP104–117, PLP139–151 and PLP178–191, to specific T cell hybridomas and Th1 lines. This timing is roughly equivalent to the initial appearance of T cells specific for the immunodominant PLP139–151 epitope. However, a temporal analysis of the appearance of various myelin epitopes (Katz-Levy *et al.*, 1999b) (Table II) indicated that some of these epitopes were available on CNS APCs well before peripheral T cell responses to most of these epitopes were demonstrable (Table I). Endogenous antigen presentation by CNS

TABLE II

TEMPORAL ABILITY OF TMEV AND PROTEOLIPID PROTEIN EPITOPES TO BE ENDOGENOUSLY PRESENTED BY CNS APCS IN THEILER'S VIRUS-INFECTED MICE

| | Activation of T Cell Lines/Hybridomas Specific for TMEV and Myelin Epitopes | | | | | |
| | TMEV Epitopes | | Myelin Protein Epitopes | | | |
Harvest of CNS APCs: Days Postinfection	$VP2_{70-86}$	$VP3_{24-37}$	PLP 56–70	PLP 104–117	PLP 139–151	PLP 178–191
35–50	+[a]	+	—	nd[b]	—	nd
60–80	+	+	—	—	+[c]	—
88–100	+	+	+	+	+	+
130–150	+	+	+	nd	+	+

[a] + indicates that CNS APCs harvested during the indicated time frame could endogenously activate a T cell line and/or hybridoma specific for the designated virus or PLP peptide.

[b] nd, Not determined.

[c] Endogneous presentation required high numbers of CNS APCs.

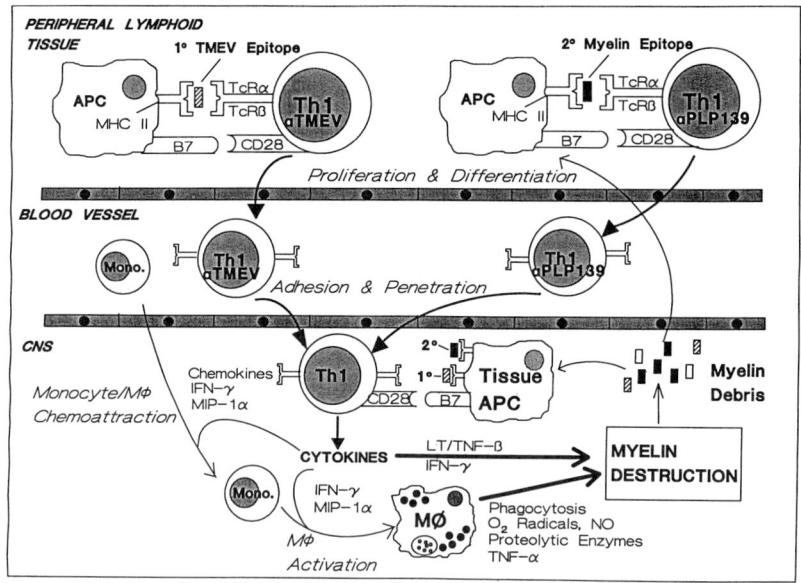

Fig 3. Model of epitope spreading in TMEV-infected SJL mice. APC, Antigen-presenting cell; IFN, interferon; LT, lymphotoxin; Mφ, macrophage; MIP, macrophage inflammatory protein; mono, monocyte; TcR, T cell receptor; TNF, tumor necrosis factor.

APCs was blocked by anti-I-A[s] and CTLA-4 Ig, indicating that the microglia/macrophages in the CNS express conventional MHC class II and costimulatory molecules. Phenotypic analysis indicates that the majority of these F4/80+, I-A[s+] CNS APCs express B7-1 and B7-2. Based on CD45 expression, the CNS APCs are roughly 60% microglia (CD45[dim]) and 30% infiltrating macrophages (CD45[bright]).

These results support a model of epitope spreading (Fig. 3) wherein localized virus-specific T cell-mediated inflammatory processes lead to the recruitment/activation of CNS-resident APCs that can serve both as effector cells for myelin destruction and as APCs that efficiently process/present endogenous self epitopes to autoreactive T cells. Regardless of the specificity of the autoreactive T cells (myelin peptides in R-EAE or TMEV epitopes in TMEV-IDD) responsible for initiating myelin destruction, epitope spreading plays an important contributory role in the chronic disease process in the genetically susceptible SJL mouse.

VIII. Summary

Epidemiological studies indicate that host immunogenetics and history of infection, particularly by viruses, may be a necessary cofactor for the induction of a variety of autoimmune diseases. To date, however, there is no clear-cut evidence, either in experimental animal models or in human autoimmune disease, that supports either molecular mimicry (Wucherpfennig and Strominger, 1995; Fujinami and Oldstone, 1985) or a role for superantigens (Scherer *et al.*, 1993) in the initiation of T cell-mediated autoimmunity. In contrast, the current data provide compelling evidence in support of a major role for epitope spreading in the induction of myelin-specific autoimmunity in mice persistently infected with TMEV. It is significant that two picornaviruses closely related to TMEV, coxsackievirus (Rose and Hill, 1996) and encephalomyocarditis virus (EMCV) (Kyu *et al.*, 1992), have been similarly shown to persist (either the viral RNA or the infectious virus) in their target organs and have been associated with the development of chronic autoimmune diseases, including myocarditis and diabetes. Thus, inflammatory responses induced by viruses that trigger proinflammatory Th1 responses, and have the ability to persist in genetically susceptible hosts, may lead to chronic organ-specific autoimmune disease via epitope spreading.

Epitope spreading has important implications for the design of antigen-specific therapies for the potential treatment of MS and other autoimmune diseases. This process indicates that autoimmune diseases are evolving entities and that the specificity of the effector autoantigen-specific T cells varies during the chronic disease process. Our experiments employing tolerance in R-EAE clearly indicate that antigen-specific treatment of ongoing disease is possible for preventing disease relapses, provided the proper relapse-associated epitope is targeted (Vanderlugt *et al.*, 1999). However, the ability to identify relapse-associated epitopes in humans will be a difficult task because immunodominance will vary in every individual. The use of costimulatory antagonists that can induce anergy without requiring prior knowledge of the exact epitopes (Miller *et al.*, 1995b), or the use of therapies that induce bystander suppression (Nicholson *et al.*, 1997; Brocke *et al.*, 1996), may thus be more practical current alternative therapies for the treatment of human autoimmune disease.

REFERENCES

Allegretta, M., Nicklas, J. A., Sriram, S., and Albertini, R. J. (1990). T cells responsive to myelin basic protein in patients with multiple sclerosis. *Science* **247:**718–721.
Barbano, R. L. and Dal Canto, M. C. (1984). Serum and cells from Theiler's virus-infected mice fail to injure myelinating cultures or to produce in vivo transfer of disease. The pathogenesis of Theiler's virus-induced demyelination appears to differ from that of EAE. *J. Neurol. Sci.* **66:**283–293.

Begolka, W. S. and Miller, S. D. (1998). Cytokines as intrinsic and exogenous regulators of pathogenesis in experimental autoimmune encephalomyelitis. *Res. Immunol.* **149:**771–781.

Begolka, W. S., Vanderlugt, C. L., Rahbe, S. M., and Miller, S. D. (1998). Differential expression of inflammatory cytokines parallels progression of central nervous system pathology in two clinically distinct models of multiple sclerosis. *J. Immunol.* **161:**4437–4446.

Bernard, C. C. and de Rosbo, N. K. (1991). Immunopathological recognition of autoantigens in multiple sclerosis. *Acta Neurologica* **13:**171–178.

Borrow, P., Tonks, P., Welsh, C. J. R., and Nash, A. A. (1992). The role of CD8+ T cells in the acute and chronic phases of Theiler's murine encephalomyelitis virus-induced disease in mice. *J. Gen. Virol.* **73:**1861–1865.

Brahic, M., Stroop, W. G., and Baringer, J. R. (1981). Theiler's virus persists in glial cells during demyelinating disease. *Cell* **26:**123–128.

Brocke, S., Gijbels, K., Allegretta, M., Ferber, I., Piercy, C., Blankenstein, T. Martin, R., Utz, U., Karin, N., Mitchell, D., Veromaa, T., Waisman, A., Gaur, A, Conlon, P., Ling, N., Fairchild, P. J., Wraith, D. C., O'Garra, A., Fathman, C. G., and Steinman, L. (1996). Treatment of experimental encephalomyelitis with a peptide analogue of myelin basic protein. *Nature* **379:**343–346.

Challoner, P. B., Smith, K. T., Parker, J. D., MacLeod, D. L., Coulter, S. N., Rose, T. M., Shultz, E. R., Bennett, J. L., Garber, R. L., Chang, M., Schad, P. A., Stewart, P. M., Nowinski, R. C., Brown, J. P., and Burmer, G. C. (1995). Plaque-associated expression of human herpesvirus 6 in multiple sclerosis. *Proc. Natl. Acad. Sci. U.S.A.* **92:**7440–7444.

Clatch, R. J., Lipton, H. L., and Miller, S. D. (1986). Characterization of Theiler's murine encephalomyelitis virus (TMEV)-specific delayed-type hypersensitivity responses in TMEV-induced demyelinating disease: correlation with clinical signs. *J. Immunol.* **136:**920–927.

Clatch, R. J., Lipton, H. L., and Miller, S. D. (1987a). Class II-restricted T cell responses in Theiler's murine encephalomyelitis virus (TMEV)-induced demyelinating disease. II. Survey of host immune responses and central nervous system virus titers in inbred mouse strains. *Microb. Pathogen.* **3:**327–337.

Clatch, R. J., Melvold, R. W., Dal Canto, M. C., Miller, S. D., and Lipton, H. L. (1987b). The Theiler's murine encephalomyelitis virus (TMEV) model for multiple sclerosis shows a strong influence of the murine equivalents of HLA-A, B, and C. *J. Neuroimmunol.* **15:**121–135.

Clatch, R. J., Melvold, R. W., Miller, S. D., and Lipton, H. L. (1985). Theiler's murine encephalomyelitis virus (TMEV)-induced demyelinating disease in mice is influenced by the H-2D region: correlation with TMEV-specific delayed-type hypersensitivity. *J. Immunol.* **135:**1408–1414.

Clatch, R. J., Miller, S. D., Metzner, R., Dal Canto, M. C., and Lipton, H. L. (1990). Monocytes/macrophages isolated from the mouse central nervous system contain infectious Theiler's murine encephalomyelitis virus (TMEV). *Virol.* **176:**244–254.

Dal Canto, M. C. and Lipton, H. L. (1975). Primary demyelination in Theiler's virus infection. An ultrastructural study. *Lab. Invest.* **33:**626–637.

Dal Canto, M. C. and Lipton, H. L. (1979). Recurrent demyelination in chronic central nervous system infection produced by Theiler's murine encephalomyelitis virus. *J. Neurol. Sci.* **42:**391–405.

Dal Canto, M. C. and Lipton, H. L. (1980). Schwann cell remyelination and recurrent demyelination in the central nervous system of mice infected with attenuated Theiler's virus. *Am. J. Pathol.* **98:**101–122.

Dal Canto, M. C. and Lipton, H. L. (1982). Ultrastructural immunohistochemical localization of virus in acute and chronic demyelinating Theiler's virus infection. *Am. J. Pathol.* **106**:20–29.

Ebers, G. C., Sadovnick, A. D., and Risch, N. J. (1995). A genetic basis for familial aggregation in multiple sclerosis. *Nature* **377**:150–151.

Fujinami, R. S. and Oldstone, M. B. (1985). Amino acid homology between the encephalitogenic site of myelin basic protein and virus: mechanism for autoimmunity. *Science* **230**:1043–1045.

Gerety, S. J., Clatch, R. J., Lipton, H. L., Goswami, R. G., Rundell, M. K., and Miller, S. D. (1991). Class II-restricted T cell responses in Theiler's murine encephalomyelitis virus-induced demyelinating disease. IV. Identification of an immunodominant T cell determinant on the N-terminal end of the VP2 capsid protein in susceptible SJL/J mice. *J. Immunol.* **146**:2401–2408.

Gerety, S. J., Rundell, M. K., Dal Canto, M. C., and Miller, S. D. (1994a). Class II-restricted T cell responses in Theiler's murine encephalomyelitis virus (TMEV)-induced demyelinating disease. VI. Potentiation of demyelination with and characterization of an immunopathologic CD4+ T cell line specific for an immunodominant VP2 epitope. *J. Immunol.* **152**:919–929.

Gerety, S. J., Karpus, W. J., Cubbon, A. R., Goswami, R. G., Rundell, M. K., Peterson, J. D., and Miller, S. D. (1994b). Class II-restricted T cell responses in Theiler's murine encephalomyelitis virus (TMEV)-induced demyelinating disease. V. Mapping of a dominant immunopathologic VP2 T cell epitope in susceptible SJL/J mice. *J. Immunol.* **152**:908–918.

Hoffman, L. M., Fife, B. T., Begolka, W. S., Miller, S. D., and Karpus, W. J. (1999). Central nervous system chemokine expression during Theiler's virus-induced demyelinating disease. *J. Neurovirol.* In Press

Karpus, W. J., Peterson, J. D., and Miller, S. D. (1994). Anergy *in vivo*: Down-regulation of antigen-specific CD4+ Th1 but not Th2 cytokine responses. *Int. Immunol.* **6**:721–730.

Karpus, W. J., Pope, J. G., Peterson, J. D., Dal Canto, M. C., and Miller, S. D. (1995). Inhibition of Theiler's virus-mediated demyelination by peripheral immune tolerance induction. *J. Immunol.* **155**:947–957.

Katz-Levy, Y., Neville, K. L., Girvin, A. M., Vanderlugt, C. L., Pope, J. G., Tan, L. J., and Miller, S. D. (1999a). Endogenous processing of self myelin epitopes by CNS-resident APCs in Theiler's virus-infected mice. *J. Clin. Invest.* **104**:599–610.

Katz-Levy, Y., Vanderlugt, C. L., Neville, K. L., Girvin, A. M., Rahbe, S. M., Padilla, J., and Miller, S. D. (1999b). Temporal development of epitope spreading in Theiler's virus-infected mice: from viral to self myelin epitopes. *J. Immunol.* **165**:5304–5314.

Kaufman, D. L., Clare-Salzler, M., Tian, J., Forsthuber, T., Ting, G. S. P., Robinson, P., Atkinson, M. A., Sercarz, E. E., Tobin, A. J., and Lehmann, P. V. (1993). Spontaneous loss of T cell tolerance to glutamic acid decarboxylase in murine insulin-dependent diabetes. *Nature* **366**:69–72.

Kennedy, M. K., Dal Canto, M. C., Trotter, J. L., and Miller, S. D. (1988). Specific immune regulation of chronic-relapsing experimental allergic encephalomyelitis in mice. *J. Immunol.* **141**:2986–2993.

Kennedy, M. K., Tan, L. J., Dal Canto, M. C., Tuohy, V. K., Lu, Z. J., Trotter, J. L., and Miller, S. D. (1990). Inhibition of murine relapsing experimental autoimmune encephalomyelitis by immune tolerance to proteolipid protein and its encephalitogenic peptides. *J. Immunol.* **144**:909–915.

Kurtzke, J. F. (1993). Epidemiologic evidence for multiple sclerosis as an infection. *Clin. Microbiol. Rev.* **6**:382–427.

Kyu, B., Matsumori, A., Sato, Y., Okada, I., Chapman, N. M., and Tracy, S. (1992). Cardiac persistence of cardioviral RNA detected by polymerase chain reaction in a murine model of dilated cardiomyopathy. *Circulation* **86:**522–530.

Lehrich, J. R., Arnason, B. G. W., and Hochberg, F. (1976). Demyelinative myelopathy in mice induced by the DA virus. *J. Neurol. Sci.* **29:**149–160.

Link, H., Baig, S., Olsson, O., Yu-Ping, J., Höjeberg, B., and Olsson, T. (1990). Persistent antimyelin basic protein IgG antibody response in multiple sclerosis cerebrospinal fluid. *J. Neuroimmunol.* **28:**237–248.

Lipton, H. L. (1975). Theiler's virus infection in mice: An unusual biphasic disease process leading to demyelination. *Infect. Immun.* **11:**1147–1155.

Lipton, H. L. (1980). Persistent Theiler's murine encephalomyelitis virus infection in mice depends on plaque size. *J. Gen. Virol.* **46**169–177.

Lipton, H. L. and Dal Canto, M. C. (1976). Theiler's virus-induced demyelination: prevention by immunosuppression. *Science* **192:**62–64.

Lipton, H. L. and Dal Canto, M. C. (1977). Contrasting effects of immunosuppression on Theiler's virus infection in mice. *Infect. Immun.* **15:**903–909.

Lipton, H. L. and Friedmann, A. (1980). Purification of Theiler's murine encephalomyelitis virus and analysis of the structural virion polypeptides: correlation of the polypeptide profile with virulence. *J. Virol.* **33:**1165–1172.

Lipton, H. L. and Melvold, R. (1984). Genetic analysis of susceptibility to Theiler's virus-induced demyelinating disease in mice. *J. Immunol.* **132:**1821–1825.

McCarron, R. M., Fallis, R. J., and McFarlin, D. E. (1990). Alterations in T cell antigen specificity and class II restriction during the course of chronic relapsing experimental allergic encephalomyelitis. *J. Neuroimmunol.* **29:**73–79.

McRae, B. L., Vanderlugt, C. L., Dal Canto, M. C., and Miller, S. D. (1995). Functional evidence for epitope spreading in the relapsing pathology of experimental autoimmune encephalomyelitis. *J. Exp. Med.* **182:**75–85.

Melvold, R. W., Jokinen, D. M., Knobler, R. L., and Lipton, H. L. (1987). Variations in genetic control of susceptibility to Theiler's murine encephalomyelitis virus (TMEV)-induced demyelinating disease. I. Differences between susceptible SJL/J and resistant BALB/c strains map near the T cell beta-chain constant gene on chromosome 6. *J. Immunol.* **138:**1429–1433.

Melvold, R. W., Jokinen, D. M., Miller, S. D., Dal Canto, M. C., and Lipton, H. L. (1990). Identification of a locus on mouse chromosome 3 involved in differential susceptibility to Theiler's murine encephalomyelitis virus-induced demyelinating disease. *J. Virol.* **64:**686–690.

Miller, S. D., Clatch, R. J., Pevear, D. C., Trotter, J. L., and Lipton, H. L. (1987). Class II-restricted T cell responses in Theiler's murine encephalomyelitis virus (TMEV)-induced demyelinating disease. I. Cross-specificity among TMEV substrains and related picornaviruses, but not myelin proteins. *J. Immunol.* **138:**3776–3784.

Miller, S. D. and Gerety, S. J. (1990). Immunologic aspects of Theiler's murine encephalomyelitis virus (TMEV)-induced demyelinating disease. *Semin. Virol.* **1:**263–272.

Miller, S. D., Gerety, S. J., Kennedy, M. K., Peterson, J. D., Trotter, J. L., Tuohy, V. K., Waltenbaugh, C., Dal Canto, M. C., and Lipton, H. L. (1990). Class II-restricted T cell responses in Theiler's murine encephalomyelitis virus (TMEV)-induced demyelinating disease. III. Failure of neuroantigen-specific immune tolerance to affect the clinical course of demyelination. *J. Neuroimmunol.* **26:**9–23.

Miller, S. D. and Karpus, W. J. (1994). The immunopathogenesis and regulation of T-cell mediated demyelinating diseases. *Immunol. Today* **15:**356–361.

Miller, S. D., McRae, B. L., Vanderlugt, C. L., Nikcevich, K. M., Pope, J. G., Pope, L., and Karpus, W. J. (1995a). Evolution of the T cell repertoire during the course of experimental autoimmune encephalomyelitis. *Immunol. Rev.* **144**:225–244.

Miller, S. D., Vanderlugt, C. L., Lenschow, D. J., Pope, J. G., Karandikar, N. J., Dal Canto, M. C., and Bluestone, J. A. (1995b). Blockade of CD28/B7-1 interaction prevents epitope spreading and clinical relapses of murine EAE. *Immunity* **3**:739–745.

Miller, S. D., Vanderlugt, C. L., Begolka, W. S., Pao, W., Yauch, R. L., Neville, K. L., Katz-Levy, Y., Carrizosa, A., and Kim, B. S. (1997). Persistent infection with Theiler's virus leads to CNS autoimmunity via epitope spreading. *Nature Med.* **3**:1133–1136.

Nicholson, L. B., Murtaza, A., Hafler, B. P., Sette, A., and Kuchroo, V. K. (1997). A T cell receptor antagonist peptide induces T cells that mediate bystander suppression and prevent autoimmune encephalomyelitis induced with multiple myelin antigens. *Proc. Natl. Acad. Sci. USA* **94**:9279–9284.

Nicholson, S. M., Dal Canto, M. C., Miller, S. D., and Melvold, R. W. (1996). Adoptively transferred CD8[+] lymphocytes provide protection against TMEV-induced demyelinating disease in BALB/c mice. *J. Immunol.* **156**:1276–1283.

Nicholson, S. M., Peterson, J. D., Miller, S. D., Dal Canto, M. C., and Melvold, R. W. (1994). BALB/c substrain differences in susceptibility to Theiler's murine encephalomyelitis virus (TMEV)-induced demyelinating disease. *J. Neuroimmunol.* **52**:19–24.

Olsberg, C., Pelka, A., Miller, S. D., Waltenbaugh, C., Creighton, T. M., Dal Canto, M. C., Lipton, H., and Melvold, R. (1993). Induction of Theiler's murine encephalomyelitis virus (TMEV)-induced demyelinating disease in genetically resistant mice. *Reg. Immunol.* **5**:1–10.

Ota, K., Matsui, M., Milford, E. L., Mackin, G. A., Weiner, H. L., and Hafler, D. A. (1990). T-cell recognition of an immunodominant myelin basic protein epitope in multiple sclerosis. *Nature* **346**:183–187.

Paterson, P. Y. and Day, E. D. (1981). Current perspectives of neuroimmunologic disease: multiple sclerosis and experimental allergic encephalomyelitis. *Clin. Immunol. Rev.* **1**:581–697.

Pelka, A., Olsberg, C., Miller, S. D., Waltenbaugh, C., Creighton, T. M., Dal Canto, M. C., and Melvold, R. (1993). Effects of irradiation on development of Theiler's murine encephalomyelitis virus (TMEV)-induced demyelinating disease in genetically resistant mice. *Cell. Immunol.* **152**:440–455.

Perry, L. L., Barzaga-Gilbert, E., and Trotter, J. L. (1991). T cell sensitization to proteolipid protein in myelin basic protein-induced experimental allergic encephalomyelitis. *J. Neuroimmunol.* **33**:7–16.

Perry, L. L. and Barzaga, M. E. (1987). Kinetics and specificity of T and B cell responses in relapsing experimental allergic encephalomyelitis. *J. Immunol.* **138**:1434–1441.

Peterson, J. D., Karpus, W. J., Clatch, R. J., and Miller, S. D. (1993). Split tolerance of Th1 and Th2 cells in tolerance to Theiler's murine encephalomyelitis virus. *Eur. J. Immunol.* **23**:46–55.

Peterson, J. D., Waltenbaugh, C., and Miller, S. D. (1992). IgG subclass responses to Theiler's murine encephalomyelitis virus infection and immunization suggest a dominant role for Th1 cells in susceptible mouse strains. *Immunol.* **75**:652–658.

Pope, J. G., Karpus, W. J., Vanderlugt, C. L., and Miller, S. D. (1996). Flow cytometric and functional analyses of CNS-infiltrating cells in SJL/J mice with Theiler's virus-induced demyelinating disease: Evidence for a CD4[+] T cell-mediated pathology. *J. Immunol.* **156**:4050–4058.

Pope, J. G., Vanderlugt, C. L., Lipton, H. L., Rahbe, S. M., and Miller, S. D. (1998). Characterization of and functional antigen presentation by central nervous system mononuclear cells from mice infected with Theiler's murine encephalomyelitis virus. *J. Virol.* **72:**7762–7771.

Rodriguez, M. and David, C. S. (1985). Demyelination induced by Theiler's virus: influence of the H-2 haplotype. *J. Immunol.* **135:**2145–2148.

Rodriguez, M., Leibowitz, J. L., and Lampert, P. W. (1983). Persistent infection of oligodendrocytes in Theiler's virus-induced encephalomyelitis. *Ann. Neurol.* **13:**426–433.

Roos, R. P., Firestone, S., Wollmann, R., Variakojis, D., and Arnason, B. G. (1982). The effect of short-term and chronic immunosuppression on Theiler's virus demyelination. *J. Neuroimmunol.* **2:**223–234.

Rose, N. R. and Hill, S. L. (1996). The pathogenesis of postinfectious myocarditis. *Clin. Immunol. Immunopathol.* **80:**S92–99.

Scherer, M. T., Ignatowicz, L., Winslow, G. M., Kappler, J. W., and Marrack, P. (1993). Superantigens: bacterial and viral proteins that manipulate the immune system. *Ann. Rev. Cell Biol.* **9:**101–128.

Sun, J. B., Olsson, T., Wang, W. Z., Xiao, B. G., Kostulas, V., Fredrikson, S., Ekre, H. P., and Link, H. (1991). Autoreactive T and B cells responding to myelin proteolipid protein in multiple sclerosis and controls. *Eur. J. Immunol.* **21:**1461–1468.

Tan, L.-J., Kennedy, M. K., Dal Canto, M. C., and Miller, S. D. (1991). Successful treatment of paralytic relapses in adoptive experimental autoimmune encephalomyelitis via neuroantigen-specific tolerance. *J. Immunol.* **147:**1797–1802.

Theiler, M. (1937). Spontaneous encephalomyelitis of mice, a new virus disease. *J. Exp. Med.* **65:**705–719.

Tisch, R., Yang, X. D., Singer, S. M., Liblau, R. S., Fugger, L., and McDevitt, H. O. (1993). Immune response to glutamic acid decarboxylase correlates with insulitis in non-obese diabetic mice. *Nature* **366:**72–75.

Tuohy, V. K., Yu, M., Weinstock-Guttman, B., and Kinkel, R. P. (1997). Diversity and plasticity of self recognition during the development of multiple sclerosis. *J. Clin. Invest.* **99:**1682–1690.

Vanderlugt, C. L., Eagar, T. N., Neville, K. L., Nikcevich, K. M., Bluestone, J. A., and Miller, S. D. (1999). Pathologic role and temporal appearance of newly emerging autoepitopes in relapsing EAE. *J. Immunol.* **164:**670–678.

Vanderlugt, C. L., Karandikar, N. J., Bluestone, J. A., and Miller, S. D. (1998). The functional significance of epitope spreading and its regulation by costimulatory interactions. *Immunol. Rev.* **164:**63–72.

Waksman, B. H. (1995). Multiple sclerosis: More genes versus environment. *Nature* **377:**105–106.

Wekerle, H. (1991). Immunopathogenesis of multiple sclerosis. *Acta Neurologica* **13:**197–204.

Welsh, C. J., Tonks, P., Nash, A. A., and Blakemore, W. F. (1987). The effect of L3T4 T cell depletion on the pathogenesis of Theiler's murine encephalomyelitis virus infection in CBA mice. *J. Gen. Virol.* **68:**1659–1667.

Wucherpfennig, K. W. and Strominger, J. L. (1995). Molecular mimicry in T cell-mediated autoimmunity: viral peptides activate human T cell clones specific for myelin basic protein. *Cell* **80:**695–705.

Yauch, R. L., Kerekes, K., Saujani, K., and Kim, B. S. (1995). Identification of a major T cell epitope within VP3(24–37) of Theiler's virus in demyelination-susceptible SJL/J mice. *J. Virol.* **69:**7315–7318.

Yauch, R. L. and Kim, B. S. (1994). A predominant viral epitope recognized by T cells from the periphery and demyelinating lesions of SJL/J mice infected with Theiler's virus is located within VP1$_{(233–244)}$. *J. Immunol.* **153:**4508–4519.

SELECTION OF AND EVASION FROM CYTOTOXIC T CELL RESPONSES IN THE CENTRAL NERVOUS SYSTEM

Stanley Perlman[*,§] and Gregory F. Wu[†]

*Departments of Pediatrics and Microbiology, †Interdisciplinary Program in Neuroscience, §Interdisciplinary Program in Immunology, University of Iowa, Iowa City, Iowa 52242

I. Objectives of This Review

Cytotoxic CD8 T lymphocytes (CTLs) are critical for the clearance of noncytopathic viruses from infected cells (Zinkernagel, 1996). Once it became evident that CTLs were indeed important for virus clearance, it was recognized that virus persistence must include the ability to evade this arm of the immune response. Several mechanisms by which viruses are able to diminish, delay, or prevent recognition by CTLs have been described (Oldstone, 1991; Ploegh, 1998). This review will concentrate on one mechanism used by viruses to persist—namely, the selection of a variant virus in which changes in the sequence of a CTL epitope abrogate recognition. In the first part of this review, the unique features of cytotoxic CD8 T cell function in the central nervous system (CNS) will be discussed. In the second part, the role of CTL escape mutants in viral evasion of the immune system and subsequent disease progression in non-CNS infections will be summarized. Finally, evidence showing that CTL escape mutants play a key role in virus amplification and the development of clinical disease in at least one model of virus-induced demyelination will be described in detail.

II. Introduction

A. Aspects of Immunological Response in the CNS

The immune response in the CNS is similar to the response in extra-neural tissue, but several aspects of activation of the immune response, cellular trafficking, and antigen presentation are unique to the CNS (reviewed in Lassmann, 1997; Lassmann *et al.*, 1991; Matyszak, 1998; Williams and Hickey, 1995)). The function of cytotoxic CD8 T cells in the inflamed CNS and the major histocompatibility complex (MHC) class I antigen expression in this tissue are most pertinent for the topic of this review and will be preferentially discussed in this section.

Although the CNS has classically been considered a site of immune privilege, surveillance of the normal CNS by circulating, activated lymphocytes occurs, with a limited number of lymphocytes being present in the normal CNS at any given time (Hickey, Hsu, and Kimura, 1991). However, considerable data suggest that the immune response in the CNS is initiated not in the parenchyma but, rather, in draining lymph nodes located in the deep cervical tissue (Cserr and Knopf, 1992). Dendritic cells, critical for antigen presentation, are virtually absent from the normal CNS, suggesting that virus antigen must spread to the draining lymph node tissue before activation can occur (Hart and Fabre, 1981; Matyszak, 1998; Matyszak and Perry, 1996). In support of this contention, influenza virus inoculated directly into the CNS parenchyma does not induce an immune response until virus spreads to the cerebrospinal fluid (CSF) and, soon thereafter, to the draining lymph nodes (Stevenson *et al.*, 1997a; Stevenson *et al.*, 1997b). Once the immune response has been initiated, breakdown of the blood–brain barrier, presence of plasma proteins, and recruitment of leukocytes characterize the CNS inflammatory response. Notably, cells with the phenotype of dendritic cells are detected as part of the cellular infiltrate in inflammatory processes and are believed to contribute to sustaining the immune response during chronic inflammation (Matyszak, 1998; Matyszak and Perry, 1996).

Within the CNS, CD8 and CD4 T cell effector functions are virtually the same as in peripheral sites of inflammation. CD8 T cells are responsible for the majority of cytotoxic activity and exhibit three effector mechanisms: perforin/granzyme-mediated killing; Fas/Fas ligand-mediated killing; and cytokine secretion (Liu, Young, and Young, 1996; Smyth and Trapani, 1998). In one well-documented example, perforin is critical for the CD8 T cell response to LCMV, both peripherally and in the CNS (Kagi *et al.*, 1994). Fas/FasL interactions have not been documented to be critical for CD8 T cell effector function in the CNS, but FasL expression by infiltrating CD4 T cells has been impli-

cated in oligodendrocyte death in inflammatory settings such as experimental allergic encephalomyelitis (EAE) and multiple sclerosis (MS)
(D'Souza et al., 1996; Sabelko et al., 1997; Waldner et al., 1997). Secretion of interferon-γ (IFN-γ) and of other cytokines are of critical importance in orchestrating multiple components of the cellular and
humoral immune responses. Furthermore, in some cases, cytokines
are capable of directly inhibiting viral replication. For example, IFN-γ
appears to be critical for virus clearance from infected oligodendrocytes in mice infected with mouse hepatitis virus (Parra et al., 1999).
In another example, genetic disruption of IFN-γ or its receptor inhibits
virus clearance in strains of mice normally resistant to infection with
Theiler's murine encephalomyelitis virus (TMEV) (Fiette et al., 1995).
CTLs also release chemokines through granule exocytosis, thereby
sustaining the immune response by contributing to the influx of other
inflammatory cells (reviewed in Baggiolini, Dewald, and Moser, 1997;
Luster, 1998). The influx of both antigen-specific and nonspecific cells
may contribute to virus clearance but also has the potential for
increasing damage to nearby uninfected cells (bystander damage).
This mechanism has been postulated to be significant in the pathogenesis of several CNS diseases with inflammatory components, including
demyelination in animals with virus-induced or autoimmune demyelination (Houtman and Fleming, 1996b; Lassmann, 1997).

One major difference between the CNS and extraneural tissue is found
in MHC class I antigen expression. In extraneural tissue, MHC class I
antigen is expressed constitutively by most cells. However, in the unin-
flamed CNS, constitutive expression of MHC class I and II antigen is
infrequent and is confined to a subset of endothelial cells (MHC class I)
and macrophages/microglia (MHC class II) (reviewed in Lampson, 1995;
Shrikant and Benveniste, 1996). Expression of these molecules by astrocytes, oligodendrocytes, or neurons is not believed to occur in the normal
CNS. However, this conclusion may need to be modified since, in a recent
study, the coordinated expression of MHC class I antigen (heavy chain
and β_2-microglobulin) and of a T cell receptor component (CD3ζ) by neurons during development was postulated to be important for synapse formation (Corriveau, Huh, and Shatz, 1998).

MHC class I and class II expression are upregulated in the CNS in
multiple pathological settings with infectious or noninfectious etiologies
(also reviewed in Lampson, 1995; Shrikant and Benveniste, 1996). As
discussed above, MHC class I expression by antigen-presenting cells
resident in the CNS is not involved in initiation of the CTL response but
is likely to be critical for its perpetuation. Expression of MHC class I
molecules by infected cells is also required for recognition and lysis by
activated antigen-specific CD8 T cells. The most convincing data show

that MHC class I and class II antigens are upregulated on microglia/macrophages in inflammatory diseases of the CNS, including MS and TMEV- and MHV-induced demyelinating encephalomyelitis. T cell costimulatory molecules such as B7–1 and B7–2 are also upregulated on these cells and, in some cases, activated microglia/macrophages isolated from the CNS can directly stimulate antigen-specific T cells proliferation (Bo *et al.*, 1994; Matyszak, 1998; Pope *et al.*, 1998; Xue *et al.*, 1999). MHC class I and class II expression by astrocytes and oligodendrocytes has been demonstrated *in vitro*, although, in some cases, only after exposure to proinflammatory cytokines, such as interferon-γ (Lampson, 1995; Shrikant and Benveniste, 1996). Whether astrocytes and oligodendrocytes upregulate MHC class I and class II expression in the inflamed CNS remains quite controversial, with expression of these molecules being reported in some studies (Lampson, 1995; Shrikant and Benveniste, 1996). In most recent studies, MHC antigen expression by astrocytes and oligodendrocytes in inflamed tissue was not detected (e.g., Pope *et al.*, 1998; Xue *et al.*, 1999), but these negative conclusions must be tempered because a biologically significant level of expression might not be detectable with presently available assays. MHC class I antigen upregulation on neurons has not been reported in pathological conditions, although a recent set of elegant studies demonstrated the expression of these molecules following inhibition of electrical activity (Neumann *et al.*, 1995; Neumann *et al.*, 1997). If verified in the infected CNS, these results would suggest that only neurons sufficiently damaged by infection with a virus or another infectious agent would be recognized by antigen-specific CTLs. Furthermore, this mechanism would minimize destruction of infected, but still functionally active, neurons.

Although there is no agreement on the specifics, these studies clearly demonstrate that MHC class I and II antigens are expressed on CNS-derived cells *in vitro* and *in vivo* during inflammatory processes. As a consequence, they provide a framework for interpreting experiments that describe the selection of CTL escape mutants in the CNS of virus-infected mice.

B. Evidence for Selection of CTL Escape Mutants in Infections Occurring Outside the CNS

The first experimental evidence for virus escape from CTL recognition by mutation of a CD8 T cell epitope was provided by Pircher *et al.*, (1990). In mice that were transgenic for a T cell receptor (TCR) specific for lymphocytic choriomeningitis virus (LCMV), infection with LCMV resulted in the rapid appearance of virus mutated in the target CTL epitope. Infection of TCR transgenic mice, but not of immunocompetent B6 mice, with mutant LCMV resulted in a delay in the kinetics of virus clearance. Similar results were obtained when wild-type mice were infected with LCMV variants in which some or all of the appropriate CTL epitopes were mutated (Lewicki et al., 1995; Moskophidis and Zinkernagel, 1995). Notably, LCMV escape mutants are not generated in immunocompetent mice after infection with wild-type virus. In these mice, CTLs recognizing multiple epitopes are detected, suggesting that an immune response directed against several epitopes precluded the selection of CTL escape mutants, at least in settings in which virus clearance was relatively efficient.

A subsequent study suggested that CTL escape mutants may emerge and become the predominant virus in human populations in which a single HLA allele is overrepresented, and in which an immunodominant epitope is restricted by that allele. The strain of Epstein-Barr virus (EBV) circulating in humans residing in the coastal regions of Papua New Guinea contains a mutation in an immunodominant CD8 T cell epitope that diminished binding to the HLA-All molecule (De Campos-Lima et al., 1994). HLA-A11 is present at a high frequency in this population, suggesting that the selection of virus encoding this mutation was immune-driven. However, this conclusion may not be warranted since the same mutation is found in virus circulating in the highland populations of Papua New Guinea, even though the HLA-A11 allele is not overrepresented in that group (Burrows et al., 1996).

The studies of LCMV-infected mice suggested that efficient virus clearance precluded the selection of CTL escape mutants. Conversely, less efficient virus clearance would be predicted to facilitate the emergence of these mutants. Therefore, the potential contribution of CTL mutants to pathogenesis was examined next in several persistent infections, including human and nonhuman primates chronically infected with the human immunodeficiency virus (HIV-1), hepatitis B virus (HBV), and hepatitis C virus. The emergence of CTL escape mutants during the course of persistence was documented in each of these infections (Franco et al., 1995; McMichael and Phillips, 1997; Weiner et al., 1995). The most convincing data suggesting a role for these variants in disease progression comes from studies in which the emergence of CTL escape mutants are documented early during the infection. Selection at early times postinfection is consistent with a role in delaying the elimination of virus from the infected host. In one study, Borrow et al. (1997) described an HIV-infected patient in which the initial CTL response was directed at a single HLA B44-restricted epitope. Early in the course of the infection, a mutation in a presumptive anchor residue was detected, leading to a loss of recognition by the patient's CTLs. Shortly thereafter, the CTL response spread to additional, presumably less immunodominant, epitopes. These responses were able to control the virus for a short period, but were, ultimately, inadequate because the patient exhibited a rapid rate of disease progression with a fatal outcome. In another study, mutations resulting in CTL escape were identified in the *nef* gene in a patient during primary HIV infection. These mutations abrogated recognition by *nef*-specific CTLs harvested from the patient, suggesting that their selection was also immune-driven and that it contributed to virus persistence (Price *et al.*, 1997).

In still another study, CTL escape mutants were identified in HIV-infected patients several years after the primary infection, coincident with increased virus replication and decreased CD4 T cell counts (Goulder *et al.*, 1997). It is uncertain, in these cases, whether CTL escape caused disease progression or whether it was a consequence of increasingly ineffectual control of virus replication by other components of the immune system, such as antibodies and CD4 T cells. CTL escape mutants that emerge only after immune surveillance has diminished may still contribute to disease progression in HIV-infected patients, even if they are not the primary cause for the loss of control by the immune system. CTL escape mutants were also selected in an HIV-infected individual treated with CTLs directed against a single epitope (Koenig *et al.*, 1995), thereby showing that, as in LCMV-infected TCR transgenic mice, a strong, monospecific CTL response facilitated their appearance.

CTL escape variants are occasionally detected in patients chronically infected with HBV, usually in the context of a narrowly focused, strong immune response (Bertoletti *et al.*, 1994a; Bertoletti *et al.*, 1994b). A more common finding in patients infected chronically with HBV is the lack of a significant CTL response (Rehermann *et al.*, 1995), suggesting that variation in HBV-specific T cell epitopes is important in only a minority of patients. Similarly, CTL escape mutants were detected in a chimpanzee chronically infected with hepatitis C virus (Weiner *et al.*, 1995), but persistence appears to be associated, most commonly, with a weak CTL response (Rehermann *et al.*, 1996).

C. Mechanisms of CTL Escape

Mutations in an epitope or in its flanking sequences can result in escape from surveillance by CD8 T cells, by affecting any one of several steps along the pathway used to generate peptides for presentation by MHC class I molecules. Antigen processing involves proteolytic cleavage by proteosomes in the cytoplasm and transport of the resulting peptides to the endoplasmic reticulum for binding to the MHC class I molecule. Mutations in sequences flanking a CTL epitope may affect proteolytic cleavage or transport into the endoplasmic reticulum. Only a few examples of mutations affecting either of these processing steps have been described (Eisenlohr, Yewdell, and Bennink, 1992; Ossendorp *et al.*, 1996). In one example, comparison of the sequence of an immunodominant CTL epitope encoded by two related murine leukemia virus families (AKV/MCF and Friend/Moloney/Rauscher [FMR]) showed that a single amino acid change present in the FMR strains abrogated CD8 T cell recognition (Ossendorp *et al.*, 1996). The mutation did not affect binding of peptide to the MHC class I molecule or immunogenicity of a peptide corresponding to the variant epitope. Rather, the loss of recognition occurred only after endogenous processing of antigen and resulted from alterations in proteosomal processing, which were shown using purified 20S proteosomes *in vitro*. These conclusions assume that processing *in vitro* mimics what occurs in the cell and that other differences in the amino acid sequence between the two viruses did not contribute to the observed differences. Mutations in sequences flanking a CTL epitope are difficult to detect, and therefore may be more important in CTL escape than is appreciated at present.

More commonly, mutations in CTL epitopes that affect binding to the MHC molecule, or affect recognition by the TCR, have been identified. Some of the mutations described above, in HIV-1 infected individuals, diminish binding to the MHC class I molecule or result in increased rates of dissociation from the MHC molecule (reviewed in McMichael and Phillips, 1997). Mutations that impair binding to the MHC molecule are the simplest to interpret and, depending on the diminution of binding, will completely abrogate recognition by epitope-specific CD8 T cells. Mutations that affected binding to TCRs were first reported in studies describing selection of LCMV CTL escape mutants in mice transgenic for an LCMV-specific TCR (Pircher *et al.*, 1990). Peptides mutated in residues that directly contact the TCR have also been described in virus isolated from HIV-infected patients (McMichael and Phillips, 1997). These mutations may result in partial or complete loss of recognition by autologous CTLs.

During the course of these studies on viral variation, mutations were identified in viral RNA and DNA isolated from patients persistently infected with HIV-1 or HBV that not only diminished recognition by epitope-specific CD8 T cells, but also inhibited recognition of wild-type epitope if both epitopes were present in the infected cell (Bertoletti *et al.*, 1994b; Klenerman *et al.*, 1994). The mechanism of this phenomenon (TCR antagonism) is not completely understood, but is believed to involve delivery of an altered signal to the CD8 T cell by the mutated peptide/MHC class I complex, resulting in nonresponsiveness or diminished responsiveness to the target agonist peptide (Sette *et al.*, 1996). Although TCR antagonism has only been demonstrated by using CTL clones, it has also been documented to occur when variant and wild-type epitopes are presented by different cells, a scenario that presumably mimics the situation in the infected host (Meier *et al.*, 1995). Another mechanism similar to antagonism is interference with CTL priming (Plebanski *et al.*, 1999). In this case, a variant epitope does not inhibit CTL effector function but, rather, the generation of activated antigen-specific CTLs from naive precursor CD8 T cells. This mechanism, identified in humans and mice infected with malaria, has not yet been described in any viral disease.

III. Selection of CTL Escape Mutants in Viral Encephalomyelitis

One of the best examples of a pathological setting in which CTL escape mutants contribute to virus persistence and the development of disease is that of mice infected with the neurotropic coronavirus, mouse hepatitis virus—strain JHM (MHV-JHM). Selection of escape mutants was not anticipated because immune selective pressure in the CNS, a site that is relatively protected from immune surveillance, might be predicted to be less than in the periphery.

Neurotropic coronaviruses, including MHV-JHM, the A59 strain of MHV (MHV-A59) and MHV-3, cause acute and chronic infections of the CNS in susceptible mice and rats (reviewed in Houtman and Fleming, 1996b; Lane and Buchmeier, 1997; Stohlman, Bergmann, and Perlman, 1998). MHV-JHM is highly neurovirulent and infection of susceptible mice results in widespread infection of neurons with a rapidly fatal outcome. However, more generally interesting are those model systems in which MHV-JHM and MHV-A59 are induced to cause either acute or chronic demyelination. General strategies for modifying the acute infection have been recently reviewed (Perlman, 1998; Stohlman, Bergmann, and Perlman, 1998), and two general approaches will be briefly described here. In the first, mice are infected with attenuated virus. Attenuated virus was cloned from pools of virus harvested from suckling mouse brain (Stohlman *et al.*, 1982). Alternatively, viral mutants were selected by chemical mutagenesis or by treatment with neutralizing antibody directed against the surface (S) glycoprotein (Dalziel *et al.,* 1986; Fleming *et al.,* 1986). In each case, the variant virus causes acute encephalitis in only a small percentage of mice but is able to persist and cause demyelination in most survivors. In mice infected with these variants, infectious virus is, in general, cleared by 2–3 weeks after inoculation. Clinical symptoms are most evident within the same time frame. Mice that survive the infection slowly recover, although evidence of ongoing demyelination can be detected for several months after the acute infection has resolved.

In the second approach, mice are infected with wild-type MHV-JHM and protected from the acute encephalitis by administration of anti-MHV antibody or MHV-specific CD4 or CD8 T cells (Buchmeier *et al.,* 1984; Stohlman *et al.,* 1986; Yamaguchi *et al.,* 1991). This intervention also results in decreased infection of neurons, but does not prevent infection of cells in the white matter or demyelination. Virus clearance is enhanced by treatment with protective antibodies or T cells in some studies (Buchmeier *et al.,* 1984; Yamaguchi *et al.,* 1991), but not all (Stohlman *et al.,* 1986).

In a variation of the latter approach, suckling C57BL/6 (B6) mice are protected from acute encephalitis with nursing by dams previously immunized to MHV-JHM (Perlman *et al.,* 1987). The suckling mice are protected from acute encephalitis and remain asymptomatic for 3–8 weeks. At that time, a variable percentage (30–90%) develop histological evidence of demyelination and clinical signs of hind-limb paralysis. Immunization of the dams may be accomplished either by active immunization with infectious virus (Perlman *et al.,* 1987), or by passive infusion of neutralizing antibody (Pewe, Xue, and Perlman, 1997), and with similar outcomes. The temporal course of this disease is very different from that observed in other models of MHV-JHM-induced demyelination, because mice are asymptomatic for several weeks before developing disease, and because infectious virus is not cleared. In a series of reports, Pewe *et al* showed that escape from the cytotoxic CD8 T cell response was a major factor in the disease progression in these mice (Pewe *et al.,* 1996; Pewe, Xue, and Perlman, 1997; Pewe, Xue, and Perlman, 1998). CTL escape mutations were detected in all mice that developed clinical disease several weeks postinfection in this model. In B6 mice, two CD8 T cell epitopes, encompassing residues 510–518 (S-510–518) (H-2D^b-restricted) and 598–605 (S-598–605 (H-2K^b-restricted), of the S glycoprotein are recognized (Bergmann *et al.,* 1996; Castro and Perlman, 1995). These two epitopes are located within a region of the protein that is prone to both mutation and deletion (termed the "hypervariable region") (Banner, Keck, and Lai, 1990;

Parker, Gallagher, and Buchmeier, 1989). The location of the epitopes within a region that can tolerate mutation, without loss of function, most likely contributes to the selection of CTL escape mutants. Mutations have only been detected in epitope S-510–518, the immunodominant epitope of the two (Castro and Perlman, 1996), and not in epitope S-598–605. Mutations are not detected in regions of the genome flanking epitope S-510–518 or in most MHV-specific CD4 T cell epitopes. The one possible exception is that a mutation is detected in a CD4 epitope encompassing residues 328–347 of the S glycoprotein (S-328–347) in several mice, although wild-type epitope is also detected in the same animals. The significance of this mutation is not known, because it does not appear to decrease recognition by CD4 T cells specific for the epitope (Xue and Perlman, 1997).

Mutations have been detected in amino acids at positions 2–7 of epitope S-510–518, with only a single variant generally being detected in any individual infected mouse (Pewe, Xue, and Perlman, 1997). A summary of mutations detected thus far is shown in Fig. 1. Based on the crystallographic structure of the H-2D^b molecule and on mutagenesis studies (Hudrisier et al., 1996; Young et al., 1994), some of these mutations are predicted to affect binding to the MHC class I molecule, whereas others inhibit binding to the TCR of the CD8 T lymphocyte. Mutations in epitope S-510–518 are selected within 10–12 days postinfection, a time at which virus is not cleared in other MHV infections (Houtman and Fleming, 1996a; Lin et al., 1999). Detection of CTL escape mutants at a time before virus clearance is expected to be complete is consistent with a role for these mutants in virus persistence. Notably, epitope S-510–518-specific CTL activity can be detected in the CNS as early as 5 days postinfection, prior to the detection of CTL escape mutants (unpublished observations). Mutations are also not detected in infected mice with severe combined immunodeficiency (SCID mice), suggesting that their selection is CD8 T cell-driven (Pewe, Xue, and Perlman, 1997). In this model, mice that do not develop hind-limb paralysis by 60–80 days postinfection remain asymptomatic. Mutations are only occasionally detected in asymptomatic mice at 60–80 days postinfection suggesting that escape from CD8 T cell surveillance correlates well with virus amplification in B6 mice but is not sufficient for the development of clinical disease.

As discussed above, the role of CTL escape mutants in viral pathogenesis is controversial. These mutants are not expected to be relevant to pathogenesis if the CTL response is directed at several CTL epitopes, as is believed to occur in many human infections. It has also been postulated that, even if a single epitope is immunodominant, a T cell

Position within epitope	1	2	3	4	5	6	7	8	9
		*M	M	T	M	T	M/T		M
Detected					S				
		F	R		T				
		Y	F		D	R	L		
		P	(P)	G	H	V	H		
Wild type epitope:	C	S	L	W	N	G	P	H	L
Not detected		C̲	H̲	S̲		W̲	R̲		W̲
				L̲		E̲			P̲
				C̲		A̲			
				R̲					

FIG 1. Summary of mutations detected in epitope S-510–518 in virus isolated from mice with hind-limb paralysis. A panel of peptides resulting from single nucleotide changes in the sequence of the wild-type epitope were tested in cytotoxicity assays. Only those changes resulting in a 20-fold decrement in activity are included in the figure. The amino acids listed above the wild-type sequence have been detected in virus isolates, whereas those listed below the wild-type sequence (underlined) have not yet been detected. The L to P change in position 3 results in only a 4-fold decrease in recognition but has been detected in a minority of cDNA clones in single mice with hind-limb paralysis (marked with parentheses). *MHC (M) or TCR (T) contact-based on published reports (Hudrisier *et al.*, 1996; Hudrisier *et al.*, 1995; Young *et al.*, 1994).

response that is comprised of many CD8 T cell clonotypes would preclude CTL escape, because a single mutation would affect recognition by only a subset of the epitope-specific T cells (Ishikawa *et al.*, 1998). To determine the biological significance of the mutations that were detected in MHV-infected mice, several additional experiments were performed. First, a panel of variant epitopes were analyzed in cytotoxicity assays, using either lymphocytes derived from the CNS of mice with acute encephalitis directly *ex vivo* or bulk populations of splenocytes cultured *in vitro* for 2 weeks with peptide S-510–518 (Pewe and Perlman, 1999; Pewe *et al.*, 1996). This panel comprised all the peptides resulting from single nucleotide changes in epitope positions 2–7 and 9. These assays showed that nearly all of the mutations that were selected *in vivo* resulted in a complete or substantial loss of recognition by either population of lymphocytes. Several additional mutations also resulted in a nearly complete loss of recognition by these lymphocytes, but have not yet been detected *in vivo* (Fig. 1) Strikingly, some mutations in position 9, a secondary anchor for binding to the MHC molecule, resulted in loss of recognition, but were never selected in infected mice. Whether this apparent selectivity reflects a sampling error, or is in fact biologically driven (i.e., results in detrimental changes in RNA or protein structure), remains to be determined.

The biological significance of these mutations was also addressed by infecting maternal antibody-protected suckling C57BL/6 mice with virus mutated in epitope S-510–518. Mice infected with this virus do not recognize epitope S-510–518 but still respond to epitope S-598–605. Variant virus-infected mice have significantly greater mortality and morbidity when compared to mice infected with wild-type virus (Pewe, Xue, and Perlman, 1998). These results suggest that the CD8 T cell response to epitope S-510–518 is an important component of the host response to MHV-JHM, and support the idea that escape from this response is beneficial to the virus and results in increased virus replication and disease progression.

These results suggest that the CTL response to epitope S-510–518 is monospecific, because single mutations in residues important for binding to either MHC antigen or the TCR result in substantial loss of activity in cytotoxicity assays. The monospecificity of the response is consistent with a monoclonal or an oligoclonal response to the epitope. To directly assess the diversity of the T cell response, TCR Vβ element usage and the complexity of the complementarity determining region 3 (CDR3) were determined. The TCR is a heterodimer consisting of an α and a β chain. The great diversity in the T cell response results from the large number of different V, D, and J elements in the germ line, coupled with imprecise joining at the V-D and D-J junctions (β chains) or V-J junction (α chain). The CDR3 encompasses the junctional region of the α and β chains and makes direct contact with the MHC/peptide complex. Analysis of a monoclonal or oligoclonal T cell response should reveal usage of only one or a small number of different Vβ and Vα elements and minimal heterogeneity of the CDR3. Sequence analyses of the CDR3 were facilitated by the use of soluble MHC/peptide tetramers specific for epitope S-510–518 (tetramer S-510). Soluble MHC class I/tetramers were first described by Altman *et al.,* and have been shown, in several subsequent studies, to detect antigen-specific T cells with high sensitivity and specificity (e.g., Altman *et al.,* 1996; Flynn *et al.,* 1998; Murali-Krishna *et al.,* 1998).

Epitope S-510–518-specific CD8 T cells were isolated directly from the CNS of mice with acute encephalitis by sorting with a fluorescent-activated cell sorter (FACS) after staining with anti-CD8 antibody and tetramer S-510. The tetramer S-510-positive T cells were analyzed initially for Vβ usage. The results showed usage of a large number of different Vβ elements by CD8 T cells responding to the epitope, with preferential usage of some Vβ elements occurring, specifically Vβ8 and Vβ13 (Pewe *et al.*, 1999). Analysis of individual CDR3 sequences from a subpopulation of tetramer S-510-positive cells, those expressing the Vβ13 element, revealed substantial heterogeneity. The CDR3 sequences were shown to fit, approximately, a logarithmic distribution (Fisher, Corbet, and Williams, 1943; Pewe *et al.*, 1999; Taylor, Kempton, and Woiwod, 1976). From this distribution, it was possible to calculate that approximately 300–500 epitope S-510–518-specific CD8 T cell clonotypes were present in the CNS of a mouse with acute encephalitis. Similar measurements using lymphocytes harvested from the CNS of mice with chronic demyelination revealed that approximately 100–900 different clonotypes were present in these animals. Previous studies suggested that either 3,000 (Butz and Bevan, 1998) or 600 (Bousso *et al.*, 1998) precursor CD8 T cells for any single epitope were present in an individual mouse. Our calculation of the number of T cells recognizing epitope S-510–518 agrees more closely with the lower estimate.

Notably, none of the wild-type epitope S-510–518 RNA, or little of it, is present in the CNS of mice with chronic demyelination, but CD8 T cells specific for the epitope are still detectable. Whether this represents stimulation by residual wild-type antigen (which may be cleared more slowly than viral RNA (Knopf *et al.*, 1998; Zhang *et al.*, 1992), or by variant sequence remains to be determined. Furthermore, many of the β chain CDR3 sequences were detected in more than one mouse. In the case of mice with chronic demyelination, very few CDR3 sequences were unique to a single animal. Thus, unlike other experimental systems (Bousso *et al.*, 1998), the repertoire of CD8 T cell TCRs recognizing epitope S-510–518 is not unique to each naive animal but is, in large part, common among all C57BL/6 mice. This lack of variability may facilitate the emergence of CTL escape mutants in a large fraction of infected mice.

These results suggest that CTL escape mutants are critical in virus persistence and in the development of clinical disease in mice infected with MHV-JHM. These results also show that CTL escape mutants are selected in the presence of a polyclonal, but monospecific, CD8 T cell response. As already mentioned, the importance of CTL escape mutants in other infections is much less certain, suggesting that the following features, present in the MHV-JHM-infected CNS, must predispose to the selection of these mutants:

1. CTL escape mutants are observed in MHV-infected rodents only when B6 mice are inoculated at the suckling stage with virulent wild-type virus and protected from acute disease by nursing by immunized dams. Inoculation with virulent MHV-JHM and use of B6 mice are essential for disease to develop. Chronic demyelination does not develop if suckling mice are inoculated with the less neurovirulent A59 strain of MHV (unpublished observations). In most other models of MHV persistence, older mice are inoculated with attenuated strains. Persistence is manifested by the presence of viral RNA and ongoing demyelination in the absence of infectious virus (Adami et al., 1995; Houtman and Fleming, 1996a; Rowe et al., 1997). CTL escape mutants have not been detected in these models, possibly because clearance of infectious virus is so rapid (Bergmann et al., 1998). A normal immune system is also required, however, because even when clearance is delayed in mice infected with attenuated virus, as occurs in immunodeficient mice, CTL escape mutants are not commonly selected (Lin et al., 1999).

2. Infection with wild-type virus at the suckling stage (10–14 days old) is also crucial in this model. Three-week-old B6 mice inoculated with wild-type virus and protected from acute encephalitis by passive administration of small amounts of neutralizing antibody are protected from acute encephalitis, but they do not develop hind-limb paralysis at later times (Sun and Perlman, 1995). Demyelination can be detected in the spinal cords of these mice at 2 months postinfection. These mice have not formally been assessed for the presence of CTL escape mutations at late times after infection, but they clearly do not develop a clinical disease consistent with the selection of CTL escape mutants. These results suggest that inoculation of mice with virulent virus at the suckling stage is essential for the selection of CTL escape mutants or, as a minimum, for their clinical manifestation. This requirement for inoculation of suckling mice may reflect the immaturity of the immune system. A consequence of infecting mice at 10 days of life may be suboptimal clearance of virus. Consistent with this possibility, after inoculation at the suckling stage, low levels of infectious virus can be detected at 40 days postinfection in BALB/c mice (unpublished observations). BALB/c mice never develop evidence of hind-limb paralysis or other clinical disease at any time and develop completely normally (Castro et al., 1994; Perlman, Schelper, and Ries, 1987).

3. Another consequence of inoculation of mice with an immature immune system may be an aberrantly polarized immune response. A

striking feature of the immune response in most C57BL/6 mice infected with MHV at the suckling stage is the lack of an appreciable antibody response, when they are assayed by ELISA or in neutralizing assays (Jacobsen and Perlman, 1990; Perlman *et al.*, 1987; Perlman, Schelper, and Ries, 1987). The CD4 and CD8 T cell response in maternal antibody-protected mice that develop chronic demyelination is proinflammatory and readily detected (Castro *et al.*, 1994; Pewe *et al.*, 1999). This strongly polarized, cell-mediated immune response in the presence of a defective humoral response may also predispose to the selection of CTL escape mutants. In support of this, BALB/c mice inoculated at the suckling stage in the maternal antibody model mount a significant antibody response and do not develop hind-limb paralysis (Castro *et al.*, 1994; Perlman, Schelper, and Ries, 1987).

4. MHV infection of congenic B10.A(18R) ($K^bD^dL^d$) mice results in the development of hind-limb paralysis, albeit at a lower frequency than in either B6 or C57BL/10 mice (Castro *et al.*, 1994). Two CTL epitopes, S-598–605 and an L^d-restricted epitope encompassing residues 318–326 of the nucleocapsid (N) protein (N-318–326), are recognized in these mice. The immunodominant epitope recognized in B6 mice, epitope S-510–518, is not recognized in B10.A(18R) mice since these mice do not encode the H-2D^b allele. Virus isolated from B10.A(18R) mice with hind-limb paralysis was evaluated for the presence of mutations in epitopes S-598–605 and N-318–326, and none were detected. The development of hind-limb paralysis in B10.A(18R) mice, in the absence of mutations in either CTL epitope, suggests that a factor expressed on the C57BL/6 background (decreased production of antibody?) may contribute to the increased amount of virus replication observed in these mice. The outgrowth of CTL escape mutants in B6 mice, in conjunction with this other putative factor, results in a much higher rate of clinical disease than is observed in B10.A(18R) mice in which CTL escape mutants are not selected.

5. A CTL response directed at a single immunodominant epitope enhances the likelihood that CTL escape mutants will be selected. In B6 mice, two MHV-specific epitopes are recognized. Cytotoxicity assays using CNS-derived lymphocytes from mice with acute encephalitis, directly ex vivo (Castro and Perlman, 1995), and limiting dilution assays using splenocytes harvested from mice intraperitoneally immunized with MHV (Castro and Perlman, 1996), suggested, in both cases, that the two epitopes were recognized by similar numbers of CD8 T cells. More recent measurements, using MHC class I peptide/tetramers to detect antigen-specific cells or methods to detect interferon-γ production after stimulation with peptide, confirmed

these results (unpublished observations). However, much more peptide S-598–605 is required to sensitize target cells for lysis. Thus, the
CTL response in these mice may be functionally directed at a single
epitope. Conversely, a CTL response to a single immunodominant epitope is not sufficient for CTL escape mutants to be selected. The CTL
response in BALB/c mice is directed at a single epitope (N-318–326),
yet these mice do not develop hind-limb paralysis in the maternal anti-
body-protection model (Castro et al., 1994), and CTL escape mutants
are not selected (unpublished observations). The ability of the virus to
tolerate mutations in the part of the protein that contains the target
CD8 T cell epitope is also important. Epitope S-510–518 is in a region
of the S protein prone to deletion and mutation, while the BALB/c epitope is located in a highly conserved region of the N protein. As mentioned above, mutations in this epitope are not detected in B10.A(18R)
mice, even though the background of this strain may be favorable for
the selection of CTL escape mutants.

6. MHV-JHM is neurotropic and the infection caused by this agent is
confined, for the most part, to the CNS. As discussed above, the
processes of initiation of an inflammatory response, trafficking to the
site of inflammation, and antigen presentation within the CNS differ
in several aspects from similar processes occurring at extraneural
sites. In particular, MHC class I expression is minimal in the normal
CNS, and while clearly upregulated within the infected CNS, determining the cellular sites of expression remains an area of active
research. Therefore, while it is speculative at present, it is formally
possible that infection within the CNS results in a modified immune
response that increases the likelihood that CTL escape mutants will be
selected during the course of an infection.

IV. Conclusions and Future Directions.

What can be concluded from these studies about the role of CTL
escape mutants in the pathogenesis of viral infections in the CNS? The
following conclusions may also be applicable for CTL escape mutants
arising in extraneural tissue.

First, CTL escape mutants are not commonly selected in most infections. One explanation for this is that the selection of CTL escape
mutants is uncommon if the CD8 T cell response is directed against
several epitopes. However, a CD8 T cell response directed against one
epitope or a few, has been observed in several human and experimental infections (Cole, Hogg, and Woodland, 1994; Flynn et al., 1998;

Lehner *et al.*, 1995; Moss *et al.*, 1991; Murali-Krishna *et al.*, 1998; Wallace *et al.*, 1999)—suggesting that an immunodominant response is not unusual. Nevertheless, CTL escape mutants are not detected in most experimental settings even when only one epitope is immunodominant, or a few are.

Second, the recognition of an epitope by a diverse population of CD8 T cell clonotypes does not prevent emergence of these variants. What does appear to be true, however, is that the diversity of the response matters less than whether the different CD8 T cell clonotypes recognize the same or different parts of the MHC/peptide complex. A response narrowly focused on one part of the complex will facilitate the selection of CTL escape mutants. The CTL response to epitope S-510–518 is very sensitive to changes in residues 4 and 6 (Pewe and Perlman, 1999; Pewe *et al.*, 1996), two residues important for binding to the TCR (Young *et al.*, 1994). Strongly focused recognition of central residues of an immunodominant H-2K^b-restricted epitope, which is recognized in Sendai virus-infected B6 mice, has also been reported (Cole, Hogg, and Woodland, 1995).

Third, in mice infected with MHV and in some humans persistently infected with HIV-1, HBV, or hepatitis C virus, CTL escape mutants play an important role in virus amplification and disease progression. In most cases, however, it is likely that there would not have been sufficient time for selection of CTL escape mutants if virus clearance were rapid and complete. Thus, selection of CTL escape mutants in the absence of any other factor is probably not sufficient to prevent rapid virus clearance from an infected host. In MHV-infected B6 mice, their selection appears to be facilitated by a suboptimal humoral immune response. This deficient antibody response may be a consequence of infection of suckling mice in the presence of maternally derived antibody. Maternal immunization has been shown to impair immune responses in other infections (Wang *et al.*, 1998), although it has never been shown to affect selectively humoral, and not cellular, immunity.

Fourth, an interesting aspect of this model system is the number of questions that it raises about the dynamics of MHV growth and cellular tropism in the CNS. Unlike the situation in mice chronically infected with TMEV, in which macrophages are the main reservoir for virus (Lipton, Twaddle, and Jelachich, 1995), no single type of CNS cell has been identified as the predominant target in mice persistently infected with MHV. The selection of CTL escape mutants in the CNS of MHV-infected mice demonstrates that MHV infection of a cell expressing MHC class I antigen is a critical part of the pathogenic process. Astrocytes, macrophages/microglia, and oligodendrocytes are all infected in these animals (Perlman and Ries, 1987; Xue et al., 1999). Of all of these cells, only macrophages/microglia express detectable levels of MHC class 1 antigen in MHV-infected mice with chronic demyelination (Xue et al., 1999). One interpretation of these results is that macrophages/microglia are in fact the primary site of productive virus infection, with replication in oligodendrocytes or astrocytes being of secondary importance or abortive. Infection of the latter two cells might be crucial for the development of clinical disease, but might be less important for propagation of the virus. Alternatively, astrocytes or oligodendrocytes might express MHC class I antigen, albeit at levels below the detection limits of assays presently available. Escape from CD8 T cell surveillance might occur in these cells, thereby resulting, simultaneously, in increased virus replication and disease progression. Infection with MHV affects MHC class I antigen expression by astrocytes *in vitro* and in newborn mice but this has not been demonstrated *in vivo* in older mice (Correale et al., 1995; Gilmore, Correale, and Weiner, 1994; Suzumura et al., 1988; Suzumura et al., 1986). Conversely, proof that astrocytes or oligodendrocytes were the most important sites of MHV-JHM replication and, therefore, the sites where selection of CTL escape mutants occurred, would provide evidence that these cells express MHC class I antigen.

ACKNOWLEDGMENTS

This work was supported in part by grants from the National Institutes of Health (NIH) (NS 36592 and AI43497) and from the National Multiple Sclerosis Society. G.F.W. was supported in part by a National Research Service Award from the NIH.

REFERENCES

Adami, C., Pooley, J., Glomb, J., Stecker, E., Fazal, F., Fleming, J. O., and Baker, S. C. (1995). Evolution of mouse hepatitis virus (MHV) during chronic infection: Quasi-species nature of the persisting MHV RNA. *Virology* **209:**337–346.

Altman, J., Moss, P., Goulder, P., Barouch, D., McHeyzer-Williams, M., Bell, J., McMichael, A., and Davis, M. (1996). Phenotypic analysis of antigen-specific T lymphocytes. *Science* **274:**94–96.

Baggiolini, M., Dewald, B., and Moser, B. (1997). Human chemokines: an update. *Ann. Rev. Immunol.* **15:**675–705.

Banner, L., Keck, J. G., and Lai, M. M. C. (1990). A clustering of RNA recombination sites adjacent to a hypervariable region of the peplomer gene of murine coronavirus. *Virology* **175:**548–555.

Bergmann, C., Dimacali, E., Stohl, S., Wei, W., Lai, M. M. C., Tahara, S., and Marten, N. (1998). Variability of persisting MHV RNA sequences constituting immune and replication-relevant domains. *Virology* **244:**563–572.

Bergmann, C. C., Yao, Q., Lin, M., and Stohlman, S. A. (1996). The JHM strain of mouse hepatitis virus induces a spike protein-specific D^b-restricted CTL response. *J. Gen. Virol.* **77**:315–325.

Bertoletti, A., Costanzo, A., Chisari, F. V., Levero, M., Artini, M., Sette, A., Penna, A., Giuberti, T., Fiaccadori, F., and Ferrari, C. (1994a). Cytotoxic T lymphocyte response to a wild type hepatitis B virus epitope in patients chronically infected by variant viruses carrying substitutions within the epitope. *J. Exp. Med.* **180**:933–943.

Bertoletti, A., Sette, A., Chisari, F. V., Penna, A., Levrero, M., De Carli, M., Fiaccadori, F., and Ferrari, C. (1994b). Natural variants of cytotoxic epitopes are T-cell receptor antagonists for antiviral cytotoxic T cells. *Nature* **369**:407–410.

Bo, L., Mork, S., Kong, P., Nyland, H., Pardo, C., and Trapp, B. D. (1994). Detection of MHC class II-antigens on macrophages and microglia, but not on astrocytes and endothelia in active multiple sclerosis lesions. *J. Neuroimmunol.* **51**:135–146.

Borrow, P., Lewicki, H., Wei, X., Horwitz, M., Peffer, N., Meyers, H., Nelson, J. A., Gairin, J., Hahn, B., Oldstone, M. B. A., and Shaw, G. (1997). Antiviral pressure exerted by HIV-1-specific cytotoxic T lymphocytes (CTLs) during primary infection demonstrated by rapid selection of CTL escape mutants. *Nature Med.* **3**:205–211.

Bousso, P., Casrouge, A., Altman, J. D., Haury, M., Kanellopoulos, J., Abastado, J.-P., and Kourilsky, P. (1998). Individual variations in the murine T cell response to a specific peptide reflect variability in naive repertoires. *Immunity* **9**:169–178.

Buchmeier, M. J., Lewicki, H. A., Talbot, P. J., and Knobler, R. L. (1984). Murine hepatitis virus-4 (strain JHM)-induced neurologic disease is modulated in vivo by monoclonal antibody. *Virology* **132**:261–270.

Burrows, J. M., Burrows, S. R., Poulsen, L. M., Sculley, T. B., Moss, D. J., and Khanna, R. (1996). Unusually high frequency of Epstein-Barr virus genetic variants in Papua New Guinea that can escape cytotoxic T-cell recognition: implications for virus evolution. *J. Virol.* **70**:2490–6.

Butz, E. A., and Bevan, M. J. (1998). Massive expansion of antigen-specific CD8+ T cells during an acute virus infection. *Immunity* **8**:167–175.

Castro, R. F., Evans, G. D., Jaszewski, A., and Perlman, S. (1994). Coronavirus-induced demyelination occurs in the presence of virus-specific cytotoxic T cells. *Virology* **200**:733–743.

Castro, R. F., and Perlman, S. (1995). CD8+ T cell epitopes within the surface glycoprotein of a neurotropic coronavirus and correlation with pathogenicity. *J. Virol.* **69**:8127–8131.

Castro, R. F., and Perlman, S. (1996). Differential antigen recognition by T cells from the spleen and central nervous system of coronavirus-infected mice. *Virology* **222**:247–251.

Cole, G., Hogg, T., and Woodland, D. (1995). T cell recognition of the immunodominant Sendai virus NP324–332/Kb epitope is focused on the center of the peptide. *J. Immunol.* **155**:2841–2848.

Cole, G. A., Hogg, T. L., and Woodland, D. L. (1994). The MHC class I-restricted T cell response to Sendai virus infection in C57BL/6 mice: a single immunodominant epitope elicits an extremely diverse repertoire of T cells. *Int. Immunol.* **6**:1767–1775.

Correale, J., Li, S., Weinter, L., and Gilmore, W. (1995). Effect of persistent mouse hepatitis virus infection on MHC class I expression in murine astrocytes. *J. Neurosci. Res.* **40**:10–21.

Corriveau, R. A., Huh, G. S., and Shatz, C. J. (1998). Regulation of class I MHC gene expression in the developing and mature CNS by neural activity. *Neuron* **21**:505–20.

Cserr, H. F., and Knopf, P. M. (1992). Cervical lymphatics, the blood-brain barrier and the immunoreactivity of the brain: a new view. *Immunol. Today* **13**:507–512.

D'Souza, S., Bonetti, B., Balasingam, V., Cashman, N., Barker, P., Troutt, A., Raine, C. S., and Antel, J. P. (1996). Multiple sclerosis: Fas signaling in oligodendrocyte cell death. *J. Exp. Med.* **184:**2361–2370.

Dalziel, R. G., Lampert, P. W., Talbot, P. J., and Buchmeier, M. J. (1986). Site-specific alteration of murine hepatitis virus type 4 peplomer glycoprotein E2 results in reduced neurovirulence. *J. Virol.* **59:**463–471.

De Campos-Lima, P. O., Levitsky, V., Brooks, J., Lee, S. P., Hu, L., Rickinson, A. B., and Masucci, M. G. (1994). T cell responses and virus evolution: loss of HLA A11-restricted CTL epitopes in Epstein-Barr virus isolates from highly A11-positive populations by selective mutation of anchor residues. *J. Exp. Med.* **179:**1297–1305.

Eisenlohr, L. C., Yewdell, J. W., and Bennink, J. R. (1992). Flanking sequences influence the presentation of an endogenously synthesized peptide to cytotoxic T lymphocytes. *J. Exp. Med.* **175:**481–487.

Fiette, L., Aubert, C., Muller, U., Huang, S., Aguet, M., Brahic, M., and Bureau, J. F. (1995). Theiler's virus infection of 129Sv mice that lack the interferon alpha/beta or interferon gamma receptors. *J. Exp. Med.* **181:**2069–76.

Fisher, R. A., Corbet, A. S., and Williams, C. B. (1943). The relationship between the number of species and the number of individuals in a random sample of an animal population. *J. Anim. Ecol.* **12:**42–58.

Fleming, J. O., Trousdale, M. D., El-Zaatari, F., Stohlman, S. A., and Weiner, L. P. (1986). Pathogenicity of antigenic variants of murine coronavirus JHM selected with monoclonal antibodies. *J. Virol.* **58:**869–875.

Flynn, K. J., Belz, G., Altman, J. D., Ahmed, R., Woodland, D. L., and Doherty, P. C. (1998). Virus-specific CD8+ T cells in primary and secondary influenza pneumonia. *Immunity* **8:**683–691.

Franco, A., Ferrari, C., Sette, A., and Chisari, F. V. (1995). Viral mutations, TCR antagonism and escape from the immune response. *Curr. Opin. Immunol.* **7:**524–531.

Gilmore, W., Correale, J., and Weiner, L. (1994). Coronavirus induction of class I major histocompatibility complex expression in murine astrocytes is virus strain specific. *J. Exp. Med.* **180:**1013–1023.

Goulder, P., Phillips, R., Colbert, R., McAdam, S., Ogg, G., Nowak, M., Giangrande, P., Luzzi, G., Morgan, B., Edwards, A., McMichael, A. J., and Rowland-Jones, S. (1997). Late escape from an immunodominant cytotoxic T-lymphocyte response associated with progression to AIDS. *Nature Med.* **3:**212–217.

Hart, D. N. J., and Fabre, J. W. (1981). Demonstration and characterization of Ia-positive dendritic cells in the interstitial connective tissues of rat heart and other tissues, but not brain. *J. Exp. Med.* **154:**347–361.

Hickey, W. F., Hsu, b. L., and Kimura, H. (1991). T-lymphocyte entry into the central nervous system. *J. Neurosci. Res.* **28:**254–260.

Houtman, J. J., and Fleming, J. O. (1996a). Dissociation of demyelination and viral clearance in congenitally immunodeficient mice infected with murine coronavirus JHM. *J. Neurovirol.* **2:**101–110.

Houtman, J. J., and Fleming, J. O. (1996b). Pathogenesis of mouse hepatitis virus-induced demyelination. *J. Neurovirol.* **2:**361–376.

Hudrisier, D., Mazarguil, H., Laval, F., Oldstone, M. B. A., and Gairin, J. E. (1996). Binding of viral antigens to major histocompatibility complex class I H-2Db molecules is controlled by dominant negative elements at peptide non-anchor residues. *J. Biol. Chem.* **271:**17,829–17,836.

Hudrisier, D., Mazarguil, H., Oldstone, M. B. A., and Gairin, J. E. (1995). Relative implication of peptide residues in binding to major histocompatibility complex class I H-

2D[b]: Application to the design of high-affinity, allele-specific peptides. *Mol. Immunol.* **32:**895–907.

Ishikawa, T., Kono, D., Chung, J., Fowler, P., Theofilopouos, A., Kakumu, S., and Chisari, F., V. (1998). Polyclonality and multispecificity of the CTL response to a single viral epitope. *J. Immunol.* **161:**5843–5850.

Jacobsen, G., and Perlman, S. (1990). Localization of virus and antibody response in mice infected persistently with MHV-JHM. *Adv. Exp. Med. and Biol.* **276:**573–578.

Kagi, D., Ledermann, B., Burki, K., Seiler, P., Odermatt, B., Olsen, K. J., Podack, E. R., Zinkernagel, R. M., and Hengartner, H. (1994). Cytotoxicity mediated by T cells and natural killer cells is greatly impaired in perforin-deficient mice. *Nature* **369:**31–37.

Klenerman, P., Rowland-Jones, S., McAdam, S., Edwards, J., Daenke, S., Lalloo, D., Koppe, B., Rosenberg, W., Boyd, D., Edwards, A., Glangrande, P., Phillips, R. E., and McMichael, A. J. (1994). Cytotoxic T-cell activity antagonized by naturally occurring HIV-1 gag variants. *Nature* **369:**403–407.

Knopf, P. M., Harling-Berg, C. J., Cserr, H. F., Basu, D., Sirulnick, E., Nolan, S., Park, J., Keir, G., Thompson, E., and Hickey, W. F. (1998). Antigen-dependent intrathecal antibody synthesis in the normal rat brain: Tissue entry and local retention of antigen-specific B cells. *J. Immunol.* **161:**692–701.

Koenig, S., Conley, A. J., Brewah, Y. A., Jones, G. M., Leath, S., Boots, L. J., Davey, V., Pantaleo, G., Demarest, J. F., and Carter, C. (1995). Transfer of HIV-1-specific cytotoxic T lymphocytes to an AIDS patient leads to selection for mutant HIV variants and subsequent disease progression [see comments]. *Nature Med.* **1:**330–336.

Lampson, L. A. (1995). Interpreting MHC class I expression and class I/class II reciprocity in the CNS: Reconciling divergent findings. *Microscopy Res. Tech.* **32:**267–285.

Lane, T. E., and Buchmeier, M. J. (1997). Murine coronavirus infection: a paradigm for virus-induced induced demyelinating disease. *Trends Microbiol.* **5:**9–14.

Lassmann, H. (1997). Basic mechanisms of brain inflammation. *J. Neural Transmission. Suppl.* **50:**183–90.

Lassmann, H., Zimprich, F., Rossler, K., and Vass, K. (1991). Inflammation in the nervous system. Basic mechanisms and immunological concepts. *Rev. Neurol. (Paris)* **147:**763–781.

Lehner, P. J., Wang, E. C., Moss, P. A., Williams, S., Platt, K., Friedman, S. M., Bell, J. I., and Borysiewicz, L. K. (1995). Human HLA-A0201-restricted cytotoxic T lymphocyte recognition of influenza A is dominated by T cells bearing the V beta 17 gene segment. *J. Exp. Med.* **181:**79–91.

Lewicki, H., Von Herrath, M., Evans, C., Whitton, J. L., and Oldstone, M. (1995). CTL escape viral variants. II. Biologic activity in vivo. *Virology* **211:**443–450.

Lin, M. T., Hinton, D. R., Marten, N. W., Bergmann, C. C., and Stohlman, S. A. (1999). Antibody prevents virus reactivation within the central nervous system. *J. Immunol.* **162:**7358–7368.

Lipton, H. L., Twaddle, G., and Jelachich, M. L. (1995). The predominant virus antigen burden is present in macrophages in Theiler's murine encephalomyelitis virus-induced demyelinating disease. *J. Virol.* **69:**2525–2533.

Liu, C.-C., Young, Y., and Young, J. D. (1996). Lymphocyte-mediated cytolysis and disease. *New Engl. J. Med.* **335:**1651–1659.

Luster, A. (1998). Chemokines—chemotactic cytokines that mediate inflammation. *New Engl. J. Med.* **338**(7):436–445.

Matyszak, M. K. (1998). Inflammation in the CNS: balance between immunological privilege and immune responses. *Prog. Neurobiol.* **56:**19–35.

Matyszak, M. K., and Perry, V. H. (1996). The potential role of dendritic cells in immune-mediated inflammatory diseases in the central nervous system. *Neuroscience* **74**:599–608.

McMichael, A. J., and Phillips, R. E. (1997). Escape of human immunodeficiency virus from immune control. *Annu. Rev. Immunol.* **15**:271–296.

Meier, U., Klenerman, P., Griffin, P., James, W., Koppe, B., Larder, B., McMichael, A., and Phillips, R. (1995). Cytotoxic T lymphocyte lysis inhibited by viable HIV mutants. *Science* **270**:1360–1362.

Moskophidis, D., and Zinkernagel, R. M. (1995). Immunobiology of cytotoxic T-cell escape mutants of lymphocytic choriomeningitis virus. *J. Virol.* **69**:2187–2193.

Moss, P. A., Moots, R. J., Rosenberg, W. M., Rowland-Jones, S. J., Bodmer, H. C., McMichael, A. J., and Bell, J. I. (1991). Extensive conservation of alpha and beta chains of the human T-cell antigen receptor recognizing HLA-A2 and influenza A matrix peptide. *Proc. Natl. Acad. Sci. U.S.A.* **88**:8987–8990.

Murali-Krishna, K., Altman, J. D., Suresh, M., Sourdive, D., Zajac, A., Miller, J., Slansky, J., and Ahmed, R. (1998). Counting antigen-specific CD8 T cells: A reevaluation of bystander activation during viral infection. *Immunity* **8**:177–187.

Neumann, H., Cavalie, A., Jenne, D., and Wekerle, H. (1995). Induction of MHC class I genes in neurons. *Science* **269**:549–552.

Neumann, H., Schmidt, H., Cavalie, A., Jenne, D., and Wekerle, H. (1997). Major histocompatibility complex (MHC) class I gene expression in single neurons of the central nervous system: differential regulation by interferon (IFN)-gamma and tumor necrosis factor (TNF)-alpha. *J. Exp. Med.* **185**:305–16.

Oldstone, M. B. A. (1991). Molecular anatomy of viral persistence. *J. Virol.* **65**:6381–6386.

Ossendorp, F., Eggers, M., Neisig, A., Rupport, T., Groettrup, M., Kloetzel, P., Neefjes, J., Koszinowski, U., and Melief, C. (1996). A single residue exchange within a viral CTL epitope alters proteasome-mediated degradation resulting in lack of antigen presentation. *Immunity* **5**:115–124.

Parker, S. E., Gallagher, T. M., and Buchmeier, M. J. (1989). Sequence analysis reveals extensive polymorphism and evidence of deletions within the E2 glycoprotein gene of several strains of murine hepatitis virus. *Virology* **173**:664–673.

Parra, B., Hinton, D., Marten, N., Bergmann, C., Lin, M. T., Yang, C. S., and Stohlman, S. A. (1999). IFN-γ is required for viral clearance from central nervous system oligodendroglia. *J. Immunol.* **162**:1641–1647.

Perlman, S. (1998). Pathogenesis of coronavirus-induced infections: Review of pathological and immunological aspects. *Adv. Expt. Med. Biol.* **440**:503–513.

Perlman, S., and Ries, D. (1987). The astrocyte is a target cell in mice persistently infected with mouse hepatitis virus, strain JHM. *Microb. Pathog.* **3**:309–314.

Perlman, S., Schelper, R., Bolger, E., and Ries, D. (1987). Late onset, symptomatic, demyelinating encephalomyelitis in mice infected with MHV-JHM in the presence of maternal antibody. *Microb. Pathog.* **2**:185–194.

Perlman, S., Schelper, R., and Ries, D. (1987). Maternal antibody-modulated MHV-JHM infection in C57BL/6 and BALB/c mice. *Adv. Exp. Med. and Biol.* **218**:297–305.

Pewe, L., Heard, S. B., Bergmann, C. C., Dailey, M. O., and Perlman, S. (1999). Selection of CTL escape mutants in mice infected with a neurotropic coronavirus: Quantitative estimate of TCR diversity in the infected CNS. *J. Immunol.* **163**:6106–6113.

Pewe, L., and Perlman, S. (1999). Immune response to the immunodominant epitope of mouse hepatitis virus is polyclonal, but functionally monospecific in C57BL/6 mice. *Virology* **255**:106–116.

Pewe, L., Wu, G., Barnett, E. M., Castro, R., and Perlman, S. (1996). Cytotoxic T cell-resistant variants are selected in a virus-induced demyelinating disease. *Immunity* **5**:253–262.

Pewe, L., Xue, S., and Perlman, S. (1997). Cytotoxic T cell-resistant variants arise at early times after infection in C57BL/6 but not in SCID mice infected with a neurotropic coronavirus. *J. Virol.* **71:**7640–7647.

Pewe, L., Xue, S., and Perlman, S. (1998). Infection with cytotoxic T-lymphocyte escape mutants results in increased mortality and growth retardation in mice infected with a neurotropic coronavirus. *J. Virol.* **72:**5912–5918.

Pircher, H., Moskophidis, D., Rohrer, U., Burki, K., Hengartner, H., and Zinkernagel, R. (1990). Viral escape by selection of cytotoxic T cell-resistant virus variants in vivo. *Nature* **346:**629–633.

Plebanski, M., Lee, E., Hannan, C., Flanagan, K. L., Gilbert, S. C., Gravenor, M. B., and Hill, A. V. S. (1999). Altered peptide ligands narrow the repertoire of cellular immune responses by interfering with T-cell priming. *Nature Med.* **5:**565–571.

Ploegh, H. L. (1998). Viral strategies of immune evasion. *Science* **280:**248–253.

Pope, J. G., Vanderlugt, C. L., Rahbe, S. M., Lipton, H. L., and Miller, S. D. (1998). Characterization of and functional antigen presentation by central nervous system mononuclear cells from mice infected with Theiler's murine encephalomyelitis virus. *J. Virol.* **72:**7762–7771.

Price, D., Goulder, P., Klenerman, P., Sewell, A., Easterbrook, P., Troop, M., Bangham, C. R., and Phillips, R. E. (1997). Positive selection of HIV-1 cytotoxic T lymphocyte escape variants during primary infection. *Proc. Natl. Acad. Sci* **94:**1890–1895.

Rehermann, B., Chang, K., McHutchinson, J. G., Kokka, R., Houghton, M., and Chisari, F. V. (1996). Quantitative analysis of the peripheral blood cytotoxic T lymphocyte response in patients with chronic hepatitis C virus infection. *J. Clin. Invest.* **98:**1432–1440.

Rehermann, B., Pasquinelli, C., Mosier, S., and Chisari, F. (1995). Hepatitis B virus (HBV) sequence variation in cytotoxic T lymphocyte epitopes is not common in patients with chronic HBV infection. *J. Clin. Invest.* **96:**1527–1534.

Rowe, C. L., Baker, S. C., Nathan, M. J., and Fleming, J. O. (1997). Evolution of mouse hepatitis virus: Detection and characterization of S1 deletion variants during persistent infection. *J. Virol.* **71:**2959–2967.

Sabelko, K., Kelly, K., Nahm, M., Cross, A., and Russell, J. (1997). Fas and fas ligand enhance the pathogenesis of experimental allergic encephalomyelitis, but are not essential for immune privilege in the central nervous system. *J. Immunol.* **159:**3096–3099.

Sette, A., Alexander, J., Snoke, K., and Grey, H. M. (1996). Antigen analogs as tools to study T-cell activation function and activation. *Sem. in Immunol.* **8:**103–8.

Shrikant, P., and Benveniste, E. N. (1996). The central nervous system as an immunocompetent organ: role of glial cells in antigen presentation. *J. Immunol.* **157:**1819–22.

Smyth, M. J., and Trapani, J. A. (1998). The relative role of lymphocyte granule exocytosis versus death receptor-mediated cytotoxicity in viral pathophysiology. *J. Virol.* **72:**1–9.

Stevenson, P., Freeman, S., Bangham, C. R. M., and Hawke, S. (1997a). Virus dissemination through the brain parenchyma without immunologic control. *J. Immunol.* **159:**1876–1884.

Stevenson, P. G., Hawke, S., Sloan, D. J., and Bangham, C. R. M. (1997b). The immunogenicity of intracerebral virus infection depends on anatomical site. *J. Virol.* **71:**145–151.

Stohlman, S. A., Bergmann, C. C., and Perlman, S. (1998). Persistent infection by mouse hepatitis virus. *In* "Persistent Viral Infections" (R. Ahmed and I. Chen, eds.), pp. 537–557. John Wiley & Sons, New York.

Stohlman, S. A., Fleming, J. O., Brayton, P. R., Weiner, L. P., and Lai, M. M. C. (1982). Murine coronaviruses: isolation and characterization of two plaque morphology variants of the JHM neurotropic strain. *J. Gen. Virol.* **63:**265–275.

Stohlman, S. A., Matsushima, G. K., Casteel, N., and Weiner, L. P. (1986). In vivo effects of coronavirus-specific T cell clones: DTH inducer cells prevent a lethal infection but do not inhibit virus replication. *J. Immunol.* **136:**3052–3056.

Sun, N., and Perlman, S. (1995). Spread of a neurotropic coronavirus to spinal cord white matter via neurons and astrocytes. *J. Virol.* **69:**633–641.

Suzumura, A., Lavi, E., Bhat, S., Murasko, D., Weiss, S., and Silberberg, D. (1988). Induction of glial cell MHC antigen expression in neurotropic coronavirus infections. *J. Immunol.* **140:**2068–2072.

Suzumura, A., Lavi, E., Weiss, S. R., and Silberberg, D. H. (1986). Coronavirus infection induces H-2 antigen expression on oligodendrocytes and astrocytes. *Science* **232:**991–993.

Taylor, L. P., Kempton, R. A., and Woiwod, I. P. (1976). Diversity statistics and the log-series model. *J. Anim. Ecol.* **45:**255–272.

Waldner, H., Sobel, R., Howard, E., and Kuchroo, V. (1997). Fas- and FasL-deficient mice are resistant to induction of autoimmune encephalomyelitis. *J. Immunol.* **159:**3100–3103.

Wallace, M., Keating, R., Health, W., and Carbone, F. (1999). The cytotoxic T-cell response to herpes simplex virus type 1 infection of C57BL/6 mice is almost entirely directed against a single immunodominant determinant. *J. Virol.* **73:**7619–7626.

Wang, Y., Xiang, Z., Pasquini, S., and Ertl, H. C. (1998). Effect of passive immunization or maternally transferred immunity on the antibody response to a genetic vaccine to rabies virus. *J. Virol.* **72:**1790–6.

Weiner, A., Erickson, A., Kanospon, J., Crawford, K., Muchmore, E., Hughes, A., Houghton, M., and Walker, C. M. (1995). Persistent hepatitis C virus infection in a chimpanzee is associated with emergence of a cytotoxic T lymphocyte escape variant. *Proc. Natl. Acad. Sci. U.S.A.* **92:**2755–2759.

Williams, K. C., and Hickey, W. F. (1995). Traffic of hematogenous cells through the central nervous system. *Curr. Top. Microbiol. & Immunol.* **202:**221–245.

Xue, S., and Perlman, S. (1997). Antigen specificity of CD4 T cell response in the central nervous system of mice infected with mouse hepatitis virus. *Virology* **238:**68–78.

Xue, S., Sun, N., van Rooijen, N., and Perlman, S. (1999). Depletion of blood-borne macrophages does not reduce demyelination in mice infected with a neurotropic coronavirus. *J. Virol.* **73:**6327–6334.

Yamaguchi, K., Goto, N., Kyuwa, S., Hayami, M., and Toyoda, Y. (1991). Protection of mice from a lethal coronavirus infection in the central nervous system by adoptive transfer of virus-specific T cell clones. *J. Neuroimmunol.* **32:**1–9.

Young, A., Zhang, W., Sacchettini, J., and Nathenson, S. (1994). The three-dimensional structure of H-2D at 2.4 A resolution: Implications for antigen-determinant selection. *Cell* **76:**39–50.

Zhang, E. T., Richards, H. K., Kida, S., and Weller, R. O. (1992). Directional and compartmentalised drainage of interstitial fluid and cerebrospinal fluid from the rat brain. *Acta Neuropathol.* **83:**233–239.

Zinkernagel, R. M. (1996). Immunology taught by viruses. *Science* **271:**173–178.

ADVANCES IN VIRUS RESEARCH, VOL 56

DNA IMMUNIZATION AND CENTRAL NERVOUS SYSTEM VIRAL INFECTION

J. Lindsay Whitton* and Robert S. Fujinami[†]

*Department of Neuropharmacology, CVN-9
The Scripps Research Institute, La Jolla California 92037
†Department of Neurology
University of Utah
Salt Lake City 84132

I. Virus Infections of the Central Nervous System

Mild central nervous system (CNS) symptoms such as headache and drowsiness can result from systemically elevated cytokine levels, and therefore are common in many virus infections, even in the absence of infection of the CNS. In this chapter we shall consider only those viruses that are known to infect the CNS.

A. *Poliovirus*

Poliovirus, a member of the picornavirus family and *Enterovirus* genus, was a major scourge in the earlier part of the twentieth century. As the genus name indicates, the virus replicates in the gastrointestinal tract; as such, it is usually transmitted by the fecal–oral route. Viremia is common, but the vast majority of infections remain asymptomatic. CNS infection is quite unusual, and is initiated either as a result of the viremia or, more rarely, by neural spread. The virus infects the anterior horn motor neurons of the spinal cord, causing poliomyelitis (from the Greek *polios* plus *myelos*—"inflammation of the gray marrow"), the disease for which the virus is named. Loss of these cells results in paralysis, often of a lower limb. In some cases, the infection ascends the cord to cause paralysis of upper limbs, and in the most extreme cases the infection reaches the junction of the spinal cord and the brain, resulting in paralysis of the muscles of respiration (bulbar palsy) and requiring that the victim be placed in an "iron lung." The development, in the mid-1950s, of killed and live polio vaccines massively reduced the frequency of this infection, and of the associated disease, and a program directed by the World Health Organization aims to eradicate poliovirus within the next few years. If successful, this will be the second virus (following smallpox) to have been exterminated by vaccination. The pathogenesis of the murine "equivalent" of poliovirus—Theiler's virus—has been extensively studied, and will be described later in this chapter.

B. *Herpesviruses*

Herpesviruses are probably the commonest viruses to infect neuronal tissue. Herpes simplex virus (HSV) and varicella-zoster virus (VZV) establish latent infections in the dorsal root ganglia of the peripheral nervous system and, particularly in the immunocompromised, can reactivate and disseminate to cause encephalitis or, more commonly, vesicular eruptions in the skin area innervated by the infected neurons—leading to cold sores/fever blisters (HSV) or shingles (VZV). Epstein-Barr virus (EBV), the cause of infectious mononucleosis, is also associated with a rare encephalitis (Andersson *et al.*, 1999; Schiff *et al.*, 1982).

C. Measles Virus

Measles virus is a negative-stranded RNA virus, and a member of the morbillivirus genus. Infection by this virus most commonly results in the characteristic rash, which is immunopathological, being mediated not by direct viral cytotoxicity, but instead by the T cell response of the host. However other organs are frequently affected, and giant cell pneumonia can be lethal. CNS infection is infrequent, but meningitis and encephalitis can occur. Postinfectious encephalomyelitis occurs in ~1/1000 cases, usually occurring within weeks of infection; however, it is difficult to detect virus in the CNS, and it has been suggested that the observed perivenular demyelination is immune-mediated (Gendelman *et al.*, 1984). A rare complication (~1 in 2×10^6 cases) is subacute sclerosing panencephalitis (SSPE), in which measles virus RNA and protein persist in glial cells and neurons, leading to CNS dysfunction and death.

D. Lentiviruses

In the 1950s, the Icelandic virologist Bjorn Sigurdsson carried out epidemiological studies of the sheep diseases rida (Sigurdsson, 1954a) (more commonly known by the English term "scrapie") and visna, and suggested that they were infectious in origin, but had an incubation period much longer than that of "standard" viruses (Sigurdsson, 1954b). These led him to propose a new category of virus, the "slow virus." Slow viral diseases were observed in many other species, including humans; the CNS disease kuru, first described in Papua New Guinea, had an incubation period measured in years. For many years this curious and clinically defined category of viruses remained devoid of molecularly characterized members, but eventually it became clear that the grouping encompassed several very different agents. Some of these agents were relatively standard viruses, but other agents—including the agent of scrapie—were refractory to categorization until Stanley Prusiner's groundbreaking identification of prions, which are described in Section I,I below. One of the first-characterized slow viruses was a retrovirus that caused visna; indeed, the long incubation period gave this virus group its name (lentiviruses; from the Latin word *lentus,* "slow"). Human immunodeficiency virus (HIV) is a lentivirus and, in common with all of this group, disease often does not appear until many years postexposure. The later stages of HIV infection are frequently characterized by encephalitis and dementia (Fox et al., 1997; Moller et al., 1988; Wiley et al., 1991).

E. Rabies Virus

This virus, a member of the rhabdovirus family, kills ~40,000 people annually, and is transmitted through the saliva of infected animals, usually by bites. The virus may undergo local replication at the site of inoculation, perhaps in muscle cells, but a key feature is its subsequent centripetal spread to the CNS within neuronal axons. Viremia is not a prominent feature, and disease can be prevented by physical or chemical interruption of axonal transport (Ceccaldi et al., 1989; Tsiang, 1979). On reaching the CNS, the virus spreads within the brain (often resulting in Negri body formation, especially in the hippocampus); however the detectable histological damage is often less extensive than might be expected, given the severity of the neurological and behavioral symptoms observed (the term "rabies" is derived from the Latin term for "madness"). After CNS infection has been established, the virus may spread centrifugally to various tissues, including the salivary glands, from which it is secreted into the saliva.

F. Arenaviruses

This family includes a variety of human pathogens, some of which cause hemorrhagic fevers (Lassa, Junin, and Machupo viruses). The prototype of the family, lymphocytic choriomeningitis virus (LCMV), infects humans more frequently than is often realized. The virus is rodent-borne, and recent surveys in Baltimore indicated that 9% of house mice and 4.7% of humans were seropositive (Childs et al., 1991; Childs et al., 1992). The spectrum of human disease ranges from subclinical infection to fatal meningoencephalitis, and the virus is teratogenic, leading to hydrocephalus (Larsen et al., 1993). The immunobiology and pathogenesis of LCMV infection have been extensively studied, and the use of this model to evaluate DNA immunization will be described later in this chapter.

G. Arboviruses

Arboviruses (Arthropod-borne viruses) are important human pathogens causing, for example, yellow fever and dengue hemorrhagic fever. In the United States, the primary clinical manifestation of arboviral disease is encephalitis. Many viruses, from several different viral families, are implicated; most are mosquito-borne, but some are transmitted by ticks. The most commonly diagnosed arboviral encephalitis in the United States is that caused by the flavivirus St. Louis encephalitis (SLE) virus. However, we cannot underestimate the capacity of viruses to enter new ecological niches. An outbreak of viral encephalitis in New York in September 1999 was initially ascribed to SLE, but subsequently was attributed to West Nile virus, which had not previously been identified in this country. Other arboviral encephalitides include those caused by western equine and Venezuelan equine encephalitis viruses (alphaviruses) and the California group of encephalitis viruses (bunyaviruses).

H. Miscellaneous Viral Encephalitides

Borna disease virus (BDV) is the prototype of a new family of negative-stranded RNA virus (Briese *et al.*, 1994; de la Torre, 1994). This virus causes an immune-mediated encephalitis in many species (Planz *et al.*, 1995; Stitz *et al.*, 1991). Infectious BDV has not been identified in humans, but antibodies, proteins, and nucleic acids have been identified in the sera and/or CNS of patients suffering from certain psychiatric disorders (Bode *et al.*, 1995; de la Torre *et al.*, 1996a; de la Torre *et al.*, 1996b), raising the intriguing possibility that some psychoses may be virus-induced; a chapter of this book is devoted to BDV neurotropism and its consequences. Also described elsewhere in this volume are coronavirus infections of the CNS. Finally, viral diseases of the CNS include progressive multifocal leukoencephalopathy (PML), caused by the papillomavirus JC. More than 80% of humans carry antibodies specific for this virus, but PML usually is seen only in immunosuppressed patients; consequently, the incidence of PML has increased in parallel with HIV infection (Dorries, 1998; Gordon and Khalili, 1998; Jensen and Major, 1999; Weber and Major, 1997).

I. Prions

As mentioned above, these agents have an extremely protracted incubation period. The prion protein (PrP) was first identified as part of the protease-resistant material proposed by Prusiner as a protein-only infectious agent responsible for scrapie, and for other transmissible spongiform encephalopathies (TSEs). This heretical notion engendered justified skepticism among many experts, but it has resisted numerous challenges, and the evidence in its favor is now very strong. The PrP gene encodes a cellular protein, expressed on many cell types including neurons and cells of the immune system, whose normal function remains uncertain. Mice lacking this gene (PrPko mice) have few detectable CNS abnormalities (Bueler *et al.*, 1992; Kuwahara *et al.*, 1999). Expression of this gene is a prerequisite for host susceptibility to TSE agents, and PrPko mice are resistant to challenge (Bueler *et al.*, 1993; Prusiner *et al.*, 1993). This host protein can exist in at least two conformations, which are distinguished by their sensitivity to protease; the protease-sensitive form appears to be the "normal" conformer, present in normal hosts, and not causing disease, while the protease-resistant form appears to be infectious. Prion replication and infectivity appear to be determined by the ability of the "pathogenic" conformer to initiate conformational changes in its "normal" counterpart; this is the key to prion diseases, and underpins both the replication of the agent in an infected host, and its transmission in the absence of a nucleic acid. Such conformational changes have recently been demonstrated in tissue culture studies (Bessen *et al.*, 1995; Kocisko *et al.*, 1994; Kocisko *et al.*, 1995). The TSEs, then, are transferred to the new host when the misfolded protein uses the normal proteins of the new host as substrates for production of abnormal conformers. The genes permitting the replication of this novel form of infectious pathogen are therefore provided by the unwitting victim. They are unusual, as they contain no nucleic acid genome. Instead, prions are infectious proteins that are encoded by the host's own PrP gene. This protein can exist in at least two conformations; one is normal (and seems to serve some function in the CNS), while the other is abnormal. The abnormal conformer, which is infectious, appears able to act as a template, causing its normal siblings to convert to abnormality. The accumulation of abnormal conformers leads to spongiform encephalopathy, the histological hallmark of prion diseases. The historical relationship between "slow viruses" and prions has often resulted in the inclusion of prions in virological textbooks—indeed, they are the topic of two chapters in this volume—but their radically different (a) coding strategy (as a host gene), (b) mode of replication (by directed misfolding of a self-protein into an abnormal conformer), and (c) mechanism of infection (as an infectious protein) surely render them unique. Despite their questionable membership in the virus taxon, we mention them here because they cause CNS diseases, and because recent studies suggest that immunization may modify other CNS diseases characterized by abnormal deposition of self-proteins.

J. CNS Diseases That May (or May Not) Be of Viral Origin

The causes of certain CNS diseases remain unknown. For example, multiple sclerosis (MS) is a degenerative disease of the CNS, which is characterized by demyelination. The clinical and histological features

of MS can vary from relapsing–remitting disease to chronic progression. What role might viruses play? First, MS may be the result of a
persistent virus infection, acting to drive a chronic immune response.
Over the past several decades, various viruses have been advanced as
the cause of MS but, to date, none of these suggestions has withstood
further analysis; as a result, enthusiasm for this hypothesis has perhaps diminished. However, while it is unlikely that MS results from
persistent infection by a known virus, it is possible that it is caused by
a virus which has yet to be identified. While it may be tempting to
think that, at the millennium, we have identified all microbes and their
associated infections, it has been estimated that only ~0.4% of extant
bacteria have been cataloged, and new viruses continue to be identified.
Indeed, even entire virus families have been discovered in the past
decade (e.g., the *Bornaviridae,* mentioned above). No animal other than
humans develops MS. This is not true of other autoimmune diseases.
For example, humans and other animals develop diabetes, arthritis,
and thyroiditis. Thus, MS could be caused by a microbe whose host
range is tightly restricted to humans. However, a second, more popular,
hypothesis implicates autoimmunity; MS, like several other autoimmune diseases, it is commoner in women than in men. A number of
ideas have been advanced to explain virus-induced autoimmunity
(recently reviewed in Oldstone, 1998; and in Whitton and Fujinami,
1999); these include the release, from infected cells, of sequestered host
proteins which then act as autoantigens (see the chapter on "epitope
spreading" in this book). Consistent with this, antibodies and T cells
specific for CNS self-antigens are detectable in the CNS of MS patients,
but not of healthy individuals. It is thought that MS may be initiated
by an infection with one virus, but that subsequent infections—with
unrelated viruses—might "boost" the immune response against the
released CNS self-antigen. This proposition is supported by an apparent association between a relapse of MS and recent infection. However,
one could argue that relapses are caused instead by a transient
immunosuppression, accompanying virus infection, with a resulting
reactivation of an unidentified persistent or latent virus. Thus, while
the pathogenesis of MS remains uncertain, many researchers feel that
viruses play some role in the initiation and maintenance of the disease;
a chapter in this volume is devoted to this important topic.

II. Antiviral Immune Response

To evaluate the role of DNA immunization in protecting against
viral infection of the CNS, and against the related diseases, we must

first consider how the immune system recognizes viruses, and virus-infected cells, and how it deals with these challenges. This topic has recently been summarized (Whitton and Oldstone, 2000).

A. Overview

The immune response to virus infection is divided into two components, the innate response and the adaptive response, which are serially expressed in partially overlapping temporal phases. Soon after the host is first infected with a virus, the innate immune response is activated. Many cells secrete interferons α and β, while natural killer (NK) cells (and, rarely, activated macrophages) secrete interferon-γ (IFNγ). (IFNγ is also an important effector molecule released by T cells during the antigen-specific phase of the immune response; this is described in more detail below.) Upon exposure to these cytokines, noninfected cells are rendered resistant to virus infection; the interferons therefore limit the ability of the virus to spread locally. Meantime the NK cell population expands, usually peaking approximately 3–4 days postinfection. These cells cannot specifically detect virus-infected cells, instead being triggered by a combination of 2 factors: poorly characterized stimulatory molecules, and the absence of class I major histocompatibility complex (MHC) molecules. As the innate response wanes, so the adaptive response expands. The adaptive response differs from the innate in two key ways. First, the adaptive response is *antigen-specific;* it recognizes specific structures (usually proteins, but occasionally carbohydrates and glycolipids) on viruses or on virus-infected cells. Second, the adaptive response exhibits *memory;* antigen-specific cells are maintained long after the infection is cleared, and these memory cells permit a more rapid and elevated response if the host is reexposed to the antigen. Antigen-specific memory forms the cornerstone of vaccination; a vaccine induces memory cells specific for the appropriate antigen(s), and these cells respond rapidly, should the host encounter the related pathogen. This chapter is devoted to vaccination, and so we shall focus below on the adaptive immune response. All antigen-specific immunity relies on lymphocytes, of which there are two types: B lymphocytes (which produce antibodies) and T lymphocytes.

B. How Antibodies Recognize Viruses and Virus-Infected Cells

Antibodies recognize antigen through regions of hypervariable sequence. Crystallographic analyses of the antibody–antigen union indicate that the union is more "hand in glove"—in which components can, to some extent, alter their conformation to accommodate one another—than "lock and key," in which both elements are fixed, with each being unable to modulate to the other (Arevalo *et al.*, 1993; Rini *et al.*, 1992). As a rule, when playing their part in host immunity, antibodies recognize intact proteins. Thus, antibodies can interact with bacteria and viruses, as well as with viral proteins (most often glycoproteins) expressed on the surface of infected cells.

C. How Antibodies Control Virus Infections

Viruses are obligate intracellular parasites, but they are (usually) not transmitted in association with cells but, rather, as free infectious particles. Since antibodies can recognize free viruses, it is easy to see how antibodies can play a major role in controlling virus infection, by inactivating the virus before it can enter the cell. Antibodies play an extremely important part in antiviral immunity. Indeed, in many cases, the administration of specific antibodies can confer complete protection against subsequent challenge with the relevant virus, and passively transferred antibody remains an important component of medical treatment of patients exposed to certain viruses (e.g., rabies). Antibodies neutralize viruses in a number of ways: (1) they may bind to the part of the virus that interacts with its cell-surface receptor, preventing virus attachment to the cell; (2) they may agglutinate many infectious particles into a single "clump," thus reducing the number of cells that will become infected; (3) viruses may activate complement (directly; or indirectly, via antibodies), releasing chemotactic factors such as C5a and C3q. Note that intact antibodies are not a prerequisite for antiviral effectiveness; Fab fragments specific for Respiratory Syncytial Virus (RSV) F glycoprotein, when instilled into the lungs of infected mice, were therapeutically effective (Crowe *et al.*, 1994). Such approaches hold promise, particularly in the light of recent advances in technologies that allow the rapid production of antibodies of any desired specificity (Barbas *et al.*, 1991; Kang *et al.*, 1991).

There are five different classes of antibody, immunoglobulin (Ig)A, IgG, IgM, IgD, and IgE, each with different functional attributes. During natural infection, most viruses gain entry via respiratory or enteric mucosal surfaces. It is therefore not surprising that mucosal immunity and, in particular, secretory IgA, plays an important role in control of viral infections (Ogra and Garofalo, 1990). The pentameric, decavalent IgM molecule is produced early after virus infection, is usually independent of T cell help, and acts as the initial antibody-mediated systemic antiviral response. Later in infection, and on secondary exposure, most IgM-producing cells switch to produce IgG of the same antigen specificity. IgG_1 is the major complement-binding and opsonizing antibody in humans (Spiegelberg, 1990), and complexing of viruses with IgG will also facilitate their Fc receptor-mediated phagocytosis by monocytes macrophages and by polymorphonuclear leukocytes.

Even after cell entry, antibodies can exert effects on virus infection by interacting with viral proteins (most often glycoproteins) on the surface of infected cells, lysing the infected cell in association with complement, or modulating the intracellular viral replication (Fujinami and Oldstone, 1979). Relevant to this chapter, it has been suggested that neuronal virus infections can be eradicated by antibodies, without damaging the neurons (Levine et al., 1991); the mechanism for this remains undefined. However, many viruses delay glycoprotein expression until late in the infective cycle, when viral maturation may have occurred, and at this point the antibody-mediated effects may be biologically inconsequential. How can a host detect a virus-infected cell early in the infection process, thus maximizing its immunological advantage? Here, antibodies are less effective, being limited by their recognition requirements, whereas T cells play a critical role, as detailed below.

D. How T Cells Recognize Viruses and Virus-Infected Cells

T cells can be categorized by the surface marker proteins (CD4 or CD8) that they express. The majority of CD8$^+$ cells are cytotoxic T lymphocytes (CTLs), although some CD8$^+$ cells are nonlytic and exert their antiviral effects by cytokine release (Levy et al., 1996), while most CD4$^+$ cells are helper cells that secrete cytokines to assist B cell maturation, and perhaps to aid a developing CD8$^+$ T cell response. T cells recognize antigens via a cell surface heterodimer, the T-cell receptor (TcR). This molecule is structurally reminiscent of the Fab portion of an antibody molecule, but the nature of T cell recognition differs from that of antibody recognition in one critical aspect: while antibodies recognize antigen in isolation, T cells react to antigen in the form of a short peptide presented by a host glycoprotein encoded in the MHC. There are two major classes of MHC molecule (class I and class II), and there is a close relationship between the class of MHC/peptide complex recognized by a T cell and the surface marker (CD8 or CD4) borne by the T cell. MHC class I molecules are the "classical" molecules associated with graft rejection (the phenomenon that gave the MHC its name); they are expressed on most somatic cells, and they interact with T cells bearing the CD8 surface marker. In contrast, MHC class II molecules have a much more restricted expression, being found only on specialized antigen-presenting cells (e.g., macrophages, B lymphocytes, dendritic cells), and they interact with T cells carrying the CD4 surface marker. At the target cell surface, class I and class II molecules are similar in overall structure. The class I heterodimer comprises the class I heavy (H) chain closely complexed with a non-MHC-encoded protein, β_2-microglobulin (β_2M). The class II heterodimer consists of two similar chains, α and β. Both class I and class II form a structure graphically described as a Venus fly trap, the groove of which binds an antigenic peptide, in a sequence-specific manner, and presents it on the cell surface for the perusal of T cells (Bjorkman et al., 1987a; Bjorkman et al., 1987b). Although superficially similar, the two types of MHC/peptide complex differ in how they reach the cell surface. MHC class I is optimized to present intracellular antigen, while MHC class II presents antigen captured from the extracellular milieu. Thus, when T cells distinguish between a peptide/class I complex and a peptide/class II complex, they are really discriminating on the basis of the source of the peptide—was that peptide derived from protein made within the cell, or from protein taken up from the extracellular spaces?

The MHC class I pathway is vital for recognition of virus-infected cells. Viral proteins are synthesized and degraded within the cell, and the resulting peptides are transported to the endoplasmic reticulum, where they encounter empty MHC class I molecules. Peptides with sufficient affinity for particular MHC alleles bind in the groove; β_2M attaches to, and stabilizes, the complex; and the trimolecular structure travels to the cell membrane, to be screened by the CD8[+] T cells of the host; these cells, as the effector arm of the antiviral T cell response, therefore assume great significance in antiviral immune responses. One major advantage of this arrangement is that CD8[+] T cells can recognize almost any viral protein (as long as it contains a peptide sequence that can be presented by MHC class I). Therefore, even proteins expressed at the beginning of the viral life cycle, and limited to the cytosol, are vulnerable to degradation and MHC class I presentation. In this way, the host can identify and eradicate infected cells at a very early stage, long before viral maturation can occur. For example, the major CTL response to human cytomegalovirus (HCMV) is directed to a protein expressed immediately on infection; a similar situation exists for VZV and HSV. Any defect in the MHC class I antigen-processing pathway may result in the infected cell's being unable to present viral peptide on the cell membrane, which in turn would render the virus "invisible" to CD8[+] T cells.

E. How T Cells Control Virus Infections

Antibodies are important in limiting the number of infected cells, and in clearing virus from the host, but CD8+ T cell responses play a critical role in the control of many virus infections. These cells have been extensively characterized in animal models, and the results equate well with those obtained in human studies. Although the role of CD8+ T cells in controlling primary virus infection has been recognized for some time, the importance of these cells in vaccine-induced protective immunity is often disregarded. Many studies have shown that vaccine-induced CD8+ T cell responses, in the absence of vaccine-induced antibody responses, are sufficient to confer solid protective immunity against a subsequent virus challenge. For example, in the LCMV mouse model, recombinant vaccines containing "minigenes" that encode isolated LCMV CTL epitopes as short as 11 residues can confer protection against normally lethal doses of challenge virus, and different epitopes can be linked on a "string of beads" to protect on several MHC backgrounds (An and Whitton, 1997; An and Whitton, 1999; Whitton *et al.*, 1993). No LCMV-specific antibody responses are induced by these vaccines, which proves that protective effects can be mediated by cellular immune responses. CD8+ T cells are also important in the control of human viral diseases. EBV infects and transforms human B lymphocytes, and the control of this cell population appears to be managed in large part by virus-specific CD8+ cells. Indeed, some immunosuppressed individuals, lacking such cells, may develop EBV+ lymphomata (Rickinson *et al.*, 1992). Marked CD8+ T cell responses have also been found against influenza virus, measles virus, mumps, respiratory syncytial virus, human immunodeficiency virus, and other agents.

Two major effector mechanisms underlie the *in vivo* antiviral effects of virus-specific CD8+ T cells: cell lysis and cytokine release. Most virus-specific CD8+ T cells can lyse infected target cells, and thus justify the name CTL. CTLs contain the protein perforin (Podack *et al.*, 1988), which is released on contact with an infected cell, and self-assembles into transmembrane pores that penetrate the cytoplasmic membrane of the target cell—leading to cell death. Transgenic mice with a dysfunctional perforin gene are much less effective at controlling infection by some (though not all) viruses (Kagi *et al.*, 1994a; Kagi *et al.*, 1994b; Walsh *et al.*, 1994). Furthermore, virus-specific CD8+ T cells can induce apoptotic lysis when the Fas ligand (FasL) protein, expressed on the T cell membrane (Suda and Nagata, 1994), interacts with Fas protein on the infected cell, initiating a signaling cascade that ends in target cell death (Shresta *et al.*, 1998; Welsh *et al.*, 1990;

Zychlinsky *et al.*, 1991). CD8+ T cells release antiviral cytokines. Many CD8+ T cells release high levels of cytokines—for example, interferon-γ (IFNγ) and tumor necrosis factor-α (TNFα). Mice lacking the IFNγ receptor have increased susceptibility to several infections, despite apparently normal CTL and Th responses (Huang *et al.*, 1993). It has been cogently argued that a major role of the TcR/MHC/peptide interaction is simply to hold CD8+ T cells in the immediate proximity of virus-infected cells, thus focusing cytokines on the infected cell (Ramsay *et al.*, 1993; Ruby and Ramshaw, 1991); and convincing data from mice persistently infected with LCMV (Oldstone *et al.*, 1986; Tishon *et al.*, 1995), from hepatitis B virus (HBV) transgenic mice (Guidotti and Chisari, 1996; Guidotti *et al.*, 1996), and from HBV-infected primates (Guidotti *et al.*, 1999) have shown that viral materials can be eradicated *in vivo* from neurons (Oldstone *et al.*, 1986; Tishon *et al.*, 1995) and from hepatocytes (Guidotti and Chisari, 1996; Guidotti *et al.*, 1996), in the absence of cytolysis.

The availability of these two T cell effector mechanisms allows us to consider how virus infections might ideally be handled by the host. In the following scenarios we shall consider two interacting variables: first, the pathogenicity of the virus; and second, the resilience of the infected organ. Consider a cell infected by a highly lytic virus. Intuitively, it may seem that the host should attempt to lyse the doomed cell; after all, the cell will die soon, and early lysis may benefit the host, by destroying a virus "factory" and thus preventing release of infectious particles. In many organs, this is precisely what happens. Often, the cells that are lysed are later replaced; for example, the regenerative power of the liver is legendary (it has been estimated that 10^9 hepatocytes are produced daily to replenish cells lost during HBV infection [Nowak *et al.*, 1996]). Furthermore, even if tissue regeneration is incomplete, most host organs are sufficiently functionally redundant so as to allow the host to tolerate loss of a significant proportion of the organ mass; for example, we can tolerate loss of ~90% of kidney function before suffering signs and symptoms of renal failure. However, what of a tissue which can neither regenerate, nor function appropriately, if some of its components are lost? In such a tissue, it would not make sense for the host to lyse infected cells; it would be better to take the risk that the virus is lytic than to consign the cell to certain immunopathological death. Since the host presumably cannot foresee the lytic capacity of an infectious agent, it faces a dilemma— should it kill an infected cell (beneficial in most tissues, for both lytic and nonlytic viruses), or should it instead secrete cytokines, allowing the infected cell to survive (possibly beneficial, for nonlytic infections in organs that have minimal regenerative capacity, and in which the host cannot tolerate cell loss)? To render such a choice meaningful, the CD8+ T cell response would have to be able to mount responses that were nonlytic in nature, thus providing the capacity for a cytokine-mediated antiviral effect, while permitting survival of the infected cell. There is some evidence for the existence of nonlytic CD8+ T cells (Blackbourn et al., 1994; Levy et al., 1996).

III. Central Nervous System as a Haven for Viruses

Many DNA and RNA viruses establish infection in the CNS, and often in neurons. Why should this be the case? The most likely reason is that the CNS, and its cells, are immune privileged—they are less open to immune surveillance than most other organs or cell types. Intuitively, this makes sense. CNS neurons are nondividing cells, and have historically been considered irreplaceable. Although some recent findings indicate that neurons and their pathways may be more resilient and plastic than previously thought, it is clear nevertheless that the host can ill afford to lose such vital cells. Were neurons to be as accessible as most somatic cells following virus infection, they would be susceptible to lysis by virus-specific CTLs, if sufficient class I/peptide complexes were expressed (as discussed below). Perhaps to limit this destruction, evolution has rendered neurons less open to immune surveillance.

A. Blood–Brain Barrier

Much of the CNS resides behind the blood-brain barrier, which resists passage of most cells, and even of many molecules. As a result, the CNS is biochemically and cellularly distinct from other organs. (The blood–brain barrier is the topic of another chapter in this volume.)

B. CNS Cells are Not Easily Recognized by Antigen-Specific T Cells

When analyzed in vitro, neurons show minimal transcription or surface expression of class I MHC, although this is inducible by IFN-γ (Joly et al., 1991; Lampson et al., 1983; Lampson and Fisher, 1984; Neumann et al., 1995). Furthermore, neurons differ from most cell types in failing to express several other components of the class I antigen presentation pathway (e.g., β_2m and the TAP transporters) (Joly and Oldstone, 1992). Interestingly, IFN-γ upregulates these molecules, along with the class I heavy chain, leading to cell-surface expression of peptide/MHC complexes; in contrast, TNFα upregulates transcription of class I, but not of various accessory molecules, and therefore, there is no increase in cell surface class I/peptide expression following exposure to this extremely toxic cytokine (Neumann *et al.*, 1997). We have recently shown that different populations of virus-specific CD8+ T cells can selectively express IFNγ or TNFα (Slifka and Whitton, 2000); it is tempting to suggest that interactions between infected neurons and particular subpopulations of virus-specific T cells in the CNS might secrete TNFα, but not IFNγ, thereby permitting the eradication of virus, without causing extensive disruption of the immune recognition status of neighboring CNS cells. Furthermore, we have recently analyzed the regulation of cytokine synthesis by antigen-specific CD8+ T cells, and have shown it to be exquisitely sensitive to antigen contact (Slifka *et al.*, 1999). Even at the height of infection, cytokine synthesis is turned off in the vast majority of the virus-specific CD8+ T cells; transcription of cytokine mRNA begins immediately upon antigen contact, and cytokine production terminates instantly upon antigen disengagement. Thus, CD8+ T cells produce cytokines only when they are in direct contact with the appropriate peptide/MHC complex. One can speculate that there may be an evolutionary advantage of this arrangement in the CNS. Cytokines such as IFN-γ upregulate cell-surface expression of class I MHC, which may be disadvantageous to the host; if T cells produced IFN-γ in a promiscuous manner, this might lead to the display of MHC complexes on cells throughout the CNS, so the tight regulation of cytokine production ensures that this risk is minimized.

Of course, *in vitro* results may not reflect the normal status and responsiveness of neurons *in vivo*. However, *in vivo* studies have shown that, under normal circumstances, CNS neurons—whether they lie within or outwith the blood–brain barrier—exhibit low-to-undetectable levels of MHC class I and β2m (Lampson and Hickey, 1986; Whelan *et al.*, 1986).

C. CNS Environment May Suppress T Cell Activity

Gangliosides—glycosphingolipids—have long been thought to modulate the immune response mounted by NK and T cells (Bergelson *et al.*, 1989; Bergelson, 1993; Bergelson, 1995), and recent work (discussed in another chapter) indicates that these molecules may contribute to an immunosuppressive milieu in the CNS during virus infection (Irani *et al.*, 1996; Irani, 1998). It is therefore possible that activated virus-

258   J. LINDSAY WHITTON AND ROBERT S. FUJINAMI

specific T cells which enter the CNS are functionally impaired by inter-actions with these complex lipids, which are abundant in this tissue.

IV. Vaccinating against Virus-Induced CNS Diseases: An Introduction to Two Mouse Models

A. Vaccinating Against CNS Viral Diseases

Vaccines are designed not to prevent infection, but to diminish the frequency and severity of disease. Immunizing against virally induced CNS diseases does not necessarily require the induction of immunity within the CNS itself. Perhaps the best example is polio vaccine, which induces a strong antibody-mediated mucosal immunity. If the vaccinee ingests virally contaminated material, these antibodies either prevent enteric infection, or else radically reduce the level to which the virus can replicate in the gastrointestinal tract. This has two benefits. First, the infected individual is less likely to develop a severe viremia, which in turn greatly diminishes the risk of CNS infection and disease. Second, the infected host will excrete less virus, thus reducing the risk of infection for his or her susceptible neighbors. Thus, polio vaccines can protect the individual and the community against poliomyelitis, with-out inducing CNS-specific immune responses in the vaccinee. Indeed, none of the currently available vaccines against the diseases reviewed in Section I of this chapter are known to induce responses in the CNS; all of them work by inducing systemic immunity, which limits infection or viral replication/dissemination. In many ways this is encouraging, for it implies that, to be successful, a vaccine does not have to over-come the immune privilege present in a healthy CNS.

B. Two Mouse Models of CNS Virus Infection and Disease

Having argued that the CNS is an immune-privileged site, we must now acknowledge that this privilege is incomplete. Indeed, this is implicit in the fact that virus infections can result in encephalitis; the inflammatory response (particularly the presence of virus-specific lymphocytes) is unequivocal evidence that any immune privilege has been breached. Animal models have revealed much about the immune responses that take place in the CNS. Here we shall describe two mod-els, the LCMV and Theiler's virus, which are studied by our laborato-ries; these virus infections allow us to demonstrate different facets of the immune response in the CNS.

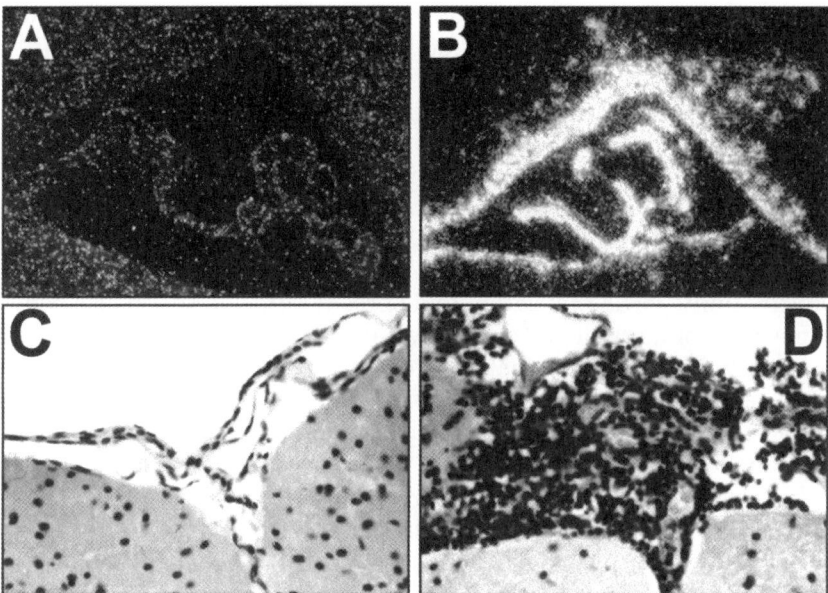

Fig 1. Lymphocytic choriomeningitis 6 days after intracranial LCMV inoculation, and virus distribution in the choriomeninges. (A and C) Brain sections from uninfected control mice; (B and D) from a BALB/c mouse 6 days after intracranial LCMV infection (infectious dose: 2 pfu). Sections were evaluated using either *in situ* hybridization to detect LCMV (A, B), or by H&E staining to evaluate inflammatory changes (C, D).

1. LCMV

This arenavirus causes aseptic meningitis in humans and mice. In the mouse model, choriomeningitis is most consistently achieved by intracranial inoculation of a low dose of virus (~0.2–2 pfu [plaque-forming units] in 20–50 μl). The mice appear essentially normal for ~5 days; on the sixth day they become ill (ruffled fur, hunched posture, reduced mobility), and they die between days 7 and 8. Analysis of the cerebrospinal fluid reveals a massive lymphocytic infiltration (as shown in Fig. 1), which is dominated by CD8+ T cells whose depletion permits the mouse to survive (Dixon *et al.*, 1987). Therefore, lethal LCM is a good example of CD8+ T cell-mediated immunopathology. Although the lethal outcome is CTL-dependent, these same cells can confer protection against infection and disease (Allan and Doherty, 1985). Indeed, as stated above, a vaccine encoding a single CTL epitope can protect against subsequent intracranial LCMV challenge (Klavinskis *et al.*, 1989; Whitton *et al.*, 1993). This apparent anomaly is explained by considering the relative kinetics of virus infection and of the immune response. In a previously naive mouse, the virus replicates in the original infected cell and, in the absence of an established CTL response, is free to disseminate throughout the choriomeninges. By the time that the virus-specific CTL response has amplified to a meaningful level, the choriomeningeal cells are heavily infected, as shown in Fig. 1B. The CTL response is therefore intense and extensive (Fig. 1D), and results in death of the host. In contrast, if the mouse has been successfully vaccinated to induce epitope-specific CTL, the accelerated CTL response quickly limits virus replication and spread; although the mouse shows some signs of morbidity around day 4 (presumably the result of a mild meningitis), the virus is cleared by day 7 and the animal makes a complete recovery. The kinetics of the immune response are crucial; if the response induced is too low, the disease may, in fact, be exacerbated (Oehen et al., 1991). Surprisingly, we still do not know precisely why the naive animals succumb to LCMV challenge. Mice lacking the perforin gene survive, despite mounting a strong virus-specific CD8+ T cell response leading to histological choriomeningitis (Kagi et al., 1994a; Walsh et al., 1994); this indicates that abrogation of the CD8+ T cells' lytic activity is sufficient to prevent death, even in the presence of an infiltrate. It is hypothesized—but not proven—that death results from perforin-mediated destruction of the choroid plexus, which leads to dysregulation at the blood/CSF interface.

2. Theiler's Virus Infection of CNS

Theiler's murine encephalomyelitis virus (TMEV) is a single-stranded positive-sense RNA virus. These viruses can be separated into two general groups, depending on neurovirulence. Highly neurovirulent strains include the GDVII and FA viruses. As little as 5 pfu injected intracranially causes a massive infection of the limbic system, particularly the hippocampus (Fig. 2A), and can kill a mouse in 7 days. Neurons die by apoptosis (Tsunoda et al., 1997). This is contrasted with infection of mice with the less neurovirulent strains, DA, WW, and BeAn viruses, which leads to an acute polioencephalomyelitis that is followed by a chronic inflammatory demyelinating disease. An interesting feature of this less virulent infection is that the CNS distribution of lesions and virus alters as the infection transits from the acute phase to the chronic phase. During the acute phase of infection in susceptible mice, viral antigens and RNA are found mostly in neurons of the gray matter. Inflammation is also exclusively present in the gray matter. In contrast, during the chronic phase, virus-infected cells, and inflammation accompanied by demyelination, are primarily detected in the white

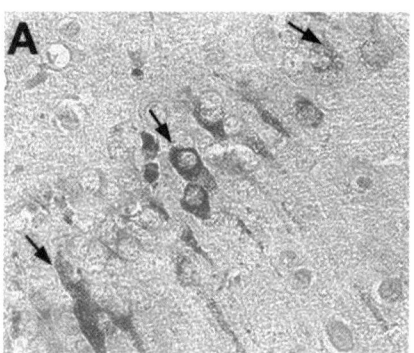 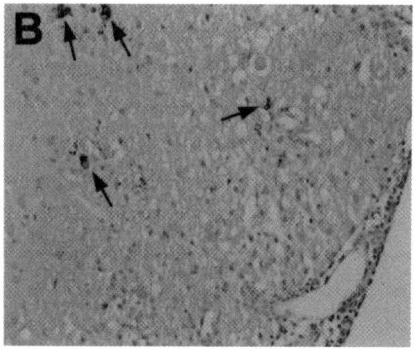

FIG 2. Acute and chronic infections with Theiler's virus. (A) Acute disease during Theiler's virus infection. Hippocampal neurons are shown to contain viral proteins. Magnification: ×200. (B) Root entry zone from a mouse chronically infected with Theiler's virus. Magnification: ×100. Inflammation and viral antigen-positive cells are present in close association.

matter of the spinal cord (Fig. 2B) (Yamada *et al.*, 1991). It is still not clear whether the astrocyte, oligodendrocyte, microglial cell, or macrophage (or a combination thereof) is the primary site of virus persistence. Resistance to chronic Theiler's virus disease maps to the MHC class I H-2D region (reviewed in Yamada *et al.*, 1991). MHC class I-restricted CD8+ CTLs, specific for VP1 and VP2 capsid proteins, are found in resistant mice, indicating that MHC class I-restricted virus-specific CD8+ CTLs are important in clearance of infection. Since MHC class I expression is upregulated in the CNS during Theiler's virus infection, it has been hypothesized that the CTL response eliminates virus during the acute phase. Neurons are infected during the early acute phase of infection, but during this phase MHC class I molecules are expressed only in glial and endothelial cells, not in neurons (Altintas *et al.*, 1993; Lindsley *et al.*, 1992). Therefore, in resistant mice, CTL may play a role in clearing virus from macrophages and/or glial cells, resulting in protection of these mice from the chronic stage. Tolerance induction of mice in regard to myelin did not alter the development of inflammatory demyelinating lesions characteristic of Theiler's mouse encephalomyelitis (Lang *et al.*, 1985). However, tolerance induction in regard to Theiler's virus prevented the development of clinical disease including inflammation and demyelination, which suggests that chronic immunopathogenic disease was directed against virus antigens persisting in the CNS (Karpus *et al.*, 1995). Some of the clinical and pathological features mimic the human demyelinating disease, MS. It appears that both CD4+ and CD8+ T cells contribute to the TMEV-induced inflammatory demyelinating disease. Therefore, infection of mice with the less neurovirulent strains of TMEV has been used as an experimental animal model for the progressive forms of MS.

V. DNA VACCINES AND CNS VIRAL INFECTIONS

DNA vaccination is a relatively new entrant in the vaccine sweep-stakes, but is viewed with optimism, for a number of reasons. This topic has been reviewed (Donnelly et al., 1997; Hassett and Whitton, 1996; Liu et al., 1997), but the following advantages of DNA vaccines should be noted. First, introduction of the encoded proteins into the MHC class I pathway induces good CD8+ T cell responses. Second, in most cases, proteins also should encounter the MHC class II pathway, and B cells, thus inducing CD4+ T cell and antibody responses. Third, the space limitation of most potential viral vectors does not apply to DNA vaccines, since many different plasmids could be contained in a single vaccine "cocktail." Fourth, it is possible to manipulate the immune response induced—for example, by directing plasmid-encoded proteins to selectively induce CD8+ T cells (Rodriguez et al., 1997; Rodriguez et al., 1998)—or to enhance induction of CD4+ T cells (Rodriguez and Whitton, unpublished data). Fifth, DNA vaccines should be safe, and easy to produce cheaply, in quantity, and at a high level of purity. These benefits have led many laboratories to evaluate DNA vaccines in a number of animal models, including several involving viruses that infect the CNS.

A. DNA Vaccines Against LCMV

DNA vaccines encoding the nucleoprotein (NP) from LCMV can confer protection against the normally lethal intracranial challenge (Yokoyama et al., 1995; Zarozinski et al., 1995), and can prevent the establishment of persistent infection (Pedroza Martins et al., 1995). Protection is CTL-mediated and does not depend on the induction of antiviral antibodies. The vehicle (saline, or lipid-associated) and the route of administration are important in determining the level of induced immunity (Yokoyama et al., 1996; Yokoyama et al., 1997). The LCMV model has allowed the demonstration of the exquisite flexibility of DNA vaccines. If the LCMV NP gene is fused to the host protein ubiquitin, the resulting protein is targeted for very rapid intracellular degradation; a plasmid encoding this ubiquitin–NP fusion induces enhanced protection against intracranial challenge (Rodriguez et al., 1997), perhaps because it increases the precursor frequency of NP-specific CTLs (Rodriguez et al., 1998). In addition, the LCMV model has been used to study neonatal DNA immunization. A single inoculation, within hours of birth, is sufficient to induce protective immunity, even in the presence of maternal antibodies (Hassett *et al.*, 1997), and these responses are long-lived and remarkably abundant; as long as 1 year post-DNA immunization, 1–2% of the animal's CD8+ T cells are NP-specific (Hassett *et al.*, 2000). Note that all of these studies employed "peripheral" immunization; none of them attempted to induce responses within the CNS. However, these data indicate that, following intracranial inoculation of virus, the DNA-vaccine-induced CTL can enter the CNS and limit LCMV replication and dissemination.

B. DNA Vaccines Against Theiler's Virus

To investigate the utility of DNA vaccines against Theiler's virus, cDNAs encoding the viral capsid proteins, VP1, VP2, and VP3, were constructed (Tolley *et al.*, 1999). Susceptible SJL/J mice were vaccinated intramuscularly one, two, or three times with the DNA vaccines. Mice were then infected with Theiler's virus, and clinical and pathological features of disease were followed. Interestingly, mice vaccinated with cDNA encoding VP2 were partially protected from clinical and pathological disease. In addition, VP3 vaccination was somewhat able to ameliorate clinical disease in infected mice. VP4 vaccination also protects mice from demyelinating disease (Tsunoda and Fujinami, unpublished). In contrast, mice vaccinated with cDNA encoding VP1 had a more severe clinical disease and enhanced histopathology as compared to nonvaccinated mice. There was no relationship between the antivirus antibody titers and the extent or course of disease. Thus, different outcomes were observed, depending on the viral antigen included in the vaccine.

C. DNA Vaccines Against Other Viruses that Cause CNS Disease

DNA vaccines have been shown to be effective against several of the agents reviewed in section I. In rabies, in a mouse model, immunization with plasmids encoding the rabies glycoprotein conferred complete protection against subsequent viral challenge (Ray *et al.*, 1997; Xiang *et al.*, 1994; Xiang *et al.*, 1995); protection was also seen in mice immunized as neonates (Wang *et al.*, 1997), confirming the efficacy of neonatal DNA immunization as demonstrated in the LCMV model. Recently, DNA immunization of Cynomolgus monkeys was shown to completely protect against subsequent challenge, and to generate levels of antibodies comparable to those induced by the standard human diploid cell vaccine

(Lodmell *et al.*, 1998). These data suggest that DNA immunization may have a future in higher primates, such as the readers of this chapter. In measles, there is ample evidence showing that DNA vaccines can induce measles-specific humoral and cell-mediated immunity (Cardoso *et al.*, 1996). Furthermore, neonates are an important target for measles vaccination, and DNA immunization at this age induces measles-specific CTL (Martinez *et al.*, 1997). Although there is no widely used small animal model for measles-induced postinfectious encephalomyelitis or SSPE, intracranial measles virus inoculation can cause encephalitis, and this disease is abrogated by prior DNA immunization with a plasmid-encoding measles nucleoprotein (Hsu *et al.*, 1998). Recently, a transgenic mouse line has been developed that expresses measles virus receptor in neurons (Rall *et al.*, 1997), and provides the opportunity to evaluate the effects of measles-specific immune responses in the CNS (Lawrence *et al.*, 1999). DNA vaccines are also effective (in animal models) in combating various arboviral encephalitides, including St. Louis encephalitis (Konishi *et al.*, 1998; Phillpotts *et al.*, 1996), Japanese encephalitis (Ashok and Rangarajan, 1999; Konishi *et al.*, 1999; Lin *et al.*, 1998), La Crosse encephalitis (Schuh *et al.*, 1999) and Murray Valley encephalitis (Colombage *et al.*, 1998).

D. DNA Vaccines Against Autoimmune Diseases of the CNS

Several virus-induced CNS diseases may be explained by their triggering of autoimmunity. Experimental autoimmune encephalomyelitis (EAE) is a well-characterized CNS disease induced by the administration of certain CNS proteins (or epitopes from these proteins). We have shown that peripheral immunization with recombinant vaccinia viruses or plasmid DNAs encoding these CNS proteins or epitopes can radically alter the susceptibility of the host to EAE (Barnett *et al.*, 1993; Barnett *et al.*, 1996; Tsunoda *et al.*, 1998; Wang *et al.*, 1999)—thus establishing the potential for vaccination against autoimmune phenomena. However, vaccination with plasmid DNA alone can potentiate both EAE and the Theiler's virus demyelinating disease, most likely due to the immunostimulatory CpG motifs contained in the bacterial DNA (Tsunoda *et al.*, 1999).

E. DNA Vaccines Against Prion Diseases

TSEs are rare in humans, and at present the major medical interest probably comes from the risk of interspecies transfer to humans; it is hypothesized that a number of unusual cases of CJD in young Britons resulted from their having been exposed to products from cattle carrying bovine spongiform encephalopathy (Will *et al.*, 1996). Although this problem may have been partially addressed by the culling of infected herds, the investigation of interspecies transfer remains important, as the pooled offal and rendering products, although no longer fed to animals directly in the human foodchain, in some cases are used to make other products to which some of us are intimately exposed (cosmetics, for example). Thus, despite the rarity of TSEs, the prospect of being able to immunize against them is exciting.

Most infectious agents stimulate antigen-specific host immune responses. Current dogma suggests that host immunity plays little or no role in the pathogenesis of TSE; indeed, since the infectious protein is host-encoded, it might be expected that no immune response would be mounted and, consistent with this, a mouse TSE agent inoculated into normal mice appears to induce a very limited immune response. Thus, it might appear pointless to pursue the idea of vaccinating against "self"-proteins. However—possibly relevant to the immunological modification of prion diseases—Alzheimer's disease may result from CNS deposition of the misfolded β-amyloid protein, and immunization with this protein's precursor fragment (which is, of course, a self-protein) slows the development of the characteristic neuropathological changes (Schenk *et al.*, 1999), raising the possibility that immune responses to PrP might alter the disease course. Furthermore, steroids appear to reduce susceptibility to TSE (Outram *et al.*, 1974), and recent findings indicate that CD8[+] T cell infiltration may occur as an early indicator of TSE (Betmouni *et al.*, 1996), although the antigen-specificity of these T cells was not defined. DNA immunization offers a promising tool for evaluating the relevance of prion-specific immune responses, because it induces CD8[+] T cells, and may even be able to overcome a "nonresponder" status of the host (Schirmbeck *et al.*, 1995). In fact, DNA immunization of PrPko mice does indeed induce anti-PrP antibodies, but T cell responses were not pursued (Krasemann *et al.*, 1996). Might DNA immunization protect against disease (a vaccine against interspecies transfer?), or might it exacerbate disease by priming for immunopathology? Such studies are under way in one of our laboratories (JLW).

Acknowledgments

We are grateful to Annette Lord and Kathleen Borick for excellent secretarial support. This work was supported by NIH grant AI-37186 (JLW) and grant AI-42525 (RSF). This is drawn from manuscript number 12617-NP from The Scripps Research Institute.

REFERENCES

Allan, J. E. and Doherty, P. C. (1985). Immune T cells can protect or induce fatal neurological disease in murine lymphocytic choriomeningitis. *Cell Immunol.* **90**:401–407.

Altintas, A., Cai, Z., Pease, L. R., and Rodriguez, M. (1993). Differential expression of H-2K and H-2D in the central nervous system of mice infected with Theiler's virus. *J. Immunol.* **151**:2803–2812.

An, L. L. and Whitton, J. L. (1997). A multivalent minigene vaccine, containing B cell, CTL, and Th epitopes from several microbes, induces appropriate responses *in vivo*, and confers protection against more than one pathogen. *J. Virol.* **71**:2292–2302.

An, L. L. and Whitton, J. L. (1999). Multivalent Minigene Vaccines Against Infectious Disease. *Curr. Opin. Molec. Ther.* **1**:16–21.

Andersson, J., Isberg, B., Christensson, B., Veress, B., Linde, A., and Bratel, T. (1999). Interferon-γ deficiency in generalized Epstein-Barr virus infection with interstitial lymphoid and granulomatous pneumonia, focal cerebral lesions, and genital ulcers: remission following IFN-gamma substitution therapy. *Clin. Infect. Dis.* **28**:1036–1042.

Arevalo, J. H., Taussig, M. J., and Wilson, I. A. (1993). Molecular basis of crossreactivity and the limits of antibody–antigen complementarity. *Nature* **365**:859–863.

Ashok, M. S. and Rangarajan, P. N. (1999). Immunization with plasmid DNA encoding the envelope glycoprotein of Japanese encephalitis virus confers significant protection against intracerebral viral challenge without inducing detectable antiviral antibodies. *Vaccine* **18**:68–75.

Barbas, C. F., Kang, A. S., Lerner, R. A., and Benkovic, S. J. (1991). Assembly of combinatorial antibody libraries on phage surfaces: the gene III site. *Proc. Natl. Acad. Sci. U.S.A.* **88**:7978–7982.

Barnett, L. A., Whitton, J. L., Wada, Y., and Fujinami, R. S. (1993). Enhancement of autoimmune disease using recombinant vaccinia virus encoding myelin proteolipid protein. *J. Neuroimmunol.* **44**:15–25.

Barnett, L. A., Whitton, J. L., Wang, L. Y., and Fujinami, R. S. (1996). Virus encoding an encephalitogenic peptide protects mice from experimental allergic encephalomyelitis. *J. Neuroimmunol.* **64**:163–173.

Bergelson, L. D. (1993). Gangliosides and antitumor immunity. *Clin. Investig.* **71**:590–594.

Bergelson, L. D. (1995). Serum gangliosides as endogenous immunomodulators. *Immunol. Today* **16**:483–486.

Bergelson, L. D., Dyatlovitskaya, E. V., Klyuchareva, T. E., Kryukova, E. V., Lemenovskaya, A. F., Matveeva, V. A., and Sinitsyna, E. V. (1989). The role of glycosphingolipids in natural immunity. Gangliosides modulate the cytotoxicity of natural killer cells. *Eur. J. Immunol.* **19**:1979–1983.

Bessen, R. A., Kocisko, D. A., Raymond, G. J., Nandan, S., Lansbury, P. T., and Caughey, B. (1995). Non-genetic propagation of strain-specific properties of scrapie prion protein. *Nature* **375**:698–700.

Betmouni, S., Perry, V. H., and Gordon, J. L. (1996). Evidence for an early inflammatory response in the central nervous system of mice with scrapie. *Neuroscience* **74**:1–5.

Bjorkman, P. J., Saper, M. A., Samraoui, B., Bennett, W. S., Strominger, J. L., and Wiley, D. C. (1987a). Structure of the human class I histocompatibility antigen HLA-A2. *Nature* **329**:506–512.

Bjorkman, P. J., Saper, M. A., Samraoui, B., Bennett, W. S., Strominger, J. L., and Wiley, D. C. (1987b). The foreign antigen binding site and T cell recognition regions of class I histocompatibility antigens. *Nature* **329**:512–518.

Blackbourn, D. J., Mackewicz, C. E., Barker, E., and Levy, J. A. (1994). Human CD8+ cell non-cytolytic anti-HIV activity mediated by a novel cytokine. *Res. Immunol.* **145**:653–658.

Bode, L., Zimmermann, W., Ferszt, R., Steinbach, F., and Ludwig, H. (1995). Borna disease virus genome transcribed and expressed in psychiatric patients. *Nat. Med.* **1:**232–236.

Briese, T., Schneemann, A., Lewis, A. J., Park, Y. S., Kim, S., Ludwig, H., and Lipkin, W. I. (1994). Genomic organization of Borna disease virus. *Proc. Natl. Acad. Sci. U.S.A.* **91:**4362–4366.

Bueler, H., Aguzzi, A., Sailer, A., Greiner, R. A., Autenried, P., Aguet, M., and Weissmann, C. (1993). Mice devoid of PrP are resistant to scrapie. *Cell* **73:**1339–1347.

Bueler, H., Fischer, M., Lang, Y., Bluethmann, H., Lipp, H. P., DeArmond, S. J., Prusiner, S. B., Aguet, M., and Weissmann, C. (1992). Normal development and behaviour of mice lacking the neuronal cell-surface PrP protein. *Nature* **356:**577–582.

Cardoso, A. I., Blixenkrone-Moller, M., Fayolle, J., Liu, M. A., Buckland, R., and Wild, T. F. (1996). Immunization with plasmid DNA encoding for the measles virus hemagglutinin and nucleoprotein leads to humoral and cell-mediated immunity. *Virol.* **225:**293–299.

Ceccaldi, P. E., Gillet, J. P., and Tsiang, H. (1989). Inhibition of the transport of rabies virus in the central nervous system. *J. Neuropathol. Exp. Neurol.* **48:**620–630.

Childs, J. E., Glass, G. E., Korch, G. W., Ksiazek, T. G., and Leduc, J. W. (1992). Lymphocytic choriomeningitis virus infection and house mouse (Mus musculus) distribution in urban Baltimore. *Am. J. Trop. Med. Hyg.* **47:**27–34.

Childs, J. E., Glass, G. E., Ksiazek, T. G., Rossi, C. A., Oro, J. G., and Leduc, J. W. (1991). Human-rodent contact and infection with lymphocytic choriomeningitis and Seoul viruses in an inner-city population. *Am. J. Trop. Med. Hyg.* **44:**117–121.

Colombage, G., Hall, R., Pavy, M., and Lobigs, M. (1998). DNA-based and alphavirus-vectored immunisation with prM and E proteins elicits long-lived and protective immunity against the flavivirus, Murray Valley encephalitis virus. *Virol.* **250:**151–163.

Crowe, J. E., Murphy, B. R., Chanock, R. M., Williamson, R. A., Barbas III, C. F., and Burton, D. R. (1994). Recombinant human RSV monoclonal antibody Fab is effective therapeutically when introduced directly into the lungs of respiratory syncytial virus-infected mice. *Proc. Natl. Acad. Sci. U.S.A.* **91:**1386–1390.

de la Torre, J. C. (1994). Molecular biology of borna disease virus: prototype of a new group of animal viruses. *J. Virol.* **68:**7669–7675.

de la Torre, J. C., Bode, L., Durrwald, R., Cubitt, B., and Ludwig, H. (1996b). Sequence characterization of human Borna disease virus. *Virus Res.* **44:**33–44.

de la Torre, J. C., Gonzalez-Dunia, D., Cubitt, B., Mallory, M., Mueller-Lantzsch, N., Grasser, F. A., Hansen, L. A., and Masliah, E. (1996a). Detection of borna disease virus antigen and RNA in human autopsy brain samples from neuropsychiatric patients. *Virol.* **223:**272–282.

Dixon, J. E., Allan, J. E., and Doherty, P. C. (1987). The acute inflammatory process in murine lymphocytic choriomeningitis is dependent on Lyt-2+ immune T cells. *Cell Immunol.* **107:**8–14.

Donnelly, J. J., Ulmer, J. B., and Liu, M. A. (1997). DNA vaccines. *Life Sci.* **60:**163–172.

Dorries, K. (1998). Molecular biology and pathogenesis of human polyomavirus infections. *Dev. Biol. Stand.* **94:**71–79.

Fox, L., Alford, M., Achim, C., Mallory, M., and Masliah, E. (1997). Neurodegeneration of somatostatin-immunoreactive neurons in HIV encephalitis. *J. Neuropathol. Exp. Neurol.* **56:**360–368.

Fujinami, R. S. and Oldstone, M. B. A. (1979). Antiviral antibody reacting on the plasma membrane alters measles virus expression inside the cell. *Nature* **279:**529–530.

Gendelman, H. E., Wolinsky, J. S., Johnson, R. T., Pressman, N. J., Pezeshkpour, G. H., and Boisset, G. F. (1984). Measles encephalomyelitis: lack of evidence of viral invasion of the central nervous system and quantitative study of the nature of demyelination. *Ann. Neurol.* **15:**353–360.

Gordon, J. and Khalili, K. (1998). The human polyomavirus, JCV, and neurological diseases (review). *Int. J. Mol. Med.* **1:**647–655.

Guidotti, L. G. and Chisari, F. V. (1996). To kill or to cure: options in host defense against viral infection. *Current Opinion in Immunology* **8:**478–483.

Guidotti, L. G., Ishikawa, T., Hobbs, M. V., Matzke, B., Schreiber, R., and Chisari, F. V. (1996). Intracellular inactivation of the hepatitis B virus by cytotoxic T lymphocytes. *Immunity.* **4:**25–36.

Guidotti, L. G., Rochford, R., Chung, J., Shapiro, M., Purcell, R., and Chisari, F. V. (1999). Viral Clearance Without Destruction of Infected Cells During Acute HBV Infection. *Science* **284:**825–829.

Hassett, D. E. and Whitton, J. L. (1996). DNA Immunization. *Trends in Microbiol.* **4:**307–312.

Hassett, D. E., Zhang, J., Slifka, M. K., and Whitton, J. L. (2000). Immune responses following neonatal DNA immunization are long-lived, abundant, and qualitatively similar to those induced by conventional vaccination. *J. Virol.* **74:**2620–2627.

Hassett, D. E., Zhang, J., and Whitton, J. L. (1997). Neonatal DNA immunization with an internal viral protein is effective in the presence of maternal antibodies and protects against subsequent viral challenge. *J. Virol.* **71:**7881–7888.

Hsu, S. C., Obeid, O. E., Collins, M., Iqbal, M., Chargelegue, D., and Steward, M. W. (1998). Protective cytotoxic T lymphocyte responses against paramyxoviruses induced by epitope-based DNA vaccines: involvement of IFN-γ. *Int. Immunol.* **10:**1441–1447.

Huang, S., Hendriks, W., Althage, A., Hemmi, S., Bluethmann, H., Kamijo, R., Vilcek, J., Zinkernagel, R. M., and Aguet, M. (1993). Immune response in mice that lack the interferon-γ receptor. *Science* **259:**1742–1745.

Irani, D. N. (1998). The susceptibility of mice to immune-mediated neurologic disease correlates with the degree to which their lymphocytes resist the effects of brain-derived gangliosides. *J. Immunol.* **161:**2746–2752.

Irani, D. N., Lin, K. I., and Griffin, D. E. (1996). Brain-derived gangliosides regulate the cytokine production and proliferation of activated T cells. *J. Immunol.* **157:**4333–4340.

Jensen, P. N. and Major, E. O. (1999). Viral variant nucleotide sequences help expose leukocytic positioning in the JC virus pathway to the CNS. *J. Leukoc. Biol.* **65:**428–438.

Joly, E., Mucke, L., and Oldstone, M. B. A. (1991). Viral persistence in neurons explained by lack of major histocompatibility class I expression. *Science* **253:**1283–1285.

Joly, E. and Oldstone, M. B. A. (1992). Neuronal cells are deficient in loading peptides onto MHC class I molecules. *Neuron* **8:**1185–1190.

Kagi, D., Ledermann, B., Burki, K., Seiler, P., Odermatt, B., Olsen, K. J., Podack, E. R., Zinkernagel, R. M., and Hengartner, H. (1994a). Cytotoxicity mediated by T cells and natural killer cells is greatly impaired in perforin-deficient mice. *Nature* **369:**31–37.

Kagi, D., Vignaux, F., Ledermann, B., Burki, K., Depraetere, V., Nagata, S., Hengartner, H., and Golstein, P. (1994b). Fas and Perforin Pathways as Major Mechanisms of T Cell-Mediated Cytotoxicity. *Science* **265:**528–530.

Kang, A. S., Barbas, C. F., Janda, K. D., Benkovic, S. J., and Lerner, R. A. (1991). Linkage of recognition and replication functions by assembling combinatorial antibody Fab libraries along phage surfaces. *Proc. Natl. Acad. Sci. U.S.A.* **88:**4363–4366.

Karpus, W. J., Pope, J. G., Peterson, J. D., Dal Canto, M. C., and Miller, S. D. (1995). Inhibition of Theiler's virus-mediated demyelination by peripheral immune tolerance induction. *J. Immunol.* **155:**947–957.

Klavinskis, L. S., Whitton, J. L., and Oldstone, M. B. A. (1989). Molecularly engineered vaccine which expresses an immunodominant T-cell epitope induces cytotoxic T lymphocytes that confer protection from lethal virus infection. *J. Virol.* **63:**4311–4316.

Kocisko, D. A., Come, J. H., Priola, S. A., Chesebro, B., Raymond, G. J., Lansbury, P. T., and Caughey, B. (1994). Cell-free formation of protease-resistant prion protein. *Nature* **370:**471–474.

Kocisko, D. A., Priola, S. A., Raymond, G. J., Chesebro, B., Lansbury, P. T. J., and Caughey, B. (1995). Species specificity in the cell-free conversion of prion protein to protease-resistant forms: a model for the scrapie species barrier. *Proc. Natl. Acad. Sci. U.S.A.* **92:**3923–3927.

Konishi, E., Yamaoka, M., Khin, S. W., Kurane, I., and Mason, P. W. (1998). Induction of protective immunity against Japanese encephalitis in mice by immunization with a plasmid encoding Japanese encephalitis virus premembrane and envelope genes. *J. Virol.* **72:**4925–4930.

Konishi, E., Yamaoka, M., Khin, S. W., Kurane, I., Takada, K., and Mason, P. W. (1999). The anamnestic neutralizing antibody response is critical for protection of mice from challenge following vaccination with a plasmid encoding the Japanese encephalitis virus premembrane and envelope genes. *J. Virol.* **73:**5527–5534.

Krasemann, S., Groschup, M., Hunsmann, G., and Bodemer, W. (1996). Induction of antibodies against human prion proteins (PrP) by DNA- mediated immunization of PrP0/0 mice. *J. Immunol. Methods* **199:**109–118.

Kuwahara, C., Takeuchi, A. M., Nishimura, T., Haraguchi, K., Kubosaki, A., Matsumoto, Y., Saeki, K., Matsumoto, Y., Yokoyama, T., Itohara, S., and Onodera, T. (1999). Prions prevent neuronal cell-line death. *Nature* **400:**225–226.

Lampson, L. A. and Fisher, C. A. (1984). Weak HLA and beta 2-microglobulin expression of neuronal cell lines can be modulated by interferon. *Proc. Natl. Acad. Sci. U.S.A.* **81:**6476–6480.

Lampson, L. A., Fisher, C. A., and Whelan, J. P. (1983). Striking paucity of HLA-A, B, C and beta 2-microglobulin on human neuroblastoma cell lines. *J. Immunol.* **130:**2471–2478.

Lampson, L. A. and Hickey, W. F. (1986). Monoclonal antibody analysis of MHC expression in human brain biopsies: tissue ranging from "histologically normal" to that showing different levels of glial tumor involvement. *J. Immunol.* **136:**4054–4062.

Lang, W., Wiley, C., and Lampert, P. (1985). Theiler's virus encephalomyelitis is unaffected by treatment with myelin components. *J. Neuroimmunol.* **9:**109–113.

Larsen, P. D., Chartrand, S. A., Tomashek, K. M., Hauser, L. G., and Ksiazek, T. G. (1993). Hydrocephalus complicating lymphocytic choriomeningitis virus infection. *Pediatr. Infect. Dis. J.* **12:**528–531.

Lawrence, D. M., Vaughn, M. M., Belman, A. R., Cole, J. S., and Rall, G. F. (1999). Immune response-mediated protection of adult but not neonatal mice from neuron-restricted measles virus infection and central nervous system disease. *J. Virol.* **73:**1795–1801.

Levine, B., Hardwick, J. M., Trapp, B. D., Crawford, T. O., Bollinger, R. C., and Griffin, D. E. (1991). Antibody-mediated clearance of alphavirus infection from neurons. *Science* **254:**856–860.

Levy, J. A., Mackewicz, C. E., and Barker, E. (1996). Controlling HIV pathogenesis: the role of the noncytotoxic anti-HIV response of CD8+ T cells. *Immunol. Today* **17:**217–224.

Lin, Y. L., Chen, L. K., Liao, C. L., Yeh, C. T., Ma, S. H., Chen, J. L., Huang, Y. L., Chen, S. S., and Chiang, H. Y. (1998). DNA immunization with Japanese encephalitis virus nonstructural protein NS1 elicits protective immunity in mice. *J. Virol.* **72:**191–200.

Lindsley, M. D., Patick, A. K., Prayoonwiwat, N., and Rodriguez, M. (1992). Coexpression of class I major histocompatibility antigen and viral RNA in central nervous system of mice infected with Theiler's virus: a model for multiple sclerosis. *Mayo Clin. Proc.* **67:**829–838.

Liu, M. A., McClements, W., Ulmer, J. B., Shiver, J., and Donnelly, J. (1997). Immunization of non-human primates with DNA vaccines. *Vaccine* **15:**909–912.

Lodmell, D. L., Ray, N. B., Parnell, M. J., Ewalt, L. C., Hanlon, C. A., Shaddock, J. H., Sanderlin, D. S., and Rupprecht, C. E. (1998). DNA immunization protects nonhuman primates against rabies virus. *Nat. Med.* **4:**949–952.

Martinez, X., Brandt, C., Saddallah, F., Tougne, C., Barrios, C., Wild, F., Dougan, G., Lambert, P. H., and Siegrist, C. A. (1997). DNA immunization circumvents deficient induction of T helper type 1 and cytotoxic T lymphocyte responses in neonates and during early life. *Proc. Natl. Acad. Sci. U.S.A.* **94:**8726–8731.

Moller, A. A., Gasser, T., Jager, H., and Hedl, A. (1988). Clinical course of subacute HIV encephalitis. *J. Neuroimmunol.* **20:**145–147.

Neumann, H., Cavalie, A., Jenne, D. E., and Wekerle, H. (1995). Induction of MHC class I genes in neurons. *Science* **269:**549–552.

Neumann, H., Schmidt, H., Cavalie, A., Jenne, D., and Wekerle, H. (1997). Major histocompatibility complex (MHC) class I gene expression in single neurons of the central nervous system: differential regulation by interferon (IFN)-gamma and tumor necrosis factor (TNF)-alpha. *J. Exp. Med.* **185:**305–316.

Nowak, M. A., Bonhoeffer, S., Hill, A. M., Boehme, R., Thomas, H. C., and McDade, H. (1996). Viral dynamics in hepatitis B virus infection. *Proc. Natl. Acad. Sci. U.S.A.* **93:**4398–4402.

Oehen, S., Hengartner, H., and Zinkernagel, R. M. (1991). Vaccination for disease. *Science* **251:**195–198.

Ogra, P. L. and Garofalo, R. (1990). Secretory antibody response to viral vaccines. *Prog. Med. Virol.* **37:**156–189.

Oldstone, M. B. A. (1998). Molecular mimicry and immune-mediated diseases. *FASEB J.* **12:**1255–1265.

Oldstone, M. B. A., Blount, P., Southern, P. J., and Lampert, P. W. (1986). Cytoimmunotherapy for persistent virus infection reveals a unique clearance pattern from the central nervous system. *Nature* **321:**239–243.

Outram, G. W., Dickinson, A. G., and Fraser, H. (1974). Reduced susceptibility to scrapie in mice after steroid administration. *Nature* **249:**855–856.

Pedroza Martins, L., Lau, L. L., Asano, M. S., and Ahmed, R. (1995). DNA vaccination against persistent viral infection. *J. Virol.* **69:**2574–2582.

Phillpotts, R. J., Venugopal, K., and Brooks, T. (1996). Immunisation with DNA polynucleotides protects mice against lethal challenge with St. Louis encephalitis virus. *Arch. Virol.* **141:**743–749.

Planz, O., Bilzer, T., and Stitz, L. (1995). Immunopathogenic role of T-cell subsets in Borna disease virus-induced progressive encephalitis. *J. Virol.* **69:**896–903.

Podack, E. R., Lowrey, D. M., Lichtenheld, M., and Hameed, A. (1988). Function of granule perforin and esterases in T cell-mediated reactions. Components required for delivery of molecules to target cells. *Ann. N. Y. Acad. Sci.* **532:**292–302.

Prusiner, S. B., Groth, D., Serban, A., Koehler, R., Foster, D., Torchia, M., Burton, D., Yang, S. L., and DeArmond, S. J. (1993). Ablation of the prion protein (PrP) gene in mice prevents scrapie and facilitates production of anti-PrP antibodies. *Proc. Natl. Acad. Sci. U.S.A.* **90:**10608–10612.

Rall, G. F., Manchester, M., Daniels, L. R., Callahan, E. M., Belman, A. R., and Oldstone, M. B. (1997). A transgenic mouse model for measles virus infection of the brain. *Proc. Natl. Acad. Sci. U.S.A.* **94:**4659–4663.

Ramsay, A. J., Ruby, J., and Ramshaw, I. A. (1993). A case for cytokines as effector molecules in the resolution of virus infection. *Immunol. Today* **14:**155–157.

Ray, N. B., Ewalt, L. C., and Lodmell, D. L. (1997). Nanogram quantities of plasmid DNA encoding the rabies virus glycoprotein protect mice against lethal rabies virus infection. *Vaccine* **15:**892–895.

Rickinson, A. B., Murray, R. J., Brooks, J., Griffin, H., Moss, D. J., and Masucci, M. G. (1992). T cell recognition of Epstein-Barr virus associated lymphomas. *Cancer Surv.* **13:**53–80.

Rini, J. M., Schulze-Gahmen, U., and Wilson, I. A. (1992). Structural evidence for induced fit as a mechanism for antibody- antigen recognition. *Science* **255:**959–965.

Rodriguez, F., An, L. L., Harkins, S., Zhang, J., Yokoyama, M., Widera, G., Fuller, J. T., Kincaid, C., Campbell, I. L., and Whitton, J. L. (1998). DNA immunization with minigenes: low frequency of memory CTL and inefficient antiviral protection are rectified by ubiquitination. *J. Virol.* **72:**5174–5181.

Rodriguez, F., Zhang, J., and Whitton, J. L. (1997). DNA immunization: ubiquitination of a viral protein enhances CTL induction, and antiviral protection, but abrogates antibody induction. *J. Virol.* **71:**8497–8503.

Ruby, J. and Ramshaw, I. A. (1991). The antiviral activity of immune CD8⁺ T cells is dependent on interferon-gamma. *Lymphokine Cytokine. Res.* **10:**353–358.

Schenk, D., Barbour, R., Dunn, W., Gordon, G., Grajeda, H., Guido, T., Hu, K., Huang, J., Johnson-Wood, K., Khan, K., Kholodenko, D., Lee, M., Liao, Z., Lieberburg, I., Motter, R., Mutter, L., Soriano, F., Shopp, G., Vasquez, N., Vandevert, C., Walker, S., Wogulis, M., Yednock, T., Games, D., and Seubert, P. (1999). Immunization with amyloid-beta attenuates Alzheimer-diseae-like pathology in the PDAPP mouse. *Nature* **400:**173–177.

Schiff, J. A., Schaefer, J. A., and Robinson, J. E. (1982). Epstein-Barr virus in cerebrospinal fluid during infectious mononucleosis encephalitis. *Yale J. Biol. Med.* **55:**59–63.

Schirmbeck, R., Bohm, W., Ando, K., Chisari, F. V., and Reimann, J. (1995). Nucleic acid vaccination primes hepatitis B virus surface antigen-specific cytotoxic T lymphocytes in nonresponder mice. *J. Virol.* **69:**5929–5934.

Schuh, T., Schultz, J., Moelling, K., and Pavlovic, J. (1999). DNA-based vaccine against La Crosse virus: protective immune response mediated by neutralizing antibodies and CD4⁺ T cells. *Hum. Gene Ther.* **10:**1649–1658.

Shresta, S., Pham, C. T., Thomas, D. A., Graubert, T. A., and Ley, T. J. (1998). How do cytotoxic lymphocytes kill their targets? *Curr. Opin. Immunol.* **10:**581–587.

Sigurdsson, B. (1954a). Rida, a chronic encephalitis of sheep with general remarks on infections which develop slowly and some of their special characteristics. *Br. Vet. J.* **110:**341–354.

Sigurdsson, B. (1954b). Observations on three slow infections of sheep. *Br. Vet. J.* **110:**255–270.

Slifka, M. K., Rodriguez, F., and Whitton, J. L. (1999). Rapid on/off cycling of cytokine production by virus-specific CD8⁺ T cells. *Nature* **401:**76–79.

Slifka, M. K. and Whitton, J. L. (2000). Activated and memory CD8⁺ T cells can be distinguished by thier cytokine profiles and phenotypic markers. *J. Immunol.* (in press).

Spiegelberg, H. L. (1990). The role of interleukin-4 in IgE and IgG subclass formation. *Springer Semin. Immunopathol.* **12:**365–383.

Stitz, L., Planz, O., Bilzer, T., Frei, K., and Fontana, A. (1991). Transforming growth factor-beta modulates T cell-mediated encephalitis caused by Borna disease virus. Pathogenic importance of CD8⁺ cells and suppression of antibody formation. *J. Immunol.* **147:**3581–3586.

Suda, T. and Nagata, S. (1994). Purification and characterization of the Fas-ligand that induces apoptosis. *J. Exp. Med.* **179:**873–879.

Tishon, A., Lewicki, H., Rall, G. F., von Herrath, M. G., and Oldstone, M. B. A. (1995). An essential role for type 1 interferon-gamma in terminating persistent viral infection. *Virol.* **212:**244–250.

Tolley, N. D., Tsunoda, I., and Fujinami, R. S. (1999). DNA vaccination against Theiler's murine encephalomyelitis virus leads to alterations in demyelinating disease. *J. Virol.* **73:**993–1000.

Tsiang, H. (1979). Evidence for an intraaxonal transport of fixed and street rabies virus. *J. Neuropathol. Exp. Neurol.* **38:**286–299.

Tsunoda, I., Kuang, L. Q., Tolley, N. D., Whitton, J. L., and Fujinami, R. S. (1998). Enhancement of experimental allergic encephalomyelitis (EAE) by DNA immunization with myelin proteolipid protein (PLP) plasmid DNA. *J. Neuropathol. Exp. Neurol.* **57:**758–767.

Tsunoda, I., Kurtz, C. I., and Fujinami, R. S. (1997). Apoptosis in acute and chronic central nervous system disease induced by Theiler's murine encephalomyelitis virus. *Virol.* **228:**388–393.

Tsunoda, I., Tolley, N. D., Theil, D. J., Whitton, J. L., Kobayashi, H., and Fujinami, R. S. (1999). Exacerbation of viral and autoimmune animal models for multiple sclerosis by bacterial DNA. *Brain Pathol.* **9:**481–493.

Walsh, C. M., Matloubian, M., Liu, C. C., Ueda, R., Kurahara, C. G., Christensen, J. L., Huang, M. T., Young, J. D., Ahmed, R., and Clark, W. R. (1994). Immune function in mice lacking the perforin gene. *Proc. Natl. Acad. Sci. U.S.A.* **91:**10854–10858.

Wang, L. Y., Theil, D. J., Whitton, J. L., and Fujinami, R. S. (1999). Infection with a recombinant vaccinia virus encoding myelin proteolipid protein causes suppression of chronic relapsing-remitting experimental allergic encephalomyelitis. *J. Neuroimmunol.* **96:**148–157.

Wang, Y., Xiang, Z., Pasquini, S., and Ertl, H. C. (1997). Immune response to neonatal genetic immunization. *Virol.* **228:**278–284.

Weber, T. and Major, E. O. (1997). Progressive multifocal leukoencephalopathy: molecular biology, pathogenesis and clinical impact. *Intervirology* **40:**98–111.

Welsh, R. M., Nishioka, W. K., Antia, R., and Dundon, P. L. (1990). Mechanism of killing by virus-induced cytotoxic T lymphocytes elicited in vivo. *J. Virol.* **64:**3726–3733.

Whelan, J. P., Wysocki, C. J., and Lampson, L. A. (1986). Distribution of beta 2-microglobulin in olfactory epithelium: a proliferating neuroepithelium not protected by a blood-tissue barrier. *J. Immunol.* **137:**2567–2571.

Whitton, J. L. and Fujinami, R. S. (1999). Viruses as triggers of autoimmunity: facts and fantasies. *Curr. Opin. Microbiol.* **2:**392–397.

Whitton, J. L. and Oldstone, M. B. A. (2000). The Immune Response to Viruses. In *Fields' Virology* (Fields, B. N., Knipe, D. M., and Howley, P. M., eds.), 4th ed. Lippincott Williams & Wilkins, Philadelphia.

Whitton, J. L., Sheng, N., Oldstone, M. B. A., and McKee, T. A. (1993). A "string-of-beads" vaccine, comprising linked minigenes, confers protection from lethal-dose virus challenge. *J. Virol.* **67:**348–352.

Wiley, C. A., Schrier, R. D., Morey, M., Achim, C., Venable, J. C., and Nelson, J. A. (1991). Pathogenesis of HIV encephalitis. *Acta Pathol. Jpn.* **41:**192–196.

Will, R. G., Ironside, J. W., Zeidler, M., Cousens, S. N., Estibeiro, K., Alperovitch, A., Poser, S., Pocchiari, M., Hofman, A., and Smith, P. G. (1996). A new variant of Creutzfeldt-Jakob disease in the UK. *Lancet* **347:**921–925.

Xiang, Z. Q., Spitalnik, S., Tran, M., Wunner, W. H., Cheng, J., and Ertl, H. C. (1994). Vaccination with a plasmid vector carrying the rabies virus glycoprotein gene induces protective immunity against rabies virus. *Virol.* **199:**132–140.

Xiang, Z. Q., Spitalnik, S. L., Cheng, J., Erikson, J., Wojczyk, B., and Ertl, H. C. (1995). Immune responses to nucleic acid vaccines to rabies virus. *Virol.* **209:**569–579.

Yamada, M., Zurbriggen, A., and Fujinami, R. S. (1991). Pathogenesis of Theiler's murine encephalomyelitis virus. *Adv. Virus Res.* **39:**291–320.

Yokoyama, M., Hassett, D. E., Zhang, J., and Whitton, J. L. (1997). DNA immunization can stimulate florid local inflammation, and the antiviral immunity induced varies depending on injection site. *Vaccine* **15:**553–560.

Yokoyama, M., Zhang, J., and Whitton, J. L. (1995). DNA immunization confers protection against lethal hocytic choriomeningitis virus infection. *J. Virol.* **69:**2684–2688.

Yokoyama, M., Zhang, J., and Whitton, J. L. (1996). DNA immunization: effects of vehicle and route of administration on the induction of protective antiviral immunity. *FEMS Immunol. Med. Microbiol.* **14:**221–230.

Zarozinski, C. C., Fynan, E. F., Selin, L. K., Robinson, H. L., and Welsh, R. M. (1995). Protective CTL-dependent immunity and enhanced immunopathology in mice immunized by particle bombardment with DNA encoding an internal virion protein. *J. Immunol.* **154:**4010–4017.

Zychlinsky, A., Zheng, L. M., Liu, C. C., and Young, J. D. (1991). Cytolytic lymphocytes induce both apoptosis and necrosis in target cells. *J. Immunol.* **146:**393–400.

SPONGIFORM ENCEPHALOPATHIES

TRANSMISSIBLE SPONGIFORM ENCEPHALOPATHIES AND PRION PROTEIN INTERCONVERSIONS

Byron Caughey and Bruce Chesebro

Laboratory of Persistent Viral Diseases
Rocky Mountain Laboratories
National Institute of Allergy and Infectious Diseases
National Institutes of Health
Hamilton, Montana 59840

I. Introduction

Transmissible spongiform encephalopathy (TSE) diseases, or prion diseases, are rare, fatal neurodegenerative diseases of humans and other animals. In the past decade a high awareness of TSE diseases has developed due to the appearance of bovine spongiform encephalopathy (BSE), or "mad cow disease," in the United Kingdom. Due to the potential for human infection, in Europe, BSE has influenced medical, agricultural, economic and political issues, perhaps even to a greater extent than has AIDS (acquired immunodeficiency syndrome). North America has been spared the ravages of the BSE epidemic; however, there is growing concern over the surprisingly high incidence of the cervid TSE, chronic wasting disease (CWD), in wild and captive populations of deer and elk in the western United States and Canada (Miller et al., 1998).

TSE diseases are transmissible by inoculation or ingestion of infected tissues. Incubation periods prior to clinical symptoms range from months to years and, in the case of some kuru patients, may have been as long as 40 years. Primary symptoms of TSE diseases in humans are dementia and ataxia. These diseases are usually characterized by spongiform degeneration of the brain, accompanied by appearance of activated astrocytes. Most distinctive, however, is the accumulation of abnormal protease-resistant forms of host-derived prion protein (PrP) in the central nervous system and, to a lesser extent, in lymphoreticular tissues. Much evidence suggests that abnormal forms of PrP of some sort are critical in the transmission and pathogenesis of TSE diseases (Caughey and Chesebro, 1997; reviewed in Chesebro, 1999; Weissmann, 1999). Indeed, it has been proposed, but not yet proven, that an abnormal PrP is the infectious TSE agent or prion (Prusiner, 1998). TSE-associated forms of PrP have been termed PrP^{Sc}, PrP^{CJD}, PrP^{BSE}, etc., according to the particular TSE involved or, more operationally, PrP-res, for protease-resistant PrP. However, diversity in the structures and properties of various abnormal PrP molecules has made the precise definitions and use of these terms problematic (see the discussion below). Nonetheless, with that caveat, we will use the generic term PrP-res, unless referring specifically to the abnormal PrP forms associated with a particular TSE disease. In scrapie, PrP-res is formed from the normal PrP by an apparent change in conformation and aggregation state. Normally, PrP is a protease-sensitive sialoglycoprotein that is anchored to membranes via glycosylphosphatidylinositol (GPI) (Caughey and Chesebro, 1997; reviewed in Prusiner, 1998; Weissmann, 1999). Here we will use the term PrP^C to refer to PrP in its normal structure and conformation and the term PrP-sen to refer generically to protease-sensitive forms of PrP, whether normal (i.e., PrP^C) or not (e.g., various recombinant forms).

This chapter will begin with descriptions of important natural human and animal TSE diseases and various experimental models. This will be followed by a summary of PrP biochemistry and cell biology, especially as it relates to the TSE-associated conversion of PrP^C to PrP-res. Finally, we will consider several of the fundamental puzzles that still face the TSE field, including the nature of the infectious agent and the potential for therapeutics.

II. TRANSMISSIBLE SPONGIFORM ENCEPHALOPATHIES IN HUMANS AND ANIMALS

A. Human TSE Diseases

TSE diseases in humans can be divided into three groups: sporadic, familial, and iatrogenic (or infectious) (Table I) (Brown *et al.*, 1994). Sporadic Creutzfeldt-Jakob disease (CJD) is not associated with any known mutations and occurs worldwide at an incidence of around one per 2 million. The source of this disease is completely unknown. Familial TSE diseases are all associated with different mutations in the PrP gene (Fig. 1), and include familial CJD, Gerstmann-Sträussler-Scheinker (GSS) syndrome and fatal familial insomnia (FFI). Infectious or iatrogenic TSE diseases include kuru, which was spread by ritual cannibalism among New Guinea tribesmen; CJD, spread by transplantation or inoculation with brain tissues or extracts from unsuspected CJD patients; and variant CJD, apparently due to infection of humans with the agent of BSE (Hill *et al.*, 1997; Bruce *et al.*, 1997).

B. Natural and Experimental Scrapie in Sheep

Scrapie has been recognized as a disease in sheep for over two centuries, and was the first TSE disease to be shown as experimentally transmissible. Sheep scrapie thus provides an unusual opportunity to compare natural and experimental TSE disease processes. Although in animals there are no known genetic cases of TSE disease comparable

TABLE I
HUMAN TSE DISEASES

Type	Comments
Sporadic	
	No clear exposure to infectious TSE agent
	Creutzfeldt-Jakob disease(CJD); sporadic fatal familial insomnia (FFI)
	No PrP mutation
	1:2,000,000 incidence worldwide
Familial/genetic	
	No clear exposure to infectious TSE agent
	Familial CJD (dementia); GSS (ataxia); FFI (sleep abnormalities)
	Associated with PrP mutations
Infectious/iatrogenic	
	Kuru; variant CJD
	Neurosurgery, corneal transplant, growth hormone therapy

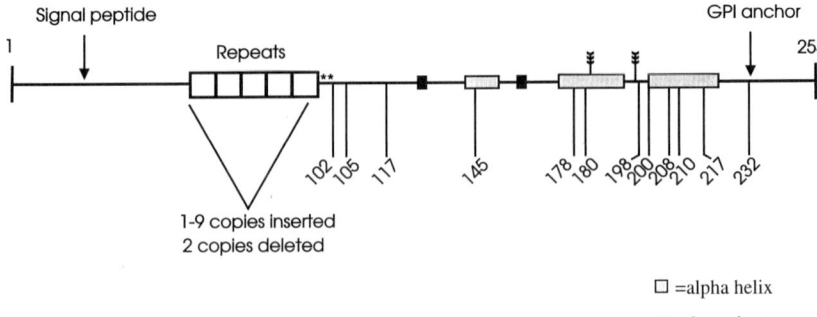

FIG 1. Mutations in human PrP associated with familial TSE. The numbered sites are positions of single-residue mutations or of insertion or deletion mutations. Arrows show the sites of cleavages, occurring during biosynthesis, that remove the signal peptide and the C-terminal GPI anchor signal sequence. The triple-headed arrows show the N-linked glycosylation sites. Asterisks indicate the approximate beginning of the PK-resistant region occurring in most PrP-res molecules. Superimposed on the molecule are the secondary structures derived from the high-resolution NMR structure of mouse PrP (Riek *et al.*, 1996). Courtesy of Suzette A. Priola.

to those seen in humans, allelic variations in the sheep PrP sequence do occur, and variation at several residues in the PrP sequence influences susceptibility to both natural and experimental scrapie infection (Goldmann *et al.*, 1994; Smits *et al.*, 1997). Although the mechanism of these effects is not certain, these susceptibility of these allelic variants to scrapie correlates with the relative efficiencies with which the respective PrP^C molecules convert to PrP-res in *in vitro* systems (see the discussion below) (Raymond *et al.*, 1997; Bossers *et al.*, 1997; Bossers *et al.*, 2000). In any case, it is clear that PrP is an important susceptibility factor for TSE diseases. A similar effect may also occur in humans where variation at PrP codon 129 does not induce a familial TSE disease, but instead appears to influence susceptibility to sporadic CJD (Palmer *et al.*, 1991) as well as the clinical phenotype of at least one familial TSE disease (Goldtarb *et al.*, 1992).

Although the infectious nature of sheep scrapie has been long recognized, the mode of transmission within a flock is not clear. Direct physical contact between individuals is not required, and there are reports of transmission by pasturing sheep on ground previously occupied by a flock with a high scrapie incidence (Greig, 1940; Palsson, 1979). Such pastures appear to be contaminated by scrapie infectivity perhaps derived from placental tissue, fetal membranes, or decomposing carcasses (Race *et al.*, 1998).

C. Bovine Spongiform Encephalopathy

In the 1990s, the bovine spongiform encephalopathy (BSE) epidemic in the United Kingdom has brought international attention to the TSE family of diseases; for a review, see Collee and Bradley, (1997). BSE was spread by the feeding of protein supplements contaminated with the rendered tissues of BSE-positive cattle. Several laboratory tests, including PrP-res banding patterns in protein gels and comparative titration in various mouse strains, identify similarities in BSE from all sources tested, as compared to most commonly known isolates of sheep scrapie. However, it remains unclear whether BSE originated by adaptation from an unusual strain of sheep scrapie or from an unrecognized bovine TSE case (Hope et al., 1999).

BSE has also been transmitted to other species by the feeding of contaminated meat and bone meal to ungulates and large felines in zoos and probably also to domestic cats. Transmission to humans has also been suggested by the appearance of variant CJD (vCJD) in over 84 younger humans primarily in the United Kingdom. The similarity in pathology and in laboratory tests lends support to this interpretation (Bruce et al., 1997; Hill et al., 1997). At present one cannot be certain whether the vCJD cases represent the beginning of a much larger epidemic in humans or whether they are unusual cases of transmission across a rather resistant species barrier. At the molecular level, at least, the interactions between PrP^{BSE} and human PrP^C that lead to human PrP-res formation are rather inefficient (Raymond et al., 1997). However, this apparent "molecular species barrier" is clearly only one of multiple factors that are likely to influence the extent of BSE transmissions to humans. In contrast to BSE, there is no evidence of spread of sheep scrapie to humans. Therefore, there may be important fundamental differences between scrapie and BSE in their interactions with different hosts.

D. Chronic Wasting Disease

Chronic wasting disease (CWD) in deer and Rocky Mountain elk is another example of a TSE disease of unknown origin. CWD is now recognized to be a serious problem in many "game farms" in the western United States and Canada. Its spread appears to be enhanced by the abnormal population densities found in such facilities, although the actual mechanism of transmission is unknown (Miller et al., 1998; Williams et al., 2000). Although first recognized in captive animals, CWD has now also been detected at incidences of up to >15% in wild deer and elk populations in southeastern Wyoming and northern

Colorado. The fact that CWD is found in wild ruminants on the same range as that of domestic cattle raises concern over the possibility that CWD could be transmitted to cattle and possibly might also pose a risk for human infection similar to BSE. Furthermore, it remains unclear whether there is a risk of direct transmission of CWD from deer or elk to humans by the handling or consumption of meat or other tissues derived from affected animals.

E. Transmissible Mink Encephalopathy

Transmissible mink encephalopathy (TME) is a TSE disease believed to be acquired by the feeding of animals with tissues from scrapie sheep or TSE-infected cattle. TME has been found in several mink ranches in the United States (Marsh et al., 1991). Although thought by many to be a mink-adapted form of sheep scrapie, there is anecdotal evidence to suggest that it might have arisen from a cattle TSE disease. However, TME has been readily transmitted to hamsters (Kimberlin and Marsh, 1975), whereas this has not been observed with BSE from Europe. Therefore, TME and BSE appear to be distinct TSE agents.

F. Experimental Animal Models

TSE diseases have been studied experimentally in several laboratory species including mice, rats, hamsters, and nonhuman primates. In general, TSE diseases show a preference for transmission to the species of origin or to a closely related species. Most noteworthy was the original demonstration of the transmissibility of CJD and kuru from humans to chimpanzees (Gajdusek et al., 1966; Gajdusek and Gibbs, 1971). Transmission to a less closely related species is also possible and appears to involve a progressive adaptation during serial passage in the new host. For example, scrapie from sheep or goats, and BSE from cattle, have produced typical TSE disease in mice, and mouse and hamster-adapted agents from these and other sources have been used extensively for pathogenesis studies and characterization of the agents.

Early experiments identified the Sinc gene as important in host susceptibility to scrapie (Dickinson et al., 1968), and subsequently Sinc was found to be the gene encoding PrP. More recent studies showed that expression of PrP-sen is required for susceptibility to TSE diseases, and that propagation of infectivity is eliminated or drastically reduced in the absence of the PrP gene (Bueler et al., 1993). This has been interpreted to imply that PrP is either a receptor for the infectious agent or an integral component of the agent.

In addition to the role of PrP as a necessary susceptibility factor for infection with homologous TSE agents, expression of PrP has also been noted as being necessary for in vivo persistence of foreign TSE agents, even in the absence of clinical disease (Race and Chesebro, 1998). This finding could be quite important when considering the possibility that nonbovine species fed BSE-contaminated feed might be able to harbor BSE infectivity without evidence of disease.

Experiments with PrP transgenic mice, where expression of hamster PrP was found to render transgenic mice susceptible to hamster-specific scrapie strains, also identified PrP as a susceptibility factor for cross-species transmission (Scott et al., 1989). In molecular studies the use of chimeric PrP molecules in transgenic mice has identified the importance of amino acid residues in the central portion of PrP in species-specific interactions between the inoculated TSE agent and the host animal. Other studies extended these results using scrapie-infected mouse neuroblastoma cells and cell-free systems, where differences at PrP amino acid residues in this region were identified as being critical for cross-species resistance between mice and hamsters as measured by PrP-res formation (Priola et al., 1994; Kocisko et al., 1995; Priola and Chesebro, 1995; Horiuchi et al., 2000). Interestingly, one of these polymorphic residues (138) is homologous to a polymorphic residue at position 142 in goat PrP that was previously found to influence resistance to BSE and to certain sheep scrapie strains in vivo (Goldmann et al., 1996).

Similar studies have shown that transgenic mice expressing human PrP have increased susceptibility to many human TSE disease isolates (Telling et al., 1994; Collinge et al., 1995; Telling et al., 1995; Asante and Collinge, 2001). This has broadened the possibilities for studying human isolates in less expensive and more rapid rodent models suitable for screening of possible therapeutic drugs. However, in spite of extensive knowledge of PrP sequences from a variety of species, the extent of species-specific resistance to TSE diseases remains impossible to predict solely by analysis of PrP sequences. This is of critical importance in the matter of human susceptibility to BSE.

Transgenic mice expressing PrP with the use of tissue-specific promoters have been studied to analyze the roles of different PrP expressing cells in susceptibility to TSE disease. PrP expression in either neurons or astrocytes is sufficient to allow the induction of clinical disease and typical scrapie neuropathology by scrapie infection (Race et al., 1995; Raeber et al., 1997). This would appear to imply that generation of a toxic PrP-res or peptide moiety in the brain may be independent of the brain cell type producing the normal PrP. Testing of mice expressing PrP in microglia or oligodendroglia will be required to examine this matter more thoroughly. In contrast to the above results, PrP expression in liver,

T lymphocytes, and B lymphocytes did not induce scrapie susceptibility (Raeber *et al.*, which 1999), suggests that perhaps the toxic PrP product must be produced in the brain to have a pathogenic effect.

G. Transgenic Mice Models of Familial Human TSE Disease

None of the animal TSE diseases mentioned above serve as models for the familial forms of human TSE diseases that are strongly associated with different PrP mutations. However, PrP with the mutation from proline to leucine at position 102, found in human GSS patients, has been expressed in transgenic mice (Hsiao *et al.*, 1990). More recently, human PrP with extra amino acid octarepeat regions, and also associated with familial TSE disease, has been generated in transgenic mice (Chiesa *et al.*, 1998). For both the Leu-102 PrP mutant and the extra amino acid octarepeat PrP mutant, mice expressing high amounts of the mutant PrP develop a fatal neurological disease with neuropathology similar, but not identical, to TSE disease. However, in neither model is there generation of abundant PrP-res with the high degree of protease-resistance that is found in the human counterparts of these models. Furthermore, the transmissibility of the diseases produced in these mouse models remains questionable. For the octarepeat mutant there are no transmission data yet available. For the Leu-102 mutant, the disease cannot be transmitted to normal PrP (Pro-102) mice, but can be transmitted to transgenic mice expressing Leu-102 PrP at levels too low to induce disease spontaneously (Hsiao *et al.*, 1994). This result has been interpreted by the authors as evidence for transmission, but it clearly does not mimic the transmission of known TSE diseases including the human familial Leu-102 PrP disease, which is in fact transmissible to monkeys and mice expressing only Pro-102 PrP (Tateishi *et al.*, 1992; Brown *et al.*, 1994). Thus, in this mouse model, the brain disease induced lacks the two critical hallmarks of TSE disease, PrP-res and transmissibility, and may in fact be a disease due to overexpression of a mutant protein, rather than being a true TSE disease.

To avoid artifacts due to an abnormal transgene copy number and abnormal integration sites, PrP with the Pro102Leu mutation has been recently substituted for the normal mouse PrP gene by homologous recombination (Manson *et al.*, 1999). In contrast to the previous transgenic mice, such recombinant mice fail to develop spontaneous central nervous system (CNS) disease. However, they do have an increased susceptibility to infection by a human GSS isolate. These results suggest that mutant Pro102Leu PrP may be an important susceptibility factor rather than a direct cause of GSS.

III. PRION PROTEIN: CELLULAR AND MOLECULAR ASPECTS

A. *Normal Function of PrP^C*

Although PrP is expressed in a wide variety of tissues and cell types (Bendheim *et al.*, 1992; Caughey *et al.*, 1988) and is subject to developmental regulation (Manson *et al.*, 1992), its normal function is not clear. Its ability to bind copper has suggested that PrP may play a role in copper homeostasis (Brown *et al.*, 1997; Viles *et al.*, 1999). PrP is not essential for viability as some lines of PrP knockout mice are fertile and neurologically normal, although some sleep abnormalities have been observed (Tobler *et al.*, 1997; Tobler *et al.*, 1996). Other lines of PrP knockout mice have shown cerebellar degeneration, but this difference may be due to disruption of adjacent genes.

B. *Scrapie-Associated Changes in PrP*

No consistent difference in covalent structure is known to distinguish PrP^C and PrP-res. However, the two isoforms can be readily discriminated in other ways (reviewed in Caughey and Chesebro, 1997; Prusiner, 1998; Weissmann, 1999). In scrapie-infected cells, PrP-res is formed slowly from mature PrP^C after it reaches the plasma membrane, and has a much slower turnover rate than PrP^C. PrP^C is fully and rapidly digested by proteinase K (PK), whereas PK usually removes only roughly 67 of the ~210 total residues from the N terminus of most PrP-res molecules. PrP^C is generally soluble in mild detergents, whereas PrP-res is much less soluble and tends to assemble into amorphous aggregates or amyloid fibril-like structures (scrapie-associated fibrils or prion rods).

An analysis of secondary structure of PrP^C by circular dichroism (CD) spectroscopy revealed high α-helical content and little β sheet (Pan *et al.*, 1993). High-resolution nuclear magnetic resonance (NMR) studies of various recombinant PrP-sen molecules identified three α helices and a short two-stranded β-sheet structure in PrP-sen in a C-terminal domain of mouse PrP-sen (residues 121–231) (Hornemann *et al.*, 1997; Riek *et al.*, 1997; Riek *et al.*, 1996; Liu *et al.*, 1999; Donne *et al.*, 1997). Subsequent studies with full-length recombinant PrP-sen molecules (residues ~23–231) or PrP-sen fragments corresponding to the usual protease-resistant core of PrP-res (residues ~90–231) have revealed that residues lying outside the 121–231 domain are so flexible as to have little regular structure by NMR.

In contrast to PrP^C, scrapie PrP-res, (PrP^{Sc}) is predominantly β sheet with lower α-helix content, according to a variety of low-resolution methods of analysis (Caughey *et al.*, 1991; Pan *et al.*, 1993; Safar *et al.*, 1993; Nguyen *et al.*, 1995; Caughey *et al.*, 1998). The physical properties of PrP-res have made high-resolution methods of analyses extremely difficult. Nonetheless, the available evidence indicates that the transition of α-helix and/or disordered secondary structures to β sheet is likely to be the key to the conversion of PrP^C to PrP-res.

C. Diversity in Abnormal TSE-Associated PrP Structures

The terms PrP-res, PrP^{Sc}, etc., are often used as if they represent a single conformational state or entity. However, this is an oversimplification. For instance, abnormal TSE-associated PrP is known to vary in many parameters including resistance to proteolysis (Safar *et al.*, 1991; Bessen and Marsh, 1994; Tagliavini *et al.*, 1994; Bessen *et al.*, 1995; Caughey *et al.*, 1998); insolubility in detergents (Bessen and Marsh, 1994; Muramoto *et al.*, 1996); secondary structure (Fig. 2) (Caughey *et al.*, 1998); glycoform ratios (Kascsak *et al.*, 1986; Monari *et al.*, 1994; Collinge *et al.*, 1996; Somerville *et al.*, 1997; Hope *et al.*, 1999); exposure of epitopes by denaturants (Safar *et al.*, 1998); multispectral ultraviolet fluorescence (Rubenstein *et al.*, 1998); and ultrastructure (McKinley *et al.*, 1991; Giaccone *et al.*, 1992; Jeffrey *et al.*, 1994). Another source of complexity is the fact that PrP^{Sc} can be purified from scrapie brain as a mixture of both partially PK-resistant and fully PK-sensitive molecules (Caughey *et al.*, 1995). While there is evidence that the PK-sensitive molecules are not required for infectivity and autocatalytic activity (Caughey *et al.*, 1997), it is possible that they are important in neuropathogenesis. Finally, as detailed below, manipulations with various recombinant and mutant PrP molecules have generated numerous forms of PrP that appear to share some of the properties of bona fide TSE-associated PrP isolated from infected tissues. A current challenge is to understand which abnormal states of PrP play roles in various aspects of TSE transmission and/or pathogenesis. The requirements may be different for infectious-versus-neurotoxic forms of PrP. Accordingly, Charles Weissmann (1991) has proposed the use of the term PrP* for the putative infectious form of PrP, whatever it may be.

D. Mechanistic Models for PrP-res Formation

Although the detailed mechanism for PrP-res formation is far from clear, numerous mechanisms and permutations of mechanisms have been proposed to account for PrP-res formation. Two of the most com-

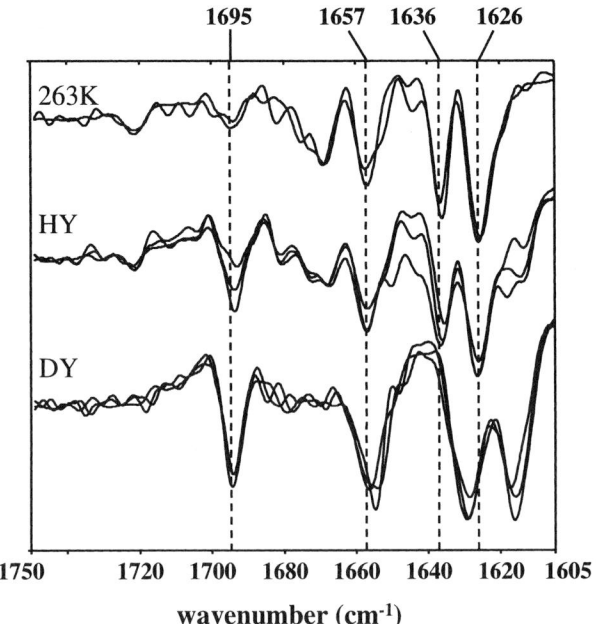

FIG 2. Strain-dependent differences in beta sheet conformations of PrP-res. Second-derivative Fourier transform infrared spectra of PrP-res isolated from hamsters with either the 263K, HY, or DY TSE strains. Although the PrP-res in these samples are derived from the same hamster PrP^C sequence, major differences are observed between the DY strains and the other two in the beta sheet region of the spectrum (~1616–1640 cm^{-1}). The data are consistent with the concept that different PrP-res conformations may account for the existence and individual properties of TSE strains. Adapted from Caughey *et al.* (1998).

monly considered are the heterodimer or template assistance-type model (Prusiner, 1998; Bolton and Bendheim, 1988; Griffith, 1967) and the nucleation (seed)-dependent polymerization model (Griffith, 1967; Gadjusek, 1988; Jarrett and Lansbury, Jr. 1993; Lansbury, Jr. and Caughey, 1995) (Fig. 3). In its simplest form, the heterodimer model proposes that PrP-res exists in a stable monomeric state that can bind PrP^C, forming a heterodimer, and catalyze a conformational change in PrP^C so as to form a homodimer of PrP-res. The PrP-res homodimer then splits apart to give two PrP-res monomers. In this model, PrP-res is more stable thermodynamically than PrP^C, but conversion of PrP^C to PrP-res is rare unless it is catalyzed by a preexisting PrP-res template.

The most fundamental aspect of the nucleated polymerization type of model is that oligomerization/polymerization of PrP is necessary to stabi-

Heterodimer model

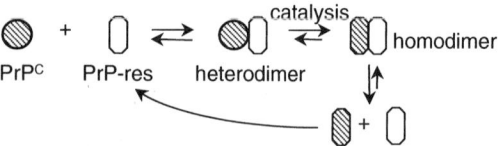

Autocatalytic nucleated polymerization

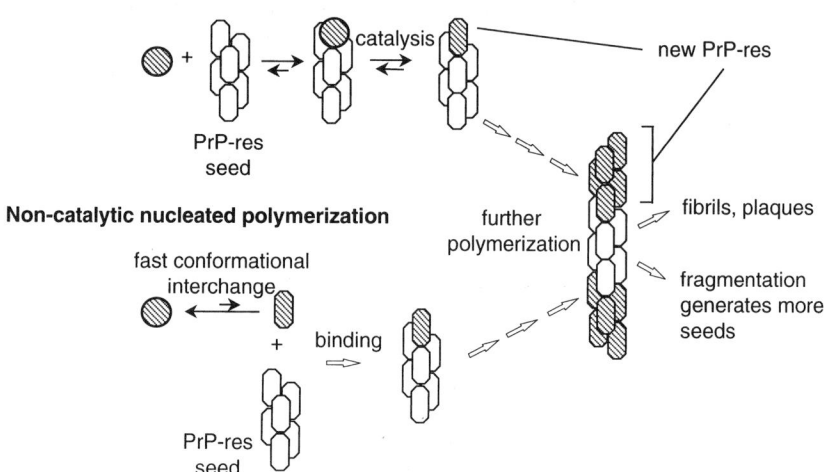

FIG 3. Heterodimer and autocatalytic and noncatalytic nucleated (seeded) polymerization models for formation of PrP-res from PrP^C. In the heterodimer and autocatalytic nucleated polymerization models, the conformational conversion of PrP^C to PrP-res is presumed to be extremely rare unless catalyzed by contact with monomeric or polymeric PrP-res, respectively. In the noncatalytic nucleated polymerization model, the conformational interchange is relatively rapid, but the PrP-res conformer is poorly populated unless stabilized by binding to a preformed, stable PrP-res polymer. See the text for further explanations.

lize PrP-res, to allow its accumulation to biologically relevant levels. Initial formation of nuclei or seeds of PrP-res is rare because of the weakness of monovalent interactions between PrP^C molecules and/or the rarity of polymerizable conformers. However, once formed, oligomeric or polymeric seeds are stabilized by multivalent interactions (Jarrett and Lansbury, Jr., 1993). Recruitment of new PrP molecules would also be favored by multivalent interactions at seeding surfaces.

Some crossover can occur between the two types of models in their various renditions. For instance, autocatalysis of the conformational change in PrP^C by PrP-res can be an element of either the template-assisted or nucleated polymerization models (Fig. 3). However, the nucleated polymerization model could also include a scenario in which PrP^C rapidly interchanges between PrP^C-like and PrP^{Sc}-like conformers, with the latter being poorly populated until they are stabilized by binding to a preexisting polymeric seed of PrP-res. In either type of model, there may be a metastable PrP^C folding intermediate that most favorably interacts with PrP-res in the conversion reaction. Such an intermediate might resemble molecules generated from recombinant PrP-sen under acidic conditions (Swietnicki *et al.*, 1997; Zhang *et al.*, 1997; Hornemann and Glockshuber, 1998; Jackson *et al.*, 1999), such as those found in endosomes or lysosomes where PrP-res may be formed (Caughey *et al.*, 1991b; McKinley *et al.*, 1991b; Borchelt *et al.*, 1992b).

Practically speaking, the critical question is what physical state of PrP-res is active in TSE disease processes *in vivo?* The nucleated polymerization model predicts that active PrP-res seeds could range in size from the minimum stable nucleus (in theory, as small as a trimer) to huge polymers. Consistent with this prediction is the frequent observation that autocatalytic activity (see the discussion below) and infectivity are associated not with monomers, but with PrP-res aggregates that range widely in size (Prusiner *et al.*, 1993; Hope, 1994; Caughey *et al.*, 1995; Caughey *et al.*, 1997). In contrast, the template assistance–type models propose discrete monomers, or perhaps small oligomers, as the active autocatalytic unit. Aggregation of PrP-res would then be considered a secondary event, albeit one that might further stabilize PrP-res in the biological milieu.

In inherited and sporadic human TSEs that are not clearly of infectious origin, the spontaneous stochastic formation of the template or seed of PrP-res might initiate the pathogenic propagation of PrP-res in the host. On the other hand, even in these TSE diseases, initiation of the process might be explained by exogenous "infection" by exposure to the template/seed or by another as yet unidentified infectious agent.

E. PrP-res-Induced Conversion of PrP^C to Protease-Resistant State

Several indirect lines of evidence have strongly implied that interactions between PrP-sen and PrP-res are required for the formation of PrP-res and TSE pathogenesis (reviewed in Priola, 1999). Direct evidence for this concept was provided by cell-free studies demonstrating that PrP^{Sc} induces ^{35}S-labeled PrP-sen to convert to a PK-resistant state characteristic of TSE-associated PrP-res (Fig. 4) (Kocisko *et al.*, 1994). This cell-free conversion reaction is highly specific and repro-

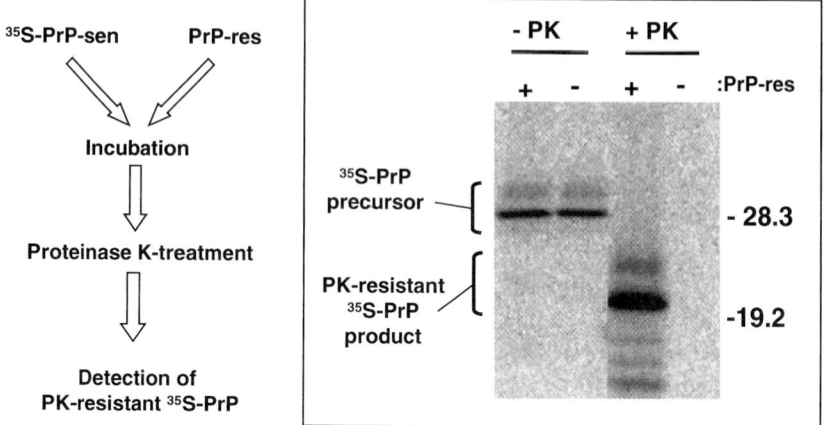

FIG 4. Cell-free reaction showing that PrP-res induces the conversion of [^{35}S]PrP-sen to the proteinase K (PK)-resistant form. The right panel is a phosphor-autoradiographic image of an SDS-PAGE gel showing only the radiolabeled hamster PrP molecules in the reaction with and without PK digestion after incubation in the presence or absence of unlabeled PrP-res. The PrP-res (PrP^{Sc}) was isolated from scrapie-infected hamster brain tissue. Details of the reaction can be found in Kocisko et al. (1994).

duces strain-specific biochemical properties of hamster-adapted TSE strains (Bessen et al., 1995). Furthermore, the PrP sequence specificities of the conversion reaction correlate with inter- and intraspecies TSE transmissions in vivo (Kocisko et al., 1995; Bossers et al., 1997; Raymond et al., 1997). Although the PrP-res-associated activity that induces the conversion reaction correlates with scrapie infectivity in guanidine hydrochloride denaturation studies (Caughey et al., 1997), it is not clear whether the products of cell-free conversion reactions are themselves infectious (Hill et al., 1999; and Raymond and Caughey, unpublished observations). Unfortunately, there are major technical difficulties associated with discriminating any newly formed infectivity from that already associated with the input PrP-res that drives the conversion reaction. In any case, the strain and species specificities of the cell-free conversion reactions suggest that this experimental system is relevant to TSE biology and a powerful tool for further analyses of the direct interactions between PrP-res and PrP-sen under defined conditions. In addition, the cell-free conversion reactions have been used for practical purposes such as predicting possible risks of cross-species transmission of TSE (Raymond et al., 1997) and testing potential anti-TSE compounds (Caughey et al., 1998; Chabry et al., 1998; Demaimay et al., 1998).

Although the cell-free conversion reaction was initially observed under decidedly nonphysiological conditions with high guanidine hydrochloride concentrations, more physiological conditions have now been established. For instance, an *in situ* conversion reaction has been shown to occur within the context of intact brain tissue slices (Bessen *et al.,* 1997). The elimination of denaturants such as guanidine hydrochloride also has allowed analyses of the impact of biological factors in the conversion of PrP-sen to PrP-res (DebBurman *et al.,* 1997; Horiuchi *et al.,* 1999; Saborio *et al.,* 1999). For instance, the chaperone proteins GroEL and Hsp104 can stimulate the conversion reaction (DebBurman *et al.,* 1997). Kinetic analyses under such conditions showed that the conversion process can be separated into two stages— first, the highly specific binding of PrP-sen to PrP^{Sc} (Fig. 5); and, then, a slower conversion of the bound PrP-sen to a protease-resistant form

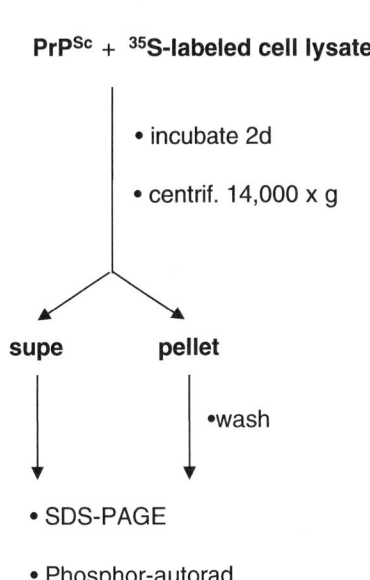 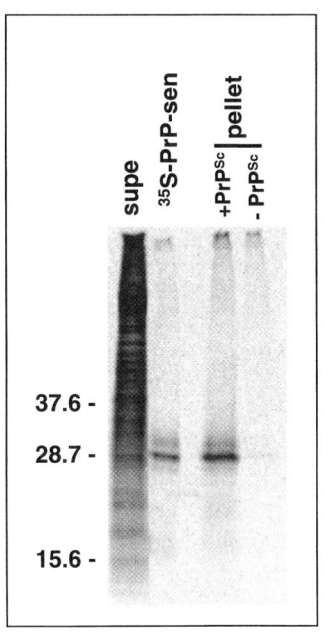

FIG 5. Highly specific binding between PrP-sen and PrP^{Sc}. PrP^{Sc} isolated from scrapie-infected brain tissue was incubated with detergent cell lysates (postnuclear supernatants) from cells metabolically labeled with [^{35}S]methionine. ^{35}S-labeled proteins that were bound to the PrP^{Sc} aggregates were isolated by the protocol shown. The PrP^{Sc}-bound ^{35}S-labeled proteins were compared to immunopurified [^{35}S]PrP-sen, the total labeled proteins in the lysate, and to pellets obtained in the absence of PrP^{Sc} by SDS-PAGE and phosphor autoradiography. Adapted from Horiuchi *et al.* (1999).

(Bessen *et al.*, 1997; DebBurman *et al.*, 1997; Horiuchi *et al.*, 1999). Anti-PrP antibodies have been used to probe the domain(s) on the PrP-sen involved in the initial binding between PrP-sen and PrP^{Sc} (Horiuchi *et al.*, 1999). The experiments, when considered with the NMR structure of PrP-sen, suggested the likely involvement of sites close in space to the C terminus, such as the extended chain of residues ~119–140, helical residues ~206–223; and the loop of residues ~165 to 174 between the second β-strand and second α-helix (Horiuchi *et al.*, 1999). The importance of residues in the 119–140 region has also been implicated by the inhibition of the conversion reaction by peptide fragments from this region (Chabry *et al.*, 1998; Chabry *et al.*, 1999).

As noted above, cell-free conversion systems have been used to dissect the conversion process into initial binding and conversion-to-PrP-res steps by kinetic means (DebBurman *et al.*, 1997; Horiuchi *et al.*, 1999). More recently, conversion has also been segregated from the initial binding in cell-free studies using mouse and hamster PrP isoforms. Consistent with the difficulties of transmitting mouse scrapie agent to hamsters, or vice versa, conversion of PrP-sen with the PrP^{Sc} from the other species is inefficient (Kocisko *et al.*, 1995). However, binding of PrP-sen to heterologous PrP^{Sc} was found to occur efficiently (Horiuchi *et al.*, 2000). This suggests that the species specificity of PrP-res formation is determined more at the conversion step than at the initial binding step.

F. Unfolding and Refolding of PrP-res

Although the high-resolution structure of PrP-res has not been determined, some conflicting clues to PrP-res folding have been obtained from analyses of the effects of denaturants. Studies of hamster PrP^{Sc} unfolding and disaggregation concluded that guanidine hydrochloride and acid (<pH 2) treatments irreversibly disaggregate and denature PrP^{Sc} via a monomeric molten globule intermediate (Safar *et al.*, 1994; Safar *et al.*, 1993), with the formation of the dissociated folded monomer occurring fully at 1.5 *M* guanidine hydrochloride. However, these results are at odds with more recent studies showing that without PK-treatment, PrP^{Sc} preparations contain a mixture of partially PK-resistant and fully protease-sensitive molecules (Kocisko *et al.*, 1996; Caughey *et al.*, 1995). The latter are solubilized by ~2.5–3 *M* guanidine hydrochloride, but the protease-resistant molecules remain polymerized and only partially and reversibly unfolded at a domain containing residues ~90–115. It is the polymerized, protease-resistant PrP fraction that confractionates with PrP converting activity and scrapie infectivity (Caughey *et al.*, 1995; Kocisko *et al.*, 1996;

Caughey *et al.*, 1997). These unfolding studies suggest that PrP^{Sc} has at least three domains: (1) the highly PK-sensitive domain of residues 23–~90 containing the octapeptide repeats; (2) the normally PK-resistant domain of residues ~90–~115 that is readily and reversibly unfolded by 2.5–3.5 M guanidine hydrochloride; and (3) the most stable, guanidine hydrochloride–resistant and PK-resistant C-terminal domain comprised roughly of residues 119–226 (Kocisko *et al.*, 1996). Unfolding of the latter domain with guanidine hydrochloride appears to be irreversible.

Attempts to reverse losses of infectivity due to denaturation of PrP-res have met with variable results. McKenzie *et al.* (1998) concluded that the addition of copper ions aids in the refolding of PrP^{Sc} after partial unfolding in guanidine hydrochloride and the apparent loss of TSE infectivity. However, others have not been able to restore scrapie infectivity lost to treatments with urea, chaotropic salts, and SDS (Prusiner *et al.*, 1993; Riesner *et al.*, 1996; Post *et al.*, 1998; Wille and Prusiner, 1999). Indeed, even aggregated and PK-resistant forms of PrP have been generated from denatured PrP^{Sc} that has little or no infectivity (Riesner *et al.*, 1996; Post *et al.*, 1998; Shaked *et al.*, 1999). Collectively, these and other studies (e.g., Hill *et al.*, 1999) indicate that PK resistance and insolubility alone are not sufficient for PrP to be infectious.

G. Studies with Synthetic PrP Peptides

Short synthetic peptide fragments of the PrP amino acid sequence have been used to model the fibril formation and the conformational conversion of PrP-sen to PrP-res (Gasset *et al.*, 1992; Come *et al.*, 1993; Goldfarb *et al.*, 1993; Selvaggini *et al.*, 1993; Come and Lansbury, Jr. 1994; Nguyen *et al.*, 1995). These studies revealed that, as is the case with numerous other small peptides, certain PrP peptides with disordered or helical structures in solution can polymerize into high-β sheet fibrils. This occurs via a nucleated polymerization mechanism (Come *et al.*, 1993). Vast stoichiometric excesses of hamster PrP synthetic peptides 90–145 and 109–141 have been shown to enhance the PK-resistance of PrP-sen (Kaneko *et al.*, 1995). However, total protease resistance of PrP-sen in the peptide mixtures was not the characteristic partial PK resistance exhibited by bona fide PrP-res. Other PrP peptides such as residues 119–136 have been shown to inhibit PrP-res formation in both cell-free conversion reactions and scrapie-infected neuroblastoma cells (Chabry *et al.*, 1998; Chabry *et al.*, 1999). Although peptide studies have revealed important mechanistic possibilities, how well they approximate the structural transitions of whole PrP molecules remains to be determined.

H. Folding Studies with Recombinant PrP Molecules

More extended portions of the PrP sequence have been expressed in bacteria and used in PrP structure and folding studies. C-terminal mouse PrP residues 121–231 act as an "autonomous folding unit" that undergoes a cooperative and reversible unfolding/refolding transition in the presence of guanidine hydrochloride (Hornemann and Glockshuber, 1996). Stopped flow studies indicated that the folding of this PrP fragment was possibly the fastest protein-folding reaction known at that time, without detectable intermediates at neutral pH (Wildegger et al., 1999). However, at an acidic pH, a high β sheet equilibrium folding intermediate of this fragment is populated (Hornemann and Glockshuber, 1998). A similar scenario was observed in studies of a larger fragment of recombinant human PrP (residues ~90–231) (Swietnicki et al., 1997). At a nearly neutral pH, a cooperative two-state guanidine hydrochloride–induced unfolding was observed. However, at a more acidic pH, a high β sheet folding intermediate, with exposed hydrophobic surfaces, was stabilized. A similar fragment of hamster PrP has been shown to assume two different α-helical states, one obtained directly from an acid environment and another refolded at a nearly neutral pHs (Zhang et al., 1997). Only the latter was soluble at a physiological pH. The high α-helical state required a disulfide bond between the two cysteine residues in PrP. On the other hand, insoluble, high β sheet forms were obtained with or without the intramolecular disulfide bond. More recently, a monomeric, high β sheet, PK-resistant and fibrillogenic form of human PrP residues 91–231 was generated reversibly by reducing the disulfide bond and dropping the pH to 4 (Jackson et al., 1999). Since PrP-res formation may occur in endosomes or lysosomes in scrapie-infected cells (Caughey et al., 1991; Caughey and Raymond, 1991; Borchelt et al., 1992), it is possible that acidification and redox changes in those compartments could influence the PrP^C-to-PrP^{Sc} transformation. However, the following observations are inconsistent with the idea that reduction of the disulfide bond is critical in PrP^{Sc} formation: (1) both PrP^C and PrP^{Sc} have intact disulfide bonds (Turk et al., 1988); (2) substitution of Ala for Cys-178 prevented the PrP-res formation in scrapie-infected cells (Muramoto et al., 1996); and (3) reduction of the disulfide bond in PrP^C inhibits its PrP^{Sc}-induced conversion to the protease resistant form in vitro (Herrmann and Caughey, 1998).

I. Effects of Familial TSE-Associated Mutations in PrP Folding

PrP mutants found in certain inherited human TSE can be inappropriately compartmentalized within cells and can display some biochemical characteristics of PrP-res, such as detergent insolubility and relative resistance to PK digestion, albeit much less PK resistance than that shown by PrP-res in brains of TSE-infected animals (Lehmann and Harris, 1995; Lehmann and Harris, 1996; Petersen *et al.*, 1996; Singh *et al.*, 1997; Priola and Chesebro, 1998). Although no infectivity has been demonstrated for PrP-res-like aggregates generated in tissue culture cells, these results suggest that at least some mutant PrP molecules can have some predisposition to form PrP-res-like aggregates.

Biophysical studies have directly addressed the question of whether amino acid changes destabilize and/or alter the secondary structure of PrP molecules. In circular dichroism spectroscopy experiments, recombinant human PrP (residues 90–231) containing Pro102Leu or Glu200Lys mutation (Swietnicki *et al.*, 1998), and eight recombinant mouse PrPs (residues 121–231) containing various point mutation (Liemann and Glockshuber, 1999), showed α-helical content essentially identical to wild-type PrP and no increased β sheet content. These results suggest that these mutations do not affect the secondary structure of PrP-sen. However, others have reported that recombinant mouse PrP (residues 23–231) containing Pro102Leu mutation displayed decreased α-helical content (Cappai *et al.*, 1999). The different result may be due to the difference in length and/or species of the PrP fragment used. In addition, some, but not all, of the recombinant mouse PrP containing mutations (Asp178Asn; Thr183Ala; Phe198Ser; or Gln217Arg) showed lower thermodynamic stability than the wild-type PrP (Liemann and Glockshuber, 1999). Thus, it appears that, although mutation-induced destabilization might explain the mechanism of the PrP-res formation for particular mutations (such as Asp-178-Asn), it is not a general mechanism underlying stochastic PrP-res formation in inherited human TSE (Swietnicki *et al.*, 1998; Liemann and Glockshuber, 1999). Some mutations might instead increase stochastic conversion by stabilizing the transition state, thereby lowering the activation energy barrier.

IV. Current Issues in TSE research

A. Nature of Infectious Agent

In the 1990s, there was a massive increase in knowledge concerning many aspects of the TSE diseases, largely due to the discovery of PrP. Nevertheless, there continues to be a paucity of information concerning the structure and composition of the infectious agent. Early ultrafiltration studies suggested that the infectious particle was small and might be a virus. However, the infectivity showed a strong resistance to sterilization by heat and chemicals that led Griffith (in 1967) to propose that the agent might be a self-replicating protein. Following the discovery of scrapie-associated fibrils (Merz *et al.*, 1981) and the identification of PrP as a major component of infectious fractions, the "protein-only" hypothesis was refined into the "prion" hypothesis (Prusiner, 1982). Although the protein-only/prion hypothesis has gained considerable popularity, it has been difficult to definitively discriminate between a protein-only agent and other conceivable infectious entities.

1. Biophysical Inactivation and Viral Hypothesis

Inactivation studies with irradiation, heat, and chemicals have led to conflicting conclusions regarding the uniqueness of TSE infectivity (Rohwer, 1991). Using heat or hypochlorite, the majority of infectivity actually shows kinetics of inactivation and a small resistant fraction (0.1%) that is similar to known viral examples, such as bacteriophage fd (Rohwer, 1984b; Rohwer, 1984b). This minor resistant fraction should not be used to infer unique properties for the majority of the infectivity.

The spectrum for inactivation of scrapie infectivity by ultraviolet (UV) irradiation suggested that the critical target was neither protein nor nucleic acid, but that it instead appeared to be lipid in nature (Alper *et al.*, 1978). However, in past virological experiments using nonpenetrating radiation such as UV, shielding of the critical target molecule of the infectious agent by other molecules in the mixture, or attached to the agent, has been known to influence the results. In fact, the unusual inactivation spectrum for scrapie was similar to intact tobacco mosaic virus, a well-characterized RNA virus, whereas the isolated RNA from this virus had peak inactivation at a wavelength predicted for a typical nucleic acid (Kleczkowski and McLaren, 1967). Thus, in view of these issues and the difficulty involved in purification of the scrapie agent, UV studies may not provide definitive information as to the nature of the scrapie infectivity.

Scrapie infectivity has also been studied using more penetrating radiation, such as X-rays, where shielding or blocking of the radiation is not an issue. Many experiments have resulted in similar inactivation rate constants; however, different groups have varied markedly in their interpretation of these results (Rohwer, 1991). Using target theory calculations, some workers have concluded that the maximum genome size would be very small (Alper, 1985; Bellinger-Kawahara *et al.*, 1988). In contrast, others making empirical comparisons to viruses with known genomes have arrived at a genome size consistent with a small

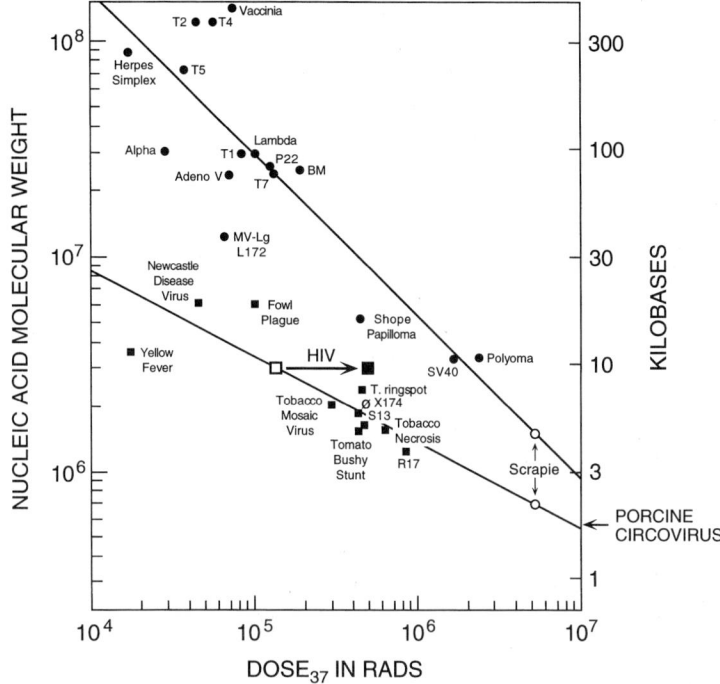

FIG 6. Comparison of viral nucleic acid molecular weight to dose of X-irradiation required for 37% inactivation (D_{37}) for a variety of viruses and for the scrapie agent. Adapted from Rohwer (1991). The solid circles are double-stranded viruses, and the solid squares are single-stranded viruses. The best-fit lines for each of these groups are shown. The open square shows the D_{37} value predicted for HIV, based on its genome size, and the adjacent solid square shows the actual value observed experimentally. Based on radiation inactivation experiments, the scrapie agent might be predicted to have a genome size of 2 to 4kb which is larger than the genome size for one of the smallest known viruses, porcine circovirus (1759 nucleotides), which is shown for comparison.

(2–4kb) virus (Fig. 6) (Rohwer, 1991; Rohwer, 1984). However, both these interpretations might underestimate the genome size if the TSE agent had a means of repairing damaged nucleic acid during replication. Such a situation occurs with retroviruses in which the two RNA genomes in each particle can be partially damaged and then repaired during reverse transcription, thus giving a higher resistance to X-irradiation than is predicted by genome size alone (see HIV in Fig. 6).

These data are only indirect evidence that may be consistent with the viral hypothesis. In spite of many efforts, there are still no data supporting any candidate viruses. Furthermore, although small nucleic acid molecules have been found in purified infectious scrapie samples, efforts to identify an intact nucleic acid molecule of potential genome size have met with failure (Kellings *et al.*, 1994). Therefore, if such a genome exists, it would have to be capable of regeneration from small fragments by a copy–choice mechanism during transcription, as described above. In addition, the role of mutant and nonmutant PrP, in the case of a viral etiology, remains hypothetical. To explain the very high correlation of TSE disease in persons with certain PrP mutations, one would have to speculate that the mutant PrP might serve as an efficient susceptibility factor or a receptor for a viral agent that might be relatively common in the population. In contrast, the normal nonmutant PrP would have to be much less efficient than mutant PrP in this role in order to account for the extremely low incidence of sporadic CJD. The recent experiments showing an increased sensitivity to human GSS in mice expressing the Pro102Leu mutation might be an example of this situation (Manson *et al.*, 1999). Similar effects probably occur for many viral diseases in humans where there is a high incidence of infection as compared to the incidence of clinical disease. Examples include human T cell leukemia virus (HTLV) I and B19 parvovirus. However, at present this possibility remains speculative as applied to TSE diseases in the absence of additional supportive data using actual candidate viruses.

2. Purification of Infectious Agent and Protein-Only Hypothesis

Although the discovery of PrP has led to a vast increase in knowledge concerning the role of PrP in susceptibility and pathogenesis of TSE diseases, the question of whether PrP-res is an integral component of the infectious agent remains unresolved. The most important evidence supporting this concept is the finding that PrP is the predominant macromolecule found in fractions of the purified infectious agent. However, two caveats persist regarding this matter. First, because of the presence of aggregated PrP, the agent is difficult to purify, and the best fractions still contain detectable nucleic acid molecules. These fractions might also conceivably contain other components relevant to infectivity. Second, in purified fractions the ratio of PrP molecules to infectious units is extremely high (approximately 100,000:1), and this fact led to the speculation that only a subfraction of the PrP-res, i.e., PrP*, is the infectious form (Weissmann, 1991). However, it remains unclear as to how to identify or biochemically distinguish the proposed infectious and noninfectious forms of the protease-resistant PrP. Thus, although considerable circumstantial evidence suggests that an abnormal form of PrP alone may be the culprit, a measure of skepticism is warranted until a pure, noninfectious form of PrP is converted *de novo* to an agent that causes TSE disease when it is injected into animals.

Among the most puzzling aspects of the protein-only hypothesis is the comparison of TSE diseases to other amyloid diseases. In all these diseases protein misfolding is a prominent part of the pathogenesis, and in many amyloid diseases interactions between the normal and abnormal proteins can lead to formation of additional abnormal protein. This is similar to the situation with TSE diseases. However, only the TSE diseases appear to be easily transmissible experimentally. This suggests that the protein interactions common to all amyloid diseases probably do not explain the unique transmissibility of TSE diseases (Chesebro, 1998). The existence of this dilemma does not imply that the protein-only or the viral hypotheses are incorrect but, rather, that we are lacking in some crucial information for explaining the differences between TSE diseases and the nontransmissible amyloid diseases.

B. Origin of Sporadic CJD

There is still only limited information regarding the origin of sporadic CJD in humans. The various animal models appear to be similar to the human diseases in the accidental or iatrogenic infection group (Table I); however, animal TSE diseases have so far not adequately modeled sporadic human TSE disease. For sporadic TSE disease the protein-only hypothesis has proposed rare spontaneous misfolding as a possible primary event. As noted above, PrP molecules with familial TSE-associated mutations have exhibited some abnormal properties *in vitro* that are somewhat reminiscent of PrP-res, but no such PrP molecules have generated any infectivity *de novo*. Furthermore, in Australia and New Zealand, where scrapie has been eradicated, there is no evidence of spontaneous recurrence of sheep scrapie (Hunter and Cairns, 1998). In addition, in humans the peak age incidence of sporadic CJD is found in those 55–60 years old, and if spontaneous misfolding were the primary event, one might expect a continuously increasing incidence with age, since more time might allow more opportunity for rare misfolding events.

C. TSE Strains

The existence of biologically different scrapie strains in inbred animals with a single type of PrP gene remains an interesting enigma (Bruce and Fraser, 1991). Disease induced by scrapie strains can differ in the clinical symptoms produced, the regions of brain affected and the incubation period prior to clinical onset. These differences might be explained by mutations in a nucleic acid genome according to the viral hypothesis, but no genomes have yet been identified. In contrast, the existence of strains may be difficult to explain by the protein-only hypothesis. However, structural variations in PrP-res might "encode" strain-specific properties, and recent data suggest that PrP-res structures might be capable of conferring such properties on newly formed PrP-res in a template-like fashion (Bessen et al., 1995; Telling et al., 1996). This possibility is further supported by studies that have indicated that different strain-associated forms of PrP of the same amino acid sequence can differ in susceptibility to PK (Bessen and Marsh, 1994; Hill et al., 1997), secondary structure (Fig. 2) (Caughey et al., 1998), and in other conformationally sensitive parameters (Rubenstein et al., 1998; Safar et al., 1998). Nonetheless, it is unclear as to how strain-specific properties or conformations might be preserved during passage between species where numerous PrP amino acid differences exist (Kimberlin et al., 1989), unless distinct PrP-res conformations can exist that propagate themselves independently of certain changes in primary sequence.

D. Routes of Neuroinvasion

In mouse scrapie, functional B lymphocytes are necessary for neuroinvasion after infection at peripheral sites, but PrP expression on B lymphocytes is not required. Previous work suggested that follicular dendritic cells (FDCs) are important sites of agent replication and/or PrP-res accumulation (Fraser and Farquhar, 1987; Muramoto et al., 1993). In recent experiments, PrP-negative B lymphocytes may have indirectly influenced neuroinvasion by allowing the development of mature spleen follicular dendritic cells (Klein et al., 1998). Although in mice, replication of scrapie in spleen may be required to amplify titers after peripheral scrapie infection, splenic infection was by itself unable to induce transport of infectivity from spleen to brain (Blattler et al., 1997). The importance of PrP-positive peripheral nerves in neuroinvasion was recently demonstrated when transgenic mice expressing PrP only in neurons were shown to develop scrapie after oral or intraperitoneal (IP) infection with high doses of agent (Race et al., 2000). Bloodborne transport of infectivity to the brain remains difficult to exclude completely; however, in most model systems infectivity levels in blood are extremely low, and transmission by blood transfusion in animals has only rarely been effective (Brown et al., 1998). Nevertheless, when considering the issue of blood safety for human transfusion, it is important to remember that different animal species and TSE strains might differ in infectivity levels of blood.

E. Therapeutic Strategies

A major challenge that remains is the development of effective therapeutics for TSE diseases. Several classes of compounds have proven to have prolonged the lives of scrapie-infected rodents if given near the time of inoculation with scrapie. These classes include polyanions (e.g., pentosan polysulfate, dextran sulfate, carrageenan, and heteropolyanion 23 (Kimberlin and Walker, 1983; Ehlers and Diringer, 1984; Kimberlin and Walker, 1986; Farquhar and Dickinson, 1986)); sulfonated dyes (Congo red [Ingrosso et al., 1995]); polyene antifungals (e.g., Amphotericin B and MS8209 (Demaimay et al., 1999; Adjou et al., 1998; Demaimay et al., 1994); an anthracycline (Tagliavini et al., 1998); and cyclic tetrapyrroles (porphyrins and phthalocyanines) (Priola et al., 2000). All but the amphotericins and anthracycline are known to inhibit PrP-res formation, which is their likely mechanism of action (Caughey et al., 1998; Caughey, 1994; Caughey and Raymond, 1993). The therapeutic utility of all but the amphotericins is severely limited by their failure to be effective when treatment is initiated long after the initial infection or in the clinical phase of disease. The amphotericins, on the other hand, are generally too toxic for anti-TSE applications in humans. A likely problem with polyanions, sulfonated dyes, and tetrapyrroles that have been tested may be impermeability to the blood–brain barrier. Thus, a fruitful approach may be to find members of these groups of potent inhibitors of PrP-res formation that can penetrate the central nervous system. Another therapeutic strategy is suggested by the studies showing that specific synthetic PrP peptides can inhibit PrP-res formation in *in vitro* systems (Chabry et al., 1999; Chabry et al., 1998). The peptides themselves, or compounds that mimic the inhibitory peptides, might have therapeutic effects if a suitable means of delivery to infected hosts were discovered. Therapeutics might also be based on other recently described inhibitors of PrP-res formation such as antimalarial lysosomotropic amines, cysteine protease inhibitors (Doh-Ura et al., 2000), and branched polyamines (Supattopone et al., 1999)

Finally, it may be possible to slow the neuropathogenesis by blocking undesirable responses of glial cells to TSE infections. For instance, there is evidence that microglial responses mediate the cytotoxicity of PrP-res and neurotoxic PrP peptides (Giese et al., 1998). Furthermore, changes in cytokine expression have been noted in TSE-infected brain tissue (Campbell et al., 1994) and in microglia exposed to neurotoxic PrP peptides (Peyrin et al., 1999), making it possible that manipulation of brain cytokine levels could influence the course of disease.

REFERENCES

Adjou, K. T., Deslys, J.-P., Demaimay, R., Seman, M., and Dormont, D. (1998). Prospects for the pharmacological treatment of human prion diseases. *CNS Drugs* **10**:83–89.

Alper, T., Haig, D. A., and Clarke, M. C. (1978). The scrapie agent: evidence against its dependence for replication on intrinsic nucleic acid. *J. Gen. Virol.* **41**:503–516.

Alper, T. (1985). Scrapie agent unlike viruses in size and susceptibility to inactivation by ionizing or ultraviolet radiation. *Nature* **317**:750

Asante, E. A, and Collinge, J. (2001). Transgenic studies of the influence of the PrP structure on TSE diseases. *Adv. Prot. Chem* (in press).

Bellinger-Kawahara, C. G., Kempner, E., Groth, D., Gabizon, R., and Prusiner, S. B. (1988). Scrapie prion liposomes and rods exhibit target sizes of 55,000 Da. *Virology* **164**:537–541.

Bendheim, P. E., Brown, H. R., Rudelli, R. D., Scala, L. J., Goller, N. L., Wen, G. Y., Kascsak, R. J., Cashman, N. R., and Bolton, D. C. (1992). Nearly ubiquitous tissue distribution of the scrapie agent precursor protein. *Neurology* **42**:149–156.

Bessen, R. A., Kocisko, D. A., Raymond, G. J., Nandan, S., Lansbury, P. T., Jr., and Caughey, B. (1995). Nongenetic propagation of strain-specific phenotypes of scrapie prion protein. *Nature* **375**:698–700.

Bessen, R. A., Raymond, G. J., and Caughey, B. (1997). *In situ* formation of protease-resistant prion protein in transmissible spongiform encephalopathy-infected brain slices. *J. Biol. Chem.* **272**:15227–15231.

Bessen, R. A. and Marsh, R. F. (1994). Distinct PrP properties suggest the molecular basis of strain variation in transmissible mink encephalopathy. *J. Virol.* **68**:7859–7868.

Blattler, T., Brandner, S., Raeber, A. J., Klein, M. A., Voigtlander, T., Weissmann, C., and Aguzzi, A. (1997). PrP-expressing tissue required for transfer of scrapie infectivity from spleen to brain. *Nature* **389**:69–73.

Bolton, D. C. and Bendheim, P. E. (1988). *Novel Infectious Agents and the Central Nervous System* (Bock, G., and Marsh, J., eds.), John Wiley & Sons, Chichester. 164–181.

Borchelt, D. R., Taraboulos, A., and Prusiner, S. B. (1992). Evidence for synthesis of scrapie prion protein in the endocytic pathway. *J. Biol. Chem.* **267**:16188–16199.

Bossers, A., Belt, P. B. G. M., Raymond, G. J., Caughey, B., de Vries, R., and Smits, M. A. (1997). Scrapie susceptibility-linked polymorphisms modulate the *in vitro* conversion of sheep prion protein to protease-resistant forms. *Proc. Natl. Acad. Sci. U.S.A.* **94**:4931–4936.

Bossers, A., de Vries, R., and Smits, M. (2000). Susceptibility of sheep for scrapie as assessed by in vitro conversion of nine naturally occurring variants of PrP. *J. Virol.* **74**:1407–141.

Brown, D. R., Qin, K., Herms, J. W., Madlung, A., Manson, J., Strome, R., Fraser, P. E., Kruck, T., von Bohlens, A., Schulz-Schaeffer, W., Giese, A., Westaway, D., and Kretzschmar, H. (1997). The cellular prion protein binds copper in vivo. *Nature* **390**:684–687.

Brown, P., Gibbs, C. J., Jr., Rodgers-Johnson, P., Asher, D. M., Sulima, M. P., Bacote, A., Goldfarb, L. G., and Gajdusek, D. C. (1994). Human spongiform encephalopathy: the National Institutes of Health series of 300 cases of experimentally transmitted disease. *Ann. Neurol.* **35**:513–529.

Brown, P., Rohwer, R. G., Dunstan, B. C., MacAuley, C., Gajdusek, D. C., and Drohan, W. N. (1998). The distribution of infectivity in blood components and plasma derivatives in experimental models of transmissible spongiform encephalopathy. *Transfusion* **38**:810–816.

Bruce, M. E., Will, R. G., Ironside, J. W., McConnell, I., Drummond, D., Suttie, A., McCardle, L., Chree, A., Hope, J., Birkett, C., Cousens, S., Fraser, H., and Bostock, C. J. (1997). Transmissions to mice indicate that `new variant' CJD is caused by the BSE agent. *Nature* **389**:498–501.

Bruce, M. E., and Fraser, H. (1991). Scrapie strain variation and its implications. *Curr. Top. Microbiol. Immunol.* **172:**125–138.

Bueler, H., Aguzzi, A., Sailer, A., Greiner, R.-A., Autenried, P., Aguet, M., and Weissmann, C. (1993). Mice devoid of PrP are resistant to scrapie. *Cell* **73:**1339–1347.

Campbell, I. L., Eddleston, M., Kemper, P., Oldstone, M. B. A., and Hobbs, M. V. (1994). Activation of cerebral cytokine gene expression and its correlation with onset of reactive astrocyte and acute-phase response gene expression in scrapie. *J. Virol.* **68:**2383–2387.

Cappai, R., Stewart, L., Jobling, M. F., Thyer, J. M., White, A. R., Beyreuther, K., Collins, S. J., Masters, C. L., and Barrow, C. J. (1999). Familial prion disease mutation alters the secondary structure of recombinant mouse prion protein: implications for the mechanism of prion formation. *Biochemistry* **38:**3280–3284.

Caughey, B., Race, R. E., and Chesebro, B. (1988). Detection of prion protein mRNA in normal and scrapie-infected tissues and cell lines. *J. Gen. Virol.* **69:**711–716.

Caughey, B., Raymond, G. J., Ernst, D., and Race, R. E. (1991). N-terminal truncation of the scrapie-associated form of PrP by lysosomal protease(s): implications regarding the site of conversion of PrP to the protease-resistant state. *J. Virol.* **65:**6597–6603.

Caughey, B. (1994). Scrapie-associated PrP accumulation and agent replication: effects of sulphated glycosaminoglycan analogues. *Phil. Trans. R. Soc. Lond. B* **343:**399–404.

Caughey, B., Kocisko, D. A., Raymond, G. J., and Lansbury, P. T. (1995). Aggregates of scrapie associated prion protein induce the cell-free conversion of protease-sensitive prion protein to the protease-resistant state. *Chem. & Biol.* **2:**807–817.

Caughey, B., Raymond, G. J., Kocisko, D. A., and Lansbury, P. T., Jr. (1997). Scrapie infectivity correlates with converting activity, protease resistance, and aggregation of scrapie-associated prion protein in guanidine denaturation studies. *J. Virol.* **71:**4107–4110.

Caughey, B., Raymond, G. J., and Bessen, R. A. (1998). Strain-dependent differences in beta-sheet conformations of abnormal prion protein. *J. Biol. Chem.* **273:**32230–32235.

Caughey, B., and Chesebro, B. (1997). Prion protein and the transmissible spongiform encephalopathies. *Trends Cell. Biol.* **7:**56–62.

Caughey, B., and Raymond, G. J. (1991). The scrapie-associated form of PrP is made from a cell surface precursor that is both protease- and phospholipase-sensitive. *J. Biol. Chem.* **266:**18217–18223.

Caughey, B., and Raymond, G. J. (1993). Sulfated polyanion inhibition of scrapie-associated PrP accumulation in cultured cells. *J. Virol.* **67:**643–650.

Caughey, B. W., Dong, A., Bhat, K. S., Ernst, D., Hayes, S. F., and Caughey, W. S. (1991). Secondary structure analysis of the scrapie-associated protein PrP 27–30 in water by infrared spectroscopy. *Biochemistry* **30:**7672–7680.

Caughey, W. S., Raymond, L. D., Horiuchi, M., and Caughey, B. (1998). Inhibition of protease-resistant prion protein formation by porphyrins and phthalocyanines. *Proc. Natl. Acad. Sci. U.S.A.* **95:**12117–12122.

Chabry, J., Caughey, B., and Chesebro, B. (1998). Specific inhibition of in vitro formation of protease-resistant prion protein by synthetic peptides. *J. Biol. Chem.* **273:**13203–13207.

Chabry, J., Priola, S. A., Wehrly, K., Nishio, J., Hope, J., and Chesebro, B. (1999). Species-Independent Inhibition of Abnormal Prion Protein (PrP) Formation by a Peptide Containing a Conserved PrP Sequence. *J. Virol.* **73:**6245–6250.

Chesebro, B. (1998). Prion diseases-BSE and prions: Uncertainties about the agent. *Science* **279:**42–43.

Chesebro, B. (1999). Prion protein and the transmissible spongiform encephalopathy diseases. *Neuron* **24:**503–506.

Chiesa, R., Piccardo, P., Ghetti, B., and Harris, D. A. (1998). Neurological illness in transgenic mice expressing a prion protein with an insertional mutation. *Neuron* **21:**1339–1351.

Collee, J. G. and Bradley, R. (1997). BSE: a decade on—Part 2. *Lancet* **349:**715–721.

Collinge, J., Palmer, M. S., Sidle, K. C., Hill, A. F., Gowland, I., Meads, J., Asante, E., Bradley, R., Doey, L. J., Lantos, P. L. (1995). Unaltered susceptibility to BSE in transgenic mice expressing human prion protein. (1995). *Nature* **378:**779–783.

Collinge, J., Sidle, K. C. L., Meads, J., Ironside, J., and Hill, A. F. (1996). Molecular analysis of prion strain variation and the aetiology of "new variant" CJD. *Nature* **383:**685–690.

Come, J. H., Fraser, P. E., and Lansbury, P. T., Jr. (1993). A kinetic model for amyloid formation in the prion diseases: importance of seeding. *Proc. Natl. Acad. Sci. U.S.A.* **90:**5959–5963.

Come, J. H. and Lansbury, P. T., Jr. (1994). Predisposition of prion protein homozygotes to Creutzfeldt-Jakob disease can be explained by a nucleation-dependent polymerization mechanism. *J. Am. Chem. Soc.* **116:**4109–4110.

DebBurman, S. K., Raymond, G. J., Caughey, B., and Lindquist, S. (1997). Chaperone-supervised conversion of prion protein to its protease-resistant form. *Proc. Natl. Acad. Sci. U.S.A.* **94:**13938–13943.

Demaimay, R., Adjou, K., Lasmezas, C., Lazarini, F., Cherifi, K., Seman, M., Deslys, J. P., and Dormont, D. (1994). Pharmacological studies of a new derivative of amphotericin B, MS-8209, in mouse and hamster scrapie. *J. Gen. Virol.* **75:**2499–2503.

Demaimay, R., Harper, J., Gordon, H., Weaver, D., Chesebro, B., and Caughey, B. (1998). Structural aspects of Congo red as an inhibitor of protease-resistant prion protein formation. *J. Neurochem.* **71:**2534–2541.

Demaimay, R., Race, R., and Chesebro, B. (1999). Effectiveness of polyene antibiotics in treatment of transmissible spongiform encephalopathy in transgenic mice expressing Syrian hamster PrP only in neurons. *J. Virol.* **73:**3511–3513.

Dickinson, A. G., Meikle, V. M. H., and Fraser, H. G. (1968). Identification of a gene which controls the incubation period of some strains of scrapie agent in mice. *J. Comp. Pathol.* **78:**293–299.

Doh-Ura, K., Iwaki, T. and Caughey, B. (2000) Lysosomotropic agents and cysteine protease inhibitors inhibit accumulation of scrapie-associated prion protein. *J. Virol.* **74:**4894–4897.

Donne, D. G., Viles, J. H., Groth, D., Mehlhorn, I., James, T. L., Cohen, F. E., Prusiner, S. B., Wright, P. E., and Dyson, H. J. (1997). Structure of the recombinant full-length hamster prion protein PrP(29–231): the N terminus is highly flexible [see comments]. *Proc. Natl. Acad. Sci. U.S.A.* **94:**13452–13457.

Ehlers, B. and Diringer, H. (1984). Dextran sulphate 500 delays and prevents mouse scrapie by impairment of agent replication in spleen. *J. Gen. Virol.* **65:**1325–1330.

Farquhar, C. F. and Dickinson, A. G. (1986). Prolongation of scrapie incubation period by an injection of dextran sulphate 500 within the month before or after infection. *J. Gen. Virol.* **67:**463–473.

Fraser, H. and Farquhar, C. F. (1987). Ionising radiation has no influence on scrapie incubation period in mice. *Vet. Microbiol.* **13:**211–223.

Gadjusek, D. C. (1988). Transmissible and nontransmissible amyloidoses: Autocatalytic post-translational conversion of host precursor proteins to beta-pleated configurations. *J. Neuroimmunol.* **20:**95–110.

Gajdusek, D. C., Gibbs, C. J., Jr., and Alpers, M. (1966). Experimental transmission of a kuru-like syndrome to chimpanzees. *Nature* **209:**794–796.

Gajdusek, D. C. and Gibbs, C. J. J. (1971). Transmission of the two subacute spongiform encephalopathies of man (kuru and Creutzfeldt-Jakob disease) to New World monkeys. *Nature* **230:**588–591.

Gasset, M., Baldwin, M. A., Lloyd, D. H., Gabriel, J., Holtzman, D. M., Cohen, F., Fletterick, R., and Prusiner, S. B. (1992). Predicted α-helical regions of the prion protein when synthesized as peptides form amyloid. *Proc. Natl. Acad. Sci. U.S.A.* **89**:10940–10944.

Giaccone, G., Verga, L., Bugiani, O., Frangione, B., Serban, D., Prusiner, S. B., Farlow, M. R., Ghetti, B., and Tagliavini, F. (1992). Prion protein preamyloid and amyloid deposits in Gerstmann-Straussler-Scheinker disease, Indiana kindred [published erratum appears in Proc. Natl. Acad. Sci. U.S.A. 1993, Jan 1;90(1):302]. *Proc. Natl. Acad. Sci. U.S.A.* **89**:9349–9353.

Giese, A., Brown, D. R., Groschup, M. H., Feldmann, C., Haist, I., and Kretzschmar, H. A. (1998). Role of microglia in neuronal cell death in prion disease. *Brain Pathol.* **8**:449–457.

Goldfarb, L. G., Petersen, R. B., Tabaton, M., Brown, P., LeBlanc, A. C., Montagna, P., Cortelli, P., Julien, J., Vital, C., Pendelbury, W. W., Haltia, M., Wills, P. R., Hauw, J. J., McKeever, P. E., Monari, L., Schrank, B., Swergold, G. D., Autilio-Gambetti, L., Gadjusek, D. C., Lugaresi, E. and Gambetti, P. (1992). Fatal familial insomnia and familial Creutzfeldt-Jakob disease: Disease phenotype determined by a DNA polymorphism. *Science* **258**:806–808.

Goldfarb, L. G., Brown, P., Haltia, M., Ghiso, J., Frangione, B., and Gajdusek, D. C. (1993). Synthetic peptides corresponding to different mutated regions of the amyloid gene in familial Creutzfeldt-Jakob disease show enhanced in vitro formation of morphologically different amyloid fibrils. *Proc. Natl. Acad. Sci. U.S.A.* **90**:4451–4454.

Goldmann, W., Hunter, N., Smith, G., Foster, J., and Hope, J. (1994). PrP genotype and agent effects in scrapie: change in allelic interaction with different isolates of agent in sheep, a natural host of scrapie. *J. Gen. Virol.* **75**:989–995.

Goldmann, W., Martin, T., Foster, J., Hughes, S., Smith, G., Hughes, K., Dawson, M., and Hunter, N. (1996). Novel polymorphisms in the caprine PrP gene: a codon 142 mutation associated with scrapie incubation period. *J. Gen. Virol.* **77**:2885–2891.

Greig, J. R. (1940). Scrapie: observations on the transmission of the disease by mediate contact. *Veterinary Journal* **96**:203–206.

Griffith, J. S. (1967). Self-replication and scrapie. *Nature* **215**:1043–1044.

Herrmann, L. M. and Caughey, B. (1998). The importance of the disulfide bond in prion protein conversion. *Neuroreport* **9**:2457–2461.

Hill, A. F., Desbruslais, M., Joiner, S., Sidle, K. C. L., Gowland, I., Collinge, J., Doey, L. J., and Lantos, P. (1997). The same prion strain causes vCJD and BSE. *Nature* **389**:448–450.

Hill, A. F., Antoniou, M., and Collinge, J. (1999). Protease-resistant prion protein produced in vitro lacks detectable infectivity. *J. Gen. Virol.* **80**:11–14.

Hope, J. (1994). The nature of the scrapie agent: the evolution of the virino. *Ann. N.Y. Acad. Sci.* **724**:282–289.

Hope, J., Wood, S. C., Birkett, C. R., Chong, A., Bruce, M. E., Cairns, D., Goldmann, W., Hunter, N., and Bostock, C. J. (1999). Molecular analysis of ovine prion protein identifies similarities between BSE and an experimental isolate of natural scrapie, CH1641. *J. Gen. Virol.* **80**:1–4.

Horiuchi, M., Chabry, J., and Caughey, B. (1999). Specific binding of normal prion protein to the scrapie form via a localized domain initiates its conversion to the protease-resistant state. *EMBO J.* **18**:3193–3203.

Horicuchi, M., Prida, S. A., Chabry, J., and Caughey, B. (2000). Interactions between heterologous forms of prion protein: Binding, inhibition of conversion, and species barriers. *Proc. Natl. Acad. Sci. U.S.A.* **97**:5836–5841.

Hornemann, S., Korth, C., Oesch, B., Riek, R., Wider, G., Wuthrich, K., and Glockshuber, R. (1997). Recombinant full-length murine prion protein, mPrP(23–231): purification and spectroscopic characterization. *FEBS Lett.* **413**:277–281.

Hornemann, S. and Glockshuber, R. (1996). Autonomous and reversible folding of a soluble amino-terminally truncated segment of the mouse prion protein. *J. Mol. Biol.* **261**:614–619.

Hornemann, S. and Glockshuber, R. (1998). A scrapie-like unfolding intermediate of the prion protein domain PrP(121–231) induced by acidic pH. *Proc. Natl. Acad. Sci. U.S.A.* **95**:6010–6014.

Hsiao, K. K., Scott, M., Foster, D., Groth, D. F., DeArmond, S. J., and Prusiner, S. B. (1990). Spontaneous neurodegeneration in transgenic mice with mutant prion protein. *Science* **250**:1587–1590.

Hsiao, K. K., Groth, D., Scott, M., Serban, H., Rapp, D., Foster, D., Torchia, M., DeArmond, S. J., and Prusiner, S. B. (1994). Serial transmission in rodents of neurodegeneration from transgenic mice expressing mutant prion protein. *Proc. Natl. Acad. Sci. U.S.A.* **91**:9126–9130.

Hunter, N. and Cairns, D. (1998). Scrapie-free Merino and Poll Dorset sheep from Australia and New Zealand have normal frequencies of scrapie-susceptible PrP genotypes. *J. Gen. Virol.* **79**:2079–2082.

Ingrosso, L., Ladogana, A., and Pocchiari, M. (1995). Congo red prolongs the incubation period in scrapie-infected hamsters. *J. Virol.* **69**:506–508.

Jackson, G. S., Hosszu, L. L., Power, A., Hill, A. F., Kenney, J., Saibil, H., Craven, C. J., Waltho, J. P., Clarke, A. R., and Collinge, J. (1999). Reversible conversion of monomeric human prion protein between native and fibrilogenic conformations. *Science* **283**:1935–1937.

Jarrett, J. T. and Lansbury, P. T. Jr. (1993). Seeding "One-Dimensional Crystallization" of Amyloid: A Pathogenic Mechanism in Alzheimer's Disease and Scrapie? *Cell* **73**:1055–1058.

Jeffrey, M., Goodsir, C. M., Bruce, M. E., McBride, P. A., and Farquhar, C. (1994). Morphogenesis of amyloid plaques in 87V murine scrapie. *Neuropathology and Applied Neuropathology* **20**:535–542.

Kaneko, K., Peretz, D., Pan, K., Blockberger, T. C., Wille, H., Gabizon, R., Griffith, O. H., Cohen, F. E., Baldwin, M. A., and Prusiner, S. B. (1995). Prion protein (PrP) synthetic peptides induce cellular PrP to acquire properties of the scrapie isoform. *Proc. Natl. Acad. Sci. U.S.A.* **92**:11160–11164.

Kascsak, R. J., Rubenstein, R., Merz, P. A., Carp, R. I., Robakis, N. K., Wisniewski, H. M., and Diringer, H. (1986). Immunological comparison of scrapie-associated fibrils isolated from animals infected with four different scrapie strains. *J. Virol.* **59**:676–683.

Kellings, K., Prusiner, S. B., and Riesner, D. (1994). Nucleic acids in prion preparations: unspecific background or essential component? *Philos. Trans. R. Soc. Lond. B. Biol. Sci.* **343**:425–430.

Kimberlin, R. H., Walker, C. A., and Fraser, H. (1989). The genomic identity of different strains of mouse scrapie is expressed in hamsters and preserved on reisolation in mice. *J. Gen. Virol.* **70**:2017–2025.

Kimberlin, R. H. and Marsh, R. F. (1975). Comparison of scrapie and transmissible mink encephalopathy in hamsters. I. Biochemical studies of brain during development of disease. *J. Infect. Dis.* **131**:97–103.

Kimberlin, R. H. and Walker, C. A. (1983). The antiviral compound HPA-23 can prevent scrapie when administered at the time of infection. *Arch. Virol.* **78**:9–18.

Kimberlin, R. H. and Walker, C. A. (1986). Suppression of scrapie infection in mice by heteropolyanion 23, dextran sulfate, and some other polyanions. *Antimicrob. Agents Chemother.* **30**:409–413.

Kleczkowski, A. and McLaren, A. D. (1967). Inactivation of infectivity of RNA of tobacco mosaic virus during ultraviolet-irradiation of the whole virus at two wavelengths. *J. Gen. Virol.* **1**:441–448.

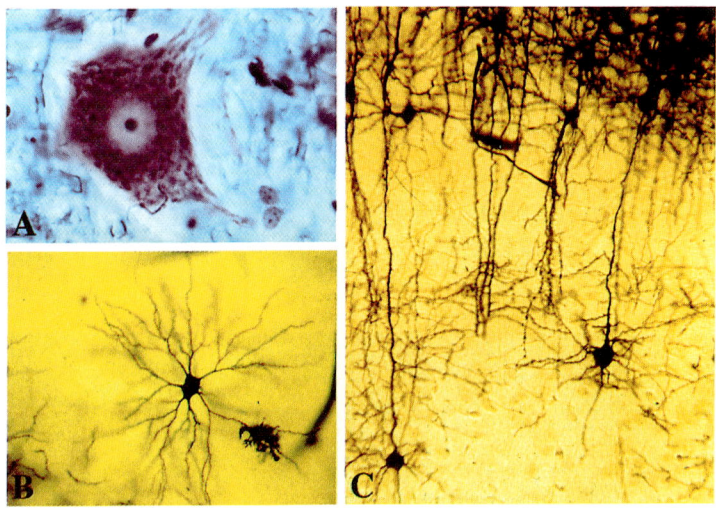

FIG 1. The characteristic morphological diversity of neurons in the central nervous system as revealed by Nissl stains (A) and by silver impregnation methods (B, C) is illustrated. The differential staining of ribosomes, rough endoplasmic reticulum, and RNA by the basophilic Nissl stain provides a beautiful presentation of neuronal perikarya but does not reveal the extensive process networks characteristically associated with the neuron. Nevertheless, this is the routine stain used to demonstrate cytoarchitecture in histological preparations of sections through the brain (see Fig. 3). Silver impregnation, or Golgi preparations, provide an exceptional illustration of the polarized nature of neurons. Note the differences in the organization of the dendritic arbors of the local circuit neuron (B) and the cortical pyramidal neurons (C). Whereas the dendrites of the local circuit neuron (B) extend radially from the cell body, the dendrites arising from the pyramidal neurons are polarized and far more extensive. [For text discussion, see p. 41 in article by Card, this volume.]

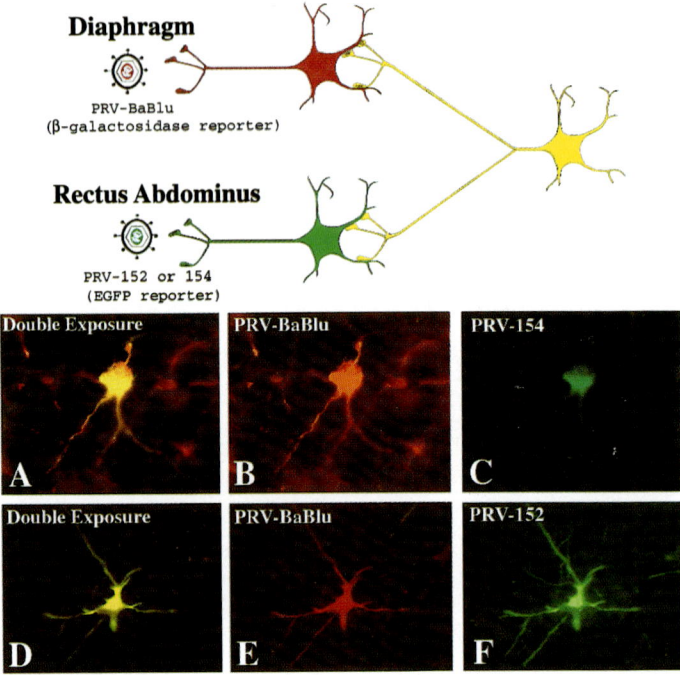

FIG 7. The utility of using antigenically distinct recombinant viruses for transneuronal analysis is illustrated. In this experiment, PRV recombinants constructed by L. W. Enquist and colleagues were used to identify neurons in the CNS that branch to synapse on motor neurons that control the activity of the diaphragm and rectus abdominus muscles of the ferret (see Billig *et al.* [1999a] and Yates *et al.* [1999] for a definition of the circuits in single injection viral studies). The genomic organization of each recombinant is illustrated in Fig. 2. Simultaneous inoculation of the two recombinants into separate muscles as shown in the schematic diagram led to retrograde infection of the motor neurons that innervate them and to transsynaptic infection of neurons whose axons branched to synapse on both populations of motor neurons. Examples of dual-infected neurons using immunofluorescence localization of the unique reporters (β-galactosidase and enhanced green fluorescent protein) are shown in (A)–(F). PRV-BaBlu and PRV-152 produce extensive staining of the somatodendritic compartment of infected neurons (B, E, F), whereas the EGFP-Us9 fusion protein reporter of PRV-154 is differentially transported to the trans-Golgi network (C). See the text for further details and also Billig *et al.*, 2000. [For text discussion, see p. 62 in article by Card, this volume.]

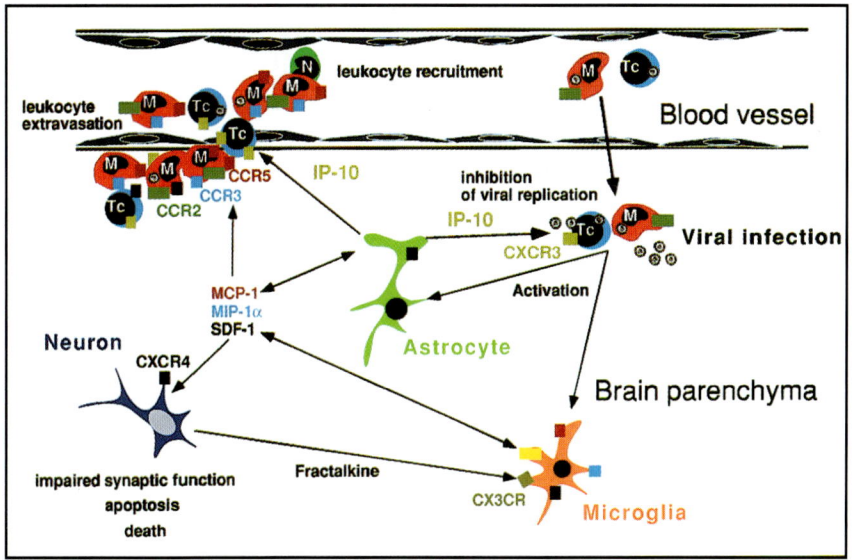

Fig 3. Chemokines are potentially multifunctional effectors in CNS viral infections. In this schema, initial infection of the CNS with a virus results in the localized production of chemokines by reactive astrocytes and microglia. Some chemokines such as SDF-1 and fractalkine are produced constitutively in the CNS and their production may also be altered by viral infection. Increased chemokine production may then promote the further recruitment and extravasation of antiviral immune effector cells from the periphery. Chemokine receptors are widely expressed by neuronal and glial cells and additional functions of the chemokines are likely. These may include direct antiviral actions (for example, by IP-10), and intercellular communication (for example, neuronally derived fractalkine can stimulate microglia). In addition, chemokines may exert detrimental actions; for example, SDF-1 can impair neuronal activity and stimulate apoptosis of these cells. M, macrophage; Tc, T lymphocyte; N, neutrophil. [For text discussion, see p. 158 in article by Arsensio and Campbell, this volume.]

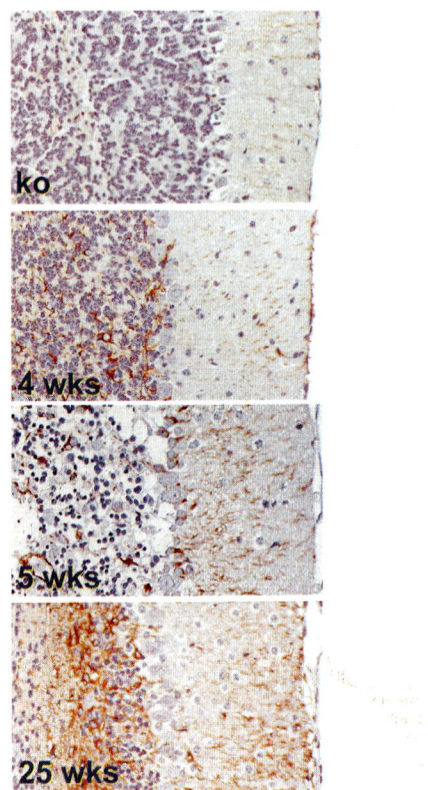

FIG 3. Progressive degeneration of the granule cell layer in mice expressing D32–134 truncated PrP. Development of the cerebellum proceeds normally until early postnatal life, and leads to formation of a molecular layer, a Purkinje cell layer, and a granule cell layer (right to left in the figure) of normal thickness. However, at 4 weeks, some degree of pathological astrogliosis can already be discerned. At 5 weeks, massive degeneration of granule cells by apoptosis is ongoing. Note strong gliosis affecting also the molecular layer. At the end stage of disease, mice suffer from a profound cerebellar syndrome and the thickness of the granule cell layer is considerably reduced (Shmerling *et al.*, 1998). In some areas of the cerebellar cortex, granule cells disappear completely (this is not shown). [For text discussion, see p. 323 in article by Aguzzi *et al.*, this volume.]

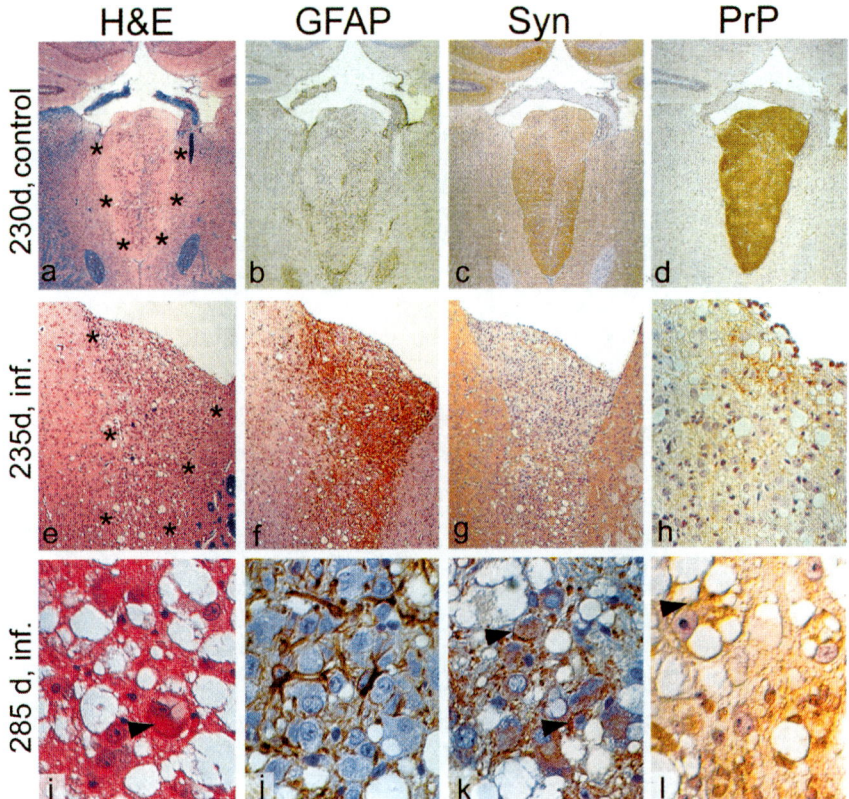

FIG 4. Noninfected and scrapie-infected neural grafts in brains of *Prnp^{o/o}* mice. *Upper row* (a, b, c, d): healthy control graft 230 days after mock inoculation. The graft is located in the third ventricle of the recipient mouse (a, see asterisks, hematoxylin–eosin), and shows no spongiform change; little gliosis (b: immunostain for GFAP); and strong expression of synaptophysin (c) and of PrP^C (d). *Middle row* (e, f, g, h): scrapie-infected graft 235 days after inoculation with increased cellularity (e), brisk gliosis (f); and a significant loss of synaptophysin (g) and of PrP (h) staining intensity is shown. *Bottom row:* high magnification of a similar graft shows characteristic pathological changes in a chronically infected graft: (i) appearance of large vacuoles and ballooned neurons (arrow). In the GFAP immunostain (j), astrocytes appear wrapped around densely packed neurons. Granular deposits and intracytoplasmic accumulation of synaptophysin (k) and PrP immunoreactivity (l) in the cytoplasm of neurons. [For text discussion, see p. 333 in article by Aguzzi *et al.,* this volume.]

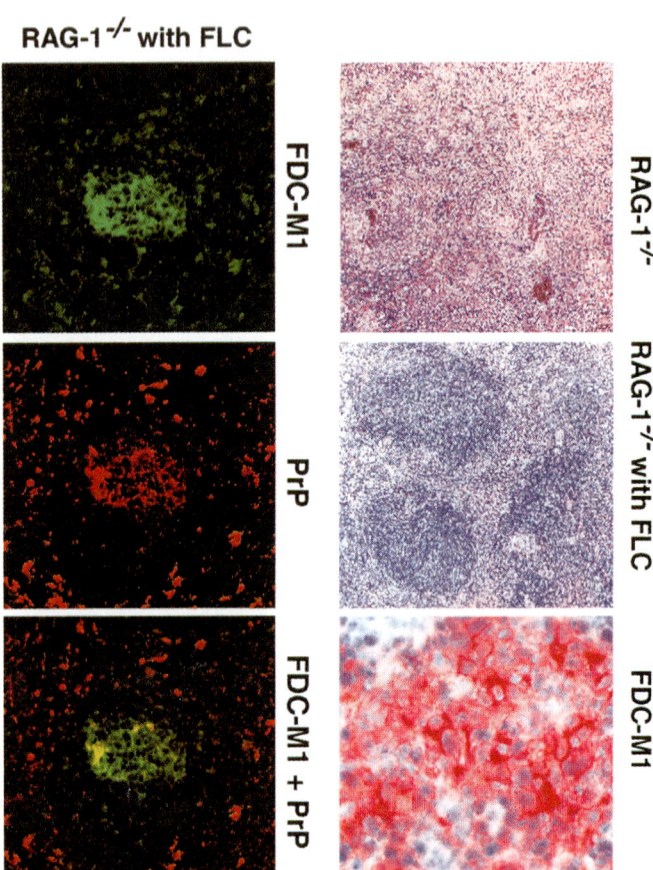

FIG 7. Spleen histology of Rag-1$^{-/-}$ mice reconstituted with $Prnp^{o/o}$ fetal liver cells. *Upper row:* Paraffin sections stained with hemalaun before FLC transfer (left). No B cell follicles and germinal centers were discernible in *Rag-1$^{-/-}$* mice. Restoration of organized B cell follicles and germinal centers after FLC reconstitution (middle). Magnification: ×200. Frozen sections immunostained with antibody FDC-M1 revealed formation of prominent FDC clusters within germinal centers after FLC transfer (right). Magnification: ×250. *Lower row:* Confocal double-color immunofluorescence analysis of splenic germinal centers in *Rag-1$^{-/-}$* mice reconstituted with $Prnp^{o/o}$ FLCs after inoculation with RML prions. Sections were stained with antibody FDC-M1 to follicular dendritic cells (green, left) and with antiserum R340 to PrP (red, middle). Regions in which both signals are detectable appear yellow in superimposed images (right). Magnification: ×250. Most of the PrP signal in germinal centers appeared to colocalize with the FDC-network. [For text discussion, see p. 340 in article by Aguzzi *et al.*, this volume.]

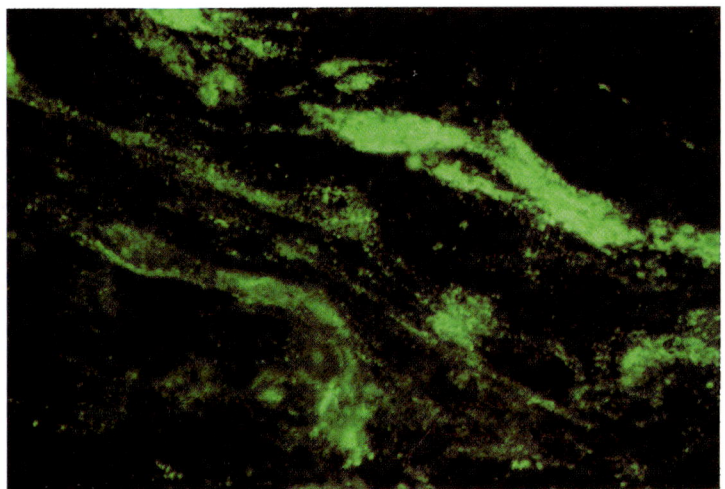

FIG 2. HIV productively infects brain MVEC *in vitro*. Positive indirect immunofluorescence for viral p24 in HIV-infected MVEC. [For text discussion, see p. 364 in article by Strelow *et al.*, this volume.]

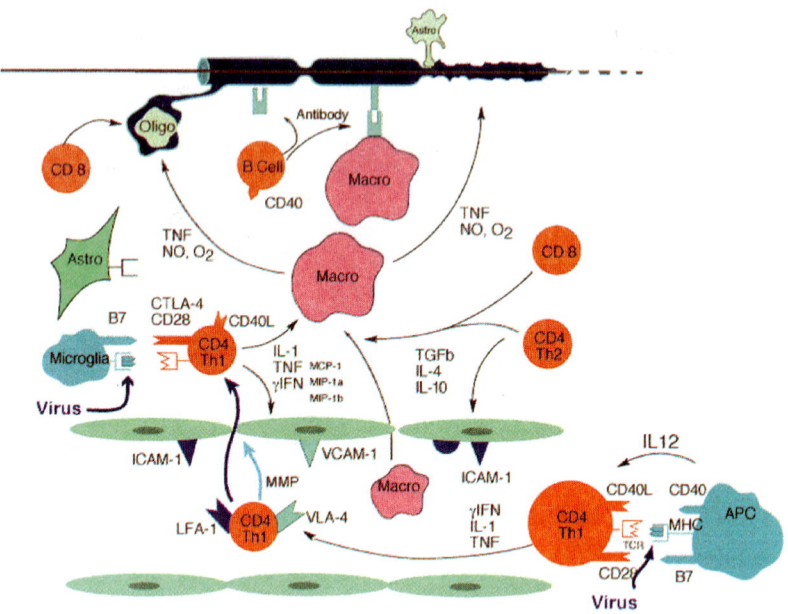

FIG. 2. Elements of the MS lesion. A diagrammatic representation of the complex immunological mechanisms that are proposed to be involved in the initiation and maintenance of the developing MS lesion. Viruses may play a role in this model by initiating a virus-specific immune response in the periphery (lower right corner) that crosses the blood–brain barrier and encounters antigen in the CNS. Antigen may be presented on virus-infected glial cells or by resident CNS antigen-presenting cells (microglia). This interaction may result in a cascade of cytokines and chemokines that are associated with lesion development. Alternatively, autoreactive cells either in the periphery or the CNS may cross-react with viral antigens (molecular mimicry) that lead to activation of these T cells and subsequent CNS damage. (We thank H. McFarland for the use of this figure.) [For text discussion, see p. 521 in article by Soldan and Jacobson, this volume.]

Klein, M. A., Frigg, R., Raeber, A. J., Flechsig, E., Hegyi, I., Zinkernagel, R. M., Weissmann, C., and Aguzzi, A. (1998). PrP expression in B lymphocytes is not required for prion neuroinvasion [see comments]. *Nat. Med.* **4:**1429–1433.

Kocisko, D. A., Come, J. H., Priola, S. A., Chesebro, B., Raymond, G. J., Lansbury, P. T., and Caughey, B. (1994). Cell-free formation of protease-resistant prion protein. *Nature* **370:**471–474.

Kocisko, D. A., Priola, S. A., Raymond, G. J., Chesebro, B., Lansbury, P. T., Jr., and Caughey, B. (1995). Species specificity in the cell-free conversion of prion protein to protease-resistant forms: a model for the scrapie species barrier. *Proc. Natl. Acad. Sci. U.S.A.* **92:**3923–3927.

Kocisko, D. A., Lansbury, P. T., Jr., and Caughey, B. (1996). Partial unfolding and refolding of scrapie-associated prion protein: evidence for a critical 16-kDa C-terminal domain. *Biochemistry* **35:**13434–13442.

Lansbury, P. T., Jr. and Caughey, B. (1995). The chemistry of scrapie infection: implications of the `ice 9' metaphor. *Chem. & Biol.* **2:**1–5.

Lehmann, S. and Harris, D. A. (1995). A mutant prion protein displays an aberrant membrane association when expressed in cultured cells. *J. Biol. Chem.* **270:**24589–24597.

Lehmann, S. and Harris, D. A. (1996). Mutant and infectious prion proteins display common biochemical properties in cultured cells. *J. Biol. Chem.* **271:**1633–1637.

Liemann, S. and Glockshuber, R. (1999). Influence of amino acid substitutions related to inherited human prion diseases on the thermodynamic stability of the cellular prion protein. *Biochemistry* **38:**3258–3267.

Liu, H., Farr-Jones, S., Ulyanov, N. B., Llinas, M., Marqusee, S., Groth, D., Cohen, F. E., Prusiner, S. B., and James, T. L. (1999). Solution structure of Syrian hamster prion protein rPrP(90–231). *Biochemistry* **38:**5362–5377.

Manson, J., West, J. D., Thomson, V., McBride, P., Kaufman, M. H., and Hope, J. (1992). The prion protein gene: a role in mouse embryogenesis? *Development* **115:**117–122.

Manson, J. C., Jamieson, E., Baybutt, H., Tuzi, N. L., Barron, R., McConnell, I., Somerville, R., Ironside, J., Will, R., Sy, M. S., Melton, D. W., Hope, J., and Bostock, C. (1999). A single amino acid alteration (101L) introduced into murine PrP dramatically alters incubation time of transmissible spongiform encephalopathy. *EMBO J.* **18:**6855–6864.

Marsh, R. F., Bessen, R. A., Lehmann, S., and Hartsough, G. R. (1991). Epidemiological and experimental studies on a new incident of transmissible mink encephalopathy. *J. Gen. Virol.* **72:**589–594.

McKenzie, D., Bartz, J., Mirwald, J., Olander, D., Marsh, R., and Aiken, J. (1998). Reversibility of scrapie inactivation is enhanced by copper. *J. Biol. Chem.* **273:**25545–25547.

McKinley, M. P., Meyer, R. K., Kenaga, L., Rahbar, F., Cotter, R., Servan, A., and Prusiner, S. B. (1991a). Scrapie prion rod formation in vitro requires both detergent extraction and limited proteolysis. *J. Virol.* **65:**1340–1351.

McKinley, M. P., Taraboulos, A., Kenaga, L., Serban, D., Stieber, A., DeArmond, S. J., Prusiner, S. B., and Gonatas, N. (1991b). Ultrastructural localization of scrapie prion proteins in cytoplasmic vesicles of infected cultured cells. *Lab. Invest.* **65:**622–630.

Merz, P. A., Somerville, R. A., Wisniewski, H. M., and Iqbal, K. (1981). Abnormal fibrils from scrapie-infected brain. *Acta Neuropathol.* **54:**63–74.

Miller, M. W., Wild, M. A., and Williams, E. S. (1998). Epidemiology of chronic wasting disease in captive Rocky Mountain elk. *J. Wildl. Dis.* **34:**532–538.

Monari, L., Chen, S. G., Brown, P., Parchi, P., Petersen, R. B., Mikol, J., Gray, F., Cortelli, P., Montagna, P., Ghetti, B., Goldfarb, L. G., Gajdusek, D. C., Lugaresi, A., Gambetti, P., and Autilio-Gambetti, L. (1994). Fatal familial insomnia and familial

Creutzfeldt-Jakob disease: different prion proteins determined by a DNA polymorphism. *Proc. Natl. Acad. Sci. U.S.A.* **91**:2939–2842.

Muramoto, T., Kitamoto, T., Hoque, M. Z., Tateishi, J., and Goto, I. (1993). Species barrier prevents an abnormal isoform of prion protein from accumulating in follicular dendritic cells of mice with Creutzfeldt-Jakob disease. *J. Virol.* **67**:6808–6810.

Muramoto, T., Scott, M., Cohen, F. E., and Prusiner, S. B. (1996). Recombinant scrapie-like prion protein of 106 amino acids is soluble. *Proc. Natl. Acad. Sci. U.S.A.* **93**:15457–15462.

Nguyen, J. T., Inouye, H., Baldwin, M. A., Fletterick, R. J., Cohen, F. E., Prusiner, S. B., and Kirschner, D. A. (1995). X-ray diffraction of scrapie prion rods and PrP peptides. *J. Mol. Biol.* **252**:412–422.

Palmer, M. S., Dryden, A. J., Hughes, J. T., and Collinge, J. (1991). Homozygous prion protein genotype predisposes to sporadic Creutzfeldt-Jakob disease. *Nature* **352**:340–342.

Palsson, P. A. (1979). *Slow transmissible diseases of the nervous system* (Prusiner, S. B. and Hadlow, W. J., eds.), 357–366. Academic Press, New York.

Pan, K.-P., Baldwin, M., Nguyen, J., Gasset, M., Serban, A., Groth, D., Mehlhorn, I., Huang, Z., Fletterick, R. J., Cohen, F. E., and Prusiner, S. B. (1993). Conversion of alpha-helices into beta-sheets features in the formation of the scrapie prion protein. *Proc. Natl. Acad. Sci. U.S.A.* **90**:10962–10966.

Petersen, R. B., Parchi, P., Richardson, S. L., Urig, C. B., and Gambetti, P. (1996). Effect of the D178N mutation and the codon 129 polymorphism on the metabolism of the prion protein. *J. Biol. Chem.* **271**:12661–12668.

Peyrin, J. M., Lasmezas, C. I., Haik, S., Tagliavini, F., Salmona, M., Williams, A., Richie, D., Deslys, J. P., and Dormont, D. (1999). Microglial cells respond to amyloidogenic PrP peptide by the production of inflammatory cytokines. *Neuroreport.* **10**:723–729.

Post, K., Pitschke, M., Schafer, O., Wille, H., Appel, T. R., Kirsch, D., Mehlhorn, I., Serban, H., Prusiner, S. B., and Riesner, D. (1998). Rapid acquisition of beta-sheet structure in the prion protein prior to multimer formation. *Biol. Chem.* **379**:1307–1317.

Priola, S. A., Caughey, B., Race, R. E., and Chesebro, B. (1994). Heterologous PrP molecules interfere with accumulation of protease-resistant PrP in scrapie-infected murine neuroblastoma cells. *J. Virol.* **68**:4873–4878.

Priola, S. A. (1999). Prion protein and species barriers in the transmissible spongiform encephalopathies. *Biomed. Pharmacother.* **53**:27–33.

Priola, S. A. and Chesebro, B. (1995). A single hamster amino acid blocks conversion to protease-resistant PrP in scrapie-infected mouse neuroblastoma cells. *J. Virol.* **69**:7754–7758.

Priola, S. A. and Chesebro, B. (1998). Abnormal properties of prion protein with insertional mutations in different cell types. *J. Biol. Chem.* **273**:11980–11985.

Priola, S. A., Raines, A., and Caughey, W. S. (2000). Porphyrin and phthalocyanine antiscrapie compounds. *Science* **287**:1503–1506.

Prusiner, S. B. (1982). Novel proteinaceous infectious particles cause scrapie. *Science* **216**:136–144.

Prusiner, S. B., Groth, D., Serban, A., Stahl, N., and Gabizon, R. (1993). Attempts to restore scrapie prion infectivity after exposure to protein denaturants. *Proc. Natl. Acad. Sci. U.S.A.* **90**:2793–2797.

Prusiner, S. B. (1998). Prions. *Proc. Natl. Acad. Sci. U.S.A.* **95**:13363–13383.

Race, R., Jenny, A., and Sutton, D. (1998). Scrapie infectivity and proteinase K-resistant prion protein in sheep placenta, brain, spleen, and lymph node: implications for transmission and antemortem diagnosis. *J. Infect. Dis.* **178**:949–953.

Race, R., Oldstone, M. B. A., and Chesebro, B. (2000). Entry versus blockade of brain infection following oral or intraperitoneal scrapie administration: role of PrP expression in peripheral nerves and spleen. *J. Virol.* (in press).

Race, R. and Chesebro, B. (1998). Scrapie infectivity found in resistant species. *Nature* **392:**770.

Race, R. E., Priola, S. A., Bessen, R. A., Ernst, D., Dockter, J., Rall, G. F., Mucke, L., Chesebro, B., and Oldstone, M. B. A. (1995). Neuron-specific expression of a hamster prion protein minigene in transgenic mice induces susceptibility to hamster scrapie agent. *Neuron* **15:**1183–1191.

Raeber, A. J., Race, R. E., Brandner, S., Priola, S. A., Sailer, A., Bessen, R. A., Aguzzi, A., Oldstone, M. B. A., Weissmann, C., and Chesebro, B. (1997). Astrocyte-specific expression of hamster prion protein (PrP) renders PrP knockout mice susceptible to hamster scrapie. *EMBO J.* **16:**6057–6065.

Raeber, A. J., Sailer, A., Hegyi, I., Klein, M. A., Rulicke, T., Fischer, M., Brandner, S., Aguzzi, A., and Weissmann, C. (1999). Ectopic expression of prion protein (PrP) in T lymphocytes or hepatocytes of PrP knockout mice is insufficient to sustain prion replication. *Proc. Natl. Acad. Sci. U.S.A.* **96:**3987–3992.

Raymond, G. J., Hope, J., Kocisko, D. A., Priola, S. A., Raymond, L. D., Bossers, A., Ironside, J., Will, R. G., Chen, S. G., Petersen, R. B., Gambetti, P., Rubenstein, R., Smits, M. A., Lansbury, P. T., Jr., and Caughey, B. (1997). Molecular assessment of the transmissibilities of BSE and scrapie to humans. *Nature* **388:**285–288.

Riek, R., Hornemann, S., Wider, G., Billeter, M., Glockshuber, R., and Wuthrich, K. (1996). NMR structure of the mouse prion protein domain PrP(121–231). *Nature* **382:**180–182.

Riek, R., Hornemann, S., Wider, G., Glockshuber, R., and Wuthrich, K. (1997). NMR characterization of the full-length recombinant murine prion protein, mPrP(23–21). *FEBS Lett.* **413:**282–288.

Riesner, D., Kellings, K., Post, K., Wille, H., Serban, H., Groth, D., Baldwin, M. A., and Prusiner, S. B. (1996). Disruption of prion rods generates 10-nm spherical particles having high alpha-helical content and lacking scrapie infectivity. *J. Virol.* **70:**1714–1722.

Rohwer, R. G. (1984a). Scrapie infectious agent is virus-like in size and susceptibility to inactivation. *Nature* **308:**658–661.

Rohwer, R. G. (1984b). Virus-like sensitivity of the scrapie agent to heat inactivation. *Science* **223:**600–602.

Rohwer, R. G. (1991). The scrapie agent: "a virus by any other name." *Curr. Top. Microbiol. Immunol.* **172:**195–232.

Rubenstein, R., Gray, P. C., Wehlburg, C. M., Wagner, J. S., and Tisone, G. C. (1998). Detection and discrimination of PrPSc by multi-spectral ultraviolet fluorescence. *Biochem. Biophys. Res. Commun.* **246:**100–106.

Saborio, G. P., Soto, C., Kascsak, R. J., Levy, E., Kascsak, R., Harris, D. A., and Frangione, B. (1999). Cell-lysate conversion of prion protein into its protease-resistant isoform suggests the participation of a cellular chaperone. *Biochem. Biophys. Res. Commun.* **258:**470–475.

Safar, J., Ceroni, M., Gajdusek, D. C., and Gibbs, C. J., Jr. (1991). Differences in the membrane interaction of scrapie amyloid precursor proteins in normal and scrapie- or Creutzfeldt-Jakob disease-infected brains. *J. Infect. Dis.* **163:**488–494.

Safar, J., Roller, P. P., Gajdusek, D. C., and Gibbs, C. J., Jr. (1993). Conformational transitions, dissociation, and unfolding of scrapie amyloid (prion) protein. *J. Biol. Chem.* **268:**20276–20284.

Safar, J., Roller, P. P., Gajdusek, D. C., and Gibbs, C. J., Jr. (1994). Scrapie amyloid (prion) protein has the conformational characteristics of an aggregated molten globule folding intermediate. *Biochemistry* **33:**8375–8383.

Safar, J., Wille, H., Itri, V., Groth, D., Serban, H., Torchia, M., Cohen, F. E., and Prusiner, S. B. (1998). Eight prion strains have PrP(Sc) molecules with different conformations [see comments]. *Nat. Med.* **4:**1157–1165.

Scott, M., Foster, D., Mirenda, C., Serban, D., Coufal, F., Walchli, M., Torchia, M., Groth, D., Carlson, G., DeArmond, S. J., Westaway, D., and Prusiner, S. B. (1989). Transgenic mice expressing hamster prion protein produce species-specific scrapie infectivity and amyloid plaques. *Cell* **59:**847–857.

Selvaggini, C., DeGioia, L., Cantu, L., Ghibaudi, E., Diomede, L., Passerini, F., Forloni, G., Bugiani, O., Tagliavini, F., and Salmona, M. (1993). Molecular characteristics of a protease-resistant, amyloidogenic and neurotoxic peptide homologous to residues 106–126 of the prion protein. *Biochem. Biophys. Res. Commun.* **194:**1380–1386.

Shaked, G. M., Fridlander, G., Meiner, Z., Taraboulos, A., and Gabizon, R. (1999). Protease-resistant and Detergent-insoluble Prion Protein Is Not Necessarily Associated with Prion Infectivity. *J. Biol. Chem.* **274:**17981–17986.

Singh, N., Zanusso, G., Chen, S. G., Fujioka, H., Richardson, S., Gambetti, P., and Petersen, R. B. (1997). Prion protein aggregation reverted by low temperature in transfected cells carrying a prion protein gene mutation. *J. Biol. Chem.* **272:**28461–28470.

Smits, M. A., Bossers, A., and Schreuder, B. E. (1997). Prion protein and scrapie susceptibility. *Vet. Q.* **19:**101–105.

Somerville, R. A., Chong, A., Mulqueen, O. U., Birkett, C. R., Wood, S. C., and Hope, J. (1997). Biochemical typing of scrapie strains [letter; comment]. *Nature* **386:**564.

Supattapone, S., Nguyen, H. O., Cohen, F. E., Prusiner, S. B., Scott, M. R. (1999). Elimination of prions by branched chain polyamines and implications for therapeutics. *Proc. Natl. Acad. Sci. U.S.A.* **96:**14529–14534

Swietnicki, W., Petersen, R., Gambetti, P., and Surewicz, W. K. (1997). pH-dependent stability and conformation of the recombinant human prion protein PrP(90–231). *J. Biol. Chem.* **272:**27517–27520.

Swietnicki, W., Petersen, R. B., Gambetti, P., and Surewicz, W. K. (1998). Familial mutations and the thermodynamic stability of the recombinant human prion protein. *J. Biol. Chem.* **273:**31048–31052.

Tagliavini, F., Prelli, F., Porro, M., Rossi, G., Giaccone, G., Farlow, M. R., Dlouhy, S. R., Ghetti, B., Bugiani, O., and Frangione, B. (1994). Amyloid fibrils in Gerstmann-Straussler-Scheinker disease (Indiana and Swedish kindreds) express only PrP peptides encoded by the mutant allele. *Cell* **79:**695–703.

Tagliavini, F., McArthur, R. A., Canciani, B., Giaccone, G., Porro, M., Bugiani, M., Lievens, P. M.-J., Bugiani, O., Peri, E., Dall'Ara, P., Rocchi, M., Poli, G., Forloni, G., Bandiera, T., Varasi, M., Suarato, A., Cassutti, P., Cervini, M. A., Lansen, J., Salmona, M., and Post, C. (1998). Effectiveness of anthracycline against experimental prion disease in Syrian hamsters. *Science* **276:**1119–1122.

Tateishi, J., Doh-ura, K., Kitamoto, T., Tanchant, C., Steinmetz, G., Warter, J. M., and Boellaard, J. W. (1992). *Prion Diseases of Humans and Animals* (Prusiner, S. B., Collinge, J., Powell, J., and Anderton, B., eds.), 129–138. Ellis Horwood, Chichester.

Telling, G. C., Parchi, P., DeArmond, S. J., Cortelli, P., Montagna, P., Gabizon, R., Mastrianni, J., Lugaresi, E., Gambetti, P., and Prusiner, S. B. (1996). Evidence for the conformation of the pathologic isoform of the prion protein enciphering and propagating prion diversity. *Science* **274:**2079–2082.

Telling, G. C., Scott, M., Hsiao, K. K., Foster, D., Tochia, M., Sidle, K. C., Collinge, J., DeArmond, S. J., Prusiner, S. B. (1994). Transmission of Crutzfeldt-Jakob disease from humans to transgenic mice expressing chimeric human-mouse prion protein. *Proc. Natl. Acad. Sci. U.S.A.* **91:**9936–9940.

Telling, G. C., Scott, M., Mastrianni, J., Gabizon, R., Torchia, M., Cohen, F. E., DeArmond, S. J., and Prusiner, S. B. (1995). Prion propagation in mice expressing human and chimeric PrP transgenes implicates the interaction of cellular PrP with another protein. *Cell* **83:**79–90.

Tobler, I., Gaus, S. E., Deboer, T., Achermann, P., Fischer, M., Rulicke, T., Moser, M., Oesch, B., McBride, P. A., and Manson, J. C. (1996). Altered circadian activity rhythms and sleep in mice devoid of prion protein. *Nature* **380:**639–642.

Tobler, I., Deboer, T., and Fischer, M. (1997). Sleep and sleep regulation in normal and prion protein-deficient mice. *J. Neurosci.* **17:**1869–1879.

Turk, E., Teplow, D. B., Hood, L. E., and Prusiner, S. B. (1988). Purification and properties of the cellular and scrapie hamster prion proteins. *Eur. J. Biochem.* **176:**21–30.

Viles, J. H., Cohen, F. E., Prusiner, S. B., Goodin, D. B., Wright, P. E., and Dyson, H. J. (1999). Copper binding to the prion protein: structural implications of four identical cooperative binding sites. *Proc. Natl. Acad. Sci. U.S.A.* **96:**2042–2047.

Weissmann, C. (1991). A unified theory of prion propagation. *Nature* **352:**679–683.

Weissmann, C. (1999). Molecular genetics of transmissible spongiform encephalopathies. *J. Biol. Chem.* **274:**3–6.

Wildegger, G., Liemann, S., and Glockshuber, R. (1999). Extremely rapid folding of the C-terminal domain of the prion protein without kinetic intermediates. *Nat. Struct. Biol.* **6:**550–553.

Wille, H. and Prusiner, S. B. (1999). Ultrastructural studies on scrapie prion protein crystals obtained from reverse micellar solutions. *Biophys. J.* **76:**1048–1062.

Williams, E. S., Kirkwood, J. K., and Miller, M. W. (2000). *Infectious Diseases in Wild Mammals* (Williams, E. S., and Barkis, I. K., eds.), 3rd ed. Iowa State University Press, Iowa City.

Zhang, H., Stockel, J., Mehlhorn, I., Groth, D., Baldwin, M. A., Prusiner, S. B., James, T. L., and Cohen, F. E. (1997). Physical studies of conformational plasticity in a recombinant prion protein. *Biochemistry* **36:**3543–3553.

ADVANCES IN VIRUS RESEARCH, VOL 56

SPONGIFORM ENCEPHALOPATHIES: INSIGHTS FROM TRANSGENIC MODELS

Adriano Aguzzi, Sebastian Brandner, Michael B. Fischer, Hisako Furukawa, Markus Glatzel, Cynthia Hawkins, Frank L. Heppner, Fabio Montrasio, Beatriz Navarro, Petra Parizek, Vladimir Pekarik, Marco Prinz, Alex J. Raeber, Christiane Röckl, and Michael A. Klein

Institute of Neuropathology
Department of Pathology
Schmelzbergstrasse 12, University Hospital, 8091
Zurich, Switzerland

I. Introduction

The prion was defined by Stanley B. Prusiner as the infectious agent that causes transmissible spongiform encephalopathies (TSEs) A pathological protein accumulating in the brain of scrapie-infected hamsters was isolated in 1982 and termed prion protein (PrP^{Sc}). Its gene, *Prnp*, was identified more than a decade ago by Charles Weissmann, and was shown to encode the host protein, PrP^C. Since the latter discovery, transgenic mice have contributed many important

313

insights into the field of prion biology, including the understanding of the molecular basis of the species barrier for prions. By disrupting the *Prnp* gene, it was shown that an organism that lacks PrP^C is resistant to infection by prions. Introduction of mutant PrP genes into PrP-deficient mice was used to investigate the structure–activity relationship of the PrP gene with regard to scrapie susceptibility. Ectopic expression of PrP in PrP knockout mice proved a useful tool for the identification of host cells competent for prion replication. Finally, the availability of PrP knockout mice and transgenic mice overexpressing PrP allows selective reconstitution experiments aimed at expressing PrP in neurografts or in specific populations of hemato- and lymphopoietic cells. The latter studies have allowed us to clarify some of the mechanisms of prion spread and disease pathogenesis.

A. Recent History of Prion Research

Prion diseases, or transmissible spongiform encephalopathies, are neurological disorders caused by transmissible pathogens termed prions. While the prototype of all prion diseases, scrapie in sheep and goats, has been known for more than two centuries, a recent form of animal prion disease, designated as bovine spongiform encephalopathy (BSE), has, since its first recognition in 1986, developed into an epizootic (Anderson *et al.*, 1996; Weissmann and Aguzzi, 1997; Wilesmith *et al.*, 1992). The emergence of a new variant form of Creutzfeldt-Jakob disease (vCJD) in young people in the United Kingdom has raised the possibility that BSE has spread to humans by dietary exposure (Chazot *et al.*, 1996; Will *et al.*, 1996). This scenario has recently been supported by experimental evidence claiming that the agent causing BSE is indistinguishable from the vCJD agent (Aguzzi, 1996; Aguzzi and Weissmann, 1996b; Bruce *et al.*, 1997; Hill *et al.*, 1997a; Will *et al.*, 1999).

Some of the milestones in prion research were the purification of the infectious agent and the discovery of the prion protein (PrP) (Bolton *et al.*, 1982; Prusiner, 1982), which were followed by the molecular cloning of the PrP cDNA (Chesebro *et al.*, 1985; Oesch *et al.*, 1985) and the gene (Basler *et al.*, 1986). Mutations in the host-encoded PrP gene were shown to be genetically linked to human prion diseases (Hsiao *et al.*, 1989). In addition, the PrP gene controls many features of prion diseases, such as incubation time, species barrier, and strain specificity (Aguzzi and Weissmann, 1997; Prusiner, 1997; Raeber *et al.*, 1998; Weissmann, 1996).

Since the discovery of the PrP gene, transgenetic investigations of prion diseases have become a fruitful area of research on these still enigmatic disorders. In this chapter, we summarize some of the transgenic mouse models that contributed to the current understanding of the pathogenesis of transmissible spongiform encephalopathies.

B. Prion Protein and Molecular Biology of Prions

Although the exact physical nature of the transmissible agent is still controversial, a very large body of experimental data supports the "protein only" hypothesis, which postulates that the agent is devoid of nucleic acid and consists solely of an abnormal conformer of the cellular prion protein, PrP^C (Griffith, 1967; Prusiner, 1982). Accumulation of an abnormal isoform (PrP^{Sc} or PrP-res) of the host-encoded prion protein (PrP^C or PrP-sen) (McKinley et al., 1983; Oesch et al., 1985; Prusiner et al., 1983) in the central nervous system (CNS) is a hallmark of prion diseases. It has been proposed that the partially protease-resistant and detergent-insoluble PrP^{Sc} is congruent with the infectious agent (Prusiner, 1982)—experimental verification of this hypothesis is, however, still amiss. Two distinct models have been postulated to explain the mechanism by which a misfolded form of PrP could induce the refolding of "native," normal PrP molecules into the abnormal conformation: the template assistance model, and the nucleation–polymerization model (Fig. 1). In the first model, which was proposed by Prusiner, a PrP^{Sc} monomer promotes the conformational conversion of PrP^C, or of a partially destabilized intermediate, to the PrP^{Sc} conformation. In this model, PrP^{Sc} is inherently more stable than PrP^C, but kinetically inaccessible (Prusiner, 1991). In the second model, the formation of PrP^{Sc} is initiated by an aggregate of PrP^{Sc} acting as a seed in a nucleation-dependent polymerization process. In contrast to the template assistance model, the PrP^{Sc} monomer is less stable than PrP^C but is stabilized on binding to the PrP^{Sc} aggregate (Gajdusek, 1988; Jarrett and Lansbury, 1993). Consistent with the latter model, cell-free conversion studies indicate that PrP^{Sc} aggregates are able to convert PrP^C into a protease-resistant PrP isoform (Caughey et al., 1997; Chabry et al., 1998; Horiuchi and Caughey, 1999; Raymond et al., 1997).

Because of its location at the outer surface of cells, anchored by a phosphatidylinositol glycolipid (Stahl et al., 1987), PrP is a candidate for signaling, cell adhesion, or perhaps even for some transport functions. PrP is expressed on many cell types, including neurons (Kretzschmar et al., 1986), astrocytes and possibly oligodendrocytes (Moser et al., 1995), and lymphocytes (Cashman et al., 1990), and appears to be developmentally regulated during mouse embryogenesis (Manson

a   Template-directed refolding

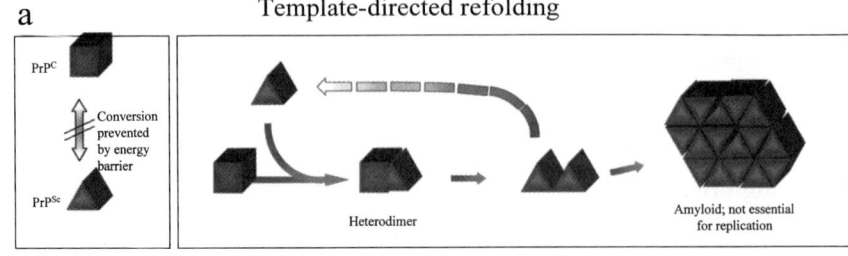

b   Seeding (nucleated crystallization)

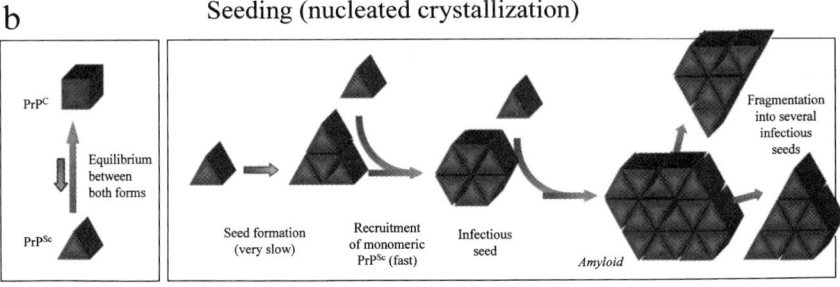

FIG 1. Models for the conformational conversion of PrP^C into PrP^{Sc}. (a) The "template-directed refolding model" postulates an interaction between exogenously introduced PrP^{Sc} and endogenous PrP^C, which is induced to transform itself into further PrP^{Sc}. A high-energy barrier may prevent spontaneous conversion of PrP^C into PrP^{Sc}. Sporadic CJD may arise when spontaneous mutations in single cells lead to the conversion of PrP^C to PrP^{Sc} and give rise to manifold subsequent conversions. (b) The seeding, or nucleation, model proposes that PrP^C and PrP^{Sc} are in a reversible thermodynamic equilibrium. Only if several monomeric PrP^{Sc} molecules are mounted up to a highly ordered seed can further monomeric PrP^{Sc} be recruited and eventually aggregate to form amyloid. The likelihood of spontaneous formation of a seed is a function of the local PrP^{Sc} concentration (which may be modulated by PrP^{Sc}-binding proteins) and is inversely dependent on the number of monomers needed to form a protoseed. Within such a crystal-like seed, PrP^{Sc} becomes stabilized. Fragmentation of PrP^{Sc} aggregates increases the number of nuclei that can recruit further PrP^{Sc} and therefore results in apparent replication of the agent.

et al., 1992). Although PrP is predominantly found in brain tissue, high levels are also present in the heart, skeletal muscle, and the kidney, whereas it is barely detectable in the liver (Bendheim *et al.*, 1992). Several candidate proteins that bind PrP^C have been reported. Among them are the amyloid precursor-like protein 1 (APLP1) (Yehiely *et al.*, 1997), the human laminin receptor precursor (Rieger *et al.*, 1997), and an uncharacterized 66-kDa membrane protein (Martins *et al.*, 1997). Evidence that any of these interactions are physiologically significant is, however, still missing.

II. TRANSGENIC MODELS FOR HUMAN HEREDITARY PRION DISEASES

Human prion diseases are characterized by extended incubation periods ranging from several months to decades, which are followed by a progressive clinical phase presenting with severe dementia and ataxia (Weber and Aguzzi, 1997). Clinical disease is always lethal: death can occur within as short a period as a few weeks, but occasionally in a period of up to a few years. The new variant of Creutzfeldt-Jakob disease (CJD), which is suspected to be caused by infection with BSE prions, tends to display a protracted clinical course (Aguzzi, 1996; Aguzzi and Weissmann, 1996b; Bruce et al., 1997; Hill et al., 1997a). Neuropathological features include profound astrocytic gliosis, spongiosis, and neuronal cell loss. A typical, albeit not invariable, finding in certain forms of human prion diseases is the presence of amyloid plaques consisting, at least in part, of PrP^{Sc}.

Human prion diseases come in three types: sporadic, genetic, and infectious. Most cases of CJD occur sporadically with a frequency of $1:10^6$–$1:10^7$ per year worldwide. A compilation of Swiss epidemiological data of the last three years has yielded an incidence in excess of $2:10^6$ per year (Hegyi and Aguzzi, unpublished data). In these patients, mutations of the PrP coding sequence are not found. The genetic forms of prion diseases include Gerstmann-Sträussler-Scheinker disease (GSS), the familial forms of CJD, and fatal familial insomnia (FFI) (Aguzzi and Weissmann, 1996a; Prusiner, 1994). All of the genetic forms are transmitted as autosomal dominant traits, albeit with somewhat variable penetrance. The first mutation in the PrP gene that was found to be genetically linked to hereditary forms of GSS was the codon 102 Pro>Leu mutation (Hsiao et al., 1989) that has since been found in many GSS families throughout the world (Goldgaber et al., 1989; Kretzschmar et al., 1991). By now, all genetic forms of prion disease have been linked to mutations in the human PrP gene (Aguzzi and Brandner, 1999). FFI has been linked to a pathogenic Prnp mutation, in conjunction with a common polymorphism (Aguzzi and Weissmann, 1996a) (Fig. 2). Most human prion diseases can be transmitted to experimental animals, including mice.

It has been speculated that mutations in the PrP gene give rise to an unstable PrP^C protein that can spontaneously convert into the abnormal conformer, PrP^{Sc}. Sporadic forms of the disease would have to be explained by a very rarely occurring spontaneous transition of PrP^C into PrP^{Sc} or by somatic mutations in the PrP gene. Once conversion has started, it would be followed by "autocatalytic" propagation—the latter term is enclosed by quotation marks because, in our view, it

Mutations and polymorphisms of the human prion gene

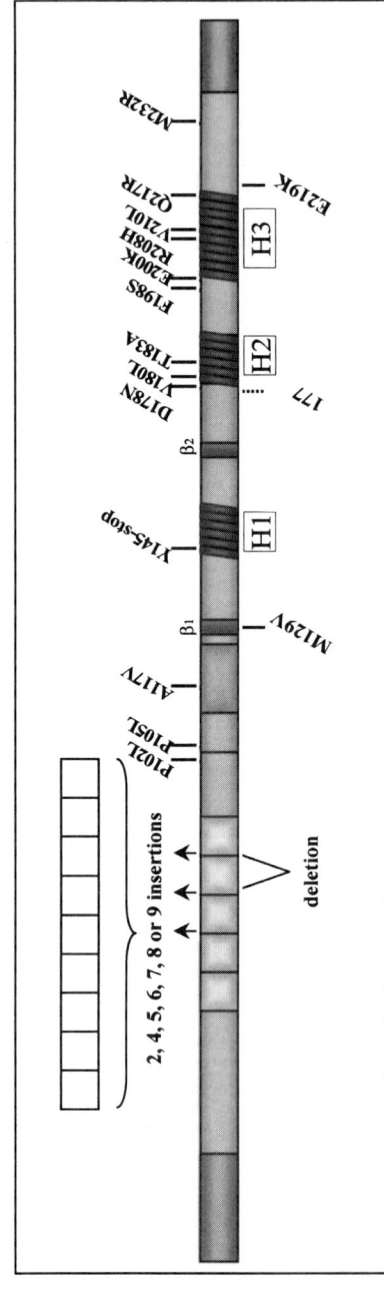

FIG 2. Schematic drawing of the coding region of the human *PRNP* gene. Mutations that segregate with inherited prion diseases are shown above, as well as nonpathogenic polymorphisms (red numbers) and silent mutations (blue). The N- and C-terminal domains are signal peptides that are cleaved off during maturation of PrP^C. Octarepeat regions are represented by yellow boxes, and pathogenic octarepeat insertions of 16, 32, 40, 48, 56, 64, and 72 amino acids are shown below. Deletion of one octarepeat stretch does not segregate with a neurodegenerative disorder (Laplanche *et al.*, 1990). The light green box indicates a conserved region and β sheet domains are drawn in red. There is an apparent clustering of pathogenic mutations associated with a CJD phenotype around the α-helical domains (H1, H2, H3, dark blue), while amino acid exchanges associated with a GSS phenotype are located further upstream around the conserved region.

remains to be established whether the conversion process resembles the chemical process of catalysis.

Prusiner and colleagues introduced the leucine substitution corresponding to the human GSS mutation at codon 102 into codon 101 of the murine *Prnp* gene. One transgenic line harboring this modified transgene developed a spontaneous neurological disease with spongiform degeneration of the brain (Hsiao *et al.*, 1990). Most intriguingly, the disease could be transmitted by inoculating brain extracts of these mice into transgenic animals harboring the same mutant PrP transgene but expressing it at lower levels, which did not develop spontaneous illness (Telling *et al.*, 1996a). An inherent problem with these studies, however, is the fact that the mutant protein was overexpressed severalfold compared to wild-type mice. Considering that the overexpression of wild-type PrP by itself from a cosmid PrP transgene causes a spontaneous disease phenotype in transgenic mice (Westaway *et al.*, 1994), it was desirable to study the mutation in the context of the endogenous PrP gene. Using a two-step gene targeting strategy in embryonic stem cells, the Pro-101Leu mutation (equivalent to codon 102 in the human PrP gene) was introduced into the endogenous murine PrP gene (Moore *et al.*, 1995). Mice either homozygous or heterozygous for the codon 101 mutation were generated from gene-targeted embryonic stem cells. However, these mice remained healthy for more than 650 days, showing no signs of spontaneous CNS disease and no abnormal pathology (Moore and Melton, 1997). This shows that the presence of the mutated PrP at physiological levels is not sufficient to cause spontaneous neurodegeneration in mice.

III. Mice as Transgenic Models

A. Phenotypes of PrP Knockout Mice

According to the "protein only" hypothesis, PrP^C is a substrate for the PrP^{Sc}-mediated conversion of PrP^C into new PrP^{Sc} molecules. An important corollary to this hypothesis is that an organism lacking PrP^C should be resistant to scrapie and unable to propagate the infectious agent. Büeler and colleagues (1992) generated mice with a targeted disruption of the *Prnp* gene by using homologous recombination in embryonic stem cells. In the disrupted *Prnp* allele, 184 codons of the *Prnp* coding region (which consists of 254 codons) were replaced by a drug-resistance gene as a selectable marker. A second line of PrP knockout mice was generated by Manson and co-workers (1994a) by inserting a selectable marker into a unique *KpnI* site of the PrP open reading frame, thereby disrupting—

but not deleting—the *Prnp* coding region. Sakaguchi and colleagues generated a third line of PrP knockout mice by replacing the whole PrP open reading frame (ORF) and, in addition, about 250 bp of the 5′ intron, and 452 bp of 3′ untranslated sequences, with a drug-resistance gene (Sakaguchi *et al.*, 1995). Both the Büeler and Sakaguchi mice were on a mixed genetic (129/Sv × C57BL) background, whereas the Manson mice were bred on a pure 129/Ola background. According to the terminology which has become customary in the literature, and which we adopt in this chapter, the Zurich mice have been designated as $Prnp^{o/o}$, while the Edinburgh and the Japanese mice are termed $Prnp^{-/-}$.

Although it was proposed that PrP, which is a ubiquitously expressed neuronal protein, may have a housekeeping function (Basler *et al.*, 1986), homozygous PrP knockout mice generated by Büeler and by Manson were viable with no behavioral impairment, and showed no overt phenotypic abnormalities, which suggests that PrP^C does not play a crucial role in the development or function of the nervous system (Büeler *et al.*, 1992; Manson *et al.*, 1994a). However, electrophysiological defects such as weakened γ-aminobutyric acid (GABA)-A receptor-mediated fast inhibition and impaired long-term potentiation in the hippocampus were reported for these two lines of PrP knockout mice when they were compared to their corresponding wild-type counterparts, indicating that PrP might play a role in synaptic plasticity (Collinge *et al.*, 1994). Interestingly, this electrophysiological phenotype could be rescued by a transgene encoding human PrP (Whittington *et al.*, 1995). In contrast, no electrophysiological abnormalities were found by others in the hippocampus (Lledo *et al.*, 1996) or in the cerebellum (Herms *et al.*, 1995), at least when the line of PrP null mice generated by Büeler *et al.* (1992) was used. Altered sleep patterns and rhythms of circadian activity have been reported in the Büeler and the Manson mice (Tobler *et al.*, 1996).

The $Prnp^{-/-}$ mice generated by Sakaguchi *et al.* (1995) presented with the most severe phenotype, which consisted of progressive ataxia from 70 weeks of age. Analysis of the brains of affected animals revealed an extensive loss of cerebellar Purkinje cells (Sakaguchi *et al.*, 1996). Because no such phenotype was observed in the other two lines of PrP knockout mice, it seems likely that this phenotype is not due to the lack of PrP but, rather, to the deletion of flanking sequences. Interestingly, a Purkinje cell-specific enhancer was proposed to be contained within the second intron of *Prnp* (Fischer *et al.*, 1996). However, the report that expression of a *Prnp* transgene can rescue this phenotype argues against the hypothesis that the phenotype was caused by deletion of a regulatory element, rather than of the *Prnp* reading frame (Nishida *et al.*, 1999)

B. Prnp$^{o/o}$ Mice Resistant to Scrapie

All three lines of PrP null mice were identical with regard to their resistance to scrapie. The Prnp$^{o/o}$ mice generated by Büeler et al., were inoculated with the RML isolate of mouse-adapted prions, and they remained healthy for their whole life span, whereas all wild-type (129/Sv × C57BL/6) mice developed clinical scrapie symptoms at 158 days ± 11, and died of scrapie at 171 days ± 11, after inoculation (Büeler et al., 1993). Similar results were obtained with the other two lines of PrP null mice challenged with the ME7 isolate (Manson et al., 1994b) and with the mouse-adapted Fukuoka-1 strain of CJD prions (Sakaguchi et al., 1995). All three lines of PrP null mice showed a complete lack of scrapie-typical neuropathology following inoculation with prions.

An additional consequence of the findings mentioned above is that prion propagation should not occur in PrP knockout mice. Indeed, animals lacking both Prnp alleles were unable to propagate prions in brain and spleen, whereas prion levels in brain and spleen of Prnp$^{+/+}$ mice increased to about 8.6 and 6.9 log LD$_{50}$ units/ml, respectively, by 20 weeks postinfection (Büeler et al., 1993; Sailer et al., 1994). Mice hemizygous for the disrupted Prnp gene (Prnp$^{0/+}$) also showed partial resistance to scrapie infection, as manifested by prolonged incubation times of about 290 days, as compared to about 160 days in the case of Prnp$^{+/+}$ mice. Although times to onset of symptoms and rate of disease progression seem to correlate with the steady-state levels of PrP^{C} in the host, final severity of the disease, as assessed by the extent of scrapie pathology and levels of prion infectivity, were not dependent on the PrP^{C} level (Büeler et al., 1994; Manson et al., 1994b). This suggests that the amount of PrP^{C} protein in the brain is a rate-limiting step in the development of the disease, and that therapeutic efforts aimed at reducing the production of the normal PrP^{C} isoform may be effective. Interestingly, Prnp$^{0/+}$ mice harbored high levels of infectivity and of PrP^{Sc} by 140 days after inoculation but survived thereafter for at least another 140 days without showing severe clinical symptoms (Büeler et al., 1994). These findings suggest that despite high levels of prion infectivity and PrP^{Sc} in the brain, clinical disease may not manifest in an organism. Whether this important finding is also true for humans and cattle remains to be seen.

C. Restoration of Scrapie Susceptibility in Prnp$^{o/o}$ Mice

Mice devoid of PrP are viable and resistant to scrapie. To unequivocally show that this phenotype is due to the disruption of the PrP gene, it was necessary to demonstrate that reintroduction of PrP into Prnp$^{o/o}$ mice is able to restore scrapie susceptibility. This was achieved by crossing $Prnp^{o/o}$ mice with transgenic mice expressing a Syrian hamster PrP gene. The resulting double-transgenic mice were very susceptible to hamster prions, with average incubation times of about 56 days, but much less susceptible to mouse prions (300 days incubation time) (Büeler et al., 1993). Further, murine Prnp genes introduced by transgenesis into $Prnp^{o/o}$ mice were shown to restore the scrapie susceptibility of $Prnp^{o/o}$ mice in a dose-dependent fashion (Fischer et al., 1996). Two lines of transgenic mice (tga19 and tga20), which contained about 30 copies of a Prnp transgene lacking the 10 kb intron 2 ("half-genomic" Prnp: phgPrP), and which overexpressed PrP in the brain by 3–4-fold (tga19) and 6–7-fold (tag20) as compared with wild-type brain, showed susceptibility to scrapie, with incubation times of 87 days ± 13 (tga19) and 64 days ± 9 (tga20) (Fischer et al., 1996). This confirmed an inverse relationship between the steady-state level of PrP^C in the brain and the incubation time for scrapie, as reported earlier (Prusiner et al., 1990). However, accurate titration of the Rocky Mountain Laboratory standard prion inoculum into wild-type and tga20 mice demonstrated that the actual sensitivity of each strain to prions, when exposed to limiting dilutions of the inoculum, was very similar (Brandner et al., 1996a).

While overexpression of PrP from a cosmid Prnp transgene caused a lethal neurologic disease associated with scrapie-like spongiform degeneration in the brain, demyelination of the sciatic nerve, and muscle degeneration in old transgenic mice (Westaway et al., 1994), no such phenotype was observed in $Prnp^{o/o}$ mice overexpressing PrP from the half-genomic Prnp transgene (Fischer et al., 1996). Since PrP expression levels were similar in these transgenic mice, it seems unlikely that overexpression of PrP^C is sufficient to account for this novel phenotype. Alternatively, different expression patterns due to the constructs used are most likely to be responsible for the observed neurologic syndrome. Incidentally, it was found that transgenic mice generated with the half-genomic Prnp transgene showed no detectable PrP RNA in Purkinje cells. This suggests that one or more control elements responsible for Purkinje cell-specific expression are absent from the half-genomic construct (Fischer et al., 1996).

IV. Structure–Function Studies on PrP Gene

Limited proteolysis truncates the N-terminus of PrP^C to form PrP^{27-30} without loss of infectivity—this argues that at least 60 amino proximal residues of PrP^{Sc} are not required for infectivity (Hope et al.,

1988; McKinley *et al.*, 1983). It had been shown that PrP^C lacking residues 23–88 could be converted into protease-resistant PrP in scrapie-infected neuroblastoma cells (Rogers *et al.*, 1993). The question then arose as to whether N-terminally truncated PrP molecules can support prion replication in mice. For that purpose, PrP transgenes harboring N-terminal deletions were introduced into PrP-deficient mice by transgenesis. Indeed, mutant PrP with amino-proximal deletions of residues 32–80 (lines *tgd11* and *tgd12*) (Fischer *et al.*, 1996) and of residues 32–93 (lines C4 and C15) (Shmerling *et al.*, 1998), which correspond to truncations of 49 and 63 residues, restore scrapie susceptibility, prion replication, and formation of truncated PrP^{Sc}. These experiments demonstrate that the octapeptide region encompassing residues 51–90 of murine PrP is dispensable for scrapie pathogenesis. This is remarkable in view of the fact that additional octapeptide repeats, instead of the normal 5, segregate with affected individuals in families with inherited CJD (Goldfarb *et al.*, 1991), and that expression of a mutant PrP with a pathological number of octarepeats induces a neurodegenerative disease in transgenic mice (Chiesa *et al.*, 1998).

A. *Spontaneous Phenotype in Mice Expressing Truncated PrP*

NMR structure determinations of full-length mature PrP have revealed a highly flexible amino-terminal tail that lacks ordered secondary structures extending from residue 23 to 121 (Donne *et al.*, 1997; Riek *et al.*, 1997). The carboxy-terminal part of PrP consists of a stably folded globular domain (James *et al.*, 1997; Riek *et al.*, 1996). The flexible tail, part of which is protease-sensitive in PrP^{Sc}, comprises the most conserved region of PrP across all species examined (Schatzl *et al.*, 1995). It was proposed that the flexible tail may play a role in the conformational transition of PrP^C to PrP^{Sc} by initiating the structural rearrangements from α-helices to β-sheets (Peretz *et al.*, 1997; Riek *et al.*, 1997). To further analyze the importance of the flexible tail with regard to scrapie susceptibility, amino-proximal deletions of residues 32–121 and 32–134 were generated and the transgenes introduced into PrP-deficient mice. Mice overexpressing these truncated PrP transgenes (lines E11, F11, and F35) developed severe ataxia and neuronal death limited to the granular layer of the cerebellum, as early as 1–3 months of age (Fig. 3, also see color section). No pathological phenotype was observed in transgenic mice with shorter deletions encompassing residues 32–80, 32–93, and 32–106, respectively. The selective degeneration of granule cells of

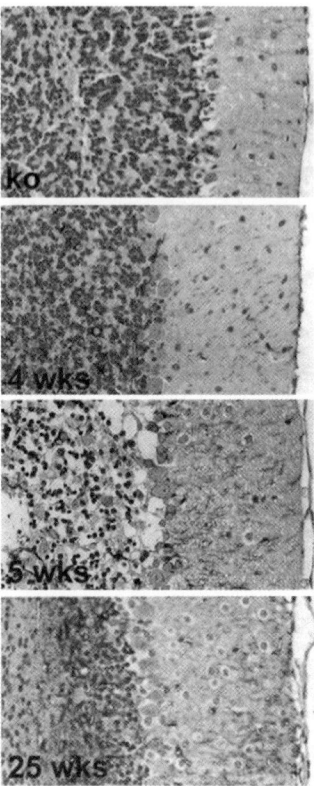

FIG 3. [For color reproduction, see color section.] Progressive degeneration of the granule cell layer in mice expressing D32–134 truncated PrP. Development of the cerebellum proceeds normally until early postnatal life, and leads to formation of a molecular layer, a Purkinje cell layer, and a granule cell layer (right to left in the figure) of normal thickness. However, at 4 weeks, some degree of pathological astrogliosis can already be discerned. At 5 weeks, massive degeneration of granule cells by apoptosis is ongoing. Note strong gliosis affecting also the molecular layer. At the end stage of disease, mice suffer from a profound cerebellar syndrome and the thickness of the granule cell layer is considerably reduced (Shmerling et al., 1998). In some areas of the cerebellar cortex, granule cells disappear completely (this is not shown).

the cerebellum argues against a nonspecific toxic effect elicited by the truncated PrP. This is further supported by the fact that neurons in the cortex and elsewhere express truncated PrP at similar levels but do not undergo cell death by apoptosis. The pathological phenotype was abolished by the introduction of one copy of a wild-type *Prnp* allele. These results are consistent with a model in which truncated PrP acts as the dominant negative inhibitor of a functional homologue of PrP, with both competing for the same putative PrP ligand (Shmerling *et al.*, 1998).

Although the functional role of PrP remains elusive, there is recent evidence, *in vitro* and *in vivo*, that the octarepeats in the flexible tail of PrP^C exhibit binding sites for copper (Brown *et al.*, 1997; Hornshaw *et al.*, 1995). The cerebellar defects are only apparent in transgenic mice with PrP deletions encompassing the whole flexible tail. Therefore, it is conceivable that the flexible tail is involved in a signal transduction pathway that may be modulated by copper.

A different spontaneous neurologic phenotype was reported in mice carrying PrP transgenes with internal deletions corresponding to either of the two carboxy-proximal α-helices. Two transgenic mouse lines generated on the *Prnp*$^{o/o}$ background expressing mutant PrP with deletions of residues 23–88, and either residues 177–200 or 201–217, developed CNS dysfunction and neuropathological changes characteristic of a neuronal storage disease (Muramoto *et al.*, 1997). Because deletion of residues 23–88 alone did not lead to a spontaneous phenotype, it was concluded that ablation of either of the two carboxy-terminal helices is sufficient to cause this novel CNS illness. Ultrastructural studies indicated extensive proliferation of the endoplasmic reticulum and revealed accumulation of mutant PrP within cytoplasmic inclusions in enlarged neurons. Because both Asn-linked glycosylation sites are located within residues 177–200, it is conceivable that aberrant glycosylation affects processing of the mutant PrP. However, it is unlikely that altered glycosylation of PrP is sufficient to account for neuronal storage disease, because transgenic mice expressing hamster PrP with point mutations that block Asn-linked glycosylation did not show this spontaneous disease phenotype (DeArmond *et al.*, 1997).

V. SPECIES BARRIER

Transmission of prions from one species to another is usually accompanied by a prolongation of the incubation period in the first passage and incomplete penetrance of the disease. Subsequent passage in the same species occurs with high frequency and shortened incubation times (Pattison, 1966). This so-called species barrier can be overcome by introducing into the recipient host PrP transgenes derived from the prion donor. Thus, transgenic mice harboring Syrian hamster (Sha) PrP transgenes developed hamster scrapie with a latency of 75 days following inoculation with hamster prions, while wild-type littermates failed to show symptoms after more than 500 days. Brains of hamster prion-infected Tg(SHaPrP) mice contained high levels of hamster prions, hamster PrPSc, and a distribution of the pathological lesions characteristic of hamster scrapie. This finding demonstrated that the PrP gene profoundly influences the species specificity of prions and consequently modulates scrapie susceptibility, incubation times, and neuropathology (Scott et al., 1989). In a following experiment, several lines of Tg(SHaPrP) mice with various transgene copy numbers and PrP expression levels were analyzed for their susceptibility to hamster scrapie. These studies showed that the length of the incubation time after inoculation with hamster prions is inversely correlated with the steady-state level of HaPrP^C in the brains of these mice. Moreover, inoculation of Tg(SHaPrP) mice with hamster prions led to the production of about 10^9 LD$_{50}$ units of hamster prions and <10 LD$_{50}$ units of mouse prions. Similarly, inoculation with mouse prions resulted in the accumulation of high levels of mouse prions in the brains of sick animals with scrapie and no detectable hamster infectivity. These results suggest that the prion inoculum dictates which prions are synthesized in a mouse containing both murine and hamster PrP genes. To explain these data within the framework of the "protein only" hypothesis, Prusiner proposed that efficient interaction of PrPSc with host-derived PrP^C requires homology between the two (Prusiner et al., 1990).

A further series of experiments by the Prusiner group was aimed at defining the regions of the PrP molecule involved in determining species specificity. Chimeric PrP genes composed of portions from the Syrian hamster and mouse PrP genes were introduced into transgenic mice (Scott et al., 1992). These transgenic mice allowed for a detailed mapping of the molecular domains of the PrP gene responsible for homotypic interactions between PrPSc in the inoculum and host PrP^C. The brains of sick transgenic mice contained prions with an artificial host range favoring propagation in mice that express the corresponding chimeric PrP and were also transmissible, albeit at reduced efficiency, to nontransgenic mice and hamsters (Scott et al., 1993b). These findings add considerable strength to the notion that homotypic interactions between cellular and pathological isoforms of PrP are a fundamental pathogenetic principle in spongiform encephalopathies.

Crossing the species barrier for transmission of prions from humans to mice turned out to be much more difficult than crossing between hamsters and mice. This might be explained by the sequence differences of the corresponding PrP genes. While hamster PrP shows only 16 amino acid differences as compared with mouse PrP, human PrP differs from mouse PrP at 28 of 254 amino acids. Human prions were found to infect wild-type mice with an efficiency of only about 10% and with incubation times exceeding 500 days (Telling *et al.*, 1994). Introduction of human PrP transgenes into wild-type mice was not sufficient to abrogate resistance to human prions (Telling *et al.*, 1994; Telling *et al.*, 1995). Introduction of the transgene array into a PrP null background by crossing Tg(HuPrP) mice with *Prnp*^{o/o} mice abrogated the species barrier to human prions, resulting in incubation times of 250–270 days. Moreover, mice expressing chimeric human-mouse PrP (MHu2M) transgenes on a *Prnp*^{+/+} background were susceptible to human prions with incubation times of 210–240 days. Surprisingly, crossing the transgene array into the MoPrP gene-ablated background rendered the mice only slightly more susceptible to human prions with incubation times of 180–220 days. It was concluded that endogenous mouse PrP^C interferes with the conversion of human PrP^C to human PrP^{Sc}, and this was said to be due to a species-specific factor X interacting with the COOH-terminal domain of PrP (Telling *et al.*, 1995).

Similar transgenetic studies were applied to the investigation of characteristics of transmission of BSE prions to mice. Unexpectedly, mice expressing a bovine PrP transgene on the *Prnp*^{o/o} background were susceptible to BSE prions, with incubation times ranging from 230 to 360 days but Tg(MBo2M) mice harboring a chimeric bovine–mouse PrP transgene were resistant to BSE prions. This puzzling result may perhaps be explained by amino acid differences in the C-terminus between mouse, bovine, and human PrP that constitute an epitope that was said to modulate interspecies transmission of prions (Scott *et al.*, 1997b).

In summary, a wealth of transgenetic studies support an important role for PrP in the transmission of prions between different species. In addition, recent data suggest that the PrP gene might be required for the long-term persistence of prions in a host that is normally resistant to foreign prions. In that particular study, it was demonstrated that very low levels of hamster prions persist for more than 700 days in the brain and spleen of wild-type mice inoculated intracerebrally with hamster prions, but that the mice failed to develop clinical disease. Surprisingly, hamster prions were undetectable in brain and spleen of *Prnp*^{o/o} mice, therefore arguing that PrP^C is required for the long-term persistence of prions (Race and Chesebro, 1998). This might be explained if wild-type, but not *Prnp*^{o/o}, mice were immune tolerant to PrP^{Sc} from a different species. In fact, *Prnp*^{o/o} mice are able to mount an immune response to PrP (Brandner *et al.*, 1996b; Korth *et al.*, 1997; Prusiner *et al.*, 1993) and therefore might be able to clear hamster prions efficiently (Aguzzi and Weissmann, 1998). This observation raises the possibility that BSE prions might also persist in various resistant species, including humans, after exposure to BSE-contaminated, cattle-derived products (Collee and Bradley, 1997).

VI. Prion Strains

Distinct isolates, or "strains," of prions were first recognized in goats with scrapie in cases where two different clinical manifestations, described originally as "scratching" and "drowsy" were identified (Pattison and Millson, 1961). Prion strains were originally characterized by their incubation time and the distribution of vacuolar lesions in the brain. Studies of scrapie in inbred mice led to the identification of a single autosomal gene, $Sinc$ or $Prni,$ controlling incubation time of mice infected with mouse prion strains (Carlson $et\ al.,$ 1986; Dickinson $et\ al.,$ 1968; Hunter $et\ al.,$ 1987). Short and long incubation times for the ME7 strain of murine prions segregated with two different alleles of the $Prnp$ gene, $Prnp^a$ and $Prnp^b,$ encoding prion proteins that differ at two amino acid residues (Westaway $et\ al.,$ 1987). $Prnp^a$ is characterized by residues Leu-108 and Thr-189, and $Prnp^b$ encodes Phe-108 and Val-189. In contrast to the ME7 strain, the mouse-adapted BSE strain 301V shows long incubation times for the $Prnp^a$ allele and short incubation times for the $Prnp^b$ allele (Bruce $et\ al.,$ 1994). Classical genetic studies and study of transgenic mice failed to resolve the question as to whether $Sinc/Prni$ and $Prnp$ were closely linked or congruent (Carlson $et\ al.,$ 1994; Hunter $et\ al.,$ 1992; Westaway $et\ al.,$ 1991).

Using a two-step, double-replacement, gene-targeting strategy in embryonic stem cells (Moore $et\ al.,$ 1995), the $Prnp^a$ allele was changed to encode the $Prnp^b$-specific residues Phe-108 and Val-189, within the context of the $Prnp^a$ allele. Following a challenge with 301V prions, mice homozygous for the targeted $Prnp^{a[108F189V]}$ allele had an average incubation time of 133 days, while wild-type 129/Ola mice $(Prnp^a)$ developed disease after about 244 days. The profound shortening of the incubation time in gene-targeted mice with only two amino acid differences at codon 108 and 189 demonstrates unequivocally that $Sinc/Prni$ and the PrP gene are congruent (Moore $et\ al.,$ 1998).

Although these studies clearly establish an important role for the endogenous PrP gene in the host response to a particular prion strain, they cannot explain why different prion strains can be propagated, without changing their properties, in the same mouse strain homozygous for the PrP gene. It was argued that such a phenomenon could best be accounted for by the existence of an additional component within the infectious agent (Bruce and Dickinson, 1987; Dickinson and

Outram, 1988; Kimberlin and Walker, 1978a; Weissmann, 1991). To explain prion strains by the protein-only hypothesis, it was suggested that different chemical or conformational modifications of PrP^{Sc} might be responsible for the strain properties. The amino acid sequence of PrP^{Sc} is likely to be the same for strains derived from the same host, whereas the strain-specific differences might be encoded by the asparagine-linked oligosaccharides of PrP^{Sc}. Studies with transgenic mice expressing PrP mutated at one or both of the glycosylation consensus sequences indicate that prion strains are not encoded by the sugars (DeArmond et al., 1997). Alternatively, it was proposed that PrP^{Sc} derived from the same precursor might assume different conformations that can be stably propagated by a nongenetic mechanism.

Evidence that PrP^{Sc} acquires strain-specific properties, which manifest themselves in differential susceptibility to protease digestion, has been reported for two different strains of transmissible mink encephalopathy in hamsters (Bessen and Marsh, 1994), and for various CJD isolates (Parchi et al., 1996). Protease-teated PrP^{Sc} displays three distinct bands corresponding to different glycoforms of the same protein. Based on the fragment size and on the relative abundance of the individual bands, three distinct patterns (PrP^{Sc} types 1–3) were defined for sporadic and iatrogenic CJD cases. In contrast, all cases of vCJD exhibited a novel pattern, designated as the type-4 pattern. Moreover, extracts from the brains of BSE-infected cattle, cats, or kudu that were thought to have acquired BSE, and macaques that were infected experimentally with BSE, all showed the type-4 pattern (Collinge et al., 1996). Even more intriguingly, transmission of BSE or vCJD to mice produced mouse PrP^{Sc} with a type-4 pattern indistinguishable from the original inoculum (Hill et al., 1997a). These findings strongly support the view that vCJD is the human counterpart of BSE (Bruce et al., 1997). Furthermore, transmission of CJD and FFI isolates to transgenic mice expressing chimeric human-mouse PrP genes have shown that the PrP^{Sc} strain phenotype is preserved on passage to the new host, thereby providing further support for the hypothesis that strains are encoded by the tertiary structure of PrP^{Sc} (Telling et al., 1996b). It has been shown that molecular strain typing by Western blot analysis can be used in the differential diagnosis of vCJD (Hill et al., 1997b). However, some concern about the general applicability of glycoform ratio analysis for strain typing has been expressed (Somerville et al., 1997). More powerful methods for resolution of various glycoforms, perhaps in conjunction with a PrP^{Sc}-specific reagent, may eventually serve as a diagnostic tool for prion strains—one promising approach utilizes the differential affinity of antibodies to

PrP^{Sc} of different strains (Safar *et al.*, 1998) and provides further evidence for the conjecture that strain specificities are encrypted within the physical structure of PrP (Aguzzi, 1998b).

But how does a strain emerge on its passage from the original host through a variety of intermediate hosts? More than 20 different strains of mouse prions are known (Dickinson and Meikle, 1971), and it has been suggested that new strains arise by mutation and selection of a putative nucleic acid within the infectious particle (Bruce and Dickinson, 1987; Kimberlin *et al.*, 1989). New evidence regarding this issue has been provided by using transgenic mice expressing a chimeric mouse-hamster (MH2M) PrP transgene. By using serial transmission to Tg(MH2M) mice as an intermediate host, it was shown that two prion strains, Me7 and Sc237, derived from completely different primary sources converged to yield identical strains with respect to incubation time and pathology. These striking results suggest that prion strain characteristics change, depending on the sequence of PrP encoded by the host during multiple serial transmissions. These studies further imply that prion strain diversity is limited to a finite and highly restricted number of conformations of PrP^{Sc} that can be adopted by the sequence of PrP encoded by the host (Scott *et al.*, 1997a).

VII. Ectopic Expression of PrP in *Prnp*-Ablated Mice

The finding that PrP null mice are unable to replicate prions shows that PrP is a necessary host factor for prion replication. However, is PrP also sufficient for prion replication, or are additional, perhaps cell- or tissue-specific, factors required for prion propagation? This question was addressed by generating transgenic mice that express PrP ectopically in distinct cell types.

Because neurons, astrocytes, oligodendrocytes, and probably also microglia express PrP, it is not clear as to which cell type in the CNS is capable of generating infectivity. It was reported that transgenic mice expressing HaPrP under the control of the neuron-specific enolase (NSE) promoter are highly susceptible to hamster prions. In these mice, HaPrP expression was found exclusively in neurons and not in glial cells or cells within the spleen or lymph nodes. Thus, neuron-specific PrP expression is sufficient to sustain scrapie infection, and PrP expression in nonneuronal cells—in particular, astrocytes and cells of the lymphoreticular system—is not required, at least in the case of intracerebral inoculation (Race *et al.*, 1995).

The possibility that astrocytes might also contribute to the natural disease process is suggested not only by the finding that astrocytes express PrP, but also by the fact that in at least one model they are the earliest site of PrP^{Sc} accumulation in the brain (Diedrich *et al.*, 1991). In addition, astrocytic activation occurs very early in the disease process, leading to physiological effects such as impairment of the blood–brain barrier (Chung *et al.*, 1995; Wisniewski *et al.*, 1983). To study the role of astrocytes in prion-elicited pathogenesis, *Prnp*$^{o/o}$ mice expressing HaPrP, under the control of the glial fibrillary acidic protein (GFAP) promoter, were generated. These mice expressed HaPrP only in astrocytes and not in neurons. After inoculation with hamster prions, these mice developed neurologic disease and accumulated, in their brains, infectivity and hamster PrP^{Sc} to high levels (Raeber *et al.*, 1997). These findings demonstrate that not only neurons, but also astrocytes, are capable of prion replication. Interestingly, scrapie neuropathology in transgenic animals was strikingly similar to that in wild-type mice. How astrocytes are involved in the pathogenesis of prion diseases, and whether indirect effects, perhaps mediated by cytokines (Campbell *et al.*, 1994), play a role in the disease process, remain to be elucidated. Growing evidence also incriminates microglial cells in prion-elicited pathogenesis. To what extent oligodendrocyte-specific expression of PrP^C can contribute to disease progression remains to be elucidated. Recently, it was demonstrated that in a tissue culture model the neuronal damage elicited by a fragment of PrP is dependent on PrP^C and on microglial cells (Brown *et al.*, 1996). It will be interesting to learn whether transgenic mice expressing PrP exclusively in microglial cells are susceptible to scrapie, and whether lineage ablation of microglia may protect against some of the histopathological signs of scrapie.

Prion replication in cells of the central nervous system seems to be the cardinal event in scrapie pathogenesis. Although accumulation of the infectious agent in lymphoid organs always precedes invasion of the brain (Eklund *et al.*, 1967; Fraser and Dickinson, 1970), the lymphoreticular system (LRS) is not essential for the development of disease after intracerebral inoculation. However, peripheral uptake of prions is epidemiologically more relevant than intracerebral administration of prions. In particular, BSE and vCJD, for which a common agent has been demonstrated (Bruce *et al.*, 1997; Hill *et al.*, 1997a), but also scrapie, are due to oral transmission of prions. Which cell types in the lymphoreticular system are targets for the scrapie agent is not known. Involvement of follicular dendritic cells has been postulated, based on immunohistochemical detection of PrP^{Sc} in these cells (Kitamoto *et al.*, 1991).

Transgenic *Prnp*$^{o/o}$ mice harboring a PrP transgene, driven by the heterologous IRF1-promoter/Eμ-enhancer, expressed high levels of

PrP on B and T lymphocytes and only low levels in brain. Following intraperitoneal inoculation with mouse scrapie, these mice propagated prions early on in spleen and thymus at a level similar to that in wild-type mice, but infectivity was below detectability in the brain at six months postinoculation (Raeber *et al.*, 1999b). Transgenic *Prnp$^{o/o}$* mice containing a PrP transgene controlled by the *lck* promoter overexpressed PrP on T cells about 100-fold as compared to wild-type T cells but lacked PrP expression on B cells. Surprisingly, these mice were resistant to scrapie, following intraperitoneal inoculation with mouse prions, and were unable to propagate prions in thymus, spleen, and brain (Raeber *et al.*, 1999b). These findings show that PrP expression is not sufficient for prion replication in T cells, and that perhaps cell-specific factors such as a prion receptor, a chaperone, or a protein X are required (Telling *et al.*, 1995).

VIII. Prions and the Central Nervous System

A. Neurografts in Prion Research

Because *Prnp$^{o/o}$* mice show normal development and behavior (Büeler *et al.*, 1992; Manson *et al.*, 1994a), it has been argued that scrapie pathology may come about because PrP^{Sc} deposition is neurotoxic (Forloni *et al.*, 1993), rather than being caused by depletion of cellular PrP^{C} (Collinge *et al.*, 1994). In the latter case, lack of PrP^{C} might result in embryonic or perinatal lethality, especially because PrP^{C} is encoded by a unique gene for which no related family members have been found. However, acute depletion of PrP^{C} may be much more deleterious than its lack throughout development because the organism may then not have the time to enable compensatory mechanisms.

To address the question of neurotoxicity, we exposed brain tissue of *Prnp$^{o/o}$* mice to a continuous source of PrP^{Sc}. To this end, we grafted neural tissue overexpressing PrP into the brain of PrP-deficient mice by using protocols established earlier (Isenmann *et al.*, 1996a; Isenmann *et al.*, 1996b; Isenmann *et al.*, 1996c). After intracerebral inoculation with scrapie prions, grafts accumulated high levels of PrP^{Sc} and infectivity and developed severe histopathological changes characteristic of scrapie. Substantial amounts of graft-derived PrP^{Sc} migrated into the host brain, and even in areas distant from the grafts, substantial amounts of infectivity were detected (Brandner *et al.*, 1996a; Fischer *et al.*, 1996). Nonetheless, even 16 months after transplantation and infection with prions, no pathological changes were detected in the

PrP-deficient tissue, not even in the immediate vicinity of the grafts or the PrP deposits. These results suggest that PrP^{Sc} is inherently non-toxic and that PrP^{Sc} plaques found in spongiform encephalopathies may be an epiphenomenon rather than a cause of neuronal damage (Aguzzi, 1998a). It is conceivable that PrP^{Sc} is only toxic when it is formed and accumulated within the cell, but not when it is presented from outside. On the other hand, it was reported that synthetic amyloidogenic fragments corresponding to the central domain of the prion protein can evoke cellular responses when they are exposed to microglia *in vitro* (Combs *et al.,* 1999).

Because the host mice harboring a chronically scrapie-infected neural graft did not develop any sign of disease, they not only enabled us to study the effects of prions on the surrounding tissue but were also an ideal medium to assess all changes occurring during the progression of scrapie disease in neuroectodermal tissue. With increasing length of the incubation time, grafts underwent progressive astrogliosis and spongiosis that were accompanied by loss of neuronal processes within the grafts and subsequent destruction of the neuropil (Fig. 4, see also color section). The terminal stage of the disease (435 days after inoculation) was characterized by an increase in cellular density in the graft, probably due to astroglial proliferation and a complete loss of neurons. Intriguingly, by *in vivo* imaging with magnetic resonance imaging, using gadolinium as a contrast enhancing medium, a progressive disruption of the blood–brain barrier in scrapie-infected grafts was detected during the course of the disease (Brandner *et al.,* 1998). These findings confirmed several predictions about the pathogenesis of spongiform encephalopathies, mainly that scrapie leads to selective neuronal loss, while astrocytes and perhaps other neuroectodermal cells, although being affected by the disease, can survive and maintain their phenotypic characteristics for very long periods of time.

Disruption of blood–brain barrier function is a finding that has been reported for experimental hamster scrapie (Chung *et al.,* 1995), but it was not found in human spongiform encephalopathies. The localized blood–brain barrier disruption in chronically infected grafts might contribute to the spread of prions from grafts to the surrounding brain, as described previously (Brandner *et al.,* 1996a). It may also account for the accumulation pattern of protease-resistant PrP within the white matter and in brain areas surrounding the grafts. Another explanation might be that vasogenic diffusion from the affected graft toward the host brain is a mechanism contributing to prion spread within the central nervous system, if the blood–brain barrier is impaired due to astrocytic activation and/or damage during prion disease.

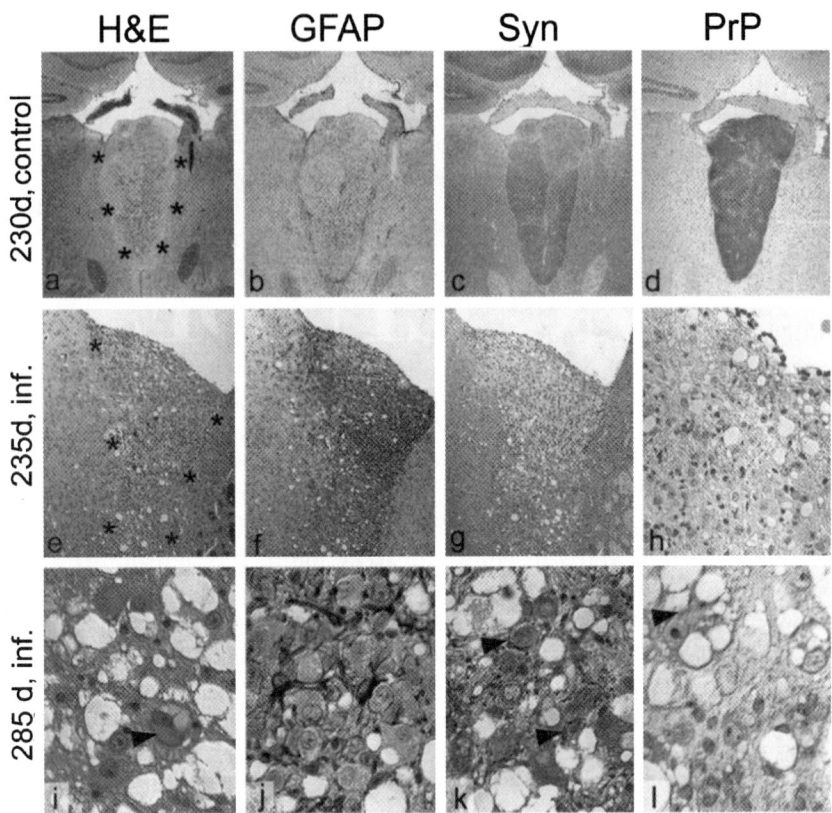

FIG 4. [For color reproduction, see color section.] Noninfected and scrapie-infected neural grafts in brains of *Prnp⁰ᐟ⁰* mice. *Upper row* (a, b, c, d): healthy control graft 230 days after mock inoculation. The graft is located in the third ventricle of the recipient mouse (a, see asterisks, hematoxylin–eosin), and shows no spongiform change; little gliosis (b: immunostain for GFAP); and strong expression of synaptophysin (c) and of PrP^C (d). *Middle row* (e, f, g, h): scrapie-infected graft 235 days after inoculation with increased cellularity (e), brisk gliosis (f); and a significant loss of synaptophysin (g) and of PrP (h) staining intensity is shown. *Bottom row:* high magnification of a similar graft shows characteristic pathological changes in a chronically infected graft: (i) appearance of large vacuoles and ballooned neurons (arrow). In the GFAP immunostain (j), astrocytes appear wrapped around densely packed neurons. Granular deposits and intracytoplasmic accumulation of synaptophysin (k) and PrP immunoreactivity (l) in the cytoplasm of neurons.

B. Spread of Prions in the Central Nervous System

Intracerebral inoculation of scrapie-infected brain homogenate into suitable recipients is the most effective method for transmission of spongiform encephalopathies and may even facilitate circumvention of the species barrier. However, prion diseases can also be initiated by feeding (Anderson et al., 1996; Kimberlin and Wilesmith, 1994; Wells et al., 1987), by intravenous and intraperitoneal injection (Kimberlin and Walker, 1978b), as well as from the eye by conjunctival instillation (Scott et al., 1993a), corneal grafts (Duffy et al., 1974), and intraocular injection (Fraser, 1982). The latter method has proved particularly useful for studying neural spread of the agent, since the retina is a part of the central nervous system (CNS), and since intraocular injection does not produce direct physical trauma to the brain, which might disrupt the blood–brain barrier and impair other aspects of brain physiology. The assumption that spread of prions occurs axonally rests mainly on the demonstration of diachronic spongiform changes along the retinal pathway, following intraocular infection (Fraser, 1982).

It has been repeatedly shown that expression of PrP^C is required for prion replication (Büeler et al., 1993; Sailer et al., 1994) and also for neurodegenerative changes to occur (Brandner et al., 1996a). To investigate whether spread of prions within the CNS is dependent on PrP^C expression in the visual pathway, PrP-producing neural grafts were used as sensitive indicators of the presence of prion infectivity in the brain of an otherwise PrP-less host.

Following prion inoculation into the eye of grafted $Prnp^{o/o}$ mice, none of the grafts showed signs of spongiosis, gliosis, synaptic loss, or PrP^{Sc}. In one instance, the graft of an intraocularly inoculated mouse was assayed and found to be devoid of infectivity. Therefore, it was concluded that infectivity administered to the eye of PrP-deficient hosts cannot induce scrapie in a PrP-expressing brain graft (Brandner et al., 1996b).

Engraftment of $Prnp^{o/o}$ mice with PrP^C-producing tissue might lead to an immune response to PrP (Prusiner et al., 1993) and possibly to neutralization of infectivity. Indeed, analysis of sera from grafted mice revealed significant anti-PrP antibody titers (Brandner et al., 1996b), and it was shown that PrP^C presented by the intracerebral graft (rather than the inoculum or graft-borne PrP^{Sc}) was the offending antigen. In order to definitively rule out the possibility that prion transport was disabled by a neutralizing immune response, the experiments were repeated in mice tolerant to PrP—namely, the $Prnp^{o/o}$ mice transgenic for the PrP coding sequence under the control of the lck-promoter described above. These mice overexpressed PrP on T lymphocytes, but were resistant to scrapie and did not replicate prions in brain, spleen, and thymus after intraperitoneal inoculation with scrapie prions (Raeber et al., 1999b). Engraftment of these mice with PrP-overexpressing neuroectoderm did not lead to the development of antibodies to PrP after intracerebral or intraocular inoculation, presumably due to clonal deletion of PrP-immunoreactive lymphocytes. As described before, intraocular inoculation with prions did not provoke scrapie in the graft, which supported the conclusion that lack of PrP^C, rather than an immune response to PrP, prevented prion spread (Brandner et al., 1996b). Therefore, PrP^C appears to be necessary for the spread of prions along the retinal projections and within the intact CNS.

These results indicate that intracerebral spread of prions is based on a PrP^C-paved chain of cells, perhaps because they are capable of supporting prion replication. When such a chain is interrupted by interposed cells that lack PrP^C, as in the case described here, no propagation of prions to the target tissue can occur. Perhaps prions require PrP^C for propagation across synapses—PrP^C is present in the synaptic region (Fournier et al., 1995), and certain synaptic properties are altered in Prnp$^{o/o}$ mice (Collinge et al., 1994; Whittington et al., 1995). Perhaps transport of prions within (or on the surface of) neuronal processes is PrP^C-dependent. Within the framework of the protein-only hypothesis (Griffith, 1967; Prusiner, 1989), these findings may be accommodated by a "domino-stone" model in which spreading of scrapie prions in the CNS occurs per continuitatem through conversion of PrP^C by adjacent PrP^{Sc} (Aguzzi, 1997).

C. Spread of Prions from Extracerebral Sites to the CNS

Oral uptake of prions may be epidemiologically more relevant than intracerebral transmission, because it is thought to be responsible for the BSE epidemic and for transmission of BSE to a variety of species including humans (Bruce et al., 1997; Hill et al., 1997a). Prions can find their way through the body to the brain of their host, yet histopathological changes have not been identified in organs other than the CNS. But the prions may multiply silently in "reservoirs" during the incubation phase of the disease. One such reservoir may be the immune system, and many studies point to the importance of the prion replication in lymphoid organs that always precedes prion replication in the CNS, even if infectivity is administered intracerebrally (Eklund et al., 1967). Infectivity can accumulate in all components of the lymphoreticular system (LRS), including lymph nodes and intestinal Peyer's patches, where prions replicate almost immediately after oral administration of prions to mice (Kimberlin and Walker, 1989a). Recently, it was shown that vCJD prions accumulate in the lymphoid tissue of tonsils in such large amounts that PrP^{Sc} can easily be detected with antibodies on histological sections (Hill et al., 1997c).

Although a wealth of early studies point to the importance of prion replication in lymphoid organs, little is known about which cells support prion propagation in the lymphoreticular system. Whole-body ionizing radiation studies in mice, after intraperitoneal infection, have suggested that the critical cells are long-lived (Fraser and Farquhar, 1987). The follicular dendritic cell (FDC) would be a prime candidate and, indeed, PrP^{Sc} accumulates in such cells of wild-type and nude mice (which have a selective T-cell defect) (Kitamoto et al., 1991). Moreover, intraperitoneal infection does not lead to replication of prions in the spleen, nor to cerebral scrapie in mice with severe combined immunodeficiency (SCID) whose FDCs are thought to be functionally impaired (Muramoto et al., 1993). Reconstitution of SCID mice with wild-type spleen cells restores susceptibility to scrapie after peripheral infection (Lasmezas et al., 1996). These findings suggest that components of the immune system are required for efficient transfer of prions from the site of peripheral infection to the CNS.

Using a panel of immune-deficient mice inoculated intraperitoneally with prions, we found that defects affecting T cells had no apparent effect, but that all mutations that disrupted the differentiation of B cells prevented the development of clinical scrapie (Klein et al., 1997) (Fig. 5). These results argue for a crucial role of B cells in the development of scrapie after peripheral infection. But do B cells suffice for transporting prions all the way from the periphery to the CNS? This is unlikely, because lymphocytes do not normally cross the blood–brain barrier unless they have a specific reason to do so (e.g., during an inflammatory reaction). Moreover, up to 30% of B cell-deficient mice contain prions in their brains, despite no signs of clinical disease (Frigg et al., 1999). How then might prions spread in the body? Perhaps, prions administered to peripheral sites are first brought to lymphatic organs by mobile immune cells such as B cells. Once infection has been established in the LRS, prions invade peripheral nerve endings (Groschup et al., 1996; Kimberlin and Walker, 1989b) and reach the CNS, where further spread occurs transsynaptically and along fiber tracts (Kimberlin and Walker, 1980; Kimberlin and Walker, 1986). Is it possible to interfere with this chain of events without resorting to ablation of a functional immune system, such as in the case of SCID mice? Again, PrP^C may offer an intriguing handle. PrP^C is crucial for prion spread within the CNS (Brandner et al., 1996b), and it is conceivable that PrP^C is also required for spread of prions from peripheral sites to the CNS. Indeed, PrP-expressing neurografts in Prnp$^{o/o}$ mice did not develop scrapie histopathology after intraperitoneal or intravenous inoculation with prions, and no infectivity was detectable in the spleen. Following reconstitution of the host lymphohemopoietic system with PrP-expressing cells, prion titers in the spleen were restored to wild-type levels but, surprisingly, PrP-express-

a

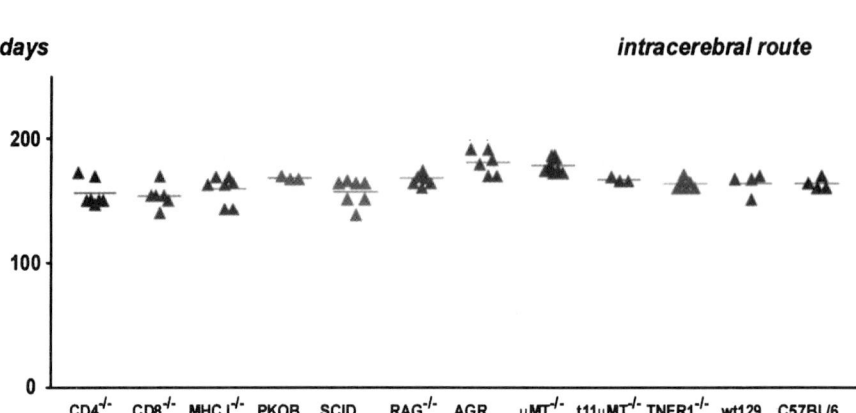

b

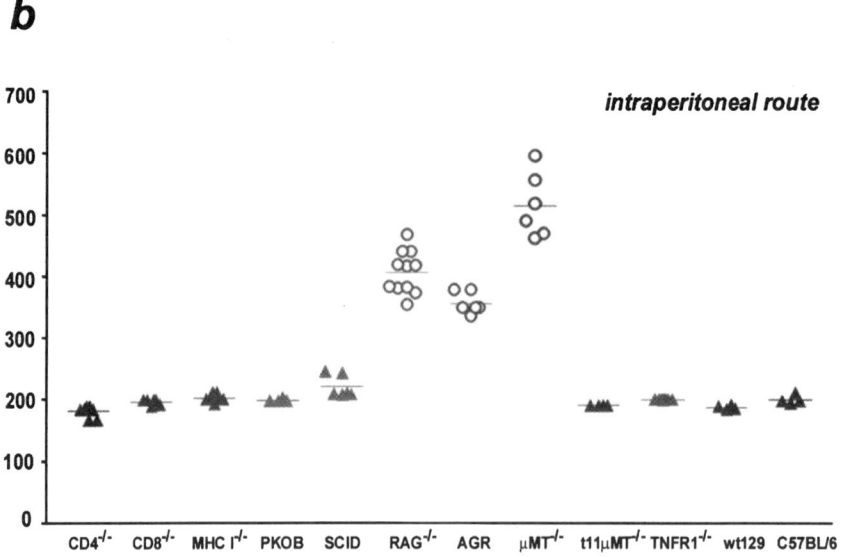

FIG 5. Latency of scrapie in various strains of immunodeficient mice. All mice developed spongiform encephalopathy after intracerebral inoculation (Fig. 1a; closed triangles), regardless of their immune status. In contrast, B-cell deficient mice remained healthy after intraperitoneal inoculation of RML scrapie prions with 100µl of a 10^{-1} or of a 10^{-4} dilution of a scrapie-infected brain homogenate. (Fig. 1b; open triangles). The horizontal lines represent mean values.

ing grafts failed to develop scrapie on intraperitoneal or intravenous infection with prions (Blättler *et al.*, 1997). These findings suggest that transfer of infectivity from the spleen to the CNS is crucially dependent on the expression of PrP in a tissue compartment, interposed between the LRS and the CNS, that cannot be reconstituted by bone marrow transfer. Indirect evidence suggests that this compartment may comprise part of the peripheral nervous system (Beekes *et al.*, 1996; Kimberlin *et al.*, 1983).

D. PrP^C, B Lymphocytes, and Neuroinvasion

Because the replication of prions (Büeler *et al.*, 1993) and their transport from the periphery to the CNS (Blättler *et al.*, 1997) are dependent on expression of PrP^C, we examined whether expression of PrP^C by B cells was necessary to support neuroinvasion. Mice with various immune defects were repopulated by adoptive transfer of hematopoietic stem cells, which expressed, or lacked expression of, PrP^C.

Adoptive transfer of either $Prnp^{+/+}$ or $Prnp^{o/o}$ fetal liver cells (FLCs) induced formation of germinal centers in spleens of recipient mice and differentiation of FDCs, as visualized by staining with antibody FDC-M1 (Klein *et al.*, 1998). However, no FDCs were found in B and T cell-deficient mice reconstituted with FLCs from μMT embryos (B cell deficient), consistent with the notion that B cells or products thereof are required for FDC maturation.

Reconstituted mice were challenged intraperitoneally with scrapie prions. Surprisingly, all mice that received FLCs of either genotype, $Prnp^{+/+}$ or $Prnp^{o/o}$, from immunocompetent donors, succumbed to scrapie after inoculation with a high dose of prions, and most mice succumbed after a low dose (Fig. 6). Transfer of FLCs from μMT donors, as well as omission of the adoptive transfer procedure, did not restore susceptibility to disease in any of the immune-deficient mice challenged with the low dose of prions. We also confirmed that by using a high dose of inoculum, susceptibility to scrapie could be restored even in the absence of B cells and FDCs. However, reconstituted mice that received bone marrow from TCRα^{-/-} donors, which possess B cells and lack all T cells, but those that express TCRγδ receptors, regained susceptibility to scrapie, again confirming the dependency of infectibility on the presence of B cells (Fig. 6). By transmitting individual samples of brain and spleen from the scrapie-inoculated bone marrow chimeras, we observed restoration of infectious titers and PrP^Sc deposition in spleens and brains of recipient mice carrying either $Prnp^{+/+}$or $Prnp^{o/o}$ donor cells (Klein *et al.*, 1998).

Development of scrapie in bone marrow reconstituted mice

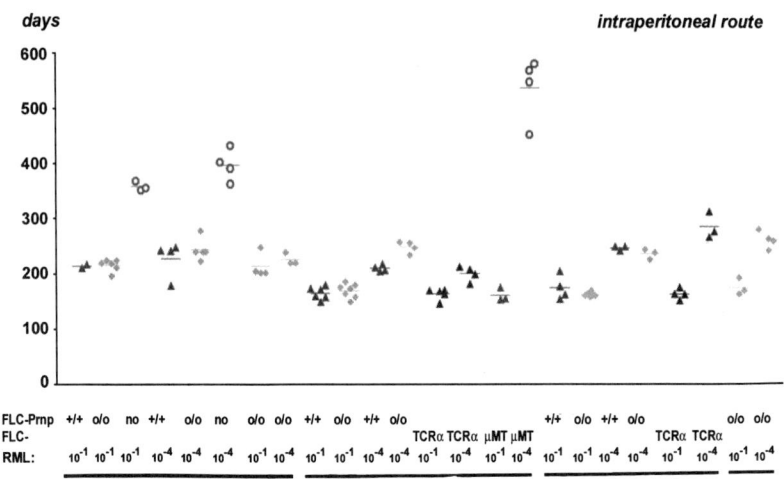

FIG 6. Latency of clinical scrapie on intraperitoneal inoculation of RML prions in various bone marrow-reconstituted mice. Horizontal bars: average incubation time. Transfer of fetal liver cells derived from $Prnp^{+/+}$ or from $Prnp^{o/o}$ mice (closed triangles), but not from μMT fetal liver cells (open triangles), restored infectibility of immune-deficient mice on intraperitoneal inoculation with 3–4 $\log LD_{50}$ scrapie prions. An analogous trend was seen when 7–8 $\log LD_{50}$ were inoculated, although resistance to CNS disease of immunodeficient mice was often overcome by this high dose.

Although B cells are clearly a cofactor in peripheral prion pathogenesis, the identity of those cells in which prions actually replicate within lymphatic organs is uncertain. In a further step to clarify this issue, we investigated whether spleen PrP^{Sc} was associated with FDCs in repopulated mice. Double-color immunofluorescence confocal microscopy revealed deposits of PrP-immunoreactive material in germinal centers which appeared largely co-localized with the follicular dendritic network in spleens of reconstituted mice (Fig. 7, also see color section).

Collectively, these findings are compatible with the hypothesis that cells whose maturation depends on B cells are responsible for accumulation of prions in lymphoid tissue such as the spleen. FDCs, although their origin remains rather obscure, are a likely candidate for the site of prion replication because their maturation correlates with the presence of B cells and their products. However, it is still possible that the follicular dendritic network serves merely as a reservoir for the accumulation of prions and that other B cell-dependent processes are involved in the transport of the infectious agent. Prions may be transported on or within B cells directly as they cross peripheral lymphoid tissue and

RAG-1⁻/⁻ with FLC

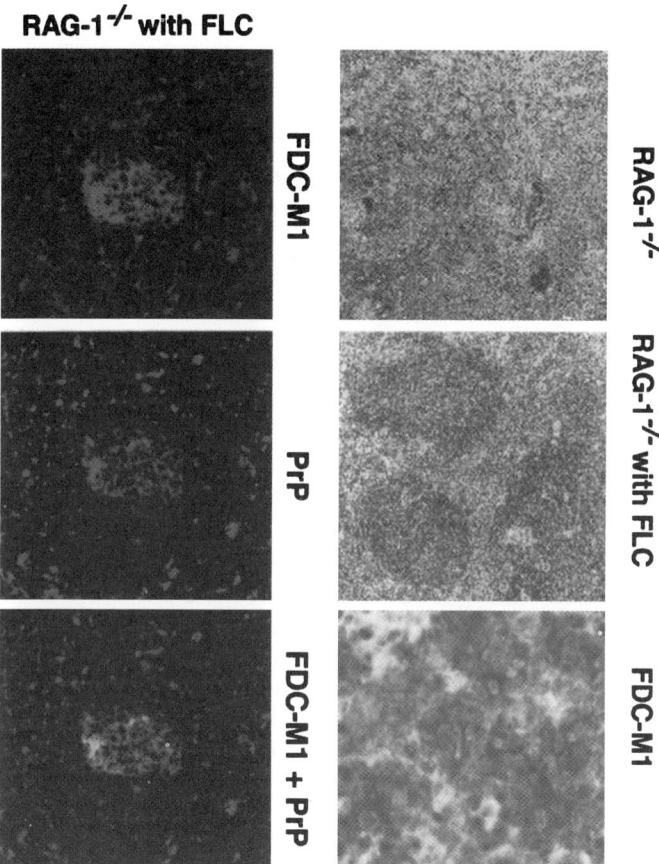

FIG 7. [For color reproduction, see color section.] Spleen histology of Rag-1⁻/⁻ mice reconstituted with *Prnp*ᵒ/ᵒ fetal liver cells. *Upper row:* Paraffin sections stained with hemalaun before FLC transfer (left). No B cell follicles and germinal centers were discernible in *Rag-1*⁻/⁻ mice. Restoration of organized B cell follicles and germinal centers after FLC reconstitution (middle). Magnification: ×200. Frozen sections immunostained with antibody FDC-M1 revealed formation of prominent FDC clusters within germinal centers after FLC transfer (right). Magnification: ×250. *Lower row:* Confocal double-color immunofluorescence analysis of splenic germinal centers in *Rag-1*⁻/⁻ mice reconstituted with *Prnp*ᵒ/ᵒ FLCs after inoculation with RML prions. Sections were stained with antibody FDC-M1 to follicular dendritic cells (green, left) and with antiserum R340 to PrP (red, middle). Regions in which both signals are detectable appear yellow in superimposed images (right). Magnification: ×250. Most of the PrP signal in germinal centers appeared to colocalize with the FDC-network.

localize in autonomic nerve terminals. Indeed, recent investigations have demonstrated that prion infectivity is mainly associated with B and T lymphocytes and less with a stromal fraction containing FDCs (Raeber *et al.*, 1999a). Alternatively, antibodies or other B cell factors may bind prions and fulfill this role. This is particularly likely because PrP^{Sc} can be detected by immunohistochemistry in the germinal center area of lymphatic organs, where immune complexes are deposited.

IX. CONCLUSION

Peripheral prion pathogenesis and ultimately neuroinvasion are dependent on components of the host immune system. Collectively, these processes require either B cells per se or their products. At least one B cell-dependent event is the acquisition of a functional FDC network within the germinal centers of peripheral lymphoid tissue. These cells are the major sites of extraneuronal PrP^C expression and probably the principal sites of PrP^{Sc} accumulation. The mechanism by which prions accumulate within lymphoid tissue remains to be established. An attractive hypothesis is that prions bind to antibodies that localize onto the surface of FDCs as prion–antibody complexes in a manner analogous to the normal function of the FDC.

The second phase of neuroinvasion appears to be the progression of prions from lymphoid tissue to nerve endings of the sympathetic nervous system. This may occur via the direct transport of prions into peripheral lymphoid tissue or from reservoirs of infectivity associated with FDCs, although no PrP^{Sc} has been detected so far in the autonomic peripheral nervous system. It is worthwhile noting that the innervation of lymphoid tissue is at least in part controlled by lymphocytes themselves as both T and B cells secrete nerve growth factor and, vice versa, nerve terminals secrete a variety of factors to stimulate the immune system (Straub *et al.*, 1998). These factors may play a critical role in the neuroinvasion process and represent a critical site for modulation of disease progression. For example, drugs that act on lymphocytes or at the synaptic innervation of lymphoid tissue, or those that prevent cytokine release or block neurotransmission, may have a strong influence on the immune modulation and might represent useful tools for studying the cellular and molecular basis of prion neuroinvasion. A thorough understanding of the role of the immune system in peripheral prion pathogenesis is of immediate importance in assessing the risk of iatrogenic transmission of prions via exposure to blood or tissues from individuals suffering from preclinical prion disease.

ACKNOWLEDGMENTS

We thank Marianne König, Petra Schwarz, and Norbert Wey for technical help; Dr. C. Weissmann for support; Dr. K. Rajewsky for μMT mice; and R. Chase for administrative help. This work is supported by the Canton of Zurich, the Bundesämter für Gesundheit, Veterinärwesen, Bildung, und Wissenschaft, and by grants from the Swiss National Research Program NFP38/NFP38+, and from the Abbott and Baxter. FLH is a fellow of the Human Frontiers Science Foundation, and BN is an EMBO fellow. MP and MAK were supported by the Deutsche Forschungsgemeinschaft.

REFERENCES

Aguzzi, A. (1996). Between cows and monkeys. *Nature* **381**:734.

Aguzzi, A. (1997). Neuro-immune connection in spread of prions in the body? *The Lancet* **349**:742–743.

Aguzzi, A. (1998a). Grafting mouse brains: from neurocarcinogenesis to neurodegeneration. *Embo J* **17**:6107–6114.

Aguzzi, A. (1998b). Protein conformation dictates prion strain. *Nat Med* **4**:1125–1126.

Aguzzi, A., and Brandner, S. (1999). The genetics of prions—a contradiction in terms? *Lancet* **354 Suppl 1**:S122–S125.

Aguzzi, A., and Weissmann, C. (1996a). Sleepless in Bologna: transmission of fatal familial insomnia. *Trends Microbiol* **4**:129–131.

Aguzzi, A., and Weissmann, C. (1996b). Spongiform encephalopathies: a suspicious signature. *Nature* **383**:666–667.

Aguzzi, A., and Weissmann, C. (1997). Prion research: the next frontiers. *Nature* **389**:795–798.

Aguzzi, A., and Weissmann, C. (1998). Spongiform encephalopathies. The prion's perplexing persistence. *Nature* **392**:763–764.

Anderson, R. M., Donnelly, C. A., Ferguson, N. M., Woolhouse, M. E., Watt, C. J., Udy, H. J., MaWhinney, S., Dunstan, S. P., Southwood, T. R., Wilesmith, J. W., Ryan, J. B., Hoinville, L. J., Hillerton, J. E., Austin, A. R., and Wells, G. A. (1996). Transmission dynamics and epidemiology of BSE in British cattle. *Nature* **382**:779–788.

Basler, K., Oesch, B., Scott, M., Westaway, D., Walchli, M., Groth, D. F., McKinley, M. P., Prusiner, S. B., and Weissmann, C. (1986). Scrapie and cellular PrP isoforms are encoded by the same chromosomal gene. *Cell* **46**:417–428.

Beekes, M., Baldauf, E., and Diringer, H. (1996). Sequential appearance and accumulation of pathognomonic markers in the central nervous system of hamsters orally infected with scrapie. *J. Gen. Virol.* **77**:1925–1934.

Bendheim, P. E., Brown, H. R., Rudelli, R. D., Scala, L. J., Goller, N. L., Wen, G. Y., Kascsak, R. J., Cashman, N. R., and Bolton, D. C. (1992). Nearly ubiquitous tissue distribution of the scrapie agent precursor protein. *Neurology* **42**:149–156.

Bessen, R. A., and Marsh, R. F. (1994). Distinct PrP properties suggest the molecular basis of strain variation in transmissible mink encephalopathy. *J. Virol.* **68**:7859–7868.

Blättler, T., Brandner, S., Raeber, A. J., Klein, M. A., Voigtländer, T., Weissmann, C., and Aguzzi, A. (1997). PrP-expressing tissue required for transfer of scrapie infectivity from spleen to brain. *Nature* **389**:69–73.

Bolton, D. C., McKinley, M. P., and Prusiner, S. B. (1982). Identification of a protein that purifies with the scrapie prion. *Science* **218**:1309–1311.

Brandner, S., Isenmann, S., Kuhne, G., and Aguzzi, A. (1998). Identification of the end stage of scrapie using infected neural grafts. *Brain Pathol.* **8**:19–27.

Brandner, S., Isenmann, S., Racber, A., Fischer, M., Sailer, A., Kobayashi, Y., Marino, S., Weissmann, C., and Aguzzi, A. (1996a). Normal host prion protein necessary for scrapie-induced neurotoxicity. *Nature* **379**:339–343.

Brandner, S., Raeber, A., Sailer, A., Blattler, T., Fischer, M., Weissmann, C., and Aguzzi, A. (1996b). Normal host prion protein (PrPC) is required for scrapie spread within the central nervous system. *Proc. Natl. Acad. Sci. U.S.A.* **93**:13148–13151.

Brown, D. R., Qin, K., Herms, J. W., Madlung, A., Manson, J., Strome, R., Fraser, P. E., Kruck, T., von Bohlen, A., Schulz-Schaeffer, W., Giese, A., Westaway, D., and Kretzschmar, H. (1997). The cellular prion protein binds copper in vivo. *Nature* **390**:684–687.

Brown, D. R., Schmidt, B., and Kretzschmar, H. A. (1996). Role of microglia and host prion protein in neurotoxicity of a prion protein fragment. *Nature* **380**:345–347.

Bruce, M., Chree, A., McConnell, I., Foster, J., Pearson, G., and Fraser, H. (1994). Transmission of bovine spongiform encephalopathy and scrapie to mice: strain variation and the species barrier. *Philos. Trans. R. Soc. Lond. B. Biol. Sci.* **343**:405–411.

Bruce, M. E., and Dickinson, A. G. (1987). Biological evidence that scrapie agent has an independent genome. *J. Gen. Virol.* **68**:79–89.

Bruce, M. E., Will, R. G., Ironside, J. W., McConnell, I., Drummond, D., Suttie, A., McCardle, L., Chree, A., Hope, J., Birkett, C., Cousens, S., Fraser, H., and Bostock, C. J. (1997). Transmissions to mice indicate that `new variant' CJD is caused by the BSE agent [see comments]. *Nature* **389**:498–501.

Büeler, H., Raeber, A., Sailer, A., Fischer, M., Aguzzi, A., and Weissmann, C. (1994). High prion and PrPSc levels but delayed onset of disease in scrapie-inoculated mice heterozygous for a disrupted PrP gene. *Mol. Med.* **1**:19–30.

Büeler, H. R., Aguzzi, A., Sailer, A., Greiner, R. A., Autenried, P., Aguet, M., and Weissmann, C. (1993). Mice devoid of PrP are resistant to scrapie. *Cell* **73**:1339–1347.

Büeler, H. R., Fischer, M., Lang, Y., Bluethmann, H., Lipp, H. P., DeArmond, S. J., Prusiner, S. B., Aguet, M., and Weissmann, C. (1992). Normal development and behaviour of mice lacking the neuronal cell-surface PrP protein. *Nature* **356**:577–582.

Campbell, I. L., Eddleston, M., Kemper, P., Oldstone, M. B., and Hobbs, M. V. (1994). Activation of cerebral cytokine gene expression and its correlation with onset of reactive astrocyte and acute-phase response gene expression in scrapie. *J. Virol.* **68**:2383–2387.

Carlson, G. A., Ebeling, C., Yang, S. L., Telling, G., Torchia, M., Groth, D., Westaway, D., DeArmond, S. J., and Prusiner, S. B. (1994). Prion isolate specified allotypic interactions between the cellular and scrapie prion proteins in congenic and transgenic mice. *Proc. Natl. Acad. Sci. U.S.A.* **91**:5690–5694.

Carlson, G. A., Kingsbury, D. T., Goodman, P. A., Coleman, S., Marshall, S. T., DeArmond, S., Westaway, D., and Prusiner, S. B. (1986). Linkage of prion protein and scrapie incubation time genes. *Cell* **46**:503–511.

Cashman, N. R., Loertscher, R., Nalbantoglu, J., Shaw, I., Kascsak, R. J., Bolton, D. C., and Bendheim, P. E. (1990). Cellular isoform of the scrapie agent protein participates in lymphocyte activation. *Cell* **61**:185–192.

Caughey, B., Raymond, G. J., Kocisko, D. A., and Lansbury, P. T., Jr. (1997). Scrapie infectivity correlates with converting activity, protease resistance, and aggregation of scrapie-associated prion protein in guanidine denaturation studies. *J. Virol.* **71**:4107–4110.

Chabry, J., Caughey, B., and Chesebro, B. (1998). Specific inhibition of in vitro formation of protease-resistant prion protein by synthetic peptides. *J. Biol. Chem.* **273**:13203–13207.

Chazot, G., Broussolle, E., Lapras, C., Blattler, T., Aguzzi, A., and Kopp, N. (1996). New variant of Creutzfeldt-Jakob disease in a 26-year-old French man [letter]. *Lancet* **347:**1181.

Chesebro, B., Race, R., Wehrly, K., Nishio, J., Bloom, M., Lechner, D., Bergstrom, S., Robbins, K., Mayer, L., Keith, J. M., and *et al.* (1985). Identification of scrapie prion protein-specific mRNA in scrapie-infected and uninfected brain. *Nature* **315:**331–333.

Chiesa, R., Piccardo, P., Ghetti, B., and Harris, D. A. (1998). Neurological illness in transgenic mice expressing a prion protein with an insertional mutation. *Neuron* **21:**1339–1351.

Chung, Y. L., Williams, A., Beech, J. S., Williams, S. C., Bell, J. D., Cox, I. J., and Hope, J. (1995). MRI assessment of the blood–brain barrier in a hamster model of scrapie. *Neurodegeneration* **4:**203–207.

Collee, J. G., and Bradley, R. (1997). BSE: a decade on—Part I. *Lancet* **349:**636–641.

Collinge, J., Sidle, K. C., Meads, J., Ironside, J., and Hill, A. F. (1996). Molecular analysis of prion strain variation and the aetiology of `new variant' CJD. *Nature* **383:**685–690.

Collinge, J., Whittington, M. A., Sidle, K. C., Smith, C. J., Palmer, M. S., Clarke, A. R., and Jefferys, J. G. (1994). Prion protein is necessary for normal synaptic function. *Nature* **370:**295–297.

Combs, C. K., Johnson, D. E., Cannady, S. B., Lehman, T. M., and Landreth, G. E. (1999). Identification of microglial signal transduction pathways mediating a neurotoxic response to amyloidogenic fragments of beta-amyloid and prion proteins. *J. Neurosci.* **19:**928–939.

DeArmond, S. J., Sanchez, H., Yehiely, F., Qiu, Y., Ninchak-Casey, A., Daggett, V., Camerino, A. P., Cayetano, J., Rogers, M., Groth, D., Torchia, M., Tremblay, P., Scott, M. R., Cohen, F. E., and Prusiner, S. B. (1997). Selective neuronal targeting in prion disease. *Neuron* **19:**1337–1348.

Dickinson, A. G., and Meikle, V. M. (1971). Host-genotype and agent effects in scrapie incubation: change in allelic interaction with different strains of agent. *Mol. Gen. Genet.* **112:**73–79.

Dickinson, A. G., Meikle, V. M., and Fraser, H. (1968). Identification of a gene which controls the incubation period of some strains of scrapie agent in mice. *J. Comp. Pathol.* **78:**293–299.

Dickinson, A. G., and Outram, G. W. (1988). Genetic aspects of unconventional virus infections: the basis of the virino hypothesis. *Ciba. Found Symp.* **135:**63–83.

Diedrich, J. F., Bendheim, P. E., Kim, Y. S., Carp, R. I., and Haase, A. T. (1991). Scrapie-associated prion protein accumulates in astrocytes during scrapie infection. *Proc. Natl. Acad. Sci. U.S.A.* **88:**375–379.

Donne, D. G., Viles, J. H., Groth, D., Mehlhorn, I., James, T. L., Cohen, F. E., Prusiner, S. B., Wright, P. E., and Dyson, H. J. (1997). Structure of the recombinant full-length hamster prion protein PrP(29–231): the N terminus is highly flexible. *Proc. Natl. Acad. Sci. U.S.A.* **94:**13452–13457.

Duffy, P., Wolf, J., Collins, G., DeVoe, A. G., Streeten, B., and Cowen, D. (1974). Possible person-to-person transmission of Creutzfeldt-Jakob disease. *N. Engl. J. Med.* **290:**692–693.

Eklund, C. M., Kennedy, R. C., and Hadlow, W. J. (1967). Pathogenesis of scrapie virus infection in the mouse. *J. Infect. Dis.* **117:**15–22.

Fischer, M., Rülicke, T., Raeber, A., Sailer, A., Moser, M., Oesch, B., Brandner, S., Aguzzi, A., and Weissmann, C. (1996). Prion protein (PrP) with amino-proximal deletions restoring susceptibility of PrP knockout mice to scrapie. *EMBO J.* **15:**1255–1264.

Forloni, G., Angeretti, N., Chiesa, R., Monzani, E., Salmona, M., Bugiani, O., and Tagliavini, F. (1993). Neurotoxicity of a prion protein fragment. *Nature* **362:**543–546.

Fournier, J. G., Escaig Haye, F., Billette de Villemeur, T., and Robain, O. (1995). Ultrastructural localization of cellular prion protein (PrPc) in synaptic boutons of normal hamster hippocampus. *C. R. Acad. Sci. III* **318**:339–344.

Fraser, H. (1982). Neuronal spread of scrapie agent and targeting of lesions within the retino-tectal pathway. *Nature* **295**:149–150.

Fraser, H., and Dickinson, A. G. (1970). Pathogenesis of scrapie in the mouse: the role of the spleen. *Nature* **226**:462–463.

Fraser, H., and Farquhar, C. F. (1987). Ionising radiation has no influence on scrapie incubation period in mice. *Vet. Microbiol.* **13**:211–223.

Frigg, R., Klein, M. A., Hegyi, I., Zinkernagel, R. M., and Aguzzi, A. (1999). Scrapie pathogenesis in subclinically infected B cell deficient mice. *J. Virol.*

Gajdusek, D. C. (1988). Transmissible and non-transmissible amyloidoses: autocatalytic posttranslational conversion of host precursor proteins to beta-pleated sheet configurations. *J. Neuroimmunol.* **20**:95–110.

Goldfarb, L. G., Brown, P., McCombie, W. R., Goldgaber, D., Swergold, G. D., Wills, P. R., Cervenakova, L., Baron, H., Gibbs, C. J., Jr., and Gajdusek, D. C. (1991). Transmissible familial Creutzfeldt-Jakob disease associated with five, seven, and eight extra octapeptide coding repeats in the PRNP gene. *Proc. Natl. Acad. Sci. U.S.A.* **88**:10926–10930.

Goldgaber, D., Goldfarb, L. G., Brown, P., Asher, D. M., Brown, W. T., Lin, S., Teener, J. W., Feinstone, S. M., Rubenstein, R., Kascsak, R. J., and et al. (1989). Mutations in familial Creutzfeldt-Jakob disease and Gerstmann-Straussler-Scheinker's syndrome. *Exp. Neurol.* **106**:204–206.

Griffith, J. S. (1967). Self-replication and scrapie. *Nature* **215**:1043–1044.

Groschup, M. H., Weiland, F., Straub, O. C., and Pfaff, E. (1996). Detection of Scrapie Agent in the Peripheral Nervous System of a Diseased Sheep. *Neurobiology of Disease* **3**:191–195.

Herms, J. W., Kretzchmar, H. A., Titz, S., and Keller, B. U. (1995). Patch-clamp analysis of synaptic transmission to cerebellar purkinje cells of prion protein knockout mice. *Eur. J. Neurosci.* **7**:2508–2512.

Hill, A. F., Desbruslais, M., Joiner, S., Sidle, K. C., Gowland, I., Collinge, J., Doey, L. J., and Lantos, P. (1997a). The same prion strain causes vCJD and BSE [letter] [see comments]. *Nature* **389**:448–450.

Hill, A. F., Will, R. G., Ironside, J., and Collinge, J. (1997b). Type of prion protein in UK farmers with Creutzfeldt-Jakob disease [letter]. *Lancet* **350**:188.

Hill, A. F., Zeidler, M., Ironside, J., and Collinge, J. (1997c). Diagnosis of new variant Creutzfeldt-Jakob disease by tonsil biopsy. *Lancet* **349**:99.

Hope, J., Multhaup, G., Reekie, L. J., Kimberlin, R. H., and Beyreuther, K. (1988). Molecular pathology of scrapie-associated fibril protein (PrP) in mouse brain affected by the ME7 strain of scrapie. *Eur. J. Biochem.* **172**:271–277.

Horiuchi, M., and Caughey, B. (1999). Specific binding of normal prion protein to the scrapie form via a localized domain initiates its conversion to the protease-resistant state [in process citation]. *Embo. J.* **18**:3193–3203.

Hornshaw, M. P., McDermott, J. R., and Candy, J. M. (1995). Copper binding to the N-terminal tandem repeat regions of mammalian and avian prion protein. *Biochem. Biophys. Res. Commun.* **207**:621–629.

Hsiao, K., Baker, H. F., Crow, T. J., Poulter, M., Owen, F., Terwilliger, J. D., Westaway, D., Ott, J., and Prusiner, S. B. (1989). Linkage of a prion protein missense variant to Gerstmann-Straussler syndrome. *Nature* **338**:342–345.

Hsiao, K. K., Scott, M., Foster, D., Groth, D. F., DeArmond, S. J., and Prusiner, S. B. (1990). Spontaneous neurodegeneration in transgenic mice with mutant prion protein. *Science* **250:**1587–1590.

Hunter, N., Dann, J. C., Bennett, A. D., Somerville, R. A., McConnell, I., and Hope, J. (1992). Are Sinc and the PrP gene congruent? Evidence from PrP gene analysis in Sinc congenic mice. *J. Gen. Virol.* **73:**2751–2755.

Hunter, N., Hope, J., McConnell, I., and Dickinson, A. G. (1987). Linkage of the scrapie-associated fibril protein (PrP) gene and Sinc using congenic mice and restriction fragment length polymorphism analysis. *J. Gen. Virol.* **68:**2711–2716.

Isenmann, S., Brandner, S., and Aguzzi, A. (1996a). Neuroectodermal grafting: a new tool for the study of neurodegenerative diseases. *Histol. Histopathol.* **11:**1063–1073.

Isenmann, S., Brandner, S., Kuhne, G., Boner, J., and Aguzzi, A. (1996b). Comparative in vivo and pathological analysis of the blood–brain barrier in mouse telencephalic transplants. *Neuropathol. Appl. Neurobiol.* **22:**118–128.

Isenmann, S., Brandner, S., Sure, U., and Aguzzi, A. (1996c). Telencephalic transplants in mice: characterization of growth and differentiation patterns. *Neuropathol. Appl. Neurobiol.* **21:**108–117.

James, T. L., Liu, H., Ulyanov, N. B., Farr-Jones, S., Zhang, H., Donne, D. G., Kaneko, K., Groth, D., Mehlhorn, I., Prusiner, S. B., and Cohen, F. E. (1997). Solution structure of a 142-residue recombinant prion protein corresponding to the infectious fragment of the scrapie isoform. *Proc. Natl. Acad. Sci. U.S.A.* **94:**10086–10091.

Jarrett, J. T., and Lansbury, P. T., Jr. (1993). Seeding "one-dimensional crystallization" of amyloid: a pathogenic mechanism in Alzheimer's disease and scrapie? *Cell* **73:**1055–1058.

Kimberlin, R. H., Hall, S. M., and Walker, C. A. (1983). Pathogenesis of mouse scrapie. Evidence for direct neural spread of infection to the CNS after injection of sciatic nerve. *J. Neurol. Sci.* **61:**315–325.

Kimberlin, R. H., and Walker, C. A. (1978a). Evidence that the transmission of one source of scrapie agent to hamsters involves separation of agent strains from a mixture. *J. Gen. Virol.* **39:**487–496.

Kimberlin, R. H., and Walker, C. A. (1978b). Pathogenesis of mouse scrapie: effect of route of inoculation on infectivity titres and dose-response curves. *J. Comp. Pathol.* **88:**39–47.

Kimberlin, R. H., and Walker, C. A. (1980). Pathogenesis of mouse scrapie: evidence for neural spread of infection to the CNS. *J. Gen. Virol.* **51:**183–187.

Kimberlin, R. H., and Walker, C. A. (1986). Pathogenesis of scrapie (strain 263K) in hamsters infected intracerebrally, intraperitoneally or intraocularly. *J. Gen. Virol.* **67:**255–263.

Kimberlin, R. H., and Walker, C. A. (1989a). Pathogenesis of scrapie in mice after intragastric infection. *Virus. Res.* **12:**213–220.

Kimberlin, R. H., and Walker, C. A. (1989b). The role of the spleen in the neuroinvasion of scrapie in mice. *Virus. Res.* **12:**201–211.

Kimberlin, R. H., Walker, C. A., and Fraser, H. (1989). The genomic identity of different strains of mouse scrapie is expressed in hamsters and preserved on reisolation in mice. *J. Gen. Virol.* **70:**2017–2025.

Kimberlin, R. H., and Wilesmith, J. W. (1994). Bovine spongiform encephalopathy. Epidemiology, low dose exposure and risks. *Ann. N.Y. Acad. Sci.* **724:**210–220.

Kitamoto, T., Muramoto, T., Mohri, S., Doh ura, K., and Tateishi, J. (1991). Abnormal isoform of prion protein accumulates in follicular dendritic cells in mice with Creutzfeldt-Jakob disease. *J. Virol.* **65:**6292–6295.

Klein, M. A., Frigg, R., Flechsig, E., Raeber, A. J., Kalinke, U., Bluethmann, H., Bootz, F., Suter, M., Zinkernagel, R. M., and Aguzzi, A. (1997). A crucial role for B cells in neuroinvasive scrapie. *Nature* **390**:687–690.

Klein, M. A., Frigg, R., Raeber, A. J., Flechsig, E., Hegyi, I., Zinkernagel, R. M., Weissmann, C., and Aguzzi, A. (1998). PrP expression in B lymphocytes is not required for prion neuroinvasion. *Nat. Med.* **4**:1429–1433.

Korth, C., Stierli, B., Streit, P., Moser, M., Schaller, O., Fischer, R., Schulz-Schaeffer, W., Kretzschmar, H., Raeber, A., Braun, U., Ehrensperger, F., Hornemann, S., Glockshuber, R., Riek, R., Billeter, M., Wuthrich, K., and Oesch, B. (1997). Prion (PrPSc)-specific epitope defined by a monoclonal antibody. *Nature* **390**:74–77.

Kretzschmar, H. A., Honold, G., Seitelberger, F., Feucht, M., Wessely, P., Mehraein, P., and Budka, H. (1991). Prion protein mutation in family first reported by Gerstmann, Straussler, and Scheinker [letter]. *Lancet* **337**:1160.

Kretzschmar, H. A., Prusiner, S. B., Stowring, L. E., and DeArmond, S. J. (1986). Scrapie prion proteins are synthesized in neurons. *Am. J. Pathol.* **122**:1–5.

Laplanche, J. L., Chatelain, J., Launay, J. M., Gazengel, C., and Vidaud, M. (1990). Deletion in prion protein gene in a Moroccan family. *Nucleic. Acids Res.* **18**:6745.

Lasmezas, C. I., Cesbron, J. Y., Deslys, J. P., Demaimay, R., Adjou, K. T., Rioux, R., Lemaire, C., Locht, C., and Dormont, D. (1996). Immune system-dependent and -independent replication of the scrapie agent. *J. Virol.* **70**:1292–1295.

Lledo, P. M., Tremblay, P., Dearmond, S. J., Prusiner, S. B., and Nicoll, R. A. (1996). Mice Deficient For Prion Protein Exhibit Normal Neuronal Excitability and Synaptic Transmission in the Hippocampus. *Proceedings of the National Academy of Sciences of the United States of America* **93**:2403–2407.

Manson, J., West, J. D., Thomson, V., McBride, P., Kaufman, M. H., and Hope, J. (1992). The prion protein gene: a role in mouse embryogenesis? *Development* **115**:117–122.

Manson, J. C., Clarke, A. R., Hooper, M. L., Aitchison, L., McConnell, I., and Hope, J. (1994a). 129/Ola mice carrying a null mutation in PrP that abolishes mRNA production are developmentally normal. *Mol. Neurobiol.* **8**:121–127.

Manson, J. C., Clarke, A. R., McBride, P. A., McConnell, I., and Hope, J. (1994b). PrP gene dosage determines the timing but not the final intensity or distribution of lesions in scrapie pathology. *Neurodegeneration* **3**:331–340.

Martins, V. R., Graner, E., Garcia-Abreu, J., de Souza, S. J., Mercadante, A. F., Veiga, S. S., Zanata, S. M., Neto, V. M., and Brentani, R. R. (1997). Complementary hydropathy identifies a cellular prion protein receptor [see comments]. *Nat. Med.* **3**:1376–1382.

McKinley, M. P., Bolton, D. C., and Prusiner, S. B. (1983). A protease-resistant protein is a structural component of the scrapie prion. *Cell* **35**:57–62.

Moore, R. C., Hope, J., McBride, P. A., McConnell, I., Selfridge, J., Melton, D. W., and Manson, J. C. (1998). Mice with gene targetted prion protein alterations show that Prnp, Sinc and Prni are congruent. *Nat. Genet.* **18**:118–125.

Moore, R. C., and Melton, D. W. (1997). Transgenic analysis of prion diseases. *Mol. Hum. Reprod.* **3**:529–544.

Moore, R. C., Redhead, N. J., Selfridge, J., Hope, J., Manson, J. C., and Melton, D. W. (1995). Double replacement gene targeting for the production of a series of mouse strains with different prion protein gene alterations. *Biotechnology (New York)* **13**:999–1004.

Moser, M., Colello, R. J., Pott, U., and Oesch, B. (1995). Developmental expression of the prion protein gene in glial cells. *Neuron* **14**:509–517.

Muramoto, T., DeArmond, S. J., Scott, M., Telling, G. C., Cohen, F. E., and Prusiner, S. B. (1997). Heritable disorder resembling neuronal storage disease in mice expressing prion protein with deletion of an alpha-helix. *Nat. Med.* **3**:750–735.

Muramoto, T., Kitamoto, T., Hoque, M. Z., Tateishi, J., and Goto, I. (1993). Species barrier prevents an abnormal isoform of prion protein from accumulating in follicular dendritic cells of mice with Creutzfeldt-Jakob disease. *J. Virol.* **67:**6808–6810.

Nishida, N., Tremblay, P., Sugimoto, T., Shigematsu, K., Shirabe, S., Petromilli, C., Erpel, S. P., Nakaoke, R., Atarashi, R., Houtani, T., Torchia, M., Sakaguchi, S., DeArmond, S. J., Prusiner, S. B., and Katamine, S. (1999). A mouse prion protein transgene rescues mice deficient for the prion protein gene from purkinje cell degeneration and demyelination. *Lab. Invest.* **79:**689–697.

Oesch, B., Westaway, D., Walchli, M., McKinley, M. P., Kent, S. B., Aebersold, R., Barry, R. A., Tempst, P., Teplow, D. B., Hood, L. E., and Weissmann, C. (1985). A cellular gene encodes scrapie PrP 27–30 protein. *Cell* **40:**735–746.

Parchi, P., Castellani, R., Capellari, S., Ghetti, B., Young, K., Chen, S. G., Farlow, M., Dickson, D. W., Sima, A. A. F., Trojanowski, J. Q., Petersen, R. B., and Gambetti, P. (1996). Molecular Basis of Phenotypic Variability in Sporadic Creutzfeldt-Jakob Disease. *Annals of Neurology* **39:**767–778.

Pattison, I. H. (1966). The relative susceptibility of sheep, goats and mice to two types of the goat scrapie agent. *Res. Vet. Sci.* **7:**207–212.

Pattison, I. H., and Millson, G. C. (1961). Experimental transmission of scrapie to goats and sheep by the oral route. *J. Comp. Pathol.* **71:**171–176.

Peretz, D., Williamson, R. A., Matsunaga, Y., Serban, H., Pinilla, C., Bastidas, R. B., Rozenshteyn, R., James, T. L., Houghten, R. A., Cohen, F. E., Prusiner, S. B., and Burton, D. R. (1997). A conformational transition at the N terminus of the prion protein features in formation of the scrapie isoform. *J. Mol. Biol.* **273:**614–622.

Prusiner, S. B. (1982). Novel proteinaceous infectious particles cause scrapie. *Science* **216:**136–144.

Prusiner, S. B. (1989). Scrapie prions. *Annu. Rev. Microbiol.* **43:**345–374.

Prusiner, S. B. (1991). Molecular biology of prion diseases. *Science* **252:**1515–1522.

Prusiner, S. B. (1994). Inherited prion diseases. *Proc. Natl. Acad. Sci. U.S.A.* **91:**4611–4614.

Prusiner, S. B. (1997). Prion diseases and the BSE crisis. *Science* **278:**245–251.

Prusiner, S. B. Groth, D., Serban, A., Koehler, R., Foster, D., Torchia, M., Burton, D., Yang, S. L., and DeArmond, S. J. (1993). Ablation of the prion protein (PrP) gene in mice prevents scrapie and facilitates production of anti-PrP antibodies. *Proc. Natl. Acad. Sci. U.S.A.* **90:**10608–10612.

Prusiner, S. B., McKinley, M. P., Bowman, K. A., Bolton, D. C., Bendheim, P. E., Groth, D. F., and Glenner, G. G. (1983). Scrapie prions aggregate to form amyloid-like birefringent rods. *Cell* **35:**349–358.

Prusiner, S. B., Scott, M., Foster, D., Pan, K. M., Groth, D., Mirenda, C., Torchia, M., Yang, S. L., Serban, D., Carlson, G. A., and *et al.* (1990). Transgenetic studies implicate interactions between homologous PrP isoforms in scrapie prion replication. *Cell* **63:**673–686.

Race, R., and Chesebro, B. (1998). Long-term persistence of scrapie infectivity in brain and spleen tissue of a clinically resistant species: implications for control of BSE. *Nature.*

Race, R. E., Priola, S. A., Bessen, R. A., Ernst, D., Dockter, J., Rall, G. F., Mucke, L., Chesebro, B., and Oldstone, M. B. (1995). Neuron-specific expression of a hamster prion protein minigene in transgenic mice induces susceptibility to hamster scrapie agent. *Neuron* **15:**1183–1191.

Raeber, A. J., Brandner, S., Klein, M. A., Benninger, Y., Musahl, C., Frigg, R., Roeckl, C., Fischer, M. B., Weissmann, C., and Aguzzi, A. (1998). Transgenic and knockout mice in research on prion diseases. *Brain Pathol.* **8:**715–733.

Raeber, A. J., Klein, M. A., Frigg, R., Flechsig, E., Aguzzi, A., and Weissmann, C. (1999a). PrP-dependent association of prions with splenic but not circulating lymphocytes of scrapie-infected mice. *EMBO J.* **18:**2702–2706.

Raeber, A. J., Race, R. E., Brandner, S., Priola, S. A., Sailer, A., Bessen, R. A., Mucke, L., Manson, J., Aguzzi, A., Oldstone, M. B., Weissmann, C., and Chesebro, B. (1997). Astrocyte-specific expression of hamster prion protein (PrP) renders PrP knockout mice susceptible to hamster scrapie. *Embo. J.* **16:**6057–6065.

Raeber, A. J., Sailer, A., Hegyi, I., Klein, M. A., T. R., Fischer, M., Brandner, S., Aguzzi, A., and Weissmann, C. (1999b). Ectopic expression of prion protein (PrP) in T lymphocytes or hepatocytes of PrP knockout mice is insufficient to sustain prion replication. *Proc. Natl. Acad. Sci. U.S.A.* **96:**3987–3992.

Raymond, G. J., Hope, J., Kocisko, D. A., Priola, S. A., Raymond, L. D., Bossers, A., Ironside, J., Will, R. G., Chen, S. G., Petersen, R. B., Gambetti, P., Rubenstein, R., Smits, M. A., Lansbury, P. T., Jr., and Caughey, B. (1997). Molecular assessment of the potential transmissibilities of BSE and scrapie to humans [see comments]. *Nature* **388:**285–288.

Rieger, R., Edenhofer, F., Lasmezas, C. I., and Weiss, S. (1997). The human 37-kDa laminin receptor precursor interacts with the prion protein in eukaryotic cells [see comments]. *Nat. Med.* **3:**1383–1388.

Riek, R., Hornemann, S., Wider, G., Billeter, M., Glockshuber, R., and Wuthrich, K. (1996). NMR structure of the mouse prion protein domain PrP(121–321). *Nature* **382:**180–182.

Riek, R., Hornemann, S., Wider, G., Glockshuber, R., and Wüthrich, K. (1997). NMR characterization of the full-length recombinant murine prion protein, mPrP(23–231). *FEBS Lett.* **413:**282–288.

Rogers, M., Yehiely, F., Scott, M., and Prusiner, S. B. (1993). Conversion of truncated and elongated prion proteins into the scrapie isoform in cultured cells. *Proc. Natl. Acad. Sci. U.S.A.* **90:**3182–3186.

Safar, J., Wille, H., Itri, V., Groth, D., Serban, H., Torchia, M., Cohen, F. E., and Prusiner, S. B. (1998). Eight prion strains have PrP(Sc) molecules with different conformations [see comments]. *Nat. Med.* **4:**1157–1165.

Sailer, A., Büeler, H., Fischer, M., Aguzzi, A., and Weissmann, C. (1994). No propagation of prions in mice devoid of PrP. *Cell* **77:**967–968.

Sakaguchi, S., Katamine, S., Nishida, N., Moriuchi, R., Shigematzu, K., Sugimoto, T., Nakatni, A., Kataoka, Y., Houtani, H., Shirabe, S., Okada, H., Hasegawa, S., Myamoto, T., and Noda, T. (1996). Loss of cerebellar Purkinje cells in aged mice homozygous for a disrupted PrP gene. *Nature* **380:**528–531.

Sakaguchi, S., Katamine, S., Shigematsu, K., Nakatani, A., Moriuchi, R., Nishida, N., Kurokawa, K., Nakaoke, R., Sato, H., Jishage, K., and *et al.* (1995). Accumulation of proteinase K-resistant prion protein (PrP) is restricted by the expression level of normal PrP in mice inoculated with a mouse-adapted strain of the Creutzfeldt-Jakob disease agent. *J. Virol.* **69:**7586–7592.

Schatzl, H. M., Da Costa, M., Taylor, L., Cohen, F. E., and Prusiner, S. B. (1995). Prion protein gene variation among primates. *J. Mol. Biol.* **245:**362–374.

Scott, J. R., Foster, J. D., and Fraser, H. (1993a). Conjunctival instillation of scrapie in mice can produce disease. *Vet. Microbiol.* **34:**305–309.

Scott, M., Foster, D., Mirenda, C., Serban, D., Coufal, F., Waelchli, M., Torchia, M., Groth, D., Carlson, G., DeArmond, S. J., Westaway, D., and Prusiner, S. B. (1989). Transgenic mice expressing hamster prion protein produce species-specific scrapie infectivity and amyloid plaques. *Cell* **59:**847–857.

Scott, M., Groth, D., Foster, D., Torchia, M., Yang, S. L., DeArmond, S. J., and Prusiner, S. B. (1993b). Propagation of prions with artificial properties in transgenic mice expressing chimeric PrP genes. *Cell* **73:**979–988.

Scott, M. R., Groth, D., Tatzelt, J., Torchia, M., Tremblay, P., DeArmond, S. J., and Prusiner, S. B. (1997a). Propagation of prion strains through specific conformers of the prion protein. *J. Virol.* **71:**9032–9044.

Scott, M. R., Kohler, R., Foster, D., and Prusiner, S. B. (1992). Chimeric prion protein expression in cultured cells and transgenic mice. *Protein Sci.* **1:**986–997.

Scott, M. R., Safar, J., Telling, G., Nguyen, O., Groth, D., Torchia, M., Koehler, R., Tremblay, P., Walther, D., Cohen, F. E., DeArmond, S. J., and Prusiner, S. B. (1997b). Identification of a prion protein epitope modulating transmission of bovine spongiform encephalopathy prions to transgenic mice. *Proc. Natl. Acad. Sci. U.S.A.* **94:**14279–14284.

Shmerling, D., Hegyi, I., Fischer, M., Blattler, T., Brandner, S., Gotz, J., Rulicke, T., Flechsig, E., Cozzio, A., von Mering, C., Hangartner; C., Aguzzi, A., and Weissmann, C. (1998). Expression of amino-terminally truncated PrP in the mouse leading to ataxia and specific cerebellar lesions. *Cell* **93:**203–214.

Somerville, R. A., Chong, A., Mulqueen, O. U., Birkett, C. R., Wood, S. C., and Hope, J. (1997). Biochemical typing of scrapie strains [letter]. *Nature* **386:**564.

Stahl, N., Borchelt, D. R., Hsiao, K., and Prusiner, S. B. (1987). Scrapie prion protein contains a phosphatidylinositol glycolipid. *Cell* **51:**229–240.

Straub, R. H., Westermann, J., Scholmerich, J., and Falk, W. (1998). Dialogue between the CNS and the immune system in lymphoid organs. *Immunol. Today* **19:**409–413.

Telling, G. C., Haga, T., Torchia, M., Tremblay, P., Dearmond, S. J., and Prusiner, S. B. (1996a). Interactions Between Wild-Type and Mutant Prion Proteins Modulate Neurodegeneration Transgenic Mice. *Genes & Development* **10:**1736–1750.

Telling, G. C., Parchi, P., DeArmond, S. J., Cortelli, P., Montagna, P., Gabizon, R., Mastrianni, J., Lugaresi, E., Gambetti, P., and Prusiner, S. B. (1996b). Evidence for the conformation of the pathologic isoform of the prion protein enciphering and propagating prion diversity [see comments]. *Science* **274:**2079–2082.

Telling, G. C., Scott, M., Hsiao, K. K., Foster, D., Yang, S. L., Torchia, M., Sidle, K. C., Collinge, J., DeArmond, S. J., and Prusiner, S. B. (1994). Transmission of Creutzfeldt-Jakob disease from humans to transgenic mice expressing chimeric human-mouse prion protein. *Proc. Natl. Acad. Sci. U.S.A.* **91:**9936–9940.

Telling, G. C., Scott, M., Mastrianni, J., Gabizon, R., Torchia, M., Cohen, F. E., DeArmond, S. J., and Prusiner, S. B. (1995). Prion propagation in mice expressing human and chimeric PrP transgenes implicates the interaction of cellular PrP with another protein. *Cell* **83:**79–90.

Tobler, I., Gaus, S. E., Deboer, T., Achermann, P., Fischer, M., Rülicke, T., Moser, M., Oesch, B., McBride, P. A., and Manson, J. C. (1996). Altered circadian activity rhythms and sleep in mice devoid of prion protein. *Nature* **380:**639–642.

Weber, T., and Aguzzi, A. (1997). The spectrum of transmissible spongiform encephalopathies. *Intervirology* **40:**198–212.

Weissmann, C. (1991). A `unified theory' of prion propagation. *Nature* **352:**679–683.

Weissmann, C. (1996). Molecular biology of transmissible spongiform encephalopathies. *FEBS Lett.* **389:**3–11.

Weissmann, C., and Aguzzi, A. (1997). Bovine spongiform encephalopathy and early onset variant Creutzfeldt-Jakob disease. *Curr. Opin. Neurobiol.* **7:**695–700.

Wells, G. A., Scott, A. C., Johnson, C. T., Gunning, R. F., Hancock, R. D., Jeffrey, M., Dawson, M., and Bradley, R. (1987). A novel progressive spongiform encephalopathy in cattle. *Vet. Rec.* **121:**419–420.

Westaway, D., DeArmond, S. J., Cayetano Canlas, J., Groth, D., Foster, D., Yang, S. L., Torchia, M., Carlson, G. A., and Prusiner, S. B. (1994). Degeneration of skeletal muscle, peripheral nerves, and the central nervous system in transgenic mice overexpressing wild-type prion proteins. *Cell* **76:**117–129.

Westaway, D., Goodman, P. A., Mirenda, C. A., McKinley, M. P., Carlson, G. A., and Prusiner, S. B. (1987). Distinct prion proteins in short and long scrapie incubation-period mice. *Cell* **51:**651–662.

Westaway, D., Mirenda, C. A., Foster, D., Zebarjadian, Y., Scott, M., Torchia, M., Yang, S. L., Serhan, H., DeArmond, S. J., Ebeling, C., and *et al.* (1991). Paradoxical shortening of scrapie incubation times by expression of prion protein transgenes derived from long incubation-period mice. *Neuron* **7:**59–68.

Whittington, M. A., Sidle, K. C., Gowland, I., Meads, J., Hill, A. F., Palmer, M. S., Jefferys, J. G., and Collinge, J. (1995). Rescue of neurophysiological phenotype seen in PrP null mice by transgene encoding human prion protein. *Nat. Genet.* **9:**197–201.

Wilesmith, J. W., Ryan, J. B., Hueston, W. D., and Hoinville, L. J. (1992). Bovine spongiform encephalopathy: epidemiological features, 1985 to 1990. *Vet. Rec.* **130:**90–94.

Will, R., Cousens, S., Farrington, C., Smith, P., Knight, R., and Ironside, J. (1999). Deaths from variant Creutzfeldt-Jakob disease. *Lancet* **353:**9157–9158.

Will, R., Ironside JW, Zeidler M, Cousens SN, Estibeiro K, Alperovitch A, Poser S, Pocchiari M, Hofman A, and Smith (1996). A new variant of Creutzfeldt-Jakob disease in the UK. *Lancet* **347:**921–925.

Wisniewski, H. M., Lossinsky, A. S., Moretz, R. C., Vorbrodt, A. W., Lassmann, H., and Carp, R. I. (1983). Increased blood–brain barrier permeability in scrapie-infected mice. *J. Neuropathol. Exp. Neurol.* **42:**615–626.

Yehiely, F., Bamborough, P., Da Costa, M., Perry, B. J., Thinakaran, G., Cohen, F. E., Carlson, G. A., and Prusiner, S. B. (1997). Identification of candidate proteins binding to prion protein. *Neurobiol. Dis.* **3:**339–355.

HIV: HUMAN IMMUNODEFICIENCY VIRUS

ADVANCES IN VIRUS RESEARCH, VOL 56

THE BLOOD–BRAIN BARRIER AND AIDS

Lisa I. Strelow,* Damir Janigro,† and Jay A. Nelson*,§

*Oregon Health Sciences University
Department of Molecular Microbiology and Immunology
Portland, Oregon 97201
†Cleveland Clinic Foundation
Cerebrovascular Research
Cleveland, Ohio 44195
§Vaccine and Gene Therapy Institute
Portland, Oregon 97201

I. INTRODUCTION

Human immunodeficiency virus (HIV) infection of the central nervous system (CNS) is currently one of the more challenging aspects of HIV-induced disease, despite the recent relative success of highly active antiretroviral therapy (HAART). While it is now clear that HIV infects the brain, HIV infection of the central nervous system was not initially recognized in AIDS (acquired immunodeficiency syndrome) patients. HIV-infected patients often present with neurological complications, but these were not initially thought to be outside the realm of what had been previously seen in other immunosuppressed patients. The social apathy and withdrawal that we now know to be part of an HIV-induced syndrome were initially perceived to be merely symptoms of depression. It was not until the mid-1980s that researchers began to make an association between productive HIV brain infection and neurological impairment in AIDS patients. In 1985, HIV viral nucleic acids were detected in the brains of AIDS patients by *in situ* and

355

Southern analyses (Shaw et al., 1985). Additionally, HIV was isolated from both the brain and the cerebrospinal fluid (CSF) of AIDS patients with neurological dysfunction (Levy et al., 1985; Ho et al., 1985). These results, taken together with studies demonstrating retrovirus-induced neurological disease in animals (Haase 1986), led to further studies aimed at elucidating the mechanisms and consequences of HIV infection of the central nervous system (reviewed in Spencer and Price, 1992; and in Glass and Johnson, 1996).

Although HIV-induced pathological changes in the brain have since been well documented in the absence of opportunistic infections, the mechanisms of viral entry into the brain are little understood, as are the sequelae following virus entry. While HIV has now been clearly shown to infect the CNS, neural elements—including neurons and oligodendrocytes—surprisingly do not appear to be directly infected by virus. Rather, the brain cells susceptible to HIV infection include, predominantly, macrophages and microglia, with fewer but significant numbers of microvascular endothelial cells (MVECs) and possibly astrocytes (especially in pediatric patients) also demonstrating immunohistochemical evidence of infection. The toxic effects of HIV infection must therefore be indirect, in contrast to a variety of other CNS infectious agents. For example, viral meningitis is associated with direct infection of ependymal and meningeal cells, and JC virus infects oligodendrocytes. Poliovirus infects motor neurons, thereby leading to poliomyelitis. Currently, how HIV mediates CNS disease through macrophage infiltration or microglia infection is a matter of continuing debate.

The relationship between disease and HIV infection of the brain is confounded by further paradoxical observations. While virus can enter the brain very early after initial infection and virus can be isolated from the brain and/or CSF throughout infection, the majority of patients do not suffer from neurological symptoms until late in the disease course, most often after the onset of clinical immunosuppression. There is, additionally, a striking disparity between the amount of virus and virally infected cells detected in the brain and the severity of both the clinical disease and the neuropathological changes that are seen (Pumarola-Sune et al., 1987, Navia et al., 1986b, Glass et al., 1993, Gray et al., 1988). Thus, the severity of clinical disease rarely correlates with the amount of HIV present in the brain, although some investigators have demonstrated a correlation between peripheral plasma viral load and the development of dementia (Childs et al., 1999). In some cases, however, virus cannot be detected in the brains of patients with CNS disease. Paradoxes such as these limit our understanding of the mechanisms and consequences of HIV infection of the brain.

The blood–brain barrier (BBB) is an anatomical barrier between the general circulation and the brain parenchyma. The BBB is composed of specialized MVECs in contact with astrocytes. The contact between brain MVECs and astrocyte foot processes has been postulated to be critical for the formation of the barrier (Beck *et al.*, 1984; Joo, 1985; Tontsch and Bauer, 1991). Some of the enabling properties of the BBB are limited pinocytotic vesicular transport, the presence of tight junctions between endothelial cells, and a selective permeability to physiological ions. Thus, while the BBB functions to exclude most molecules from entering the brain, the barrier must also function to selectively transport necessary ions from the blood. As the critical interface between the circulation and the brain, the BBB must not only be encountered, but also crossed, before access to the brain parenchyma can be achieved. This chapter will therefore focus on how HIV and simian immunodeficiency virus (SIV) are proposed to interact with and cross the BBB, and to effect CNS damage. In addition, the CNS as a reservoir of virus, and the BBB as a limitation to the eradication of HIV, will be discussed. Lastly, animal and tissue culture models for the study of virus interactions with the BBB will be reviewed.

II. AIDS Dementia versus HIV Encephalitis

Although the two terms have often been used interchangeably, AIDS dementia complex (ADC) clearly involves a separate clinical disease course from the pathobiological changes of HIV encephalitis (HIVE). AIDS dementia complex is characterized by a clinical syndrome of motor, cognitive, and behavioral abnormalities (outlined in Table I). Originally described by Navia *et al.* (1986a), ADC presents as a subcortical dementia involving predominantly subcortical (white matter, basal ganglia) pathologies, with symptoms that are similar to those of Parkinson's and Huntington's diseases. This syndrome differs from the cortical dementias, which involve language and perception difficulties, and are best exemplified by Alzheimer's disease and Creutzfeldt-Jakob disease. For the diagnosis of AIDS dementia, patients must exhibit a progressive decline in at least two of the following areas: motor speed, nonverbal memory, and frontal lobe tasks (Tross *et al.*, 1988). The progressive nature of ADC is one of the hallmarks of this syndrome, with the prevalence and severity of ADC increasing as the immune system fails (Price *et al.*, 1988).

Pathologically, there is a generalized atrophy of the brain in ADC patients, with some neuronal dropout being demonstrated (reviewed

TABLE I.

SYMPTOMS ASSOCIATED WITH AIDS DEMENTIA COMPLEX (ADC) VERSUS PATHOBIOLOGY
OF HIV ENCEPHALITIS (HIVE)

ADC	HIVE
Motor	
Psychomotor slowing and loss of precision	Astrogliosis
	Microglial nodules
Ataxia	Multiple foci of multinucleated giant cells
Tremor	Regions of demyelination
Hyperreflexia	Diffuse white matter pallor
Hypertonia	
Release signs	
Impaired rapid movements	
Behavioral	
Lethargy	
Social withdrawal, withdrawal from routine activities	
Depressive symptoms	
Apathy	
Cognitive	
Cognitive changes	
Impaired memory, attention, and concentration	
Diffuse cerebral atrophy and white matter rarefaction	
Dementia	

in Spencer and Price, 1992). Only about half of AIDS dementia cases are associated with encephalitis, however, and some dementia patients do not exhibit neuropathological evidence of HIV encephalitis (Rosenblum, 1990; Budka, 1991; Navia *et al.*, 1986; Wiley *et al.*, 1994; Glass *et al.*, 1993). Rather, diffuse white matter pallor is predominant in ADC, perhaps as a consequence of indirect damage to the brain. Magnetic resonance imaging (MRI) results often demonstrate hyperintensities on T2-weighted images in ADC patients, which is indicative of an abnormal retention of water, possibly as a result of BBB disruption. In support of this hypothesis, diffuse myelin pallor and MRI abnormalities are associated with BBB perturbation in other disease states (e.g., metastatic tumors, as reviewed in Power *et al.*, 1993). The cause of white matter pallor has been postulated to be due either to demyelination or to an altered BBB or altered microvasculature. A study by Power *et al.* (1993) suggested that the BBB is perturbed without evidence of widespread demyelination. A diffuse

BBB leak is also observed in approximately 50% of all patients with AIDS and may be seen in the absence of HIVE (Petito and Cash, 1992). Increased CSF–albumin ratios, an indication of BBB damage in neurological diseases, occur in HIV infections and become abnormally high in AIDS dementia patients (McArthur et al., 1993). Serum proteins are normally excluded by an intact BBB. Marked serum protein accumulation is found in the brains of AIDS dementia patients, however, raising the possibility that BBB alterations may contribute to AIDS dementia (Power et al., 1993; Rhodes, 1991). Supporting this hypothesis, a study by Power et al. (1993) found a statistically significant association between diffuse myelin pallor, serum protein immunoreactivity, and the clinical expression of AIDS dementia. This study also found serum protein reactivity in the brains of some non-demented AIDS patients, which suggests that abnormalities of the BBB are chronic and precede the onset of AIDS dementia. An increased cerebrovascular permeability in AIDS dementia, coupled with the presence of increased numbers of perivascular macrophages, may have a role in perturbing the BBB (reviewed in Power et al., 1993), and in leading to a chronic, slowly progressing breach of the BBB. ADC also occurs in HIV-infected children, who may demonstrate a loss of developmental milestones in addition to other symptoms of ADC (Epstein et al., 1985; Belman et al., 1985). Children are rarely affected by opportunistic infections—an observation that helped to confirm the role of HIV in AIDS dementia. In addition, the symptoms of ADC can occur in the absence of opportunistic infections in adult patients (Power et al., 1993).

HIV encephalitis (HIVE) is an inflammatory process, characterized by focal regions of intense macrophage infiltration. The neuropathological changes characteristic of HIVE are outlined in Table I. The lesions of HIVE are found most frequently in the deep white matter of cerebral hemispheres, basal ganglia, and the brainstem and are often adjacent to blood vessels. The hallmark of HIVE is the multinucleated giant cell (MNGC), which is commonly believed to be the result of cell fusion of HIV-infected macrophages. In contrast to ADC, focal lesions such as those seen in HIVE are associated with BBB disruption only when tissue necrosis is also present (Petito and Cash, 1992).

Thus, the two separate manifestations of HIV infection of the brain-the clinical symptoms of ADC and the pathological effects of HIVE-appear to have a differential effect on the BBB. While ADC is almost universally associated with a diffuse breakdown of the BBB, as demonstrated by leakage of serum proteins, HIVE results in a perturbation of the BBB only in the presence of concomitant tissue necrosis.

III. Timing of Viral Entry into the CNS

As discussed above, HIV enters the brain soon after infection (Bell *et al.*, 1993; reviewed in Spencer and Price, 1992). Virus can also be isolated from CSF at early times during infection (Michaels *et al.*, 1988; Ho *et al.*, 1985). Several studies have detected the presence of virus in HIV-infected patient brains at early times, when patients are still asymptomatic (Sinclair *et al.*, 1992; An *et al.*, 1996; Wilkinson *et al.*, 1997). One of these studies demonstrated that brain spectroscopic abnormalities also occur early during HIV infection, before the onset of severe immunosuppression (Wilkinson *et al.*, 1997). In a case study of an accidental patient inoculation with HIV-infected blood cells, virus could be isolated from the recipient's brain at 15 days postinoculation, which was one day after virus was first isolated from blood (Davis *et al.*, 1992). The recovery of virus from brain occurred despite the administration of antiretroviral therapy (ART) within 45 minutes of the accidental inoculation. Thus, even when ART is initiated at or before seroconversion, virus gained access to the brain. This finding has important implications for the brain as a CNS reservoir of virus, as well as for the prevention of reservoir formation.

The early timing of HIV entry into brain is also paralleled in an animal model. Studies with SIV infection of macaques demonstrate that virus enters the brain soon after intravenous inoculation, and can be detected in the brain as early as 7 days postintravenous inoculation (Chakrabarti *et al.*, 1991; Sharer *et al.*, 1991; Lackner *et al.*, 1991b; Smith *et al.*, 1995). In one experimental SIV infection of pig-tailed macaques, neurovirulent virus administered orally was detectable in the brain as early as 5 days postinoculation (Novembre *et al.*, 1998). Clearly, peripheral virus is capable of gaining access to the brain at very early times postinfection. Such early infection of the CNS is paradoxical, given that HIV-infected humans and SIV-infected macaques do not display symptoms of neurological disease until much later during the disease course, usually after the onset of clinical immunosuppression. The question then arises as to what is the role of virus in the brain at early times, and why symptoms of neuropathology do not manifest until late times during the course of disease.

IV. Mechanisms and Models of Viral Entry into the Brain

Several theoretical models have been proposed for the *in vivo* transmigration of HIV across the BBB into the brain parenchyma (see Fig. 1). In one model, virus could gain access to the brain parenchyma via

"Trojan Horse" model

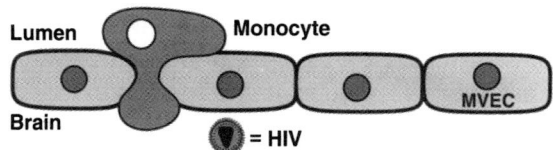

Transcytosis model

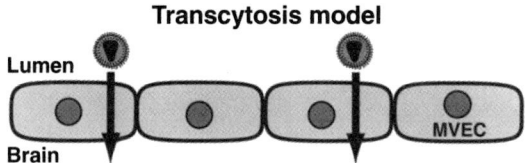

Direct viral infection of MVEC model

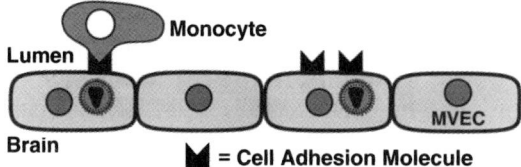

BBB disruption model

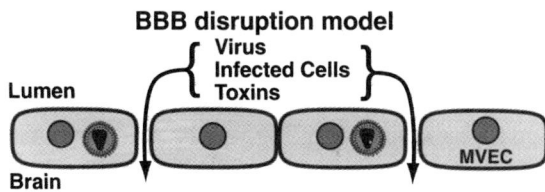

Fig 1. Models for HIV-1 infection of the brain. Trojan horse: diapedesis of HIV-infected cell through the MVEC monolayer and subsequent brain infection. Transcytosis: direct passage of HIV through MVEC monolayer by adsorption and subsequent budding to abluminal side. Direct infection of MVEC: chemoattraction and adhesion of uninfected monocytes to HIV-infected, adhesion molecule-expressing MVECs with subsequent diapedesis of monocyte through MVEC monolayer. Infection of monocytes would result from passage through the infected MVEC monolayer, and spread of virus into the brain parenchyma would result from the now perivascular, HIV-infected monocytes. Blood–brain barrier disruption: any or all of the models proposed may result in a transient disruption of tight junctions, allowing free virus, virus-infected cells, and neurotoxic molecules access to the brain parenchyma.

direct transcytosis of virus through the MVEC layer of the BBB. Transcytosis would result from virus absorption on the lumenal side of the BBB and from budding of virus into the abluminal, brain parenchyma side. The most popular theory is that HIV latently infected monocytes/macrophages traverse the BBB, resulting in the activation and release of virus in the brain parenchyma (Trojan horse model; Peluso *et al.*, 1986). One shortcoming of this model is that inflammatory cells do not usually pass through an intact BBB unless there is circulation of brain-specific activated T cells, or there is an enhanced expression of endothelial adhesion molecules (Hickey, 1991; Lassman *et al.*, 1991). This observation leads to another model, in which uninfected monocytes migrate through the BBB, becoming infected as they pass through an adhesion molecule-expressing, infected MVEC monolayer (diagrammed in Fig. 1). Thus, two of the models for monocyte migration are as follows: (1) infected monocytes induce the expression of adhesion molecules on MVECs, adhere to MVECs and transmigrate through the BBB; (2) infected or otherwise activated MVECs express adhesion molecules, to which monocytes bind and thereby pass through the BBB.

While there is more *in vitro* evidence for the former model, the two models need not be mutually exclusive. Thus, MVECs express elevated levels of adhesion molecules during conditions of inflammation, such as would be found during viral infection (Beekhuizen *et al.*, 1993; Beilke, 1989). Specifically, HIV-infected MVECs were found to express intercellular cell adhesion molecule (ICAM-1) (Moses and Nelson, 1994). Additionally, macrophages in direct contact with MVECs demonstrate an upregulation of HIV-1 replication (Gilles *et al.*, 1995). Nottet *et al.* (1996) found that cell–cell contact between MVECs and HIV-infected or otherwise immune-activated monocytes resulted in the upregulation of E-selectin and vascular cell adhesion molecule (VCAM-1) on MVECs. During an animal infection with SIV, leukocytes were recruited to the CNS by the expression of VCAM-1 on MVECs (Sasseville *et al.*, 1995). These data indicate that MVECs are capable of recruiting macrophages via expression of adhesion molecules, and that HIV-infected macrophages can cause MVECs to express adhesion molecules. It is interesting to note that immune-activated macrophages are capable of inducing adhesion molecule expression on MVECs, obviating the need for productive viral infection of monocytes as an initiating factor in adherence to MVECs.

Several studies have elucidated additional potential mechanisms whereby infected monocytes traffic into the CNS through the endothelium of the BBB. Weiss *et al.* (1999) observed that the HIV-1 tat protein induces MCP-1 expression in astrocytes, leading to the recruitment of infected monocytes into the CNS. A study by Dhawan *et al.* (1995) demonstrated that HIV-infected monocytes express elevated levels of the lymphocyte function-associated antigen-1 (LFA-1, for which ICAM is a ligand), induce endothelial monolayer permeability and express elevated levels of gelatinase, leading to an extravasation of monocytes through an endothelial monolayer. Monocyte migration through the endothelium was also shown to result from immune activation, resulting in an altered MVEC morphology and concomitant formation of holes in the MVEC monolayer (Persidsky et al., 1997). These studies describe a common pathway for HIV entry into brain via interactions between MVECs and macrophages. While macrophages are certainly infected during the course of HIV disease, it is likely that infected MVECs contribute to macrophage–MVEC interactions and to the spread of virus across the BBB and into the CNS.

Direct infection of the BBB MVECs (discussed below) could also contribute to BBB dysfunction (reviewed in Moses and Nelson, 1995). HIV infection of MVECs could result in tight junction leakage, leading to altered barrier properties, which could allow trafficking of additional infected cells into the brain parenchyma, and subsequent viral spread. Alternatively, the tight junction leakage may allow cytokines and other toxic metabolites to gain access to the brain. For example, interleukin (IL)-1α has been found to alter tight junctions and to increase pinocytosis in MVECs (Quagliarello et al., 1991; Ramilo et al., 1990), while cytokine dysregulation has been implicated in other studies of HIV infection (Tyor et al., 1992). Additional toxic molecules and metabolites that might gain access to the normally privileged CNS via disruption of BBB include TNF–α, quinolinic acid, arachidonic acid metabolites, nitric oxide, and other toxic factors (Tyor et al., 1992; Wesselingh et al., 1993; Saukkonen et al., 1990; Heyes et al., 1991; Genis et al., 1992; Wahl et al., 1989; Giulian et al., 1990, Giulian et al., 1993). Viral proteins such as gp120 (Kaiser et al., 1990) and tat (Weiss et al., 1999) can also contribute to CNS dysfunction, once they get across the BBB. Altered BBB integrity might also directly contribute to CNS dysfunction. Additionally, HIV-infected MVECs might contribute to the abnormal production of neurotoxic or cytotoxic factors or to endothelial cell-induced astrocytosis. Tight junction and/or BBB disruption—via infection of MVECs, generalized immune activation, or elevated serum cytokine levels (Rosenberg and Fauci, 1990)—may result in a more permeable BBB. The resulting chronic CNS edema could lead to gliosis (Okazaki, 1989), an altered ionic environment, and an altered function of excitable membranes (Power et al., 1993). Thus, direct infection of MVECs could contribute to BBB disruption through permeability changes as well as through altered cytokine profiles.

One of the hallmarks of HIV-induced CNS damage is that, while the brain harbors virus early during the disease course, neurologic damage occurs late. It is interesting to note that the temporal changes of neurologic symptoms in HIV-infected individuals may be explained by chronic, persistent, but low-level damage to the BBB, consistent with persistent, nonlytic MVEC infection, as has been seen *in vitro* (discussed below). Additionally, azidothymidine (AZT) and diadenosine may temporarily reverse AIDS dementia in infected patients (Pizzo *et al.*, 1988; Brouwers *et al.*, 1997), suggesting that neurons are not damaged irreversibly by a virus-induced pathologic cascade. In addition, SIV infection of macaques results in electrophysiological abnormalities in only certain portions of evoked potentials, suggesting that virus selectively affects only certain regions or functions of CNS neurons (Horn *et al.*, 1998). These studies present evidence for HIV-induced selective and chronic damage to CNS functions, and are consistent with a persistent and chronic infection of brain microvascular endothelial cells.

A. Infection of Microvascular Endothelial Cells

Within the infected brain, macrophages and microglial cells are the major targets for HIV-1 and SIV infection (Takahashi *et al.*, 1996; Wiley *et al.*, 1986; Budka *et al.*, 1991; Vazeux *et al.*, 1987; Simon *et al.*, 1992). With the advent of more sensitive detection methods, however, it has become increasingly clear that other brain cell types are infected *in vivo*, including astrocytes (Tornatore *et al.*, 1994; Saito *et al.*, 1994; An *et al.*, 1999) and MVECs (Wiley *et al.*, 1986; Koenig *et al.*, 1986; Gabuzda *et al.*, 1986; Ward *et al.*, 1987; Bagasra *et al.*, 1996; An *et al.*, 1999). A recent study by An *et al.* (1999) utilized double-labeling methods (*in situ* PCR [polymerase chain reaction] for viral genes and immunohistochemistry for cellular markers) to unequivocally identify virus localization in MVECs as well as in astrocytes isolated from HIV-positive patient brains. The localization of virus signal specifically to MVECs of the brain vasculature is an important distinction from the perivascular localization reported for a subset of HIV-infected macrophages. Importantly, MVECs and astrocytes are the cell types comprising the BBB.

In vitro infections of MVECs have yielded controversial results (see Table II; Figs. 2 and 3). Several groups have clearly demonstrated productive HIV and SIV infection of various types of MVECs (see Figs. 2 and 3), while others have been unable to show productive infection. The disparity between these two findings may have to do with the cell type in which infection was attempted, the activation state of the cell, the strain of virus used, and/or the method(s) used for viral detection. Thus, macrovascular endothelial cells (such as those from aorta and venous sources) are much more refractory to HIV infection than microvascular endothelial cells. In addition, certain strains of HIV

TABLE II.

STUDIES OF *IN VITRO* ENDOTHELIAL CELL INFECTION BY HIV AND SIV

Type of Endothelial Cell	Reference
Susceptible to infection	
Adipose tissue	Re *et al.*, 1991
Adipose tissue	Cenacchi *et al.*, 1992
Renal glomeruli	Green *et al.*, 1992
Brain microvascular	Moses *et al.*, 1993
Brain microvascular	Poland *et al.*, 1995
Liver sinusoids	Steffan *et al.*, 1992
Human umbilical vein (HUVEC)	Conaldi *et al.*, 1995
Human umbilical vein	Corbeil *et al.*, 1995
Simian brain microvascular	Mankowski *et al.*, 1994
Simian brain microvascular	Strelow *et al.*, 1998
Not susceptible to infection	
Continuous endothelial cell line	Ades *et al.*, 1993
Human aortic	Teitel *et al.*, 1989
Human umbilical vein	Lafon *et al.*, 1992
Human umbilical vein	Gilles *et al.*, 1995
Brain microvascular	Gilles *et al.*, 1995
Saphenous vein	Scheglovitova *et al.*, 1993

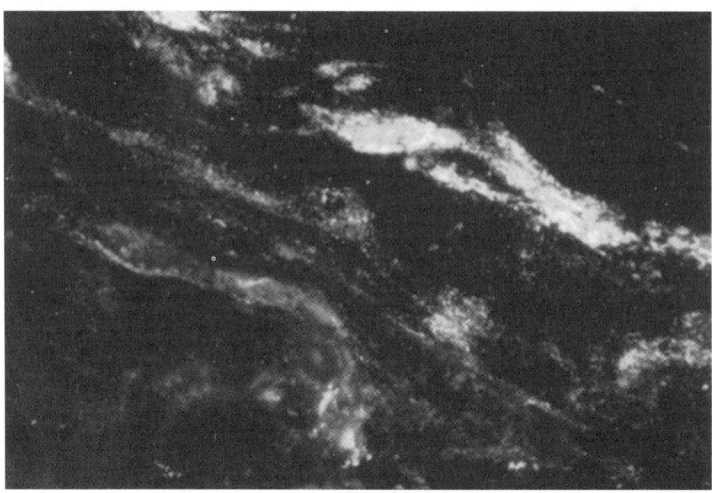

FIG 2. [For color reproduction, see the color section.] HIV productively infects brain MVEC *in vitro*. Positive indirect immunofluorescence for viral p24 in HIV-infected MVEC.

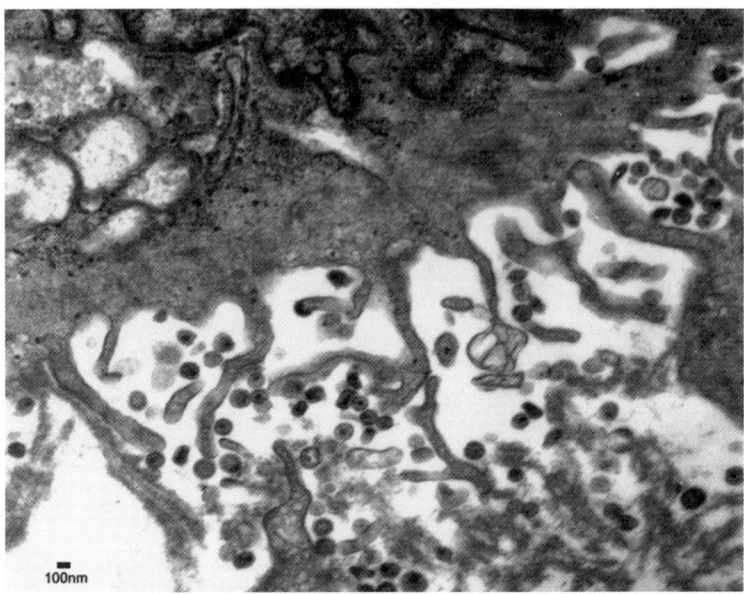

FIG 3. SIV productively infects brain MVEC *in vitro*. Electron microscopy of SIV-infected simian MVEC at day 16 postinfection. Note budding virion in center of micrograph. Original magnification: × 20,000.

readily infect brain MVECs while others do not (Moses *et al.*, 1996a). The sequences regulating MVEC infection do not appear to correlate with sequences known to be involved with either T cell line or macrophage tropism (Moses *et al.*, 1996a), suggesting that there are independent mechanisms for MVEC tropism. Fig. 2 (see also color section) demonstrates the presence of HIV p24 antigen in a culture of human brain MVECs exposed to HIV *in vitro*.

When SIV-infected brain MVECs are monitored for viral p27 (core protein) production over time, peak levels of virus are routinely seen at days 6–8 postinfection (Mankowski *et al.*, 1994; Strelow *et al.*, 1998). Productive infection of simian brain MVECs can also be demonstrated by the presence of viral particles via electron microscopy (Fig. 3). However, results from our laboratory indicate that some MVECs may initially appear to be refractory to SIV infection. Thus, replication-competent virus has been obtained after as many as 40 days postinfection, with virus previously undetectable (Strelow and Nelson, unpublished results, 1999). Clearly, this time frame is not amenable to most *in vitro* studies. However, the delay in productive viral infection may explain the disparity observed by different groups.

Although the source of CNS-tropic strains of virus is controversial, the bone marrow may play an important role in the development of these strains. Most primary retroviral infections of murine and feline species target the bone marrow early in the development of disease (Lander *et al.*, 1999; Hayes *et al.*, 1995; Beebe *et al.*, 1992). Early studies that centered on the development of CNS–tropic SIV strains utilized intracerebral inoculation of rhesus macaques with the bone marrow of SIV-infected animals (Sharma *et al.*, 1992). Serial passage of this virus resulted in the derivation of a neuropathogenic virus that exemplified MVEC tropism (Mankowski *et al.*, 1997). We have also observed that HIV obtained from bone marrow stroma also efficiently infects brain MVECs. Interestingly, the crucial cell type infected in the stroma is the bone marrow MVEC (Moses *et al.*, 1996b). Taken together, these studies suggest that virus derived from infected bone marrow MVECs may have an enhanced capacity to enter the brain. We propose a model (Fig. 4) in which HIV/SIV infects the bone marrow stroma early during the disease and results in the development of MVEC-tropic strains capable of infecting and penetrating the BBB.

B. Viral Entry into Microvascular Endothelial Cells

Early studies of HIV and SIV infection of brain MVECs demonstrated that MVEC infection occurs via a CD4-independent mecha-

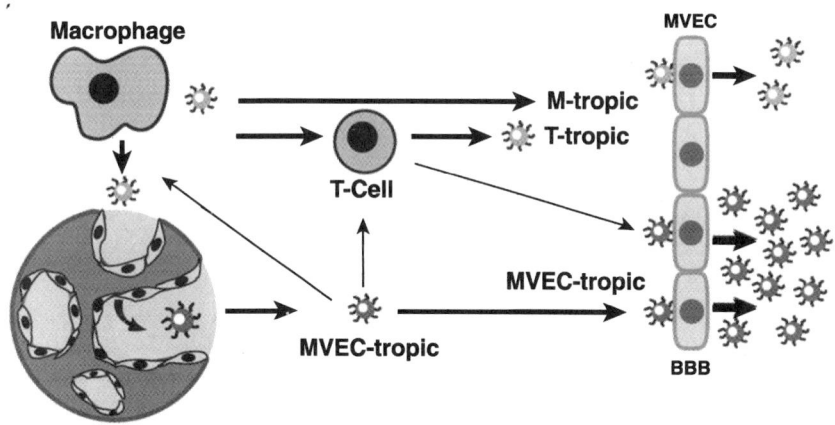

FIG 4. Model for the generation of MVEC-tropic HIV within infected bone marrow. HIV infects bone marrow early during the disease course, resulting in the derivation of MVEC-tropic strains that then preferentially seed into brain by penetrating the BBB.

nism (Re et al., 1991; Moses et al., 1993; Mankowski et al., 1994), raising the question of how these cells were capable of infection. The discovery that the family of seven-transmembrane receptors can function as coreceptors for HIV entry led several investigators to look for the expression of chemokine receptors on CD4-negative (CD4⁻) cells. These studies have identified the presence of the chemokine receptors CCR1, CCR3, and CCR5, as well as of the orphan seven-transmembrane receptor STRL33, on brain MVECs (Rottman et al., 1996; Sanders et al., 1998; Edinger et al., 1997). Another orphan seven-transmembrane receptor, APJ, has been shown to be present in brain tissues (Matsumoto et al., 1996). Furthermore, the presence of the chemokine receptors CXCR4 and CCR5 has been demonstrated in the brains of children with AIDS (Vallat et al., 1998). These studies suggested that certain chemokine receptors could be the primary receptor for the entry of HIV and SIV into MVECs. In support of this hypothesis, a study by Edinger et al. (1997) demonstrated that CCR5 could serve as a receptor for several different strains of SIV. In addition, the study showed that rhesus brain MVECs could be infected by a neurovirulent SIV strain through a CD4-independent, CCR5-dependent mechanism. A further study confirmed the presence of the APJ receptor in brain, and demonstrated that APJ can function as a receptor for HIV and SIV (Edinger et al., 1998). Thus, the ability of CCR5 and other chemokine receptors to function as the primary receptor for HIV and SIV entry into MVECs has elucidated a CD4-independent pathway for viral entry into cells.

V. The Central Nervous System as A Viral Reservoir

Despite the efficacy of highly active antiretroviral therapy in suppressing plasma viremia, under certain conditions, HIV may be isolated from latently infected pools of resting CD4⁺ T cells after months to years of undetectable plasma viremia (Finzi et al., 1997; Wong et al., 1997; Finzi et al., 1999; Chun et al., 1997; Zhang et al., 1999; Chun et al., 1999). This pool of latently infected T cells has an extremely long half-life in vivo, and can be induced in vitro to yield infectious, replication-competent virus. Early studies indicated that virus obtained from these patients does not show evidence for the evolution of drug resistance mutations, indicating that the latent pool is not likely to be replicating to high levels in patients (Finzi et al., 1997; Wong et al., 1997). More recent studies have observed very low levels of persistent viral replication in the face of undetectable plasma viremia (Chun et al., 1997; Finzi et al., 1999). The CD4⁺ resting T cell reservoir is established early, as evidenced by the detection of this reservoir population even when patients are started on HAART before or soon after sero-conversion (Finzi *et al.*, 1997; Finzi *et al.*, 1999). Additionally, the formation of an HIV reservoir in the brain is not preventable by current standards of care and antiretroviral regimens, as shown by the presence of virus in the CNS despite the use of HAART therapy. While CD4$^+$ T cells are considered to be the main reservoir for long-term persistence of HIV under HAART, most investigators acknowledge the probability of the existence of other stable reservoirs. In studies demonstrating viral rebound following cessation of HAART, replication-competent HIV was undetectable in resting CD4$^+$ T cells despite the subsequent recurrence of plasma viremia, suggesting that other tissue reservoirs may have been responsible for the rebound (Chun *et al.*, 1999). In another study, HIV-related polyneuropathy and myelopathy not only persisted but also worsened in a patient undergoing successful HAART (as measured by plasma viremia), suggesting that the CNS can be a reservoir for HIV (Pialoux *et al.*, 1997). With an increase in the sensitivity of detection methods, HIV provirus and RNA have been shown to be present in many of the CNS cell types, raising the possibility that the CNS may be a sanctuary for persistence of HIV (Bagasra *et al.*, 1996; An *et al.*, 1999).

A mild amelioration of HIV-associated cognitive motor complex has been seen in adult patients treated with azidothymidine (also known as zidovudine [ZDV] or as Retrovir) (Yarchoan *et al.*, 1987; Schmitt *et al.*, 1988; Portegies *et al.*, 1989; Sidtis *et al.*, 1993). However, the neuroprotective effects of azidothymidine appear to be limited to approximately 18 months, possibly due to viral resistance variants (Baldeweg *et al.*, 1998). One of the challenges of treating HIV brain disease is the limited ability of most antiretrovirals to cross the BBB and thereby gain access to infected CNS tissues. Although AZT appears to weakly penetrate the BBB, active efflux of AZT occurs within the BBB (Takasawa *et al.*, 1997). Thus, infected cells of the CNS may act as a viral reservoir, because they are sequestered from the successful antiretroviral effects of the HAART drugs currently in use.

Cells that are potential reservoirs of HIV must possess the following characteristics: (1) the cell type must be susceptible to infection by HIV; (2) the infection must be either persistent or latent, as well as noncytolytic; (3) the cell type must exhibit an extended half-life *in vivo;* (4) the cell type must be inaccessible to conventional antiretrovirals, or be resistant by virtue of maintaining a quiescent state.

Several different cell types in the CNS would fulfill most or all of these criteria. In one study, sensitive viral detection methods using DNA and reverse transcriptase–*in situ* PCR assays, in combination with immunohistochemistry for identifying infected cell types in the CNS (Bagasra et al., 1996), have shown HIV infection of microglia and macrophages, as had been previously demonstrated (Wiley et al., 1986; Koenig et al., 1986; Vazeux et al., 1987; Budka et al., 1991; Takahashi et al., 1996). The study by Bagasra et al. (1996) also demonstrated convincing evidence for the infection, at low levels, of neurons, oligodendrocytes, and astrocytes. In addition, HIV-infected MVECs were observed in 19 out of 22 brains. Not only did MVECs express viral DNA, but a significant proportion of cells also expressed HIV mRNA. HIV-1 infection of MVECs has been demonstrated by other authors, both in vivo (Wiley et al., 1986; Koenig et al., 1986; Gabuzda et al., 1986; Ward et al., 1987; An et al., 1999), and in vitro (Moses et al., 1993). Therefore, any or all of these CNS cell types may harbor virus in a quiescent state and therefore act as a CNS viral reservoir.

Brain MVECs are attractive candidates for a viral CNS reservoir because these cells have been shown to be susceptible to HIV infection in vivo and in vitro. Cultures of human brain MVECs infected with various strains of HIV-1 and HIV-2 have been maintained in vitro for over a year (A. V. Moses, unpublished data). The cells do not produce significant amounts of virus, as measured by p24 ELISA (enzyme-linked immunosorbent assay) and/or indirect immunofluorescence for viral p24 protein, but are capable of yielding infectious virus when cocultured with 174×CEM cells (a T cell–B cell hybrid cell line permissive for various strains of HIV and SIV). These results are paralleled in the SIV-macaque model, where persistent infections of simian MVECs with diverse strains of SIV have been demonstrated (Strelow and Nelson, manuscript in preparation). Our unpublished results indicate that infection of MVECs in vitro is noncytolytic. However, work by Adamson et al. (1996) indicates that SIV in vivo is capable of inducing apoptosis in simian MVECs. The appearance of multinucleated giant cells was first noted at a low level in persistently SIV-infected simian MVEC cultures at late times postinfection (over a year in culture), which is suggestive of ongoing viral replication with mutation to a more virulent phenotype (Strelow and Nelson, unpublished results). In spite of the persistent production of virus throughout the course of the in vitro infection (as measured by an s-MAGI assay for infectious titer, as well as by p27 ELISA), the simian MVECs remained morphologically constant (i.e., there was no evidence of overt cell death) until this point. Taken together, these results indicate that simian brain MVECs are capable of a noncytolytic, persistent SIV infection.

MVECs in vitro are a long-lived cell population. In an in vitro model of the BBB (the DIV-BBB, outlined later), persistent, long-term (>100 days) infection of simian brain MVECs by SIV was readily accomplished under conditions of flow (Janigro *et al.*, 1998). In this dynamic BBB model, MVECs show a decrease in glucose consumption after approximately 200 days in culture, but maintain a low but measurable and consistent rate of consumption. This result suggests that MVECs in dynamic culture enter a relatively quiescent state. The viability of cells was determined by the glucose consumption, and the presence of a functional barrier was measured by Evan's blue dye exclusion and by a reduced permeability to radiolabeled compounds (data are not published; Strelow and Nelson). When passaged, MVECs retain their characteristic morphology for up to 16 *in vitro* passages, after which point the cells express actin stress fibers and the growth rate slows considerably. Simian MVECs have been similarly cultured *in vitro* without passaging for up to 385 days (Strelow and Nelson, manuscript in preparation). These results suggest that MVECs have a long half-life in culture as well as *in vivo*, thereby satisfying another criterion of a CNS viral reservoir.

The antiretroviral compound zidovudine (AZT) crosses the CNS's BBB initially, but quickly undergoes active efflux from the brain via a probenecid-sensitive transporter system in the brain endothelium (Takasawa *et al.*, 1997). The inability of antiretrovirals to gain access to the CNS through the BBB therefore raises the potential for the CNS to act as a reservoir for virus, even in patients undergoing successful HAART to reduce plasma viremia (Pialoux *et al.*, 1997; Cohen, 1998). A CNS viral reservoir has the potential to be established quite early during infection. As detailed above, the accidental infection of a patient with HIV-infected white blood cells resulted in the presence of virus in the brain after 15 days, despite the administration of ART within 45 minutes of the infusion error (Davis *et al.*, 1992). The presence of virus in the brain at early time points indicates that the CNS has the potential to become an early site of viral persistence.

Thus, MVECs must be considered a potential viral reservoir in the CNS. MVECs are infectible with a variety of HIV and SIV strains; demonstrate persistence of viral infection; are long-lived *in vitro* and *in vivo;* and are refractory to the effects of antiretrovirals due to their function as a part of the BBB.

VI. ANIMAL MODELS

Simian immunodeficiency virus is a member of the primate lentivirus family of retroviruses, and causes infections in macaques that share many of the pathological changes and clinical disease characteristics of human HIV infection (Desrosiers, 1990; Letvin et al., 1985; Ringler et al., 1988). Experimental SIV infection of macaques results in a progressive immunodeficiency syndrome that parallels human AIDS (Desrosiers, 1990; Kestler et al., 1990; Kindt et al., 1992), with CD4+ T cells and tissue macrophages being the predominant virus-infected cell types. SIV-infected macaques, similar to HIV-infected humans, demonstrate the characteristic decrease in levels of CD4+ T cells, concomitant with an increased susceptibility to opportunistic infections over the disease course. Studies seeking to characterize neuropathologic changes in SIV-infected macaques have identified neurologic abnormalities paralleling those seen in HIV-infected brains (Simon et al., 1992). The characteristic histopathologic findings in SIV-infected brains consist of multinucleated giant cells, perivascular cuffing, and focal glial nodules (Letvin et al., 1985; Sharer et al., 1988; Lackner et al., 1991a, Sharer et al., 1991). SIV-infected macaques are not routinely imaged; therefore, BBB perturbations have been difficult to measure except at termination. However, calculation of CSF/serum albumin ratios in vivo indicates that transient disruptions of BBB may occur with SIV infection (Horn et al., 1998).

Several recent studies have shown that SIV-infected rhesus macaques also develop physiologic abnormalities such as motor problems, abnormalities in sensory-evoked potentials, and cognitive anomalies, similar to what has been observed in HIV-infected patients (Murray et al., 1991; Sharer et al., 1988; Prospero-Garcia et al., 1996; Gold et al., 1998). The SIV-macaque model therefore provides a workable paradigm for studies of retrovirus-induced neuropathogenesis. Although virus may hematogenously disseminate into the brain parenchyma, in the context of this discussion "neuroinvasive" refers to virus that is capable of entering the brain after inoculation of virus into nonneuronal sites. "Neurotropic" virus represents virus capable of replicating in brain tissues, while "neurovirulent" refers to virus that is capable of causing neurologic disease as measured by histopathology and/or neurophysiological abnormalities.

Several groups have derived neurovirulent clones or strains of SIV in order to better model neuropathogenesis in macaques. In one model, a neurotropic virus was derived from lymphocyte–tropic parental SIV after multiple animal passages of infected cells (Sharma et al., 1992; Mankowski et al., 1997). A viral stock of cloned SIVmac239, a lymphocyte-tropic, pathogenic strain of SIVmac, was inoculated intracerebrally into the brain of a rhesus macaque (the first animal passage). As this animal did not show signs of neurologic disease, despite peripheral viremia, brain homogenate from this animal was inoculated intracerebrally into a second macaque. Brain homogenate as well as bone marrow cells from the second passage were inoculated intracerebrally into a third animal. The third animal passage resulted in macrophage infection in the periphery, but not in brain infection. A fourth animal passage was therefore attempted, with infected bone marrow cells being inoculated intracerebrally into two macaques, and this resulted in SIV encephalitis in both animals. A fifth animal passage, utilizing brain cells from the fourth animal passage, resulted in a virus strain termed SIV/17E-Br (for brain) which was macrophage-tropic and caused encephalitis. Intracerebral inoculation of six macaques with SIV/17E-Br resulted in the development of CNS disease in three of the animals (Adamson et al., 1996). Analysis of histopathological specimens from diseased macaques with SIV encephalitis demonstrated evidence of apoptosis in brain MVECs, utilizing terminal transferase dUTP, nicked-end labeling (TUNEL). This result suggests that the BBB may have been compromised in these three SIV/17E-Br-infected animals. Consistent with this hypothesis, a loss of glucose transporter-1 staining was also observed in the SIV encephalitic brains (Adamson et al., 1996). Several molecular clones of SIVmac239-17E-Br have been generated from uncloned stock. One clone, SIV/17E-Fr, caused neurologic disease in six of nine intravenously inoculated animals (Mankowski et al., 1997).

During the course of neuroadaptation of a lymphocyte-tropic virus in the above studies, several interesting results were observed during the course of experiments. First, macrophage-tropic virus that developed in the periphery following primary inoculation with SIVmac239 was incapable of crossing the BBB to gain access to brain (Sharma et al., 1992). Thus, macrophage tropism alone is not sufficient to confer neuroinvasiveness. This finding was confirmed in a study by Mankowski et al. (1997), in which viral determinants of macrophage tropism and neurovirulence were genetically distinct. Second, macrophage-tropic and lymphocyte-tropic viruses were frequently found within the same animal, although in different tissues. This partitioning of viral genotypes has also been seen in HIV infection (Koyanagi et al., 1987), and may result from tissue-specific selective pressures, especially from those present in the CNS. Third, in an additional study, a correlation was found between the lesions that occur with neuroinvasive strains of SIV and the level of host immunosuppression (Zink et al., 1997). Therefore, in a dual SIV infection, the rate of neurological deterioration paralleled that of immunosuppression. In the same dual-SIV-infection study with SIV/DeltaB670 (a dual-tropic SIV swarm) and the neurovirulent molecular clone SIV/17E-Fr, SIV/17E-Fr preferentially entered the CNS and replicated there (Zink *et al.*, 1997). These studies suggest that immunosuppression contributes to neurological manifestations of SIV disease and, further, that neurovirulent strains can selectively partition to, and become established in, the CNS.

Further animal studies utilizing SIV/17E-Fr demonstrated that, in addition to the development of SIV encephalitis in these animals, viral RNA was detectable in macaque brains by both *in situ* hybridization and reverse transcriptase–*in situ* PCR (Mankowski *et al.*, 1994). By double-label immunohistochemistry and *in situ* hybridization, brain cells staining for von Willebrand factor VIII-related antigen, an endothelial cell marker, were shown to be virally infected. SIV-infected cells lining the lumens of small brain vessels, suggestive of endothelial cells, were also observed in these tissues. The lumenal staining pattern was distinct from the perivascular staining of macrophages.

The finding that endothelial cells were infected *in vivo* was followed by *in vitro* studies demonstrating productive viral replication in MVECs of brain origin (Mankowski *et al.*, 1994). Brain MVECs may therefore have a role in the selection of neurovirulent viruses, since MVECs were not seen to be infected by nonneurovirulent, nonneuroinvasive strains of SIVmac (Mankowski *et al.*, 1994). Flaherty *et al.* (1997) subsequently mapped determinants of MVEC tropism. Specific amino acid changes in the transmembrane (TM) portion of the *env* gene, as well as the presence of a full-length *nef* gene, were found to be absolutely required for SIV replication in MVEC. Thus, the ability to replicate productively in brain MVECs differs from macrophage tropism. Indeed, not all macrophage-tropic strains of SIV are neurovirulent (Mankowski *et al.*, 1997). The derivation of a neuroinvasive and neurovirulent strain of SIVmac from a strictly lymphocyte-tropic strain in the studies outlined above required multiple animal passages of SIV-infected tissues, including bone marrow and brain. The SIV17E-Br strain and resulting clones induce significant neuropathology in most, but not all, macaques inoculated with virus. Additionally, the derivation strategy involved serial intracerebral inoculations, which may have selected for neurotropic variants but not for neuroinvasive ones.

Another group of investigators derived a neuroinvasive, neurotropic, and neurovirulent SIVmac strain after only two animal passages of infected brain cells (Watry *et al.*, 1995). Unlike the studies described above, these investigators utilized intravenous rather than intracerebral inoculation of an SIVmac251 swarm, which contains variants that are naturally neuroinvasive. Initial infection of one macaque resulted in peripheral viremia, without signs of neurologic dysfunction, when the animal was sacrificed 15 months postinoculation. In order to select for neuroinvasive quasispecies of virus, the investigators then isolated fresh (i.e., not cultured) microglia-enriched populations of brain cells for intravenous inoculation (the first animal passage). This animal developed widespread macrophage infiltration, microglial nodules, and MNGCs in the brain after only 4 months (Watry *et al.*, 1995). Microglia-enriched brain cells were isolated from this animal and cryopreserved for subsequent infections. Intravenous inoculation of an additional three animals, with the cryopreserved microglia derived from the first animal passage, resulted in neuropathology in all three newly inoculated animals (the second animal passage). Therefore, all four animals inoculated with microglia-enriched brain cells developed significant neuropathological lesions, demonstrating that neuropathogenic virus can partition into the brain during the course of a natural SIV infection.

In subsequent experiments, four additional animals were infected intravenously with microglia from the first animal passage. These animals uniformly developed infiltrating T cells and macrophages, as well as perivascular macrophages, in their brains, in addition to developing neurophysiologic abnormalities (Prospero-Garcia *et al.*, 1996). One of the four animals was tested for BBB integrity via Evan's blue dye exclusion, and was found to have BBB leakage in the temporal lobe. Cell-free virus (as opposed to virus-infected microglia) derived from this microglia-passaged SIVmac251 also resulted in neuropathologic lesions in four intravenously inoculated animals (Horn *et al.*, 1998). Two of the four SIV-infected animals tested in these experiments demonstrated a transient disruption of the BBB, as evidenced by a higher-than-normal albumin quotient (CSF/serum albumin concentration).

Infection with the microglia-passaged virus also resulted in neurophysiological deficits in infected macaques (Prospero-Garcia *et al.*, 1996; Horn *et al.*, 1998). Thus, in four SIV-infected animals evaluated for electrophysiological criteria, certain portions of sensory-evoked responses were significantly delayed as compared to those of uninfected control animals. SIV-infected monkeys also displayed a significant behavioral impairment indicative of frontal lobe function and consistent with findings in HIV-infected humans (Prospero-Garcia *et al.*, 1996). These changes in evoked potentials noted in SIV-infected animals may be due to virally induced cortical hyperexcitability. Furthermore, SIV infection with microglia-passaged virus resulted in distinct changes in body temperature and motor activity in five infected animals within a month of inoculation (Horn *et al.*, 1998), mimicking episodes of fever and motor impairment seen in HIV-infected humans.

The body temperature increases may affect generalized brain function, since many brain biochemical processes are temperature sensitive. Motor impairment has been seen consistently in other studies of SIV infection (Murray *et al.*, 1992). Thus, infection with a brain-derived, microglia-passaged variant of SIVmac resulted not only in significant brain pathology in all infected animals, but also in reproducible signs of physiological dysfunction that were indicative of motor and cognitive impairments.

A final selection scheme for neuropathogenic virus involved the derivation of a viral strain pathogenic for pig-tailed macaques from a naturally SIV- and STLV-infected sooty mangabey (Novembre *et al.*, 1998). Blood from the original animal (termed FGb), a sooty mangabey, was transfused into six monkeys: two rhesus macaques, two pig-tailed macaques, and two sooty mangabeys. Of these six animals, one rhesus macaque and both pig-tailed macaques developed SIV encephalitis and had SIV RNA present in the CNS, as measured by *in situ* hybridization. Neither sooty mangabey developed neurologic disease. Cell-free virus derived from infected pig-tailed macaque lymph node cells and infected rhesus macaque CSF was inoculated intravenously into six rhesus and six pig-tailed macaques. Notably, all six pig-tailed macaques receiving this second animal passage virus had SIV-positive cells within the brain, as measured by *in situ* hybridization. However, the rhesus macaques did not exhibit neurologic signs of infection. Therefore, the derivation of this sooty mangabey SIV strain (SIVsmm-FGb) resulted in a virus that is neuropathogenic for pig-tailed macaques only. Several recent studies have demonstrated that pig-tailed macaques *(M. nemestrina)* are more susceptible to SIV-induced neuropathology than are other macaque species such as rhesus *(M. mulatta)* and cynomolgus *(M. fascicularis)* (Novembre *et al.*, 1998; Zink *et al.*, 1997). The use of pig-tailed macaques with the SIVsmm strain did result in a virus that was 100% neurovirulent in pig-tailed macaques. Infected lymph node cells from the first animal passage were further cultured with peripheral blood mononuclear cells (PBMCs) and subsequently used to isolate an infectious molecular clone of FGb, termed PGm5.3. At day 5 postoral inoculation with PGm5.3, two pig-tailed macaques were sacrificed and found to have virus in their brains, indicating that the molecular clone of SIVsmm is also neuroinvasive for pig-tailed macaques (Novembre *et al.*, 1998). Although SIVsmm isolates are much less effectively characterized than SIVmac isolates in terms of genetic determinants of pathogenesis and of macrophage and MVEC tropism, this SIVsmm model of neuropathogenesis will also be useful in investigating the basis of retrovirus-induced neurologic disease.

VII. Tissue Culture Models of Blood–Brain Barrier

In order to study BBB properties *in vitro,* several different models of BBB have been developed (reviewed in Janigro *et al.,* 1999, and in Grant *et al.,* 1998). Commonly, the models involve a static coculture of endothelial cells with astrocytes and are characterized by a relatively high sucrose permeability and the lack of specific transporters demonstrated by the BBB *in vivo.* In transwell models of the BBB, endothelial cells of varying origins are usually cultured on the apical surface of an insert consisting of a permeable membrane. Astrocytes or astrocytic lines are then cultured in the well below the insert, or, alternatively, on the basolateral surface of the insert (see Fig. 5). These types of studies have the advantage that the cells involved can be readily imaged and that transmigration studies can be easily performed. Studies utilizing transwell inserts have elucidated many of the mechanisms involved in monocyte–MVEC interactions and adherence (outlined above). One disadvantage of the transwell system is that the BBB properties demonstrated by the static coculture of endothelial cells with astrocytes do not approach those seen *in vivo.* Additionally, cell lines or nonbrain endothelial cells have often been used in lieu of genuine brain MVECs, as these cells are easier to obtain and maintain.

A dynamic, *in vitro* model of the BBB, called the DIV–BBB model, has recently been developed (Stanness *et al.,* 1996, Stanness *et al.,* 1997). In the DIV–BBB model, MVEC are cocultured with astrocytes under pulsatile flow, which has been shown to induce ultrastructural differentiation of endothelial cells (Ballerman and Ott, 1995). The DIV–BBB model is characterized by a low permeability to sucrose, the expression of a BBB-like glucose transporter, the presence of tight interendothelial junctions, and the presence of stereospecific amino acid transporters (Stanness *et al.,* 1996; Stanness *et al.,* 1997). A schematic comparing the DIV–BBB model to a transwell insert is shown in Fig. 5.

For studies of SIV interactions with the BBB, primary simian brain MVECs were inoculated into the lumenal compartment and normal human astrocytes were seeded into the extracapillary space of DIV–BBB cartridges. Two DIV–BBB cartridges were infected with the SIVmac251 molecular clone and monitored over time, by using several different criteria to detect viral infection (Strelow and Nelson, manuscript in preparation). SIV infection was persistent throughout the time course of infection (>100 days). Virus was first detected at day 2 postinfection, the first time point sampled, and continued to be detectable until day 116 postinfection, at which point the DIV–BBB cartridges were terminated. Although viral titers and p27 values decreased over time, infectious virus was present both in culture

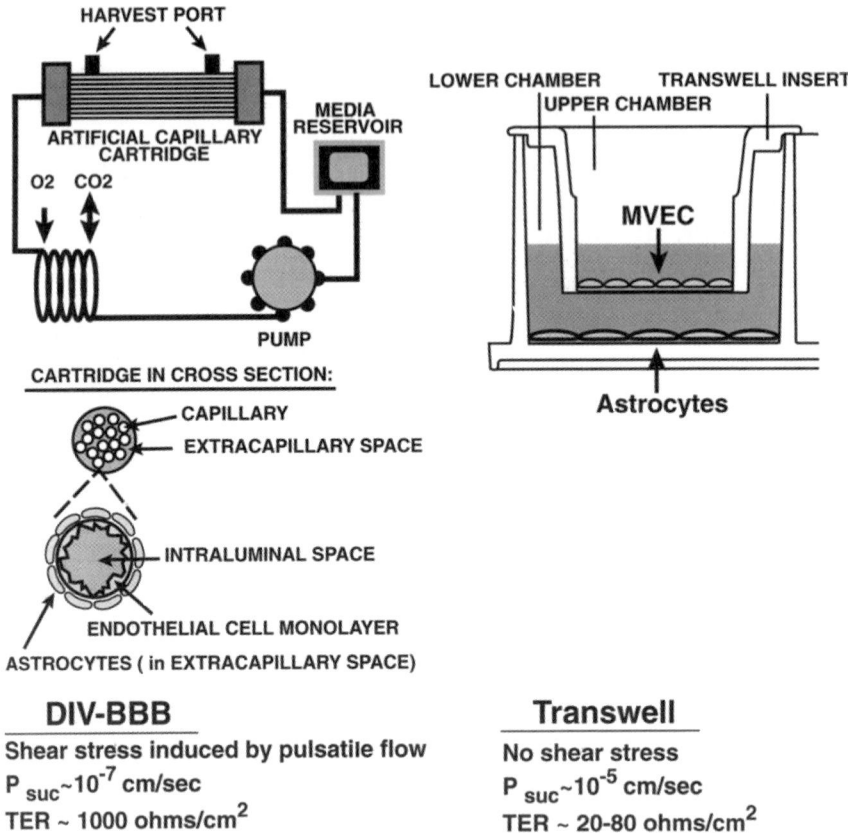

FIG 5. Comparison of *in vitro* BBB models: DIV–BBB versus transwell model.

supernatants and in infected MVECs shed into the medium at late times postinfection. The viral phenotype of SIVmac251, grown through the DIV–BBB, changed over time from non-syncytium-inducing (NSI) to syncytium-inducing (SI). The significance of the viral phenotypic change is under investigation. Sequence analysis of the *env* and *nef* regions of the input virus, versus virus isolated late during MVEC infection, revealed several amino acid changes in both genes. The detection of virus at all times postinfection demonstrates that the DIV–BBB model is permissive for persistent, productive infection by SIV. Furthermore, the demonstration of a phenotypic change in virus grown through the DIV–BBB suggests the presence of active viral replication concomitant with mutation in the MVECs of this model.

The DIV–BBB model has several features that make it a more physiologic model as compared to other models of BBB. First, the cell types utilized in the DIV–BBB are those found within the BBB *in situ,* namely, brain MVECs and astrocytes. Second, culturing cells within the context of the DIV–BBB resulted in a barrier to sucrose that approached that seen in the *in vivo* BBB. The reduced permeability to sucrose is a significant improvement over current models, most of which have P_{suc} values in the range of 10^{-3} cm/sec^2. Third, SIV infection of the BBB was not toxic to cells and could be established in the presence of a forming barrier. Additionally, SIV infection did not appear to prevent the formation of a barrier. In this regard, it is interesting to note that pseudotyped HIV infection of a polarized epithelial cell monolayer resulted in a transient delay in the re-formation of EGTA-disrupted tight junctions as compared to mock-infected cells (A. V. Moses, unpublished results, 1999), and that HIV is capable of causing significant disruptions in tight junctions (Dallasta *et al.,* 1999). In the DIV–BBB, one of the limitations of the model is that no such transient effects are measurable. Thus, while SIV infection of MVECs did not appear to adversely affect barrier formation, transient effects on the barrier or on BBB formation cannot be ruled out. Fourth, a productive and persistent infection of the DIV–BBB was established, in which virus was capable of traversing the MVEC monolayer into the extracapillary space (astrocyte layer), suggesting that SIV is able to cross the BBB from the lumenal to the ablumenal compartments.

Preliminary evidence from SIV infection of the DIV-BBB indicates that cell–cell contact (as in MVECs with T cells or monocytes) may be important to viral dissemination. Thus, cell-free culture supernatants taken from adherent MVECs from the DIV-BBB, were able to induce syncytia in CEM × 174 indicator cells after 11 days of coculture, whereas CEM × 174 cells added directly to similar cultures of adherent MVEC demonstrated syncytia in as little as 3 days. Several studies have outlined the importance, in HIV and SIV infections, of MVEC-induced monocyte adherence (Dhawan *et al.,* 1995; Nottet *et al.,* 1996; Weiss *et al.,* 1999; Persidsky *et al.,* 1997; Sasseville *et al.,* 1995; Gilles *et al.,* 1995). It will now be possible to examine cell trafficking through the BBB. Infected macrophages can be pulsed through the MVEC-lined lumen of the DIV-BBB and their trafficking patterns followed over time.

The DIV-BBB model demonstrates properties that closely approximate the *in vivo* BBB. As such, the model will be useful for examining mechanisms of viral persistence and activation, as well as for examining viral entry into the CNS through the BBB. In addition, the DIV-BBB can be used to study viral interference with either tight junction formation or transporter function associated with a functional BBB. Finally, the DIV-

BBB model can be used to test the delivery of antiretroviral drugs across the BBB in the presence or absence of HIV/SIV infection. Additionally, the DIV-BBB can be utilized to test the effectiveness of antiretroviral therapy and of other prophylactic therapeutics on HIV/SIV persistence in both the BBB itself and within cells of the CNS.

REFERENCES

Adamson, D. C., Dawson, T. M., Zink, M. C., Clements, J. E., and Dawson V. L. (1996). Neurovirulent simian immunodeficiency virus infection induces neuronal, endothelial, and glial apoptosis. *Mol. Med.* **2:**417–428.

Ades, E. W., Comans, T. W., Nicholson, J. K. A., and Browning, S. W. (1993). Lack of evidence that human immunodeficiency virus can infect human endothelial cells *in vitro*. *J. Acquir. Immune Defic. Syndr.* **6:**104.

An, S. F., Giometto, B., and Scaravilli, F. (1996). HIV-1 DNA in brains in AIDS and pre-AIDS: correlation with the stage of disease. *Ann. Neurol.* **40:**611–617.

An, S. F., Groves, M., Gray, F., and Scaravilli, F. (1999). Early entry and widespread cellular involvement of HIV-1 DNA in brains of HIV-1 positive asymptomatic individuals. *J. Neuropathol. Exp. Neurol.* **58:**1156–1162.

Bagasra, O., Lavi, E., Bobroski, L., Khalili, K., Pestaner, J. P., Tawadros, R., and Pomerantz, R. J. (1996). Cellular reservoirs of HIV-1 in the central nervous system of infected individuals: identification by the combination of *in situ* polymerase chain reaction and immunohistochemistry. *AIDS* **10:**573–585.

Baldeweg, T., Catalan, J., and Gazzard, B. (1998). Risk of HIV dementia and opportunistic brain disease in AIDS and zidovudine therapy. *J. Neurology, Neurosurgery and Psychiatry* **65:**34–41.

Ballerman, B. J., and Ott, M. J. (1995). Adhesion and differentiation of endothelial cells by exposure to chronic shear stress: A vascular graft model. *Blood Purif.* **13:**125–134.

Beck, D. W., Vinters, H. V., Hart, M. N., and Cancilla, P. A. (1984). Glial cells influence polarity of the blood–brain barrier. *J. Neuropathol. Exp. Neurol.* **43:**219–224.

Beebe, A. M., Gluckstern, T. G., George, J., Pedersen, N. C., and Dandekar, S. (1992). Detection of feline immunodeficiency virus infection in bone marrow of cats. *Vet. Immunol. Immunopathol.* **35:**37–49.

Beekhuizen, H., and vanFurth, R. (1993). Monocyte adherence to human vascular endothelium. *J. Leukocyte Biol.* **54:**363–378.

Beilke, M. (1989). Vascular endothelium in immunology and infectious disease. *Reviews of Infectious Disease* **11:**273–283.

Bell, J. E., Busuttil, A., Ironside, J. W., Rebus, S., Donaldson, Y. K., Simmonds, P., and Peutherer, J. F. (1993). Human immunodeficiency virus and the brain: investigations of virus load and neuropathologic changes in pre-AIDS subjects. *J. Infectious Disease* **168:**818–824.

Belman, A. L., Ultmann, M. H., Horoupian, D., Novick, B., Spiro, A. J., Rubenstein, A., Kurtzburg, D., and Cone-Wesson, B. (1985). Neurologic complications in infants and children with acquired immunodeficiency syndrome. *Ann. Neurol.* **18:**560–566.

Brouwers, P., Hendricks, M., Lietzau, J. A., Mitsuya, H., Broder, S., and Yarchoan, R. (1997). Effect of combination therapy with zidovudine and didanosine on neuropsychological functioning in patients with symptomatic HIV disease: a comparison of simultaneous and alternating regimens. *AIDS* **11:**59–66.

Budka, H. (1991). Neuropathology of human immunodeficiency virus infection. *Brain Pathol.* **1:**163–175.

Budka, H., Wiley, C. A., Kleihues, P., Artigas, J., Asbury, A. K., Cho, E. S., Cornblath, D. R., Dal Canto, M. C., DeGirolani, U., Dickson, D. *et al.* (1991). HIV-associated disease of the central nervous system: review of nomenclature and proposal for neuropathology-based terminology. *Brain Pathol.* **1:**143–152.

Catalan, J. (1998). Update on HIV-1-associated dementia and related disorders. *AIDS Targeted Information* **12:**R63–R65.

Cenacchi, G., Re, M. C., Preda, P., Pasquinelli, G., Furlini, G., Apkarian, R. P., La Place, M., and Martinelli, G. M. (1992). Human immunodeficiency virus type-1 (HIV-1) infection of endothelial cells *in vitro:* a virological, ultrastructural and immunocytochemical approach. *J. Submicrosc. Cytol. Pathol.* **24:**155–161.

Chakrabarti, L., Hurtrel, M. I., Maire, M.-A., Vazeux, R., Dormont, D., Montagnier, L., and Hurtrel, B. (1991). Early viral replication in the brain of SIV-infected rhesus monkeys. *Am. J. Pathol.* **139:**1273–1280.

Childs, E. A., Lyles, R. H., Selnes, O. A., Chen, B., Miller, E. N., Cohen, B. A., Becker, J. T., Mellors, J., and McArthur, J. C. (1999). Plasma viral load and CD4 lymphocytes predict HIV-associated dementia and sensory neuropathy. *Neurology* **52:**607–613.

Chun, T.-W., Stuyver, L., Mizell, S. B., Ehler, L. A., Mican, J. M., Baseler, M., Lloyd, A. L., Nowak, M. A., and Fauci, A. S. (1997). Presence of an inducible HIV-1 latent reservoir during highly active antiretroviral therapy. *Proc. Natl. Acad. Sci. U.S.A.* **94:**13193–13197.

Chun, T.-W., Davey, R. T. Jr, Engel, D., Lane, H. C., and Fauci, A. S. (1999). Reemergence of HIV after stopping therapy. *Nature* **401:**874–875.

Cohen, J. (1998). Exploring how to get at—and eradicate—hidden HIV. *Science* **279:**1854–1855.

Conaldi, P. G., Serra, C., Dolei, A., Basolo, F., Falcone, V., Mariani, G., Speziale, P., and Toniolo, A. (1995). Productive HIV-1 infection of human vascular endothelial cells requires cell proliferation and is stimulated by combined treatment with interleukin-1α plus tumor necrosis factor-α. *J. Med. Virology* **47:**355–363.

Corbeil, J., Evans, L. E., McQueen, P. W., Vasak, E., Edward, P. D., Richman, D. D., Penny, R., and Cooper, D. A. (1995). Productive *in vitro* infection of human umbilical vein endothelial cells and three colon carcinoma cell lines with HIV-1. *Immunology and Cell Biology* **73:**140–145.

Dallasta, L. M., Pisarov, L. A., Esplen, J. E., Werley, J. V., Moses, A. V., Nelson, J. A., and Achim, C. L. (1999). Blood-brain barrier tight junction disruption in human immunodeficiency virus encephalitis. *Am. J. Pathol.* **155:**1915–1927.

Davis, L. E., Hjelle, B. L., Miller, V. E., Palmer, D. L., Llewellyn, A. L., Merlin, T. L., Young, S. A., Mills, R. G., Wachsman, W., and Wiley, C. A. (1992). Early viral brain invasion in iatrogenic human immunodeficiency virus infection. *Neurol.* **42:**1736–1739.

Desrosiers, R. C. (1990). The simian immunodeficiency viruses. *Annu. Rev. Immunol.* **8:**557–578.

Dhawan, S., Weeks, B. S., Soderland, C., Schnapper, H. W., Toro, L. A., Asthana, S. P., Hewlett, I. K., Stetler-Stevenson, W. G., Yamada, S. S., Yamada, K. M., et al. (1995). HIV-1 infection alters monocyte interactions with human microvascular endothelial cells. *J. Immunol.* **154:**422–432.

Edinger, A. L., Mankowski, J. L., Dorantz, B. J., Margulies, B. J., Lee, B., Rucker, J., Sharron, M., Hoffman, T. L., Berson, J. F., Zink, M. C., Hirsch, V. M., Clements, J. E., and Doms, R. W. (1997). CD4-independent, CCR5-dependent infection of brain capillary endothelial cells by a neurovirulent simian immunodeficiency virus strain. *Proc. Natl. Acad. Sci. U.S.A.* **94:**14742–14747.

Edinger, A. L., Hoffman, T. L., Sharron, M., Lee, B., Yi, Y., Choe, W., Kolson, D. L., Mitrovic, B., Zhou, Y., Faulds, D., Collman, R. G., Hesselgesser, J., Horuk, R., and Doms, R. W. (1998). An orphan seven-transmembrane domain receptor expressed widely in the brain functions as a coreceptor for human immunodeficiency virus type 1 and simian immunodeficiency virus. *J. Virol.* **72:**7934–7940.

Epstein, L. G., Sharer, L. R., Joshi, V. V., Fojas, M., Koenigsberger, M. R., and Oleske, J. M. (1985). Progressive encephalopathy in children with acquired immunodeficiency syndrome. *Ann. Neurol.* **17:**488–496.

Finzi, D., Hermankova, M., Pierson, T. Carruth, L. M., Buck, C., Chaisson, R. E., Quinn, T. C., Chadwick, K., Margolick, J., Brookmeyer, R., Gallant, J., Markowitz, M., Ho, D. D., Richman, D. D., and Siciliano, R. F. (1997). Identification of a reservoir for HIV-1 in patients on highly active antiretroviral therapy. *Science* **278:**1295–1300.

Finzi, D., Blankson, J., Siciliano, J. D., Margolick, J. B., Chadwick, K., Pierson, T., Mith, K., Lisziewicz, J., Lori, F., Flexner, C., Quinn, T. C., Chaisson, R. E., Rosenberg, E., Walker, B., Gange, S., Gallant, J., and Siciliano, R. F. (1999). Latent infection of CD4+ T cells provides a mechanism for lifelong persistence of HIV-1, even in patients on effective combination therapy. *Nature Med.* **5:**512–517.

Flaherty, M. T., Hauer, D. A., Mankowski, J. L., Zink, M. C., and Clements, J. E. (1997). Molecular and biological characterization of a neurovirulent molecular clone of simian immunodeficiency virus. *J. Virol.* **71:**5790–5798.

Gabuzda, D. H., Ho, D. D., de la Monte, S. M., Hirsch, M. S., Rota, T. R., and Sobel, R. A. (1986). Immunohistochemical identification of HTLV III antigen in brains of patients with AIDS. *Ann. Neurol.* **20:**289–295.

Genis, P., Jett, M., Bernton, E. W., Boyle, T., Gelbard, H. A., Dzenko, K., Keane, R. W., Resnick, L., Mizrachi, Y., Volsky, D. J., *et al.* (1992). Cytokines and arachadonic metabolites produced during human immunodeficiency virus (HIV)-infected macrophage-astroglia interactions: implications for the neuropathogensis of HIV disease. *J. Exp. Med.* **176:**1703–1718.

Gilles, P. N., Lathey, J. L., and Spector, S. A. (1995). Replication of macrophage-tropic and T-cell-tropic strains of human immunodeficiency virus type 1 is augmented by macrophage-endothelial cell contact. *J. Virol.* **69:**2133–2139.

Giulian, D., Vaca, K., and Noonan, C. A. (1990). Secretion of neurotoxins by mononuclear phagocytes infected with HIV-1. *Science* **250:**1593–1596.

Giulian, D., Vaca, K., and Corpuz, M. (1993). Brain glia release factors with opposing actions upon neuronal survival. *J. Neurosci.* **13:**29–37.

Glass, J. D., Wesselingh, S. L., Selnes, O. A., and McArthur, J. C. (1993). Clinical-neuropathologic correlation in HIV-associated dementia. *Neurology* **43:**2230–2237.

Glass, J. D., and Johnson, R. T. (1996). Human immunodeficiency virus and the brain. *Annual Rev. Neurosci.* **19:**1–26.

Gold, L. H., Fox, H. S., Henriksen, S. J., Buchmeier, M. J., Weed, M. R., Taffe, M. A., Huitron-Resendez, S., Horn, T. F. W., and Bloom, F. E. (1998). Longitudinal analysis of behavioral neurophysiological viral and immunological effects of SIV infection in rhesus monkeys. *J. Med. Primatol.* **27:**104–112.

Grant, G. A., Abbott, N. J., and Janigro, D. (1998). Understanding the physiology of the blood–brain barrier: *In vitro* models. *News Physiol. Sci.* **13:**287–293.

Gray, F., Gherardi, R., and Scaravilli, F. (1988). The neuropathology of the acquired immunodeficiency syndrome (AIDS). A review. *Brain* **111:**245–266.

Green, D. F., Resnick, L., and Bourgoignie, J. J. (1992). HIV infects glomerular endothelial and mesangial but not epithelial cells *in vitro*. *Kidney International* **41:**956–960.

Haase, A. T. (1986). Pathogenesis of lentivirus infections. *Nature* **322:**130–136.

Hayes, K. A., Wilkinson, J. G., Frick, R., Francke, S., and Mathes, L. E. (1995). Early suppression of plasma viremia by ZDV does not alter the spread of feline immunodeficiency virus infection in cats. *J. Acquir. Immune Defic. Syndr. Hum. Retrovirol.* **9:**114–122.

Heyes, M. P., Brew, B. J., Martin, A., Price, R. W., Salazar, A. M., Sidtis, J. J., Yergey, J. A., Mouradian, M. M., Sadler, A. E., Keilp, J. *et al.* (1991). Quinolinic acid in cerebrospinal fluid and serum in HIV-1 infection: relationship to clinical and neurological status. *Ann. Neurol.* **29:**202–209.

Hickey, W. F. (1991). Migration of hematogenous cells through the blood–brain barrier and the initiation of CNS inflammation. *Brain Pathol.* **1:**97–105.

Ho, D. D., Rota, T. R., Schooley, R. T., Kaplan, J. C., Allan, J. D., Groopman, J. E., Resnick, L., Felsenstein, D., Andrews, C. A., and Hirsch, M. S. (1985). Isolation of HTLV-III from CSF and neural tissues of patients with AIDS-related neurologic syndromes. *New Engl. J. Med.* **313:**1493–1497.

Horn, T. F., Huitron-Resendiz, S., Weed, M. R., Henriksen, S. J., and Fox, H. S. (1998). Early physiological abnormalities after simian immunodeficiency virus infection. *Proc. Natl. Acad. Sci. U.S.A.* **95:**15072–15077.

Janigro, D., Strelow, L., Grant, G., and Nelson, J. A. (1998). An *in vitro* blood brain barrier model for HIV-induced CNS disease. *NeuroAIDS* **1:** issue 4.

Janigro, D., Leaman, S. M., and Stanness, K. A. (1999). Dynamic *in vitro* modeling of the blood–brain barrier: a novel tool for studies of drug delivery to the brain. *Pharmaceutical Science and Technology Today* **2:**7–12.

Joo, F. (1985). The blood–brain barrier *in vitro:* Ten years of research on microvessels isolated from brain. *Neurochem. Intl.* **7:**1–25.

Kaiser, P. K., Offerman, J. T., and Lipton, S. A. (1990). Neuronal injury due to HIV-1 envelope protein is blocked by anti-gp120 antibodies but not by anti-CD4 antibodies. *Neurology* **40:**1757–1761.

Kestler, H., Kodama, T., Ringler, D., Martas, M., Pedersen, N., Lackner, A., Regier, D., Sehgal, P., Daniel, M. King, N., and Desrosiers, R. (1990). Induction of AIDS in rhesus monkeys by molecularly cloned simian immunodeficiency virus. *Science* **248:**1109–1112.

Kindt, T., Hirsch, V., Johnson, P., and Sawadikosol, S. (1992). Animal models for acquired immunodeficiency syndrome. *Adv. Immunol.* **52:**425–473.

Koenig, S., Gendelman, H. E., Orenstein, J. M., Del Canto, M. C., Pezeshkpour, G. M., Yungbluth, M., Jenotta, F., Akasmit, A., Martin, M. A., and Fauci, A. S. (1986). Detection of AIDS virus in macrophages in brain tissue from AIDS patients with encephalopathy. *Science* **233:**1089–1093.

Koyanagi, Y., Miles, S., Mitsuyasu, R. T., Merrill, J. E., Vinters, H. V., and Chen, I. S. (1987). Dual infection of the central nervous system by AIDS viruses with distinct cellular tropisms. *Science* **236:**819–822.

Kure, K., Llena, J. F., Lyman, W. D., Soeiro, R., Weidenheim, K. M., Hirano, A., and Dickson, D. W. (1991). Human immunodeficiency virus-1 infection of the nervous system: an autopsy study of 268 adult, pediatric, and fetal brains. *Hum. Pathol.* **22:**700–710.

Lackner, A. A., Dandekar, S., and Gardner, M. B. (1991a). Neurobiology of simian and feline immunodeficiency virus infections. *Brain Pathol.* **1:**202–212.

Lackner, A., Smith, M., Munn, R., Martfield, D., Gardner, M., Marx, P. and Dandekar, S. (1991b). Localization of simian immunodeficiency virus in the central nervous system of rhesus monkeys. *Am. J. Pathol.* **139:**609–621.

Lafon, M. E., Steffan, A. M., Gendrult, J. L., Klein-Soyer, C., Gloeckler-Tondre, L., Royer, C., and Kirn, A. (1992). Interaction of human immunodeficiency virus with human macrovascular endothelial cells *in vitro. AIDS Res. Hum. Retroviruses* **8:**1567–1569.

Lander, J. K., Chesebro, B., and Fan, H. (1999). Appearance of mink cell focus-inducing recombinants during *in vivo* infection by moloney murine leukemia virus (M-MuLV) or the Mo+PyF101 M-MuLV enhancer variant: implications for sites of generation and roles in leukemogenesis. *J. Virol.* **73:**5671–5680.

Lane, T. E., Buchmeier, M. J., Watry, D. D., Jakubowski, D. B., and Fox, H. S. (1995). Serial passage of microglial SIV results in selection of homogeneous *env* quasispecies in the brain. *Virology* **212:**458–465.

Lassman, H., Zimprich, F., Rossler, K., and Vass, K. (1991). Inflammation in the nervous system. *Rev. Neurol.* **147:**763–781.

Lazarini, F., Seilhean, D., Rosenblum, O., Suarez, S., Conquy, L., Uchihara, T., Sazdovich, V., Mokhtari, K., Maisonobe, T., Boussin, F., Katlama, C., Bricaire, F., Duyckaerts, C., and Hauw, J.-J. (1997). Human immunodeficiency virus type 1 DNA and RNA load in brains of demented and non-demented patients with acquired immunodeficiency syndrome. *J. NeuroVirol.* **3:**299–303.

Letvin, N. L., Daniel, M. D., Sehgal, P. K., Desrosiers, R. C., Hunt, R. D., Waldron, L. M., MacKey, J. J., Schmidt, D. K., Chalifoux, L. V., and King, N. W. (1985). Induction of AIDS-like disease in macaque monkeys with T-cell tropic retrovirus STLV-III. *Science* **230:**71–73.

Levy, J. A., Shimabukuro, J., Hollander, H., Mills, J., and Kaminsky, L. (1985). Isolation of AIDS-associated retrovirus from cerebrospinal fluid and brain of patients with neurological symptoms. *Lancet* **2:**586–588.

Mankowski, J. L., Spelman, J. P., Ressetar, H. G., Strandberg, J. D., Laterra, J., Carter, D. L., Clements, J. E., and Zink, M. C. (1994). Neurovirulent simian immunodeficiency virus replicates productively in endothelial cells of the central nervous system in vivo and in vitro. *J. Virol.* **68:**8202–8208.

Mankowski, J. L., Flaherty, M. L., Spelman, J. P., Hauer, D. A., Didier, P. J., Amedee, A. M., Murphey-Corb, M., Kirstein, L. M., Munoz, A., Clements, J. E., and Zink, M. C. (1997). Pathogenesis of simian immunodeficiency virus encephalitis: viral determinants of neurovirulence. *J. Virol.* **71:**6055–6060.

Matsumoto, M., Hidaka, K., Akiho, H., Tada, S., Okada, M., and Yamaguchi, T. (1996). Low stringency hybridization study of the dopamine D4 receptor revealed D4-like mRNA distribution of the orphan seven-transmembrane receptor, APJ, in human brain. *Neurosci. Letters* **219:**119–122.

McArthur, J. C., Nance-Sproson, T. E., Griffin, D. E., Hoover, D., Selnes, O. A., Miller, E. N., Margolick, J. B., Cohen, B. A., Farzadegan, H., and Saah, A. (1993). The diagnostic utility of elevation in cerebrospinal fluid beta-2-microglobulin in HIV-1 dementia. Multicenter AIDS cohort study. *Neurology* **42:**1707–1712.

Michaels, J., Sharer, L. R., and Epstein, L. G. (1988). Human immunodeficiency virus type 1 (HIV-1) infection of the central nervous system: A review. *Immunodefic. Rev.* **1:**71–104.

Moses, A. V., Bloom, F. E., Pauza, C. D., and Nelson, J. A. (1993). HIV infection of human brain capillary endothelial cells occurs via a CD4/galactosylceramide-independent mechanism. *Proc. Natl. Acad. Sci. U.S.A.* **90:**10474–10478.

Moses, A. V., and Nelson, J. A. (1994). HIV infection of human brain capillary endothelial cells—implications for dementia. *Advances in Neuroimmunology* **4:**239–247.

Moses, A. V., Stenglein, S. G., Strussenberg, J. G., Wehrly, K., Chesebro, B., and Nelson, J. A. (1996a). Sequences regulating tropism of human immunodeficiency virus type 1 for brain capillary endothelial cells map to a unique region on the viral genome. *J. Virol.* **70:**3401–3406.

Moses, A. V., Williams, S., Heneveld, M. L., Strussenberg, J., Rarick, M., Loveless, M., Bagby, G., and Nelson, J. A. (1996b). Human immunodeficiency virus infection of bone marrow endothelium reduces induction of stromal hematopoietic growth factors. *Blood* **87:**919–925.

Murray, E. A., Rausch, D. M., Lendvay, J., Sharer, L. R., and Eiden, L. E. (1991). Cognitive and motor impairments associated with SIV infection in rhesus monkeys. *Science* **255**:1246–1249.

Navia, B., Jordan, B. D., and Price, R. W. (1986a). The AIDS dementia complex: I. Clinical features. *Ann. Neurol.* **19**:517–524.

Navia, B. A., Cho, E.-S., Petito, C. K., and Price, R. W. (1986b). The AIDS dementia complex: II. Neuropathology. *Ann. Neurol.* **19**:525–535.

Nottet, H. S., Persidsky, Y., Sasseville, V. G., Nukuna, A. N., Bock, P., Zhai, Q. H., Sharer, L. R., McComb, R. D., Swindells, S., Soderland, C., and Gendleman, H. E. (1996). Mechanisms for the transendothelial migration of HIV-1-infected monocytes into brain. *J. Immunol.* **156**:1284–1295.

Novembre, F. J., De Rosayro, J., O'Neil, S. P., Anderson, D. C., Klumpp, S. A., and McClure, H. M. (1998). Isolation and characterization of a neuropathogenic simian immunodeficiency virus derived from a sooty mangabey. *J. Virol.* **72**:8841–8851.

Okazaki, O. (1989). "Fundamentals of neuropathology: morphologic basis of neurologic disorders," 2nd ed. Igaku-Shoin, New York.

Peluso, R., Haase, A., Stowring, L., Edwards, M., and Ventura, P. (1986). A trojan horse mechanism for the spread of visna virus in monocytes. *Virology* **147**:231–236.

Persidsky, Y., Stins, M., Way, D., Witte, M. H., Weinand, M., Kim, K. S., Bock, P., Gendleman, H. E., and Fiala, M. (1997). A model for monocyte migration through the blood-brain barrier during HIV-1 encephalitis. *J. Immunol.* **158**:3499–3510.

Petito, C. K., and Cash, K. S. (1992). Blood–brain barrier abnormalities in the acquired immunodeficiency syndrome: immunohistochemical localization of serum proteins in postmortem brain. *Ann. Neurol.* **32**:658–666.

Pialoux, G., Fournier, S., Moulingnier, A., Poveda, J.-D., Clavel, F., and DuPont, B. (1997). Central nervous system as a sanctuary for HIV-1 infection despite treatment with zidovudine, lamivudine and indinavir. *AIDS* **11**:1302–1303.

Pizzo, P. A., Eddy, J., Falloon, J., Balis, F. M., Murphy, R. F., Moss, H., Wolters, P., Brouwers, P., Jarosinki, P., Rubin, M. et al. (1988). Effect of continuous intravenous infusion of zudovudine (AZT) in children with symptomatic HIV infection. *New Engl. J. Med.* **319**:889–896.

Poland, S. D., Rice, G. P. A., and Dekaban, G. A. (1995). HIV-1 infection of human brain-derived microvascular endothelial cells in vitro. *J. Acquir. Immune Defic. Syndr. Hum. Retroviruses* **8**:437–445.

Pomerantz, R. J., Kuritzkes, D. R., de la Monte, S. J., Rota, T. R., Baker, A. S., Albert, D., Bor, D. H., Feldman, E. L., Schooley, R. T., and Hirsch, M. S. (1987). Infection of the retina by human immunodeficiency virus type 1. *New Engl. J. Med.* **317**:1643–1647.

Portegies, P., de Gans, J., Langer, J., Derix, M., Speelman, H., and Bakker, R. (1989). Declining incidence of ADC after the introduction of AZT treatment. *British Med. J.* **299**:819–821.

Power, C., Kong, P.-A., Crawford, T. O., Wesselingh, S., Glass, J. D., McArthur, J. C., and Trapp, B. D. (1993). Cerebral white matter changes in acquired immunodeficiency syndrome dementia: alterations of the blood–brain barrier. *Ann. Neurol.* **34**:339–350.

Power, C., and Johnson, R. T. (1995). HIV-1-associated dementia: clinical features and pathogenesis. *Can. J. Neurol. Sci.* **22**:92–100.

Price, R. W., Brew, B., Sidtis, J., Rosenblum, M., Scheck, A. C., and Cleary, P. (1988). The brain in AIDS: central nervous system HIV-1 infection and AIDS dementia complex. *Science* **239**:586–591.

Prospero-Garcia, O., Gold, L. H., Fox, H. S., Polis, I., Koob, G. F., Bloom, F. E., and Henriksen, S. J. (1996). Microglia-passaged simian immunodeficiency virus induces neurophysiological abnormalities in monkeys. *Proc. Natl. Acad. Sci. U.S.A.* **93**:14158–14163.

Pumarola-Sune. T., Navia, B. A., Cordon-Cardo, C., Cho, E.-S., and Price, R. W. (1987). HIV antigen in the brains of patients with the AIDS dementia complax. *Ann. Neurol.* **21:**490–496.

Quagliarello, V. J., Wispelway, B., Long, W. J. Jr., and Scheld, W. M. (1991). Recombinant human interleukin-1 induces meningitis and blood–brain barrier injury in the rat. Characterization and comparison with tumor necrosis factor. *J. Clin. Invest.* **87:**1360–1366.

Ramilo, O., Saez-Llorens, X., Mertsola, J., Jafari, H., Olsen, K. D., Hansen, E. J., Yoshinaga, M., Ohkawara, S., Nariuchi, H., and McCracken, G. H. Jr. (1990). Tumor necrosis factor-alpha/cachetin and interleukin 1-beta initiate meningeal inflammation. *J. Exp. Med.* **172:**497–507.

Re, M. C., Furlini, G., Cennachi, G., Preda, P., and LaPlaca, M. (1991). Human immunodeficiency virus type 1 infection of endothelial cells *in vitro*. *Microbiologica* **14:**149–152.

Rhodes, R. H. (1991). Evidence of serum-protein leakage across the blood–brain barrier in the acquired immunodeficiency syndrome. *J. Neuropathol. Exp. Neurol.* **50:**171–183.

Ringler, D. J., Hunt, R. D., Desrosiers, R. C., Daniel, M. D., Chalifoux, L. V., and King, N. W. (1988). Simian immunodeficiency virus-induced meningoencephalitis: natural history and retrospective study. *Ann. Neurol.* **23:**S101–S107.

Rosenberg, Z. F., and Fauci, A. S. (1990). Immunopathogenic mechanisms of HIV infection: cytokine induction of HIV expression. *Immunol. Today* **11:**176–180.

Rosenblum, M. K. (1990). Infection of the central nervous system by the human immunodeficiency virus type 1: morphology and relation to syndromes of progressive encephalopathy and myelopathy in patients with AIDS. *Pathol. Ann.* **1:**117–169.

Rottman, J. B., Ganley, K. P., Williams, K., Wu, L., Mackay, C. R., and Ringler, D. J. (1997). Cellular localization of the chemokine receptor CCR5: correlation to cellular targets of HIV-1 infection. *Am. J. Pathol.* **151:**1341–1351.

Saito, Y., Sharer, L. R., Epstein, L. G., Michaels, J., Mintz, M., Louder, M., Golding, K., Cvetkovich, T. A., and Blumberg, B. M. (1994). Overexpression of nef as a marker for restricted HIV-1 infection of astrocytes in postmortem pediatric central nervous tissues. *Neurology* **44:**474–481.

Sanders, V. J., Pittman, C. A., White, M. G. Wang, G., Wiley, C. A., and Achim, C. L. (1998). Chemokines and receptors in HIV encephalitis. *AIDS* **12:**1021–1026.

Sasseville, V. G., Lane, L. H., Walsh, D., Ringler, D. J., and Lackner, A. A. (1995). VCAM-1 expression and leukocyte trafficking to the CNS occur early in infection with pathogenic isolates of SIV. *J. Med. Primatology* **24:**123–131.

Saukkonen, K., Sande, S., Cioffe, C., Wolpe, S., Shery, B., Cerami, A., and Tuomanen, E. (1990). The role of cytokines in the generation of inflammation and tissue damage in experimental gram-positive meningitis. *J. Exp. Med.* **171:**439–448.

Scheglovitova, O., Capobianchi, M. R., Antonelli, G., Guamnu, D., and Dianzani, F. (1993). CD4-positive lymphoid cells rescue HIV-1 replication from abortively infected primary endothelial cells. *Archives of Virology* **132:**267–280.

Schmitt, F. A., Bigley, J. W., McKinnis, R., Logue, P. E., Evans, R. W., and Drucker, J. L. (1988). Neuropsychological outcome of zidovudine (AZT) treatment of patients with AIDS and AIDS-related complex. *New Engl. J. Med.* **319:**1573–1578.

Sharer, L. R., Baskin, G. B., Cho, E.-S., Murphey-Corb, M., Blumberg, B. M., and Epstein, L. G. (1988). Comparison of simian immunodeficiency virus and human immunodeficiency virus encephalitides in the immature host. *Ann. Neurol.* **23** Suppl:108–112.

Sharer, L. R., Michaels, J., Murphey-Corb, M., Hu, F. S., Kuebler, D. J., Martin, L. N., and Baskin, G. B. (1991). Serial pathogenesis study of SIV brain infection. *J. Med. Primatol.* **20:**211–217.

Sharer, L. R. (1992). Pathology of HIV-1 infection of the central nervous system: a review. *J. Neuropathol. Exp. Neurol.* **51:**3–11.

Sharma, D. P., Zink, M. C., Anderson, M., Adams, R., Clements, J. E., Joag, S. V., and Narayan, O. (1992). Derivation of neurotropic simian immunodeficiency virus from exclusively lymphocytotropic parental virus: pathogenesis of infection in macaques. *J. Virol.* **66:**3550–3556.

Shaw, G. M., Harper, M. E., Hahn, B. H., Epstein, L. G., Gajdusek, D. C., Price, R. W., Navia, B. A., Petito, C. K., O'Hara, C. J., Groopman, J. E. *et al.* (1985). HTLV-III infection in brains of children and adults with AIDS encephalopathy. *Science* **227:**177–182.

Sidtis, J., Gatsonis, C., Price, R., Singer, E., Collier, A., and Richmond, D. (1993). Zidovudine treatment of the AIDS dementia complex: placebo-controlled trial. AIDS Clinical Trials Group. *Ann. Neurol.* **33:**343–349.

Simon, M. A., Chalifoux, L. V., and Ringler, D. J. (1992). Pathologic features of SIV-induced disease and the association of macrophage infection with disease evolution. *AIDS Res. Hum. Retroviruses* **8:**327–337.

Sinclair, E., Gray, F., and Scaravilli, F. (1992). PCR detection of HIV proviral DNA in the brain of an asymptomatic HIV-positive patient. *J. Neurol.* **239:**469–470.

Smith, M. O., Heyes, M. P., and Lackner, A. A. (1995). Early intrathecal events in rhesus macaques (Macaca mulatta) infected with pathogenic or nonpathogenic molecular clones of simian immunodeficiency virus. *Lab. Invest.* **72:**547–558.

Smith, T. W., De Girolami, U., Henin, D., Bolgert, F., and Hauw, J.-J. (1990). Human immunodeficiency virus (HIV) leukoencephalopathy and the microcirculation. *J. Neuropathol. Exp. Neurol.* **49:**357–370.

Spencer, D. C., and Price, R. W. (1992). Human immunodeficiency virus and the central nervous system. *Annu. Rev. Microbiol.* **46:**655–693.

Stanness, K. A., Guatteo, E, and Janigro, D. (1996). A dynamic model of the blood–brain barrier "in vitro." *Neuro Toxicology* **17:**481–496.

Stanness, K. A., Westrum, L. E., Fornaciari, E., Mascagni, P., Nelson, J. A., Stenglein, S. G., Myers, T., and Janigro, D. (1997). Morphological and functional characterization of an in vitro blood–brain barrier model. *Brain Res.* **771:**329–342.

Steffan, M., Lafon, M. E., Gendrault, J. L., Schwiter, C., Royer, C., Jaeck, D., Arnaud, J. P., Schmitt, M. P., Aubertin, A. M., and Kirn, A. (1992). Primary cultures of endothelial cells from the human liver sinusoid are permissive for HIV-1. *Proc. Natl. Acad. Sci. U.S.A.* **89:**1582–1586.

Strelow, L. I., Watry, D. D., Fox, H. S., and Nelson, J. A. (1998). Efficient infection of brain microvascular endothelial cells by an in vivo-selected neuroinvasive SIV$_{mac}$ variant. *J. Neurovirol.* **4:**269–280.

Takahashi, K., Wesselingh, S. L., Griffin, D. E., McArthur, J. C., Johnson, R. T., and Glass, J. D. (1996). Localization of HIV-1 in human brain using polymerase chain reaction/*in situ* hybridization and immunohistochemistry. *Ann. Neurol.* **39:**705–711.

Takasawa, K., Terasaki, T., Suzuki, H., and Sugiyama, Y. (1997). *In vivo* evidence for carrier-mediated efflux transport of 3'-azido-3'-deoxythymidine and 2', 3'-dideoxyinosine across the blood–brain barrier via a probenecid-sensitive transport system. *J. Pharmacol. Exp. Ther.* **281:**369–375.

Teitel, J. M., Shore, A., Read, S. E., and Schiavone, A. (1989). Immune function of vascular endothelial cells is impaired by HIV. *J. Inf. Dis.* **160:**551–552.

Tontsch, U., and Bauer, H. C. (1991). Glial cells and neurons induce blood–brain barrier related enzymes in cultured cerebral endothelial cells. *Brain Res.* **6:**271–283.

Tornatore, C., Chandra, R., Berger, J. R., and Major, E. O. (1994). HIV-1 infection of subcortical astrocytes in the pediatric central nervous system. *Neurology* **44:**481–487.

Tross, S., Price, R. W., Navia, B., Thaler, H. T., Gold, J., Hirsch, D. A., and Sidtis, J. J. (1988). Neuropsychological characterization of the AIDS dementia complex: a preliminary report. *AIDS* **2:**81–88.

Tyor, W. R., Glass, J. D., Griffin, J. W., Becker, P. S., McArthur, J. C., Bezman, L., and Griffin, D. E. (1992). Cytokine expression in the brain during AIDS. *Ann. Neurol.* **31:**349–360.

Vallat, A.-V., De Girolami, U., He, J., Mhashikar, A., Marasco, W., Shi, B., Gray, F., Bell, J., Keohane, C., Smith, T. W., and Gabuzda, D. (1998). Localization of HIV-1 coreceptors CCR5 and CXCR4 in the brain of children with AIDS. *Am. J. Pathol.* **152:**167–178.

Vazeux, R., Brousse, N., Jarry, A., Henin, D., Marche, C., Vendrenne, C., Mikol, J., Wolff, M., Michou, C., Rozenbaum, W., et al. (1987). AIDS subacute encephalitis: identification of HIV-infected cells. *Am. J. Pathol.* **126:**403–410.

Wahl, L. M., Corcoran, M. L., Pyle, S. W., Arthur, L. O., Harel-Bellan, A., and Farrar, W. L. (1989). Human immunodeficiency virus glycoprotein 120 (gp 120) induction of monocyte arachadonic acid metabolites and interleukin 1. *Proc. Natl. Acad. Sci. U.S.A.* **86:**621–625.

Ward, J. M., O'Leary, T. J., Baskins, G. B., Beveniste, R., Harris, C. A., Nora, P. L., and Rhodes, R. H. (1987). Immunohistochemical localization of human and immunodeficiency viral antigens in fixed tissue sections. *Am. J. Pathol.* **127:**199–205.

Watry, D., Lane, T. E., Streb, M., and Fox, H. S. (1995). Transfer of neuropathogenic simian immunodeficiency virus with naturally infected microglia. *Am. J. Pathol.* **146:**914–923.

Weiss, J. M., Nath, A., Major, E. O., and Berman, J. W. (1999). HIV-1 Tat induces monocyte chemoattractant protein-1-mediated monocyte transmigration across a model of the human blood-brain barrier and up-regulates CCR5 expression on human astrocytes. *J. Immunol.* **163:**2953–2959.

Wesselingh, S. L., Power, C., Glass, J. D., Tyor, W. R., McArthur, J. C., Farber, J. M., Griffin, J. W., and Griffin, D. E. (1993). Intracerebral cytokine mRNA expression in AIDS. *Ann. Neurol.* **33:**576–582.

Wiley, C. A., Schrier, R. D., Nelson, J. A., Lampert, P. W., and Oldstone, M. B. A. (1986). Cellular localization of human immunodeficiency virus infection within the brains of acquired immunodeficiency syndrome patients. *Proc. Natl. Acad. Sci. U.S.A.* **83:**7089–7093.

Wiley, C. A., and Achim, C. (1994). Human immunodeficiency virus encephalitis is the pathological correlate of dementia in acquired immunodeficiency syndrome. *Ann. Neurol.* **36:**673–676.

Wilkinson, I. D., Miller, R. F., Miszkiel, K. A., Paley, M. N. J., Hall-Craggs, M. A., Baldeweg, T., Williams, I. G., Carter, S., Newman, S. P., Kendall, B. E., Catalan, J., Chinn, R. J. S., and Harrison, M. J. G. (1997). Cerebral proton magnetic resonance spectroscopy in asymptomatic HIV infection. *AIDS* **11:**289–295.

Wong, J. K., Hezareh, M., Gunthard, H., Havlir, D. V., Ignacio, C. C., Spina, C. A., and Richman, D. D. (1997). Recovery of replication-competent HIV despite prolonged suppression of plasma viremia. *Science* **278:**1291–1295.

Yarchoan, R., Berg, G., Brouwers, P., Fischal, M. A., Spitzer, A. R., Wichman, A., Grafman, J., Thomas, R. V., Safai, B., Brunetti, A. et al. (1987). Response of human immunodeficiency virus-associated neurological disease to 3'-azido-3'-deoxythymidine. *Lancet* **1:**132–135.

Zhang, Z.-Q., Schuler, T., Zupancic, M., Wietgrefe, S., Staskus, K. A., Reimann, K. A., Reinhart, T. A., Rogan, M., Cavert, W., Miller, C. J., Veazey, R. S., Notermans, D., Little, S., Danner, S. A., Richman, D. D., Havlir, D., Wong, J., Jordan, H. L., Schacker, T. W., Racz, P., Tenner-Racz, K., Letvin, N. L., Wolinsky, S., and Haase, A. T. (1999). Sexual transmission and propagation of SIV and HIV in resting and activated CD4+ T cells. *Science* **286:**1353–1357.

Zink, M. C., Amadee, A. M., Mankowski, J. L., Craig, L., Didier, P., Carter, D. L., Munoz, A., Murphey-Corb, M., and Clements, J. E. (1997). Pathogenesis of SIV encephalitis: selection and replication of neurovirulent SIV. *Am. J. Pathol.* **151:**793–803.

NEUROIMMUNE AND NEUROVIROLOGICAL ASPECTS OF HUMAN IMMUNODEFICIENCY VIRUS INFECTION

Christopher Power and Richard T. Johnson

Departments of Clinical Neuroscience, Microbiology, and Infectious Diseases
University of Calgary, Calgary Alberta, Canada
Department of Neurology, Johns Hopkins University
Baltimore Maryland 21287

I. INTRODUCTION

Over 40 million individuals, worldwide, have been infected by the human immunodeficiency virus type 1 (HIV-1) (Nicoll and Gill, 1999). Like all lentiviruses, HIV-1 infection is defined by multisystem involvement, including neurological disease, although several lentiviral infections are associated with profound immune suppression (Table I) (Narayan and Clements, 1989). HIV-1 also shares the property of all lentiviruses—unlike many other retroviruses—of infecting macrophages, which are assumed to be pivotal in the development of organ (neurological, respiratory, gastrointestinal) disease (Levy, 1998). Although HIV-related neurological disease is most obvious at later stages of infection, over 90% of infected individuals will manifest some neurological involvement by the time of death (Johnson, 1998). The evolution of HIV-induced immunological disease begins soon after infection; it is characterized by a mononucleosis-like syndrome that is accompanied by transient leucopenia, thrombocytopenia, and a burst

389

TABLE I

NEUROLOGICAL DISEASES ASSOCIATED WITH LENTIVIRAL INFECTIONS[a]

Virus	Systemic Disease	Blood Cells Infected	CNS Signs	Neuropathology	Neural Cells Infected
Visna-maedi	Pneumonitis, arthritis, mastitis	M	+	Leukoencephalitis and demyelination	Macrophages and microglia, ? vascular endothelia
Caprine arthritis – encephalitis	Arthritis, pneumonitis, mastitis	M	+	Leukoencephalitis and demyelination	Macrophages and microglia
Equine infectious anemia	Recurrent autoimmune hemolytic crises	M	+	Ependymitis, subependymal encephalomalacia, perivascular inflammation	Perivascular and meningeal cells
Bovine immuno-deficiency	Lymphadenopathy, lymphocytosis, and wasting	T	+	Perivascular inflammation	?
Feline immuno-deficiency	Immunodeficiency with opportunistic infections	M+T	+	Perivascular mononuclear cells, gliosis, and microglial nodules	Macrophages, astrocytes, microglia
Simian immuno-deficiency	Immunodeficiency with opportunistic infections	M+T	+	Perivascular and meningeal mononuclear cells and giant cells; minimal myelin pallor	Macrophages and microglia, vascular endothelia, astrocytes
Human immuno-deficiency	Immunodeficiency with opportunistic infections	M+T	+	Gliosis, microglial nodules, giant cells; myelin pallor; neuronal loss; vacuolar myelopathy, axonal neuropathy	Macrophages and microglia, astrocytes

[a] M, Macrophages; T, T cells. Adapted with permission from Johnson (1998).

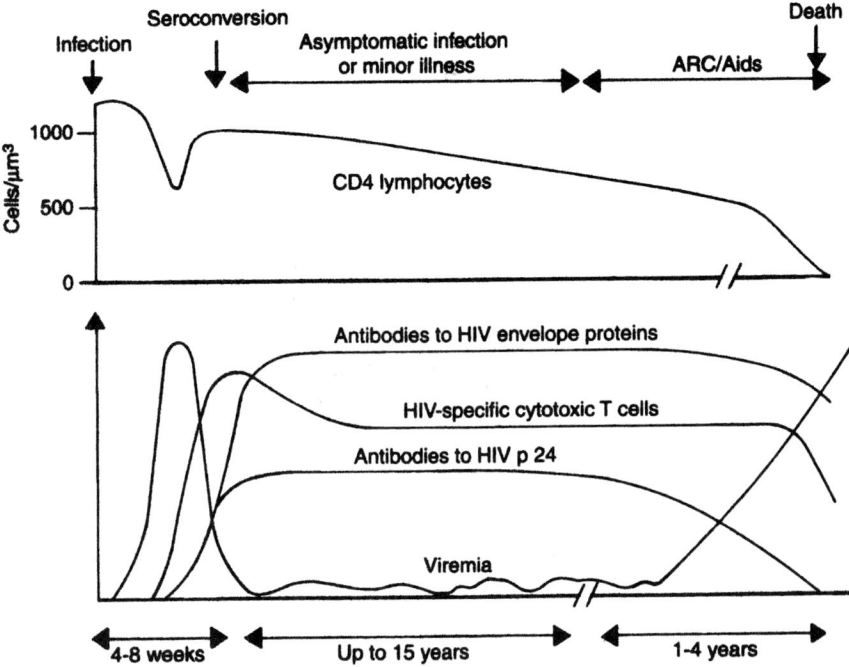

FIG. 1. Course of HIV infection displaying CD4 decline, viremia, and immunological events. Adapted with permission from Johnson (1998).

of viral replication, measured as viral RNA copies per ml (load) of plasma (Weiss, 1996). A decline in viral load and a rise in CD8 positive lymphocytes usually succeed these events, within weeks to months postcontact. Seroconversion usually occurs within 6 weeks of initial infection (Fig. 1). Thereafter, an insidious progression in viral replication occurs during a relatively symptom-free period until the eventual development of the acquired immunodeficiency syndrome (AIDS), defined by less than 200 CD4 cells/µl in blood. AIDS may present as early as 2–3 years after infection or not develop in small groups of patients, termed long-term nonprogressors. However, the onset of AIDS after initial infection in North America and Western Europe occurs in approximately 10 years (O'Brien et al., 1996). The clinical progression is accompanied by a rise in viral load and a decline in CD4 lymphocytes in blood. Opportunistic infections such as pneumocystis pneumonia, central nervous system (CNS) toxoplasmosis, or tumors such as Kaposi's sarcoma or CNS lymphoma, are usually the cause of death.

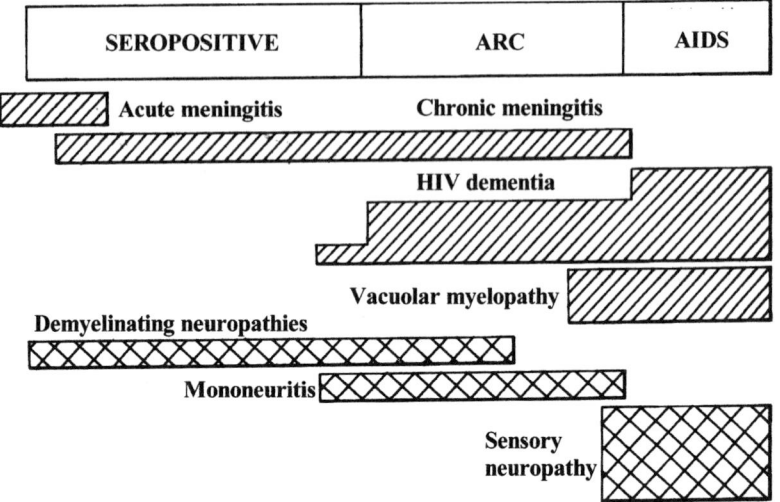

FIG. 2. Primary HIV-induced neurological disorders represented at different stages of infection and at multiple levels of the neuroaxis. Adapted with permission from Johnson (1998).

These immunological and clinical events can be arrested or retarded by antiretroviral therapies, thereby influencing the natural disease course dramatically, as recently shown by the use of highly active antiretroviral therapy (HAART) (Collier *et al.*, 1996). Throughout the course of HIV infection, different neurological syndromes may present, although individual syndromes (Johnson, 1998), such as dementia, are closely associated with the extent of immune suppression (Fig. 2). In the following discussion of HIV-related neurological disease, the clinical manifestations and neuropathogenesis will be highlighted at each stage of infection.

II. CLINICAL ASPECTS

In 1985, primary infection of the nervous system by HIV-1 was first shown by several groups (Ho *et al.*, 1985) (Levy *et al.*, 1985) that isolated virus from the cerebrospinal fluid (CSF), brain, spinal cord and peripheral nerves of patients dying with AIDS. Viral RNA was identified in microglial nodules in the brains of children, and viral DNA levels in brain were found to be higher than in lymph node and spleen (Shaw *et al.*, 1985). These findings were complemented by the demonstration of intrathecal synthesis of antibodies against HIV (Resnick *et al.,* 1985). In subsequent prospective studies, recoverable virus, pleocytosis, and elevated immunoglobulins were shown in the CSF of two-thirds of infected healthy gay men soon after seroconversion (McArthur *et al.,* 1988). These latter findings indicated that the CNS (McArthur *et al.,* 1992) was infected early after infection and that productive infection was occurring in the absence of overt neurological disease. Early on in the epidemic, it became apparent that two general categories of neurological diseases developed as HIV infection progressed; the first group includes opportunistic infections of the CNS such as toxoplasmosis, lymphoma, and cryptococcal meningitis, which are complications of systemic immune suppression; the second group includes primary HIV-induced syndromes such as dementia, myelopathies, and peripheral neuropathies (McArthur, 1987). These latter conditions may emerge soon after primary infection and include an acute demyelinating peripheral neuropathy, similar to Guillain-Barré syndrome, acute encephalopathy, and meningitis (McArthur, 1987). Other syndromes occur during the asymptomatic period of infection, such as recurrent headaches with pleocytosis or mononeuritis multiplex defined by vasculitic infarctions of multiple nerves. However, the neurological syndromes with greatest morbidity were dementia, vacuolar myelopathy, and painful axonal sensory neuropathy; these conditions have attracted the most attention over the past decade (Simpson and Tagliati, 1994; Epstein and Gendelman, 1993; Lipton and Gendelman, 1995).

A. Early HIV-associated Neurological Syndromes

Approximately 50% of infected individuals experience fever and headache at the time of seroconversion, and a subset of these individuals will develop nuchal rigidity, cranial nerve palsies, photophobia, myelopathy, radiculopathies, acute demyelinating neuropathy, and, on rare occasions, an acute encephalopathy that may lead to death (Hollander and Stringari, 1987; McArthur, 1987). A mild cognitive impairment may be recognized in HIV-infected persons prior to AIDS, which has been termed Minor Motor/Cognitive Disorder (Janssen, 1991; Marder, 1996) and may precede the development of dementia (Ellis *et al.,* 1997). The relative impact of the latter syndrome on day-to-day life remains unclear. These syndromes may be accompanied by a pleocytosis that reflects CD4:CD8 ratios in blood (McArthur, 1987). The occurrence of these neurological symptoms and signs soon after primary infection may portend an accelerated progression to AIDS (Boufassa *et al.,* 1995). Neuropathological studies of patients dying from other causes, soon after seroconversion or during the asymptomatic phase of infection, disclose microglial nodules, perivascular lymphocytic cuffs and lymphocytic meningitis, while virus was detectable in brain by PCR in about half of individuals (Bell et al., 1993). Rarely, multifocal CNS demyelination resembling postinfectious encephalomyelitis has been observed in a patient dying immediately after seroconversion (Silver et al., 1997). Several peripheral nerve syndromes have been identified during seroconversion or the asymptomatic period after infection, including syndromes simulating acute Guillain-Barré syndrome and chronic inflammatory demyelinating polyneuropathy, and a mononeuritis multiplex. Inflammatory infiltrates and demyelination characterize nerve biopsy in the former two conditions. Conversely, mononeuritides manifested as cranial neuropathies or mononeuritis multiplex have been reported, and corresponding pathological studies demonstrate a vasculitic process (Brannagan, 1998). These conditions are assumed to have an autoimmune pathogenesis, which is prompted by immune dysregulation that frequently accompanies HIV infection (Hirsch and Curran, 1996). The finding that patients appear to benefit from treatment with plasmapheresis and/or corticosteroids supports this assumption.

B. AIDS-related Neurological Syndromes

At the onset of AIDS, when CD4 counts drop and opportunistic infections occur, several well-defined neurological syndromes that are presumably HIV-induced may arise in addition to central nervous system opportunistic infections (Levy, Mackewicz, and Barker, 1996). The three prototypic HIV-induced syndromes include (1) HIV-associated dementia (HIVD) that has also been termed AIDS dementia complex or HIV encephalopathy, (2) vacuolar myelopathy (VM), and (3) a predominantly (painful) sensory neuropathy (PSN) (McArthur, 1987).

Estimates of HIVD prevalence range from 5% to 20% of patients with AIDS and an annual incidence of dementia after AIDS development of 7% per year (McArthur et al., 1993). HIVD is also associated with a significantly worsened survival prognosis (McArthur et al., 1993). HIVD dementia is characterized by progressive motor (tremor, gait instability, and loss of fine motor control), cognitive (mental slowing, forgetfulness, and poor concentration), and behavioral abnormalities (mania at the outset, apathy, emotional lability), and displays diversity in its clinical phenotype (Navia, Jordan, and Price, 1986) (Mirsattari et al., 1999) (Maher et al., 1997). The course of the dementia is variable with an abrupt decline in function among some individuals over weeks, while others display a protracted course over several years, although the

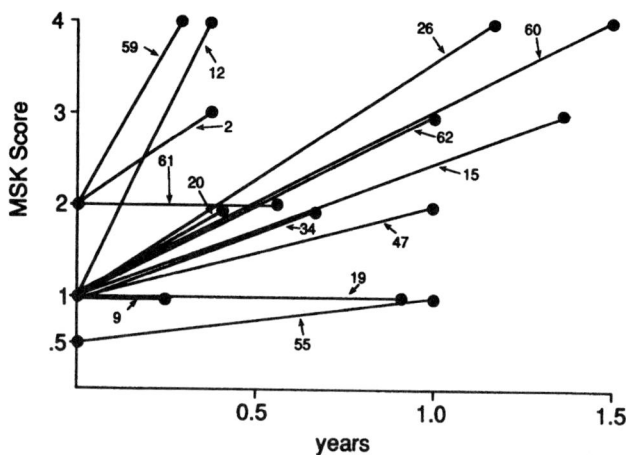

FIG. 3. Severity of dementia measured by the Memorial Sloan-Kettering Scale (MSK) over time. Adapted with permission from Power *et al.* (1994).

mean survival had been 3–6 months, prior to the advent of HAART (Fig. 3). Radiological features accompanying HIV dementia include cerebral and basal ganglia atrophy, and diffuse white matter hyperintensities on MRI T2 weighted images (Simpson and Tagliati, 1994) (Dal Pan *et al.*, 1992). More recent magnetic resonance spectroscopy studies reveal diminished N-acetyl aspartate levels in brain, implying neuronal injury or death (Chang *et al.*, 1999). The newer antiretroviral drugs may influence this variation in presentation and course (Sacktor *et al.*, 1999) (Dore *et al.*, 1999). Dementia is usually manifested after the development of AIDS, but dementia is the AIDS-defining illness in 3% of infected individuals. HIV dementia has been shown to be transiently responsive to zidovudine (AZT) therapy (Sidtis *et al.*, 1993), indicating that viral replication in the brain may participate in the development of this neurological syndrome. HIVD has attracted the most attention of all the primary HIV-induced syndromes because of its devastating effects on patients, because it heralds a poorer prognosis and it is to some extent improved with antiretroviral drugs (Tozzi *et al.*, 1999). It has been identified in all populations infected by HIV-1, regardless of viral clade (subtype) (Maj *et al.*, 1994). The majority of adult patients who develop HIVD, VM, and PSN are severely immune-suppressed, indicating that systemic immune suppression and its effects are key determinants of these syndromes.

Among children with advanced HIV infection, the prevalence of progressive HIV encephalopathy approaches 50% and is defined by devel-

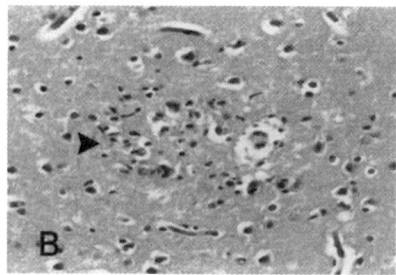

Fig. 4. Multinucleated giant cells (A) and microglial nodules (B) in an adult AIDS patient. Generously contributed by J. K. Holden, Department of Pathology, St. Paul's Hospital, Vancouver, BC, Canada.

opmental delays, impaired brain growth, and abnormalities in motor function and tone (Belman, 1997). Neuroradiological studies reveal calcification of the basal ganglia in HIV-infected children, although atrophy is also commonly observed. Like adults infected with HIV, markers of macrophage activation such as interleukin (IL)-6 are increased in the spinal fluid of children with HIV encephalopathy (Gallo *et al.*, 1989). In contrast to adults, however, myelopathies and peripheral neuropathies are infrequently encountered among HIV-infected children. HIV-infected children may show marked improvement in measures of cognition such as IQ, following treatment with antiretroviral drugs (Pizzo and Wilfert, 1994).

The neuropathological hallmarks of HIV infection in the adult brain include multinucleated giant cells, diffuse white matter pallor, and microglial nodules (Navia *et al.*, 1986) (Sharer, 1992) (Vinters, 1990)). HIV encephalitis is defined by the presence of multinucleated giant cells and microglial nodules (Fig. 4). Among HIV-infected adults, approximately 20–80% of individuals will display multinucleated giant cells (Navia *et al.*, 1986) (Glass *et al.*, 1996) (Sharer *et al.*, 1986). In cases in which multinucleated giant cells are not detectable, the presence of viral antigen has been used to confirm this diagnosis (Wiley, Baldwin, and Achim, 1996). Diffuse white matter pallor with sparing of the U fibers shows relatively preserved myelin proteins but conversely, deposition of serum proteins is observed in white matter, suggesting altered permeability of the blood–brain barrier (BBB) (Petito and Cash, 1992) (Power *et al.*, 1993). This latter finding is complemented by recent studies showing apoptotic cell death in cerebral endothelia in brains of HIV-infected patients (Shi, 1996). Perivascular CNS demyelination has also been reported and is reminiscent of postinfectious encephalomyelitis. Diffuse astrocytosis and microgliosis in the cerebral gray matter define a third neuropathological feature that has been termed diffuse poliodystrophy (Budka, 1991). The latter two neuropathological entities are not correlated with any specific clinical features, although diffuse myelin pallor and multinucleated giant cells are associated with the occurrence of HIVD (Glass et al., 1995). Similar features, including multinucleated giant cells, white matter pallor, and microglial nodules, have been observed in children with AIDS but the most prominent feature observed in pediatric brains is that of mineralizations adjacent to small blood vessels of the basal ganglia and frontal white matter (Belman et al., 1986). This latter feature may affect up to 90% of infected children coming to autopsy but its pathological significance remains obscure.

A limited correlation has been found between HIVD (a clinical entity) and HIV encephalitis (a pathological entity). Over half of adult AIDS patients with dementia do not exhibit diffuse myelin pallor or multinucleated giant cells at autopsy while microglial nodules may be present in 90% of autopsied AIDS patients (Glass et al., 1995). A correlation between viral antigen abundance and dementia has been proposed (Achim, Heyes, and Wiley, 1993). Other studies have shown that macrophage and microglia activation, particularly in the basal ganglia, is a stronger predictive marker for HIV dementia (Glass et al., 1995). Neuronal injury and death, in the cortex and deep gray matter, have also been observed in the brains of patients with AIDS (Everall, Luthert, and Lantos, 1991) (Masliah et al., 1992a; Masliah et al., 1992b). Studies reveal neuronal loss and a reduction in neuronal cell body volume in the frontal cortex of patients who are AIDS defined (Ketzler et al., 1990) and the degree of neuronal loss is correlated with the severity of cognitive impairment (Everall et al., 1999). Dendritic simplification, diminished arborization, and loss of presynaptic terminals have also been reported in the brains of HIV-infected individuals (Masliah et al., 1992b). The mechanism of neuronal death remains uncertain, although both necrosis and apoptosis have been implicated in the loss of cortical and basal ganglia neurons in both children and adults. Indeed, these neuronal abnormalities are likely responsible for the phenotypic expression of HIV dementia.

The prevalence of clinical myelopathy is approximately 15% and is characterized by progressive spasticity and weakness, loss of sensation in the legs, and bladder dysfunction (Dal Pan, Glass, and McArthur, 1994). Hyperreflexia may be present in the arms, although weakness and sensory changes are not usually detected. These clinical findings are associated with vacuolization observed pathologically in the posterior and lateral columns of the spinal cord (Dal Pan, 1997).

Vacuolization represents intralaminar edema within the myelin sheaths, with relative preservation of the axon. Macrophages may be detected in the vacuole (Tan, 1995). Within the posterior columns, increased numbers of macrophages and enhanced expression of activation markers on macrophages have been reported in VM (Tyor et al., 1993a). Viral antigen and genome detection in the spinal cord are infrequent and not correlated with the severity of pathology and clinical disease (Johnson and Chesebro, unpublished data). Focal lesions in the spinal cord, distinguished by degeneration of myelin and axons within the gracile tracts, have been reported and assumed to represent a dying back process due to injury to the dorsal root ganglia (Rance et al., 1988). A rare abnormality in which necrotic lesions with proliferation of macrophages, microglia, and multinucleated giant cells has also been rarely reported (Geny et al., 1991) and may be the spinal cord correlate of HIV encephalitis.

The most common peripheral nerve abnormality associated with HIV infection is PSN with a prevalence of 20% among AIDS patients (McArthur, 1987) (de la Monte et al., 1988) (Simpson and Tagliati, 1994). Symptoms include sensory loss in the hands and feet, complicated by painful dyesthesias and a loss of deep tendon reflexes. Pathological studies reveal distal axonal degeneration involving myelinated and unmyelinated fibers with a loss of dorsal root ganglia neurons (Brannagan, 1998). Macrophage infiltration and increased expression of activation markers on macrophages within the lesion have also been reported. Autonomic neuropathy manifested as impotence, urinary dysfunction, and orthostatic hypotension have been observed among patients with HIV infection, although the underlying pathogenesis remains obscure. Axonal neuropathy has become an increasing problem over the past several years due to the neurotoxic actions of several antiretroviral drugs including didanosine (ddl), zalcitabine (ddC), stavudine (d4T), and lamivudine (3TC) (Brannagan, 1998). These neurotoxin-induced neuropathies are dose-dependent and reversible with cessation of the drug. Although AZT does not cause a neuropathy, it is associated with a myopathy that exhibits a distinctive morphology of ragged-red fibers that is suggestive of mitochondrial dysfunction and is reversible if the drug is stopped (Grau et al., 1993)

In addition to the well-recognized syndromes cited above, other less common CNS and PNS disorders associated with HIV infection have been reported, including paroxysmal dyskinesias, parkinsonism, ocular flutter, and spinal myoclonus (Mirsattari, Power, and Nath, 1998) (Maher et al., 1997). Episodes of unilateral weakness or sensory loss suggestive of transient ischemic attacks or stroke have also been observed in patients with HIV infection and may be related to the presence of antiphospholid antibodies (Abuaf *et al.,* 1997) (Brew *et al.,* 1996). The clinical presentation and epidemiology of these primary HIV-induced syndromes is changing as newer and more effective drugs to treat HIV infection become available. For example, the incidence of HIV dementia has dropped markedly over the past 2–3 years but many patients now harbor drug-resistant viral strains and, hence, a resurgence of HIVD may occur (Dore *et al.,* 1999). In contrast, vascular complications, including strokes, have been reported and appear to arise from some of the newer antiretroviral treatments, such as the protease inhibitors, that result in rises in plasma lipids. Many fundamental clinical questions regarding HIV neuropathogenesis remain unanswered, including the existence of host susceptibility genes, long-term impact of drug abuse, gender differences, and the effects of socioeconomic status.

III. VIROLOGICAL ASPECTS

HIV-1 belongs to the lentiviral group of retroviruses, which are defined by a relatively slow disease course in their natural hosts. Like all retroviruses, the HIV-1 genome structure is defined by *gag, pol,* and *env* genes, but HIV also contains several nonstructural genes that influence splicing and transcriptional events, for a total of 10 open reading frames within approximately 10 kilobase pairs (Greene, 1991). There is extensive variation within different HIV-1 strains that is manifested both genotypically as well as phenotypically (Levy, 1998) (Korber *et al.,* 1997), and that arises because of poor fidelity during reverse transcription and recombination in the virion between the diploid RNA molecules. Many strains of HIV-1 from around the world have now been sequenced and, in some, the virus has been phenotypically characterized *in vitro.* Unlike other many retroviruses, lentiviruses, including HIV, are capable of infecting terminally differentiated cells. Viral phenotypes have been defined in terms of cell tropism: most HIV strains are either macrophage- or T cell-tropic but dual-tropic viruses have also been described (Collman *et al.,* 1992). Human-to-human transmission is usually characterized by macrophage-tropic strains, which predominate in blood early after infection (McNearney *et al.,* 1992). HIV-1 cell tropism is chiefly defined by which chemokine receptor the individual strain uses as a coreceptor, together with CD4 for infection (Deng *et al.,* 1996). In addition, different HIV strains have also been defined *in vitro* as syncytia (SI)- or nonsyncytia-inducing (NSI), although this classification is dependent on which cell type is being used in the assay (Koot *et al.*, 1992). It is clear from viral sequence analyses, *in vitro* studies, and studies of autopsied or biopsied brain that HIV infection of the CNS principally involves cells of macrophage lineage, although other cell types, including astrocytes, are infected, as outlined below.

IV. NEUROPATHOGENESIS

Like all viruses infecting the CNS, HIV-related pathogenesis can be discussed in terms of (a) neuroinvasion, or the ability to enter the CNS; (b) neurotropism, or the ability to infect brain cells; and (c) neurovirulence, or the ability to cause disease (Johnson, 1998).

A. Neuroinvasion

Neuroinvasion by HIV-1 occurs early after primary infection and is present in almost all patients at the time of death. In addition, HIV antigens and/or genome have been detected in the brains of HIV-infected patients at all stages of infection (Bell *et al.*, 1993) (Budka, 1991). Studies of patients who have died soon after infection, of other causes, exhibit viral antigen and neuropathological findings indicative of HIV infection (Davis *et al.*, 1992). The mechanism by which HIV enters the CNS is assumed to be infection of macrophages, which cross the blood–brain barrier and infect other cells of macrophage lineage including microglia and perivascular macrophages, termed the "Trojan Horse" hypothesis (Haase, 1986). HIV infection of brain endothelial cells has been demonstrated *in vitro* but not convincingly *in vivo*, to date (Poland, Rice, and Dekaban, 1995) (Moses *et al.*, 1996). An alternative pathway for entry into the CNS has been postulated to occur through the choroid plexus, and examination of viral sequences from choroid plexus suggests viruses of differing tropism including potentially macrophage or T cell-tropic strains (Falangola *et al.*, 1995). In general, definitive studies examining the route by which the virus enters the CNS have yet to be performed.

The viruses recovered from the brains of AIDS and pre-AIDS patients are macrophage-tropic viruses in terms of sequence similarities and *in vitro* tropism (Cheng-Mayer *et al.*, 1989) (Reddy *et al.*, 1996) (Power *et al.*, 1994). But monocyte/macrophage traffic through the CNS is limited, unless some injury or infection has occurred within the brain. Several chemokines, including SDF-1 and MCP-1 show enhanced expression in the brains of HIV infected patients indicating that monocyte/macrophages (HIV-infected or uninfected) may be recruited into the CNS through a chemotactic mechanism (Letendre, Lanier, and McCutchan, 1999). Additionally, upregulation of adhesion molecules, such as intercellular adhesion molecule (ICAM) on the luminal aspect of brain endothelial cells, has been demonstrated *in vivo,* which may facilitate monocyte/macrophage adherence and subsequent CNS entry (Hurwitz, Berman, and Lyman, 1994). Free virus directly entering the CNS during initial plasma viremia—this route has appeal because of the high levels of macrophage-tropic virus early in infection in the plasma and CSF. The relationship of CSF viral subtype and quantity or "load" to infection of the brain also remains unclear. Whether the virus found in the CSF influences the extent or pattern of infection of the brain is uncertain. As discussed below, viral load in CSF may be a predictor of severity of HIV dementia (Brew *et al.,* 1997).

How HIV enters the CNS is a pivotal issue because strategies to prevent CNS entry by HIV would be of immense value in preventing the development of neurological disease. These include enhancing the humoral response if free virus entry from the plasma was the chief route of entry, or blocking adherence of monocyte/macrophages if these cells were the principal vehicles for viral entry into the brain.

B. *Neurotropism*

HIV antigens and genome (DNA and RNA) are found primarily in microglia and pericytes (perivascular macrophages) (Koenig *et al.,* 1986) (Wiley *et al.,* 1986), and in astrocytes, albeit less frequently (Takahashi *et al.,* 1996). Infected cells are principally localized in the central white matter and the deep gray matter structure, including the basal ganglia, and include parenchymal microglia and macrophages and perivascular cells of macrophage lineage (Kure *et al.,* 1990). Among HIV-infected children, there appear to be greater infection of astrocytes and differential expression of antigens with enhanced expression of HIV nef (Saito *et al.,* 1994). These findings have been confirmed by several different methods, including immunocytochemistry detecting several different viral proteins including the *gag*-encoded p24 (Kure *et al.,* 1990) and the *env*-encoded transmembrane unit, gp41 (Glass *et al.,* 1996). *In situ* hybridization with and without PCR confirm these immunocytochemical findings (Shaw *et al.,* 1985) (Bagasra *et al.,* 1996; Nuovo *et al.,* 1994) (Takahashi *et al.,* 1996). To date, endothelia and oligodendrocytes have not been shown consistently to be infected *in vivo* although *in vitro* studies indicate these cell types are permissive to some strains of HIV-1

(Moses *et al.*, 1993) (Gyorkey, Melnick, and Gyorkey, 1987). The question of *in vivo* infection of neurons remains controversial: Nuovo *et al.*, (1994) have reported that viral genome in neurons is detectable by *in situ* PCR (polymerase chain reaction) and this finding has been confirmed is detectable by at least one other group (Bagasra *et al.*, 1996). However, *in vivo* productive infection of neurons has not been shown to date, although *in vitro* productive infection of neuronal cell lines has been demonstrated.

HIV infection has been shown to be mediated by CD4 as the principal receptor on lymphocytes and macrophages (Levy, 1998). However, CD4 expression is comparatively low in the brain, although it is clear that brain-derived strains of HIV utilize CD4 for infection. Several other coreceptors, such as CCR5 on macrophages and CXCR4 on lymphocytes, have been shown to exist on blood-derived cells (Feng *et al.*, 1996) (Alkhatib *et al.*, 1996). Other coreceptors have been reported (Deng *et al.*, 1996), although their roles are less well defined. In the brain, CCR5 and CCR3 have been postulated as potential receptors on microglia (He *et al.*, 1997a). This observation has been confirmed by *in vitro* studies showing that infectious recombinant HIV clones containing brain-derived envelope sequences use CCR5 and CCR3 to a lesser extent (Chan *et al.*, 1999). CXCR4 has also been demonstrated on microglia, in addition to cells such as neurons (Sanders *et al.*, 1998). The failure of T-cell tropic HIV-1 strains to establish productive infection of microglia, despite the presence of CXCR4, is enigmatic and may reflect a requirement for interaction between multiple coreceptors for infection (Kaul and Lipton, 1999). Nonetheless, the role of this coreceptor in HIV neuropathogenesis remains intriguing because it may have important implications in mediating neuronal injury and death (Zheng *et al.*, 1999). Other HIV receptors in the brain have been reported, including galactosyl ceramide (Harouse *et al.*, 1991) and a 260 kDa astrocyte cell membrane protein (Ma, Geiger, and Nath, 1994). Hence, the mechanisms of viral fusion and entry into brain cells remain uncertain and may differ from blood cells.

Cell tropism, or infectivity of a retrovirus, is determined by multiple viral genes that influence events during infection, including viral entry, reverse transcription, integration, transport of viral proteins and genome to the cell surface, and budding of virions (Levy, 1998). Hence, several genes found within HIV are likely to influence its tropism. HIV entry appears to be modulated by different regions of gp120 that are encoded by the *env* (Chesebro *et al.*, 1991) (Westervelt *et al.*, 1992) (Fig. 5). For example, the CD4 binding domain lies primarily within C4 region of the gp120 but its sequence is relatively conserved

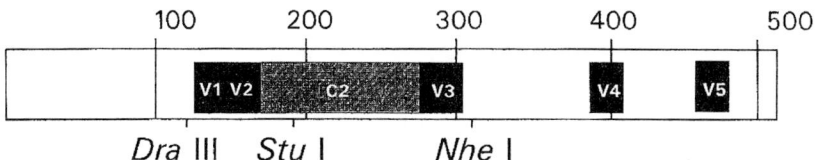

FIG. 5. HIV gp120 sequence (511 amino acids) showing regions that influence cell tropism and corresponding restriction sites. Adapted with permission from Power *et al.* (1998).

across different viral strains (Korber *et al.*, 1997). In contrast, the V3 hypervariable region of gp120 has attracted the most attention in terms of influencing tropism and it is also the virus' principal neutralizing determinant (Hwang *et al.*, 1991). The V3 region's control of viral infectivity appears to be dependent on distinct sequences and specific amino acids that are necessary for macrophage tropism and its interaction with coreceptors such as CCR5 (Cocchi *et al.*, 1996). The V1 and V2 hypervariable regions of gp120 have also been implicated in regulating tropism (Koito *et al.*, 1994) (Power *et al.*, 1998b) (Toohey *et al.*, 1995), and distinctive V2 sequences have been associated with the SI viral phenotype (Groenink *et al.*, 1993). The HIV-1 envelope leader sequence has been implicated in controlling virulence and cell tropism in different *in vitro* assays, possibly due to its markedly hydrophobic region (Korber *et al.*, 1997). The V4 and V5 regions have not received as much experimental scrutiny as other regions of the HIV envelope (Korber *et al.*, 1994) (Gartner *et al.*, 1997), although comparison of viral sequences derived from brain and other organs suggests that these domains may also influence neurotropism. Studies of other viral genes and their ability to influence neurotropism indicate that the *gag* (Morris *et al.*, 1999) and *tat* (Mayne *et al.*, 1998) sequences, derived from brain, may influence neurotropism but functional confirmatory studies are pending. How the different regions of gp120 interact with plasma membrane-bound proteins is uncertain but a conformationally dependent interaction between multiple regions of the envelope and the different receptors has been proposed (Chesebro *et al.*, 1992) (Kwong *et al.*, 1998).

HIV-1 strains isolated from brain appear to be principally macrophage-tropic (Cheng-Mayer *et al.*, 1989; Power *et al.*, 1994; Reddy *et al.*, 1996) and have also been shown to infect microglia *in vitro* (Watkins *et al.*, 1990) (Power *et al.*, 1995). The V3 region of brain-derived HIV has been shown (Jordan *et al.*, 1991) (Power *et al.*, 1995) to be critical for HIV infection of macrophages and microglia. More recently, it has

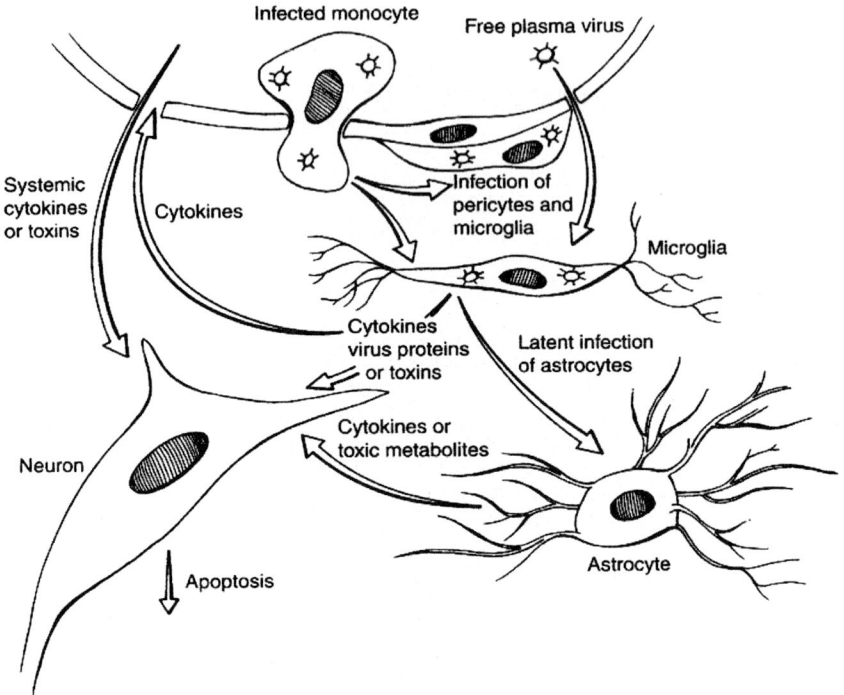

FIG. 6. Potential interactions between different cell types involved in the HIV-1 neuropathogenesis in the CNS. Virus may enter the CNS directly or via infected macrophages, resulting in infection of perivascular cells of macrophage lineage. Nonproductive infection of astrocytes may result in impaired blood–brain barrier function and diminished neuroprotection. Adapted with permission from Johnson (1998).

been demonstrated that brain-derived HIV-1 V3 sequences mediate the use of CCR5 and CCR3, but not CXCR4, as coreceptors (Chan *et al.*, 1999). Repeated passages of HIV *in vitro* can result in mutations in the envelope that resemble mutations identified in brain-derived sequences, suggesting that the virus may adapt to the brain (Strizki *et al.*, 1996). Nonetheless, other groups using transgenic mice have shown that the HIV-1 LTR may be important for expression in the brain (Corboy *et al.*, 1992). Reverse transcriptase (Wong *et al.*, 1997) and envelope (Morris *et al.*, 1999) sequences from brain, blood, and spleen cluster separately by phylogenetic analyses, suggesting that different organs may exert selective evolutionary pressures on viral replication (Fig. 6). These results are not incompatible with the above V3 findings because multiple genes in many viruses determine cell tropism.

C. NEUROVIRULENCE

1. Viral Strains

Among several lentiviruses, including simian immunodeficiency virus (SIV) (Zink *et al.*, 1997), feline immunodeficiency virus (FIV) (Power *et al.*, 1998a) and visna-maedi virus (VMV) (Andresson *et al.*, 1993), it is clear that specific viral strains cause the development of neurological disease (discussed below; reviewed by Narayan and Clements, 1989). In addition, recently reported chimeric SHIV strains also cause neurological disease in monkeys (Liu *et al.*, 1999). There are extensive *in vitro* data indicating HIV-1 strains differ in tropism and cytotoxicity. As discussed above, there appear to be distinct HIV-1 envelope sequences that influence the ability of different viruses to infect specific cell types. At the same time, characteristic strains have been reported to cause specific pathologies such as thrombocytopenia (Voulgaropoulou *et al.*, 1999). Furthermore, it has become apparent over the last 5 years that individual viral strains, defined by specific mutations in the reverse transcriptase and protease genes, are resistant to drug therapy (Perelson, Essunger, and Ho, 1997). Clinical drug resistance, with a subsequent rise in plasma viral load, is correlated with *in vitro* drug resistance (Arts and Wainberg, 1996).

By studying a well-characterized, prospective cohort of AIDS patients with and without HIVD that came to autopsy, our group has shown that specific mutations within the V3 (Power *et al.*, 1994) and V1 (Power *et al.*, 1998b) domains of brain-derived HIV-1 envelope sequences differed between AIDS patients with and without dementia. These same domains also determined the ability of infectious recombinant clones to infect and spread in macrophage and mixed glial cultures but did not replicate in T cell analog (HeLa/CD4) cultures (Power *et al.*, 1995) (Power *et al.*, 1998b). In addition, conditioned media from HIV-infected macrophages caused varying levels of neuronal death, when applied to human neuronal cultures; neurotoxicity induced by recombinant viruses derived from patients with HIVD was higher than neurotoxicity caused by viruses derived from nondemented patients (Power *et al.*, 1998b). The neurotoxicity of conditioned media was resistant to boiling and was partially blocked by the NMDA receptor antagonist, APV. Because of the immense molecular diversity within HIV-1, it has been difficult to identify strain-specific HIV-related syndromes. Indeed, other groups have not been able to identify mutations in the V3 region of the envelope associated with HIV dementia. In each case, they have used viral isolates from CSF or blood (Kuiken *et al.*, 1995) (Di Stefano *et al.*, 1996), or brain-derived sequences from patients with limited clinical inclusion criteria. It is difficult to compare these findings with our results because cells isolated from CSF tend to reflect cell profiles in blood and not brain and the clinical characterizations were not comparable. Similarly, we have not found mutations in *tat* and reverse transcriptase sequences that differentiated patients with and without HIV dementia in the same cohort, although phylogenetic comparisons showed some clustering of *tat* sequences from patients with dementia (Bratanich *et al.*, 1998). A recent study found that T cell-tropic strains of HIV-1 induced the highest levels of neuronal injury when tested in an *in vitro* model of neuronal death, possibly through activation of CXCR4 expressed on neurons (Ohagen *et al.*, 1999). What constitutes a viral strain remains uncertain because of the diversity within viral sequences that can be isolated from blood and other organs, including brain. It is assumed that the viral sequences cloned from plasma or tissues represent the predominant sequences, but the possibility that a sequence that is expressed to a lesser extent could be pathogenic should be considered. This is a particularly difficult issue in HIV neuropathogenesis of brain, in which most studies are predicated on autopsied tissues.

2. Viral Proteins

There is a rapidly expanding literature implicating several different proteins in HIV-1 neuropathogenesis (Nath and Geiger, 1998). HIV-1 gp 120 has been shown to be directly and indirectly neurotoxic *in vitro* and *in vivo* (Dreyer *et al.*, 1990) (Kaiser, Offermann, and Lipton, 1990) (Barks *et al.*, 1997). Specific domains within gp120 have been implicated as especially neuropathogenic, including the CD4 binding (Giulian *et al.*, 1993) and the V3 regions (Pattarini, Pittaluga, and Raiteri, 1998). The underlying mechanism appears to be related to accumulation of intracellular calcium in neurons following activation of glutamate receptors and voltage-operated calcium channels (Lipton and Gendelman, 1995). For example, indirect activation of the NMDA receptor, resulting in neuronal death, possibly through binding to the adjacent glycine receptor or increased free zinc concentrations (Lee, Zipfel, and Choi, 1999), has been reported and can be blocked with several different NMDA receptor antagonists such as memantine and APV (Lipton, 1992). Other neurotransmitters may be affected by gp 120-induced activation of NMDA receptors, such as impaired dopamine transport, as demonstrated in dopaminergic neurons cultured from rat midbrain (Bennett, Rusyniak, and Hollingsworth, 1995). Dawson *et al.*, (1993) have also shown that nitric oxide may be influencing gp120 neurotoxicity. Transgenic mice expressing HIV-1 gp120 in astrocytes display neuropathological findings including astrogliosis, neuronal loss, and dendritic vacuolizations, resembling HIV encephalitis (Toggas *et al.,* 1994). This model used a gp120 sequence from a T cell tropic strain of HIV-1; initial studies did not show protein expression but later studies disclosed gp120 protein detection and corresponding neurophysiological changes (Krucker *et al.,* 1998). Studies of gp120 action on glial cells suggest that it may also alter glial function through pertubation of the Na^+/H^+ ion transporter(s) in astrocytes that might contribute to neuronal dysfunction and death (Benos *et al.,* 1994); gp120 has also been shown to affect intracellular signaling that controls the expression of different cell adhesion molecules and cytokines and perhaps nitric oxide through the JAK-STAT pathway (Shrikant *et al.,* 1996). The above *in vitro* and *in vivo* findings support the hypothesis that gp120 is involved in the pathogenesis of HIV-induced neurological injury.

Other HIV proteins, including Tat, gp41, and Nef have been shown to be neurotoxic *in vitro* (Adamson *et al.,* 1996) (Nath *et al.,* 1996). The transactivating protein, Tat, has attracted extensive attention in neuropathogenesis studies because of early studies showing that it was neurotoxic, was released from infected lymphocytes, and was taken up by cells and could in turn transactivate host genes such as TNF-α (Chen *et al.,* 1997) and IL-1 (Nath *et al.,* 1999). As with gp120, several domains have been found to be especially neurotoxic, including the basic region and the RGD residues in the second exon, while other groups have shown that the entire first exon is necessary for neurotoxicity. More recently, Conant *et al.* have shown that Tat induces expression MCP-1 in astrocytes, which may influence macrophage trafficking in the CNS (Conant *et al.,* 1998). Other groups have shown that multiple glial-derived proteins, including cytokines and chemokines, can be induced by treatment with Tat through several intracellular signaling pathways.

A concern regarding studies using viral proteins to study pathogenesis is that the concentrations of protein tested in many of these *in vitro* and *in vivo* studies were extremely high. It is difficult to reconcile these concentrations with the findings that extracellular HIV proteins are difficult to detect (gp41, p24, Nef) or have not been detected in the brain (Tat, gp120). Nonetheless, the mRNAs corresponding to all of these proteins are readily detectable in brain and perhaps viral proteins are rapidly degraded (Nath *et al.,* 1999).

3. Viral Load

Recent studies indicate that viral replication and "load" in blood are extremely high in persons with HIV-1 infection (Perelson *et al.,* 1997). However, the role of viral load in brain, in relation to the development of neurological disease, is uncertain. Viral load in cerebrospinal fluid is usually several orders less than plasma but correlated to some extent with the presence of dementia (McArthur, 1987). Differing results, depending on the method of viral or proviral quantitation, have been reported from various groups examining the relationship between viral load in the central nervous system and in the development of HIV dementia. QC-PCR studies of brain-derived viral mRNA and provirus in brain indicate no significant difference in levels between AIDS patients with and without HIV dementia (Johnson et al., 1996); (Lazarini et al., 1997). In contrast, viral protein and RNA levels detected in CSF (Royal et al., 1994) (Brew 1997) and brain, detected by immunocytochemistry (Brew et al., 1995) (Glass et al., 1995), show modestly elevated levels in patients with HIV dementia, compared to nondemented AIDS controls. More recent studies suggest that CSF viral load is correlated with neuropsychological abnormalities (Ellis et al., 1997), but the relationship of plasma viral load to CSF viral load is dependent on the level of immune suppression (McArthur et al., 1997). Viral load in the brain, measured by immunostaining or quantititative molecular methods, is closely associated with the extent of pathological change accompanying HIV encephalitis (Vazeux et al., 1992). The regulation of viral replication in the brain has been difficult to study for obvious reasons. Several studies (e.g., Chao et al., 1996) indicate that opiate receptors may be an important neurotransmitter pathway that can be exploited in regulating viral replication. The lack of definitive evidence that would indicate that viral load in the brain is closely correlated with the development of HIV dementia may reflect the problem of using autopsied tissues to assess a relatively dynamic process such as viral replication. Several studies have also suggested that levels of unintegrated viral DNA in brain macrophages are associated with the development of HIV dementia (Pang et al., 1991) (Teo et al., 1997), implying that increased viral genome levels in select cell populations may be important determinants of disease in the brain.

Specific drug resistance conferring mutations in the reverse transcriptase and protease encoding genes have been identified in viruses from patients who show high viral loads despite HAART (Harrigan and Alexander, 1999) (Hirsch et al., 1998). The extent to which these mutations are present in the viruses found in the brains of patients treated with antiretroviral drugs remains uncertain. Recent studies suggest that brain-derived viruses have fewer drug-resistant than matched blood-derived HIV sequences; that may reflect poor CNS penetration of the drug and/or limited replication in the brain and, hence, a restricted potential for drug resistant mutations to emerge

(Bratanich *et al.*, 1998) (Wong *et al.*, 1997). Drug resistance is becoming an increasingly important issue and has resulted in at least one clinical trial failing to show clinical benefit among AIDS patients with dementia who received a drug with high CNS penetration.

4. Host Molecules / Neuroinflammation

Excess production of host-encoded inflammatory molecules by microglia and perhaps astrocytes has been proposed as a chief cause of damage within the brain in a number of CNS diseases, including Alzheimer's disease, stroke, and HIV dementia (Kolson and Pomerantz, 1996). This hypothesis is predicated on data derived from *in vitro,* animal models and studies of autopsied tissues. Among studies of HIV neuropathogenesis, this hypothesis has gained wide popularity because microglia and astrocytes are the principal cells infected by HIV-1.

In 1990, Giulian and colleagues showed that HIV-infected monocytoid cell lines secreted diffusible molecules that killed several different neuronal cell types by a presumed excitotoxic mechanism, mediated by NMDA receptors (Giulian, Vaca, and Noonan, 1990). Pulliam *et al.* (1991) also reported that HIV-infected macrophages produced a neurotoxic compound causing cytopathic effects in cultured cell aggregates from human fetal brain tissue. In contrast, other groups (e.g., Tardieu and Janabi, 1994) have shown that only following direct contact with neurons, could HIV-infected monocytes induce neurotoxicity (but a soluble neurotoxin could not be demonstrated). Despite the controversy surrounding this area, several potential neurotoxins have been identified and characterized *in vitro* and *in vivo* (Table II).

a. Cytokines: Multiple cytokines have been shown to be elevated in the brains and CSF of patients with HIV dementia (Tyor et al., 1992) (Wesselingh et al., 1993) (Gelbard et al., 1994); these include tumor necrosis factor-alpha (TNF-α), interleukin-1 (IL-1), interleukin-6 (IL-6), interferon-alpha (IFN-α), and tissue growth factor-beta (TGF-β). Although almost all cells in the CNS can produce cytokines, the chief sources of these small soluble molecules are activated glial cells that include macrophages, microglia, and astrocytes. TNF-α is an inflammatory cytokine that has received extensive attention for its potential neurotoxic effects in HIV infection and its ability to influence the release of other cytokines (Barna et al., 1990) (Benvieniste, 1994). It has been shown to be released by microglia and astrocytes (Tyor et al., 1992) (Wilt et al., 1995) in HIV infection and prevent uptake of glutamate (Fine et al., 1996). Several studies have shown that TNF-α mRNA and protein levels are increased in the brains and CSF of

TABLE II
POTENTIALLY NEUROTOXIC MOLECULES IMPLICATED IN HIV NEUROPATHOGENESIS[a]

Source of molecules	Molecule
Viral proteins	Tat, Nef, gp41, gp120
Macrophage factors	Low molecular weight toxic factors (Ntox)
	Quinolinic acid
	Arachidonic acid metabolites (prostaglandin and leukotriene)
	Platelet-activating factor
	Matrix metalloproteinases
	NO, superanion
	Cytokines: TNF-α, IL-1β, IL-6, GM-CSF, IFN-α
	Chemokines: MCP-1, RANTES, MIP-1α
Decreased neurotrophic factors	Neuroleukin
Autoimmune mimicry	Antibrain antibodies

[a] GM-CSF, granulocyte–macrophage colony-stimulating factor; IFN, interferon; IL, interleukin: TNF, tumor necrosis factor. Adapted with permission from Johnson (1998).

patients with HIV infection. Notably, Wesselingh et al. (1993) showed that TNF-α mRNA levels were increased in patients with HIV dementia as compared to AIDS patients without dementia or noninfected controls; furthermore, the level of mRNA was correlated with the severity of dementia. These in vivo studies are supported by in vitro studies showing neurotoxic and gliotoxic effects of TNF-α, often at low concentrations in conjunction with other potentially neurotoxic molecules (Westmoreland, Kolson, and Gonzalez-Scarano, 1996).

Conversely, other groups have demonstrated a neuroprotective effect by TNF-α in vitro (Mattson et al., 1995), or have shown increased TNF-α levels in brains from patients with other neurological disease manifesting neuropathological and clinical phenotypes other than dementia (Selmaj and Raine, 1988). These dichotomies have, to some extent, been resolved by studies showing that the activity and kinetics of TNF-α expression may be dependent on the simultaneous action of other molecules to produce neurological disease (Chao and Hu, 1994). II-6 has also been reported to be increased in the brains of patients with HIV infection (Tyor et al., 1992) (Sippy et al., 1995). Recent in vitro studies suggest that it may mediate neurotoxicity directly (Gruol 1998). TGF-β, IL-1, and NGF have been reported to be overexpressed in HIV-infected brains (Boven et al., 1999b) These three molecules have some neurotrophic properties and, hence, their increased production may reflect a host defense response to the neurotoxic actions of HIV.

It is also possible that the increased expression of inflammatory cytokines in brain may indirectly mediate neurological injury during HIV infection by enhancing the permeability of the BBB. Inflammatory cytokines such as TNF-α increase BBB permeability, as shown for other CNS diseases (Quagliarello et al., 1991). Increased BBB permeability, which has been shown to be present in HIV infection, may permit the influx of potential neurotoxins from the systemic circulation.

b. Arachidonic acid: Cells of macrophage lineage are major producers of arachidonic acid and its metabolites (Triggiani et al., 1995). Arachidonic acid is metabolized through the cyclooxygenase or lipoxygenase pathways to form prostaglandins and thromboxanes or leukotrienes, respectively. These metabolites have been shown to influence neuronal NMDA and non-NMDA receptor action and cell survival. Griffin et al. (Griffin, Wesselingh, and McArthur, 1994) reported that prostaglandins E_2 and F_{2a} and thromboxane B_2 were elevated in CSF from patients with AIDS dementia as compared to patients without dementia. Other groups have shown elevated levels of arachidonic acid metabolites in HIV-infected macrophages, although most of the products were produced through the lipoxygenase pathway (Genis et al., 1992). It may be that arachidonic acid metabolites influence neurotoxicity indirectly, such as by regulating the expression of glutamate uptake by astrocytes, as has been shown in vitro (Rothstein et al., 1993, 1996). These potential mediators of neuronal survival are likely to be involved in the neuropathogenesis of HIV infection.

c. Quinolinic acid: Quinolinic acid (QA) is produced by macrophages following different types of stimulation and as a product of tryptophan metabolism (Saito and Heyes, 1996). By binding to NMDA (N-methyl D-asparate) receptors, QA has been shown to have neurotoxic properties, following acute or chronic exposure (Moroni, 1999). The source of QA in the brain is not clear, as it has been shown to cross the blood–brain barrier. QA has been shown to be elevated in CSF and brains from patients with HIV dementia as well as in CSF from macaques infected with simian immunodeficiency virus as compared to controls (Heyes et al., 1990) (Heyes et al., 1991). In addition, QA levels in CSF appear to correlate with the severity of dementia (Sei et al., 1995) (Heyes et al., 1991). Like TNF-α, another product of macrophages, QA levels are increased in other neurological diseases in which macrophages or microglia are activated (Flanagan et al., 1995; Heyes et al., 1997). Whether QA is directly neurotoxic, or is merely a marker of macrophage activation, remains to be demonstrated among patients with HIV dementia.

Other products of macrophages and microglia have been shown to be potential neurotoxins in several neurological conditions, including

HIV infection (Gendelman et al., 1997). For example, increased levels of inducible nitric oxide synthase (iNOS) have been reported in the brains of patients with severe HIV dementia, implicating nitric oxide, in its capacity as a neurotransmitter, as a potential neurotoxin (Rostasy et al., 1999) (Boven et al., 1999a). The iNOS levels were also highly correlated with the levels of HIV gp41 expression in brain (Adamson et al., 1996). These findings are supported by recent studies showing that metabolites of NO are also increased in the brains of patients with HIV dementia as compared to nondemented controls. An intriguing report described a novel neurotoxin, Ntox, which is released by activated microglial cells, although full characterization of the molecule is pending (Giulian et al., 1996). Other molecules produced by activated macrophages include neopterin (Griffin, McArthur, and Cornblath, 1991) and β_2-microglobulin (Brew et al., 1996), and these have been shown to be increased in the CSF of patients with HIV infection. To some extent, these levels are correlated with the severity of dementia, although a direct relationship between the activity of these molecules and pathogenic mechanism is unclear at present (Brew et al., 1990) (McArthur et al., 1992). The roles of other macrophage-derived molecules, such as macrophage-related proteins 8 and 14 and matrix metalloproteinases (Conant et al., 1999), have not fully been defined in the neuropathogenesis of HIV infection, although they hold promise as additional molecules that may contribute to the complex cascade that results in neuronal injury.

d. Chemokines: With the increased understanding of the role of chemokine receptors as coreceptors for HIV infection, there has also been a concomitant expanding interest in the actions of chemokines in the nervous system in the context of HIV infection and other neurological diseases (Ransohoff and Tani, 1998). Several chemokines have been shown to be increased in the spinal fluids and brains of patients with HIV infection. (Letendre, Lanier, and McCutchan, 1999) (Sanders et al., 1998) (Kelder et al., 1998), including macrophage inflammatory protein-1α (MIP-1α) and macrophage inflammatory protein-1β (MIP-1β); regulated upon activation, normal T cell expressed and secreted (RANTES); and inflammatory protein-10. Other groups have shown that monocyte chemoattractant protein-1 (MCP-1) levels are increased in the cerebrospinal fluids and brains of patients with HIV dementia (Conant et al., 1998). The exact role of these chemokines remains uncertain because several groups have shown, in vitro, that MIP-1α and RANTES are able to block gp120-induced neuronal death (Kaul and Lipton, 1999). These studies are supported by other studies showing that different chemokines affect calcium signaling in neurons and demonstrate that chemokine receptors have shown to be expressed on neurons, thereby directly influencing neuronal survival. The exact mechanisms by which chemokines induce signaling events in macrophages and/or neurons remain uncertain.

5. Autoimmunity

Dysregulation of the immune system is a cardinal feature of HIV infection, resulting in both systemic immune suppression and activation. Within the CNS, increased expression of Class I and II antigens on glia and release of inflammatory molecules reflect immune activation. However, activation of autoimmune mechanisms may also participate in the development of HIV-induced neurological disease. These potential mechanisms include (1) the expression of superantigens, (2) molecular mimicry, and (3) polyclonal B cell activation.

Although a superantigen expressed by HIV has not been demonstrated to date, other retroviruses such as mouse mammary tumor virus (MMTV) (Acha-Orbea et al., 1999) and a human endogenous retrovirus (Conrad et al., 1997) have been shown to express superantigens that have been implicated in pathogenesis. A superantigen would explain some features of HIV infection, including activation and T cell depletion. This is supported by a limited repertoire of T cell receptor Vβ sequences among some HIV-infected individuals. Conflicting data, exhibiting MHC-restricted immune responses, have also been reported, thus leaving the role of superantigens open for further analysis. Molecular mimicry has also been proposed as a possible mechanism by which neurological disease develops during HIV infection. Autoreactive antibodies from HIV-infected individuals that recognize HIV gp41 have been shown to bind an astrocyte-derived protein, although a pathogenic role for these antibodies has not been demonstrated (Yamada et al., 1991). Regions of homology between HIV gp160 and specific HLA domains, IL-2 and human IgA have also suggested that molecular mimicry may participate in HIV pathogenesis.

Polyclonal B cell activation has been posited as the explanation of the increased levels of immunoglobulins in the blood and CSF of HIV-infected persons. These include IgG, IgA, IgM, and IgD, of which approximately 20%–50% target HIV-encoded antigens. Nonetheless, the remaining fraction are viewed as polyclonal or oligoclonal immunoglobulins that may target specific pathogens or host antigens. Antibodies directed to antigens such as phospholipids (Rubbert et al., 1994) have been reported in HIV infection and may be involved in the neuropathogenesis of HIV infection.

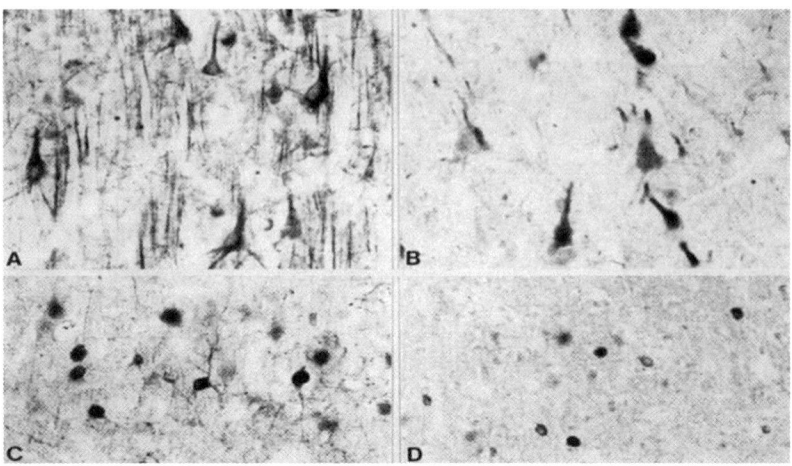

FIG. 7. Immunocytochemical detection of microtubule-associated protein-2 (A, B) and calbindin (C, D) in the frontal lobes of AIDS patients with (B, D) or without (HIVD). Neuronal injury and loss are associated with the development of HIVD.

6. Neuronal Damage

In recent years there has been an increasing interest in, and understanding of, the intracellular pathways and mechanisms by which neurons are damaged and/or killed during ontogeny and disease (Nicotera and Lipton, 1999). In the context of HIV infection, these questions are just beginning to be addressed *in vitro*. Studies of autopsied tissues indicate several distinctive patterns of neuronal loss (Masliah *et al.*, 1992a) (Everall *et al.*, 1994) (Wiley *et al.*, 1991) (Fig. 7). Select neuronal subpopulations, including larger pyramidal cells within the cortex identified by stereological methods, are at greater risk of cell death, and, similarly, neuronal populations defined by expression of certain neurotransmitters such as GABA or proteins, including parvalbumin and calbindin, are more likely to be diminished in HIV-infected brains (Masliah *et al.*, 1992a). Synaptic density is also diminished in patients with HIV-induced cognitive impairment (Everall *et al.*, 1999). In contrast, other neuronal populations expressing neuropeptides such as somatostatin are relatively resistant to HIV-induced injury (Fox *et al.*, 1997). The mechanism of cell death has been shown to be apoptotic in some studies (Gelbard *et al.*, 1995) (Adle-Biassette *et al.*, 1995), although this was not correlated with the occurrence of dementia (Adle-Biassette *et al.*, 1999). This finding may reflect similar mechanisms demonstrated in studies showing that HIV-infected macrophages mediate cell CD8 death, through an apoptotic mechanism involving gp120 binding to CXCR4 (Herbein *et al.*, 1998). However, it is unclear at present if programmed cell death is the major mechanism of neuronal loss, nor is it apparent as to the extent to which glial cell die during the course of HIV infection of the brain (He *et al.*, 1997b). *In vitro* studies have implicated different intracellular signaling pathways in neurons as potential routes to cell death. For example, *in vitro* studies show that gp 120 can induce apoptosis in human fetal neurons through the activation of JNK and ERK pathways (Lannuzel *et al.*, 1997). More recent studies suggested that gp120, derived from a T cell tropic HIV strain, induced neuronal apoptosis that was mediated p38 mitogen-activated protein kinase (Kaul, 1999). *In vitro,* studies using HIV-1 Tat indicate that the glycogen synthase kinase-3 beta and caspase 9 are also involved in neuronal death (Maggirwar *et al.*, 1999). Thus, these studies indicate that multiple pathways may determine the mechanism and frequency of HIV-related neuronal death.

V. Animal Models

Different models have been reported in which pathological findings and behavioral features resemble those observed among patients with HIV infection. Each model has different strengths and weaknesses that will be described below.

A. Lentiviral Models

The model that has received the most attention over the past 10 years is the simian immunodeficiency virus infection of rhesus macaques (Sasseville and Lackner, 1997). Using this model, investigators have shown that many of the pathological findings of HIV encephalitis can be evoked, depending on the viral strain (Flaherty *et al.*, 1997). In addition, the neurological disease develops quickly and often severe systemic immune suppression accompanies infection (Narayan and Clements, 1989). This model has many appealing aspects because of the virus' molecular homology to HIV and its induction of many similar neuropathological features. Like HIV infection, the correlation is weak between neuropathological and neurobehavioral findings in SIV infection (Murray *et al.*, 1992). This model has recently been adapted by creating a SIV/HIV chimera (SHIV) that includes the HIV envelope and several accessory genes in an SIV background (Liu *et al.*, 1999). The resulting SHIV chimera infects rhesus macaques and produces neuropathological findings similar to those of HIV infection.

The SIV model has several drawbacks, including the inability to induce disease in its natural host and the cost. For these reasons, researchers have sought other models to study lentiviral neuropathogenesis including feline immunodeficiency virus (FIV) (Podell *et al.*, 1997) (Power *et al.*, 1997); visna-maedi virus (VMV) (Bergsteinsdottir *et al.*, 1998), and caprine arthritis–encephalitis virus (Perk, 1999). VMV was a popular model for multiple sclerosis research in the past (Chebloune *et al.*, 1998). SIV and FIV have largely supplanted the sheep and goat models in HIV pathogenesis studies. FIV infection causes neurological disease in its natural host, which exhibits a slow onset and parallels the occurrence of systemic immune suppression (Phillips *et al.*, 1996). The extent of CNS disease is FIV strain-dependent (Power, 1998) but the neuropathological findings may be subtle (Hurtrel *et al.*, 1992), despite neurobehavioral and neurophysiological abnormalities (Phillips *et al.*, 1994). Both SIV and FIV have the shared appeal of using chemokine receptors as coreceptors for infection (Kovacs, Baxter, and Robinson, 1999) (Baik, Doms, and Doranz, 1999), and have similar tropisms for CNS cells.

B. Rodent Models

Two interesting strategies have emerged, using rodent models to study HIV neuropathogenesis, including (1) transgenic mice expressing HIV genes or the entire genome (Corboy *et al.*, 1992), (Toggas *et al.*, 1994) (Thomas *et al.*, 1994), and (2) the introduction of HIV-infected human brain tissue or macrophages in murine brains (Tyor *et al.*, 1993b), (Persidsky *et al.*, 1996), or the anterior chamber of the eye lens (Epstein *et al.*, 1992). Several transgenic models have been reported over the past decade beginning with the observation that differential expression by the HIV long terminal repeat (LTR) was dependent on the viral strain used and that only CNS-derived LTRs expressed in the brain (Corboy *et al.*, 1992). Similarly, when the entire HIV genome was expressed (in neurons) under the neurofilament promoter, neuronal abnormalities were observed (Thomas *et al.*, 1994). Indeed, one of the most intriguing transgenic models has shown that HIV gp120 expression in astrocytes causes neuronal injury and neurobehavioral abnormalities (Toggas *et al.*, 1994). Nonetheless, all of the above models have been subject to criticism because in humans, HIV is expressed chiefly in cells of macrophage lineage and, hence, it is difficult to interpret the findings in mice.

In contrast, the strategy in which HIV-infected human cells are introduced into murine brain offers some advantages. Tyor *et al.*, showed that HIV-infected macrophages induced increased gliosis in the brains of severe combined immunodeficient (SCID) mice that were found to be accompanied by neurobehavioral abnormalities (Avgeropoulos *et al.*, 1998). Other groups have shown that microglial activation and neuronal death may be present in the SCID mouse model (Persidsky and Gendelman, 1997). Criticisms of this model stem from the issue of the incompatibility of human HIV-infected macrophages delivered into a murine brain and how the neuropathological findings might be interpreted.

There are many murine retroviruses that cause neurological disease and show neuropathological findings that resemble HIV infection and are viral strain-dependent (Kustova *et al.*, 1998) (Portis and Lynch, 1998). To date, these models have not been exploited to their fullest potential, although they have provided a strong basis for many of the studies performed in the SIV and FIV models.

VI. Unanswered questions

Although the AIDS epidemic continues to march on with ever-increasing numbers of infected individuals, many important questions remain unanswered regarding the neuropathogenesis of this infection. The role of the brain as a protected viral reservoir that can potentially reseed the systemic circulation, after virus has become undetectable in the blood, has not been resolved. This issue bears directly on current therapeutic regimens and on the impact of drug resistance mutations encoded by the virus, found systemically and possibly also in the brain. Most HIV neuropathogenesis studies using human samples, performed to date, have focused on gay males infected with B clade viruses. The extent to which this population sample reflects HIV neuropathogenesis *in toto* is uncertain. Although the CNS disease in children is in some ways best recapitulated by the SIV model, there are few studies that have examined neuropathogenesis of HIV-infected children. Likewise, the relative neuropathogenicity of non-B clade viruses, which may undergo multiple recombination events, has not been examined. Finally, the exact function of the plethora of inflammatory molecules released by macrophages infected by, or treated with, proteins derived from HIV has yet to be defined. This latter question is of great interest because it may provide insights into the mechanisms of neuronal injury mediated by HIV infection, while at the same time providing clues to the pathogenesis of other CNS diseases characterized by glial activation and neuronal damage.

VII. Summary

Like most lentiviruses, HIV-1 causes both immune suppression and neurological disease. Neurological disease may occur at any stage of HIV infection but is most apparent with severe immune suppression. Cognitive impairment, reflected strikingly by HIV-associated dementia, has attracted intense interest since the outset of the HIV epidemic, and understanding of its pathogenesis has been spurred on by the emergence of several hypotheses outlining potential pathogenic mechanisms. The release of inflammatory molecules by HIV-infected microglia and macrophages and the concurrent neuronal damage play central roles in the conceptualization of HIV neuropathogenesis. Many inflammatory molecules appear to contribute to the pathogenic cascade and their individual roles remain undefined. At the same time, the abundance of virus in the brain and the type or strain of virus found in the brain may also be important codeterminants of neurological disease, as shown for other neurotropic viruses. Coreceptor use by HIV found in the brain appears to closely mirror what has been reported in systemic macrophages. The impact of HAART on viral genotype and phenotype found in the brain, and its relationship to clinical disease, remain uncertain. Several interesting animal models have been developed, using other lentiviruses, transgenic animals, and HIV-infected SCID mice, that may prove useful in future pathogenesis and therapeutic studies. Despite the progress in the understanding of HIV neuropathogenesis, many questions remain unanswered.

Acknowledgments

The authors thank B. Ibrahim, C. Silva, and J. B. Johnston for technical assistance in the preparation of this chapter.

References

Abuaf, N., Laperche, S., Rajoely, B., Carsique, R., Deschamps, A., Rouquette, A. M., Barthet, C., Khaled, Z., Marbot, C., Saab, N., Rozen, J., Girard, P. M., and Rozenbaum, W. (1997). Autoantibodies to phospholipids and to the coagulation proteins in AIDS. *Thromb. Haemost.* **77**(5):856–861.

Acha-Orbea, H., Finke, D., Attinger, A., Schmid, S., Wehrli, N., Vacheron, S., Xenarios, I., Scarpellino, L., Toellner, K. M., MacLennan, I. C., and Luther, S. A. (1999). Interplays between mouse mammary tumor virus and the cellular and humoral immune response. *Immunol. Rev.* **168**:287–303.

Achim, C. L., Heyes, M. P., and Wiley, C. A. (1993). Quantitation of human immunodeficiency virus, immune activation factors, and quinolinic acid in AIDS brains. *J. Clin. Invest.* **91**(6):2769–2775.

Adamson, D. C., Wildemann, B., Sasaki, M., Glass, J. D., McArthur, J. C., Christov, V. I., Dawson, T. M., and Dawson, V. L. (1996). Immunologic NO synthase: elevation in severe AIDS dementia and induction by HIV-1 gp41. *Science* **274**(5294):1917–1921.

Adle-Biassette, H., Chretien, F., Wingertsmann, L., Hery, C., Ereau, T., Scaravilli, F., Tardieu, M., and Gray, F. (1999). Neuronal apoptosis does not correlate with dementia in HIV infection but is related to microglial activation and axonal damage. *Neuropathol. Appl. Neurobiol.* **25**(2):123–133.

Adle-Biassette, H., Levy, Y., Colombel, M., Poron, F., Natchev, S., Keohane, C., and Gray, F. (1995). Neuronal apoptosis in HIV infection in adults. *Neuropathol. Appl. Neurobiol.* **21**(3):218–227.

Alkhatib, G., Combadiere, C., Broder, C. C., Feng, Y., Kennedy, P. E., Murphy, P. M., and Berger, E. A. (1996). CC CKR5: a RANTES, MIP-1alpha, MIP-1beta receptor as a fusion cofactor for macrophage-tropic HIV-1. *Science* **272**(5270):1955–1958.

Andresson, O. S., Elser, J. E., Tobin, G. J., Greenwood, J. D., Gonda, M. A., Georgsson, G., Andresdottir, V., Benediktsdottir, E., Carlsdottir, H. M., and Mantyla, E. O. (1993). Nucleotide sequence and biological properties of a pathogenic proviral molecular clone of neurovirulent visna virus. *Virology* **193**(1):89–105.

Arts, E. J., and Wainberg, M. A. (1996). Mechanisms of nucleoside analog antiviral activity and resistance during human immunodeficiency virus reverse transcription. *Antimicrob Agents Chemother* **40**(3):527–540.

Avgeropoulos, N., Kelley, B., Middaugh, L., Arrigo, S., Persidsky, Y., Gendelman, H. E., and Tyor, W. R. (1998). SCID mice with HIV encephalitis develop behavioral abnormalities. *J. Acquir. Immune. Defic. Syndr. Hum. Retrovirol.* **18**(1):13–20.

Bagasra, O., Lavi, E., Bobroski, L., Khalili, K., Pestaner, J. P., Tawadros, R., and Pomerantz, R. J. (1996). Cellular reservoirs of HIV-1 in the central nervous system of infected individuals: identification by the combination of in situ polymerase chain reaction and immunohistochemistry. *AIDS* **10**(6):573–585.

Baik, S. S., Doms, R. W., and Doranz, B. J. (1999). HIV and SIV gp120 binding does not predict coreceptor function. *Virology* **259**(2):267–273.

Barks, J. D., Liu, X. H., Sun, R., and Silverstein, F. S. (1997). gp120, a human immunodeficiency virus-1 coat protein, augments excitotoxic hippocampal injury in perinatal rats. *Neuroscience* **76**(2):397–409.

Barna, B. P., Estes, M. L., Jacobs, B. S., Hudson, S., and Ransohoff, R. M. (1990). Human astrocytes proliferate in response to tumor necrosis factor alpha. *J. Neuroimmunol.* **30**(2–3):239–243.

Bell, J. E., Busuttil, A., Ironside, J. W., Rebus, S., Donaldson, Y. K., Simmonds, P., and Peutherer, J. F. (1993). Human immunodeficiency virus and the brain: investigation of virus load and neuropathologic changes in pre-AIDS subjects. *J. Infect. Dis.* **168**(4):818–824.

Belman, A. (1997). Infants, Children and Adolescents. 2nd ed. *In* "AIDS and the Nervous System" (J. L. Berger, R. M. Levy, ed.), pp. 223–253. Lippincott-Raven Publishers, Philadelphia.

Belman, A. L., Lantos, G., Horoupian, D., Novick, B. E., Ultmann, M. H., Dickson, D. W., and Rubinstein, A. (1986). AIDS: calcification of the basal ganglia in infants and children. *Neurology* **36**(9):1192–1199.

Bennett, B. A., Rusyniak, D. E., and Hollingsworth, C. K. (1995). HIV-1 gp120-induced neurotoxicity to midbrain dopamine cultures. *Brain Res.* **705**(1–2):168–176.

Benos, D. J., McPherson, S., Hahn, B. H., Chaikin, M. A., and Benveniste, E. N. (1994). Cytokines and HIV envelope glycoprotein gp120 stimulate Na+/H+ exchange in astrocytes. *J. Biol. Chem.* **269**(19):13811–1316.

Benveniste. (1994). Cytokine circuits in brain: implications for AIDS dementia complex. *In* "HIV, AIDS, and the Brain" (R. W. Price and S. W. Perry, eds.), pp. 71–88. Raven Press, New York.

Bergsteinsdottir, K., Arnadottir, S., Torsteinsdottir, S., Agnarsdottir, G., Andresdottir, V., Petturscon, G., and Georgsson, G. (1998). Constitutive and visna virus induced expression of class I and II major histocompatibility complex antigens in the central nervous system of sheep and their role in the pathogenesis of visna lesions. *Neuropathol. Appl. Neurobiol.* **24**(3):224–232.

Boufassa, F., Bachmeyer, C., Carre, N., Deveau, C., Persoz, A., Jadand, C., Sereni, D., and Bucquet, D. (1995). Influence of neurologic manifestations of primary human immunodeficiency virus infection on disease progression. SEROCO Study Group. *J. Infect. Dis.* **171**(5):1190–1195.

Boven, L. A., Gomes, L., Hery, C., Gray, F., Verhoef, J., Portegies, P., Tardieu, M., and Nottet, H. S. (1999a). Increased peroxynitrite activity in AIDS dementia complex: implications for the neuropathogenesis of HIV-1 infection. *J. Immunol.* **162**(7):4319–4327.

Boven, L. A., Middel, J., Portegies, P., Verhoef, J., Jansen, G. H., and Nottet, H. S. (1999b). Overexpression of nerve growth factor and basic fibroblast growth factor in AIDS dementia complex. *J. Neuroimmunol.* **97**(1–2):154–162.

Brannagan, T. H., McAlarney, T., and Latov, N. (1998). Peripheral neuropathy in HIV-1 infection. *In* "Immunological and Infectious Diseases of the Peripheral Nerves" (N. W. Latov, J. H. J. Wokke, Kelly, JJ, ed.), pp. 285–307. Cambridge University Press, Cambridge.

Bratanich, A. C., Liu, C., McArthur, J. C., Fudyk, T., Glass, J. D., Mittoo, S., Klassen, G. A., and Power, C. (1998). Brain-derived HIV-1 tat sequences from AIDS patients with dementia show increased molecular heterogeneity. *J. Neurovirol.* **4**(4):387–393.

Brew, B. J., Bhalla, R. B., Paul, M., Gallardo, H., McArthur, J. C., Schwartz, M. K., and Price, R. W. (1990). Cerebrospinal fluid neopterin in human immunodeficiency virus type I infection. *Ann. Neurol.* **28**(4):556–560.

Brew, B. J., Dunbar, N., Pemberton, L., and Kaldor, J. (1996). Predictive markers of AIDS dementia complex: CD4 cell count and cerebrospinal fluid concentrations of beta 2-microglobulin and neopterin. *J. Infect. Dis.* **174**(2):294–298.

Brew, B. J., Pemberton, L., Cunningham, P., and Law, M. G. (1997). Levels of human immunodeficiency virus type 1 RNA in cerebrospinal fluid correlate with AIDS dementia stage. *J. Infect. Dis.* **175**(4):963–966.

Brew, B. J., Rosenblum, M., Cronin, K., and Price, R. W. (1995). AIDS dementia complex and HIV-1 brain infection: clinical–virological correlations [see comments]. *Ann. Neurol.* **38**(4):563–570.

Budka, H. (1991). Neuropathology of human immunodeficiency virus infection. *Brain Pathol.* **1**(3):163–175.

Chan, S. Y., Speck, R. F., Power, C., Gaffen, S. L., Chesebro, B., and Goldsmith, M. A. (1999). V3 recombinants indicate a central role for CCR5 as a coreceptor in tissue infection by human immunodeficiency virus type 1. *J. Virol.* **73**(3):2350–2358.

Chang, L., Ernst, T., Leonido-Yee, M., Walot, I., and Singer, E. (1999). Cerebral metabolite abnormalities correlate with clinical severity of HIV-1 cognitive motor complex. *Neurology* **52**(1):100–108.

Chao, C. C., Gekker, G., Hu, S., Sheng, W. S., Shark, K. B., Bu, D. F., Archer, S., Bidlack, J. M., and Peterson, P. K. (1996). kappa opioid receptors in human microglia downregulate human immunodeficiency virus 1 expression. *Proc. Natl. Acad. Sci. U.S.A.* **93**(15):8051–8056.

Chao, C. C., and Hu, S. (1994). Tumor necrosis factor-alpha potentiates glutamate neurotoxicity in human fetal brain cell cultures. *Dev. Neurosci.* **16**(3–4):172–179.

Chebloune, Y., Karr, B. M., Raghavan, R., Singh, D. K., Leung, K., Sheffer, D., Pinson, D., Foresman, L., and Narayan, O. (1998). Neuroinvasion by ovine lentivirus in infected sheep mediated by inflammatory cells associated with experimental allergic encephalomyelitis. *J. Neurovirol.* **4**(1):38–48.

Chen, P., Mayne, M., Power, C., and Nath, A. (1997). The Tat protein of HIV-1 induces tumor necrosis factor-alpha production. Implications for HIV-1-associated neurological diseases. *J. Biol. Chem.* **272**(36):22385–22388.

Cheng-Mayer, C., Weiss, C., Seto, D., and Levy, J. A. (1989). Isolates of human immunodeficiency virus type 1 from the brain may constitute a special group of the AIDS virus. *Proc. Natl. Acad. Sci. U.S.A.* **86**(21):8575–8579.

Chesebro, B., Nishio, J., Perryman, S., Cann, A., O'Brien, W., Chen, I. S., and Wehrly, K. (1991). Identification of human immunodeficiency virus envelope gene sequences influencing viral entry into CD4-positive HeLa cells, T-leukemia cells, and macrophages. *J. Virol.* **65**(11):5782–5789.

Chesebro, B., Wehrly, K., Nishio, J., and Perryman, S. (1992). Macrophage-tropic human immunodeficiency virus isolates from different patients exhibit unusual V3 envelope sequence homogeneity in comparison with T-cell-tropic isolates: definition of critical amino acids involved in cell tropism. *J. Virol.* **66**(11):6547–6554.

Cocchi, F., DeVico, A. L., Garzino-Demo, A., Cara, A., Gallo, R. C., and Lusso, P. (1996). The V3 domain of the HIV-1 gp120 envelope glycoprotein is critical for chemokine-mediated blockade of infection [see comments]. *Nat. Med.* **2**(11):1244–1247.

Collier, A. C., Coombs, R. W., Schoenfeld, D. A., Bassett, R. L., Timpone, J., Baruch, A., Jones, M., Facey, K., Whitacre, C., McAuliffe, V. J., Friedman, H. M., Merigan, T. C., Reichman, R. C., Hooper, C., and Corey, L. (1996). Treatment of human immunodeficiency virus infection with saquinavir, zidovudine, and zalcitabine. AIDS Clinical Trials Group. *N. Engl. J. Med.* **334**(16):1011–1017.

Collman, R., Balliet, J. W., Gregory, S. A., Friedman, H., Kolson, D. L., Nathanson, N., and Srinivasan, A. (1992). An infectious molecular clone of an unusual macrophage-tropic and highly cytopathic strain of human immunodeficiency virus type 1. *J. Virol.* **66**(12):7517–7521.

Conant, K., Garzino-Demo, A., Nath, A., McArthur, J. C., Halliday, W., Power, C., Gallo, R. C., and Major, E. O. (1998). Induction of monocyte chemoattractant protein-1 in HIV-1 Tat-stimulated astrocytes and elevation in AIDS dementia. *Proc. Natl. Acad. Sci. U.S.A.* **95**(6):3117–3121.

Conant, K., McArthur, J. C., Griffin, D. E., Sjulson, L., Wahl, L. M., and Irani, D. N. (1999). Cerebrospinal fluid levels of MMP-2, 7, and 9 are elevated in association with human immunodeficiency virus dementia. *Ann. Neurol.* **46**(3):391–398.

Conrad, B., Weissmahr, R. N., Boni, J., Arcari, R., Schupbach, J., and Mach, B. (1997). A human endogenous retroviral superantigen as candidate autoimmune gene in type I diabetes. *Cell* **90**(2):303–313.

Corboy, J. R., Buzy, J. M., Zink, M. C., and Clements, J. E. (1992). Expression directed from HIV long terminal repeats in the central nervous system of transgenic mice. *Science* **258**(5089):1804–1808.

Dal Pan, G. B., JR. (1997). Spinal Cord Disease in Human Immunodeficiency Virus Infection. 2nd ed. *In* "AIDS and the Nervous System" (J. L. Berger, RM, ed.):pp. 173–187. Lippincott-Raven Publishers, Philadelphia.

Dal Pan, G. J., Glass, J. D., and McArthur, J. C. (1994). Clinicopathologic correlations of HIV-1-associated vacuolar myelopathy: an autopsy-based case-control study. *Neurology* **44**(11):2159–2164.

Dal Pan, G. J., McArthur, J. H., Aylward, E., Selnes, O. A., Nance-Sproson, T. E., Kumar, A. J., Mellits, E. D., and McArthur, J. C. (1992). Patterns of cerebral atrophy in HIV-1-infected individuals: results of a quantitative MRI analysis. *Neurology* **42**(11):2125–2130.

Davis, T. H., Morton, C. C., Miller-Cassman, R., Balk, S. P., and Kadin, M. E. (1992). Hodgkin's disease, lymphomatoid papulosis, and cutaneous T-cell lymphoma derived from a common T-cell clone. *N. Engl. J. Med.* **326**(17):1115–1122.

Dawson, V. L., Dawson, T. M., Uhl, G. R., and Snyder, S. H. (1993). Human immunodeficiency virus type 1 coat protein neurotoxicity mediated by nitric oxide in primary cortical cultures. *Proc. Natl. Acad. Sci. U.S.A.* **90**(8):3256–3259.

de la Monte, S. M., Gabuzda, D. H., Ho, D. D., Brown, R. H., Jr., Hedley-Whyte, E. T., Schooley, R. T., Hirsch, M. S., and Bhan, A. K. (1988). Peripheral neuropathy in the acquired immunodeficiency syndrome. *Ann. Neurol.* **23**(5):485–492.

Deng, H., Liu, R., Ellmeier, W., Choe, S., Unutmaz, D., Burkhart, M., Di Marzio, P., Marmon, S., Sutton, R. E., Hill, C. M., Davis, C. B., Peiper, S. C., Schall, T. J., Littman, D. R., and Landau, N. R. (1996). Identification of a major co-receptor for primary isolates of HIV-1 [see comments]. *Nature* **381**(6584):661–666.

Di Stefano, M., Wilt, S., Gray, F., Dubois-Dalcq, M., and Chiodi, F. (1996). HIV type 1 V3 sequences and the development of dementia during AIDS. *AIDS Res. Hum. Retroviruses* **12**(6):471–476.

Dore, G. J., Correll, P. K., Li, Y., Kaldor, J. M., Cooper, D. A., and Brew, B. J. (1999). Changes to AIDS dementia complex in the era of highly active antiretroviral therapy. *Aids* **13**(10):1249–1253.

Dreyer, E. B., Kaiser, P. K., Offermann, J. T., and Lipton, S. A. (1990). HIV-1 coat protein neurotoxicity prevented by calcium channel antagonists [see comments]. *Science* **248**(4953):364–367.

Ellis, R. J., Hsia, K., Spector, S. A., Nelson, J. A., Heaton, R. K., Wallace, M. R., Abramson, I., Atkinson, J. H., Grant, I., and McCutchan, J. A. (1997). Cerebrospinal fluid human immunodeficiency virus type 1 RNA levels are elevated in neurocognitively impaired individuals with acquired immunodeficiency syndrome. HIV Neurobehavioral Research Center Group [see comments]. *Ann. Neurol.* **42**(5):679–688.

Epstein, L. G., Cvetkovich, T. A., Lazar, E., Dehlinger, K., Dzenko, K., del Cerro, C., and del Cerro, M. (1992). Successful xenografts of second trimester human fetal brain and retinal tissue in the anterior chamber of the eye of adult immunosuppressed rats. *J. Neural. Transplant Plast.* **3**(2–3):151–158.

Epstein, L. G., and Gendelman, H. E. (1993). Human immunodeficiency virus type 1 infection of the nervous system: pathogenetic mechanisms [see comments]. *Ann. Neurol.* **33**(5):429–436.

Everall, I. P., Glass, J. D., McArthur, J., Spargo, E., and Lantos, P. (1994). Neuronal density in the superior frontal and temporal gyri does not correlate with the degree of human immunodeficiency virus-associated dementia. *Acta. Neuropathol.* **88**(6):538–544.

Everall, I. P., Heaton, R. K., Marcotte, T. D., Ellis, R. J., McCutchan, J. A., Atkinson, J. H., Grant, I., Mallory, M., and Masliah, E. (1999). Cortical synaptic density is reduced in mild to moderate human immunodeficiency virus neurocognitive disorder. HNRC Group. HIV Neurobehavioral Research Center. *Brain Pathol.* **9**(2):209–217.

Everall, I. P., Luthert, P. J., and Lantos, P. L. (1991). Neuronal loss in the frontal cortex in HIV infection [see comments]. *Lancet* **337**(8750):1119–1121.

Falangola, M. F., Hanly, A., Galvao-Castro, B., and Petito, C. K. (1995). HIV infection of human choroid plexus: a possible mechanism of viral entry into the CNS. *J. Neuropathol. Exp. Neurol.* **54**(4):497–503.

Feng, Y., Broder, C. C., Kennedy, P. E., and Berger, E. A. (1996). HIV-1 entry cofactor: functional cDNA cloning of a seven-transmembrane, G protein-coupled receptor [see comments]. *Science* **272**(5263):872–877.

Fine, S. M., Angel, R. A., Perry, S. W., Epstein, L. G., Rothstein, J. D., Dewhurst, S., and Gelbard, H. A. (1996). Tumor necrosis factor alpha inhibits glutamate uptake by primary human astrocytes. Implications for pathogenesis of HIV-1 dementia. *J. Biol. Chem.* **271**(26):15303–15306.

Flaherty, M. T., Hauer, D. A., Mankowski, J. L., Zink, M. C., and Clements, J. E. (1997). Molecular and biological characterization of a neurovirulent molecular clone of simian immunodeficiency virus. *J. Virol.* **71**(8):5790–5798.

Flanagan, E. M., Erickson, J. B., Viveros, O. H., Chang, S. Y., and Reinhard, J. F., Jr. (1995). Neurotoxin quinolinic acid is selectively elevated in spinal cords of rats with experimental allergic encephalomyelitis. *J. Neurochem.* **64**(3):1192–1196.

Fox, L., Alford, M., Achim, C., Mallory, M., and Masliah, E. (1997). Neurodegeneration of somatostatin-immunoreactive neurons in HIV encephalitis. *J. Neuropathol. Exp. Neurol.* **56**(4):360–368.

Gallo, P., Frei, K., Rordorf, C., Lazdins, J., Tavolato, B., and Fontana, A. (1989). Human immunodeficiency virus type 1 (HIV-1) infection of the central nervous system: an evaluation of cytokines in cerebrospinal fluid. *J. Neuroimmunol.* **23**(2):109–116.

Gartner, S., McDonald, R. A., Hunter, E. A., Bouwman, F., Liu, Y., and Popovic, M. (1997). Gp120 sequence variation in brain and in T-lymphocyte human immunodeficiency virus type 1 primary isolates. *J. Hum. Virol.* **1**(1):3–18.

Gelbard, H. A., James, H. J., Sharer, L. R., Perry, S. W., Saito, Y., Kazee, A. M., Blumberg, B. M., and Epstein, L. G. (1995). Apoptotic neurons in brains from paediatric patients with HIV-1 encephalitis and progressive encephalopathy. *Neuropathol. Appl. Neurobiol.* **21**(3):208–217.

Gelbard, H. A., Nottet, H. S., Swindells, S., Jett, M., Dzenko, K. A., Genis, P., White, R., Wang, L., Choi, Y. B., Zhang, D., *et al.* (1994). Platelet-activating factor: a candidate human immunodeficiency virus type 1-induced neurotoxin. *J. Virol.* **68**(7):4628–4635.

Gendelman, H. E., Persidsky, Y., Ghorpade, A., Limoges, J., Stins, M., Fiala, M., and Morrisett, R. (1997). The neuropathogenesis of the AIDS dementia complex. *AIDS* **11**(Suppl A):S35–S45.

Genis, P., Jett, M., Bernton, E. W., Boyle, T., Gelbard, H. A., Dzenko, K., Keane, R. W., Resnick, L., Mizrachi, Y., Volsky, D. J., *et al.* (1992). Cytokines and arachidonic metabolites produced during human immunodeficiency virus (HIV)-infected macrophage-astroglia interactions: implications for the neuropathogenesis of HIV disease. *J. Exp. Med.* **176**(6):1703–1718.

Geny, C., Gherardi, R., Boudes, P., Lionnet, F., Cesaro, P., and Gray, F. (1991). Multifocal multinucleated giant cell myelitis in an AIDS patient. *Neuropathol. Appl. Neurobiol.* **17**(2):157–162.

Giulian, D., Vaca, K., and Noonan, C. A. (1990). Secretion of neurotoxins by mononuclear phagocytes infected with HIV-1. *Science* **250**(4987):1593–1596.

Giulian, D., Wendt, E., Vaca, K., and Noonan, C. A. (1993). The envelope glycoprotein of human immunodeficiency virus type 1 stimulates release of neurotoxins from monocytes. *Proc. Natl. Acad. Sci. U.S.A.* **90**(7):2769–2773.

Giulian, D., Yu, J., Li, X., Tom, D., Li, J., Wendt, E., Lin, S. N., Schwarcz, R., and Noonan, C. (1996). Study of receptor-mediated neurotoxins released by HIV-1-infected mononuclear phagocytes found in human brain. *J. Neurosci.* **16**(10):3139–3153.

Glass, J. D., Fedor, H., Wesselingh, S. L., and McArthur, J. C. (1995). Immunocytochemical quantitation of human immunodeficiency virus in the brain: correlations with dementia. *Ann. Neurol.* **38**(5):755–762.

Glass, M., Faull, R. L., Bullock, J. Y., Jansen, K., Mee, E. W., Walker, E. B., Synek, B. J., and Dragunow, M. (1996). Loss of A1 adenosine receptors in human temporal lobe epilepsy. *Brain Res.* **710**(1–2):56–68.

Grau, J. M., Masanes, F., Pedrol, E., Casademont, J., Fernandez-Sola, J., and Urbano-Marquez, A. (1993). Human immunodeficiency virus type 1 infection and myopathy: clinical relevance of zidovudine therapy [see comments]. *Ann. Neurol.* **34**(2):206–211.

Greene, W. C. (1991). The molecular biology of human immunodeficiency virus type 1 infection. *N. Engl. J. Med.* **324**(5):308–317.

Griffin, D. E., McArthur, J. C., and Cornblath, D. R. (1991). Neopterin and interferon-gamma in serum and cerebrospinal fluid of patients with HIV-associated neurologic disease. *Neurology* **41**(1):69–74.

Griffin, D. E., Wesselingh, S. L., and McArthur, J. C. (1994). Elevated central nervous system prostaglandins in human immunodeficiency virus-associated dementia. *Ann. Neurol.* **35**(5):592–597.

Groenink, M., Fouchier, R. A., Broersen, S., Baker, C. H., Koot, M., van't Wout, A. B., Huisman, H. G., Miedema, F., Tersmette, M., and Schuitemaker, H. (1993). Relation of phenotype evolution of HIV-1 to envelope V2 configuration [see comments]. *Science* **260**(5113):1513–1516.

Gyorkey, F., Melnick, J. L., and Gyorkey, P. (1987). Human immunodeficiency virus in brain biopsies of patients with AIDS and progressive encephalopathy. *J. Infect. Dis.* **155**(5):870–876.

Haase, A. T. (1986). Pathogenesis of lentivirus infections. *Nature* **322**(6075):130–136

Harouse, J. M., Bhat, S., Spitalnik, S. L., Laughlin, M., Stefano, K., Silberberg, D. H., and Gonzalez-Scarano, F. (1991). Inhibition of entry of HIV-1 in neural cell lines by antibodies against galactosyl ceramide. *Science* **253**(5017):320–323.

Harrigan, P. R., and Alexander, C. S. (1999). Selection of drug-resistant HIV [published erratum appears in trends Microbiol 1999 Jul;7(7):302]. *Trends Microbiol.* **7**(3):120–123.

He, J., Chen, Y., Farzan, M., Choe, H., Ohagen, A., Gartner, S., Busciglio, J., Yang, X., Hofmann, W., Newman, W., Mackay, C. R., Sodroski, J., and Gabuzda, D. (1997a). CCR3 and CCR5 are co-receptors for HIV-1 infection of microglia. *Nature* **385**(6617):645–649.

He, J., deCastro, C. M., Vandenbark, G. R., Busciglio, J., and Gabuzda, D. (1997b). Astrocyte apoptosis induced by HIV-1 transactivation of the c-kit protooncogene. *Proc. Natl. Acad. Sci. U.S.A.* **94**(8):3954–3959.

Herbein, G., Mahlknecht, U., Batliwalla, F., Gregersen, P., Pappas, T., Butler, J., O'Brien, W. A., and Verdin, E. (1998). Apoptosis of CD8$^+$ T cells is mediated by macrophages through interaction of HIV gp120 with chemokine receptor CXCR4 [see comments]. *Nature* **395**(6698):189–194.

Heyes, M. P., Brew, B. J., Martin, A., Price, R. W., Salazar, A. M., Sidtis, J. J., Yergey, J. A., Mouradian, M. M., Sadler, A. E., Keilp, J., and *et al.* (1991). Quinolinic acid in cerebrospinal fluid and serum in HIV-1 infection: relationship to clinical and neurological status. *Ann. Neurol.* **29**(2):202–209.

Heyes, M. P., Gravell, M., London, W. T., Eckhaus, M., Vickers, J. H., Yergey, J. A., April, M., Blackmore, D., and Markey, S. P. (1990). Sustained increases in cerebrospinal fluid quinolinic acid concentrations in rhesus macaques (Macaca mulatta) naturally infected with simian retrovirus type-D. *Brain Res.* **531**(1–2):148–158.

Heyes, M. P., Saito, K., Chen, C. Y., Proescholdt, M. G., Nowak, T. S., Jr., Li, J., Beagles, K. E., Proescholdt, M. A., Zito, M. A., Kawai, K., and Markey, S. P. (1997). Species heterogeneity between gerbils and rats: quinolinate production by microglia and astrocytes and accumulations in response to ischemic brain injury and systemic immune activation. *J. Neurochem.* **69**(4):1519–1529.

Hirsch, M., and Curran, J. (1996). Human Immunodeficiency Viruses. 3rd ed. *In* "Fields Virology" (B. Fields, D. Knipe and P. Howley, eds.), pp. 1953–1975. Lippincott-Raven Publishers, Philadelphia.

Hirsch, M. S., Conway, B., D'Aquila, R. T., Johnson, V. A., Brun-Vezinet, F., Clotet, B., Demeter, L. M., Hammer, S. M., Jacobsen, D. M., Kuritzkes, D. R., Loveday, C., Mellors, J. W., Vella, S., and Richman, D. D. (1998). Antiretroviral drug resistance testing in adults with HIV infection: implications for clinical management. International AIDS Society—USA Panel [see comments]. *Jama* **279**(24):1984–1991.

Ho, D. D., Rota, T. R., Schooley, R. T., Kaplan, J. C., Allan, J. D., Groopman, J. E., Resnick, L., Felsenstein, D., Andrews, C. A., and Hirsch, M. S. (1985). Isolation of HTLV-III from cerebrospinal fluid and neural tissues of patients with neurologic syndromes related to the acquired immunodeficiency syndrome. *N. Engl. J. Med.* **313**(24):1493–1497.

Hollander, H., and Stringari, S. (1987). Human immunodeficiency virus-associated meningitis. Clinical course and correlations. *Am. J. Med.* **83**(5):813–816.

Hurtrel, M., Ganiere, J. P., Guelfi, J. F., Chakrabarti, L., Maire, M. A., Gray, F., Montagnier, L., and Hurtrel, B. (1992). Comparison of early and late feline immunodeficiency virus encephalopathies. *AIDS* **6**(4):399–406.

Hurwitz, A. A., Berman, J. W., and Lyman, W. D. (1994). The role of the blood–brain barrier in HIV infection of the central nervous system. *Adv. Neuroimmunol.* **4**(3):249–256.

Hwang, S. S., Boyle, T. J., Lyerly, H. K., and Cullen, B. R. (1991). Identification of the envelope V3 loop as the primary determinant of cell tropism in HIV-1. *Science* **253**(5015):71–74.

Janssen, R. (1991). Nomenclature and research case definitions for neurologic manifestations of human immunodeficiency virus-type 1 (HIV-1) infection. Report of a Working Group of the American Academy of Neurology AIDS Task Force [see comments]. *Neurology* **41**(6):778–785.

Johnson, R. (1998). "Viral Infections of the Nervous System." 2nd ed. Lippincott-Raven Publishers, Philadelphia.

Johnson, R. T., Glass, J. D., McArthur, J. C., and Chesebro, B. W. (1996). Quantitation of human immunodeficiency virus in brains of demented and nondemented patients with acquired immunodeficiency syndrome. *Ann. Neurol.* **39**(3):392–395.

Jordan, C. A., Watkins, B. A., Kufta, C., and Dubois-Dalcq, M. (1991). Infection of brain microglial cells by human immunodeficiency virus type 1 is CD4 dependent. *J. Virol.* **65**(2):736–742.

Kaiser, P. K., Offermann, J. T., and Lipton, S. A. (1990). Neuronal injury due to HIV-1 envelope protein is blocked by anti-gp120 antibodies but not by anti-CD4 antibodies. *Neurology* **40**(11):1757–1761.

Kaul, M., and Lipton, S. A. (1999). Chemokines and activated macrophages in HIV gp120-induced neuronal apoptosis. *Proc. Natl. Acad. Sci. U.S.A.* **96**(14):8212–8216.

Kelder, W., McArthur, J. C., Nance-Sproson, T., McClernon, D., and Griffin, D. E. (1998). Beta-chemokines MCP-1 and RANTES are selectively increased in cerebrospinal fluid of patients with human immunodeficiency virus-associated dementia. *Ann. Neurol.* **44**(5):831–835.

Ketzler, S., Weis, S., Haug, H., and Budka, H. (1990). Loss of neurons in the frontal cortex in AIDS brains. *Acta. Neuropathol.* **80**(1):92–94.

Koenig, S., Gendelman, H. E., Orenstein, J. M., Dal Canto, M. C., Pezeshkpour, G. H., Yungbluth, M., Janotta, F., Aksamit, A., Martin, M. A., and Fauci, A. S. (1986). Detection of AIDS virus in macrophages in brain tissue from AIDS patients with encephalopathy. *Science* **233**(4768):1089–1093.

Koito, A., Harrowe, G., Levy, J. A., and Cheng-Mayer, C. (1994). Functional role of the V1/V2 region of human immunodeficiency virus type 1 envelope glycoprotein gp120 in infection of primary macrophages and soluble CD4 neutralization. *J. Virol.* **68**(4):2253–2259.

Kolson, D., and Pomerantz, R. (1996). AIDS Dementia and HIV-1-Induced Neurotoxicity: Possible Pathogenic Associations and Mechanisms. *J. Biomed. Sci.* **3**:389–414.

Koot, M., Vos, A. H., Keet, R. P., de Goede, R. E., Dercksen, M. W., Terpstra, F. G., Coutinho, R. A., Miedema, F., and Tersmette, M. (1992). HIV-1 biological phenotype in long-term infected individuals evaluated with an MT-2 cocultivation assay. *Aids* **6**(1):49–54.

Korber, B., Foley, B., Leitner, T., McCutchan, F., Hahn, B., Mellors, J., Myers, G., and Kuiken, C., eds. (1997). Human Retroviruses and AIDS 1997. Theoretical Biology and Biophysics, Los Alamos.

Korber, B. T., Kunstman, K. J., Patterson, B. K., Furtado, M., McEvilly, M. M., Levy, R., and Wolinsky, S. M. (1994). Genetic differences between blood-and brain-derived viral sequences from human immunodeficiency virus type 1-infected patients: evidence of conserved elements in the V3 region of the envelope protein of brain-derived sequences. *J. Virol.* **68**(11):7467–7481.

Kovacs, E. M., Baxter, G. D., and Robinson, W. F. (1999). Feline peripheral blood mononuclear cells express message for both CXC and CC type chemokine receptors. *Arch. Virol.* **144**(2):273–285.

Krucker, T., Toggas, S. M., Mucke, L., and Siggins, G. R. (1998). Transgenic mice with cerebral expression of human immunodeficiency virus type-1 coat protein gp120 show divergent changes in short- and long-term potentiation in CA1 hippocampus. *Neuroscience* **83**(3):691–700.

Kuiken, C. L., Goudsmit, J., Weiller, G. F., Armstrong, J. S., Hartman, S., Portegies, P., Dekker, J., and Cornelissen, M. (1995). Differences in human immunodeficiency virus type 1 V3 sequences from patients with and without AIDS dementia complex. *J. Gen. Virol.* **76**(Pt 1):175–180.

Kure, K., Lyman, W. D., Weidenheim, K. M., and Dickson, D. W. (1990). Cellular localization of an HIV-1 antigen in subacute AIDS encephalitis using an improved double-labeling immunohistochemical method. *Am. J. Pathol.* **136**(5):1085–1092.

Kustova, Y., Ha, J. H., Espey, M. G., Sei, Y., Morse, D., and Basile, A. S. (1998). The pattern of neurotransmitter alterations in LP-BM5 infected mice is consistent with glutamatergic hyperactivation. *Brain Res.* **793**(1–2):119–126.

Kwong, P. D., Wyatt, R., Robinson, J., Sweet, R. W., Sodroski, J., and Hendrickson, W. A. (1998). Structure of an HIV gp120 envelope glycoprotein in complex with the CD4 receptor and a neutralizing human antibody [see comments]. *Nature* **393**(6686):648–659.

Lannuzel, A., Barnier, J. V., Hery, C., Huynh, V. T., Guibert, B., Gray, F., Vincent, J. D., and Tardieu, M. (1997). Human immunodeficiency virus type 1 and its coat protein gp120 induce apoptosis and activate JNK and ERK mitogen-activated protein kinases in human neurons. *Ann. Neurol.* **42**(6):847–856.

Lazarini, F., Seilhean, D., Rosenblum, O., Suarez, S., Conquy, L., Uchihara, T., Sazdovitch, V., Mokhtari, K., Maisonobe, T., Boussin, F., Katlama, C., Bricaire, F., Duyckaerts, C., and Hauw, J. J. (1997). Human immunodeficiency virus type 1 DNA and

RNA load in brains of demented and nondemented patients with acquired immunodeficiency syndrome. *J. Neurovirol.* **3**(4):299–303.

Lee, J., Zipfel, G., and Choi, D. (1999). The changing landscape of ischaemic brain injury mechanisms. *Nature* **399**(Supp):A7–A14.

Letendre, S. L., Lanier, E. R., and McCutchan, J. A. (1999). Cerebrospinal fluid beta chemokine concentrations in neurocognitively impaired individuals infected with human immunodeficiency virus type 1. *J. Infect. Dis.* **180**(2):310–319.

Levy, J. (1998). "HIV and the Pathogenesis of AIDS." 2nd ed. American Society of Microbiology, Washington, D.C.

Levy, J. A., Mackewicz, C. E., and Barker, E. (1996). Controlling HIV pathogenesis: the role of the noncytotoxic anti-HIV response of CD8+ T cells. *Immunol. Today* **17**(5):217–224.

Levy, J. A., Shimabukuro, J., Hollander, H., Mills, J., and Kaminsky, L. (1985). Isolation of AIDS-associated retroviruses from cerebrospinal fluid and brain of patients with neurological symptoms. *Lancet* **2**(8455):586–588.

Lipton, S. A. (1992). Memantine prevents HIV coat protein-induced neuronal injury in vitro [see comments]. *Neurology* **42**(7):1403–1405.

Lipton, S. A., and Gendelman, H. E. (1995). Dementia associated with the acquired immunodeficiency syndrome. *N. Engl. J. Med.* **332**:934–940.

Liu, Z. Q., Muhkerjee, S., Sahni, M., McCormick-Davis, C., Leung, K., Li, Z., Gattone, V. H., 2nd, Tian, C., Doms, R. W., Hoffman, T. L., Raghavan, R., Narayan, O., and Stephens, E. B. (1999). Derivation and biological characterization of a molecular clone of SHIV(KU-2) that causes AIDS, neurological disease, and renal disease in rhesus macaques. *Virology* **260**(2):295–307.

Ma, M., Geiger, J. D., and Nath, A. (1994). Characterization of a novel binding site for the human immunodeficiency virus type 1 envelope protein gp120 on human fetal astrocytes. *J. Virol.* **68**(10):6824–6828.

Maggirwar, S. B., Tong, N., Ramirez, S., Gelbard, H. A., and Dewhurst, S. (1999). HIV-1 Tat-mediated activation of glycogen synthase kinase-3beta contributes to Tat-mediated neurotoxicity. *J. Neurochem.* **73**(2):578–586.

Maher, J., Choudhri, S., Halliday, W., Power, C., and Nath, A. (1997). AIDS dementia complex with generalized myoclonus. *Mov. Disord.* **12**(4):593–597.

Maj, M., Satz, P., Janssen, R., Zaudig, M., Starace, F., D'Elia, L., Sughondhabirom, B., Mussa, M., Naber, D., Ndetei, D., and et al. (1994). WHO Neuropsychiatric AIDS study, cross-sectional phase II. Neuropsychological and neurological findings. *Arch. Gen. Psychiatry.* **51**(1):51–61.

Marder, K. (1996). Clinical confirmation of the American Academy of Neurology algorithm for HIV-1-associated cognitive/motor disorder. The Dana Consortium on Therapy for HIV Dementia and Related Cognitive Disorders. *Neurology* **47**(5):1247–1253.

Masliah, E., Ge, N., Achim, C. L., Hansen, L. A., and Wiley, C. A. (1992a). Selective neuronal vulnerability in HIV encephalitis. *J. Neuropathol. Exp. Neurol.* **51**(6):585–593.

Masliah, E., Ge, N., Morey, M., DeTeresa, R., Terry, R. D., and Wiley, C. A. (1992b). Cortical dendritic pathology in human immunodeficiency virus encephalitis [see comments]. *Lab. Invest.* **66**(3):285–291.

Mattson, M. P., Cheng, B., Baldwin, S. A., Smith-Swintosky, V. L., Keller, J., Geddes, J. W., Scheff, S. W., and Christakos, S. (1995). Brain injury and tumor necrosis factors induce calbindin D-28k in astrocytes: evidence for a cytoprotective response. *J. Neurosci. Res.* **42**(3):357–370.

Mayne, M., Bratanich, A. C., Chen, P., Rana, F., Nath, A., and Power, C. (1998). HIV-1 tat molecular diversity and induction of TNF-alpha: implications for HIV-induced neurological disease. *Neuroimmunomodulation* **5**(3–4):184–192.

McArthur, J. C. (1987). Neurologic manifestations of AIDS. *Medicine (Baltimore)* **66**(6):407–437.

McArthur, J. C., Cohen, B. A., Farzedegan, H., Cornblath, D. R., Selnes, O. A., Ostrow, D., Johnson, R. T., Phair, J., and Polk, B. F. (1988). Cerebrospinal fluid abnormalities in homosexual men with and without neuropsychiatric findings. *Ann. Neurol.* **23**(Suppl):S34=S37.

McArthur, J. C., Hoover, D. R., Bacellar, H., Miller, E. N., Cohen, B. A., Becker, J. T., Graham, N. M., McArthur, J. H., Selnes, O. A., Jacobson, L. P., *et al.* (1993). Dementia in AIDS patients: incidence and risk factors. Multicenter AIDS Cohort Study. *Neurology* **43**(11):2245–2252.

McArthur, J. C., McClernon, D. R., Cronin, M. F., Nance-Sproson, T. E., Saah, A. J., St Clair, M., and Lanier, E. R. (1997). Relationship between human immunodeficiency virus-associated dementia and viral load in cerebrospinal fluid and brain [see comments]. *Ann. Neurol.* **42**(5):689–698.

McArthur, J. C., Nance-Sproson, T. E., Griffin, D. E., Hoover, D., Selnes, O. A., Miller, E. N., Margolick, J. B., Cohen, B. A., Farzadegan, H., and Saah, A. (1992). The diagnostic utility of elevation in cerebrospinal fluid beta 2-microglobulin in HIV-1 dementia. Multicenter AIDS Cohort Study. *Neurology* **42**(9):1707–1712.

McNearney, T., Hornickova, Z., Markham, R., Birdwell, A., Arens, M., Saah, A., and Ratner, L. (1992). Relationship of human immunodeficiency virus type 1 sequence heterogeneity to stage of disease. *Proc. Natl. Acad. Sci. U.S.A.* **89**(21):10247–10251.

Mirsattari, S. M., Berry, M. E., Holden, J. K., Ni, W., Nath, A., and Power, C. (1999). Paroxysmal dyskinesias in patients with HIV infection. *Neurology* **52**(1):109–114.

Mirsattari, S. M., Power, C., and Nath, A. (1998). Parkinsonism with HIV infection. *Mov. Disord.* **13**(4):684–689.

Moroni, F. (1999). Tryptophan metabolism and brain function: focus on kynurenine and other indole metabolites. *Eur. J. Pharmacol.* **375**(1–3):87–100.

Morris, A., Marsden, M., Halcrow, K., Hughes, E. S., Brettle, R. P., Bell, J. E., and Simmonds, P. (1999). Mosaic structure of the human immunodeficiency virus type 1 genome infecting lymphoid cells and the brain: evidence for frequent in vivo recombination events in the evolution of regional populations. *J. Virol.* **73**(10):8720–8731.

Moses, A. V., Bloom, F. E., Pauza, C. D., and Nelson, J. A. (1993). Human immunodeficiency virus infection of human brain capillary endothelial cells occurs via a CD4/galactosylceramide-independent mechanism. *Proc. Natl. Acad. Sci. U.S.A.* **90**(22):10474–10478.

Moses, A. V., Stenglein, S. G., Strussenberg, J. G., Wehrly, K., Chesebro, B., and Nelson, J. A. (1996). Sequences regulating tropism of human immunodeficiency virus type 1 for brain capillary endothelial cells map to a unique region on the viral genome. *J. Virol.* **70**(6):3401–3406.

Murray, E. A., Rausch, D. M., Lendvay, J., Sharer, L. R., and Eiden, L. E. (1992). Cognitive and motor impairments associated with SIV infection in rhesus monkeys. *Science* **255**(5049):1246–1249.

Narayan, O., and Clements, J. E. (1989). Biology and pathogenesis of lentiviruses. *J. Gen. Virol.* **70**(Pt 7):1617–1639.

Nath, A., Conant, K., Chen, P., Scott, C., and Major, E. O. (1999). Transient exposure to HIV-1 Tat protein results in cytokine production in macrophages and astrocytes. A hit and run phenomenon. *J. Biol. Chem.* **274**(24):17098–17102.

Nath, A., and Geiger, J. (1998). Neurobiological aspects of human immunodeficiency virus infection: neurotoxic mechanisms. *Prog. Neurobiol.* **54**(1):19–33.

Nath, A., Psooy, K., Martin, C., Knudsen, B., Magnuson, D. S., Haughey, N., and Geiger, J. D. (1996). Identification of a human immunodeficiency virus type 1 Tat epitope that is neuroexcitatory and neurotoxic. *J. Virol.* **70**(3):1475–1480.

Navia, B. A., Cho, E. S., Petito, C. K., and Price, R. W. (1986). The AIDS dementia complex: II. Neuropathology. *Ann. Neurol.* **19**(6):525–535.

Navia, B. A., Jordan, B. D., and Price, R. W. (1986). The AIDS dementia complex: I. Clinical features. *Ann. Neurol.* **19**(6):517–524.

Nicoll, A., and Gill, O. N. (1999). The global impact of HIV infection and disease. *Commun. Dis. Public Health* **2**(2):85–95.

Nicotera, P., and Lipton, S. A. (1999). Excitotoxins in neuronal apoptosis and necrosis. *J. Cereb. Blood Flow Metab.* **19**(6):583–591.

Nuovo, G. J., Gallery, F., MacConnell, P., and Braun, A. (1994). In situ detection of polymerase chain reaction-amplified HIV-1 nucleic acids and tumor necrosis factor-alpha RNA in the central nervous system. *Am. J. Pathol.* **144**(4):659–666.

O'Brien, W. A., Hartigan, P. M., Martin, D., Esinhart, J., Hill, A., Benoit, S., Rubin, M., Simberkoff, M. S., and Hamilton, J. D. (1996). Changes in plasma HIV-1 RNA and CD4+ lymphocyte counts and the risk of progression to AIDS. Veterans Affairs Cooperative Study Group on AIDS [see comments]. *N. Engl. J. Med.* **334**(7):426–431.

Ohagen, A., Ghosh, S., He, J., Huang, K., Chen, Y., Yuan, M., Osathanondh, R., Gartner, S., Shi, B., Shaw, G., and Gabuzda, D. (1999). Apoptosis induced by infection of primary brain cultures with diverse human immunodeficiency virus type 1 isolates: evidence for a role of the envelope. *J. Virol.* **73**(2):897–906.

Pang, S., Vinters, H. V., Akashi, T., O'Brien, W. A., and Chen, I. S. (1991). HIV-1 env sequence variation in brain tissue of patients with AIDS-related neurologic disease. *J. Acquir. Immune. Defic. Syndr.* **4**(11):1082–1092.

Pattarini, R., Pittaluga, A., and Raiteri, M. (1998). The human immunodeficiency virus-1 envelope protein gp120 binds through its V3 sequence to the glycine site of N-methyl-D-aspartate receptors mediating noradrenaline release in the hippocampus. *Neuroscience* **87**(1):147–157.

Perelson, A. S., Essunger, P., Cao, Y., Vesanen, M., Hurley, A., Saksela, K., Markowitz, M., and Ho, D. D. (1997). Decay characteristics of HIV-1-infected compartments during combination therapy [see comments]. *Nature* **387**(6629):188–191.

Perelson, A. S., Essunger, P., and Ho, D. D. (1997). Dynamics of HIV-1 and CD4+ lymphocytes in vivo. *Aids* **11**(Suppl A):S17–S24.

Perk, K. (1999). Concealed locations of lentiviruses in caprine arthritis encephalitis system. *Virology* **253**(1):8–9.

Persidsky, Y., and Gendelman, H. E. (1997). Development of laboratory and animal model systems for HIV-1 encephalitis and its associated dementia. *J. Leukocyte. Biol.* **62**:100–106.

Persidsky, Y., Limoges, J., McComb, R., Bock, P., Baldwin, T., Tyor, W., Patil, A., Nottet, H. S., Epstein, L., Gelbard, H., Flanagan, E., Reinhard, J., Pirruccello, S. J., and Gendelman, H. E. (1996). Human immunodeficiency virus encephalitis in SCID mice [see comments]. *Am. J. Pathol.* **149**(3):1027–1053.

Petito, C. K., and Cash, K. S. (1992). Blood-brain barrier abnormalities in the acquired immunodeficiency syndrome: immunohistochemical localization of serum proteins in postmortem brain. *Ann. Neurol.* **32**(5):658–666.

Phillips, T. R., Prospero-Garcia, O., Puaoi, D. L., Lerner, D. L., Fox, H. S., Olmsted, R. A., Bloom, F. E., Henriksen, S. J., and Elder, J. H. (1994). Neurological abnormalities associated with feline immunodeficiency virus infection. *J. Gen. Virol.* **75**(Pt 5):979–987.

Phillips, T. R., Prospero-Garcia, O., Wheeler, D. W., Wagaman, P. C., Lerner, D. L., Fox, H. S., Whalen, L. R., Bloom, F. E., Elder, J. H., and Henriksen, S. J. (1996). Neurologic dysfunctions caused by a molecular clone of feline immunodeficiency virus, FIV-PPR. *J. Neurovirol.* **2**(6):388–396.

Pizzo, P. A., and Wilfert, C. M., eds. (1994). Pediatric AIDS: the challenge of HIV infection in infants, children and adolescents. Williams and Wilkins, New York.

Podell, M., Hayes, K., Oglesbee, M., and Mathes, L. (1997). Progressive encephalopathy associated with CD4/CD8 inversion in adult FIV-infected cats. *J. Acquir. Immune. Defic. Syndr. Hum. Retrovirol.* **15**(5):332–340.

Poland, S. D., Rice, G. P., and Dekaban, G. A. (1995). HIV-1 infection of human brain-derived microvascular endothelial cells in vitro. *J. Acquir. Immune. Defic. Syndr. Hum. Retrovirol.* **8**(5):437–445.

Portis, J. L., and Lynch, W. P. (1998). Dissecting the determinants of neuropathogenesis of the murine oncornaviruses. *Virology* **247**(2):127–136.

Power, C., Buist, R., Johnston, J. B., Del Bigio, M. R., Ni, W., Dawood, M. R., and Peeling, J. (1998a). Neurovirulence in feline immunodeficiency virus-infected neonatal cats is viral strain specific and dependent on systemic immune suppression. *J. Virol.* **72**(11):9109–9115.

Power, C., Kong, P. A., Crawford, T. O., Wesselingh, S., Glass, J. D., McArthur, J. C., and Trapp, B. D. (1993). Cerebral white matter changes in acquired immunodeficiency syndrome dementia: alterations of the blood-brain barrier. *Ann. Neurol.* **34**(3):339–350.

Power, C., McArthur, J. C., Johnson, R. T., Griffin, D. E., Glass, J. D., Dewey, R., and Chesebro, B. (1995). Distinct HIV-1 env sequences are associated with neurotropism and neurovirulence. *Curr. Top. Microbiol. Immunol.* **202**:89–104.

Power, C., McArthur, J. C., Johnson, R. T., Griffin, D. E., Glass, J. D., Perryman, S., and Chesebro, B. (1994). Demented and nondemented patients with AIDS differ in brain-derived human immunodeficiency virus type 1 envelope sequences. *J. Virol.* **68**(7):4643–4649.

Power, C., McArthur, J. C., Nath, A., Wehrly, K., Mayne, M., Nishio, J., Langelier, T., Johnson, R. T., and Chesebro, B. (1998b). Neuronal death induced by brain-derived human immunodeficiency virus type 1 envelope genes differs between demented and nondemented AIDS patients. *J. Virol.* **72**(11):9045–9053.

Power, C., Moench, T., Peeling, J., Kong, P. A., and Langelier, T. (1997). Feline immunodeficiency virus causes increased glutamate levels and neuronal loss in brain. *Neuroscience* **77**(4):1175–1185.

Pulliam, L., Herndier, B. G., Tang, N. M., and McGrath, M. S. (1991). Human immunodeficiency virus-infected macrophages produce soluble factors that cause histological and neurochemical alterations in cultured human brains. *J. Clin. Invest.* **87**(2):503–512.

Quagliarello, V. J., Wispelwey, B., Long, W. J., Jr., and Scheld, W. M. (1991). Recombinant human interleukin-1 induces meningitis and blood-brain barrier injury in the rat. Characterization and comparison with tumor necrosis factor. *J. Clin. Invest.* **87**(4):1360–1366.

Rance, N. E., McArthur, J. C., Cornblath, D. R., Landstrom, D. L., Griffin, J. W., and Price, D. L. (1988). Gracile tract degeneration in patients with sensory neuropathy and AIDS. *Neurology* **38**(2):265–271.

Ransohoff, R. M., and Tani, M. (1998). Do chemokines mediate leukocyte recruitment in post-traumatic CNS inflammation? *Trends Neurosci.* **21**(4):154–159.

Reddy, R. T., Achim, C. L., Sirko, D. A., Tehranchi, S., Kraus, F. G., Wong-Staal, F., and Wiley, C. A. (1996). Sequence analysis of the V3 loop in brain and spleen of patients with HIV encephalitis. *AIDS Res. Hum. Retroviruses* **12**(6):477–482.

Resnick, L., diMarzo-Veronese, F., Schupbach, J., Tourtellotte, W. W., Ho, D. D., Muller, F., Shapshak, P., Vogt, M., Groopman, J. E., Markham, P. D., and *et al.* (1985). Intra-blood-brain-barrier synthesis of HTLV-III-specific IgG in patients with neurologic symptoms associated with AIDS or AIDS-related complex. *N. Engl. J. Med.* **313**(24):1498–1504.

Rostasy, K., Monti, L., Yiannoutsos, C., Kneissl, M., Bell, J., Kemper, T. L., Hedreen, J. C., and Navia, B. A. (1999). Human immunodeficiency virus infection, inducible nitric oxide synthase expression, and microglial activation: pathogenetic relationship to the acquired immunodeficiency syndrome dementia complex. *Ann. Neurol.* **46**(2):207–216.

Rothstein, J. D., Dykes-Hoberg, M., Pardo, C. A., Bristol, L. A., Jin, L., Kuncl, R. W., Kanai, Y., Hediger, M. A., Wang, Y., Schielke, J. P., and Welty, D. F. (1996). Knockout of glutamate transporters reveals a major role for astroglial transport in excitotoxicity and clearance of glutamate. *Neuron.* **16**(3):675–686.

Rothstein, J. D., Jin, L., Dykes-Hoberg, M., and Kuncl, R. W. (1993). Chronic inhibition of glutamate uptake produces a model of slow neurotoxicity. *Proc. Natl. Acad. Sci. U.S.A.* **90**(14):6591–6595.

Royal, W., 3rd, Selnes, O. A., Concha, M., Nance-Sproson, T. E., and McArthur, J. C. (1994). Cerebrospinal fluid human immunodeficiency virus type 1 (HIV-1) p24 antigen levels in HIV-1-related dementia. *Ann. Neurol.* **36**(1):32–39.

Rubbert, A., Bock, E., Schwab, J., Marienhagen, J., Nusslein, H., Wolf, F., and Kalden, J. R. (1994). Anticardiolipin antibodies in HIV infection: association with cerebral perfusion defects as detected by 99mTc-HMPAO SPECT. *Clin. Exp. Immunol.* **98**(3):361–368.

Sacktor, N. C., Lyles, R. H., Skolasky, R. L., Anderson, D. E., McArthur, J. C., McFarlane, G., Selnes, O. A., Becker, J. T., Cohen, B., Wesch, J., and Miller, E. N. (1999). Combination antiretroviral therapy improves psychomotor speed performance in HIV-seropositive homosexual men. Multicenter AIDS Cohort Study (MACS). *Neurology* **52**(8):1640–1647.

Saito, K., and Heyes, M. P. (1996). Kynurenine pathway enzymes in brain. Properties of enzymes and regulation of quinolinic acid synthesis. *Adv. Exp. Med. Biol.* **398**:485–492.

Saito, Y., Sharer, L. R., Epstein, L. G., Michaels, J., Mintz, M., Louder, M., Golding, K., Cvetkovich, T. A., and Blumberg, B. M. (1994). Overexpression of nef as a marker for restricted HIV-1 infection of astrocytes in postmortem pediatric central nervous tissues. *Neurology* **44**(3 Pt 1):474–481.

Sanders, V. J., Pittman, C. A., White, M. G., Wang, G., Wiley, C. A., and Achim, C. L. (1998). Chemokines and receptors in HIV encephalitis. *AIDS* **12**(9):1021–1026.

Sasseville, V. G., and Lackner, A. A. (1997). Neuropathogenesis of simian immunodeficiency virus infection in macaque monkeys. *J. Neurovirol.* **3**(1):1–9.

Sei, S., Saito, K., Stewart, S. K., Crowley, J. S., Brouwers, P., Kleiner, D. E., Katz, D. A., Pizzo, P. A., and Heyes, M. P. (1995). Increased human immunodeficiency virus (HIV) type 1 DNA content and quinolinic acid concentration in brain tissues from patients with HIV encephalopathy. *J. Infect. Dis.* **172**(3):638–647.

Selmaj, K., and Raine, C. S. (1988). Tumor necrosis factor mediates myelin damage in organotypic cultures of nervous tissue. *Ann. NY. Acad. Sci.* **540**:568–570.

Sharer, L. R. (1992). Pathology of HIV-1 infection of the central nervous system. A review. *J. Neuropathol. Exp. Neurol.* **51**(1):3–11.

Sharer, L. R., Epstein, L. G., Cho, E. S., Joshi, V. V., Meyenhofer, M. F., Rankin, L. F., and Petito, C. K. (1986). Pathologic features of AIDS encephalopathy in children: evidence for LAV/HTLV-III infection of brain. *Hum. Pathol.* **17**(3):271–284.

Shaw, G. M., Harper, M. E., Hahn, B. H., Epstein, L. G., Gajdusek, D. C., Price, R. W., Navia, B. A., Petito, C. K., O'Hara, C. J., Groopman, J. E., *et al.* (1985). HTLV-III infection in brains of children and adults with AIDS encephalopathy. *Science* **227**(4683):177–182.

Shi, B. D., U; He, J; *et al.* (1996). Apoptosis induced by HIV-1 infection of the central nervous system. *J. Clin. Invest.* **98**:1979–1990.

Shrikant, P., Benos, D. J., Tang, L. P., and Benveniste, E. N. (1996). HIV glycoprotein 120 enhances intercellular adhesion molecule-1 gene expression in glial cells. Involvement of Janus kinase/signal transducer and activator of transcription and protein kinase C signaling pathways. *J. Immunol.* **156**(3):1307–1314.

Sidtis, J. J., Gatsonis, C., Price, R. W., Singer, E. J., Collier, A. C., Richman, D. D., Hirsch, M. S., Schaerf, F. W., Fischl, M. A., Kieburtz, K., and et al. (1993). Zidovudine treatment of the AIDS dementia complex: results of a placebo-controlled trial. AIDS Clinical Trials Group. *Ann. Neurol.* **33**(4):343–349.

Silver, B., McAvoy, K., Mikesell, S., and Smith, T. W. (1997). Fulminating encephalopathy with perivenular demyelination and vacuolar myelopathy as the initial presentation of human immunodeficiency virus infection. *Arch. Neurol.* **54**(5):647–650.

Simpson, D. M., and Tagliati, M. (1994). Neurologic manifestations of HIV infection [published erratum appears in *Ann. Intern. Med.* 1995, Feb 15;122(4):317] [see comments]. *Ann. Intern. Med.* **121**(10):769–785.

Sippy, B. D., Hofman, F. M., Wallach, D., and Hinton, D. R. (1995). Increased expression of tumor necrosis factor-alpha receptors in the brains of patients with AIDS. *J. Acquir. Immune. Defic. Syndr. Hum. Retrovirol.* **10**(5):511–521.

Strizki, J. M., Albright, A. V., Sheng, H., O'Connor, M., Perrin, L., and Gonzalez-Scarano, F. (1996). Infection of primary human microglia and monocyte-[???]

Vazeux, R., Lacroix-Ciaudo, C., Blanche, S., Cumont, M. C., Henin, D., Gray, F., Boccon-Gibod, L., and Tardieu, M. (1992). Low levels of human immunodeficiency virus replication in the brain tissue of children with severe acquired immunodeficiency syndrome encephalopathy. *Am. J. Pathol.* **140**(1):137–144.

Vinters, H. A., KH.[???] (1990). "Neuropathology of AIDS." CRC Press, Boca Raton, Florida.

Voulgaropoulou, F., Tan, B., Soares, M., Hahn, B., and Ratner, L. (1999). Distinct human immunodeficiency virus strains in the bone marrow are associated with the development of thrombocytopenia. *J. Virol.* **73**(4):3497–3504.

Watkins, B. A., Dorn, H. H., Kelly, W. B., Armstrong, R. C., Potts, B. J., Michaels, F., Kufta, C. V., and Dubois-Dalcq, M. (1990). Specific tropism of HIV-1 for microglial cells in primary human brain cultures. *Science* **249**(4968):549–553.

Weiss, R. A. (1996). HIV receptors and the pathogenesis of AIDS [see comments]. *Science* **272**(5270):1885–1886.

Wesselingh, S. L., Power, C., Glass, J. D., Tyor, W. R., McArthur, J. C., Farber, J. M., Griffin, J. W., and Griffin, D. E. (1993). Intracerebral cytokine messenger RNA expression in acquired immunodeficiency syndrome dementia. *Ann. Neurol.* **33**(6):576–582.

Westervelt, P., Trowbridge, D. B., Epstein, L. G., Blumberg, B. M., Li, Y., Hahn, B. H., Shaw, G. M., Price, R. W., and Ratner, L. (1992). Macrophage tropism determinants of human immunodeficiency virus type 1 in vivo. *J. Virol.* **66**(4):2577–2582.

Westmoreland, S. V., Kolson, D., and Gonzalez-Scarano, F. (1996). Toxicity of TNF alpha and platelet activating factor for human NT2N neurons: a tissue culture model for human immunodeficiency virus dementia. *J. Neurovirol.* **2**(2):118–126.

Wiley, C. A., Baldwin, M., and Achim, C. L. (1996). Expression of HIV regulatory and structural mRNA in the central nervous system [see comments]. *AIDS* **10**(8):843–847.

Wiley, C. A., Masliah, E., Morey, M., Lemere, C., DeTeresa, R., Grafe, M., Hansen, L., and Terry, R. (1991). Neocortical damage during HIV infection. *Ann. Neurol.* **29**(6):651–657.

Wiley, C. A., Schrier, R. D., Nelson, J. A., Lampert, P. W., and Oldstone, M. B. (1986). Cellular localization of human immunodeficiency virus infection within the brains of acquired immune deficiency syndrome patients. *Proc. Natl. Acad. Sci. U.S.A.* **83**(18):7089–7093.

Wilt, S. G., Milward, E., Zhou, J. M., Nagasato, K., Patton, H., Rusten, R., Griffin, D. E., O'Connor, M., and Dubois-Dalcq, M. (1995). In vitro evidence for a dual role of tumor necrosis factor-alpha in human immunodeficiency virus type 1 encephalopathy [see comments]. *Ann. Neurol.* **37**(3):381–394.

Wong, J. K., Ignacio, C. C., Torriani, F., Havlir, D., Fitch, N. J., and Richman, D. D. (1997). In vivo compartmentalization of human immunodeficiency virus: evidence from the examination of pol sequences from autopsy tissues. *J. Virol.* **71**(3):2059–2071.

Yamada, M., Zurbriggen, A., Oldstone, M. B., and Fujinami, R. S. (1991). Common immunologic determinant between human immunodeficiency virus type 1 gp41 and astrocytes. *J. Virol.* **65**(3):1370–1376.

Zheng, J., Ghorpade, A., Niemann, D., Cotter, R. L., Thylin, M. R., Epstein, L., Swartz, J. M., Shepard, R. B., Liu, X., Nukuna, A., and Gendelman, H. E. (1999). Lymphotropic virions affect chemokine receptor-mediated neural signaling and apoptosis: implications for human immunodeficiency virus type 1-associated dementia. *J. Virol.* **73**(10):8256–8267.

Zink, M. C., Amedee, A. M., Mankowski, J. L., Craig, L., Didier, P., Carter, D. L., Munoz, A., Murphey-Corb, M., and Clements, J. E. (1997). Pathogenesis of SIV encephalitis. Selection and replication of neurovirulent SIV. *Am. J. Pathol.* **151**(3):793–803.

ADVANCES IN VIRUS RESEARCH, VOL 56

SIMIAN IMMUNODEFICIENCY VIRUS MODEL OF HIV-INDUCED CENTRAL NERVOUS SYSTEM DYSFUNCTION

E.M.E. Burudi and Howard S. Fox

Department of Neuropharmacology
The Scripps Research Institute, CVN-8
La Jolla, California 92037

I. Introduction

The simian immunodeficiency virus (SIV) and the human immunodeficiency virus (HIV) are primate lentiviruses that have great genomic, structural, and virologic similarities (Desrosiers, 1990; Kindt *et al.,* 1992). SIV was first isolated from captive rhesus macaques monkeys *(Macaca mulatta).* SIVs have been isolated from different genera of nonhuman primates, and hence are designated by subscripts indicating the type of primate from which they originally were isolated—for example, SIV_{mac} from rhesus monkeys; SIV_{smm} from sooty mangabey monkeys *(Cercocebus atys),* etc. It is thought that this infection in captive animals originated from the sooty mangabey monkeys, the nat-

435

ural host of SIV, as macaques in the wild do not harbor this virus. Although SIV infection in its natural hosts, such as the sooty mangabey monkeys, does not result in any apparent disease, infection of rhesus monkeys with this virus results in an immune deficiency syndrome comparable to the acquired immunodeficiency syndrome (AIDS) caused by HIV in humans.

Infection of rhesus as well as pig-tailed *(M. nemestrina)* monkeys is the most popular nonhuman primate model of HIV disease. Progressive lymphoid changes, opportunistic infections, a wasting syndrome, and central nervous system (CNS) disease characterize the syndrome in both humans and monkeys (Desrosiers, 1990; Fox *et al.*, 1997; Lackner, 1994; Zink *et al.*, 1998). Both SIV and HIV cause persistent infections with a tropism for CD4+ T cells and cells of the macrophage lineage *in vivo*. However, the time course of the disease in monkeys is more rapid than that observed in humans, thus expediting experimental analysis.

Following intravenous infection of rhesus monkeys with SIV, the virus then seeds lymphoid organs, liver, and lungs, and is transiently cleared from plasma (Zhang *et al.*, 1999). An acute viremia ensues and the virus can be found in the brain within two weeks following infection (Chakrabarti *et al.*, 1991; Lackner *et al.*, 1994). Infected animals initially manifest fever, anorexia, and peripheral lymphadenopathy, which, after a variable time period, are followed by signs and symptoms of disease that can include lymphoid depletion, lethargy, chronic diarrhea, and weight loss (Desrosiers, 1990; Simon, Chalifoux, and Ringler, 1992).

In humans, the HIV-associated cognitive/motor disorder, also known as AIDS dementia complex (ADC) or neuroAIDS occurs in one-third of the patients (Glass *et al.*, 1993; Price *et al.*, 1988; Sacktor and McArthur, 1997). Although not all cases of HIV infections are associated with CNS parenchyma changes, the brains of patients presenting with HIV-associated CNS clinical symptoms show, more often than not, CNS pathology. At postmortem examination, only approximately 50% of patients with ADC manifest HIV encephalitis (HIVE) (Glass *et al.*, 1993). Histopathological findings include the presence of infected cells of the monocyte/macrophage lineage, loss of neuronal dendritic processes and decreased synaptic densities, and astrocytosis (Kure *et al.*, 1990; Wiley *et al.*, 1991). These changes are most prominent in subcortical regions (deep white matter, basal ganglia, brain stem), and, to a lesser extent, in the cortical regions (Wiley *et al.*, 1991). Not surprisingly, the subcortical regions are the sites of highest levels of viral antigen and RNA in HIVE (Kure *et al.*, 1990; Wiley *et al.*, 1998).

CNS infection and dysfunction also occur in SIV-infected monkeys. Thus this provides for an ideal animal model for clinical, pathological, immunological, and therapeutic studies of the CNS effects of HIV infection. This chapter examines the advances that have been made in the development and refinement of the SIV model of HIV dementia, and its contribution to the understanding and management of the ADC in humans.

II. ANIMAL MODELS

Initiation of SIV infection in rhesus monkeys is achieved by use of stocks, selected strains, or clones as sources of the viral inoculum. Thus far, only selective viral/host combinations have been used in studies of CNS dysfunction resulting from SIV infection, with a greater number having been examined for neuropathological effects. Following is an overview of some of the most commonly used SIV isolates (Fig. 1) in the monkey model of HIV dementia.

A. $SIV_{mac}251$ Stock

One of the first isolates to have been studied, $SIV_{mac}251$, is an uncloned stock. It was first isolated from a captive rhesus monkey suf-

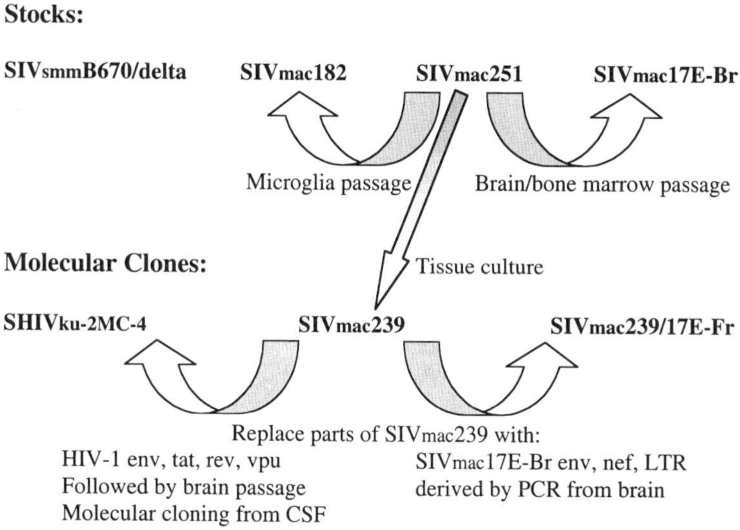

Stocks:

SIVsmmB670/delta   SIVmac182   SIVmac251   SIVmac17E-Br

Microglia passage   Brain/bone marrow passage

Molecular Clones: Tissue culture

SHIVku-2MC-4   SIVmac239   SIVmac239/17E-Fr

Replace parts of SIVmac239 with:
HIV-1 env, tat, rev, vpu   SIVmac17E-Br env, nef, LTR
Followed by brain passage   derived by PCR from brain
Molecular cloning from CSF

FIG 1. Diagrammatic representation of SIV stocks and molecular clones used in monkey studies of CNS pathogenesis. Details of the derivation of these viruses are described in the text.

fering from a lymphoma (Letvin et al., 1985). $SIV_{mac}251$ is both neuroinvasive and neurovirulent, with a tropism for macrophages as well as $CD4^+$ T cells. As with most strains and clones of SIV, the induced disease takes two courses. A rapid course of disease, with survival of less than 6 months, develops in 25–50% of animals, and is associated with a poorly developed host humoral immune response to the virus, and with a high frequency (50%) of neurological signs, including ataxia, seizures, opisthotonos, and apathy (Sopper et al., 1998). The remaining animals, exhibiting a good initial immune response against the virus, survive for greater than six months, often 1–2 years following viral inoculation, with a lower frequency (<15%) of the frank neurological signs observed in the animals with rapidly progressive disease (Sopper et al., 1998). Rhesus monkeys infected with the $SIV_{mac}251$ stock develop SIV encephalitis (SIVE) in approximately 30% of the cases. A correlation has been demonstrated between the development of SIVE and the presence of a rapid disease course (Westmoreland, Halpern, and Lackner, 1998). Several other variants have been derived from this stock, although in many cases, these have similar neuropathological effects (Czub et al., 1996; Westmoreland, Halpern, and Lackner, 1998).

The neuropathology induced by $SIV_{mac}251$ is prototypic of that found with other strains and clones of SIV. In the brain, infected cells are predominantly composed of perivascular and leptomeningeal macrophages, as well as microglia and infiltrating macrophages (Lackner, 1994; Simon, Chalifoux, and Ringler, 1992). Brain lesions include multinucleated giant cells, increased numbers of perivascular macrophages, and the presence of infiltrating macrophages. Spongiosis, vascular dilation, endothelial hypertrophy, astrogliosis, and lymphocyte infiltration are also present.

The development of overt neurological signs in rhesus monkeys infected with $SIV_{mac}251$ occurs, in some cases, even before the onset of other AIDS-defining illnesses (Sopper et al., 1998). Besides those outlined above, there are also functional deficits such as failure at gripping food. The neurological signs and functional deficits in rapid progressors are associated with the typical SIVE brain lesions described above.

B. $SIV_{smm}B670/Delta$

This stock was first isolated from asymptomatic sooty mangabey monkeys. Similar to $SIV_{mac}251$, it is both neuroinvasive and neurovirulent in rhesus monkeys and results in simian AIDS in infected animals. Following infection, rhesus monkeys develop the simian AIDS symptoms, including seizures in a few of the cases (Rausch et al., 1994). At necropsy, the typical SIV-induced neuropathology is observed, with changes being seen largely in the cerebral cortex white matter and the basal ganglia (Baskin et al., 1992; Rausch et al., 1994).

Extensive motor and cognitive testing has been performed in animals infected with $SIV_{smm}B670/Delta$. In behavioral studies designed to evaluate visual-recognition memory, recent memory, stimulus–response association, long-term memory for spatial locations, and motor tasks, infected rhesus monkeys were found to have an impairment in at least one of these functions at 2–6 months after infection (Murray et al., 1992). By 10 months following infection, most of the animals had reduced scores in motor tasks, although all these changes varied from animal to animal. However, one parameter that was always associated with motor skill impairment was the level of quinolinic acid in the cerebrospinal fluid (CSF) (Rausch et al., 1994). Hence it is deemed that quinolinic acid may be a major contributing factor in the causation of these motor deficits.

Interestingly, in clinical correlative studies in humans, inflammatory diseases, but not degenerative diseases, were associated with an accumulation of quinolinic acid and, to a lesser extent, kynurenic acid (a quinolinic acid antagonist), and L-kynurenine, within the CNS (Heyes et al., 1992). Thus quinolinic acid may be involved in development of neuronal dysfunction and nerve cell death during CNS inflammation, including lentiviral encephalitis, as opposed to neurodegenerative diseases.

Changes in several other parameters have been associated with development of motor/cognitive disorders in lentiviral infections. In the above-described study of cognitive and motor skills in infected animals, impairment of a motor skill task was the most reliable indicator of infection, and these functional deficits were most related to SIV infection of the brain, but not to inflammatory lesions at particular brain loci (Murray et al., 1992). Other pathological studies indicated that productive infection of the CNS, as determined by in situ hybridization, correlated with encephalitis in the late stage of the viral infection in rhesus monkeys (Reinhart et al., 1997).

C. $SIV_{mac}17E$-Br (Bone Marrow / Brain-Passaged) Strain

This strain was derived from the $SIV_{mac}239$ clone (see the discussion below) through neuroadaptation by sequential intracerebral inoculation of rhesus monkeys with $SIV_{mac}239$ infected bone marrow (Anderson et al., 1993). The third passage in monkey R71 produced a neurovirulent, macrophage-tropic, and neurotropic strain, designated as $SIV_{mac}239/R71$. Brain homogenate from this monkey was inoculated intracerebrally into another naive rhesus monkey resulting in a fatal neurological disease, characterized by SIVE. The viral isolate from this monkey, designated as $SIV_{mac}17E$–Br, was macrophage tropic and neurotropic. However, brain infection was only achieved when isolate 17E-Br was inoculated intracerebrally and not intravenously (Sharma et al., 1992). Interestingly, intracerebral inoculation with this virus can lead to apoptosis of neurons, endothelial cells, and glia (Adamson et al., 1996a). Subsequent studies showed that changes in the env gene of R71 and 17E-Br, as compared to the parental virus, $SIV_{mac}239$, conferred the above cellular and brain tropism to the viruses (Anderson et al., 1993).

A combination of the R71 and 17E isolates, inoculated into the bone marrow of rhesus monkeys, has been used in CNS studies. A number of experiments have been performed to attempt to discern the correlation between the functional and pathological changes in monkeys infected with this set of SIV isolates. Over the course of infection of rhesus monkeys, the motor system is affected earlier and more severely than in perpetual and decision-making systems (Marcario et al., 1999a; Marcario et al., 1999b). This model has been described to give a good correlation with HIVE in terms of neuropathological lesions, severity of clinical signs, and neurophysiological changes (Raghavan et al., 1999). There was a consistent involvement of white matter in animals infected with these isolates, regardless of the disease duration or viral strain used, suggesting a specific susceptibility to lentiviral infection of these brain regions. Sensory and motor-evoked potentials are also altered in this model (Raymond et al., 1998; Raymond et al., 1999). These neurophysiological data support our findings for $SIV_{mac}182$ infected rhesus monkeys (see below).

D. $SIV_{mac}182$ (microglia-passaged) strain

In order to achieve a reproducible CNS infection with SIV in rhesus monkeys, one of the approaches taken has been to selectively enrich for neuroinvasive and neurovirulent viral strains from a heterogeneous viral stock. Since brain macrophages and microglia are the main reservoir of SIV and HIV in the brain, they are a perfect source of potentially neurovirulent viral quasispecies. Therefore, we performed an intravenous serial passage of microglia from a chronically $SIV_{mac}251$-infected, but asymptomatic, rhesus monkey to naive rhesus monkeys, which resulted in SIV neuroinvasion with subsequent encephalitis in the recipient monkeys (Watry et al., 1995).

Microglia from the second passage group of monkeys were cultured to obtain a cell-free stock (one of which is $SIV_{mac}182$, attained from monkey 182), currently in use as a neurovirulent SIV stock. There is approximately an 80% rate of CNS pathology, which include SIV-associated encephalitis, mononuclear giant cells, macrophage infiltration, T cell infiltrates, and SIV gene expression in the brain (Fox *et al.*, 1997; Lane *et al.*, 1996; Prospero-Garcia *et al.*, 1996; Watry *et al.*, 1995) (Fig. 2). In *in vitro* studies, the virus causes infection to CD4+ T cells, macrophages, and cerebrovascular endothelial cells (Strelow *et al.*, 1998) (see the chapter by Strelow, Janigro, and Nel-

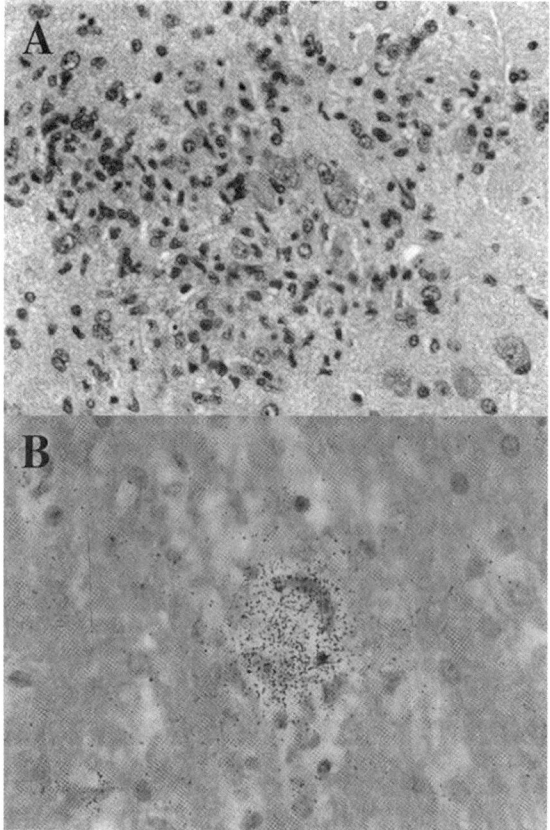

FIG 2. CNS histopathology induced by $SIV_{mac}182$ infection. (A) Photomicrograph of hematoxylin and eosin-stained section, revealing a mononuclear cell infiltrate in the brain. (B) Photomicrograph of *in situ* hybridization for TNF-α, counterstained with hematoxylin and eosin, revealing a positive signal over a multinucleate giant cell in the brain.

son in this volume), and spreads rapidly in culture (our unpublished results).

It should be noted that the original donor monkey, although asymptomatic, had no viral gene expression detectable in the brain by *in situ* hybridization, but still harbored transmissible SIV in brain macrophages and microglia. This highlights the potential of the CNS to act as a reservoir for viruses inaccessible to antiviral agents and, hence, not only sources of relapse infections, but also possible causes of continued CNS damage.

Functional studies of rhesus monkeys infected with microglia-passaged SIV have taken three paths. Cognitive and motor testing has been performed, revealing that subcortical dysfunction occurs early in the disease course, whereas impairment of higher cognitive functions occurs later (see the chapter by Weed and Gold in this volume). Electophysiological analysis has shown consistent abnormalities in sensory evoked potentials detectable throughout the course of infection (see below). Finally, radiotelemetric analysis of body temperature and movement has revealed physiological changes that can be detected early in the course of infection (see below) (Horn *et al.,* 1998). The typical findings of SIV neuropathology can be found in the brains of microglia-passaged SIV-infected monkeys, but no clear correlation to functional effects has been identified to date. We are currently utilizing this strain in functional studies of simian neuroAIDS in rhesus monkeys.

E. SIV$_{mac}$239 Molecular Clone

SIV molecular clones would be the ideal viruses for studying SIV and HIV disease *in vivo,* as this would enable standardization of the model among laboratories, and the use of molecular virologic studies. However, of the available clones, some cannot replicate in macrophages effectively, and hence cannot produce CNS disease unless mutations occur within the virus's envelope gene, a process that can take place both *in vitro* and *in vivo.* One such molecular clone is SIV$_{mac}$239. It is a T cell tropic virus, and when inoculated into rhesus monkeys, the animals develop simian AIDS, but often without detectable brain infection (Sharma *et al.,* 1992). However, *in vivo,* the virus develops several variants, some of which can effectively infect and replicate in macrophages, and can cause CNS disease (Desrosiers *et al.,* 1991). Some of these variants have been cloned (e.g., SIV-$_{mac}$239/316EM) and can themselves be inoculated into rhesus monkeys and result in AIDS, including CNS pathology (Westmoreland, Halpern, and Lackner, 1998). Detailed descriptions of functional changes induced by SIV$_{mac}$239 have so far not been reported.

F. $SIV_{mac}17E$-Fr Molecular Clone

The finding that *in vivo* passages of $SIV_{mac}239$ allowed the development of neurovirulence led to efforts to create a neurovirulent molecular clone. Initially, a chimerical clone was constructed that contained a portion of the *env* gene derived from the brain $SIV_{mac}17E$-Br stock, viruses isolated from the brain of rhesus monkeys with neurological lesions after four *in vivo* passages of $SIV_{mac}239$, on the backbone of the parent viruses $SIV_{mac}239$. In this manner the $SIV_{mac}239$ *env* was replaced with sequences encoding the entire surface glycoprotein but only a portion of the transmembrane glycoprotein from $SIV_{mac}17E$ (Anderson *et al.*, 1993). However, in two studies, infection with the virus derived from this clone, $SIV_{mac}17E$-Cl, although macrophage trophic, was not capable of causing CNS disease (Joag *et al.*, 1995; Mankowski *et al.*, 1997). When an additional recombinant was made, replacing the remaining transmembrane glycoprotein, the *nef* gene, as well as the 3' LTR (long terminal repeat) sequences, with those derived from $SIV_{mac}17E$, the virus derived from the new clone, $SIV_{mac}17E$-Fr, was still macrophage tropic, besides showing endothelial cell tropism (Flaherty *et al.*, 1997; Mankowski *et al.*, 1994). However, this new derivative was also capable of inducing mild-to-moderate CNS disease when inoculated into macaques (Mankowski *et al.*, 1997). Thus macrophage tropism, while necessary, is not sufficient to induce CNS disease in monkeys.

Since immunosuppression is a predisposing factor in the development of neurological disease as defined by encephalitis, recent studies have used a combination of an immunosuppressive virus, $SIV_{mac}B670$/Delta, with $SIV_{mac}17E$-Fr, in pig-tailed monkeys. Encephalitis resulted in more than 90% of the animals, within 6 months following infection (Zink *et al.*, 1999). In these studies, plasma and CSF were examined longitudinally to onset of AIDS, and viral load was assayed in brain tissue at postmortem. Viral peak in plasma occurred at 10–14 days following infection, and remained high throughout the infection period. Although no correlation was found between plasma viral load and SIVE, the CSF viral RNA load correlated with the development of SIVE. Furthermore, the attainment and maintenance of a high CSF viral RNA load correlated with the severity of CNS lesions. Thus, these studies argue for the use of CSF viral load measurement in the postacute phase of SIV infection as a marker of encephalitis and CNS viral replication.

Further studies with this clone have led to a new model for rapid CNS disease in monkeys (Zink *et al.*, 1997). It was found that when nine macaques of diverse genera were coinoculated with the $SIV_{mac}17E$-Fr clone and the $SIV_{mac}B670$/Delta stock, described above, seven developed

CNS disease. Four of these animals had severe SIV encephalitis, and when brain-derived virus was examined molecularly, three of the four had only SIV$_{mac}$17E-Fr sequences, whereas the fourth had a mixture of SIV$_{mac}$17E-Fr and the Delta stock. Thus, in a complex mixture of virus, neurovirulent strains can be selected for in the CNS. Clinical signs of severe neurological disease are said to occur most often in immune-compromised HIV-AIDS patients (McArthur et al., 1994; Price et al., 1988), and a correlation between the immunosuppression and the severity of encephalitis has been demonstrated in such infected rhesus monkeys (Zink et al., 1997). It was noted that pig-tailed macaques were more susceptible, and this coinfection model is now being used as a rapid, reproducible way to obtain CNS disease in monkeys.

G. SIV/HIV (SHIV) Chimeras: Stocks and a Molecular Clone

Although SIV encephalitis in rhesus monkeys is a useful model of HIV-related neurologic disease, it falls short of providing an opportunity for investigating the effects of HIV envelope glycoproteins, or other virally encoded molecules that are thought to be important in the pathogenesis of HIV-associated neurologic disease. For this reason, pathogenic SIV/HIV (SHIV) chimeras would be an ideal source of viral inocula for such studies.

The SHIV$_{ku-1}$ viral stock (Joag et al., 1996) is derived from the SHIV-4 proviral chimeric clone, containing the tat, rev, vpv, and env genes of HIV-1 HXB2, and the LTR, gag, pol, vif, vpr, vix, and nef genes of SIV$_{mac}$239 (Li et al., 1992). It is pathogenic, and leads to a latent brain infection with a rapid depletion of CD4+ T cells in pig-tailed monkeys, but does not result in lentiviral encephalitis or productive infection of the CNS (Joag et al., 1996).

A further serial passage of this chimera, using infected bone marrow and cell-free virus in rhesus monkeys, resulted in a viral stock (SHIV$_{ku-2}$) that induces both CD4+ depletion and AIDS, along with the typical SIVE lesions (Raghavan et al., 1997). Infectious CSF from one of these monkeys was used to isolate a new molecular clone, SHIV$_{ku-2mc4}$ (Liu et al., 1999). SHIV$_{ku-2mc4}$ causes a transient CSF infection in the first week following inoculation of rhesus monkeys, with typical clinical signs of AIDS. Neurological signs, which include mild muscle-power loss, ataxia, intention tremors, and reduced movement (Liu et al., 1999), are observed in 20% of the cases. The viral DNA can be detected in the brain, and other pathological features, such as perivascular and nodular mononuclear infiltration (monocytes, macrophages, microglia), T cells, multinucleated giant cells, and demyelinating lesions in cerebral and cerebellar white matter, are present. Interestingly, the disseminated white matter lesions found are made up of focal accumulations of CD3$^+$ T cells and plaque-like lesions of macrophages and astrocytes. The factors and mechanisms involved in the development of this distinctive white matter pathology induced by infection with this SHIV clone are not known.

III. BRAIN INFECTION

The infection of the brain by HIV and SIV is thought to occur by one or more of the following routes.

A. Trojan Horse Model

Infected/activated CD4$^+$ T cells and monocytes could traffic to the brain, thus carrying with them the virus (Epstein and Gendelman, 1993; Haase, 1986; Nottet et al., 1996). Capillary endothelial cells upregulate the expression of cell-adhesion molecules such as E-selectin (Nottet et al., 1996) and VCAM-1 (Sasseville et al., 1995; Sasseville et al., 1994; Sasseville et al., 1992), following HIV and SIV infection of humans and rhesus monkeys, respectively. Moreover, studies of HIV-infected patients have shown a correlation between macrophage infiltration into the brain and increased E-selectin and VCAM-1 expression by brain microvascular endothelial cells (Nottet et al., 1996). Astrocytes have also been shown to express VCAM-1 and ICAM-1 following culture with the HIV-1 Tat protein in vitro (Woodman et al., 1999). These, coupled with increased expression of chemokines in the CNS (Sanders et al., 1998; Sasseville et al., 1996), could promote the diapedesis of activated CD4$^+$ T cells and monocytes to the brain and regulate their traffic through the brain. Nitric oxide, elaborated by both endothelial cells and monocytes/macrophages, may also facilitate the transmigration of infected cells through the blood–brain barrier (BBB) by generation of the highly toxic molecule peroxynitrite, which may result in the destruction of the BBB tight-junction proteins and alteration of endothelial integrity (Boven et al., 1999). In addition, increased production of matrix metalloprotease activity, such as that induced by HIV-infected monocytes/macrophages, could contribute to the disruption of the BBB and the spread of HIV in the brain parenchyma (Sporer et al., 1998).

B. Breakdown of Blood–Brain Barrier

Loss of integrity of the BBB, as alluded to above, may occur following systemic or CNS inflammation with the resultant elaboration of high amounts of cytokines and chemokines, thus facilitating brain infection. Evidence of BBB damage in SIV and HIV infections has been demonstrated by the presence of increased serum albumin in the cerebrospinal fluid and fibrinogen and immunoglobulin in the brain matter (Horn *et al.*, 1998; Petito and Cash, 1992; Singer *et al.*, 1994; Smith, Sutjipto, and Lackner, 1994). In addition, significant apoptosis of brain-capillary endothelial cells in SIV encephalitis has been demonstrated (Adamson *et al.*, 1996a).

C. Infection of Brain Microvascular Endothelial Cells

The early descriptions of HIV in the CNS of humans included evidence of cerebrovascular endothelial cell infection (Wiley *et al.*, 1986). *In vitro* work has confirmed the infectibility of endothelial cells by HIV (Moses *et al.*, 1993; Moses *et al.*, 1996). This potential mechanism is also evidenced by the detection of virus in brain-capillary endothelial cells of rhesus monkeys following SIV inoculation and *in vitro* infection of monkey brain-capillary endothelial cells (Mankowski *et al.*, 1994; Strelow *et al.*, 1998). This route of infection of the brain provides not only a mechanism for the initial viral entry to the CNS, but also alteration of the blood–brain barrier, which could compromise its functional integrity and lead to direct access of the virus to the CNS.

IV. HOST RESPONSES

In response to the viral infection, both resident cells of the CNS and infiltrating immune cells can produce a variety of host defense molecules, including cytokines. Cytokines can be expressed by a variety of cell types in the CNS, including astrocytes, endothelial cells, and microglia, as well as infiltrating immune macrophages and T cells. Utilizing *in situ* hybridization, we have found that expression of a number of host defense molecules is present in the brains of SIV-$_{mac}$182-infected animals (Lane *et al.*, 1996). These include the cytokines interleukin (IL)-1β, tumor necrosis factor (TNF)-α and interferon (IFN)-γ. Double labeling was not performed; however, morphologically, the cells producing IL-1β and TNF-α appear to be macrophages, whereas the IFN-γ positive cells are lymphocytic. Additionally, we have identified expression of the inducible form of nitric oxide synthase (iNOS) in the brains of SIV-infected monkeys, in addition to finding stable end products of nitric oxide in the CSF (Lane *et al.*, 1996). Microglia/macrophages derived from the brains of SIV-infected animals are capable of expressing iNOS following LPS stimulation (Lane et al., 1996). Localization of iNOS expression to cells of the macrophage lineage, in the brains of monkeys with productive CNS SIV infection, has recently been reported (Li et al., 1999).

In the brains of individuals with AIDS dementia, elevated levels of cytokines (Lipton and Gendelman, 1995) and of iNOS have been found (Adamson et al., 1996b). Such expression of cytokines and iNOS, although potentially helpful to the host in combating the viral infection, may be deleterious to the brain in many ways, perhaps causing damage to neurons, glia, or compromise of the BBB. Nitric oxide, which itself can be damaging to cells, can form peroxynitrite, a potent neurotoxin, on reaction with superoxide anion, another product of activated macrophages (Lipton et al., 1993). Superoxide dismutase, which normally scavenges excess superoxide anions, has actually been shown to increase the damage mediated by peroxynitrite in neurons (Beckman and Koppenol, 1996). The degree of elevation in levels of iNOS expression in the brain at autopsy have been found to correlate with the rate of progression and severity of dementia in HIV-infected individuals (Adamson et al., 1999; Adamson et al., 1996b). Furthermore, evidence of peroxynitrite-mediated damage, iNOS, and superoxide dismutase expression has recently been reported in the brains of individuals with HIV dementia (Boven et al., 1999). Thus, oxidative stress-induced damage is another, nonexclusive candidate for mediating CNS damage in HIV infection.

CD8+ T cells play an important role in control of viral infections, through both cytotoxic and noncytotoxic mechanisms. Recent studies, using antibody treatment to deplete CD8+ T cells in monkeys infected with SIV$_{mac}$251, have confirmed their crucial role in control of the viral replication (Jin et al., 1999; Schmitz et al., 1999). We have demonstrated that CD8+ T cells are present in the brains of SIV$_{mac}$251 and microglia-passaged SIV infected monkeys, and that such cells can represent SIV-specific cytotoxic T lymphocytes (CTLs) (von Herrath, Oldstone, and Fox, 1995). In HIV infection, a vigorous CTL response has been found in the CSF of patients with AIDS dementia (Jassoy et al., 1992); no studies have been reported on the brain parenchyma, although CD8+ T-cells have been detected in the brains of SIV$_{mac}$251-infected monkeys by immunohistochemistry (Lackner et al., 1991).

Recent data from rodent systems reveal a long-term persistence of CD8+ T cells in the CNS following influenza viral infection of the CNS, even in the absence of demonstrable virus (Bergmann et al., 1999; Marten et al., 1999). The CD8+ cells expressed high levels of the early-activation marker CD69. Thus, there appears to be a continuing TCR signaling despite undetectable levels of viral protein and nucleic acid in the brain. This implies the differential regulation of CD8+ cells within the CNS as compared with those in the periphery. The presence and importance of such activated CD8+ T cells in long-term responses to viruses, such as SIV, are yet to be determined. Such data would be relevant for viruses sequestered in sites of relative immune privilege, as is the CNS.

In our original work (described above) on isolating microglia for the serial passage, fluorescence activated cell sorter (FACS) analysis of the cells isolated from the brains of SIV-infected macaques revealed a significant number of CD8+ T cells. Analysis of 10 SIV-infected animals, sacrificed at different stages of disease, revealed that a range of 3–15% of the recovered cells from the brain were CD8+. An example of such cells in the FACS analysis is shown in Fig. 3. Significant numbers of CD4+ T cells either could not be detected, or were found at greatly reduced frequency. Indeed, CD8+ T cells were found by immunohisto-chemical staining of the CNS of SIV-infected rhesus macaques, but not of uninfected animals, in one of the original descriptions of the pathology of SIV in the CNS (Lackner *et al.,* 1991). No CD4+ T cells or B cells could be found in this study in the brain parenchyma. Similar findings on the number of CD8+ T-cells in the brain of $SIV_{mac}251$-infected rhesus macaques have recently been reported (Sopper *et al.,* 1998). Interestingly, the increase was seen predominantly in the brains of slow progressors—a heightened level of CD8+ T cells was only found in 2 of 9 rapid progressors. Longitudinal studies of CSF revealed that slow

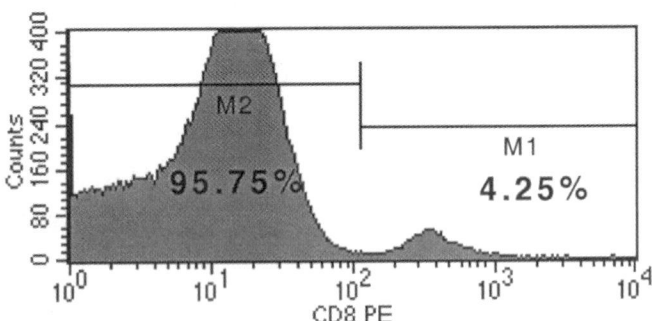

FIG 3. Histogram of FACS analysis of dissociated CNS cells. Cells were isolated from the CNS of an $SIV_{mac}182$-infected monkey, as described (Watry *et al.,* 1995), stained with a PE-conjugated anti-CD8 antibody, and analyzed by FACS. The positive cells are present in region M1.

progressors had a greater and a more sustained increase of CD8+ T cells in the CSF than did rapid progressors.

In our SIV/rhesus monkey studies (von Herrath, Oldstone, and Fox, 1995), CTLs capable of recognizing autologous, but not allogeneic, target cells infected with vaccinia virus recombinants, expressing *gag, env, nef* or *pol* of SIV, are found as early as 1 week following viral inoculation in lymphocytes obtained from blood and CSF of infected monkeys. CSF CTLs could display different activities and/or recognized different viral proteins than did contemporaneous CTLs from peripheral blood. Similarly, at autopsy, brain CTL specificities could be identified that were different from the those in CSF and the periphery, suggesting that viral-immune interactions in the brain parenchyma may be independent of those identified in the CSF and blood.

Although crucial for the control of viral infection, the presence and activity of CTLs in the CNS may cause neuronal dysfunction by lysis of functionally important cells in the CNS; by induction of other cells, such as astrocytes and microglia, to produce and release deleterious agents; and/or by production by the CTLs themselves of numerous potentially neurotoxic substances. Activated CTLs and natural killer cells induce apoptosis by degranulation of toxin-carrying granules at exocytosis, Fas/Apo-mediated pathways, or by TNF-dependent processes (Braun *et al.*, 1996; Kagi *et al.*, 1994; Lowin *et al.*, 1994). These toxins include perforin, which polymerizes into transmembrane pores, leading to intracellular delivery of other toxic agents such as granzymes (Masson and Tschopp, 1985). Alternatively, following CTL adhesion to the target cell in the process of antigen presentation, the serine proteases (granzymes) may bind to high-affinity binding sites and enter the target cell independent of perforin (Froelich *et al.*, 1996). This means that although CNS neurons are more resistant to perforin (Keane, Tallent, and Podack, 1992), they are subject to the apoptotic effects caused by the alternative mechanisms of granzyme toxicity. This underscores the potential importance of granzymes in the pathogenesis of brain ailments that involve neuronal loss, such as HIVE and SIVE. Granzyme A, produced by many CTLs, is reported to have CNS pathogenic properties (Suidan *et al.*, 1994). *In vitro* studies have demonstrated that granzyme A, when released on stimulation of cytotoxic T lymphocytes, activates the thrombin receptor on neuronal cells and astrocytes, resulting in profound morphological changes, and in invariably impaired function. Granzyme B, also released by CTL, may be similarly toxic. The extent of the specific roles these serine proteases play in HIVE and SIVE is a subject of future work.

In addition to T cells and macrophages, B cells represent an important component of the immune system. We were unable to detect B cells in our isolated cells from the brain by FACS analysis; others have found similar results with FACS (Sopper *et al.,* 1998) and with immunohistochemical staining (Lackner *et al.,* 1991). However, others have revealed that plasma cells, mature antibody-secreting cells that do not express the CD20 antigen commonly used in FACS analysis, could be found in the brain. Indeed, plasma cells secreting antibody directed against the SIV envelope protein could not only be found in the brain, but in fact, from 4 months following viral inoculation onward, constituted a high percentage (10%) of the total antibody-secreting cells in the brain (Sopper *et al.,* 1998). The continued presence of SIV-specific plasma cells in the brain, in addition to the SIV-specific CTLs, argues for a continued antigenic exposure, and for a chronic viral–host interaction in the CNS of SIV-infected monkeys.

In HIV patients, neurocognitive/neuropsychological impairment is related to the level of dendritic injury as defined by loss and vacuolization of dendritic processes (Masliah *et al.,* 1997). In the severe dementia associated with HIV infection, there is increased presence of the multinucleated giant cells and diffuse myelin pallor in the brain that are associated with the infection, but only a limited correlation with the presence of viral proteins (Glass *et al.,* 1995; Glass *et al.,* 1993; Navia *et al.,* 1986; Wiley *et al.,* 1991). Our current findings in studies of the neurophysiological and neuropsychological changes associated with SIV infection in rhesus monkeys show that the presence or absence of SIV-related neuropathology at the time of sacrifice is not predictive of CNS functional impairments (Fox *et al.,* 1997); these findings are supported by previous observations (Rausch *et al.,* 1994).

In HIV infection, the presence of HIV RNA/protein, although important for the development of CNS functional and pathologic changes, is not quantitatively linked to the severity of the CNS changes (Budka *et al.,* 1991; Masliah *et al.,* 1997). There are reports of a correlation between CSF viral levels and the severity of AIDS-related dementia (Ellis *et al.,* 1997; Wiley *et al.,* 1998), although this is yet to be ascertained with the advent of quantitative RNA-based viral load assays. Other parameters that may have a positive correlation with ADC include brain macrophage number (Glass *et al.,* 1995) and CSF monocyte chemoattractant protein-1 (MCP-1) levels (Conant *et al.,* 1998; Kelder *et al.,* 1998). These latter two parameters support the notion that macrophages, by generating harmful products, are a great contributing factor in CNS pathology.

It has been argued that one major means by which SIV and HIV may induce CNS and neuronal damage is by their infection of microglia and

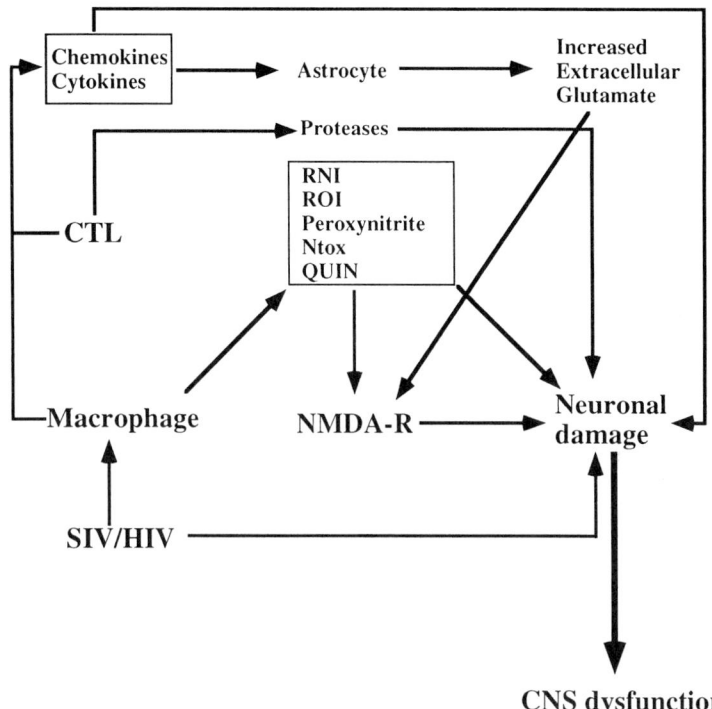

FIG 4. The interplay between host and viral factors in the pathogenesis of CNS dysfunction following SIV/HIV infection of the CNS. The central role played by cells of the macrophage lineage, and the potential convergence on neuronal NMDA receptors, are highlighted. CTL, cytotoxic lymphocyte; RNI, reactive nitrogen intermediate; ROI, reactive oxygen intermediate; QUIN, quinolinic acid; NMDA-R, N-methyl-D-aspartate receptor.

brain macrophages, which leads to release of viral coat proteins, cytokines, reactive oxygen intermediates, nitric oxide, arachidonic acid metabolites, quinolinic acid, and proteases, as described above (Fig. 4). Indirect effects on astrocytes and disruption of the BBB may also be a contributing factor (Lane *et al.*, 1996; Lipton and Gendelman, 1995; Sacktor and McArthur, 1997). A potential model for CNS damage by infected and/or activated macrophages is via N-methyl-D-aspartate (NMDA) receptors, a subtype of glutamate receptors (Lipton and Gendelman, 1995). Overstimulation of such receptors can lead to a cascade of increased neuronal intracellular calcium, increased glutamate release, and overexcitation of neurons, which in turn lead to neuronal damage and death.

V. Chemokines and Their Receptors

It is now clear that HIV and SIV can use certain chemokine receptors as coreceptors, including strains that infect brain microglia (Albright *et al.*, 1999; Ghorpade *et al.*, 1998; He *et al.*, 1997; Shieh *et al.*, 1998). These findings helped initiate new lines of inquiry into the role that chemokines and their receptors play in the modulation of brain disease. Because the envelope protein of the virus can bind to, and in some cases stimulate, these receptors (Davis *et al.*, 1997; Weissman *et al.*, 1997), the presence of chemokine receptors on neurons leads to an additional potential pathway of neuronal damage by viral proteins or the host response (i.e., the viral envelope and the chemokine themselves, respectively; see below) (Lavi *et al.*, 1997; Rottman *et al.*, 1997; Westmoreland *et al.*, 1998). The expression of chemokines has been demonstrated in HIVE/SIVE tissues. Chemokines found to be expressed in human brain include the α-chemokines IP-10 and IL-8; and the β-chemokines MCP-1, MIP-1α, and RANTES (Sanders *et al.*, 1998). Studies of SIV-infected rhesus monkeys have shown that encephalitic brain tissue from these animals has elevated expression of IP-10, MIP-1α, MIP-1β, MCP-3, and RANTES (Sasseville *et al.*, 1996). A large array of chemokine receptors (CRs) are expressed by CNS cells, including neurons, astrocytes, and macrophages, in both normal and SIV/HIV-infected tissue (Albright *et al.*, 1999; Galasso, Harrison, and Silverstein, 1998; Ghorpade *et al.*, 1998; He *et al.*, 1997; Hesselgesser *et al.*, 1998; Klein *et al.*, 1999; Lavi *et al.*, 1997; Meucci *et al.*, 1998; Westmoreland *et al.*, 1998; Xia and Hyman, 1999; Zhang *et al.*, 1998; Zheng *et al.*, 1999b).

Although the specific functions of CRs in the brain are the subject of much work in a rapidly growing field, genetically deleting the gene encoding CXCR4 in mice leads to abnormal cerebellum development, implying an important role in normal neural development for CXCR4 and for its monogamous natural ligand, SDF-1α (Ma *et al.*, 1998; Zou *et al.*, 1998). Study of the CR distribution is important because the natural role and effect of disease-related production of chemokines *in vivo* are reflective of the functional expression of their respective receptors. Interestingly, the normal expression pattern may change with CNS injury (Galasso, Harrison, and Silverstein, 1998).

Several studies have recently investigated the roles of specific chemokines in the effect of AIDS on the CNS. In studies combining use of an *in vitro* BBB model and an HIVE SCID mouse model, microglia activation by HIV infection and LPS stimulation, and/or through inter-cellular interaction with astrocytes, was found to induce β-chemokine-mediated monocyte migration (Persidsky *et al.*, 1999). These findings were corroborated by retrospective immunohistopathological analyses of human HIVE brain tissues. This process could thus lead to an expansion of the brain viral reservoir and to an exacerbation of HIV-related dementia.

Other experiments, utilizing an *in vitro* BBB model consisting of fetal umbilical cord endothelial cells and astrocytes, revealed that HIV-1 Tat, a viral protein required for efficient viral gene expression, induces MCP-1 production by astrocytes. In addition to stimulating chemotaxis through CCR2, this chemokine upregulates CCR5 expression by monocytes, thereby increasing the susceptibility of these cells to infection by HIV (He *et al.*, 1997) and promoting their migration to the brain following stimulation of the CCR5 by the MIP-1β produced by infected brain cells (Sasseville *et al.*, 1996; Weiss *et al.*, 1998). Thus, Tat can facilitate monocyte recruitment to the CNS, thereby contributing to the pathogenesis of HIVE and dementia. In these studies, Tat had no effect on either CXCR4 or CCR2 expression by monocytes.

CXCR4 is a chemokine receptor expressed on lymphocytes, macrophages, microglia, and neurons, and, like CCR3 and CCR5, is a coreceptor for HIV-1, besides having its role in normal brain development. Neurons expressing CXCR4 or other CRs could bind the viral envelope proteins, which, when followed by signaling and eventual neuronal apoptosis, would result in dementia. Indeed, transgenic mice overexpressing the HIV envelope protein gp120 display features of HIV-1 related pathophysiology (Krucker *et al.*, 1998; Toggas *et al.*, 1994), although mediation of these *in vivo* effects by CXCR4 has not yet been examined.

Studies have shown that lymphotropic (i.e., CXCR4-utilizing) HIV-1 virions inhibit cAMP, activate inositol 1,4,5-triphosphate, and induce apoptosis in neurons and astrocytes more significantly than macrophage-tropic (i.e., CCR5-utilizing) virions (Zheng *et al.*, 1999a). These effects are mediated in part via CXCR4, and are CD4 independent. Additionally, SDF-1α is upregulated in astrocytes exposed to HIV-1 or conditioned media from activated macrophages. These conditioned media, HIV-1 or SDF-1α, induce changes in G-protein-coupled signaling that include increased intracellular Ca^{2+}, with a high correlation with increased caspase-3 activity and apoptosis (Zheng *et al.*, 1999b). In other studies, it has been shown that certain gp120 isolates share some aspects of the signaling pathway induced by the CXCR4 ligand SDF-1α in the induction of apoptosis. Inhibition of the p38 mitogen-activated protein kinase ameliorates both gp120- and SDF-1α induced neuronal apoptosis, but the gp120 effect on neurons was found to be an indirect effect mediated by CRs on macrophages/microglia, whereas the SDF-1α effect was mediated directly on neurons or astrocytes (Kaul and Lipton, 1999).

Such findings may help explain the role of CRs in HIV-induced dementia; however, the demonstration that CXCR4-utilizing viruses are present in the brains of individuals with dementia remains to be shown. Whether the same mechanisms apply for SIV motor/cognitive deficits observed in infected monkeys, given that the SIV strains used in most studies cannot utilize CXCR4 as a coreceptor, is not yet clear. Still, the concept that the signaling produced in neurons, following the binding of CR by gp120, might be variant to that produced by chemokines themselves, when binding to the same CR, and thus leading to adverse effects on the neurons, is attractive in helping to explain the etiopathogenesis of CNS dysfunction induced by HIV.

VI. CENTRAL NERVOUS SYSTEM DYSFUNCTION

It is quite challenging to functionally examine "dementia" in rhesus monkeys. We and others have developed means to examine other functional CNS deficits in SIV-infected rhesus monkeys, including neurophysiological and motor abnormalities (Horn et al., 1998; Marcario et al., 1999b; Murray et al., 1992; Prospero-Garcia et al., 1996). Studies utilizing tests measuring specific cognitive and motor functions are reviewed in the chapter by Weed and Gold in this volume. Studies on physiological systems mediated by the CNS are discussed below.

A. Electrophysiological Studies

Electrophysiological analysis of evoked potentials, which measures the electrical activity produced by groups of neurons following stimulation, has been demonstrated to reveal abnormalities consequent to HIV infection. Both sensory-evoked potentials, in which response to stimulation, such as by sound or light, and event-related potentials, which require recognition of a rare event in the stimulation pattern, have been reported to be altered in both asymptomatic and symptomatic HIV-infected individuals (Coburn et al., 1992; Evers et al., 1998; Fein, Biggins, and MacKay, 1995; Messenheimer et al., 1992; Pagano et al., 1992; Schroeder et al., 1994; Somma-Mauvais and Farnarier, 1992).

In the rhesus monkey, electrophysiological measurement of brainstem auditory-evoked potentials (BSAEPs) have been successfully utilized to examine models of several human neurological conditions, including conductive hearing loss, bilirubin toxicity, and lead toxicity (Doyle and Webster, 1991; Lilienthal and Winneke, 1996; Wennberg et al., 1993). We have used electrophysiological measurement of sensory-evoked potentials, including BSAEPs, cortical auditory-evoked and visually evoked potentials (AEPs and VEPs) in the rhesus monkey/SIV system to document abnormalities in the CNS due to viral infection (Horn *et al.,* 1998; Phillips *et al.,* 1994; Prospero-Garcia *et al.,* 1996). Recordings of evoked potentials provide an elegant, noninvasive method for testing general CNS function. In contrast to EEG recordings, which measure spontaneous activity, evoked potentials are responses to a defined external stimulus (such as aural and visual stimuli), resulting in a local change in the electrical field of a neural structure. Alterations in the evoked potentials indicate perturbations in CNS neuronal function and activity. We have found that the electrophysiological data collected under our experimental conditions are extremely stable over time (Horn *et al.,* 1998). Electrophysiological studies have the additional advantage whereby, with trained personnel, obtaining the data in the monkeys, without requiring extensive animal training and testing, is relatively straightforward.

Our initial studies involved a comparison of chronically (11-month) SIV-infected monkeys with uninfected control animals. Analysis of the evoked potentials revealed marked changes in all modalities examined: in the relatively short-latency BSAEPs and the longer-latency AEPs and VEPs (Prospero-Garcia *et al.,* 1996). A longitudinal study was then performed, beginning before viral inoculation and continuing at monthly intervals after SIV infection. Interestingly, as early as one to three months following viral inoculation, a significant prolongation of the latencies of BSAEP waves P4 and P5, as well as the cortically generated AEP wave N2, were found (Horn *et al.,* 1998).

Delays in these late components of the BSAEPs are also found in another animal model for neuroAIDS—cats infected with the feline immunodeficiency virus (Phillips *et al.,* 1994)—as well as in HIV-infected people (Pagano *et al.,* 1992; Serafini *et al.,* 1998; Somma-Mauvais and Farnarier, 1992). Abnormalities in evoked potentials in SIV-infected monkeys have also been confirmed by others (Raymond *et al.,* 1998). We find that measurement of sensory-evoked potentials are sensitive probes of neuronal circuitry, and represent a crucial tool in determining the effects of viral infection–initiated damage in the CNS.

B. Control of Temperature and Activity

We have recently utilized radiotelemetric monitoring of rhesus monkeys to noninvasively measure the effects of SIV infection on the animals' temperature and movement as parameters of disease progression (Horn *et al.,* 1998). Before infection, animals manifest a consistent circadian temperature rhythm, with temperature being higher in the daytime

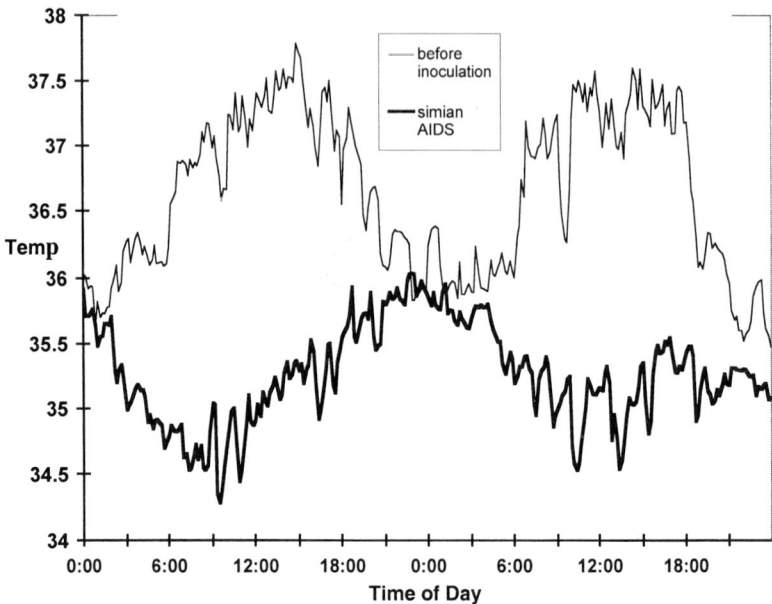

FIG 5. Temperature traces of an SIV$_{mac}$182-infected monkey. Continuous monitoring of body temperature was obtained by radiotelemetry, as described (Horn *et al.*, 1998). Data shown represent a 48-hour period before SIV inoculation, and a similar period following the development of simian AIDS.

than at night, as well as a distinct day/night pattern of motor activity. Following infection, an acute fever develops in the second week after inoculation, persists for 3–4 days, and then largely regresses. Interestingly, the average body temperature can then become slightly elevated (by up to 0.5°C), largely due to an increase in the nighttime temperature nadir. Concomitant with the acute fever, a reduction in movement activity is found, which then quickly recovers. However, by three months in the infection, a reduction of 50% is found in overall motor activity (Horn *et al.*, 1998). Later in the infection, we have observed hypothermia as well as hyperthermia, in addition to loss of circadian temperature and motor rhythms (our unpublished data; see Fig. 5).

These alterations in body temperature and motor activity might be an indicator of altered biochemical profiles, such as increased cytokine production (Kluger *et al.*, 1995; Moltz, 1993). Other host response–related changes in the CNS, and/or effects of viral proteins—all resulting in altered neuronal function—may also be responsible. These abnormal findings, affecting physiological homeostasis, may also contribute to other physiological changes, such as sleep and fatigue problems, as well as to the motor and cognitive abnormalities reported in HIV/SIV infections.

VII. CHEMOTHERAPY AND PROPHYLAXIS

Therapy for the CNS consequences of HIV infection includes treatment directed at the virus itself, and at the pathogenic mechanisms induced by viral infection. Although clinical trials are proceeding on the basis of the latter, with various strategies being used to protect the CNS from damage, studies of effects in monkeys have not been reported. Two studies to date have examined the effects of antivirals in monkeys on the CNS. In the first, infection of newborn rhesus monkeys with $SIV_{smm}B670$/Delta was followed by zidovudine treatment (Rausch et al., 1995). Zidovudine increased survival time after inoculation, as well as lowering virus in the blood and CSF. Motor abnormalities did not occur in treated animals, as compared to 2 of 5 control, infected animals. The second study utilized 9-(2-phosphonylmethoxypropyl)-adenine (PMPA) treatment of juvenile rhesus monkeys, instituted two months following $SIV_{mac}182$ infection (Fox et al., 2000). Electrophysiological analysis revealed that the BSAEPs, abnormally delayed before the institution of therapy, normalized with the PMPA-induced reduction of viral load. Withdrawal of therapy led to the return of BSAEP abnormalities. However, the SIV-induced decline in motor activity and elevation of body temperature, measured by radiotelemetry, was insensitive to treatment-induced viral load reduction.

Data are now emerging on the effects of highly active antiviral therapy (HAART)—which combines reverse transcriptase inhibitors with a protease inhibitor—on the CNS. Still, most of the data to date do not examine the CNS, but encouragingly reveal that successfully treated subjects have greatly reduced viral load in the blood and lymphoid organs (Cavert et al., 1997; Chun et al., 1997; Perelson et al., 1997). However, most of these therapeutic agents are not thought to have significant penetration of the BBB. Still, treated individuals often have quite low levels of HIV in the CSF, reflecting either activity of the agents in the CSF, or a dependence of CSF viral levels on blood levels (Eggers et al., 1999). At this time, the ability of antivirals to affect virus in the CNS, and their potential to prevent or ameliorate CNS symptoms, especially in the long term, are unknown (Lipton, 1997).

Indeed, the dramatic lowering of viral load with HAART has had an important impact on HIV-1-associated dementia, and the incidence of dementia has declined by approximately one-half (Michaels, Clark, and Kissinger, 1998; Moore and Chaisson 1999). Also encouraging are the reports of the effects of therapy on dementia and on potential inflammatory neurotoxins (Gendelman *et al.*, 1998). In contrast, the effects of HAART on the prevalence of AIDS-related dementia, as well as on the HIV-induced minor cognitive and motor disorders, remain uncertain; it is even conceivable that the prevalence may increase, given that people with AIDS are living longer. Recent studies examining the incidence of the different AIDS-defining illnesses (ADIs) have indicated that while the incidence of other AIDS-related diseases has decreased, the incidence of dementia has increased (Dore *et al.*, 1999). Also worrisome is the increase in the median CD4+ T cell count now found at the time of ADC diagnosis (Dore *et al.*, 1999). Both these findings suggest that HAART has a lesser impact on ADC than on the other ADIs, with the poor brain penetration of many antiretroviral agents being a possible explanation. Furthermore, HAART is not effective in up to 50% of HIV-infected individuals (Volberding and Deeks, 1998), and requires strict adherence to dosing regimens. Since many HIV-infected individuals are being treated early in the course of infection, the development of viral resistance is also a concern.

It is important to mention that because no antiviral therapy has so far proven foolproof for eradication of HIV or SIV *in vivo*, the SIV model still remains the outstanding tool for evaluating the efficacy of future drug development. However, one of the limiting factors in the eradication of HIV infection is the presence of HIV viral reservoirs. These are in the form of latently infected CD4+ T cells (Chun and Fauci, 1999), or are within various anatomical locations such as the brain, intestine, bone marrow, and testis. Future studies will have to address this question with a view to understanding the pathological significance of the CNS as a reservoir and its potential contribution to rebound infections that occur following termination of antiviral therapy.

VIII. CONCLUSION

The etiology of CNS dysfunction resulting from HIV and SIV infections likely depends on host as well as viral factors. The SIV/nonhuman primate model of the HIV-associated cognitive/motor disorders provides a splendid system for studying this interplay. Given the high cognitive function of rhesus monkeys and the intimate similarity between SIV and HIV, use of the SIV model provides a fundamental pathogenic and therapeutic understanding of the HIV-associated dementia. The SIV/nonhuman primate model provides the opportunity for studies of pathogenesis, evaluation of treatment and prophylaxis, invasive investigation of tissues in living animals, and comprehensive postmortem examination at any stage of the disease course. This model has been enhanced by the development of reproducibly neurovirulent SIV strains/clones; the advent of efficient cognitive, motor, and neurophysiological evaluations in rhesus monkeys; the use of methods to accurately and quantitatively assess the virus and the host response; and the development of novel brain-accessible therapeutic agents as well as palliative approaches in the management of CNS disorders. Due to all these factors, progress in this field should continue and enable the understanding, prevention, and treatment of this condition.

ACKNOWLEDGMENTS

We thank Dr. Cecilia Marcondes for the FACS data (Fig. 3), Dr. Thomas Lane for data on *in situ* hybridization (Fig. 2), Dr. Thomas Horn for telemetry traces (Fig. 5), and all the members of the laboratory for their work on the SIV/monkey neuroAIDS model. This work is supported by grants from the NIH: MH59468, MH61224, MH55836, and MH47680. This is manuscript #13231-NP from The Scripps Research Institute.

REFERENCES

Adamson, D. C., Dawson, T. M., Zink, M. C., Clements, J. E., and Dawson, V. L. (1996a). Neurovirulent simian immunodeficiency virus infection induces neuronal, endothelial, and glial apoptosis. *Mol. Med.* **2:**417–428.

Adamson, D. C., McArthur, J. C., Dawson, T. M., and Dawson, V. L. (1999). Rate and Severity of HIV-Associated Dementia (HAD): Correlations with Gp41 and iNOS. *Mol. Med.* **5:**98–109.

Adamson, D. C., Wildemann, B., Sasaki, M., Glass, J. D., McArthur, J. C., Christov, V. I., Dawson, T. M., and Dawson, V. L. (1996b). Immunologic NO synthase: elevation in severe AIDS dementia and induction by HIV-1 gp41. *Science* **274:**1917–1921.

Albright, A. V., Shieh, J. T. C., Itoh, T., Lee, B., Pleasure, D., O'Connor, M. J., Doms, R. W., and Gonzalez-Scarano, F. (1999). Microglia Express CCR5, CXCR4, and CCR3, but of These, CCR5 Is the Principal Coreceptor for Human Immunodeficiency Virus Type 1 Dementia Isolates. *J. Virol.* **73:**205–213.

Anderson, M. G., Hauer, D., Sharma, D. P., Joag, S. V., Narayan, O., Zink, M. C., and Clements, J. E. (1993). Analysis of envelope changes acquired by $SIV_{mac}239$ during neuroadaption in rhesus macaques. *Virology* **195:**616–626.

Baskin, G. B., Murphey-Corb, M., Roberts, E. D., Didier, P. J., and Martin, L. N. (1992). Correlates of SIV encephalitis in rhesus monkeys. *J. Med. Primatol.* **21:**59–63.

Beckman, J. S., and Koppenol, W. H. (1996). Nitric oxide, superoxide, and peroxynitrite: the good, the bad, and ugly. *Am. J. Physiol.* **271:**C1424–C1437.

Bergmann, C. C., Altman, J. D., Hinton, D., and Stohlman, S. A. (1999). Inverted immunodominance and impaired cytolytic function of CD8+ T cells during viral persistence in the central nervous system. *J. Immunol.* **163:**3379–3387.

Boven, L. A., Gomes, L., Hery, C., Gray, F., Verhoef, J., Portegies, P., Tardieu, M., and Nottet, H. S. (1999). Increased Peroxynitrite Activity in AIDS Dementia Complex: Implications for the Neuropathogenesis of HIV-1 Infection. *J. Immunol.* **162:**4319–4327.

Braun, M. Y., Lowin, B., French, L., Acha-Orbea, H., and Tschopp, J. (1996). Cytotoxic T cells deficient in both functional fas ligand and perforin show residual cytolytic activity, yet lose their capacity to induce lethal acute graft-versus-host disease. *J. Exp. Med.* **183:**657–661.

Budka, H., Wiley, C. A., Kleihues, P., Artigas, J., Asbury, A. K., Cho, E. S., Cornblath, D. R., Dal Canto, M. C., DeGirolami, U., Dickson, D., and et al. (1991). HIV-associated disease of the nervous system: review of nomenclature and proposal for neuropathology-based terminology. *Brain Pathol.* **1:**143–152.

Cavert, W., Notermans, D. W., Staskus, K., Wietgrefe, S. W., Zupancic, M., Gebhard, K., Henry, K., Zhang, Z. Q., Mills, R., McDade, H., Schuwirth, C. M., Goudsmit, J., Danner, S. A., and Haase, A. T. (1997). Kinetics of response in lymphoid tissues to antiretroviral therapy of HIV-1 infection. *Science* **276:**960–964.

Chakrabarti, L., Hurtrel, M., Maire, M. A., Vazeux, R., Dormont, D., Montagnier, L., and Hurtrel, B. (1991). Early viral replication in the brain of SIV-infected rhesus monkeys. *Am. J. Pathol.* **139:**1273–1280.

Chun, T. W., Carruth, L., Finzi, D., Shen, X., DiGiuseppe, J. A., Taylor, H., Hermankova, M., Chadwick, K., Margolick, J., Quinn, T. C., Kuo, Y. H., Brookmeyer, R., Zeiger, M. A., Barditch-Crovo, P., and Siliciano, R. F. (1997). Quantification of latent tissue reservoirs and total body viral load in HIV-1 infection. *Nature* **387:**183–188.

Chun, T. W., and Fauci, A. S. (1999). Latent reservoirs of HIV: obstacles to the eradication of virus. *Proc. Natl. Acad. Sci. U.S.A.* **96:**10958–10961.

Coburn, K. L., Moore, N. C., Katner, H. P., Tucker, K. A., Pritchard, W. S., and Duke, D. W. (1992). HIV and the brain: evidence of early involvement and progressive damage. *Neuroreport* **3:**539–541.

Conant, K., Garzino-Demo, A., Nath, A., McArthur, J. C., Halliday, W., Power, C., Gallo, R. C., and Major, E. O. (1998). Induction of monocyte chemoattractant protein-1 in HIV-1 Tat-stimulated astrocytes and elevation in AIDS dementia. *Proc. Natl. Acad. Sci. U.S.A.* **95:**3117–3121.

Czub, S., Muller, J. G., Czub, M., and Muller-Hermelink, H. K. (1996). Impact of various simian immunodeficiency virus variants on induction and nature of neuropathology in macaques. *Res. Virol.* **147:**165–170.

Davis, C. B., Dikic, I., Unutmaz, D., Hill, C. M., Arthos, J., Siani, M. A., Thompson, D. A., Schlessinger, J., and Littman, D. R. (1997). Signal transduction due to HIV-1 envelope interactions with chemokine receptors CXCR4 or CCR5. *J. Exp. Med.* **186:**1793–1798.

Desrosiers, R. C. (1990). The simian immunodeficiency viruses. *Annu. Rev. Immunol.* **8:**557–578.

Desrosiers, R. C., Hansen-Moosa, A., Mori, K., Bouvier, D. P., King, N. W., Daniel, M. D., and Ringler, D. J. (1991). Macrophage-tropic variants of SIV are associated with specific AIDS-related lesions but are not essential for the development of AIDS. *Am. J. Pathol.* **139:**29–35.

Dore, G. J., Correll, P. K., Li, Y., Kaldor, J. M., Cooper, D. A., and Brew, B. J. (1999). Changes to AIDS dementia complex in the era of highly active antiretroviral therapy. *AIDS* **13:**1249–1253.

Doyle, W. J., and Webster, D. B. (1991). Neonatal conductive hearing loss does not compromise brainstem auditory function and structure in rhesus monkeys. *Hear Res.* **54:**145–151.

Eggers, C. C., van Lunzen, J., Buhk, T., and Stellbrink, H. J. (1999). HIV infection of the central nervous system is characterized by rapid turnover of viral RNA in cerebrospinal fluid. *J. Acquir. Immune. Defic. Syndr. Hum. Retrovirol.* **20:**259–264.

Ellis, R. J., Hsia, K., Spector, S. A., Nelson, J. A., Heaton, R. K., Wallace, M. R., Abramson, I., Atkinson, J. H., Grant, I., and McCutchan, J. A. (1997). Cerebrospinal fluid human immunodeficiency virus type 1 RNA levels are elevated in neurocognitively impaired individuals with acquired immunodeficiency syndrome. HIV Neurobehavioral Research Center Group. *Ann. Neurol.* **42:**679–688.

Epstein, L. G., and Gendelman, H. E. (1993). Human immunodeficiency virus type 1 infection of the nervous system: pathogenetic mechanisms. *Ann. Neurol.* **33:**429–436.

Evers, S., Grotemeyer, K. H., Reichelt, D., Luttmann, S., and Husstedt, I. W. (1998). Impact of antiretroviral treatment on AIDS dementia: a longitudinal prospective event-related potential study. *J. Acquir. Immune. Defic. Syndr. Hum. Retrovirol.* **17**:143–148.

Fein, G., Biggins, C. A., and MacKay, S. (1995). Delayed latency of the event-related brain potential P3A component in HIV disease. Progressive effects with increasing cognitive impairment. *Arch. Neurol.* **52**:1109–1118.

Flaherty, M. T., Hauer, D. A., Mankowski, J. L., Zink, M. C., and Clements, J. E. (1997). Molecular and biological characterization of a neurovirulent molecular clone of simian immunodeficiency virus. *J. Virol.* **71**:5790–5798.

Fox, H. S., Gold, L. H., Henriksen, S. J., and Bloom, F. E. (1997). Simian immunodeficiency virus: a model for neuroAIDS. *Neurobiol. Dis.* **4**:265–274.

Fox, H. S., Weed, M. R., Huitron-Resendiz, S., Baig, J., Horn, T. F. W., Dailey, P. J., Bischofberger, N., and Henriksen, S. J. (2000). Normalization of CNS Evoked Potential but not movement abnormalities by antiviral treatment in SIV-infected monkeys. *J. Clin. Invest.* **106**:37–45.

Froelich, C. J., Orth, K., Turbov, J., Seth, P., Gottlieb, R., Babior, B., Shah, G. M., Bleackley, R. C., Dixit, V. M., and Hanna, W. (1996). New paradigm for lymphocyte granule-mediated cytotoxicity. Target cells bind and internalize granzyme B, but an endosomolytic agent is necessary for cytosolic delivery and subsequent apoptosis. *J. Biol. Chem.* **271**:29073–29079.

Galasso, J. M., Harrison, J. K., and Silverstein, F. S. (1998). Excitotoxic brain injury stimulates expression of the chemokine receptor CCR5 in neonatal rats. *Am. J. Pathol.* **153**:1631–1640.

Gendelman, H. E., Zheng, J., Coulter, C. L., Ghorpade, A., Che, M., Thylin, M., Rubocki, R., Persidsky, Y., Hahn, F., Reinhard, J., Jr., and Swindells, S. (1998). Suppression of inflammatory neurotoxins by highly active antiretroviral therapy in human immunodeficiency virus-associated dementia. *J. Infect. Dis.* **178**:1000–1007.

Ghorpade, A., Xia, M. Q., Hyman, B. T., Persidsky, Y., Nukuna, A., Bock, P., Che, M., Limoges, J., Gendelman, H. E., and Mackay, C. R. (1998). Role of the beta-chemokine receptors CCR3 and CCR5 in human immunodeficiency virus type 1 infection of monocytes and microglia. *J. Virol.* **72**:3351–3361.

Glass, J. D., Fedor, H., Wesselingh, S. L., and McArthur, J. C. (1995). Immunocytochemical quantitation of human immunodeficiency virus in the brain: correlations with dementia. *Ann. Neurol.* **38**:755–762.

Glass, J. D., Wesselingh, S. L., Selnes, O. A., and McArthur, J. C. (1993). Clinical–neuropathologic correlation in HIV-associated dementia. *Neurology* **43**:2230–2237.

Haase, A. T. (1986). Pathogenesis of lentivirus infections. *Nature* **322**:130–136.

He, J., Chen, Y., Farzan, M., Choe, H., Ohagen, A., Gartner, S., Busciglio, J., Yang, X., Hofmann, W., Newman, W., Mackay, C. R., Sodroski, J., and Gabuzda, D. (1997). CCR3 and CCR5 are co-receptors for HIV-1 infection of microglia. *Nature* **385**:645–669.

Hesselgesser, J., Taub, D., Baskar, P., Greenberg, M., Hoxie, J., Kolson, D. L., and Horuk, R. (1998). Neuronal apoptosis induced by HIV-1 gp120 and the chemokine SDF-1 alpha is mediated by the chemokine receptor CXCR4. *Curr. Biol.* **8**:595–598.

Heyes, M. P., Saito, K., Crowley, J. S., Davis, L. E., Demitrack, M. A., Der, M., Dilling, L. A., Elia, J., Kruesi, M. J., Lackner, A., *et al.* (1992). Quinolinic acid and kynurenine pathway metabolism in inflammatory and non-inflammatory neurological disease. *Brain* **115**:1249–1273.

Horn, T. F. W., Huitron-Resendiz, S., Weed, M. R., Henriksen, S. J., and Fox, H. S. (1998). Early physiological abnormalities after simian immunodeficiency virus infection. *Proc. Natl. Acad. Sci. U.S.A.* **95**:15072–15077.

Jassoy, C., Johnson, R. P., Navia, B. A., Worth, J., and Walker, B. D. (1992). Detection of a vigorous HIV-1-specific cytotoxic T lymphocyte response in cerebrospinal fluid from infected persons with AIDS dementia complex. *J. Immunol.* **149:**3113–3119.

Jin, X., Bauer, D. E., Tuttleton, S. E., Lewin, S., Gettie, A., Blanchard, J., Irwin, C. E., Safrit, J. T., Mittler, J., Weinberger, L., Kostrikis, L. G., Zhang, L., Perelson, A. S., and Ho, D. D. (1999). Dramatic Rise in Plasma Viremia after CD8(+) T Cell Depletion in Simian Immunodeficiency Virus-Infected Macaques. *J. Exp. Med.* **189:**991–998.

Joag, S. V., Li, Z., Foresman, L., Stephens, E. B., Zhao, L. J., Adany, I., Pinson, D. M., McClure, H. M., and Narayan, O. (1996). Chimeric simian/human immunodeficiency virus that causes progressive loss of CD4+ T cells and AIDS in pig-tailed macaques. *J. Virol.* **70:**3189–3197.

Joag, S. V., Stephens, E. B., Galbreath, D., Zhu, G. W., Li, Z., Foresman, L., Zhao, L. J., Pinson, D. M., and Narayan, O. (1995). Simian immunodeficiency virus SIV$_{mac}$ chimeric virus whose *env* gene was derived from SIV-encephalitic brain is macrophage-tropic but not neurovirulent. *J. Virol.* **69:**1367–1369.

Kagi, D., Ledermann, B., Burki, K., Seiler, P., Odermatt, B., Olsen, K. J., Podack, E. R., Zinkernagel, R. M., and Hengartner, H. (1994). Cytotoxicity mediated by T cells and natural killer cells is greatly impaired in perforin-deficient mice. *Nature* **369:**31–37.

Kaul, M., and Lipton, S. A. (1999). Chemokines and activated macrophages in HIV gp120-induced neuronal apoptosis. *Proc. Natl. Acad. Sci. U.S.A.* **96:**8212–8216.

Keane, R. W., Tallent, M. W., and Podack, E. R. (1992). Resistance and susceptibility of neural cells to lysis by cytotoxic lymphocytes and by cytolytic granules. *Transplantation* **54:**520–526.

Kelder, W., McArthur, J. C., Nance-Sproson, T., McClernon, D., and Griffin, D. E. (1998). Beta-chemokines MCP-1 and RANTES are selectively increased in cerebrospinal fluid of patients with human immunodeficiency virus-associated dementia. *Ann. Neurol.* **44:**831–835.

Kindt, T. J., Hirsch, V. M., Johnson, P. R., and Sawasdikosol, S. (1992). Animal models for acquired immunodeficiency syndrome. *Adv. Immunol.* **52:**425–474.

Klein, R. S., Williams, K. C., Alvarez-Hernandez, X., Westmoreland, S., Force, T., Lackner, A. A., and Luster, A. D. (1999). Chemokine receptor expression and signaling in macaque and human fetal neurons and astrocytes: implications for the neuropathogenesis of AIDS. *J. Immunol.* **163:**1636–1646.

Kluger, M. J., Kozak, W., Leon, L. R., Soszynski, D., and Conn, C. A. (1995). Cytokines and fever. *Neuroimmunomodulation* **2:**216–223.

Krucker, T., Toggas, S. M., Mucke, L., and Siggins, G. R. (1998). Transgenic mice with cerebral expression of human immunodeficiency virus type-1 coat protein gp120 show divergent changes in short- and long-term potentiation in CA1 hippocampus. *Neuroscience* **8:**691–700.

Kure, K., Weidenheim, K. M., Lyman, W. D., and Dickson, D. W. (1990). Morphology and distribution of HIV-1 gp41-positive microglia in subacute AIDS encephalitis. Pattern of involvement resembling a multisystem degeneration. *Acta. Neuropathol.* **80:**393–400.

Lackner, A. A. (1994). Pathology of simian immunodeficiency virus-induced disease. *Curr. Top. Microbiol. Immunol.* **188:**35–64.

Lackner, A. A., Smith, M. O., Munn, R. J., Martfeld, D. J., Gardner, M. B., Marx, P. A., and Dandekar, S. (1991). Localization of simian immunodeficiency virus in the central nervous system of rhesus monkeys. *Am. J. Pathol.* **139:**609–621.

Lackner, A. A., Vogel, P., Ramos, R. A., Kluge, J. D., and Marthas, M. (1994). Early events in tissues during infection with pathogenic (SIV$_{mac}$239) and nonpathogenic (SIV$_{mac}$1A11) molecular clones of simian immunodeficiency virus. *Am. J. Pathol.* **145:**428–439.

Lane, T. E., Buchmeier, M. J., Watry, D. D., and Fox, H. S. (1996). Expression of inflammatory cytokines and inducible nitric oxide synthase in brains of SIV-infected rhesus monkeys: applications to HIV-induced central nervous system disease. *Mol. Med.* **2:**27–37.

Lavi, E., Strizki, J. M., Ulrich, A. M., Zhang, W., Fu, L., Wang, Q., O'Connor, M., Hoxie, J. A., and Gonzalez-Scarano, F. (1997). CXCR-4 (Fusin), a co-receptor for the type 1 human immunodeficiency virus (HIV-1), is expressed in the human brain in a variety of cell types, including microglia and neurons. *Am. J. Pathol.* **151:**1035–1042.

Letvin, N. L., Daniel, M. D., Sehgal, P. K., Desrosiers, R. C., Hunt, R. D., Waldron, L. M., MacKey, J. J., Schmidt, D. K., Chalifoux, L. V., and King, N. W. (1985). Induction of AIDS-like disease in macaque monkeys with T cell-tropic retrovirus STLV-III. *Science* **230:**71–73.

Li, J., Lord, C. I., Haseltine, W., Letvin, N. L., and Sodroski, J. (1992). Infection of cynomolgus monkeys with a chimeric HIV-1/SIV$_{mac}$ virus that expresses the HIV-1 envelope glycoproteins. *J. Acquir. Immune. Defic. Syndr.* **5:**639–646.

Li, Q., Eiden, L. E., Cavert, W., Reinhart, T. A., Rausch, D. M., Murray, E. A., Weihe, E., and Haase, A. T. (1999). Increased expression of nitric oxide synthase and dendritic injury in simian immunodeficiency virus encephalitis. *J. Hum. Virol.* **2:**139–145.

Lilienthal, H., and Winneke, G. (1996). Lead effects on the brain-stem auditory-evoked potential in monkeys during and after the treatment phase. *Neurotoxicol. Teratol.* **18:**17–32.

Lipton, S. A. (1997). Treating AIDS dementia. *Science* **276:**1629–1630.

Lipton, S. A., Choi, Y. B., Pan, Z. H., Lei, S. Z., Chen, H. S., Sucher, N. J., Loscalzo, J., Singel, D. J., and Stamler, J. S. (1993). A redox-based mechanism for the neuroprotective and neurodestructive effects of nitric oxide and related nitroso-compounds. *Nature* **364:**626–632.

Lipton, S. A., and Gendelman, H. E. (1995). Seminars in medicine of the Beth Israel Hospital, Boston. Dementia associated with the acquired immunodeficiency syndrome. *N. Engl. J. Med.* **332:**934–940.

Liu, Z. Q., Muhkerjee, S., Sahni, M., McCormick-Davis, C., Leung, K., Li, Z., Gattone, V. H., 2nd, Tian, C., Doms, R. W., Hoffman, T. L., Raghavan, R., Narayan, O., and Stephens, E. B. (1999). Derivation and biological characterization of a molecular clone of SHIV(KU-2) that causes AIDS, neurological disease, and renal disease in rhesus macaques. *Virology* **260:**295–307.

Lowin, B., Hahne, M., Mattmann, C., and Tschopp, J. (1994). Cytolytic T cell cytotoxicity is mediated through perforin and Fas lytic pathways. *Nature* **370:**650–652.

Ma, Q., Jones, D., Borghesani, P. R., Segal, R. A., Nagasawa, T., Kishimoto, T., Bronson, R. T., and Springer, T. A. (1998). Impaired B-lymphopoiesis, myelopoiesis, and derailed cerebellar neuron migration in CXCR4- and SDF-1-deficient mice. *Proc. Natl. Acad. Sci. U.S.A.* **95:**9448–9453.

Mankowski, J. L., Flaherty, M. T., Spelman, J. P., Hauer, D. A., Didier, P. J., Amedee, A. M., Murphey-Corb, M., Kirstein, L. M., Munoz, A., Clements, J. E., and Zink, M. C. (1997). Pathogenesis of simian immunodeficiency virus encephalitis: viral determinants of neurovirulence. *J. Virol.* **71:**6055–6060.

Mankowski, J. L., Spelman, J. P., Ressetar, H. G., Strandberg, J. D., Laterra, J., Carter, D. L., Clements, J. E., and Zink, M. C. (1994). Neurovirulent simian immunodeficiency virus replicates productively in endothelial cells of the central nervous system in vivo and in vitro. *J. Virol.* **68:**8202–8208.

Marcario, J. K., Raymond, L. A., McKiernan, B. J., Foresman, L. L., Joag, S. V., Raghavan, R., Narayan, O., and Cheney, P. D. (1999a). Motor skill impairment in SIV-infected rhesus macaques with rapidly and slowly progressing disease. *J. Med. Primatol.* **28**:105–117.

Marcario, J. K., Raymond, L. A., McKiernan, B. J., Foresman, L. L., Joag, S. V., Raghavan, R., Narayan, O., Hershberger, S., and Cheney, P. D. (1999b). Simple and choice reaction time performance in SIV-infected rhesus macaques. *AIDS Res. Hum. Retroviruses* **15**:571–583.

Marten, N. W., Stohlman, S. A., Smith-Begolka, W., Miller, S. D., Dimacali, E., Yao, Q., Stohl, S., Goverman, J., and Bergmann, C. C. (1999). Selection of CD8⁺ T cells with highly focused specificity during viral persistence in the central nervous system. *J. Immunol.* **162**:3905–3914.

Masliah, E., Heaton, R. K., Marcotte, T. D., Ellis, R. J., Wiley, C. A., Mallory, M., Achim, C. L., McCutchan, J. A., Nelson, J. A., Atkinson, J. H., and Grant, I. (1997). Dendritic injury is a pathological substrate for human immunodeficiency virus-related cognitive disorders. HNRC Group. The HIV Neurobehavioral Research Center. *Ann. Neurol.* **42**:963–972.

Masson, D., and Tschopp, J. (1985). Isolation of a lytic, pore-forming protein (perforin) from cytolytic T-lymphocytes. *J. Biol. Chem.* **260**:9069–9072.

McArthur, J. C., Selnes, O. A., Glass, J. D., Hoover, D. R., and Bacellar, H. (1994). HIV dementia. Incidence and risk factors. *Res. Publ. Assoc. Res. Nerv. Ment. Dis.* **72**:251–272.

Messenheimer, J. A., Robertson, K. R., Wilkins, J. W., Kalkowski, J. C., and Hall, C. D. (1992). Event-related potentials in human immunodeficiency virus infection. A prospective study. *Arch. Neurol.* **49**:396–400.

Meucci, O., Fatatis, A., Simen, A. A., Bushell, T. J., Gray, P. W., and Miller, R. J. (1998). Chemokines regulate hippocampal neuronal signaling and gp120 neurotoxicity. *Proc. Natl. Acad. Sci. U.S.A.* **95**:14500–14505.

Michaels, S. H., Clark, R., and Kissinger, P. (1998). Declining morbidity and mortality among patients with advanced human immunodeficiency virus infection. *N. Engl. J. Med.* **339**:405–406.

Moltz, H. (1993). Fever: causes and consequences. *Neurosci. Biobehav. Rev.* **17**:237–269.

Moore, R. and Chaisson, R. E. (1999). Natural history of HIV infection in the era of highly active antiretroviral therapy. *AIDS* **14**:1249–1253.

Moses, A. V., Bloom, F. E., Pauza, C. D., and Nelson, J. A. (1993). Human immunodeficiency virus infection of human brain capillary endothelial cells occurs via a CD4/galactosylceramide-independent mechanism. *Proc. Natl. Acad. Sci. U.S.A.* **90**:10474–10478.

Moses, A. V., Williams, S., Heneveld, M. L., Strussenberg, J., Rarick, M., Loveless, M., Bagby, G., and Nelson, J. A. (1996). Human immunodeficiency virus infection of bone marrow endothelium reduces induction of stromal hematopoietic growth factors. *Blood* **87**:919–925.

Murray, E. A., Rausch, D. M., Lendvay, J., Sharer, L. R., and Eiden, L. E. (1992). Cognitive and motor impairments associated with SIV infection in rhesus monkeys. *Science* **255**:1246–1249.

Navia, B. A., Cho, E. S., Petito, C. K., and Price, R. W. (1986). The AIDS dementia complex: II. Neuropathology. *Ann. Neurol.* **19**:525–535.

Nottet, H. S., Persidsky, Y., Sasseville, V. G., Nukuna, A. N., Bock, P., Zhai, Q. H., Sharer, L. R., McComb, R. D., Swindells, S., Soderland, C., and Gendelman, H. E. (1996). Mechanisms for the transendothelial migration of HIV-1-infected monocytes into brain. *J. Immunol.* **156**:1284–1295.

Pagano, M. A., Cahn, P. E., Garau, M. L., Mangone, C. A., Figini, H. A., Yorio, A. A., Dellepiane, M. C., Amores, M. G., Perez, H. M., and Casiro, A. D. (1992). Brainstem auditory-evoked potentials in human immunodeficiency virus-seropositive patients with and without acquired immunodeficiency syndrome. *Arch. Neurol.* **49:**166–169.

Perelson, A. S., Essunger, P., Cao, Y., Vesanen, M., Hurley, A., Saksela, K., Markowitz, M., and Ho, D. D. (1997). Decay characteristics of HIV-1-infected compartments during combination therapy. *Nature* **387:**188–191.

Persidsky, Y., Ghorpade, A., Rasmussen, J., Limoges, J., Liu, X. J., Stins, M., Fiala, M., Way, D., Kim, K. S., Witte, M. H., Weinand, M., Carhart, L., and Gendelman, H. E. (1999). Microglial and astrocyte chemokines regulate monocyte migration through the blood-brain barrier in human immunodeficiency virus-1 encephalitis. *Am. J. Pathol.* **155:**1599–1611.

Petito, C. K., and Cash, K. S. (1992). Blood-brain barrier abnormalities in the acquired immunodeficiency syndrome: immunohistochemical localization of serum proteins in postmortem brain. *Ann. Neurol.* **32:**658–666.

Phillips, T. R., Prospero-Garcia, O., Puaoi, D. L., Lerner, D. L., Fox, H. S., Olmsted, R. A., Bloom, F. E., Henriksen, S. J., and Elder, J. H. (1994). Neurological abnormalities associated with feline immunodeficiency virus infection. *J. Gen. Virol.* **75:**979–87.

Price, R. W., Brew, B., Sidtis, J., Rosenblum, M., Scheck, A. C., and Cleary, P. (1988). The brain in AIDS: central nervous system HIV-1 infection and AIDS dementia complex. *Science* **239:**586–92.

Prospero-Garcia, O., Gold, L. H., Fox, H. S., Polis, I., Koob, G. F., Bloom, F. E., and Henriksen, S. J. (1996). Microglia-passaged simian immunodeficiency virus induces neurophysiological abnormalities in monkeys. *Proc. Natl. Acad. Sci. U.S.A.* **93:**14158–63.

Raghavan, R., Cheney, P. D., Raymond, L. A., Joag, S. V., Stephens, E. B., Adany, I., Pinson, D. M., Li, Z., Marcario, J. K., Jia, F., Wang, C., Foresman, L., Berman, N. E., and Narayan, O. (1999). Morphological correlates of neurological dysfunction in macaques infected with neurovirulent simian immunodeficiency virus. *Neuropathol. Appl. Neurobiol.* **25:**285–94.

Raghavan, R., Stephens, E. B., Joag, S. V., Adany, I., Pinson, D. M., Li, Z., Jia, F., Sahni, M., Wang, C., Leung, K., Foresman, L., and Narayan, O. (1997). Neuropathogenesis of chimeric simian/human immunodeficiency virus infection in pig-tailed and rhesus macaques. *Brain Pathol.* **7:**851–61.

Rausch, D. M., Heyes, M. P., Murray, E. A., and Eiden, L. E. (1995). Zidovudine treatment prolongs survival and decreases virus load in the central nervous system of rhesus macaques infected perinatally with simian immunodeficiency virus. *J. Infect. Dis.* **172:**59–69.

Rausch, D. M., Heyes, M. P., Murray, E. A., Lendvay, J., Sharer, L. R., Ward, J. M., Rehm, S., Nohr, D., Weihe, E., and Eiden, L. E. (1994). Cytopathologic and neurochemical correlates of progression to motor/cognitive impairment in SIV-infected rhesus monkeys. *J. Neuropathol. Exp. Neurol.* **53:**165–75.

Raymond, L. A., Wallace, D., Berman, N. E., Marcario, J., Foresman, L., Joag, S. V., Raghavan, R., Narayan, O., and Cheney, P. D. (1998). Auditory brainstem responses in a Rhesus Macaque model of neuro-AIDS. *J. Neurovirol.* **4:**512–20.

Raymond, L. A., Wallace, D., Marcario, J. K., Raghavan, R., Narayan, O., Foresman, L. L., Berman, N. E., and Cheney, P. D. (1999). Motor-evoked potentials in a rhesus macaque model of neuro-AIDS. *J. Neurovirol.* **5:**217–31.

Reinhart, T. A., Rogan, M. J., Huddleston, D., Rausch, D. M., Eiden, L. E., and Haase, A. T. (1997). Simian immunodeficiency virus burden in tissues and cellular compartments during clinical latency and AIDS. *J. Infect. Dis.* **176:**1198–1208.

Rottman, J. B., Ganley, K. P., Williams, K., Wu, L., Mackay, C. R., and Ringler, D. J. (1997). Cellular localization of the chemokine receptor CCR5. Correlation to cellular targets of HIV-1 infection. *Am. J. Pathol.* **151:**1341–51.

Sacktor, N., and McArthur, J. (1997). Prospects for therapy of HIV-associated neurologic diseases. *J. Neurovirol.* **3:**89–101.

Sanders, V. J., Pittman, C. A., White, M. G., Wang, G., Wiley, C. A., and Achim, C. L. (1998). Chemokines and receptors in HIV encephalitis. *AIDS* **12:**1021–26.

Sasseville, V. G., Lane, J. H., Walsh, D., Ringler, D. J., and Lackner, A. A. (1995). VCAM-1 expression and leukocyte trafficking to the CNS occur early in infection with pathogenic isolates of SIV. *J. Med. Primatol.* **24:**123–31.

Sasseville, V. G., Newman, W., Brodie, S. J., Hesterberg, P., Pauley, D., and Ringler, D. J. (1994). Monocyte adhesion to endothelium in simian immunodeficiency virus-induced AIDS encephalitis is mediated by vascular cell adhesion molecule-1/alpha 4 beta 1 integrin interactions. *Am. J. Pathol.* **144:**27–40.

Sasseville, V. G., Newman, W. A., Lackner, A. A., Smith, M. O., Lausen, N. C., Beall, D., and Ringler, D. J. (1992). Elevated vascular cell adhesion molecule-1 in AIDS encephalitis induced by simian immunodeficiency virus. *Am. J. Pathol.* **141:**1021–30.

Sasseville, V. G., Smith, M. M., Mackay, C. R., Pauley, D. R., Mansfield, K. G., Ringler, D. J., and Lackner, A. A. (1996). Chemokine expression in simian immunodeficiency virus-induced AIDS encephalitis. *Am. J. Pathol.* **149:**1459–67.

Schmitz, J. E., Kuroda, M. J., Santra, S., Sasseville, V. G., Simon, M. A., Lifton, M. A., Racz, P., Tenner-Racz, K., Dalesandro, M., Scallon, B. J., Ghrayeb, J., Forman, M. A., Montefiori, D. C., Rieber, E. P., Letvin, N. L., and Reimann, K. A. (1999). Control of viremia in simian immunodeficiency virus infection by CD8+ lymphocytes. *Science* **283:**857–60.

Schroeder, M. M., Handelsman, L., Torres, L., Dorfman, D., Rinaldi, P., Jacobson, J., Wiener, J., and Ritter, W. (1994). Early and late cognitive event-related potentials mark stages of HIV-1 infection in the drug-user risk group. *Biol. Psychiatry* **35:**54–69.

Serafini, G., Stagni, G., Chiarella, G., Brizi, S., and Simoncelli, C. (1998). ABR and HIV-induced impairment of the central nervous system. *Rev. Laryngol. Otol. Rhinol.* **119:**87–90.

Sharma, D. P., Zink, M. C., Anderson, M., Adams, R., Clements, J. E., Joag, S. V., and Narayan, O. (1992). Derivation of neurotropic simian immunodeficiency virus from exclusively lymphocytotropic parental virus: pathogenesis of infection in macaques. *J. Virol.* **66:**3550–56.

Shieh, J. T., Albright, A. V., Sharron, M., Gartner, S., Strizki, J., Doms, R. W., and Gonzalez-Scarano, F. (1998). Chemokine receptor utilization by human immunodeficiency virus type 1 isolates that replicate in microglia. *J. Virol.* **72:**4243–49.

Simon, M. A., Chalifoux, L. V., and Ringler, D. J. (1992). Pathologic features of SIV-induced disease and the association of macrophage infection with disease evolution. *AIDS Res. Hum. Retroviruses* **8:**327–37.

Singer, E. J., Syndulko, K., Fahy-Chandon, B., Schmid, P., Conrad, A., and Tourtellotte, W. W. (1994). Intrathecal IgG synthesis and albumin leakage are increased in subjects with HIV-1 neurologic disease. *J. Acquir. Immune. Defic. Syndr.* **7:**265–71.

Smith, M. O., Sutjipto, S., and Lackner, A. A. (1994). Intrathecal synthesis of IgG in simian immunodeficiency virus (SIV)-infected rhesus macaques (Macaca mulatta). *AIDS Res. Hum. Retroviruses* **10:**81–89.

Somma-Mauvais, H., and Farnarier, G. (1992). Evoked potentials in HIV infection. *Neurophysiol. Clin.* **22:**369–84.

Sopper, S., Sauer, U., Hemm, S., Demuth, M., Muller, J., Stahl-Hennig, C., Hunsmann, G., ter Meulen, V., and Dorries, R. (1998). Protective role of the virus-specific immune response for development of severe neurologic signs in simian immunodeficiency virus-infected macaques. *J. Virol.* **72:**9940–47.

Sporer, B., Paul, R., Koedel, U., Grimm, R., Wick, M., Goebel, F. D., and Pfister, H. W. (1998). Presence of matrix metalloproteinase-9 activity in the cerebrospinal fluid of human immunodeficiency virus-infected patients. *J. Infect. Dis.* **178:**854–57.

Strelow, L. I., Watry, D. D., Fox, H. S., and Nelson, J. A. (1998). Efficient infection of brain microvascular endothelial cells by an in vivo-selected neuroinvasive SIV$_{mac}$ variant. *J. Neurovirol.* **4:**269–80.

Suidan, H. S., Bouvier, J., Schaerer, E., Stone, S. R., Monard, D., and Tschopp, J. (1994). Granzyme A released upon stimulation of cytotoxic T lymphocytes activates the thrombin receptor on neuronal cells and astrocytes. *Proc. Natl. Acad. Sci. U.S.A.* **91:**8112–16.

Toggas, S. M., Masliah, E., Rockenstein, E. M., Rall, G. F., Abraham, C. R., and Mucke, L. (1994). Central nervous system damage produced by expression of the HIV-1 coat protein gp120 in transgenic mice. *Nature* **367:**188–93.

Volberding, P. A., and Deeks, S. G. (1998). Antiretroviral therapy for HIV infection: promises and problems. *Jama* **279:**1343–44.

von Herrath, M., Oldstone, M. B., and Fox, H. S. (1995). Simian immunodeficiency virus (SIV)-specific CTL in cerebrospinal fluid and brains of SIV-infected rhesus macaques. *J. Immunol.* **154:**5582–89.

Watry, D., Lane, T. E., Streb, M., and Fox, H. S. (1995). Transfer of neuropathogenic simian immunodeficiency virus with naturally infected microglia. *Am. J. Pathol.* **146:**914–23.

Weiss, J. M., Downie, S. A., Lyman, W. D., and Berman, J. W. (1998). Astrocyte-derived monocyte-chemoattractant protein-1 directs the transmigration of leukocytes across a model of the human blood–brain barrier. *J. Immunol.* **161:**6896–903.

Weissman, D., Rabin, R. L., Arthos, J., Rubbert, A., Dybul, M., Swofford, R., Venkatesan, S., Farber, J. M., and Fauci, A. S. (1997). Macrophage-tropic HIV and SIV envelope proteins induce a signal through the CCR5 chemokine receptor. *Nature* **389:**981–85.

Wennberg, R. P., Gospe, S. M., Jr., Rhine, W. D., Seyal, M., Saeed, D., and Sosa, G. (1993). Brainstem bilirubin toxicity in the newborn primate may be promoted and reversed by modulating PCO2. *Pediatr. Res.* **34:**6–9.

Westmoreland, S. V., Halpern, E., and Lackner, A. A. (1998). Simian immunodeficiency virus encephalitis in rhesus macaques is associated with rapid disease progression. *J. Neurovirol.* **4:**260–68.

Westmoreland, S. V., Rottman, J. B., Williams, K. C., Lackner, A. A., and Sasseville, V. G. (1998). Chemokine receptor expression on resident and inflammatory cells in the brain of macaques with simian immunodeficiency virus encephalitis. *Am. J. Pathol.* **152:**659–65.

Wiley, C. A., Masliah, E., Morey, M., Lemere, C., DeTeresa, R., Grafe, M., Hansen, L., and Terry, R. (1991). Neocortical damage during HIV infection. *Ann. Neurol.* **29:**651–57.

Wiley, C. A., Schrier, R. D., Nelson, J. A., Lampert, P. W., and Oldstone, M. B. (1986). Cellular localization of human immunodeficiency virus infection within the brains of acquired immune deficiency syndrome patients. *Proc. Natl. Acad. Sci. U.S.A.* **83:**7089–93.

Wiley, C. A., Soontornniyomkij, V., Radhakrishnan, L., Masliah, E., Mellors, J., Hermann, S. A., Dailey, P., and Achim, C. L. (1998). Distribution of brain HIV load in AIDS. *Brain Pathol.* **8:**277–84.

Woodman, S. E., Benveniste, E. N., Nath, A., and Berman, J. W. (1999). Human immunodeficiency virus type 1 TAT protein induces adhesion molecule expression in astrocytes. *J. Neurovirol.* **5:**678–84.

Xia, M. Q., and Hyman, B. T. (1999). Chemokines/chemokine receptors in the central nervous system and Alzheimer's disease. *J. Neurovirol.* **5**:32–41.

Zhang, L., Dailey, P. J., He, T., Gettie, A., Bonhoeffer, S., Perelson, A. S., and Ho, D. D. (1999). Rapid Clearance of Simian Immunodeficiency Virus Particles from Plasma of Rhesus Macaques. *J. Virol.* **73**:855–60.

Zhang, L., He, T., Talal, A., Wang, G., Frankel, S. S., and Ho, D. D. (1998). In vivo distribution of the human immunodeficiency virus/simian immunodeficiency virus coreceptors: CXCR4, CCR3, and CCR5. *J. Virol.* **72**:5035–45.

Zheng, J., Ghorpade, A., Niemann, D., Cotter, R. L., Thylin, M. R., Epstein, L., Swartz, J. M., Shepard, R. B., Liu, X., Nukuna, A., and Gendelman, H. E. (1999a). Lymphotropic virions affect chemokine receptor-mediated neural signaling and apoptosis: implications for human immunodeficiency virus type 1-associated dementia. *J. Virol.* **73**:8256–67.

Zheng, J., Thylin, M. R., Ghorpade, A., Xiong, H., Persidsky, Y., Cotter, R., Niemann, D., Che, M., Zeng, Y. C., Gelbard, H. A., Shepard, R. B., Swartz, J. M., and Gendelman, H. E. (1999b). Intracellular CXCR4 signaling, neuronal apoptosis and neuropathogenic mechanisms of HIV-1 associated dementia. *J. Neuroimmunol.* **98**:185–200.

Zink, M. C., Amedee, A. M., Mankowski, J. L., Craig, L., Didier, P., Carter, D. L., Munoz, A., Murphey-Corb, M., and Clements, J. E. (1997). Pathogenesis of SIV encephalitis. Selection and replication of neurovirulent SIV. *Am. J. Pathol.* **151**:793–803.

Zink, M. C., Spelman, J. P., Robinson, R. B., and Clements, J. E. (1998). SIV infection of macaques—modeling the progression to AIDS dementia. *J. Neurovirol.* **4**:249–59.

Zink, M. C., Suryanarayana, K., Mankowski, J. L., Shen, A., Piatak, M., Jr., Spelman, J. P., Carter, D. L., Adams, R. J., Lifson, J. D., and Clements, J. E. (1999). High viral load in the cerebrospinal fluid and brain correlates with severity of simian immunodeficiency virus encephalitis. *J. Virol.* **73**:10480–88.

Zou, Y. R., Kottmann, A. H., Kuroda, M., Taniuchi, I., and Littman, D. R. (1998). Function of the chemokine receptor CXCR4 in haematopoiesis and in cerebellar development. *Nature* **393**:595–99.

NEUROENDOCRINE-IMMUNE INTERACTIONS DURING VIRAL INFECTIONS

Brad D. Pearce,* Christine A. Biron,† and Andrew H. Miller*

*Emory University School of Medicine
Department of Psychiatry and Behavioral Sciences
Atlanta, Georgia 30322
†Department of Molecular Microbiology and Immunology
Division of Biology and Medicine
Brown University
Providence, Rhode Island, 02912

I. Introduction

On a fundamental level, the usual consequence of a viral infection is the disruption of the normal physiological equilibrium of the host. Restoration of homeostasis in complex hosts (e.g., mammals) requires that multiple organ systems be coordinated through the action of diffusible mediators such as cytokines, hormones, and neurotransmitters. Virologists tend to think of cytokines as the primary soluble molecules coordinating the response to a viral challenge. However, in a broader context, the immune system interacts with stress-activated endocrine hormones, and the mortality of the host hangs in the balance between these two powerful homeostatic systems. In this chapter we review the physiological and pathophysiological interplay between the immune and neuroendocrine systems, and emphasize the reciprocal regulation that occurs between endogenous adrenal steroids and cytokines during viral infections.

II. Hypothalamic–Pituitary–Adrenal (HPA) Axis

A. Functional Anatomy

Glucocorticoids are immunomodulatory steroid hormones produced by the adrenal cortex under both physiological and pathological conditions (Besedovsky and del Rey, 1996; McEwen *et al.*, 1997). Cortisol is the major endogenous glucocorticoid in humans while corticosterone is its functional equivalent in mice and rats (Turnbull and Rivier, 1999). In this review we use the term "CORT" to refer generically to either cortisol or corticosterone, and both these hormones are similarly regulated within the framework of the hypothalamic–pituitary–adrenal (HPA) axis

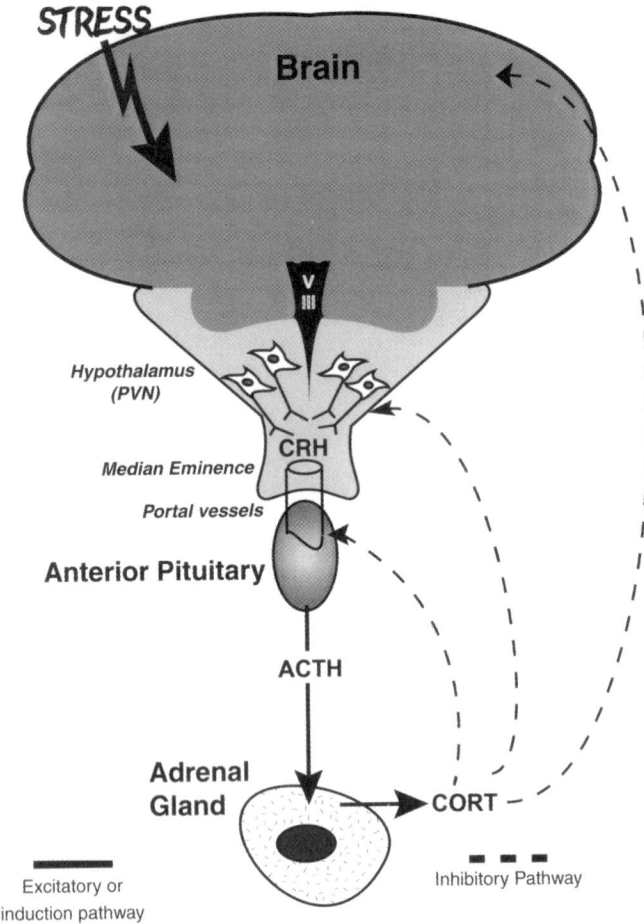

Fig 1. HPA-axis regulation

(Fig. 1). The adrenal cortex releases CORT in response to adreno-corticotropin (ACTH), a peptide hormone derived from the anterior lobe of the pituitary (Plotsky, 1991). The production of ACTH is under the regulatory control of the hypothalamus, by virtue of neurons in the paraventricular nucleus that release the peptide secretagogue, corticotrophin-releasing factor (CRF) into the hypophyseal portal circulation that leads to the anterior pituitary (Plotsky, 1991). Although CRF is the main neurohormone responsible for stimulating ACTH release, other hypothalamic peptides such as arginine vasopressin play an ancillary role (Plotsky, 1991; Sawchenko, Swanson, and Vale, 1984).

The HPA axis can be activated at multiple levels, yet the process is commonly conceptualized as beginning with a surge of CRF and culminating in the production and release of corticosteroids by the adrenal glands. Depending on the stimulus, the entire process can take only minutes to occur (McEwen et al., 1997; Turnbull and Rivier, 1999). However, this simplistic view of HPA axis regulation is belied by a plethora of studies demonstrating that the activity of the HPA is orchestrated by an intricate web of interdependent soluble mediators, which include neuropeptides, hormones, neurotransmitters, and cytokines (Besedovsky and del Rey, 1996; Dunn and Vickers, 1994; McEwen et al., 1997; Schobitz, De Kloet, and Holsboer, 1994; Turnbull and Rivier, 1999). The multitude of regulatory pathways for the HPA axis is reflected in its anatomical organization. The hypothalamus is situated neuroanatomically to integrate a variety of psychological, sensory, and physical stressors into a common hormonal message (e.g., CRF) (Plotsky, 1991). Like most of the brain, the hypothalamus is sequestered behind the blood–brain barrier, and consequently it is somewhat protected from pernicious soluble molecules and infectious agents in the peripheral circulation (Turnbull and Rivier, 1999; Watkins, Maier, and Goehler, 1995). Even so, there are regions of the hypothalamus where the blood–brain barrier is leaky or incomplete, and the pituitary gland, like the adrenals, lies outside the blood–brain barrier. Thus, compared to the hypothalamus, the pituitary and adrenal glands are more receptive to peripheral signals and also potentially more vulnerable to blood-borne infections or immunological mediators. Furthermore, the HPA axis is subject to regulatory influences at each of its anatomically separated levels, and a disseminated viral infection inducing high circulating levels of cytokines would be capable of acting at multiple locations of the HPA axis. Moreover, the impact of a viral infection on the HPA axis can be dependent on tissue tropisms. For example, a minor localized infection in or near the adrenal gland might influence CORT production (e.g., through paracrine action of cytokines) without acting on other HPA tissues.

Superimposed on the response of the HPA axis to a viral infection is the normal diurnal pattern of glucocorticoid release, with highest levels occurring on commencement of an animal's physically active period. Diurnal fluctuations in glucocorticoids are considerable, typically 10- to 15-fold for rodents (Turnbull and Rivier, 1999). However, these cyclic changes pale in contrast to the surge of glucocorticoids elicited by extreme stress or infection, whereby plasma concentrations can soar to 100-fold above baseline (Dunn and Vickers, 1994; Plotsky, 1991). The pattern of glucocorticoid fluctuation is governed by feedback regulation, which occurs at each of the sites of the HPA axis and is mediated in part by CORT binding to its receptors (Plotsky, 1991; Turnbull and Rivier, 1999). In addition, brain regions that are not considered part of the HPA axis, such as the hippocampus, participate in feedback inhibition mediated by CORT (Plotsky, 1991). As discussed below, cytokines are potent modulators of the HPA axis, and hence form one of the critical links between the immune and neuroendocrine systems.

B. Role of HPA Axis in Stress-Induced Immune Alterations

A large body of data collected from laboratory animal and human studies has demonstrated that stress is capable of leading to a multitude of immune changes involving virtually every aspect of the immune response. Some of the earliest studies on the immunologic effects of stress took place over 40 years ago and examined the influence of a range of stressful manipulations on the expression of viral disease and virus-induced tumors. Studies by Rasmussen and colleagues demonstrated that repeatedly exposing mice to a shock-avoidance paradigm, as well as to physical restraint and high-intensity sound, resulted in increased susceptibility to herpes simplex virus (Rasmussen, Marsh, and Brill, 1957); poliomyelitis virus (Johnsson and Rasmussen, 1965); coxsackie B virus (Johnsson et al., 1963); vesicular stomatitis virus (Jensen and Rasmussen, 1963); and polyomavirus infection (Rasmussen, 1969). Based on these early reports, considerable interest developed in discovering the mechanism of these effects, and attention was directed toward the link between stressful stimuli and the induction of immunoregulatory hormones, particularly glucocorticoids.

The effects of glucocorticoids on the immune system have been extensively characterized. In 1950, Philip Hench won the Nobel Prize in medicine for his discovery of the role of glucocorticoids in modulating the expression of autoimmune disorders including rheumatoid arthritis and allergic responses (Hench, 1952). Since then, a massive literature has developed documenting the multiple immunologic effects of glucocorticoids and the biochemical and molecular mechanisms involved (Besedovsky and del Rey, 1996; McEwen et al., 1997; Miller et al., 1997; Munck and Guyr, 1991; Sheridan, 1998). In brief, glucocorticoids influence the immune response by (1) modulating the trafficking of immune cells throughout the body (Dhabhar et al., 1995; Dhabhar et al., 1996; Miller et al., 1994); (2) inhibiting cytokine production and function through interaction of glucocorticoid receptors with transcription factors (e.g., AP-1, NF-κB), which in turn regulate cytokine gene expression and/or the expression of cytokine-inducible genes (Auphan et al., 1995; Vacca et al., 1992); (3) inhibiting the generation of products of the arachidonic acid pathway that mediate inflammation (Schleimer, Claman, and Oronsky, 1989); (4) inhibiting cytotoxicity (Nair and Schwartz, 1984); and (5) modulating cell death pathways in immature and mature cell types (Gonzalo et al., 1993; McEwen et al., 1997).

In terms of the effects of stress on the immune system, glucocorticoids have been shown to mediate the effects of acute restraint stress on peripheral-blood immune cell distribution in rodents and have been found to mediate inhibitory effects of stress on lymph node cellularity during viral infections in mice (Dhabhar et al., 1996; Hermann, Beck, and Sheridan, 1995). In addition, glucocorticoids have been shown to mediate the impact of stress on proliferative responses of peripheral-blood lymphocytes (McEwen et al., 1997). Finally, glucocorticoids appear to mediate some of the effects of CRF on immune responses.

Although much emphasis has been placed on the immunosuppressive effects of glucocorticoids during stress, recent studies have suggested that the acute influence of stress-induced CORT may involve facilitating protective immune responses in peripheral immune compartments such as the skin, whereas more prolonged exposure to CORT, such as during chronic stress, may reduce immune responses in all immune tissues. Taken together, these data support the notion that glucocorticoids are immunomodulators, and the ultimate consequences of stress-induced elevations in glucocorticoids is a function of the context within which they occur (Dhabhar and McEwen, 1999).

Another classic mediator of stress-induced immune changes is the sympathetic nervous system (SNS). For example, activation of SNS pathways by CRF has been shown to inhibit natural killer (NK) cell activity in the spleen of rats. Blockade of this pathway at the level of CRF or SNS innervation of the spleen is capable of disrupting the immunosuppressive effects of stress in this tissue. Thus, because physical and psychological stressors are processed by the brain, and converge on CRF, this molecule serves as a focal point for the integration of responses to stress and hence could be considered a master-modulator of immune alterations associated with various stressful stimuli, including infections. Nevertheless, there clearly are multiple avenues of bidirectional communication between the CNS and the immune system, and the discovery of a bewildering array of receptors for neurotransmitters and hormones on leukocytes serves to further demonstrate that the immune and nervous systems are deeply intertwined (Besedovsky and del Rey, 1996). In this review we focus primarily on the molecules involved in the regulation and activity of glucocorticoids because of the profound and indisputable effects of these hormones on immunity.

III. GLUCOCORTICOID INDUCTION DURING VIRAL INFECTIONS: MECHANISMS AND SIGNALING PATHWAYS

A number of studies have documented that viral infections are capable of activating the HPA axis and ultimately the release of glucocorticoids (Ben Hur et al., 1996; Ben-Hur et al., 1995; Besedovsky et al., 1986; Besedovsky and del Rey, 1989; Christeff et al., 1997; Dunn, Powell, and Gaskin, 1987; Dunn et al., 1989; Dunn et al., 1987; Dunn and Vickers, 1994; Hermann, Beck, and Sheridan, 1995; Price et al., 1996; Smith, Meyer, and Blalock, 1982). Early studies of humans detected alterations in adrenal cortical output in measles (Lorenz and Rossipal, 1965) and varicella infections (Sherman, Michaels, and Kenny, 1965; Zeitoun et al., 1973). Experimental models focused initially on virus-type stimuli such as Newcastle disease virus (NDV), which is not infectious to rodents but stimulates a pattern of cytokine production similar to the synthetic viral analog, polyinosinic–polycytidylic acid [poly(I:C)] (Guha-Thakurta and Majde, 1997). Elevations in glucocorticoids, in response to NDV, were found to be relatively acute (minutes to hours) and robust (Besedovsky et al., 1986; Besedovsky and del Rey, 1989; Dunn, Powell, and Gaskin, 1987; Dunn et al., 1987; Dunn and Vickers, 1994; Miller et al., 1997; Price et al., 1996; Smith, Meyer, and Blalock, 1982). Subsequent studies examining a range of replicating viruses in humans and laboratory animals have further demonstrated that viral infection can be associated with induction of endogenous glucocorticoids, although, not surprisingly, compared to NDV and poly(I:C), the kinetics are typically delayed with peak responses occurring days after the initial infection (Ben Hur et al., 1996; Ben-Hur et al., 1995; Christeff et al., 1997; Dunn et al., 1989; Hermann, Beck, and Sheridan, 1995; Price et al., 1996; Ruzek et al., 1997b). Using different murine viral infections, work conducted in our laboratories has demonstrated that not all viral infections are associated with increased glucocorticoid release (Miller et al., 1997), thus raising the question regarding the conditions and mediators required for HPA axis activation. Furthermore, chronic viral infections may influence multiple components of the HPA axis and, in some cases, lead to deficits in the amount or activity of glucocorticoids. The mechanisms underlying HPA axis regulation during viral infections are only beginning to be defined, and are considered in more detail in the following sections.

A. Cytokines and Immune Response to Viral Infection

Cytokines are well established as important intermediates in communicating conditions of stress or a bacterial infection to the HPA axis (Turnbull and Rivier, 1999). A substantial literature supports the contention that the cytokines that are associated with viral infections can have profound effects on various components of the HPA axis (Besedovsky and del Rey, 1996; Dunn and Wang, 1995; McEwen et al., 1997; Schobitz, De Kloet, and Holsboer, 1994; Turnbull and Rivier, 1999; Watkins, Maier, and Goehler, 1995). Nevertheless, the vast majority of such studies have examined neuroendocrine responses to purified cytokines or chemical immunostimulants such as lipopolysaccaride (LPS) or poly(I:C), and hence their findings must be cautiously extrapolated to viral infections in which endogenous cytokines are elaborated in a naturally orchestrated pattern over a sustained period of days or weeks. Despite the utility of purified cytokines and synthetic immunostimulants in revealing fundamental homeostatic mechanisms, the typical host response to a pathogen involves neither a transient surge of a single cytokine, nor the abrupt mobilization of a conglomerate of cytokines that occurs following an injection of LPS (Biron, 1994; Ahmed and Biron, 1998; McEwen et al., 1997). Furthermore, although all immune responses share certain characteristics in common, the complex of various soluble and cellular components that comprise the immune response are ultimately pathogen specific. Thus, before considering detailed mechanisms by which the immune system, and cytokines in particular, influence the HPA axis during a viral infection, we will review the fundamental elements of the host antiviral immune response.

Immune responses are generally elicited in cascades with multiple amplification and regulation pathways as well as with simultaneous activation of different cytokine and cellular components (Ahmed and Biron, 1998). In addition to mediating antiviral defense, these components can contribute to pathology if induced to high concentrations and/or at inappropriate times. As a result, they must be tightly regulated. Glucocorticoids are potent modulators of immunopathology (see below), and the HPA axis secretory activity can be under the control of multiple interacting regulatory loops involving antiviral cytokines, cellular immunity, and feedback from HPA hormones (McEwen *et al.*, 1997; Turnbull and Rivier, 1999). Thus, the fate of the host can be dependent on communication between the antiviral immune response and HPA axis activity. Work in our own and other laboratories is dissecting these regulatory loops, and defining the precarious balance maintained by glucocorticoids between cytokine antiviral effects and their capacity to cause tissue damage and death (Ruzek *et al.*, 1997a; Ruzek *et al.*, 1999). A thorough review of the cytokines induced during viral infections is beyond the scope of this chapter, and the reader is referred to recent reviews (Ahmed and Biron, 1998; Biron and Sen, 2000). To illustrate the principles most pertinent to the HPA axis, we will concentrate on studies examining infections of mice with strains of the arenavirus, lymphocytic choriomeningitis virus (LCMV), and a member of the herpes group of viruses—murine cytomegalovirus (MCMV).

Generally, host responses to first or primary viral infections can be divided into early and late phases, with early-phase responses having components contributing both to defense and to the regulation of downstream immune responses (Ahmed and Biron, 1998; McEwen *et al.*, 1997). Thus, the late or adaptive immune effector functions and development of long-term immunity are influenced by the early or innate immune responses (Fig. 2). The phases can be associated with different and overlapping mechanisms of protection. They are not absolutely divided in time but occur in waves. During the early phase, innate mech-

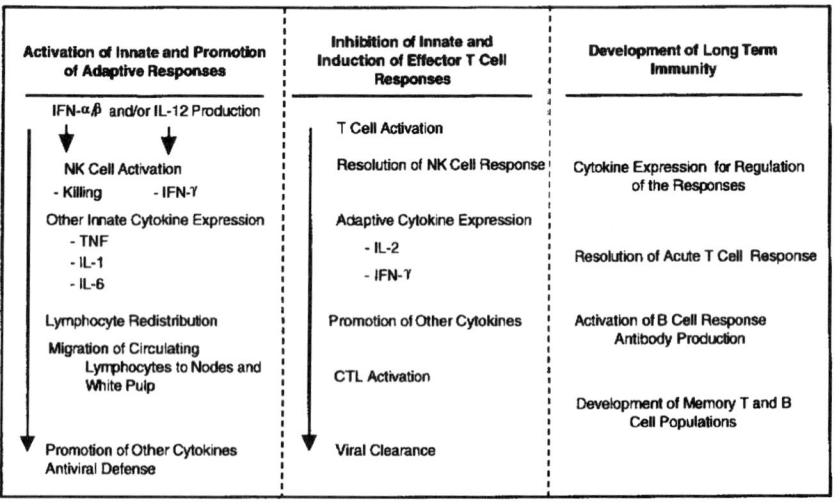

FIG 2. General features of immune-response phases during viral infections.

anisms of immune protection become activated. As an example, the type I interferons, i.e., interferons α/β (IFN-α/β), are produced by virus-infected cells, and induce biochemical pathways that render uninfected cells resistant to infection (Chong, Gresser, and Mims, 1983; Muller *et al.*, 1994; van den Broek *et al.*, 1995). High circulating levels of these type I IFNs are found in many viral infections, including those of LCMV and MCMV (Grundy *et al.*, 1982; Welsh, 1978). Type 1 IFNs also activate NK cells to mediate killing of certain virus-infected cells (Biron, 1994; Biron, 1997; Bukowski and Welsh, 1986; Bukowski *et al.*, 1983; Grundy *et al.*, 1982; Orange and Biron, 1996b; Stein-Streilein, 1988; Trinchieri, 1989; Welsh, 1978). The evidence supporting the importance of NK cell-mediated killing in antiviral responses remains limited, but NK cell-mediated cytotoxicity may be important in resistance to viral infection in the spleen but less important in the liver during defense against MCMV (Tay and Welsh, 1997).

In addition to induction of IFN-α/β and activation of NK cell cytotoxic activity, early responses to some, but not all, viral infections also result in expression of interleukin-12 (IL-12) (Biron and Orange, 1995; Coutelier, Van Broeck, and Wolf, 1995; Kanangat *et al.*, 1996; Orange and Biron, 1996a; Ruzek *et al.*, 1997b). Although NK cell expression of interferon gamma (IFN-γ) mRNA is observed during a variety of viral infections (Biron, Young, and Kasaian, 1990; Orange and Biron, 1996a; Salazar-Mather, Ishikawa, and Biron, 1996), detectable NK cell protein production requires IL-12 induction. IFN-γ can mediate direct antiviral effects and promote conditions supporting inflammation in tissues (Salazar-Mather, Hamilton, and Biron, 2000; Orange and Biron, 1996a). The role of NK cell-produced IFN-γ in defense has been demonstrated conclusively during MCMV infections (Orange *et al.*, 1995). In addition to IFN-α/β, IL-12, and IFN-γ, some viral infections induce early high levels of tumor necrosis factor (TNF), also having the potential to mediate direct antiviral effects (Campbell *et al.*, 1994; Hennet *et al.*, 1992; Orange and Biron, 1996b; Ruzek *et al.*, 1997b). Taken together, these observations indicate that early responses elicited during viral infections include innate effector mechanisms that protect the host from the spread of infection. These effector functions are critical to the host during this period and occur prior to development of adaptive protective responses, but early responses also may facilitate induction and shaping of the late-phase adaptive immune responses (Biron, 1999; Biron and Sen, 2000; Cousens *et al.*, 1999).

Later-phase immune responses are associated with CD4+ and CD8+T cell activation (Allan *et al.*, 1990; Biron, 1994; Cousens, Orange, and Biron, 1995; Karanth, Lyson, and McCann, 1994; Kasaian and Biron, 1989; Kasaian and Biron, 1990; Kasaian, Leite-Morris, and Biron, 1991; Pfizenmaier *et al.*, 1977; Sinicco *et al.*, 1993; Su *et al.*, 1994; Su *et al.*, 1998). Activated T cells can produce the potent T cell growth factor, interleukin-2 (IL-2), as well as IFN-γ (Biron, 1994; Gessner *et al.*, 1990; Kasaian and Biron, 1989; Lynch, Doherty, and Ceredig, 1989; Sinicco *et al.*, 1993; Su *et al.*, 1994). Although not completely understood, the conditions at this time support the expansion of CD8+ T cells; activation of virus-specific, CD8+ CTLs; and induction of CD8+ T cells expressing IFN-γ (Cousens, Orange, and Biron, 1995; Gessner *et al.*, 1990; Gessner, Moskophidis, and Lehmann-Grube, 1989; Kasaian and Biron, 1989; Kasaian and Biron, 1990; Kasaian, Leite-Morris, and Biron, 1991; Pfizenmaier *et al.*, 1977; Sinicco *et al.*, 1993; Su *et al.*, 1998; Zinkernagel and Hengartner, 1994). CD8+ T cells play major roles in clearing many viral infections, and, although of less importance in certain viral infections, (Kagi and Hengartner, 1996), the perforin-dependent, CD8+ T cell, cytotoxic pathway is a major mediator in control of primary LCMV infections (Kagi *et al.*, 1994; Walsh *et al.*, 1994). During the later periods of the late phase, acute effector responses subside and long-term immunity develops. Activated CD8+ T cell responses are turned off and memory T cells become apparent (Laye *et al.*, 1994; Razvi and Welsh, 1993), B cell responses are elevated, and virus-specific antibody production is detectable (Buchmeier *et al.*, 1980; Huang *et al.*, 1993). A number of nonviral infections have been associated with responses that involve predominantly or exclusively CD4+ T cells, either of the Th1 subset making IL-2 and IFN-γ, or of the T helper type 2 (Th2) subset making interleukin-4 (IL-4) and interleukin-5 (IL-5) (Biron and Gazzinelli, 1995; Heinzel *et al.*, 1989; Seder and Paul, 1994). This dichotomy does not appear to be in place during the majority of well-characterized viral infections. Under the conditions of a variety of viral challenges, IL-4 and IL-10 can be demonstrated at times coinciding with, or overlapping with, IL-2 and IFN-γ expression (Graziosi *et al.*, 1996; Graziosi *et al.*, 1994; Niemialtowski and Rouse, 1992; Sarawar and Doherty, 1994; Su *et al.*, 1998; Wesselingh *et al.*, 1994).

Thus, a number of cytokines are induced during the early and late phases of viral immunity, and most of these cytokines (when injected in purified form) have been demonstrated to influence HPA secretory activity. While the majority of such studies have focused on classic proinflammatory cytokines (IL-1, IL-6, TNF-α), there is substantial evidence that IL-2 and IFN-α also stimulate the HPA axis (Turnbull and Rivier, 1999), but the effects of IL-4 and IFNγ are less clear, and in some cases these may even dampen adrenocortical responses (Harbuz *et al.*, 1992; Vankelecom *et al.*, 1990). Whatever the case, these studies focusing predominantly on exogenous cytokines can serve only as guideposts for exploring the influence of replicating viruses on the HPA axis, because, as mentioned above, the cytokine response to such an infection occurs in waves, and cannot be mimicked by a bolus of an immunologically active molecule such as a purified cytokine or poly (I:C).

To investigate the specificity of endogenous glucocorticoid responses to viral infection, kinetic studies examining endogenous glucocorticoid induction during several experimental murine viral infections have been conducted in our laboratories (Ruzek et al., 1997b). A diagrammatic illustration of published and unpublished results is presented in Fig. 3. Times of known cytokine, NK cell, and T cell responses to infection were chosen for examination. Infections included MCMV, wild-type LCMV clone E350, and the LCMV clone 13b. MCMV is

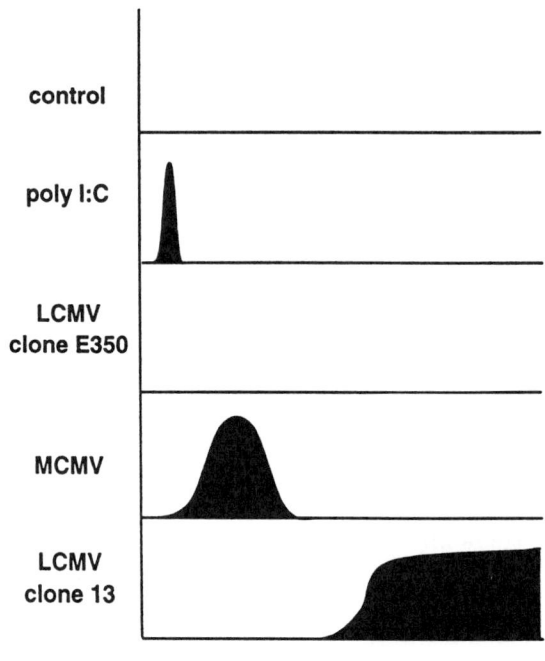

Time After Challenge in Days

Fig 3. Representation of differences in kinetics of endogenously induced glucocorticoid responses following challenges with the chemical analog for viral nucleic acid, poly (I:C), and particular viral infections. Schematic representation of results from studies of mice that are presented in Miller et al., 1997 and Ruzek et al., 1997. Corticosterone responses to poly (I:C) are rapid, peaking within hours. In contrast, some viral infections fail to induce detectable endogenous corticosterone (LCMV clone E350); others (MCMV) induce responses at times of innate immune responses, i.e., 36 hours after infection; and still others (LCMV clone 13) induce responses at times of adaptive immune responses, i.e., after 7 days of infection.

hepatotropic, whereas the liver is not a significant target of LCMV clone E350. In contrast, LCMV clone 13b is associated with tropisms for macrophages and dendritic cells (Borrow, Evans, and Oldstone, 1995; Matloubian *et al.*, 1993), as well with as impaired development of protective T cell responses and generalized immunosuppression with viral persistence (Wu-Hsieh, Howard, and Ahmed, 1988).

We compared morning (A.M.) glucocorticoid (corticosterone) responses following infection with those observed after treatment with the synthetic viral analog, poly(I:C). As previously described (Miller *et al.*, 1997), poly(I:C) induced a peak corticosterone response soon after injection, reaching serum levels of 200–400 ng/ml about 2 hours after intraperitoneal administration (Fig. 3). Infection with LCMV clone E350 elicits little or no glucocorticoid response. In MCMV-infected mice the kinetics of virus-induced corticosterone occurred later and were more protracted than seen with poly (I:C). The MCMV-induced CORT rise was evident on days 2 and 3 after infection. In contrast, studies after intravenous injection with the immunosuppressive LCMV clone 13 demonstrate that clone 13 induces a robust corticosterone response that occurs later in infection, corresponding with a loss of T cell responses and viral persistence. Other studies (data are not shown) indicate that glucocorticoid elevations following LCMV clone 13 infection are still detectable on day 9 postinfection, but resolve by day 11. In unpublished data from our groups, we found glucocorticoid responses similar to LCMV clone 13 with the hepatotrophic LCMV variant, WE, which is not associated with reduced T cell responses. Thus, based on these studies using three different conditions of viral infection (one eliciting a sharp early peak, a second failing to induce, and a third inducing late endogenous glucocorticoid responses), HPA-immune interactions appear to be both virus-specific and phase (of the immune response)-specific.

Relatively little is known concerning the cytokine determinants of HPA axis responses to viral infection. Although there has been considerable progress in defining the mechanism by which cytokines impact the HPA axis in bacterial infections, our investigations, using the models described above, suggest that glucocorticoid induction in viral infections may proceed along qualitatively different pathways involving distinct cytokine interactions (Ruzek *et al.*, 1997b). For example, we have previously demonstrated that, following infection with murine cytomegalovirus, there is induction of early circulating levels of IL-12, IFNγ, and TNF (Orange and Biron, 1996a, 1996b). More recently, we have extended characterization of these responses to include IL-1 and IL-6, and to define the role of

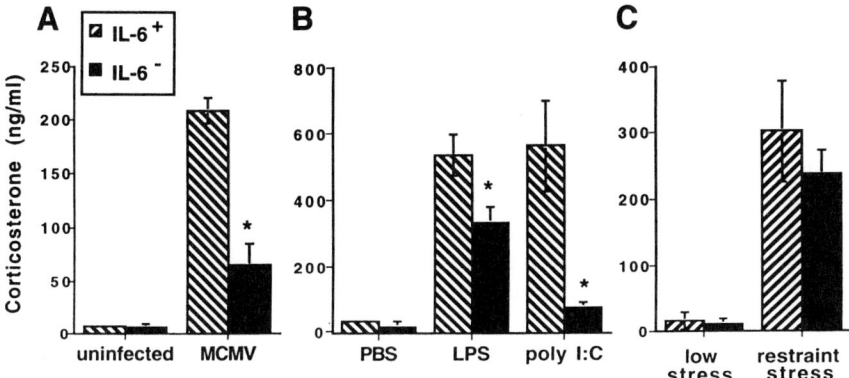

FIG 4. A major and unique role for IL-6 in induction of glucocorticoid responses to viral infections. Serum corticosterone levels were measured at peak times following 36-hour infections with MCMV (A), 2-hour challenges with LPS or poly (I:C) (B), or after 30 minutes of restraint stress (C). Effects of endogenous IL-6 were determined by comparing responses in IL-6-deficient mice to IL-6-replete mice. * p < 0.01. Reproduced from Ruzek *et al.*, *J. Exp. Med.* **185**:1185–92. (1997).

these cytokines in the induction of glucocorticoids in MCMV-infected mice. Our results demonstrate that IL-12 p40, IFNγ, TNF, IL-1α, and IL-6 are increased, but that IL-1β is undetectable, in serum of MCMV-infected mice (Ruzek *et al.*, 1997b). Studies using mice with cytokine deficiencies, or neutralized cytokine function, demonstrate that IL-6 is the pivotal mediator of the glucocorticoid response, with IL-1 (likely IL-1α) contributing to IL-6 production (Ruzek *et al.*, 1997b). The IL-6 requirement appears to be specific for some other virus-type stimuli, such as poly (I:C) (Fig. 4B). Specifically, MCMV and poly (I:C) induce IL-6-dependent glucocorticoid release (Fig. 4A, B), but a physical restraint stressor elicits IL-6-independent corticosterone responses (Ruzek *et al.*, 1997b) (Fig. 4C). The activation of the HPA axis by NDV is IL-1-dependent, although the role of IL-1 induction of IL-6 in NDV has not been clarified (Dunn and Vickers, 1994). In any case, IL-6 is a primary mediator of glucocorticoid induction in MCMV infection, and represents a specific pathway of interaction between the immune and neuroendocrine systems during viral infection.

We have additionally found that IL-2 synergizes with TNF to mediate profound glucocorticoid elevations in mice infected with LCMV E350 and treated with high doses of IL-12 (Orange *et al.*, 1997). IL-2 also has been shown to have direct effects on pituitary corticotrophs

(Kanangat *et al.*, 1996; Smith, Brown, and Blalock, 1989). These findings suggest that IL-2 may contribute to glucocorticoid induction in certain viral infections, especially in those infections in which T cell responses (with associated IL-2 release) predominate.

Besides studies using MCMV and LCMV, there have been only a few other models in which indices of HPA activation and cytokine induction have been measured contemporaneously in response to infection by a replicating virus (Ben Hur *et al.*, 1996; Ben-Hur *et al.*, 1996; Carr, 1998; Dobbs *et al.*, 1996; Espey and Basile, 1999; Trgovcich *et al.*, 1997). In general, these models have not yielded data that would identify an obligatory cytokine for CORT induction, and multiple overlapping mechanisms of HPA axis activation could be involved. For example, influenza infection of mice has been shown to induce a biphasic glucocorticoid response (Hermann *et al.*, 1994). The initial surge of CORT occurs at 2–3 days postinfection, concurrent with the peak of viral titers. In the lungs at this time, there is local production of proinflammatory cytokines that have been suggested to be a stimulus for the CORT response, although this has not been determined definitively (Hennet *et al.*, 1992; Hermann *et al.*, 1994). A second rise in CORT occurs at 7–8 days postinfection, coinciding with peak mononuclear cell infiltration and respiratory distress (Hennet *et al.*, 1992; Hermann *et al.*, 1994). At this time, cells from regional lymphatic tissue are activated, as evidenced by the detection of several virus-stimulated cytokines including IL-2, IL-6, IL-10, and IFN-γ (Dobbs *et al.*, 1996). However, respiratory distress, rather than virus-induced cytokines, could be responsible for this second rise in CORT because the HPA axis is exquisitely sensitive to both physical and psychological stressors, and in a viral infection in which pain or physical distress is elicited, the line of demarcation between these stressors is obscure (McEwen *et al.*, 1997). Thus pain or debilitation can give rise to psychological distress, potentially invoking a different set of chemical or neuronal pathways than would be elicited by a purely immunological stimulus. As discussed further in Section III,C, these pathways are poorly understood, and even nonimmune stressors can induce cytokine expression in the brain (Minami *et al.*, 1991; Shizuya *et al.*, 1997; Suzuki *et al.*, 1997; Tagoh *et al.*, 1995). The pathophysiological picture becomes more complex for viruses infecting the nervous system because in this case the region-specific expression of cytokines in the brain is a function of viral neurotropism, as well as of the character of the immune response and the ensuing physical and psychological distress (Ben-Hur *et al.*, 1996; Carr, 1998; Trgovcich *et al.*, 1997).

In many respects the HPA axis is self-regulating—e.g., the dampening of proinflammatory cytokines by CORT and feedback inhibition at multiple levels of the axis (Plotsky, 1991). Nevertheless, there is evidence that in chronic viral diseases such as AIDS, there is a long-term disruption of the harmonious balance between immune mediators and components of the HPA axis (Norbiato *et al.*, 1997). Few models have been established to examine the pertinent interactions between cytokines and the HPA axis in chronic viral diseases. Recently, Epsey and Basile reported that chronic infection of mice with the LP-BM retrovirus mixture resulted in latent activation of the HPA axis, as evidenced by increased ACTH and corticosterone at 14 to 16 weeks after the infection (Espey and Basile, 1999). This activation was not causally linked to increases in IL-1B, IL-6, or TNFα, but corresponded temporally with a shift from a TH-1 cytokine profile dominated by IFNγ, to a TH2-type pattern with increased levels of IL-4 and IL-10. Nevertheless, this shift was not believed to be responsible for the virus-associated elevations in ACTH and CORT, because attempts to mimic these elevations, by injecting IL-4 or IL-10, did not bring about an HPA axis response (Espey and Basile, 1999).

Thus while the ability of specific virus-induced cytokines to modulate the HPA axis has just recently begun to be revealed, the signaling pathways through which these cytokine messages are conveyed at various levels of the HPA axis have been examined almost exclusively in nonviral models. Furthermore, as mentioned above, the tissue tropism or site infection can play a role in the ultimate impact of a viral infection on the HPA axis. Thus in the next section we examine the consequences of a direct viral infection of the glands and tissues that comprise the HPA axis.

B. Infection of Endocrine Tissue

Perhaps the most straightforward mechanism by which viruses can influence glucocorticoids is by infection of the various endocrine tissues comprising the HPA axis. Numerous pathogens are capable of infecting the adrenal glands (Lack and Kozakewich, 1990). The herpesviruses are well represented among this group—indeed, experimental and clinical studies have demonstrated adrenal infections by several herpesviruses including cytomegalovirus (CMV), herpes simplex, varicella zoster, and Epstein-Barr virus (Lack and Kozakewich, 1990). The clinical importance of adrenal infections is illustrated by the high prevalence of CMV adrenalitis in patients with AIDS (Glasgow *et al.*, 1985; Rotterdam and Dembitzer, 1993). The adrenal medulla appears to be particularly susceptible, although the cortex is also frequently involved, and CMV-induced cortisol insufficiency has been described (Glasgow *et al.*, 1985; Rotterdam and Dembitzer, 1993).

Indeed, the role of cortisol abnormalities in the pathogenesis of HIV is an area of intense interest and debate, as will be discussed in more detail below. The contribution of CMV adrenalitis to cortisol abnormalities in AIDS has not been definitively determined, nor has there been a thorough investigation of the mechanism by which CMV can cause adrenal insufficiency. However, CMV clearly induces necrolytic and thrombotic injury to the adrenals, and in some cases the adrenal glands are the only organ containing detectable CMV (Rotterdam and Dembitzer, 1993).

Interestingly, adrenalitis has also been demonstrated following experimental infection of mice with MCMV (Price *et al.*, 1996). Price *et al.* reported that despite infected cells and inflammatory infiltrates in the adrenal cortex, the capacity of the infected adrenals to produce corticosterone, in response to exogenous ACTH, remained relatively intact (Price *et al.*, 1996). Although these data are consistent with the idea that there is a large reserve capacity of the adrenal cortex that enables the host to resist adrenal insufficiency from destructive lesions (Lack and Kozakewich, 1990), the finding of inflammatory infiltrates in MCMV-infected adrenal cortices raises the additional possibility that soluble inflammatory mediators are poised to exert paracrine modulation of corticosterone production during the infection. Along these lines, the study by Price *et al.* does not elucidate the mechanism by which MCMV infection causes the transient increase in plasma corticosterone within 2 days of infection (as described in the previous section), although Price *et al.* found peak adrenal inflammation after the corticosterone surge, suggesting that local inflammatory secretagogues were not entirely responsible for the rise (Price *et al.*, 1996). Thus the rise in CORT may be mediated at higher levels of the HPA axis, which is consistent with our recent study in which we found elevations in ACTH occurring within 28 hours of MCMV infection (Ruzek *et al.*, 1997b).

The ability of a virus to alter ACTH secretion by directly infecting the corticotroph cells in the pituitary gland has not been well studied, although the anterior lobe of the pituitary is known to support infection by several viruses including CMV in patients with AIDS (Ferreiro and Vinters, 1988; Mosca *et al.*, 1992). Besides ACTH, the anterior lobe of the pituitary produces several hormones including the macrophage migration inhibitory factor (a putative antiglucocorticoid) (Bucala, 1994), and growth hormone (Horvath and Kovacs, 1988). In a series of studies, Oldstone and colleagues investigated the pathophysiological consequences of pituitary infection in growth hormone production by using a model in which mice were infected with various strains of

LCMV (Oldstone *et al.,* 1985; Rodriguez *et al.,* 1983). Although the virus was found to target cells primarily producing growth hormone (as opposed to ACTH), this model illustrates two fundamental concepts of virus-associated endocrinopathy. First, viral tropism for specific endocrine cell types (and consequential disease expression) is dictated by both the strain of the virus and the host genotype. This finding predicts that primary pituitary dysfunction might be observed in only a small subset of virus-infected individuals, depending on the specific virus strain and host genome. Second, LCMV was shown to impair hormone production without causing evident cell injury (Oldstone *et al.,* 1985; Rodriguez *et al.,* 1983). This is potentially relevant to opportunistic viral infections in immunocompromised hosts because hormone production could be disrupted by direct glandular infection, and yet the pathological hallmarks of the virus would be absent.

The HPA axis is also vulnerable to functional perturbation by direct viral infection of the hypothalamus and interconnected brain regions. For example, hypothalamic infection of rats with a neurovirulent strain of HSV results in the loss of CRF content, despite increased plasma ACTH (Ben-Hur *et al.,* 1996). Interpreting such studies is difficult, however, because any proposed mechanism by which a CNS infection influences the HPA axis must be predicated on the incomplete and rudimentary understanding of neuronal circuits governing the axis under normal circumstances. Fortunately, viral infections of the brain parenchyma are relatively rare in humans, although the brain is nonetheless an important participant in glucocorticoid responses to most infectious agents in the periphery. Cytokines link the immune system to glucocorticoid responses, and in the following section we review the potential nervous system pathways involved, relying on evidence from nonviral studies, and on the scant relevant data available from studies investigating viral infections.

C. Translation of Immune Signals into Neuroendocrine Responses

Many of the cytokines known to be produced during viral infection are capable of influencing the HPA axis at various levels (Fig. 5). However, the first step in deciphering the actual pathways by which viral infections influence the HPA axis is to determine which viral cytokines are obligatory for inducing HPA hormones in the context of an *in vivo* infection. In this regard, the role of IL-6 and IL-1 in causing increased glucocorticoid release following viral (MCMV) infection has been discussed. Nonetheless, the mechanisms and sites of action through which these

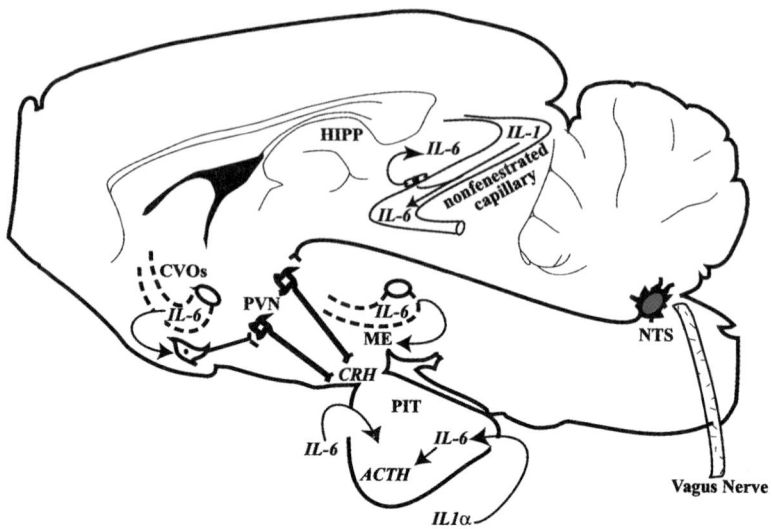

Fig 5. Schematic of hypothesized pathways for translation of IL-1α and IL-6 into neuroendocrine signals. During MCMV infection, IL-6 produced in the liver (or other abdominal viscera) may stimulate vagal afferents, thereby activating central neurons in the nucleus of the solitary tract (NTS) which in turn send catecholaminergic projections to the paraventricular nucleus of the hypothalamus. Alternatively, circulating IL-1α or IL-6 (which are confined to the vascular lumen of nonfenestrated capillaries) may act indirectly by recruiting the brain's cytokine network. In such a scenario, IL-1α could induce production of IL-6 from capillary endothelium, microvascular pericytes, or perivascular glia. Once in the brain parenchyma, IL-6 might act directly on HPA axis regulatory neurons, or, more likely, modulate HPA axis activity through intermediates such as prostaglandins. In addition, IL-6 could enter the brain parenchyma passively by diffusing through the fenestrated capillaries of circumventricular organs (CVOs). Capillaries of the median eminence (ME) are also fenestrated, allowing IL-6 to travel from the vascular lumen to nerve terminals of the PVN and consequently placing this cytokine in position to mediate CRF release. The pituitary gland is intimately associated with the brain and represents another site where MCMV-induced cytokines could mediate ACTH release. Because the anterior pituitary is outside the blood–brain barrier, IL-1α and IL-6 presumably have direct access to pituitary corticotrophs. The foregoing mechanisms are not mutually exclusive. For example, CRF could play a permissive role for the action of IL-6 at the pituitary.

and other virus-induced cytokines activate the HPA axis have not been determined. Several potential mechanisms can be considered

1. Induction of CRF Secretion

Based on a series of studies examining the mechanisms by which proinflammatory cytokines such as IL-1 and IL-6 lead to HPA axis activation, a major final common pathway involving cytokine induction of CRF in the paraventricular nucleus (PVN) has been described (Berkenbosch et al., 1987; Besedovsky and del Rey, 1996; Kovacs and Elenkov, 1995; Matta, Weatherbee, and Sharp, 1992; Naitoh et al., 1988; Rivier, 1995; Schmidt et al., 1995; Spinedi et al., 1992; Suda et al., 1990). The PVN is a major regulatory site for the HPA axis, and CRF released from this brain region into the median eminence and portal venous circulation is responsible for the release of ACTH from the pituitary (Plotsky, 1991). The ability of both lipopolysaccharide (LPS) and proinflammatory cytokines to stimulate CRF secretory neurons in the PVN has been documented by the induction of the early immediate gene, c-fos (a neuronal activation marker), as well as of CRF mRNA and protein (Ericsson, Kovacs, and Sawchenko, 1994; Rivier, 1995). Nevertheless, cytokine regulation of glucocorticoid secretion is not exclusively governed by CRF, and current evidence suggests that multiple alternate pathways may be operating concurrently or sequentially during the unfolding of the immune response. For example, cytokines elaborated during viral infection could contribute to rises in glucocorticoids by affecting steroidogenesis in the adrenal cortex or by stimulating pituitary release of ACTH in a CRF-independent manner (Besedovsky and del Rey, 1996; Turnbull and Rivier, 1999).

The increase in plasma ACTH, following MCMV infection, points to the role of peptide secretagogues in the hypothalamus or pituitary in mediating the IL-6 dependent surge of glucocorticoids that peaks at about 36 hours after infection (Besedovsky and del Rey, 1996; Callahan and Piekut, 1997; Kovacs and Elenkov, 1995; Matta, Weatherbee, and Sharp, 1992; Naitoh et al., 1988; Perlstein et al., 1993; Ruzek et al., 1997b; Spinedi et al., 1992). To further investigate the importance of CRF in mediating ACTH and corticosterone responses during viral infection, the effects of MCMV infection on HPA axis activity was recently examined in CRF-deficient mice (Pearce et al., 1999). ACTH levels were found to be no different in infected-versus-uninfected CRF-deficient mice, while infected wild-type mice had the expected increase in ACTH. These data suggest that MCMV may drive the HPA axis through CRF-dependent release of ACTH. However, further analysis indicated that this pathway may not be absolutely required for corticosterone induction because MCMV infection induced robust corticosterone increases in both CRF-deficient and wild-type mice (Pearce et al., 1999). Taken together, these data suggest that although the classic CRF-ACTH-CORT pathway may be activated during MCMV infection, glucocorticoid responses to MCMV are not strictly dependent on CRF. Thus, in the absence of CRF, other mechanisms of glucocorticoid stimulation can be revealed.

2. Paracrine Actions in Adrenal Gland

Given the importance of IL-6 in mediating MCMV-induced CORT, the various pathways by which IL-6 could facilitate adrenal corticosteroid production deserve additional consideration. Several studies have demonstrated that IL-6 can stimulate CORT production from adrenocortical cells (Path *et al.*, 1997; Salas *et al.*, 1990; Weber *et al.*, 1997). In MCMV infection there is ample opportunity for IL-6 to act as a direct secretagogue of CORT because circulating IL-6 is elevated substantially between 30 and 44 hours postinfection. Furthermore, the adrenal gland itself is a potential source of IL-6 in the infection (Ruzek *et al.*, 1997b). The local production of IL-6 by the outer layer of the adrenal cortex has been well studied (Gonzalez-Hernandez *et al.*, 1994; Judd, 1998; Judd and MacLeod, 1992), and because the release of IL-6 from the adrenal cells is potentiated by ACTH (Judd and MacLeod, 1992), a local cytokine–hormonal paracrine network may be operating in the adrenal. The capacity of other cytokines to regulate IL-6 production from the adrenal cells has not been well defined, but there is ample evidence to indicate that IL-1 can induce IL-6 production from numerous cell types (Turnbull and Rivier, 1999). Thus, IL-1-mediated production of adrenal IL-6 is one mechanism by which MCMV could be stimulating CORT in a CRF-independent manner.

3. Pathways by Which Cytokine Messages Are Conveyed

These and other studies provide emerging evidence that virus-induced cytokines can increase glucocorticoid secretion by regulating the HPA axis at several levels (Fig. 5). The underlying mechanisms by which cytokine messages are conveyed to HPA endocrine glands are distinct at each of the HPA axis levels, and are dependent on anatomical barriers and factors regulating the availability of free cytokines. For example, elevated levels of ACTH in MCMV and other viral infections point to the importance of the pituitary, and of brain regions governing CRF secretion, in making the connection between a viral infection and the glucocorticoid response. Because IL-6 and other proinflammatory cytokines do not freely cross the blood–brain barrier (BBB) under usual circumstances (i.e., in the absence of CNS infection), consideration must be given to how peripheral immune signals are transmitted to the brain to influence relevant CRF neurons (Schobitz, De Kloet, and Holsboer, 1994; Turnbull and Rivier, 1999; Watkins, Maier, and Goehler, 1995). There are data to support each of several proposed mechanisms, which are not mutually exclusive. The dominant mechanism that the host uses to restore homeostasis may depend on the type of pathogen, kinetics of the infection, and tissue compartments affected. For example, studies using predominantely a bolus of LPS or IL-1 have suggested that local (peripheral) production of proinflammatory cytokines can stimulate visceral afferents that in turn communicate with the brain through the vagus, ultimately modulating CRF secretion through interconnected neuronal circuits involving ascending catecholaminergic fibers of the nucleus of the solitary tract (NTS) that project to the parvocellular division of the PVN (Ericsson, Kovacs, and Sawchenko, 1994; Fleshner *et al.*, 1995; Kapcala *et al.*, 1996; Watkins, Maier, and Goehler, 1995). Alternatively, cytokines present in the circulation can communicate with the brain through intermediates without themselves entering the CNS parenchyma—for example, by acting on cells of the brain endothelium or choroid plexus and inducing the release of secondary messengers including prostaglandins and nitric oxide (Schobitz, De Kloet, and Holsboer, 1994; Turnbull and Rivier, 1999). In addition, it has been suggested that circulating cytokines may cross the BBB in leaky regions such as the organum vasculosum of the lamina terminalis (OVLT) and directly activate neuroendocrine pathways (Schobitz, De Kloet, and Holsboer, 1994; Turnbull and Rivier, 1999; Watkins, Maier, and Goehler, 1995). Along similar lines, the CRF-laden median eminence lies outside the BBB, and recent studies indicate that IL-6 can facilitate the release of CRF from nerve terminals located in this brain region (Matta, Weatherbee, and Sharp, 1992; Spinedi *et al.*, 1992). In addition, small quantities of proinflammatory cytokines may reach neuroendocrine regulatory circuits because of active transport mechanisms carrying these proteins across the BBB (Banks, Kastin, and Broadwell, 1995). Viral infections could disrupt the BBB and upregulate these transport pumps, thereby facilitating the communication of peripheral immune signals with CNS targets. Nonetheless, cytokines leaving the periphery and entering the brain do not become functionally disengaged from other cytokines or immune mediators. Indeed, cytokines are known to operate within a highly interconnected network, and thus a virus-induced increase in one cytokine can influence the production of another (Schobitz, De Kloet, and Holsboer, 1994) (Turnbull and Rivier, 1999). As discussed in the next section, there is considerable evidence that the brain is an intimate participant in this network.

4. CNS Cytokine Network

Although there is ample evidence for a cytokine network within the CNS, many of the details of its operation remain to be defined. Receptors for proinflammatory cytokines are located in numerous brain regions, including areas involved in HPA axis regulation, such as the hypothalamus and the hippocampus (Besedovsky and del Rey, 1996; Schobitz, De Kloet, and Holsboer, 1994). Furthermore, there is mounting evidence that several cytokines derived from neural cells may play a role as intermediaries in communicating peripheral inflammatory signals to the brain. For instance, peripheral injections of LPS (or circulating cytokines such as IL-1) have been shown to induce neural cells within the hypothalamus and other brain regions to produce proinflammatory cytokines such as IL-1, IL-6, and TNF (Gatti and Bartfai, 1993; Laye et al., 1994; Quan, Sundar, and Weiss, 1994; Spangelo et al., 1990; van Dam et al., 1992). Furthermore, recent studies suggest that such autoinduction of cytokines can be transmitted indirectly through the vagal pathways discussed above (Laye et al., 1995; Watkins, Maier, and Goehler, 1995). Thus, during an infection with a virus such as MCMV, peripherally released cytokines could either cross the BBB (e.g., at leaky sites or through active transport mechanisms) or activate peripheral afferents, and consequently recruit local neural cells to produce other cytokines such as IL-1, II-6 or TNF, thereby amplifying cytokine signals in brain areas that control the HPA axis. Among various CNS cell types, activated glia are rich sources of proinflammatory cytokines, and viruses are potent glial activators (Schobitz, De Kloet, and Holsboer, 1994). Moreover, glia are responsive to peripheral immune signals. For example, peripheral injection of LPS causes an overt induction of immunoreactivity for IL-1 in glia cells (van Dam et al., 1992). Because receptors for IL-1 and IL-6 are found at high density in some brain regions that participate in HPA axis regulation, the autoinduction of proinflammatory cytokines may be an important intermediate step in transducing or amplifying the effects of peripheral cytokines in the brain (Rivier, 1995; Schobitz, De Kloet, and Holsboer, 1994). The implications of this for relatively common viral infections could be far-reaching, yet most studies examining cytokine production from glia in response to viruses have focused on infections occurring with the CNS, which are more rare, and the impact of peripheral viral infections on neural cell cytokine responses has been relatively neglected. Because of the extensive cross-talk between many of the cytokines that have been implicated in HPA axis regulation, a number of cytokine pathways could be acting concurrently. In the context of a viral infection, such a mechanism implies, for example, that blood-borne IL-1 could induce glial production of IL-6, which would then act secondarily to stimulate the HPA axis. This would result in an increase of ACTH, while IL-1-dependent induction of IL-6 in adrenal glands might sensitize adrenal cortical cells to ACTH.

Taken together, the available data examining the stimulation of HPA axis secretory activity by cytokines during viral infections indicate that several immune–endocrine pathways can be invoked, and that elucidation of the pathways most relevant to viral pathogenesis will require a more concrete understanding of the pattern by which cytokines are elaborated in response to infection with replicating viruses. Furthermore, because adrenal steroids are themselves potent regulators of cytokine induction and activity, the outcome of infected hosts is dependent on the availability of CORT in various immune compartments, and consequently is a function of the tissue-specific effects of glucocorticoid receptor activation. Thus in the following sections we explore how neuroendocrine signals are received by target immune tissues, and examine the impact of the neuroendocrine system on specific immune responses to viral infections.

IV. IMPACT OF GLUCOCORTICOIDS ON TARGET TISSUES

A. Factors Regulating Glucocorticoid Availability

Once glucocorticoids have been induced, the relative impact of these hormones on a given target tissue is dependent on the access of free hormone to the glucocorticoid receptor (GR), and on the relative expression and function of the GR (McEwen *et al.*, 1997; Miller *et al.*, 1997). Both of these factors are potential sites where viral infections could interact with the neuroendocrine system and modulate adrenal steroid–immune interactions. The majority of circulating glucocorticoids are bound to corticosteroid binding globulin (CBG), yet only free or unbound glucocorticoids are capable of diffusing across the plasma membrane and activating the GR (Rosner, 1991; Siiteri *et al.*, 1982). Therefore, if viral infections or host immune responses alter the relative concentration of CBG, the "free" or available glucocorticoid would likely be changed accordingly, and the effectiveness of host responses would ultimately be linked to the ensuing fluctuations in GR activation at target immune tissues. In support of such a mechanism, previous studies have indicated that decreases in the circulating level of CBG are associated with evidence of increased occupation/activation of GR in the spleen of stressed rats (Spencer *et al.*, 1996). To our knowledge, only two studies have examined plasma CBG during viral infection, and no differences in plasma CBG were found in infected subjects in either case (Lambert, 1994; Miller *et al.*, 1997). However, local changes (e.g., within the microenvironment of the various lymphoid compartments) in the concentration and/or affinity of CBG during viral infection remain a consideration. Because CBG has a primary structure with marked homology to the family of serine protease inhibitors (SERPINs), it has been suggested that the local release of proteases, such as elastase, by activated neutrophils (e.g., during infection) may cleave CBG, thereby reducing its affinity for its ligand and leading to a local increase in free glucocorticoid hormone (Hammond *et al.*, 1990; Rosner, 1991).

Because the GR is an intracellular receptor, glucocorticoid ligands must pass through the cell membrane in active form before interacting with the GR in the cytoplasm (Gustafsson *et al.*, 1987). As a result, this process invokes another level at which the access of glucocorticoids can be regulated in immune tissues. Specifically, as CORT enters the cell, it can be degraded into inactive metabolites by 11-hydroxysteroid dehydrogenase (HSD) (Monder and Shackleton, 1984). Interestingly, significant differences in 11β-HSD activity have been found among immune compartments, and there is a direct correlation between 11β-HSD activity and the preferential production of Th1- versus Th2-type cytokines by T cells residing in a given tissue (Hennebold *et al.*, 1996). Inhibition of 11β-HSD activity (which would enhance the available amount of hormone to the GR) leads to a shift in the cytokine profile of activated T cells; the Th2 cytokines becoming favored over the Th1 type (Hennebold *et al.*, 1996). In addition, infection and destruction of hepatocytes that are responsible for CORT metabolism may lead to increases in circulating CORT, irrespective of HPA axis activity. Therefore, during viral infections the impact of endogenous glucocorticoids is influenced in immune as well as nonimmune tissues by the level of CBG as well as by the relative activity of 11β-HSD and other CORT metabolizing enzymes. Because HPA axis activity is dependent on CORT (through feedback inhibition), factors influencing glucocorticoid availability should be regarded as participants in HPA axis regulation (McEwen *et al.*, 1997; Plotsky, 1991).

B. Glucocorticoid Receptors in Target Tissues

The majority of the effects of glucocorticoids on immune cells are mediated through the intracellular GR, and therefore, the ultimate effect of glucocorticoids on the immune system must be established at the receptor level (Gustafsson *et al.*, 1987). There is considerable heterogeneity among immune cells and tissues in the amount of GR expressed as well as differences among immune tissues in the amount of receptor activation following a given hormone exposure. The spleen, for example, which is an important site of the immune response to infectious challenge, has been found to be relatively resistant to effects of endogenous glucocorticoids both at the level of the GR as well as the level of glucocorticoid effects on immune function (Miller *et al.*, 1992;

Miller *et al.*, 1990; Spencer *et al.*, 1993). In contrast, the thymus is exquisitely sensitive to glucocorticoids, a finding which is consistent with the high level of thymic GR expression, along with evidence of significant receptor activation in the thymus following stress-induced increases in glucocorticoids.

An important possible mechanism for GR changes during viral infection includes the potential effects of cytokines on GR number and/or function. There is a large body of data on the impact of cytokines on the GR, with over 20 studies documenting significant effects (Betancur, Borrell, and Guaza, 1995; Costas *et al.*, 1996; Hill, Stith, and McCallum, 1987; Hill, Stith, and McCallum, 1988; Hoshino *et al.*, 1994; Kam *et al.*, 1993; Krasznai *et al.*, 1986; Liu *et al.*, 1993; Mulkins and Allison, 1987; Nimmagadda *et al.*, 1997; Norbiato *et al.*, 1996; Rakasz *et al.*, 1993; Salkowski and Vogel, 1992a; Salkowski and Vogel, 1992b; Schobitz *et al.*, 1994; Sher *et al.*, 1994; Sica *et al.*, 1990; Spahn *et al.*, 1996; Stith and McCallum, 1983; Stith, McCallum, and Hill, 1989; Verheggen *et al.*, 1996). Interestingly, however, the results are divided into those studies finding an increased GR number following cytokine administration (Hoshino *et al.*, 1994; Kam *et al.*, 1993; Krasznai *et al.*, 1986; Nimmagadda *et al.*, 1997; Norbiato *et al.*, 1996; Rakasz *et al.*, 1993; Salkowski and Vogel, 1992a, 1992b; Sher *et al.*, 1994; Sica *et al.*, 1990; Spahn *et al.*, 1996; Verheggen *et al.*, 1996) and those finding a decreased GR number (Betancur, Borrell, and Guaza, 1995; Costas *et al.*, 1996; Hill, Stith, and McCallum, 1987; Hill, Stith, and McCallum, 1988; Liu *et al.*, 1993; Mulkins and Allison, 1987; Schobitz *et al.*, 1994; Stith and McCallum, 1983; Stith, McCallum, and Hill, 1989). These results appear to be dependent on how the GR was measured. For example, 8 out of 12 studies using whole-cell assay binding techniques found GR number to be increased (4 found no change) following treatment with a variety of cytokines including IL-1, IL-2, and IL-4 (Hoshino *et al.*, 1994; Kam *et al.*, 1993; Krasznai *et al.*, 1986; Nimmagadda *et al.*, 1997; Norbiato *et al.*, 1996; Rakasz *et al.*, 1993; Salkowski and Vogel, 1992a, 1992b; Sher *et al.*, 1994; Sica *et al.*, 1990; Spahn *et al.*, 1996; Verheggen *et al.*, 1996), whereas the majority of studies (6 out of 9) using cytosolic receptor binding techniques have found the GR to be decreased (2 found no change) following treatment with the similar group of cytokines (Betancur, Borrell, and Guaza, 1995; Costas *et al.*, 1996; Hill, Stith, and McCallum, 1987; Hill, Stith, and McCallum, 1988; Liu *et al.*, 1993; Mulkins and Allison, 1987; Schobitz *et al.*, 1994; Stith and McCallum, 1983; Stith, McCallum, and Hill, 1989). Results are further complicated by the fact that studies using whole-cell binding techniques tended to utilize longer incubation

494   BRAD D. PEARCE ET AL.

times (e.g., 24 hours), whereas studies using cytosolic receptor binding tended to examine shorter incubation periods (e.g., 4–6 hours). Of interest, however, is that many of the studies examining the functional consequences of cytokines (especially, proinflammatory cytokines and cytokines that mediate lymphocyte growth and differentiation—e.g., IL-2 and IL-4) on the GR have found inhibition of GR function (Miller, Pariante, and Pearce, 1999). In fact, the inhibitory impact of IL-2 and IL-4 on GR function is hypothesized to play a role in steroid-resistant asthma (Kam et al., 1993; Sher et al., 1994). Although IL-2 and IL-4 increase GR expression in steroid-resistant asthmatic patients, it should be noted that steroid resistance is associated with a reduced affinity of the GR for ligand (increased K_d). To examine potential mechanisms involved in the inhibitory effects of cytokines on GR function, a series of studies has been conducted on the effects of cytokines on the GR, through the combined application of quantitative immunocytochemistry and measurements of GR-mediated gene transcription using cells stably transfected with a chloramphenicol acetyltransferase (CAT) reporter gene construct downstream of several glucocorticoid response elements (LMCAT) (Pariante et al., 1996). Initial characterizations of cytokine–GR interactions were with IL-1 (α and β), because much of the focus on neuroendocrine–immune interactions is based on the neuroendocrine effects of these cytokines. Although IL-1α was found to increase GR protein expression in the cytoplasm, when coincubated with dexamethasone this cytokine inhibited both GR translocation from cytoplasm to nucleus and GR-mediated gene transcription. Receptor upregulation may therefore represent an attempt to overcome (or adapt to) impaired GR translocation and function. Similar results were seen with IL-1β.

Interestingly, these results might explain the mechanism underlying glucocorticoid resistance (dexamethasone nonsuppression) found in animals chronically treated with LPSs (Yirmiya, 1996). Moreover, cytokine effects on the GR may be involved in the glucocorticoid resistance described in clinical disorders of immune activation, including asthma, ulcerative colitis, rheumatoid arthritis, and HIV infection.

V. ROLE OF HPA IN SHAPING IMMUNE RESPONSE
DURING VIRAL INFECTION

There is a plethora of data demonstrating the effects of glucocorticoids on immune responses (Besedovsky and del Rey, 1996; McEwen et al., 1997; Miller, 1998; Munck and Guyr, 1991). Of particular relevance to viral infections is the ability of CORT to shape the phenotype of protective immunity as well as to contain the magnitude and/or kinetics of detrimental immune responses. As described in Section III, the immune response to a viral infection moves through different phases, each of which may be uniquely modulated by the prevailing level and activity of CORT.

The early phase of the response to viral infection primarily involves activation of innate immunity and promotion of downstream-specific (adaptive) responses (Ahmed and Biron, 1998). A dampening effect of CORT on these early events is suggested by studies showing that glucocorticoids can directly inhibit NK cell activity (Nair and Schwartz, 1984). But much of the influence glucocorticoids exert during the early phase might be associated with effects of these hormones on immune cell trafficking (Dhabhar *et al.*, 1995; Dhabhar *et al.*, 1996; Dhabhar *et al.*, 1994; Dobbs *et al.*, 1996; Miller *et al.*, 1994). Modification of cell migration is a necessary prerequisite for redistribution of mononuclear cells from peripheral sites to lymphoid compartments during early phases of the response to infection, and the delivery of activated immune effector cells to sites of infection both early and late. Although CORT has been shown to mediate some of the inhibitory effects of stress on lymph node cellularity during viral infections (see below) (Dobbs *et al.*, 1996; Hermann, Beck, and Sheridan, 1995), this endogenous glucocorticoid may also enhance the virus-induced, IFN-dependent redistribution of lymphocytes from the peripheral circulation to lymph nodes and splenic white pulp, allowing for increased immune activation (Gresser *et al.*, 1981; Ishikawa and Biron, 1993; Korngold, Blank, and Murasko, 1983; Turney, Harmsen, and Jarpe, 1986). Additionally, CORT may augment inflammation at sites of viral infection, as suggested (albeit indirectly) by a study which found that CORT could promote local inflammatory responses during site-specific antigen challenges by enhancing lymphocyte trafficking to the tissue (Dhabhar and McEwen, 1996).

Besides their action on the immune system, glucocorticoids may influence early phases of a viral infection by directly stimulating the replication of certain viruses including HIV-1, Epstein-Barr virus, herpes simplex virus, hepatitis B virus, and mouse mammary tumor virus (MMTV) (Chou *et al.*, 1992; Glaser *et al.*, 1995; Sawiris, Sydiskis, and Bashirelahi, 1994; Soudeyns *et al.*, 1993; Tanaka *et al.*, 1984; Varmus, Ringold, and Yamamoto, 1979). In this regard, the long-term repeat region of MMTV contains several glucocorticoid response elements (where activated GRs bind to DNA and regulate gene expression), as does the HIV genome (Soudeyns *et al.*, 1993; Varmus, Ringold, and

Yamamoto, 1979). Furthermore, recent findings indicate that viral proteins may directly interact with the GR to influence GR function. In particular, an accessory gene product of HIV-1, HIV-1 viral protein R (vpr), has been found to inhibit mitogen-induced lymphocyte proliferation in peripheral blood mononuclear cells taken from healthy volunteers. This inhibitory effect is reversed by glucocorticoid antagonist RU38486, suggesting that vpr may directly activate the GR to mediate its effects (Refaeli, Levy, and Weiner, 1995).

Glucocorticoids also affect the late phases of the immune response to a virus. For example, glucocorticoids can influence T cell apoptosis and inhibit cytokines that drive T cell activation. While the contribution of glucocorticoids (Gruber et al., 1994; Jondal, Okret, and McConkey, 1993; Morrissey et al., 1988; Wyllie, 1980; Zacharchuk et al., 1990; Zubiaga, Munoz, and Huber, 1992) to thymocyte apoptosis in vivo has been conclusively demonstrated (Gruber et al., 1994; Morale et al., 1995; Morrissey et al., 1988; Zacharchuk et al., 1990), the role of these hormones in peripheral T cell death is more controversial and appears to be dependent on the activation state of the immune cells (Gonzalo et al., 1993; Iwata, Hanaoka, and Sato, 1991; Zacharchuk et al., 1990; Zubiaga, Munoz, and Huber, 1992). Some studies suggest that endogenous glucocorticoids are required for early peripheral T cell depletion in response to profound immunostimulation in vivo (Gonzalo et al., 1993), while other studies imply that activation-induced and glucocorticoid-mediated death pathways can be mutually antagonistic (Iwata, Hanaoka, and Sato, 1991; Zacharchuk et al., 1990). As an example of the latter, glucocorticoids can rescue CD4+ T cells from activation-induced cell death triggered by HIV-1 infection (Lu et al., 1995). Moreover, as glucocorticoids can negatively regulate cytokines that promote activation-induced death, these hormones may indirectly inhibit apoptosis.

Glucocorticoids can act on cytokines to influence all phases of the immune response to a virus. Several cytokine genes have glucocorticoid response elements allowing for direct glucocorticoid regulation of cytokine induction (Almawi et al., 1990). Glucocorticoids are known to inhibit cytokine production and function through interaction of the activated GR with transcription factors (e.g., AP-1, NF-κB), which in turn regulate cytokine gene expression and/or the expression of cytokine-inducible genes (Almawi et al., 1990; Auphan et al., 1995; Kelso and Munck, 1984; Scheinman et al., 1995; Vacca et al., 1992). Glucocorticoid-mediated inhibition of proinflammatory cytokines (including TNF, IL-1, and IL-6), and other mediators of inflammation (including the products of the arachidonic acid pathway) early in infection, may have protective effects against potentially detrimental inflammatory responses (Schleimer, Claman, and Oronsky, 1989). Neutralization of endogenous glucocorticoid function results in enhanced pathology and mortality in animals exposed to endotoxin (e.g., LPS) and autoimmune-inducing stimuli (e.g., streptococcal cell wall antigen or myelin basic protein) (Bertini, Bianchi, and Ghezzi, 1988; Sternberg *et al.*, 1989). A similar enhanced pathology has been found to occur in the absence of glucocorticoids following viral infections, including significant mortality in mice that have been rendered glucocorticoid deficient (via adrenalectomy) prior to infection with MCMV (Price *et al.*, 1996; Ruzek *et al.*, 1997a). We recently conducted an in-depth analysis to elucidate the physiological importance of the corticosterone surge in mice infected with MCMV (Ruzek *et al.*, 1999). Mice were sham-operated or adrenalectomized to examine the biological consequences of the endogenous glucocorticoid response to MCMV infection. Removal of glucocorticoids resulted in sustained increases in serum levels of IFNγ and TNFα (Fig. 6). In addition, mRNA levels of these and other cytokines were elevated in the spleen, as determined by RNase protection assay (Fig. 7). Removal of endogenous glucocorticoids also resulted in a marked susceptibility of infected mice to MCMV-induced disease, such that the majority (80%) of mice died within 150 hours of infection (Fig. 6). Cytokine-mediated death was reversed by replicating "surge" levels of corticosterone in adrenalectomized infected mice. Studies of cytokine-deficient or cytokine-neutralized mice demonstrated that TNF α was the primary mediator of this lethality (Fig. 6). Interestingly, the log viral titer in the liver of adrenalectomized MCMV-infected mice was significantly lower than that of sham-operated, infected, control mice, indicating that increases in relevant cytokines were mediating stronger antiviral immunity. These results demonstrate that endogenous glucocorticoids maintain a delicate balance between the antiviral function of cytokines and virus-induced cytokine disease.

As more is learned about the multifarious and subtle effects of endogenous glucocorticoids on immune function, a picture emerges to suggest that glucocorticoids may help orchestrate the host response to infection—for example, by turning off immune responses that are no longer needed and by selectively promoting activation of other immune-response components. Cytokines appear to be crucial intermediates in this process. For example, glucocorticoids have been reported to inhibit IFNγ and IL-2 expression, while enhancing IL-4 (Daynes and Araneo, 1989; Daynes *et al.*, 1990; Kunicka *et al.*, 1993). This glucocorticoid-mediated shift (Rook, Hernandez-Pando, and Lightman, 1994) to preferential expression of IL-4, late in viral infec-

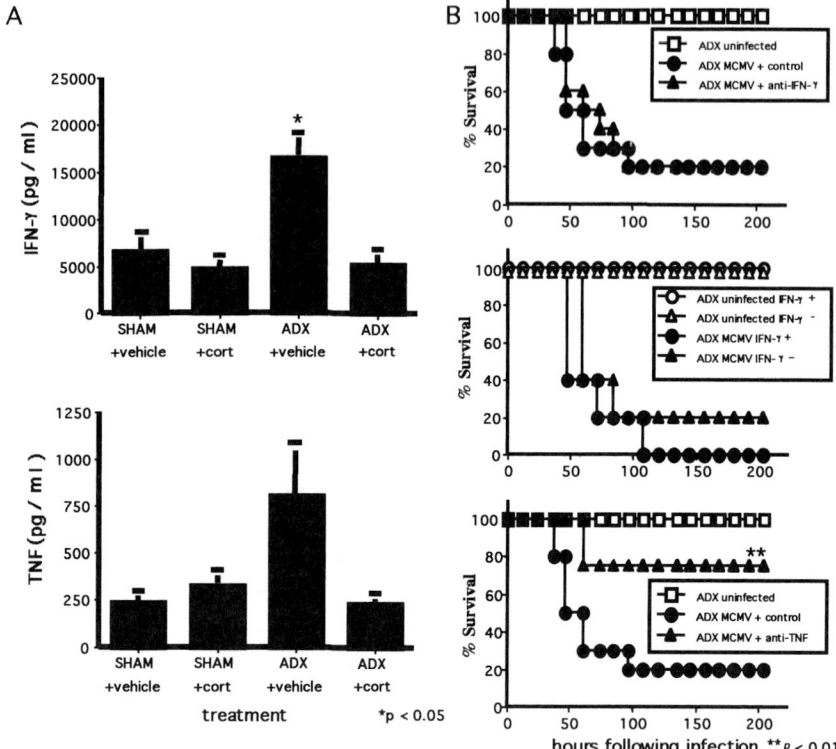

FIG 6. (A) Effects of corticosterone on cytokine responses to MCMV infection. Mice were either adrenalectomized (ADX) or SHAM-operated (by making the appropriate incisions without removing the adrenal glands). Five days following surgeries mice were infected with 5×10^4 PFU of MCMV and given either corticosterone (30 μg/ml in saline, for ADX mice; or in water, for SHAM-operated mice) or no corticosterone (vehicle) in drinking water during infection. Serum IFN-γ and TNF levels were measured 36 hours following infection. Denoted p values represent significant differences between corticosterone- and vehicle-treated ADX mice. No significant differences were observed between corticosterone-treated ADX mice and either corticosterone- or vehicle-treated SHAM-operated mice. Experimental groups contained three or four mice. (B) Cytokine effects on virus-induced mortality. *Top:* Control or anti-IFN-γ Abs were administered 8–10 hours before infection of ADX mice with 1×10^5 PFU of MCMV. *Middle:* IFN-γ⁻ (triangles) or wild-type (IFN-γ⁺) (circles) mice underwent adrenalectomy and were injected 5 days after surgery with 1×10^5 PFU of MCMV (closed symbols) or vehicle (open symbols). *Bottom:* Control or anti-TNF Abs were administered to ADX mice infected with MCMV, as in A. Experimental groups consisted of between 5 and 10 mice each. Denoted p value represents significant differences between cytokine-neutralized and control Ab-treated ADX mice infected with MCMV (reproduced from Ruzek *et al.*, *J Immunol*, 1999).

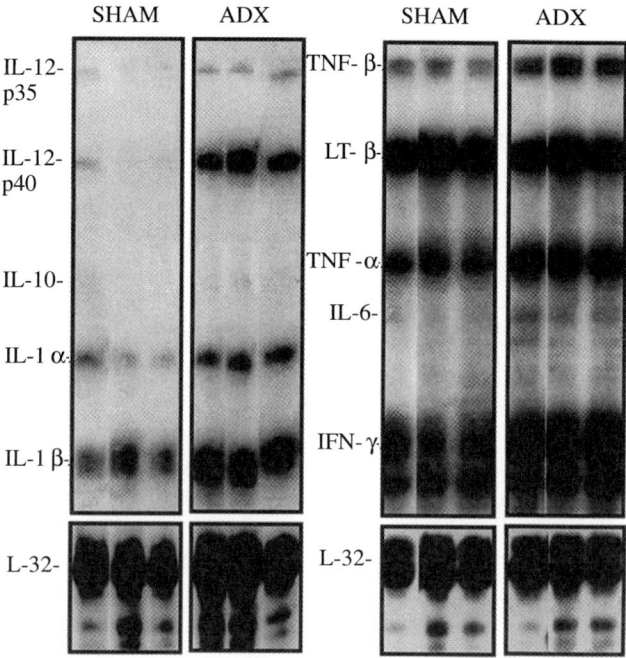

Fig 7. Effects of ADX on cytokine mRNA expression. Autoradiographic image showing RNAse protection assays for cytokine RNA isolated from spleens of SHAM-operated or ADX mice at 36 hours after infection with MCMV. Lanes show results from samples prepared from 3 individual mice (reproduced from Ruzek *et al, J Immunol,* 1999).

tion, may help drive B cell and antibody responses. The complex series of interactions between glucocorticoids and the immune response to viral infection have been elaborated into an immunoendocrinological hypothesis of AIDS, which emphasizes the role of glucocorticoids in disease progression through their inhibitory effects on cell-mediated immunity (via effects on IL-2 and IFNγ and T cell death) and their direct effects on viral replication (Clerici *et al.,* 1997). The hypothesis states that in the context of elevated glucocorticoids (e.g., through direct effects of cytokines on the HPA axis, or secondarily to environmental stimuli, including stress) viral replication is induced, T cell apoptosis is stimulated, and cytokine profiles are shifted away from protective Th1 responses to Th2 responses. However, it should be noted that this hypothesis has yet to be fully integrated with previous studies indicating glucocorticoid resistance in the later stages of HIV infection (Norbiato *et al.,* 1992, 1996).

VI. Conclusions

Interactions between the neuroendocrine and immune systems during viral infection play an important role in immune regulation and maintain a delicate balance between antiviral immune responses and potentially toxic cytokine excesses. Glucocorticoid hormones are induced under conditions of viral infection and serve to restrain cytokine responses through effects on cytokine gene expression. Consequences of glucocorticoid deficiency during viral infection include increased cytokine mRNA and protein, and cytokine-mediated toxicity and death. Although multiple pathways exist for cytokines, such as IL-6, to activate all levels of the HPA axis, the question of which pathways predominate during a given infection remains to be fully elucidated. Availability of glucocorticoid hormones to target immune cells and tissues may also be influenced by the effects of cytokines on the microenvironment of immune tissues, including cytokine effects on tissue-specific factors that bind or break down CORT and cytokine effects on glucocorticoid receptor expression and/or function. Direct infection of neuroendocrine tissues may also be involved. Given that the neuroendocrine system is exquisitely sensitive to environmental influences including exposure to psychological and physical stressors, neuroendocrine regulation of immune responses provides an important pathway for integration of the environment, as perceived by the nervous system, with that confronted by the immune system.

References

Ahmed, A., and Biron, C. A. (1998). Immunity to viruses. In "Fundamental Immunology," 4th ed. (W. E. Paul, ed.), pp. 1295–1334. Lippincott-Raven Publishers, New York.

Allan, W., Tabi, Z., Cleary, A., and Doherty, P. C. (1990). Cellular events in the lymph node and lung of mice with influenza. Consequences of depleting CD4+ T cells. *Journal of Immunology* **144**(10):3980–3986.

Almawi, W., Sewell, K., Hadro, E., Zanker, B., and Strom, T. (1990). Mode of action of the glucocorticosteroids as immunosuppressive agents. In "Molecular and Cellular Biology of Cytokines, Progress in Leukocyte Biology" (J. Oppenheim, M. Powanda, M. Kluger, and C. A. Pinarello, eds.), vol. 10A, pp. 321–326. Wiley-Liss, New York.

Auphan, N., DiDonato, J., Rosette, C., Helmberg, I., and Karin, M. (1995). Immunosuppression by glucocorticoids: Inhibition of NK-kappa B activity through induction of I kappa B synthesis. *Science* **270**:286–290.

Banks, W. A., Kastin, A. J., and Broadwell, R. D. (1995). Passage of cytokines across the blood–brain barrier. *Neuroimmunomodulation* **2**(4):241–248.

Ben Hur, T., Rosenthal, J., Itzik, A., and Weidenfeld, J. (1996). Adrenocortical activation by herpes virus: involvement of IL-1 beta and central noradrenergic system. *Neuroreport* **7** (4):927–931.

Ben-Hur, T., Conforti, N., Itzik, A., and Weidenfeld, J. (1995). Effects of HSV-1, a neurotropic virus, on the hypothalamic–pituitary–adrenocortical axis in rats. *Brain Research* **702**(1–2):17–22.

Ben-Hur, T., Rosenthal, J., Itzik, A., and Weidenfeld, J. (1996). Rescue of HSV-1 neurovirulence is associated with induction of brain interleukin-1 expression, prostaglandin synthesis and neuroendocrine responses [see comments]. *Journal of Neurovirology* **2**(4):279–288.

Berkenbosch, F., van Oers, J., del Rey, A., Tilders, F., and Besedovsky, H. (1987). Corticotropin-releasing factor-producing neurons in the rat activated by interleukin-1. *Science* **238**(4826):524–526.

Bertini, R., Bianchi, M., and Ghezzi, P. (1988). Adrenalectomy sensitizes mice to the lethal effects of interleukin 1 and tumor necrosis factor. *Journal of Experimental Medicine* **167**(5):1708–1712.

Besedovsky, H., del Rey, A., Sorkin, E., and Dinarello, C. A. (1986). Immunoregulatory feedback between interleukin-1 and glucocorticoid hormones. *Science* **233**(4764):652–654.

Besedovsky, H. O., and del Rey, A. (1989). Mechanism of virus-induced stimulation of the hypothalamus–pituitary–adrenal axis. *Journal of Steroid Biochemistry* **34**(1–6):235–239.

Besedovsky, H. O., and del Rey, A. (1996). Immune-neuroendocrine interactions: facts and hypotheses. *Endocrine Reviews* **17**(1):64–102.

Betancur, C., Borrell, J., and Guaza, C. (1995). Cytokine regulation of corticosteroid receptors in the rat hippocampus: effects of interleukin-1, interleukin-6, tumor necrosis factor and lipopolysaccharide. *Neuroendocrinology* **62**(1):47–54.

Biron, C. A. (1994). Cytokines in the generation of immune responses to, and resolution of, virus infection. *Current Opinion in Immunology* **6**(4):530–538.

Biron, C. A. (1997). Activation and function of natural killer cell responses during viral infections. *Current Opinion in Immunology* **9**(1):24–34.

Biron, C. A. (1999). Initial and innate responses to viral infections—pattern setting in immunity or disease. *Current Opinion in Microbiology* **2**(4):374–381.

Biron, C. A., and Gazzinelli, R. T. (1995). Effects of IL-12 on immune responses to microbial infections: a key mediator in regulating disease outcome. *Current Opinion in Immunology* **7**(4):485–496.

Biron, C. A., and Orange, J. S. (1995). IL12 in acute viral infectious disease. *Research in Immunology* **146**(7–8):590–600.

Biron, C. A., and Sen, G. C. (2000). Interferons and other cytokines. *In* "Fields Virology" (D. M. Knipe, P. M. Howley, D. E. Griffin, R. A. Lamb, M. A. Martin, B. Roizman, and S. E. Straus, eds.), pp. In press. Lippincott/Williams and Wilkins,.

Biron, C. A., Young, H. A., and Kasaian, M. T. (1990). Interleukin 2-induced proliferation of murine natural killer cells in vivo. *Journal of Experimental Medicine* **171**(1):173–188.

Borrow, P., Evans, C. F., and Oldstone, M. B. (1995). Virus-induced immunosuppression: immune system-mediated destruction of virus-infected dendritic cells results in generalized immune suppression. *Journal of Virology* **69**(2):1059–1070.

Bucala, R. (1994). Identification of MIF as a new pituitary hormone and macrophage cytokine and its role in endotoxic shock. *Immunology Letters* **43**(1–2):23–26.

Buchmeier, M. J., Welsh, R. M., Dutko, F. J., and Oldstone, M. B. (1980). The virology and immunobiology of lymphocytic choriomeningitis virus infection. *Advances in Immunology* **30**:275–331.

Bukowski, J. F., and Welsh, R. M. (1986). The role of natural killer cells and interferon in resistance to acute infection of mice with herpes simplex virus type 1. *Journal of Immunology* **136**(9):3481–3485.

Bukowski, J. F., Woda, B. A., Habu, S., Okumura, K., and Welsh, R. M. (1983). Natural killer cell depletion enhances virus synthesis and virus-induced hepatitis in vivo. *Journal of Immunology* 131(3):1531–1538.

Callahan, T. A., and Piekut, D. T. (1997). Differential Fos expression induced by IL-1beta and IL-6 in rat hypothalamus and pituitary gland. *Journal of Neuroimmunology* 73(1–2):207–211.

Campbell, I. L., Hobbs, M. V., Kemper, P., and Oldstone, M. B. (1994). Cerebral expression of multiple cytokine genes in mice with lymphocytic choriomeningitis. *Journal of Immunology* 152(2):716–723.

Carr, D. J. (1998). Increased levels of IFN-gamma in the trigeminal ganglion correlate with protection against HSV-1-induced encephalitis following subcutaneous administration with androstenediol. *Journal of Neuroimmunology* 89(1–2):160–167.

Chong, K. T., Gresser, I., and Mims, C. A. (1983). Interferon as a defence mechanism in mouse cytomegalovirus infection. *Journal of General Virology* 64(Pt 2):461–464.

Chou, C. K., Wang, L. H., Lin, H. M., and Chi, C. W. (1992). Glucocorticoid stimulates hepatitis B viral gene expression in cultured human hepatoma cells. *Hepatology* 16(1):13–18.

Christeff, N., Gherbi, N., Mammes, O., Dalle, M. T., Gharakhanian, S., Lortholary, O., Melchior, J. C., and Nunez, E. A. (1997). Serum cortisol and DHEA concentrations during HIV infection. *Psychoneuroendocrinology* 22 **Suppl 1**:S11–S18.

Clerici, M., Trabattoni, D., Piconi, S., Fusi, M. L., Ruzzante, S., Clerici, C., and Villa, M. L. (1997). A possible role for the cortisol/anticortisols imbalance in the progression of human immunodeficiency virus. *Psychoneuroendocrinology* 22 **Suppl 1**:S27–S31.

Costas, M., Trapp, T., Pereda, M. P., Sauer, J., Rupprecht, R., Nahmod, V. E., Reul, J. M., Holsboer, F., and Arzt, E. (1996). Molecular and functional evidence for in vitro cytokine enhancement of human and murine target cell sensitivity to glucocorticoids. TNF-alpha priming increases glucocorticoid inhibition of TNF-alpha-induced cytotoxicity/apoptosis. *Journal of Clinical Investigation* 98(6):1409–1416.

Cousens, L. P., Orange, J. S., and Biron, C. A. (1995). Endogenous IL-2 contributes to T cell expansion and IFN-gamma production during lymphocytic choriomeningitis virus infection. *Journal of Immunology* 155(12):5690–5699.

Cousens, L. P., Peterson, R., Hsu, S., Dorner, A., Altman, J. D., Ahmed, R., and Biron, C. A. (1999). Two roads diverged: interferon alpha/beta- and interleukin 12-mediated pathways in promoting T cell interferon gamma responses during viral infection. *Journal of Experimental Medicine* 189(8):1315–1328.

Coutelier, J. P., Van Broeck, J., and Wolf, S. F. (1995). Interleukin-12 gene expression after viral infection in the mouse. *Journal of Virology* 69(3):1955–1958.

Daynes, R. A., and Araneo, B. A. (1989). Contrasting effects of glucocorticoids on the capacity of T cells to produce the growth factors interleukin 2 and interleukin 4. *European Journal of Immunology* 19(12):2319–2325.

Daynes, R. A., Araneo, B. A., Dowell, T. A., Huang, K., and Dudley, D. (1990). Regulation of murine lymphokine production in vivo. III. The lymphoid tissue microenvironment exerts regulatory influences over T helper cell function. *Journal of Experimental Medicine* 171(4):979–996.

Dhabhar, F. S., and McEwen, B. S. (1996). Stress-induced enhancement of antigen-specific cell-mediated immunity. *Journal of Immunology* 156(7):2608–2615.

Dhabhar, F. S., and McEwen, B. S. (1999). Enhancing versus suppressive effects of stress hormones on skin immune function. *Proceedings of the National Academy of Sciences of the United States of America* 96(3):1059–1064.

Dhabhar, F. S., Miller, A. H., McEwen, B. S., and Spencer, R. L. (1995). Effects of stress on immune cell distribution. Dynamics and hormonal mechanisms. *Journal of Immunology* 154(10):5511–5527.

Dhabhar, F. S., Miller, A. H., McEwen, B. S., and Spencer, R. L. (1996). Stress-induced changes in blood leukocyte distribution. Role of adrenal steroid hormones. *Journal of Immunology* **157**(4):1638–1644.

Dhabhar, F. S., Miller, A. H., Stein, M., McEwen, B. S., and Spencer, R. L. (1994). Diurnal and acute stress-induced changes in distribution of peripheral blood leukocyte subpopulations. *Brain, Behavior, & Immunity* **8**(1):66–79.

Dobbs, C. M., Feng, N., Beck, F. M., and Sheridan, J. F. (1996). Neuroendocrine regulation of cytokine production during experimental influenza viral infection: effects of restraint stress-induced elevation in endogenous corticosterone. *Journal of Immunology* **157**(5):1870–1877.

Dunn, A. J., Powell, M. L., and Gaskin, J. M. (1987). Virus-induced increases in plasma corticosterone. *Science* **238**(4832):1423–1425.

Dunn, A. J., Powell, M. L., Meitin, C., and Small, P. A., Jr. (1989). Virus infection as a stressor: influenza virus elevates plasma concentrations of corticosterone, and brain concentrations of MHPG and tryptophan. *Physiology & Behavior* **45**(3):591–594.

Dunn, A. J., Powell, M. L., Moreshead, W. V., Gaskin, J. M., and Hall, N. R. (1987). Effects of Newcastle disease virus administration to mice on the metabolism of cerebral biogenic amines, plasma corticosterone, and lymphocyte proliferation. *Brain, Behavior, & Immunity* **1**(3):216–230.

Dunn, A. J., and Vickers, S. L. (1994). Neurochemical and neuroendocrine responses to Newcastle disease virus administration in mice. *Brain Research* **645**(1–2):103–112.

Dunn, A. J., and Wang, J. (1995). Cytokine effects on CNS biogenic amines. *Neuroimmunomodulation* **2**(6):319–328.

Ericsson, A., Kovacs, K. J., and Sawchenko, P. E. (1994). A functional anatomical analysis of central pathways subserving the effects of interleukin-1 on stress-related neuroendocrine neurons. *Journal of Neuroscience* **14**(2):897–913.

Espey, M. G., and Basile, A. S. (1999). Glutamate augments retrovirus-induced immunodeficiency through chronic stimulation of the hypothalamic–pituitary–adrenal axis. *Journal of Immunology* **162**(8):4998–5002.

Ferreiro, J., and Vinters, H. V. (1988). Pathology of the pituitary gland in patients with the acquired immune deficiency syndrome (AIDS). *Pathology* **20**(3):211–215.

Fleshner, M., Goehler, L. E., Hermann, J., Relton, J. K., Maier, S. F., and Watkins, L. R. (1995). Interleukin-1 beta induced corticosterone elevation and hypothalamic NE depletion is vagally mediated. *Brain Research Bulletin* **37**(6):605–610.

Gatti, S., and Bartfai, T. (1993). Induction of tumor necrosis factor-alpha mRNA in the brain after peripheral endotoxin treatment: comparison with interleukin-1 family and interleukin-6. *Brain Research* **624**(1–2):291–294.

Gessner, A., Drjupin, R., Lohler, J., Lother, H., and Lehmann-Grube, F. (1990). IFN-gamma production in tissues of mice during acute infection with lymphocytic choriomeningitis virus. *Journal of Immunology* **144**(8):3160–3165.

Gessner, A., Moskophidis, D., and Lehmann-Grube, F. (1989). Enumeration of single IFN-gamma-producing cells in mice during viral and bacterial infection. *Journal of Immunology* **142**(4):1293–1298.

Glaser, R., Kutz, L. A., MacCallum, R. C., and Malarkey, W. B. (1995). Hormonal modulation of Epstein-Barr virus replication. *Neuroendocrinology* **62**(4):356–361.

Glasgow, B. J., Steinsapir, K. D., And ers, K., and Layfield, L. J. (1985). Adrenal pathology in the acquired immune deficiency syndrome. *American Journal of Clinical Pathology* **84**(5):594–597.

Gonzalez-Hernand ez, J. A., Bornstein, S. R., Ehrhart-Bornstein, M., Spath-Schwalbe, E., Jirikowski, G., and Scherbaum, W. A. (1994). Interleukin-6 messenger ribonucleic acid expression in human adrenal gland in vivo: new clue to a paracrine or autocrine regulation of adrenal function. *Journal of Clinical Endocrinology & Metabolism* **79**(5):1492–1497.

Gonzalo, J. A., Gonzalez-Garcia, A., Martinez, C., and Kroemer, G. (1993). Glucocorti-coid-mediated control of the activation and clonal deletion of peripheral T cells in vivo. *Journal of Experimental Medicine* **177**(5):1239–1246.

Graziosi, C., Gantt, K. R., Vaccarezza, M., Demarest, J. F., Daucher, M., Saag, M. S., Shaw, G. M., Quinn, T. C., Cohen, O. J., Welbon, C. C., Pantaleo, G., and Fauci, A. S. (1996). Kinetics of cytokine expression during primary human immunodeficiency virus type 1 infection. *Proceedings of the National Academy of Sciences of the United States of America* **93**(9):4386–4391.

Graziosi, C., Pantaleo, G., Gantt, K. R., Fortin, J. P., Demarest, J. F., Cohen, O. J., Sekaly, R. P., and Fauci, A. S. (1994). Lack of evidence for the dichotomy of TH1 and TH2 predominance in HIV-infected individuals. *Science* **265**(5169):248–252.

Gresser, I., Guy-Grand, D., Maury, C., and Maunoury, M. T. (1981). Interferon induces peripheral lymphadenopathy in mice. *Journal of Immunology* **127**(4):1569–1575.

Gruber, J., Sgonc, R., Hu, Y. H., Beug, H., and Wick, G. (1994). Thymocyte apoptosis induced by elevated endogenous corticosterone levels. *European Journal of Immunology* **24**(5):1115–1121.

Grundy, J. E., Trapman, J., Allan, J. E., Shellam, G. R., and Melief, C. J. (1982). Evidence for a protective role of interferon in resistance to murine cytomegalovirus and its control by non-H-2-linked genes. *Infection & Immunity* **37**(1):143–150.

Guha-Thakurta, N., and Majde, J. A. (1997). Early induction of proinflammatory cytokine and type I interferon mRNAs following Newcastle disease virus, poly [rI:rC], or low-dose LPS challenge of the mouse. *Journal of Interferon & Cytokine Research* **17**(4):197–204.

Gustafsson, J. A., Carlstedt-Duke, J., Poellinger, L., Okret, S., Wikstrom, A. C., Bronnegard, M., Gillner, M., Dong, Y., Fuxe, K., and Cintra, A. (1987). Biochemistry, molecular biology, and physiology of the glucocorticoid receptor. *Endocrine Reviews* **8**(2):185–234.

Hammond, G. L., Smith, C. L., Paterson, N. A., and Sibbald, W. J. (1990). A role for corticosteroid-binding globulin in delivery of cortisol to activated neutrophils. *Journal of Clinical Endocrinology & Metabolism* **71**(1):34–39.

Harbuz, M. S., Stephanou, A., Sarlis, N., and Lightman, S. L. (1992). The effects of recombinant human interleukin (IL)-1 alpha, IL-1 beta or IL-6 on hypothalamo–pituitary–adrenal axis activation. *Journal of Endocrinology* **133**(3):349–355.

Heinzel, F. P., Sadick, M. D., Holaday, B. J., Coffman, R. L., and Locksley, R. M. (1989). Reciprocal expression of interferon gamma or interleukin 4 during the resolution or progression of murine leishmaniasis. Evidence for expansion of distinct helper T cell subsets. *Journal of Experimental Medicine* **169**(1):59–72.

Hench, P. S. (1952). The reversibility of certain rheumatic and non-rheumatic conditions by the use of cortisone or of the pituitry adrenocorticotropic hormone. *Ann Internal Med* **36**:1–25.

Hennebold, J. D., Ryu, S. Y., Mu, H. H., Galbraith, A., and Daynes, R. A. (1996). 11 beta-hydroxysteroid dehydrogenase modulation of glucocorticoid activities in lymphoid organs. *American Journal of Physiology* **270**(6 Pt 2):R1296–1306.

Hennet, T., Ziltener, H. J., Frei, K., and Peterhans, E. (1992). A kinetic study of immune mediators in the lungs of mice infected with influenza A virus. *Journal of Immunology* **149**(3):932–939.

Hermann, G., Beck, F. M., and Sheridan, J. F. (1995). Stress-induced glucocorticoid response modulates mononuclear cell trafficking during an experimental influenza viral infection. *Journal of Neuroimmunology* **56**(2):179–186.

Hermann, G., Tovar, C. A., Beck, F. M., and Sheridan, J. F. (1994). Kinetics of glucocorticoid response to restraint stress and/or experimental influenza viral infection in two inbred strains of mice. *Journal of Neuroimmunology* 49(1–2):25–33.

Hill, M. R., Stith, R. D., and McCallum, R. E. (1987). Monokines mediate decreased hepatic glucocorticoid binding in endotoxemia. *Journal of Leukocyte Biology* 41(3):236–241.

Hill, M. R., Stith, R. D., and McCallum, R. E. (1988). Human recombinant IL-1 alters glucocorticoid receptor function in Reuber hepatoma cells. *Journal of Immunology* 141(5):1522–1528.

Horvath, E., and Kovacs, K. (1988). Fine structural cytology of the adenohypophysis in rat and man. *Journal of Electron Microscopy Technique* 8(4):401–432.

Hoshino, J., Beckmann, G., Huser, J., and Kroger, H. (1994). Interleukin 1 beta enhances the response of rabbit synovial fibroblasts in vitro to dexamethasone injury: implication for the role of increased nuclear hypersensitive sites and the number of dexamethasone receptors. *Journal of Rheumatology* 21(4):616–622.

Huang, S., Hendriks, W., Althage, A., Hemmi, S., Bluethmann, H., Kamijo, R., Vilcek, J., Zinkernagel, R. M., and Aguet, M. (1993). Immune response in mice that lack the interferon-gamma receptor [see comments]. *Science* 259(5102):1742–1745.

Ishikawa, R., and Biron, C. A. (1993). IFN induction and associated changes in splenic leukocyte distribution. *Journal of Immunology* 150(9):3713–3727.

Iwata, M., Hanaoka, S., and Sato, K. (1991). Rescue of thymocytes and T cell hybridomas from glucocorticoid-induced apoptosis by stimulation via the T cell receptor/CD3 complex: a possible in vitro model for positive selection of the T cell repertoire. *European Journal of Immunology* 21(3):643–648.

Jensen, M., and Rasmussen, A. (1963). Stress and susceptibility to viral infection. Sound stress and susceptibility to vesicular stomatitis virus. *Journal of Immunology* 90:21–23.

Johnsson, T., Lavender, J., Hullin, E., and Rasmussen, A., Jr. (1963). The influence of avoidance-learning stress on resistance to coxsackie B virus in mice. *Journal of Immunology* 91:569–575.

Johnsson, T., and Rasmussen, A., Jr. (1965). Emotional stress and susceptibility to poliomyelitis virus infection in mice. *Archiv fur des Gesamte Virusforschung* 18:390.

Jondal, M., Okret, S., and McConkey, D. (1993). Killing of immature CD4+ CD8+ thymocytes in vivo by anti-CD3 or 5¹-(N-ethyl)-carboxamide adenosine is blocked by glucocorticoid receptor antagonist RU-486 [published erratum appears in Eur J Immunol 1993 Oct;23 (10):2734]. *European Journal of Immunology* 23(6):1246–1250.

Judd, A. M. (1998). Cytokine expression in the rat adrenal cortex. *Hormone & Metabolic Research* 30(6–7):404–410.

Judd, A. M., and MacLeod, R. M. (1992). Adrenocorticotropin increases interleukin-6 release from rat adrenal zona glomerulosa cells. *Endocrinology* 130(3):1245–1254.

Kagi, D., and Hengartner, H. (1996). Different roles for cytotoxic T cells in the control of infections with cytopathic versus noncytopathic viruses. *Current Opinion in Immunology* 8(4):472–477.

Kagi, D., Ledermann, B., Burki, K., Seiler, P., Odermatt, B., Olsen, K. J., Podack, E. R., Zinkernagel, R. M., and Hengartner, H. (1994). Cytotoxicity mediated by T cells and natural killer cells is greatly impaired in perforin-deficient mice [see comments]. *Nature* 369:31–37.

Kam, J. C., Szefler, S. J., Surs, W., Sher, E. R., and Leung, D. Y. (1993). Combination IL-2 and IL-4 reduces glucocorticoid receptor-binding affinity and T cell response to glucocorticoids. *Journal of Immunology* 151(7):3460–3466.

Kanangat, S., Thomas, J., Gangappa, S., Babu, J. S., and Rouse, B. T. (1996). Herpes simplex virus type 1-mediated up-regulation of IL-12 (p40) mRNA expression. Implications in immunopathogenesis and protection. *Journal of Immunology* 156(3):1110–1116.

Kapcala, L. P., He, J. R., Gao, Y., Pieper, J. O., and DeTolla, L. J. (1996). Subdiaphragmatic vagotomy inhibits intra-abdominal interleukin-1 beta stimulation of adrenocorticotropin secretion. *Brain Research* **728**(2):247–254.

Karanth, S., Lyson, K., and McCann, S. M. (1994). Cyclosporin A inhibits interleukin-2-induced release of corticotropin-releasing hormone. *Neuroimmunomodulation* **1**(1):82–85.

Kasaian, M. T., and Biron, C. A. (1989). The activation of IL-2 transcription in L3T4⁺ and Lyt-2⁺ lymphocytes during virus infection in vivo. *Journal of Immunology* **142**(4):1287–1292.

Kasaian, M. T., and Biron, C. A. (1990). Effects of cyclosporin A on IL-2 production and lymphocyte proliferation during infection of mice with lymphocytic choriomeningitis virus. *Journal of Immunology* **144**(1):299–306.

Kasaian, M. T., Leite-Morris, K. A., and Biron, C. A. (1991). The role of CD4⁺ cells in sustaining lymphocyte proliferation during lymphocytic choriomeningitis virus infection. *Journal of Immunology* **146**(6):1955–1963.

Kelso, A., and Munck, A. (1984). Glucocorticoid inhibition of lymphokine secretion by alloreactive T lymphocyte clones. *Journal of Immunology* **133**(2):784–791.

Korngold, R., Blank, K. J., and Murasko, D. M. (1983). Effect of interferon on thoracic duct lymphocyte output: induction with either poly I:poly C or vaccinia virus. *Journal of Immunology* **130**(5):2236–2240.

Kovacs, K. J., and Elenkov, I. J. (1995). Differential dependence of ACTH secretion induced by various cytokines on the integrity of the paraventricular nucleus. *Journal of Neuroendocrinology* **7**(1):15–23.

Krasznai, A., Aranyi, P., Feher, T., Krajcsi, P., Meszaros, K., and Horvath, I. (1986). Alterations in the number of glucocorticoid receptors of circulating lymphocytes in sepsis. *Haematologia* **19**(4):293–298.

Kunicka, J. E., Talle, M. A., Denhardt, G. H., Brown, M., Prince, L. A., and Goldstein, G. (1993). Immunosuppression by glucocorticoids: inhibition of production of multiple lymphokines by in vivo administration of dexamethasone. *Cellular Immunology* **149**(1):39–49.

Lack, E. E., and Kozakewich, P. W. (1990). Embryology, developmental anatomy, and selected aspects of non-neoplastic pathology. In "Pathology of the Adrenal Glands" (E. E. Lack, ed.), pp. 1–74. Churchill Livingstone., New York.

Lambert, M. (1994). Thyroid dysfunction in HIV infection. *Baillieres Clinical Endocrinology & Metabolism* **8**:825–835.

Laye, S., Bluthe, R. M., Kent, S., Combe, C., Medina, C., Parnet, P., Kelley, K., and Dantzer, R. (1995). Subdiaphragmatic vagotomy blocks induction of IL-1 beta mRNA in mice brain in response to peripheral LPS. *American Journal of Physiology* **268**:R1327–1331.

Laye, S., Parnet, P., Goujon, E., and Dantzer, R. (1994). Peripheral administration of lipopolysaccharide induces the expression of cytokine transcripts in the brain and pituitary of mice. *Brain Research–Molecular Brain Research* **27**(1):157–162.

Liu, L. Y., Sun, B., Tian, Y., Lu, B. Z., and Wang, J. (1993). Changes of pulmonary glucocorticoid receptor and phospholipase A2 in sheep with acute lung injury after high dose endotoxin infusion. *American Review of Respiratory Disease* **148**:878–881.

Lorenz, E., and Rossipal, E. (1965). Glucocorticoid elimination during morbilli in children. *Arch. Kinderheilk* **172**:251.

Lu, W., Salerno-Goncalves, R., Yuan, J., Sylvie, D., Han, D. S., and And rieu, J. M. (1995). Glucocorticoids rescue CD4⁺ T lymphocytes from activation-induced apoptosis triggered by HIV-1: implications for pathogenesis and therapy. *AIDS* **9**(1):35–42.

Lynch, F., Doherty, P. C., and Ceredig, R. (1989). Phenotypic and functional analysis of the cellular response in regional lymphoid tissue during an acute virus infection. *Journal of Immunology* **142**(10):3592–3598.

Matloubian, M., Kolhekar, S. R., Somasundaram, T., and Ahmed, R. (1993). Molecular determinants of macrophage tropism and viral persistence: importance of single amino acid changes in the polymerase and glycoprotein of lymphocytic choriomeningitis virus. *Journal of Virology* **67**(12):7340–7349.

Matta, S. G., Weatherbee, J., and Sharp, B. M. (1992). A central mechanism is involved in the secretion of ACTH in response to IL-6 in rats: comparison to and interaction with IL-1 beta. *Neuroendocrinology* **56**(4):516–525.

McEwen, B. S., Biron, C. A., Brunson, K. W., Bulloch, K., Chambers, W. H., Dhabhar, F. S., Goldfarb, R. H., Kitson, R. P., Miller, A. H., Spencer, R. L., and Weiss, J. M. (1997). The role of adrenocorticoids as modulators of immune function in health and disease: neural, endocrine and immune interactions. *Brain Research–Brain Research Reviews* **23**(1–2):79–133.

Miller, A. H. (1998). Neuroendocrine and immune system interactions in stress and depression. *Psychiatric Clinics of North America* **21**(2):443–463.

Miller, A. H., Pariante, C. M., and Pearce, B. D. (1999). Effects of cytokines on glucocorticoid receptor expression and function. Glucocorticoid resistance and relevance to depression. *Advances in Experimental Medicine & Biology* **461**:107–116.

Miller, A. H., Spencer, R. L., hassett, J., Kim, C., Rhee, R., Ciurea, D., Dhabhar, F., McEwen, B., and Stein, M. (1994). Effects of selective type I and II adrenal steroid agonists on immune cell distribution. *Endocrinology* **135**(5):1934–1944.

Miller, A. H., Spencer, R. L., Pearce, B. D., Pisell, T. L., Tanapat, P., Leung, J. J., Dhabhar, F. S., McEwen, B. S., and Biron, C. A. (1997). 1996 Curt P. Richter Award. Effects of viral infection on corticosterone secretion and glucocorticoid receptor binding in immune tissues. *Psychoneuroendocrinology* **22**(6):455–474.

Miller, A. H., Spencer, R. L., Pulera, M., Kang, S., McEwen, B. S., and Stein, M. (1992). Adrenal steroid receptor activation in rat brain and pituitary following dexamethasone: implications for the dexamethasone suppression test. *Biological Psychiatry* **32**(10):850–869.

Miller, A. H., Spencer, R. L., Stein, M., and McEwen, B. S. (1990). Adrenal steroid receptor binding in spleen and thymus after stress or dexamethasone. *American Journal of Physiology* **259**(3 Pt 1):E405–412.

Minami, M., Kuraishi, Y., Yamaguchi, T., Nakai, S., Hirai, Y., and Satoh, M. (1991). Immobilization stress induces interleukin-1 beta mRNA in the rat hypothalamus. *Neuroscience Letters* **123**(2):254–256.

Monder, C., and Shackleton, C. H. (1984). 11 beta-hydroxysteroid dehydrogenase: fact or fancy? *Steroids* **44**(5):383–417.

Morale, M. C., Batticane, N., Gallo, F., Barden, N., and Marchetti, B. (1995). Disruption of hypothalamic–pituitary–adrenocortical system in transgenic mice expressing type II glucocorticoid receptor antisense ribonucleic acid permanently impairs T cell function: effects on T cell trafficking and T cell responsiveness during postnatal development. *Endocrinology* **136**(9):3949–3460.

Morrissey, P. J., Charrier, K., Alpert, A., and Bressler, L. (1988). In vivo administration of IL-1 induces thymic hypoplasia and increased levels of serum corticosterone. *Journal of Immunology* **141**(5):1456–1463.

Mosca, L., Costanzi, G., Antonacci, C., Boldorini, R., Carboni, N., Cristina, S., Liverani, C., Parravicini, C., Pirolo, A., and Vago, L. (1992). Hypophyseal pathology in AIDS. *Histology & Histopathology* **7**(2):291–300.

Mulkins, M. A., and Allison, A. C. (1987). Recombinant human interleukin-1 inhibits the induction by dexamethasone of alkaline phosphatase activity in murine capillary endothelial cells. *Journal of Cellular Physiology* **133**(3):539–545.

Muller, U., Steinhoff, U., Reis, L. F., Hemmi, S., Pavlovic, J., Zinkernagel, R. M., and Aguet, M. (1994). Functional role of type I and type II interferons in antiviral defense. *Science* **264**(5167):1918–1921.

Munck, A., and Guyr, P. (1991). Glucocorticoids and Immune Function. *In* "Psychoneuroimmunology, Second Edition" R. Ader, D. Felten, and N. Cohen, eds. Academic Press, San Diego.

Nair, M. P., and Schwartz, S. A. (1984). Immunomodulatory effects of corticosteroids on natural killer and antibody-dependent cellular cytotoxic activities of human lymphocytes. *Journal of Immunology* **132**(6):2876–2882.

Naitoh, Y., Fukata, J., Tominaga, T., Nakai, Y., Tamai, S., Mori, K., and Imura, H. (1988). Interleukin-6 stimulates the secretion of adrenocorticotropic hormone in conscious, freely-moving rats. *Biochemical & Biophysical Research Communications* **155**(3):1459–1463.

Niemialtowski, M. G., and Rouse, B. T. (1992). Predominance of Th1 cells in ocular tissues during herpetic stromal keratitis. *Journal of Immunology* **149**(9):3035–3039.

Nimmagadda, S. R., Szefler, S. J., Spahn, J. D., Surs, W., and Leung, D. Y. (1997). Allergen exposure decreases glucocorticoid receptor binding affinity and steroid responsiveness in atopic asthmatics. *American Journal of Respiratory & Critical Care Medicine* **155**(1):87–93.

Norbiato, G., Bevilacqua, M., Vago, T., Baldi, G., Chebat, E., Bertora, P., Moroni, M., Galli, M., and Oldenburg, N. (1992). Cortisol resistance in acquired immunodeficiency syndrome. *Journal of Clinical Endocrinology & Metabolism* **74**(3):608–613.

Norbiato, G., Bevilacqua, M., Vago, T., and Clerici, M. (1996). Glucocorticoids and interferon-alpha in the acquired immunodeficiency syndrome. *Journal of Clinical Endocrinology & Metabolism* **81**(7):2601–2606.

Norbiato, G., Bevilacqua, M., Vago, T., Taddei, A., and Clerici. (1997). Glucocorticoids and the immune function in the human immunodeficiency virus infection: a study in hypercortisolemic and cortisol-resistant patients. *Journal of Clinical Endocrinology & Metabolism* **82**(10):3260–3263.

Oldstone, M. B., Ahmed, R., Buchmeier, M. J., Blount, P., and Tishon, A. (1985). Perturbation of differentiated functions during viral infection in vivo. I. Relationship of lymphocytic choriomeningitis virus and host strains to growth hormone deficiency. *Virology* **142**(1):158–174.

Orange, J. S., and Biron, C. A. (1996a). An absolute and restricted requirement for IL-12 in natural killer cell IFN-gamma production and antiviral defense. Studies of natural killer and T cell responses in contrasting viral infections. *Journal of Immunology* **156**(3):1138–1142.

Orange, J. S., and Biron, C. A. (1996b). Characterization of early IL-12, IFN-alpha-beta, and TNF effects on antiviral state and NK cell responses during murine cytomegalovirus infection. *Journal of Immunology* **156**(12):4746–4756.

Orange, J. S., Salazar-Mather, T. P., Opal, S. M., and Biron, C. A. (1997). Mechanisms for virus-induced liver disease: tumor necrosis factor-mediated pathology independent of natural killer and T cells during murine cytomegalovirus infection. *Journal of Virology* **71**(12):9248–9258.

Orange, J. S., Wang, B., Terhorst, C., and Biron, C. A. (1995). Requirement for natural killer cell-produced interferon gamma in defense against murine cytomegalovirus infection and enhancement of this defense pathway by interleukin 12 administration. *Journal of Experimental Medicine* **182**(4):1045–1056.

Pariante, C., Pearce, B., Pisell, T., and Miller, A. (1996).Steroid-independent regulation of glucocorticoid receptor number. *26th annual meeting of the Society for Neuroscience, Washington, D.C.*

Path, G., Bornstein, S. R., Ehrhart-Bornstein, M., and Scherbaum, W. A. (1997). Interleukin-6 and the interleukin-6 receptor in the human adrenal gland: expression and effects on steroidogenesis. *Journal of Clinical Endocrinology & Metabolism* **82**(7):2343–2349.

Pearce, B. D., Ruzek, M. C., Karalis, K., Venihaki, M., Silverman, M. N., Biron, C. A., and Miller, A. H. (1999). CRF-independent activation of corticosterone release during viral infection. *Absrtacts of the Society for Neuroscience* **29**:1447.

Perlstein, R. S., Whitnall, M. H., Abrams, J. S., Mougey, E. H., and Neta, R. (1993). Synergistic roles of interleukin-6, interleukin-1, and tumor necrosis factor in the adrenocorticotropin response to bacterial lipopolysaccharide in vivo. *Endocrinology* **132**(3):946–952.

Pfizenmaier, K., Jung, H., Starzinski-Powitz, A., Rollinghoff, M., and Wagner, H. (1977). The role of T cells in anti-herpes simplex virus immunity. I. Induction of antigen-specific cytotoxic T lymphocytes. *Journal of Immunology* **119**(3):939–944.

Plotsky, P. M. (1991). Pathways to the secretion of adrenocorticotropin: A view from the portal. *Journal of Neuroendocrinology* **3**(1):1–9.

Price, P., Olver, S. D., Silich, M., Nador, T. Z., Yerkovich, S., and Wilson, S. G. (1996). Adrenalitis and the adrenocortical response of resistant and susceptible mice to acute murine cytomegalovirus infection. *European Journal of Clinical Investigation* **26**(9):811–119.

Quan, N., Sundar, S. K., and Weiss, J. M. (1994). Induction of interleukin-1 in various brain regions after peripheral and central injections of lipopolysaccharide. *Journal of Neuroimmunology* **49**(1–2):125–134.

Rakasz, E., Gal, A., Biro, J., Balas, G., and Falus, A. (1993). Modulation of glucocorticosteroid binding in human lymphoid, monocytoid and hepatoma cell lines by inflammatory cytokines interleukin (IL)-1 beta, IL-6 and tumour necrosis factor (TNF)-alpha. *Scandinavian Journal of Immunology* **37**(6):684–689.

Rasmussen, A., Jr., Marsh, J., and Brill, N. (1957). Increased susceptibility to herpes simplex in mice subjected to avoidance-learning stress or restraint. *Proc. Soc. Exp. Biol. Medicine* **96**:183–189.

Rasmussen, A. F., Jr. (1969). Emotions and immunity. *Annals of the New York Academy of Sciences* **164**(2):458–462.

Razvi, E. S., and Welsh, R. M. (1993). Programmed cell death of T lymphocytes during acute viral infection: a mechanism for virus-induced immune deficiency. *Journal of Virology* **67**(10):5754–5765.

Refaeli, Y., Levy, D. N., and Weiner, D. B. (1995). The glucocorticoid receptor type II complex is a target of the HIV-1 vpr gene product. *Proceedings of the National Academy of Sciences of the United States of America* **92**(8):3621–3625.

Rivier, C. (1995). Influence of immune signals on the hypothalamic-pituitary axis of the rodent. *Frontiers in Neuroendocrinology* **16**:151–182.

Rodriguez, M., von Wedel, R. J., Garrett, R. S., Lampert, P. W., and Oldstone, M. B. (1983). Pituitary dwarfism in mice persistently infected with lymphocytic choriomeningitis virus. *Laboratory Investigation* **49**(1):48–53.

Rook, G. A., Hernand ez-Pand o, R., and Lightman, S. L. (1994). Hormones, peripherally activated prohormones and regulation of the Th1/Th2 balance [see comments]. *Immunology Today* **15**(7):301–303.

Rosner, W. (1991). Plasma steroid-binding proteins. *Endocrinology & Metabolism Clinics of North America* **20**(4):697–720.

Rotterdam, H., and Dembitzer, F. (1993). The adrenal gland in AIDS. *Endocrine Pathology* 4(1):4–14.

Ruzek, M., Miller, A., Pearce, B., Spencer, R., and Biron, C. (1997a). Evidence for endogenous glucocorticoid protection against MCMV-induced lethality. *27th Annual Meeting of the Society for Neuroscience, New Orleans.*

Ruzek, M. C., Miller, A. H., Opal, S. M., Pearce, B. D., and Biron, C. A. (1997b). Characterization of early cytokine responses and an interleukin (IL)-6-dependent pathway of endogenous glucocorticoid induction during murine cytomegalovirus infection. *Journal of Experimental Medicine* 185(7):1185–1192.

Ruzek, M. C., Pearce, B. D., Miller, A. H., and Biron, C. A. (1999). Endogenous glucocorticoids protect against cytokine-mediated lethality during viral infection. *Journal of Immunology* 162(6):3527–3533.

Salas, M. A., Evans, S. W., Levell, M. J., and Whicher, J. T. (1990). Interleukin-6 and ACTH act synergistically to stimulate the release of corticosterone from adrenal gland cells. *Clinical & Experimental Immunology* 79(3):470–473.

Salazar-Mather, T. P., Hamilton, T. A., and Biron, C. A. (2000). A chemokine to cytokine to chemokine cascade critical in antiviral defense. *J. Clin. Invest.* 105:985–993.

Salazar-Mather, T. P., Ishikawa, R., and Biron, C. A. (1996). NK cell trafficking and cytokine expression in splenic compartments after IFN induction and viral infection. *Journal of Immunology* 157(7):3054–3064.

Salkowski, C. A., and Vogel, S. N. (1992a). IFN-gamma mediates increased glucocorticoid receptor expression in murine macrophages. *Journal of Immunology* 148(9):2770–2777.

Salkowski, C. A., and Vogel, S. N. (1992b). Lipopolysaccharide increases glucocorticoid receptor expression in murine macrophages. A possible mechanism for glucocorticoid-mediated suppression of endotoxicity. *Journal of Immunology* 149(12):4041–4047.

Sarawar, S. R., and Doherty, P. C. (1994). Concurrent production of interleukin-2, interleukin-10, and gamma interferon in the regional lymph nodes of mice with influenza pneumonia. *Journal of Virology* 68(5):3112–3119.

Sawchenko, P. E., Swanson, L. W., and Vale, W. W. (1984). Co-expression of corticotropin-releasing factor and vasopressin immunoreactivity in parvocellular neurosecretory neurons of the adrenalectomized rat. *Proceedings of the National Academy of Sciences of the United States of America* 81(6):1883–1887.

Sawiris, G. P., Sydiskis, R. J., and Bashirelahi, N. (1994). Hormonal modulation of herpes simplex virus replication in a mouse neuroblastoma cell line. *Journal of Clinical Laboratory Analysis* 8(3):135–139.

Scheinman, R. I., Cogswell, P. C., Lofquist, A. K., and Baldwin, A. S., Jr. (1995). Role of transcriptional activation of I kappa B alpha in mediation of immunosuppression by glucocorticoids [see comments]. *Science* 270(5234):283–286.

Schleimer, R., Claman, H., and Oronsky, A. (1989). "Anti-inflammatory Steroid Action: Basic and Clinical Aspects." Academic Press, San Diego.

Schmidt, E. D., Janszen, A. W., Wouterlood, F. G., and Tilders, F. J. (1995). Interleukin-1-induced long-lasting changes in hypothalamic corticotropin-releasing hormone (CRF)—neurons and hyperresponsiveness of the hypothalamus–pituitary–adrenal axis. *Journal of Neuroscience* 15(11):7417–7426.

Schobitz, B., De Kloet, E. R., and Holsboer, F. (1994). Gene expression and function of interleukin 1, interleukin 6 and tumor necrosis factor in the brain. *Progress in Neurobiology* 44(4):397–432.

Schobitz, B., Sutanto, W., Carey, M. P., Holsboer, F., and de Kloet, E. R. (1994). Endotoxin and interleukin 1 decrease the affinity of hippocampal mineralocorticoid (type I) receptor in parallel to activation of the hypothalamic–pituitary–adrenal axis. *Neuroendocrinology* 60(2):124–133.

Seder, R. A., and Paul, W. E. (1994). Acquisition of lymphokine-producing phenotype by CD4+ T cells. *Annual Review of Immunology* **12**:635–673.

Sher, E. R., Leung, D. Y., Surs, W., Kam, J. C., Zieg, G., Kamada, A. K., and Szefler, S. J. (1994). Steroid-resistant asthma. Cellular mechanisms contributing to inadequate response to glucocorticoid therapy. *Journal of Clinical Investigation* **93**(1):33–39.

Sheridan, J. F. (1998). Norman Cousins Memorial Lecture 1997. Stress-induced modulation of anti-viral immunity. *Brain, Behavior, & Immunity* **12**(1):1–6.

Sherman, F. E., Michaels, R. H., and Kenny, F. M. (1965). Acute encephalopathy (encephalitis) complicating rubella: Report of cases with virological studies, cortisol production estimations, and observations at autopsy. *JAMA* **192**:675.

Shizuya, K., Komori, T., Fujiwara, R., Miyahara, S., Ohmori, M., and Nomura, J. (1997). The influence of restraint stress on the expression of mRNAs for IL-6 and the IL-6 receptor in the hypothalamus and midbrain of the rat. *Life Sciences* **61**(10):L135–140.

Sica, G., Lama, G., Tartaglione, R., Pierelli, L., Frati, L., della Cuna, G. R., and Marchetti, P. (1990). Effects of natural beta-interferon and recombinant alpha-2B-interferon on proliferation, glucocorticoid receptor content, and antigen expression in cultured HL-60 cells. *Cancer* **65**(4):920–925.

Siiteri, P. K., Murai, J. T., Hammond, G. L., Nisker, J. A., Raymoure, W. J., and Kuhn, R. W. (1982). The serum transport of steroid hormones. *Recent Progress in Hormone Research* **38**:457–510.

Sinicco, A., Biglino, A., Sciand ra, M., Forno, B., Pollono, A. M., Raiteri, R., and Gioannini, P. (1993). Cytokine network and acute primary HIV-1 infection. *AIDS* **7**(9):1167–1172.

Smith, E. M., Meyer, W. J., and Blalock, J. E. (1982). Virus-induced corticosterone in hypophysectomized mice: a possible lymphoid adrenal axis. *Science* **218**(4579):1311–1312.

Smith, L. R., Brown, S. L., and Blalock, J. E. (1989). Interleukin-2 induction of ACTH secretion: presence of an interleukin-2 receptor alpha-chain-like molecule on pituitary cells. *Journal of Neuroimmunology* **21**(2–3):249–254.

Soudeyns, H., Geleziunas, R., Shyamala, G., Hiscott, J., and Wainberg, M. A. (1993). Identification of a novel glucocorticoid response element within the genome of the human immunodeficiency virus type 1. *Virology* **194**(2):758–768.

Spahn, J. D., Szefler, S. J., Surs, W., Doherty, D. E., Nimmagadda, S. R., and Leung, D. Y. (1996). A novel action of IL-13: induction of diminished monocyte glucocorticoid receptor-binding affinity. *Journal of Immunology* **157**(6):2654–2659.

Spangelo, B. L., Judd, A. M., MacLeod, R. M., Goodman, D. W., and Isakson, P. C. (1990). Endotoxin-induced release of interleukin-6 from rat medial basal hypothalami. *Endocrinology* **127**(4):1779–1785.

Spencer, R. L., Miller, A. H., Moday, H., McEwen, B. S., Blanchard, R. J., Blanchard, D. C., and Sakai, R. R. (1996). Chronic social stress produces reductions in available splenic type II corticosteroid receptor binding and plasma corticosteroid binding globulin levels. *Psychoneuroendocrinology* **21**(1):95–109.

Spencer, R. L., Miller, A. H., Moday, H., Stein, M., and McEwen, B. S. (1993). Diurnal differences in basal and acute stress levels of type I and type II adrenal steroid receptor activation in neural and immune tissues. *Endocrinology* **133**(5):1941–1950.

Spinedi, E., Hadid, R., Daneva, T., and Gaillard, R. C. (1992). Cytokines stimulate the CRF but not the vasopressin neuronal system: evidence for a median eminence site of interleukin-6 action. *Neuroendocrinology* **56**(1):46–53.

Stein-Streilein, J. (1988). Natural effector cells in influenza virus infection. *In* "Functions of the Natural Immune Systems" (C. Reynolds and R. Wiltrout, eds.), pp. 67–83. Plenum Publishing Corp., New York.

Sternberg, E. M., Hill, J. M., Chrousos, G. P., Kamilaris, T., Listwak, S. J., Gold, P. W., and Wilder, R. L. (1989). Inflammatory mediator-induced hypothalamic–pituitary–adrenal axis activation is defective in streptococcal cell wall arthritis-susceptible Lewis rats. *Proceedings of the National Academy of Sciences of the United States of America* **86**(7):2374–2378.

Stith, R. D., and McCallum, R. E. (1983). Downregulation of hepatic glucocorticoid receptors after endotoxin treatment. *Infection & Immunity* **40**(2):613–621.

Stith, R. D., McCallum, R. E., and Hill, M. R. (1989). Effect of interleukin-6/interferon-beta 2 on glucocorticoid action in rat hepatoma cells. *Journal of Steroid Biochemistry* **34**(1–6):479–481.

Su, H. C., Cousens, L. P., Fast, L. D., Slifka, M. K., Bungiro, R. D., Ahmed, R., and Biron, C. A. (1998). CD4+ and CD8+ T cell interactions in IFN-gamma and IL-4 responses to viral infections: requirements for IL-2. *Journal of Immunology* **160**(10):5007–5017.

Su, H. C., Orange, J. S., Fast, L. D., Chan, A. T., Simpson, S. J., Terhorst, C., and Biron, C. A. (1994). IL-2-dependent NK cell responses discovered in virus-infected beta 2-microglobulin-deficient mice. *Journal of Immunology* **153**(12):5674–5681.

Suda, T., Tozawa, F., Ushiyama, T., Sumitomo, T., Yamada, M., and Demura, H. (1990). Interleukin-1 stimulates corticotropin-releasing factor gene expression in rat hypothalamus. *Endocrinology* **126**(2):1223–1228.

Suzuki, E., Shintani, F., Kanba, S., Asai, M., and Nakaki, T. (1997). Immobilization stress increases mRNA levels of interleukin-1 receptor antagonist in various rat brain regions. *Cellular & Molecular Neurobiology* **17**(5):557–562.

Tagoh, H., Nishijo, H., Uwano, T., Kishi, H., Ono, T., and Muraguchi, A. (1995). Reciprocal IL-1 beta gene expression in medial and lateral hypothalamic areas in SART-stressed mice. *Neuroscience Letters* **184**(1):17–20.

Tanaka, J., Ogura, T., Kamiya, S., Sato, H., Yoshie, T., Ogura, H., and Hatano, M. (1984). Enhanced replication of human cytomegalovirus in human fibroblasts treated with dexamethasone. *Journal of General Virology* **65**(Pt 10):1759–1767.

Tay, C. H., and Welsh, R. M. (1997). Distinct organ-dependent mechanisms for the control of murine cytomegalovirus infection by natural killer cells. *Journal of Virology* **71**(1):267–275.

Trgovcich, J., Ryman, K., Extrom, P., Eldridge, J. C., Aronson, J. F., and Johnston, R. E. (1997). Sindbis virus infection of neonatal mice results in a severe stress response. *Virology* **227**(1):234–238.

Trinchieri, G. (1989). Biology of natural killer cells. *Advances in Immunology* **47**:187–376.

Turnbull, A. V., and Rivier, C. L. (1999). Regulation of the hypothalamic–pituitary–adrenal axis by cytokines: actions and mechanisms of action. *Physiological Reviews* **79**(1):1–71.

Turney, T. H., Harmsen, A. G., and Jarpe, M. A. (1986). Modification of the antitumor action of Corynebacterium parvum by stress. *Physiology & Behavior* **37**(4):555–558.

Vacca, A., Felli, M. P., Farina, A. R., Martinotti, S., Maroder, M., Screpanti, I., Meco, D., Petrangeli, E., Frati, L., and Gulino, A. (1992). Glucocorticoid receptor-mediated suppression of the interleukin 2 gene expression through impairment of the cooperativity between nuclear factor of activated T cells and AP-1 enhancer elements. *Journal of Experimental Medicine* **175**(3):637–646.

van Dam, A. M., Brouns, M., Louisse, S., and Berkenbosch, F. (1992). Appearance of interleukin-1 in macrophages and in ramified microglia in the brain of endotoxin-treated rats: a pathway for the induction of non-specific symptoms of sickness? *Brain Research* **588**(2):291–296.

van den Broek, M. F., Muller, U., Huang, S., Aguet, M., and Zinkernagel, R. M. (1995). Antiviral defense in mice lacking both alpha/beta and gamma interferon receptors. *Journal of Virology* **69**(8):4792–4796.

Vankelecom, H., Carmeliet, P., Heremans, H., Van Damme, J., Dijkmans, R., Billiau, A., and Denef, C. (1990). Interferon-gamma inhibits stimulated adrenocorticotropin, prolactin, and growth hormone secretion in normal rat anterior pituitary cell cultures. *Endocrinology* **126**(6):2919–2926.

Varmus, H. E., Ringold, G., and Yamamoto, K. R. (1979). Regulation of mouse mammary tumor virus gene expression by glucocorticoid hormones. *Monographs on Endocrinology* **12**:253–278.

Verheggen, M. M., van Hal, P. T., Adriaansen-Soeting, P. W., Goense, B. J., Hoogsteden, H. C., Brinkmann, A. O., and Versnel, M. A. (1996). Modulation of glucocorticoid receptor expression in human bronchial epithelial cell lines by IL-1 beta, TNF-alpha and LPS. *European Respiratory Journal* **9**(10):2036–2043.

Walsh, C. M., Matloubian, M., Liu, C. C., Ueda, R., Kurahara, C. G., Christensen, J. L., Huang, M. T., Young, J. D., Ahmed, R., and Clark, W. R. (1994). Immune function in mice lacking the perforin gene. *Proceedings of the National Academy of Sciences of the United States of America* **91**(23):10854–10858.

Watkins, L. R., Maier, S. F., and Goehler, L. E. (1995). Cytokine-to-brain communication: a review & analysis of alternative mechanisms. *Life Sciences* **57**(11):1011–1026.

Weber, M. M., Michl, P., Auernhammer, C. J., and Engelhardt, D. (1997). Interleukin-3 and interleukin-6 stimulate cortisol secretion from adult human adrenocortical cells. *Endocrinology* **138**(5):2207–2210.

Welsh, R. M., Jr. (1978). Cytotoxic cells induced during lymphocytic choriomeningitis virus infection of mice. I. Characterization of natural killer cell induction. *Journal of Experimental Medicine* **148**(1):163–181.

Wesselingh, S. L., Levine, B., Fox, R. J., Choi, S., and Griffin, D. E. (1994). Intracerebral cytokine mRNA expression during fatal and nonfatal alphavirus encephalitis suggests a predominant type 2 T cell response. *Journal of Immunology* **152**(3):1289–1297.

Wu-Hsieh, B., Howard, D. H., and Ahmed, R. (1988). Virus-induced immunosuppression: a murine model of susceptibility to opportunistic infection. *Journal of Infectious Diseases* **158**(1):232–235.

Wyllie, A. H. (1980). Glucocorticoid-induced thymocyte apoptosis is associated with endogenous endonuclease activation. *Nature* **284**(5756):555–556.

Yirmiya, R. (1996). Endotoxin produces a depressive-like episode in rats. *Brain Research* **711**(1–2):163–174.

Zacharchuk, C. M., Mercep, M., Chakraborti, P. K., Simons, S. S., Jr., and Ashwell, J. D. (1990). Programmed T lymphocyte death. Cell activation- and steroid-induced pathways are mutually antagonistic. *Journal of Immunology* **145**(12):4037–4045.

Zeitoun, M. M., Hassan, A. I., Hussein, Z. M., Fahmy, M. S., Ragab, M., and Hussein, M. (1973). Adrenal glucocorticoid function in acute viral infections in children. *Acta Paediatrica Scand inavica* **62**(6):608–614.

Zinkernagel, R. M., and Hengartner, H. (1994). T-cell-mediated immunopathology versus direct cytolysis by virus: implications for HIV and AIDS. *Immunology Today* **15**(6):262–268.

Zubiaga, A. M., Munoz, E., and Huber, B. T. (1992). IL-4 and IL-2 selectively rescue Th cell subsets from glucocorticoid-induced apoptosis. *Journal of Immunology* **149**(1):107–112.

PRECLINICAL AND CLINICAL MODELS

ROLE OF VIRUSES IN ETIOLOGY AND PATHOGENESIS OF MULTIPLE SCLEROSIS

Samantha S. Soldan[*,†] and Steven Jacobson[*]

* Viral Immunology Section
NIH/NINDS
Bethesda, Maryland 20892
† Institute for Biomedical Sciences
Department of Genetics
George Washington University
Washington, D.C. 20052

I. Introduction

Multiple sclerosis (MS) is the most prevalent demyelinating disease of young adults, affecting an estimated 300,000 individuals in the United States alone. The natural history of the disease is unpredictable. The majority of affected individuals have a relapsing–remitting course, while a smaller subset have a more chronic–progressive presentation. Women are affected more often than men, a phenomenon associated with a number of autoimmune diseases. Although the etiology of MS is unknown, it is generally believed that genetic, immunologic, and environmental factors are involved. This chapter will highlight these issues as they suggest that exogenous factors are associated with the pathogenesis of this disorder. It has been suggested that infectious agents may comprise an environmental component of the induction and progression of this often debilitating neurological disorder. While many viruses have been investigated as potential "triggers" of MS, no virus to

517

date has been definitively associated with this disease. Recently, the human herpes virus 6 (HHV-6) has received considerable attention as an infectious agent candidate that might be associated with the pathogenesis of MS. We will focus on this agent and the data that support the role of this virus in MS disease pathogenesis. We will compare and contrast these observations with the long list of viruses that have been suggested to play a role in this disease. Additionally, we propose a model whereby in genetically susceptible individuals, multiple viruses may trigger either a virus-specific or a cross-reactive autoimmune response that results in clinical MS. Importantly, we take an open but cautious view on the role that viruses may play in the pathogenesis of a chronic, progressive neurologic disorder such as MS.

II. Etiology of Multiple Sclerosis

A. Genetic Influences

Epidemiological, familial, and molecular studies of MS have supported a strong influence of genetic background on disease susceptibility. The worldwide distribution of MS is uneven, with areas of high prevalence being found in North America, Europe, New Zealand, and Australia, and areas of lower prevalence in Asia, Africa, and South America (Kurtzke, 1995). In general, the prevalence and incidence of MS follow a north–south gradient in both hemispheres (Sadovnick and Ebers 1993; Compston, 1994; Ebers and Sadovnick, 1994). It has been suggested that the north–south gradient observed in the New World may be reflective of the propensity of individuals from regions of Europe with a high incidence of MS to migrate to the northern regions of the United States and Canada, and of individuals from regions of Europe with a lower incidence of MS to migrate to southern regions of the United States and South America (Sadovnick and Ebers, 1993; Ebers and Sadovnick, 1994). The importance of a genetic background in influencing susceptibility to MS is further supported by epidemiologic studies demonstrating different prevalences of MS among genetically disparate populations living in the same geographic area. For example, the prevalence of MS is different in Hungarians of Caucasian descent (37/100,000) as compared to Hungarian gypsies (2/100,000) (Kalman *et al.*, 1991). A similar example has been described in the United States, where the prevalence of MS among people of Japanese descent living on the Pacific Coast (6.7/100,000) is considerably lower than that of Caucasians living in California (30/100,000) (Detels *et al.*, 1977). However, it is interesting to note that people of Japanese descent living on the West Coast of the United States show a slightly higher prevalence of MS than those living in Japan (2/100,000), suggesting that environmental factors also have a significant impact on disease susceptibility (Detels *et al.*, 1977).

Family and twin studies have played an important role in establishing a genetic influence in the development of MS. It has been demonstrated that biological relatives of patients with MS have a greater likelihood of developing MS than have adoptees and that, conversely, family members of adopted individuals with MS do not have an increased risk of developing MS (Ebers *et al.*, 1995). Additionally, among biological relatives of individuals with MS, the lifetime risk of developing MS increases with closer biological relationships. The risk is greatest for siblings of affected individuals, especially sisters, and decreases in second- and third-degree relatives (Sadovnick *et al.*, 1988; Sadovnick and Ebers, 1993b). The rate of MS concordance is eight times greater in monozygotic than in dizygotic twins. However, the concordance among monozygotic twins is only 20%, which suggests that a susceptible genetic background alone is not sufficient to cause the disease (Bobowick *et al.*, 1978). Sadovnick and colleagues effectively illustrate the combined influence of genetic background and environment in the development of MS through a liability threshold model (Sadovnick *et al.*, 1999). As demonstrated in Fig. 1, the risk of acquiring MS is dependent on both genetic and environmental risk factors (load) that collectively represent an individual's "total liability." The degree of genetic and environmental loads vary on an individual basis. Once an individual's total disease liability crosses a particular threshold, clinical MS is observed (Sadovnick *et al.*, 1999).

Considerable effort has been made to ascribe MS susceptibility to various models of inheritance. However, MS has not been demonstrated to fit any of these models. In part, the inability to attribute a particular inheritance pattern to MS may arise from the difficulty in diagnosing this unpredictable disease (Lynch *et al.*, 1990; Tienari *et al.*, 1992). Additionally, because the age of high risk ranges from the late teens to the late 50s, an individual cannot be considered unaffected with certainty until they are past the age of high risk. Over the years, several genes, many of which are associated with immune function, have been tentatively associated with an increased risk of MS (McFarland *et al.*, 1997). Many of these associations have not been demonstrated consistently in different studies. However, a strong association between MS and the major histocompatibility complex (MHC) class II alleles DR2 and DQw1 has been demonstrated (Cook, 1997).

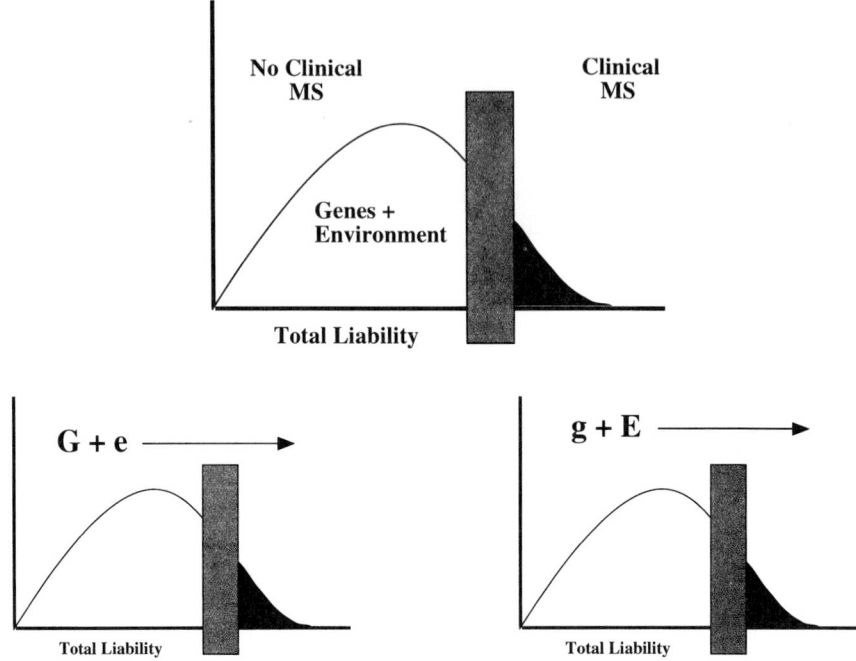

G + e = genetic load greater
g + E = non-genetic (environmental) loading greater

FIG. 1. Role of genetic and environmental influences in the lifetime risk of developing MS. Genes plus environmental factors comprise an individual's total disease liability. Below a particular threshold (gray bar), no clinical disease is evident. Above this threshold, expression of disease (black area) is present. Total disease liability in some individuals reflect a higher genetic burden (G) than environmental factors (e), relative to other patients in which environmental factors (E) are greater than genetic influences (g). Adapted from Sadovnick *et al.*, 1999.

B. Immunologic Influences

1. Immunologic Charactersitics of MS

In addition to genetic influences, it is widely accepted that T cell-mediated responses are involved in the etiology of MS. This is based on the association of MS with genes involved with the immune response, the immunopathology of the disease, the clinical response of MS patients to immunomodulatory and immunosuppressive treatments, and similarities with experimental immune-mediated demyelinating diseases in animals. As described above, the MHC class II background of an individual is an important factor in disease susceptibility. Many studies have concentrated on MHC–peptide interactions in order to determine how MHC class II alleles may confer disease susceptibility. It has been determined that the binding affinity of an antigenic peptide to an MHC allele determines T cell immunogenicity and encephalitogenicity (Greer *et al.,* 1996). The vast majority of studies concerning MHC–peptide interactions relevant to MS have focused on myelin basic protein (MBP) as the antigenic peptide. MBP-specific T cells may be demonstrated in the peripheral blood of both MS patients and normal individuals (Burns, 1983). The frequencies of MBP-specific T cells tend to be higher in MS patients than in controls (Ota *et al.,* 1990, Olson *et al.,* 1990). However, similar frequencies of MBP-specific T cells have been demonstrated in affected and unaffected family members, thereby suggesting that the frequency of MBP-specific T cells may be linked to the immunogenetic background of an individual and be a prerequisite of disease development (Joshi *et al.,* 1993). It has been suggested that molecular mimicry, a phenomenon by which environmental antigens cross-react with normal host cell components, may induce an immune response against host proteins such as MBP. Therefore, individuals with higher frequencies of MBP reactive T cells may be more likely to develop autoreactive T cell responses as a result of environmental, non-self-epitopes mimicking MBP.

A number of immune abnormalities are frequently observed in MS patients and lend support to an immunologic component of the MS disease process. As represented in Fig. 2 (see color insert) a complex series of immunological mechanisms associated with events in both the peripheral blood system and the central nervous system (CNS) has been proposed (Martin *et al.,* 1997; Brosnan *et al.,* 1997). One of the hallmarks of MS is the intrathecal secretion of oligoclonal antibodies (Kabat *et al.,* 1995; Tourtellotte, 1985). Oligoclonal bands (OCBs) are found in the CNS tissue and cerebrospinal fluid (CSF) of greater than 90% of MS patients and are used in the diagnosis of the disease. OCBs are not specific for MS as they are also found in several other chronic inflammatory CNS conditions of either infectious origin (such as CNS lyme or chronic viral and bacterial meningitis) or autoimmune origin (such as CNS lupus erythematosus). Although OCBs are not directed against a single antigen, antibody bands specific for viral and bacterial antigens, and self-antigens, have been described (Sindic *et al.,* 1994; Sriram *et al.,* 1999). Therefore, it is unclear whether the intrathecal synthesis of immunoglobulins observed in MS results from the presence of cells that are passively recruited into the CNS after the pathogenetically relevant

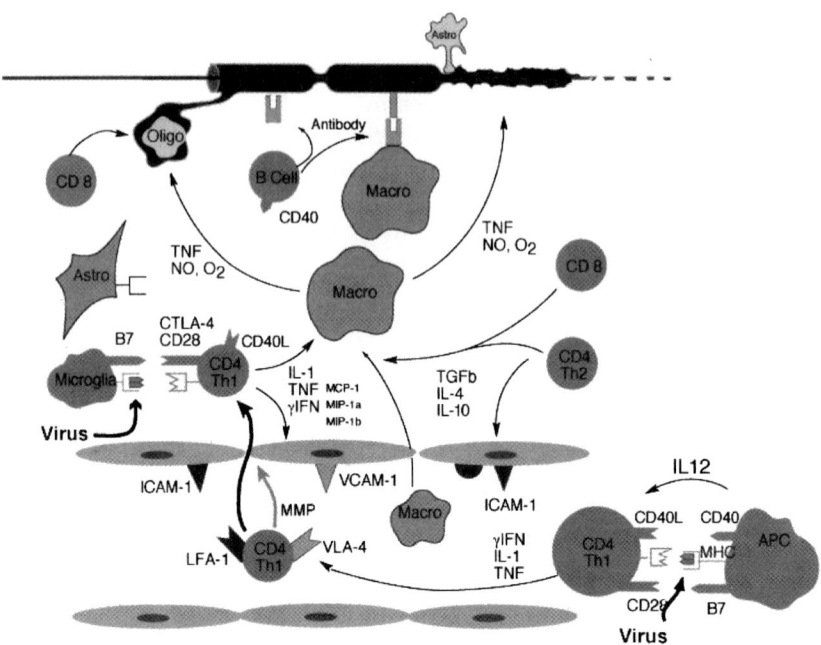

FIG. 2. [For color reproduction, see color section.] Elements of the MS lesion. A diagrammatic representation of the complex immunological mechanisms that are proposed to be involved in the initiation and maintenance of the developing MS lesion. Viruses may play a role in this model by initiating a virus-specific immune response in the periphery (lower right corner) that crosses the blood–brain barrier and encounters antigen in the CNS. Antigen may be presented on virus-infected glial cells or by resident CNS antigen-presenting cells (microglia). This interaction may result in a cascade of cytokines and chemokines that are associated with lesion development. Alternatively, autoreactive cells either in the periphery or the CNS may cross-react with viral antigens (molecular mimicry) that lead to activation of these T cells and subsequent CNS damage. (We thank H. McFarland for the use of this figure.)

cells have crossed the blood–brain barrier (BBB) or from disease-related lymphocytes. In addition to the presence of OCBs, other immunological markers of disease activity have been described in MS. The overexpression of several proinflammatory cytokines including tumor necrosis factor (TNF)-α and interferon (INF)-γ have been demonstrated in MS. Treatment of MS patients with IFN-γ resulted in a marked increase in exacerbations, which supports the model of MS as an autoimmune disease mediated by TH-1 like T cells (Panitch *et al.*, 1987). Furthermore, an increase in TNF-α expression has been found to precede relapses and inflammatory activity as measured by MRI, while the mRNA levels of inhibitory cytokines, such as interleukin (IL)-10 and transforming growth factor (TGF)-β, declined at the same time (Riekmann *et al.*, 1995). The overexpression of these cytokines may be involved in disease pathogenesis by causing the upregulation of MHC and adhesion molecule expression on endothelial and glial cells, activation of macrophages, and recruitment of TH1 cells, or by damaging oligodendroglial cells and myelin sheaths directly (Selmaj and Raine, 1988). The soluble adhesion molecules ICAM-1 and E-selectin are elevated in MS sera while soluble VCAM-1 and E-selectin are increased in the CSF of MS patients (Dore-Duffy *et al.*, 1995).

Additional support for the concept of MS as a disease with an autoimmune component is provided by the clinical improvement obtained with immunosuppressive and antiinflammatory therapies. Although corticosteroid treatment does not alter the long-term course of MS, it is used effectively in the treatment of MS exacerbations. It has been demonstrated that the administration of high-dose steroids immediately stops BBB leakage as visualized by gadolinium-enhanced MRI (Burnham *et al.*, 1991). A number of immunosuppressive and chemotherapeutic drugs including cyclophosphamide and methotrexate have been used in the treatment of MS with variable success. Currently, two immunomodulatory therapies, namely IFN-β and copolymer-1 (Cop-1), are widely used in the treatment of MS. IFN-β counters many of the effects of IFN-γ, such as the recruitment of inflammatory cells and the upregulation of MHC and adhesion molecules. Additionally, IFN-β has been shown to lower the exacerbation rate in MS patients with a relapsing–remitting course and inflammatory activity as demonstrated by MRI (Paty *et al.*, 1993; IFN-β MS Study Group, 1993; Stone *et al.*, 1995). COP-1 is a synthetic polypeptide consisting of a random sequence of four amino acids; it blocks antigen presentation by competing with antigenic peptides for the MHC binding groove. COP-1 has been demonstrated to be approximately as effective as IFN-β in early relapsing–remitting MS (Johnson, 1995). The effectiveness of these immunomodulatory therapies lends support to the presence of an autoimmune component in the pathogenesis of MS.

2. Animal Models for MS

Experimental autoimmune encephalomyelitis (EAE) models in various animals have rendered great insight into the immunopathogenesis of MS and have been especially useful in the development of immunomodulatory therapies for the disease. EAE is an acute or chronic relapsing inflammatory demyelinating disease of the CNS that is characterized by demyelinating white matter lesions and inflammation. EAE may be induced in a number of susceptible inbred animal strains by the injection of whole white matter or individual myelin proteins such as proteolipid protein (PLP), or MBP in Freund's complete adjuvant (Fritz and McFarlin, 1989). The ability to transfer EAE from an affected animal to a naive animal with cellular or humoral components demonstrates that EAE is a T cell-mediated autoimmune disease. EAE-resistant and- susceptible strains of mice, rats, and guinea pigs have been bred and, as in MS, are associated with particular MHC-class II backgrounds (Fritz and McFarlin, 1989). The clinical course and pathology of EAE varies among rodent species and strains (Stepaniak et al., 1994; Lorentzen, 1995). Animals sacrificed at various stages of the disease display lesions of various stages of inflammation, demyelination, and glial scarring reminiscent of MS plaques (Raine, 1983). EAE in primates is also useful as the inflammatory and demyelinating foci found in marmosets with EAE follow the same distribution pattern as those in MS (Massacesi et al., 1992). EAE has been used to develop treatment modalities that are broad, such as those which target the migration of encephalitogenic T cells into the CNS, and specific, such as interventions which target the trimolecular complex. The various EAE models have contributed greatly to our understanding of the immunopathogenesis and immunopathology of MS and will continue to be a great asset in the development of immunomodulatory therapies for this disease.

C. Environmental Influences

1. Epidemiological Evidence for Viruses in Multiple Sclerosis

For many years, an infectious etiology of MS has been suspected, as it fits with a number of epidemiological observations as well as the pathological characteristics of this disease. It has been widely speculated that the infectious component in the development of MS may be a virus. Data implicating a virus in the pathogenesis of MS include (1) epidemiological evidence of childhood exposure to infectious agents and an increase in disease exacerbations with viral infection (Johnson, 1994; Weinshenker, 1996); (2) geographic association of disease susceptibility with evidence of MS clustering (Haahr et al., 1997; Kurtzke, 1995); (3) evidence that migration to and from high-risk areas influences the likelihood of developing MS (Weinshenker, Alter et al., 1966); (4) abnormal immune responses to a variety of viruses (Neighbour et al., 1981; Jacobson et al., 1985); and (5) an analogy with animal models and other human diseases in which viruses can cause diseases with long incubation periods, a relapsing-remitting course, and demyelination.

As mentioned previously, the distribution of MS follows a geographic distribution, with an increased prevalence occurring in northern latitudes, which may be the result of both genetic and environmental influences. Migration studies, based chiefly on Europeans who immigrated to South Africa, Israel, and Hawaii, have also supported an infectious etiology of MS (Dean and Kurtzke, 1971; Kurtzke et al., 1970; Alter et al., 1966; Alter, 1971; Alter et al., 1978). In general, individuals who migrate from high-risk to low-risk areas after the age of 15 tend to take their risk of MS with them. However, individuals who migrate from high-risk to low-risk areas before the age of 15 acquire a lower risk. These data suggest that an environmental factor, perhaps a virus, must be presented before the age of 15 in order to influence MS susceptibility.

Reports of clusters or epidemics of MS also support a role for an infectious agent in MS. In the Faroe Islands, off the coast of Denmark, no cases of MS were reported from 1929 to 1943. After the occupation of the Faroe Islands by British troops in 1940, 20 islanders developed MS between 1940 and the end of the war. The areas of the Faroe Islands MS epidemic were found to correlate with the locations of British troop encampments after 1940. The sudden appearance of MS among the Faroese after the occupation by British troops suggests an interhuman transmission of the disease and the influence of an infectious agent (Kurtzke, 1995; Rohowsky-Kochan et al., 1995). This primary outbreak of MS was followed by three additional clusters of the disease, each being separated by 13 years. It has been proposed that the original cluster of MS was initiated by a virus that affected only the MS susceptibility of individuals between the ages of 11 and 12. The susceptible individuals harbored this virus in a latent state through adolescence, thus accounting for the 13-year interval between MS clusters on the Faroe Islands (Kurtzke, 1995). Other examples of MS epidemics have been described in the Shetland and Orkney Islands, Scotland; Key West, Florida; Mossyrock, Washington; and Mansfield, Massachusetts (Kurtzke, 1997).

2. Viruses in Demyelinating Diseases of Animals

Viruses have been implicated in a number of demyelinating diseases of the central nervous system in both animal and human subjects. The association of viruses in other demyelinating diseases further suggests a viral influence on the development of MS, by indicating that viruses are capable of inducing demyelination, and that they can persist for years in the CNS, presenting chronic diseases long after acute infection. Several of these viruses involved in demyelinating diseases of nonhumans include canine distemper virus (CDV); murine coronavirus (JHM strain); Theiler's mouse encephalomyelitis virus (TMEV); and visna virus (Appel, 1969; Kyuma and Stohlman 1990; Rodriguez *et al.*, 1987; Zink, 1992).

CDV is a member of the morbilliviruses and is related to measles and rinderpest viruses. CDV is a common infection of dogs and other members of the canine family. Acute infection with CDV is sometimes followed by a demyelinating encephalomyelitis called subacute diffuse sclerosing encephalitis. This encephalitis is characterized by tremor, paralysis, and convulsions and may not appear until weeks or months after acute infection. Lesions observed show demyelination with a sparing of axons and perivascular cuffs of lymphocytes and macrophages (Wisniewsky *et al.*, 1972). Antibodies to CDV and CNS myeline are detected in the serum and CSF of affected animals. Canine distemper demyelinating encephalomyelitis strongly resembles subacute sclerosing panencephalitis (SSPE) pathologically, virologically, and immunologically (Appel, 1969).

TMEV belongs to the family of Picornaviridae. These mouse enteroviruses are typically found in the gut. However, TMEVs are occasionally able to penetrate the CNS and cause an acute inflammation of the anterior horn cells that resembles poliomyelitis. Pathologically, this disease resembles MS in that it is characterized by demyelination with the preservation of axons. TMEV is often used as a model for MS because the pathological anomalies are limited to the CNS; infection is latent and persistent; demyelination is mediated by the immune system and occurs after a long incubation period; antibodies to myelin and proteolipid protein can be detected in diseased animals; and there are recurrences of demyelination and remyelination reminiscent of relapsing-remitting MS (Rodriguez *et al.*, 1987). Virulent and avirulent strains of TMEV have been identified. Interestingly, it is the persistent avirlulent strain that causes chronic CNS disease (Pevaar *et al.*, 1988). Susceptibility of mice to TMEV-mediated demyelinating disease is associated with MHC class I genes (Borrow *et al.*, 1992). It has been demonstrated that CD8+ T cells are critical in prevention of TMEV demyelinating disease, and that CD4+ T cells are important as helpers in the synthesis of neutralizing antiviral antibodies (Borrow *et al.*, 1992).

The murine coronavirus JHM strain is a neurotropic variant that infects small rodents. The virus readily infects oligodendrocytes and neurons and kills most animals. However, those animals that survive acute infection develop a chronic–progressive neurologic disease, while the virus establishes a persistent infection of astrocytes (Kyuwa and Stohlman 1990). These mice typically develop scattered demyelinating lesions and areas infiltrated by macrophages and lymphocytes (Kyuwa and Stohlman, 1990). As the disease progresses, lymphocytic infiltration diminishes while demyelination and astrogliosis increase. These lesions resemble the chronic plaques of MS. JHM-resistant strains have been observed in mice and rats. In JHM-infected mice, no autoimmune reaction against brain antigens has been described.

Visna virus is a member of the lentiviruses, which include human immunodeficiency viruses I and II. Infection of sheep with visna virus results in gait abnormalities followed by paraplegia and total paralysis. The disease course is variable and ranges from slowly to rapidly progressive. Neurological signs correlate with elevations in CSF protein and pleocytosis. Visna virus has a primary tropism for monocytes and macrophages and is thought to be transported to the CNS by infected monocytes that release viral particles when they differentiate into macrophages (Zink, 1989). Once released into the CNS, the virus infects microglia and leads to the recruitment and proliferation of cytotoxic T lymphocytes (CTLs). It is believed that the demyelination lesions observed in this disease may be the result of damage by CTLs specific for viral antigen or by autoantibodies that are common in chronic lentiviral diseases (Striker et al., 1987).

3. Viruses in Demyelinating Diseases of Man

Examples of viral-induced demyelinating diseases of humans include progressive multifocal leukoencephalopathy (PML), subacute sclerosing panencephalitis (SSPE), and HTLV-I-associated myelopathy/tropical spastic paraparesis (HAM/TSP). PML is a rare subacute demyelinating disease associated with JC virus, a papovavirus that is widespread in human populations worldwide. Approximately 70–50% of adult humans are seropositive for JC virus, 65% of whom are infected by the age of 14. It has been suggested that the kidneys are the site of persistent JC virus infection (Dörries and ter Meulen, 1983). Typically, PML occurs in individuals who are immunocompromised or who have defective cellular immunity. It has been reported that 3–5% of AIDS patients develop PML (Berger et al., 1987). Patients with PML present with variable symptoms, depending on the location of CNS lesions (Johnson et al., 1977). Ataxia, dementia, paralysis, and sensory abnormalities are common in PML. Pathologically, PML is typified by noninflammatory multifocal lesions scattered throughout white matter. The pathogenesis of PML is not fully understood and it is unclear why PML is such a rare disorder.

SSPE is a CNS disease of children and young adults that develops as a rare consequence of measles virus infection. The clinical course of

SSPE typically begins with subtle mental deterioration, followed by lack of coordination and other motor abnormalities (ter Meulen et al., 1983). The clinical course of SSPE may last either months or years and ultimately results in coma and death. SSPE patients have high serum and CSF antibodies to all measles virus structural proteins with the exception of the membrane (M) protein. CNS lesions in SSPE are characterized by perivascular cuffing with infiltrates of lymphocytes and plasma cells in both gray and white matter. Extensive demyelination and an increase in hypertrophic astrocytes are also observed in this disease. Cowdry type A and type B inclusion bodies containing measles virus-specific antigens are found in both neurons and glia. However, no intact measles virus particles have been observed in brain material of SSPE patients and the mechanism by which measles virus enters the CNS is unknown. The development of SSPE has not been associated with particular strains of measles virus and, therefore, it is likely that other host factors are necessary for this rare disorder to occur in measles-infected individuals.

The human T-lymphotropic virus type I (HTLV-I) is associated with a chronic, progressive neurologic disease known as HTLV-I-associated myelopathy/tropical spastic paraparesis (cited above). The clinical hallmark of HAM/TSP is a gradual onset of lower-extremity weakness, bowel and bladder dysfunction, fecal incontinence, Babinski sign, and variable sensory loss (Osame et al., 1987, 1990a, 1990b; McFarlin and Blattner, 1991). The onset of HAM/TSP is indeed gradual in most patients and the disease is clinically indistinguishable from the chronic–progressive form of MS. CSF analysis in HAM/TSP is remarkable for a mild lymphocytic pleocytosis, mild protein elevation, increased neopterin elevated IgG synthesis and IgG index, and oligoclonal bands, some of which are directed against HTLV-I (Höllsberg and Hafler, 1993; Jacobson et al., 1990). Magnetic resonance imaging has demonstrated demyelinating lesions in both the white matter and the paraventricular regions of HAM/TSP brains and swelling or atrophy in the spinal cord (Nakagawa et al., 1995). New lesions comprise mainly lymphocytes of the CD4+ lineage and monocytes, while older lesions display diffuse gliosis with macrophages and CD8+ lymphocytes in the parenchyma and perivascular spaces. Distinct plaques characteristic of MS are not observed in HAM/TSP. However, loss of myelin, with some preservation of axons, has been described. The incubation period between infection with HTLV-I and the development of HAM/TSP is typically long and, as will be described subsequently, only occurs in a minority of those infected.

III. Viruses in Multiple Sclerosis

A. *Viruses Associated with Mutiple Sclerosis*

Infectious agents have been suspected in the etiology of MS for over a century (Johnson, 1994). Over the years, several viruses have been thought to be associated with MS, and these associations are based primarily on elevated antibody titers or the isolation of a particular virus from MS material (Table I). However, none of these viruses have been definitively associated with the disease. Elevated antibody titers to several viruses, including influenza C, herpes simplex, measles, varicella-zoster, rubeola, vaccinia, Epstein-Barr, mumps, SV5, and human herpes virus 6 (HHV-6), have been reported in patients with MS, in comparison with healthy controls (Henson *et al.*, 1970, Alperovich *et al.*, 1991, Whitaker *et al.*, 1976, Ito *et al.*, 1975, Soldan *et al.*, 1997, Sumaya *et al.*, 1980). Although most of these reported agents have been discounted from consideration in the pathogenesis of MS, a few remain candidate viruses. Several bacteria have also been identified as potential etiologic agents in MS, based on the observation of increased antibody titers in MS patients as compared to controls (Salmi *et al.*, 1981, 1983; Sriram *et al.*, 1999, Vartdal *et al.*, 1980). It is not known whether elevated antibodies to infectious agents found in the CNS of MS patients represent local production of antibody in the CNS as a result of resident lymphocytes, or if they are a consequence of "spillover" of circulating serum antibodies that result from a damaged blood–brain barrier.

Of the many viruses, from a wide variety of families (Table I), that have been associated with MS, no one virus has received more consideration throughout the years than measles virus. As mentioned above, measles virus can establish persistence, both in tissue culture and *in vivo,* as is the case in SSPE, a chronic, progressive demyelinating disease of the CNS. Therefore, it has been repeatedly sought out as an etiologic agent for MS. Both humoral and cellular immune responses to measles virus differ in MS patients as compared to healthy controls. Intrathecal synthesis of measles-specific antibodies has been demonstrated in the CSF of MS patients (Norby, 1978) and, paradoxically, decreased measles-specific CTLs are found in MS patients as compared to healthy individuals (Jacobson *et al.*, 1985). Cytoplasmic tubular structures resembling measles nucleocapsids have been found in the astrocytes of one MS patient. An explant of this patient's brain tissue developed a cytopathic effect that was preventable by pretreatment with an antimeasles serum (Field, 1972). Additionally, in one study, measles virus-specific RNA has been detected by *in situ* hybridization in brain material of patients (Haase *et al.*, 1981). How-

TABLE I

Viruses Associated with MS[a]

Virus	Evidence for Association	Ref.
Coronavirus		
Coronavirus	Isolation from mice inoculated with MS brain	Burks, 1980
Herpesviruses		
HSV	Isolated in T cells from MS patients brains	Gudnadottir, 1964
	Increased CSF antibody titers in MS	Norby, 1978
	Isolated from CSF of MS patients	Bergstrom, 1989
HCMV	Isolated from chimpanzee inoculated with MS brain	Wrobleska, 1979
EBV	Higher prevalence of EBV infection in MS	Bray, 1983
HHV-6	Detection of DNA and viral protein in MS brain	Challoner, 1995
	Increased IgM and detection of serum DNA	Soldan, 1997
	Increased lymphoproliferative response to HHV-6A variant	Soldan, 2000
Flaviviruses		
Rubella	Increased antibody titers in MS	Forghani, 1978
Tick-borne encephalitis	Isolated from mice inoculated with MS blood	Vagabov, 1982
Parainfluenza viruses		
Parainfluenza virus I	Isolation in tissue culture after cell fusion of brain cells from MS patients	ter Meulen, 1972
Simian virus 5	Development of SV5 CPE in T cells after inoculation with bone marrow from an MS patient	Mitchell, 1978

Paramyxoviruses		
Measles virus	Measles RNA detected in MS brain tissue	Haase, 1991
	Impaired CTL response in MS	Jacobson, 1985
	Increased intrathecal antibody synthesis in MS	Norby, 1978
Mumps	Increased antibody titers in MS	Alpertovich, 1991
Retroviruses		
HTLV-I	Detection of retrovirus from T cells of MS	Koprowski, 1985
MSRV	Detection of retrovirus RNA in MS CSF	Peron, 1997
Retrovirus/EBV	EBV activation of retroviral-like particles in MS CSF	Munch, 1997
Rhabdovirus		
Rabies virus	Isolation from blood and CSF of two MS patients	Margulis, 1946
Undefined viral agents		
Scrapie agent	Development of scrapie in sheep after inoculation with MS brain	Mitchell, 1978
Bone marrow agent	Development of CPE in tissue culture after CSF inoculation	Mitchell, 1978

[a] Adapted from Johnson (1995).

ever, several other studies did not confirm these results (Stevens *et al.*, 1980, Hall, 1982).

Retroviruses, including HTLV-I, have been repeatedly targeted as potential agents in the pathogenesis of MS, in part due to the clinical and pathological similarities between HAM/TSP and MS. Kaprowski *et al.*, (1985) demonstrated the presence of antibodies that react with the HTLV-I gag (p24) protein in samples of serum and cerebrospinal fluid of patients with MS in Sweden and Florida. Additionally, HTLV-I sequences were reported in one-third of lymphocytes and CSF cell cultures from MS patients by *in situ* hybridization (Kaprowski *et al.*, 1985). Subsequently, it was demonstrated that there were both HTLV-I seronegative and seropositive MS patients from whom HTLV-I sequences could be amplified from the peripheral blood lymphocytes (PBLs) and CSF T cells (DeFreitas, 1995; Koprowski and DeFreitas, 1988; Redy, 1989). However, other studies have failed to confirm these results (Bangham, 1989; Richardson, 1989; Chen, 1990).

Although an association between HTLV-I and MS has not been supported, the possibility of a retroviral etiology of MS has not been excluded. In the absence of evidence for an exogenous retrovirus associated with MS, it has been suggested that human endogenous retroviruses (HERVs) could be involved (Rudge, 1991; Rassmussen *et al.*, 1993). HERVs comprise up to 1% of human DNA and have recently been suggested as "triggers" in a variety of autoimmune disorders (Krieg, 1992; Urnovitz, 1996; Nakagawa, 1997). The proposed pathogenic role for HERVs is based on the correlation of superantigen expression from the endogenous retrovirus termed $IDDMK_{1,2}22$ and insulin-dependent diabetes mellitus and the presence of autoantibodies that cross-react with HERV proteins in patients with systemic lupus erythematosus and Sjögren's syndrome (Benoist, 1997; Conrad, 1997; Bloomber, 1994; Garry, 1994).

A putative retrovirus, known as the multiple sclerosis retrovirus (MSRV), has recently joined the long list of viruses tentatively associated with MS (Peron *et al.*, 1997). MSRV *pol* (polymerase gene-encoding retroviral reverse transcriptase) sequences were isolated from retroviral particles released by leptomeningeal cells (LM7) cultured from the CSF of an MS patient (Peron *et al.*, 1997). Additionally, *pol* sequences were isolated from the serum of a significantly higher percentage of MS patients than of controls (Garson *et al.*, 1998). Interestingly, proteins from HSV-I have been demonstrated to transactivate MSRV *in vitro* (Peron *et al.*, 1997). This observation may be consistent with the correlation between MS exacerbations and viral infections (Sibley *et al.*, 1985). Sequence analysis of the MSRV *pol* gene indicates that it is virtually identical to the *pol* gene of the endogenous retrovirus –9 family that is expressed in MS and control human tissues (Brahic and Bureau, 1997). However, an extensive characterization of a new family of human endogenous retroviruses, which has been designated as human endogenous retrovirus-W (HERV-W), suggests that MSRV may be a member of the HERV-W family (Blond *et al.*, 1999; Fujinami and Libbey, 1999). Mutations in the *gag* and *pol* genes of the HERV-W family indicated that functional proteins cannot be translated. Furthermore, preliminary studies have shown that the HERV-W family is expressed solely in the placenta and fetal liver. These findings are inconsistent with the suggestion that MSRV is derived from a replication-competent endogenous retrovirus and do not provide a viable explanation for the involvement of MSRV in the pathogenesis of MS. The association of MSRV and MS is ambiguous but warrants further investigation.

B. Potential Mechanisms of Virus-Induced Demyelination in Multiple Sclerosis

There are a number of models of virus-induced demyelination in MS. All of these models attempt to explain the complex series of events (Fig. 2) that ultimately result in the MS lesion. The molecular mimicry model suggests that an immune response against viral antigens that cross-reacts with normal host cell components may contribute to the pathogenesis of MS. It has been demonstrated that several viral sequences contain part of the human MBP sequence (Jahnke *et al.*, 1985). Evidence of molecular mimicry is provided by the cross-reactivity of antibodies against proteins of herpes simplex and measles virus with human intermediate filaments (Fujinami *et al.*, 1983). Additionally, rabbits immunized with a synthetic peptide containing sequences of the hepatitis B virus polymerase developed EAE lesions. The rabbits that developed EAE as a consequence of immunization with this synthetic peptide then generated a humoral and cell-mediated immune response to both myelin and hepatitis B polymerase (Fujinami and Oldstone, 1985). Additionally, cross-reactivity between a monoclonal antibody to the VP1 protein of Theiler's murine virus and oligodendrocytes has been demonstrated (Yamada *et al.*, 1990). Demyelinating disease was observed in mice who were administered the VPI monoclonal antibody. In a recent study, amino acid homologies between immunogenic epitopes of Semliki Forest virus (SFV) and myelin autoantigens, myelin basic protein (MBP), myelin proteolipid protein, and myelin oligodendrocyte glycoprotein (MOG) were identified (Mokhtarian et al., 1999). Immunization of B6 mice with SFV proteins induced significant lymphocyte proliferation to the SFV E2 peptide as well as the MOG peptide, 18–32, but not to MBP or PLP peptides. Immunization with both MOG 18–32 and E2 115–129 induced a later-onset, chronic EAE-like disease (Mokhtarian et al., 1999). These examples of molecular mimicry support the possibility that immunological recognition of viral peptides of sufficient structural similarity to the immunodominant MBP peptide may lead to clonal expansion of MBP-reactive T cells in MS. Therefore, viruses which are known to cause latent or persistent infections, such as herpes viruses, may lead to a chronic antigenic stimulation of autoreactive T cell clones (Allegretta et al., 1990). The immunodominant region of MBP (84–102) is predominantly recognized by MS patients who are carriers of the HLA-DR2 allele. This region has been found to have sequence homology with several viruses from diverse families, thereby suggesting that multiple viral agents may "trigger" autoimmune sequelae in MS (Wucherpfennig, 1995).

A second possible mechanism of virus-induced demyelination is that of a "nonspecific bystander" effect resulting from the reaction of lymphocytes or macrophages to diverse antigens (Wisniewsky et al., 1975). In this case, oligodendrocytes or myelin sheaths could be damaged by lymphokines or proteases released by activated macrophages and immune cells in response to viral infection. The induction of inflammatory cytokines alone, such as TNF-α, has been shown to induce demyelination (Brosnan et al., 1988). This mechanism of viral stimulation of immunocompetent cells that then nonspecifically attack the myelin sheath could explain demyelination in the number of infections with diverse viruses. This mechanism for virus-induced demyelination is also proposed in the pathogenesis of HAM/TSP. It has been suggested that the recognition of HTLV-I gene products in the CNS results in the lysis of glial cells and cytokine release (Ijichi et al., 1993). This model is based on the observation that HTLV-I specific CTL restricted to immunodominant epitopes of HTLV-I gene products can be demonstrated in the PBLs and the CSF of HAM/TSP patients, and that the frequency of HTLV-I specific CTL is lower or absent in HTLV-I asymptomatic carriers. The target of the HTLV-I specific CTLs in the CNS could be either a resident glial cell (an oligodendrocyte, an astrocyte, or resident microglia), infected with HTLV-I, or an infiltrating CD4$^+$ cell. HLA class I and II are not normally expressed in the CNS, which would prevent the antigen presentation necessary for CTL activity. However, class I and class II expression are upregulated by several cytokines including IFN-γ and TNF-α, which can be induced by HTLV-

I and are known to be upregulated in HAM/TSP patients. The release of cytokine and chemokine production by HTLV-I is potentially destructive to cells of the CNS. A similar mechanism could explain the virus induction of demyelinating disease in MS.

Demyelination may also develop as a consequence of virus-induced autoimmune reaction against brain antigens. Indirect evidence in support of this theory comes from EAE in animals, and from parainfectious encephalomyelitis in humans where virus-specific CD4$^+$ lymphocytes proliferate in response to MBP (Johnson, 1985; Leibert *et al.*, 1988). It is not known how viruses break immune tolerance and force the host to mount a strong, cell-mediated immune response to brain antigens. It may be that as the virus replicates, it incorporates host antigens into its envelope and inserts, modifies, or coats itself, cellular antigens on the cell surface. It is biologically possible that these newly exposed antigens may be recognized and treated by the host as foreign (Hirsch, 1975). Söderberg-Naucler and colleagues (1996) have demonstrated the presence of CD13 in the envelope of human cytomegalovirus (HCMV). CD13 becomes associated with HCMV on budding in the Golgi-derived vacuoles during early egress (Söderberg-Nauclér *et al.*, 1996). CD13-specific antibodies were detected in the majority of patients with HCMV viremia or disease after bone marrow transplantation (Söderberg-Naucler *et al.*, 1996). These antibodies were found to cross-react with structures in normal skin biopsies (Söderberg-Nauclér *et al.*, 1996). Alternatively, lymphotropic viruses might interact with the immune regulatory system by destroying some populations of lymphocytes or stimulating the generation of autoreactive lymphocyte clones. Many lymphotropic viruses are capable of transforming infected cells and rendering them immortal (ter Meulen, 1997). It has been demonstrated *in vitro* that cells immortalized by viruses, such as EBV, are capable of secreting autoantibodies (Rosen *et al.*, 1977).

C. Ubiquity and Disease

Considerable focus has been placed on the identification of a unique virus exclusively associated with MS. However, the search for an "MS virus" (i.e., a viral infection that invariably results in MS and is not present in disease-free individuals) has been unsuccessful (Jacobson, 1998; Johnson, 1994). The inability to identify an "MS virus" could indicate that no single virus causes MS, that the putative "MS virus" has yet to be identified, or that viruses are not associated with this disease (Jacobson, 1998). Alternatively, a new paradigm of the MS disease process, which suggests that a common or ubiquitous virus may act as a trigger for MS in individuals with a genetic or an immunologic predisposition, has emerged (Jacobson, 1998). There are several examples of virus infections that lead to disease in only a subset of infected individuals. Some examples of viruses that are associated with multiple disease outcomes in different subsets of individuals include: EBV (Burkitt's lymphoma, nasopharyngeal carcinoma, mononucleosis); measles virus (SSPE); JC virus (PML); and hepatitis B and C viruses (hepatoma) (Miller, 1990; ter Meuler et al., 1983; Grinnell et al., 1983; Szmuness et al., 1978; Colombo, 1999). Perhaps the most relevant example of a virus that is common in certain populations, but results in disease only in a minority of those infected, is that of HTLV-I.

Originally identified from a T-lymphoblastoid cell line (HUT 102) of a patient diagnosed with a cutaneous T cell lymphoma, HTLV-I was the first described human retrovirus (Poiesz, 1980). In 1981, HTLV-I was established as the etiologic agent for adult T cell leukemia (ATL) (Hinuma et al., 1981), a hematological malignancy first characterized in Japan (Uchiyama et al., 1981). Since the initial description of ATL and the discovery of HTLV-I, the virus has been associated with an inflammatory, chronic, progressive neurologic disease known as HTLV-I-associated myelopathy/tropical spastic paraparesis (cited previously) and with several other inflammatory diseases (Gessain et al., 1985; Osame et al., 1986; Mochizuki et al., 1992; Nishioka et al., 1989; Morgan et al., 1989; Terada et al., 1994). While between 15 million and 25 million individuals are infected worldwide and seroprevalence rates in endemic areas can exceed 30%, the majority of individuals infected with HTLV-I are clinically asymptomatic (Gessain, 1996).

The propensity for certain individuals to develop either HAM/TSP, ATL, or other HTLV-I-associated inflammatory diseases, while at the same time, others remain clinically asymptomatic, is not fully understood. It has been suggested that host genetics and immune abnormalities influence an individual's predisposition to HAM/TSP, as they are believed to influence the likelihood of developing MS (Jacobson, 1995). Therefore, the use of HAM/TSP as a "model" of a chronic, progressive neurologic disease that occurs in only a small percentage of infected individuals is particularly germane in examining the possible involvement of a ubiquitous virus in the etiology of MS. The risk of an HTLV-I-infected individual acquiring HAM/TSP over a lifetime is estimated to be 0.25% (Osame et al., 1990c). In Japan, associations have been made between the likelihood of developing either HAM/TSP or ATL and particular HLA haplotypes (Usuku, 1988, Sonoda, 1992, Sonoda, 1996). HAM/TSP patients of Japanese descent have an increased frequency of certain HLA-Cw7,B7, and DR1 alleles represented by the A26CwB16DR9DQ3 and A24Cw7B DR1DQ1 haplotypes. In contrast, Japanese ATL patients have an increased frequency of HLA-A26, B16, and DR19 and a decreased frequency of HLA-A24 and Cw1, as compared to controls. The HLA types DRB1*0901, DQB1*0303, and DRB1*1501 in ATL patients, and the HLA types DRB1*0101, DRB1*0803, DRB1*1403 and DRB1*in HAM/TSP patients, were found to be mutually exclusive (Sonoda et al., 1996).

The neuropathology of HAM/TSP indicates that immune-mediated mechanisms are involved in the progression of this disease. Furthermore, several lines of evidence indicate that the cellular and humoral immune responses of HAM/TSP patients are altered from those of HTLV-I asymptomatic carriers and uninfected controls. The immunologic hallmarks of HAM/TSP include an increase in *ex vivo* spontaneous lymphoproliferation in the absence of antigenic stimulation or IL-2 (Kramer et al., 1989); the presence of HTLV-I-specific, CD8+ CTLs in the PBL (Jacobson et al., 1992), and an increase in antibodies to HTLV-I in sera and CSF (Gessain et al., 1985). Natural killer cells tend to be diminished in both number and activity in HAM/TSP (Kitajima et al., 1998). Although the suggestion of disease-specific HTLV-I strains has been dismissed as a factor in the determination of disease susceptibility, increased viral load has been implicated in the pathogenesis of HAM/TSP (Nagai et al., 1998). It has been suggested that increased proviral loads may be a predictor of the progression from the asymptomatic-carrier state to HAM/TSP (Jeffery et al., 1999). It has been suggested that the HLA class I allele A2*01 may confer a protective effect on the development of HAM/TSP by influencing the proviral load in infected individuals (Jeffery et al., 1999). Interestingly, the HLA A2*01 haplotype has also been shown to decrease the overall risk of MS in an HLA allele comparison study from a cohort of Swedish and Norwegian MS patients and healthy controls using polymerase chain reaction- single-strand conformation polymorphism (PCR-SSCP) (Fogdell-Hahn et al., 2000).

Several models for the immunopathogenesis of HAM/TSP have been proposed. All of these models are based on an HTLV-I-induced, immune-mediated response in the CNS to either specific viral antigens or cross-reactive self-peptides, none of which are mutually exclusive (reviewed in Kubota et al., 2000). The proposed models for the immunopathogenesis of HAM/TSP are similar to those suggested for MS. Therefore, it is hoped that insights into the pathogenesis of HAM/TSP will lead to a better understanding of MS and other neurologic disorders, such as neuro-AIDS, in which virus-mediated immunopathogenesis may occur in a subset of infected individuals.

D. Human Herpes Virus-6 and Multiple Sclerosis

One of the most recent viral candidates as an etiological agent in MS is the human herpes virus-6 (HHV-6). HHV-6 is a beta herpesvirus for which seroprevalence rates vary from 72% to 100% in healthy adults worldwide (Yamanishi, 1992; Asano and Grose, 1994; Hall *et al.*, 1994). Although HHV-6 replicates primarily in T lymphocytes, it is a pleiotropic virus that can either productively or nonproductively infect cells from several lineages including B cells, microglia, oligodendrocytes, and astrocytes (Lusso, 1988; Takahashi *et al.*, 1989; He *et al.*, 1996; Albright *et al.*, 1998). Two variants of HHV-6 (HHV-6A and HHV-6B) have been described, and are based on genomic, antigenic, and biological differences (Ablashi, 1991). The HHV-6B variant has been identified as the causative agent of exanthem subitum and accounts for the majority of symptomatic HHV-6 infections in infants. However, the HHV-6A variant has yet to be clearly associated with a particular disease (Braun *et al.*, 1997; Yaminishi *et al.*, 1988). An increased neurovirulence of the HHV-6A variant as compared to the B variant has been suggested, based on a greater detection of the HHV-6A variant than of the HHV-6 B variant in the CSF of children and adults (Hall *et al.*, 1998) Additionally, the HHV-6A variant has been isolated from the CNS of AIDS patients with areas of demyelination (Knox and Carrigan, 1995).

HHV-6 is considered to be a viable candidate as a possible etiologic agent in MS for several reasons. First, primary infection with HHV-6 usually occurs during the first few years of life, and the involvement of HHV-6 with MS is consistent with epidemiological evidence in MS suggesting exposure to an etiologic agent before puberty (Yaminishi *et al.*, 1988; Kurtzke, 1995). Second, HHV-6, particularly the HHV-6A variant, is highly neurotropic (Hall *et al.*, 1998). Primary infection with HHV-6 occasionally results in neurologic complications including meningitis and meningoencephalitis and febrile seizures (Ishiguro, 1990; Asano *et al.*, 1992). HHV-6 has been demonstrated to cause fatal encephalitis in AIDS patients and in individuals immunosuppressed as a consequence of bone marrow transplantation (Knox and Carrigan, 1994, 1995; Drobyski *et al.*, 1994). Furthermore, a neuropathogenic role for HHV-6 has been suggested, based on the development of a variety of disorders associated with active HHV-6 infection, including fulminant demyelinating encephalomyelitis, subacute leukoencephalitis, necrotizing encephalitis, progressive multifocal leukoencephalopathy, and chronic myelopathy (Kamei *et al.*, 1997; Novoa *et al.*, 1997; Carrigan *et al.*, 1996; Wagner *et al.*, 1997; Mackenzie *et al.*, 1995). Third, one of the fundamental properties of herpesviruses is their tendency to reactivate. The same factors that often lead to herpesvirus reactivation, such as stress and infection with another agent, have also been associated with MS exacerbations (Paniton, 1994). Unfortunately, the mechanisms by which HHV-6 achieves latency and reactivation are poorly understood (Yasukawa et al., 1999; Kondo, 1998). Fourth, herpesviruses are typically latent in nervous tissue and cannot be structurally identified in a latent state. Therefore, herpesviruses are not likely to be found by electron microscopy. Fifth, HHV-6 is pleiotropic and infects cells of both lymphoid and nonlymphoid origin. The pleiotropic nature of HHV-6 could be involved in the abnormalities observed in both the immune and nervous systems of patients with MS.

In 1995, Challoner and colleagues suggested a potential role for HHV-6 in MS that is based on an unbiased search for non-human DNA by representational-difference analysis (RDA). This technique is based on successive rounds of subtractive hybridization and PCR amplification, which are enriched for DNA sequences present in DNA preparations from MS disease material and control PBMCs. In this study, greater than 70 DNA fragments were analyzed. One of these fragments was found to be homologous to the MDBP gene of the HHV-6B variant Z29. HHV-6 DNA was found in 78% of MS brains and 74% of control brains. However, monoclonal antibodies against the HHV-6 101 K protein and the DNA binding protein p41 were detected in the brain tissues of MS patients, and not in controls. In MS brains, nuclear staining was found in oligodendrocytes surrounding MS plaques more frequently than in uninvolved white matter (Challoner, 1995). Additionally, a prominent cytoplasmic staining appeared in neurons in gray matter adjacent to plaques. While this study did not establish a causal link between HHV-6 and MS, it made a significant impact by suggesting the first association of a virus with MS using an unbiased technology.

The association of HHV-6 with MS has also been supported by several immunological and molecular studies. Significantly higher antibody titers against HHV-6 whole-virus preparations in MS patients as compared to normal controls, and an increase in HHV-6 DNA in the PBMCs of MS patients, determined by a polymerase chain reaction (PCR) assay, have been reported (Sola et al., 1993; Wilborn, 1994). While these preliminary studies were intriguing, they were based on methodologies that do not discriminate between latent and active infection of HHV-6. In order to distinguish between these stages of HHV-6 infection, early antibody responses to the HHV-6 p41/38 early antigen, and the presence of HHV-6 serum DNA, by a nested PCR assay, were examined (Soldan, 1997). It has been demonstrated that HHV-6 serum DNA correlates with active HHV-6 infection and is not found in healthy individuals (Secchiero et al., 1995). In this original study, a significant increase in IgM response to the p41/38 early antigen was demonstrated in patients with the relapsing–remitting form of MS, in comparison to healthy controls and individuals with other neurologic disease. An increased IgM response to the p41/38 early antigen was also observed in a group of patients with other inflammatory diseases. It is notable that of the individuals in the inflammatory disease group of increased IgM titers to HHV-6 were patients with systemic lupus erythematosus, which has been tentatively associated with active HHV-6 infection (Hoffmann, 1991; Dostal, 1997). Two additional studies have confirmed the presence of increased IgM responses to HHV-6 in patients with MS (Ablashi, 1998; Friedman, 1999), while no correlation was demonstrated in another study (Enbom, 1999).

Additionally, HHV-6 serum DNA was detected in 30% of MS patients (15 of 50) and in 0% of 47 controls consisting of healthy individuals, patients with other inflammatory diseases, and patients with other neurologic diseases (Soldan et al., 1997). This NIH cohort has been expanded to include a total of 103 MS patients and 70 controls (Fig. 3). We have continued to demonstrate the presence of HHV-6 DNA in the serum of 24% of MS patients and 0% of controls. Subsequent studies

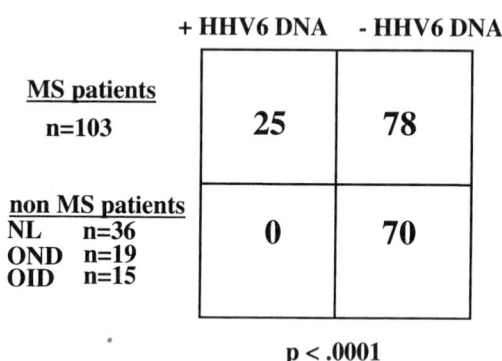

FIG. 3. An extension of a study describing the presence of HHV-6 DNA in MS patients and controls, as determined by a nested PCR assay (Soldan et al., 1997). To date, serum from 103 MS patients and 70 non-MS patients, including 36 normal donors, 19 patients with other neurological disease (OND), and 15 patients with other inflammatory diseases (OIDs) were examined for HHV-6 DNA sequences as a marker of active virus. Consistent with previous results, 24% of MS patients from the National Institutes of Health (NIH) cohort were positive, while no HHV-6 DNA was found in the serum of 70 controls. These data serve to support an association of HHV-6 in a subset of MS patients.

by a number of other groups have reexamined the presence of HHV-6 serum DNA in MS patients. Overall, the results from these studies have been equivocal. It has been suggested that discrepancies in these studies may be attributable to differences in patient selection, techniques, and reagents used (Fillet, 1998; Jacobson, 1998b). In a recent study, Locatelli and colleagues (2000) validated this theory by comparing a nested PCR assay for the HHV-6 major capsid protein region of HHV-6 (Secchiero et al., 1995) to a TaqMan quantitative PCR assay specific for the HHV-6 U67 open reading frame. The HHV-6 TaqMan assay had a greater sensitivity than the nested PCR assay had, in part due to the more efficient DNA-extraction method, which used glycogen as a carrier molecule prior to DNA precipitation (Locatelli et al., 2000). In this study, HHV-6 cell-free DNA was detected in 18% of sera and, importantly, in 34% of CSF from 29 MS patients tested. In contrast, none of the same sera and CSF were HHV-6 DNA positive by a standard nested PCR assay. This example underscores the importance of using assays of comparable sensitivity when detecting DNA of extremely low copy number. This study confirms the presence of active HHV-6 infection, as measured by serum and CSF DNA, and suggests that HHV-6 may replicate in the central nervous system of a subset of MS patients (Locatelli et al., 2000).

Lymphoproliferative responses to both the HHV-6 A and B variants have been reported in healthy adults (Wang et al., 1999; Yakushijin et al., 1991). A higher percentage of healthy adults have T cell responses to the HHV-6B variant than to the HHV-6A variant (Wang et al., 1999). The increased frequency of healthy individuals who have lymphoproliferative responses to the HHV-6B variant may reflect a higher frequency of infection with the HHV-6 B variant in the general population (Wang et al., 1999). A recent study examined the T cell lymphoproliferative responses of healthy controls and patients with MS to both variants of HHV-6 as well as to HHV-7 (Soldan et al., 2000). This study demonstrated that there was no difference in either the frequency or magnitude of proliferative responses, between healthy controls and patients with MS, to either the HHV-6B variant or HHV-7. However, a significantly higher percentage of patients with MS had proliferative responses to the HHV-6A variant (66%) than did healthy controls (33%). It is, at present, not known whether the increased frequency of lymphoproliferative responses to the HHV-6A lysate in patients with MS is the result of a higher seroprevalence of the HHV-6A variant in MS patients or of an altered host immune response (Soldan et al., 2000).

The description of an increased lymphoproliferative response to the HHV-6A variant further supports the association of HHV-6 with MS and suggests that the highly neurotropic A variant, rather than the B variant, may play a role in this disease (Soldan *et al.*, 2000). Additionally, this work emphasizes that future studies concerning the putative association of HHV-6 with MS must consider variant specific tropisms and immunology. Unfortunately, the serological methods currently available do not discriminate between the HHV-6A and HHV-6B variants, thus making it difficult to assess the seroprevalence of either variant. Presently, there is indirect evidence for the involvement of both variants in MS. HHV-6B sequences have been amplified from brain material of MS patients and controls (Challoner *et al.*, 1995; Ablashi *et al.*, 1999), and one study has demonstrated an increase in the IgM response to the HHV-6B variant in the CSF of MS patients (Ongradi *et al.*, 1999). Evidence for the involvement of the HHV-6A variant, rather than the B variant, in MS is supported by the lymphoproliferative study described and by the detection of the HHV-6 A variant in the PBMC of MS patients but not in controls (Kim *et al.*, 1999). The relationship between HHV-6 and MS remains controversial and has yet to be clearly defined. Additional serological, cellular/immune response, molecular, and clinical studies are necessary to elucidate the role, if any, of HHV-6 in the pathogenesis of MS.

E. Multiple Infectious "Triggers" in Multiple Sclerosis

It is possible that multiple viruses may be involved in the etiology of MS and that particular viruses trigger disease in different subsets of individuals. Could different viruses be associated with a disease, such as MS, through a common mechanism? A possible mechanism by which HHV-6 and, potentially, other viruses could result in MS has recently been suggested by the exciting discovery of the HHV-6 cellular receptor (Santoro *et al.*, 1999). Santoro and colleagues have clearly demonstrated that CD46, also known as the membrane cofactor protein (MCP), is the cellular receptor for HHV-6. CD46 is a member of a family of glycoproteins that are regulators of complement activation (RAC), and that prevent spontaneous activation of complement on autologous cells. CD46 is expressed on all human nucleated cells, and soluble forms can be found in plasma tears and in seminal fluid of normal individuals (Hara *et al.*, 1992). The use of a virtually ubiquitous human molecule as a surface receptor helps to explain the pleiotropism of HHV-6. Of particular interest is the fact that CD46 is also the primate-specific receptor for measles (Dorig *et al.*, 1993; Oldstone *et al.*, 1999). Significantly, HHV-6 and measles virus, which are from disparate virus families, have been associated with MS and, interestingly, use the same receptor.

Could viruses that share a receptor in common, such as HHV-6, and measles virus cause MS by a similar mechanism? It is theoretically possible that the engagement of CD46 by one or both of these viruses may result in increased activation of the complement cascade on autologous cells through downregulation of the receptor. This abnormal increase in complement could lead to widespread tissue damage through cytokine disregulation, cytolysis, and nitric oxide production (Karp *et al.*, 1996, Ghali and Schneider-Schaulies, 1998). Furthermore, an increase in soluble CD46 has been described in several autoimmune disorders, including systemic lupus erythematosus and Sjogren's syndrome and may be implicated in the pathogenesis of other autoimmune diseases (Cuida *et al.*, 1997; Kawano *et al.*, 1999). Additionally, other viruses use various members of the RAC family as cellular receptors. Epstein-Barr virus, which has also been implicated in MS, uses CD21, while CD55 is used by several echoviruses and coxsackie viruses (Yerfenol *et al.*, 1976; Bergelson *et al.*, 1994, 1995). Further studies are needed to determine whether these members of the RAC family that serve as virus receptors play a role in the pathogenesis of MS.

The potential influence of viruses that utilize members of the RAC family, such as HHV-6 and measles virus, on the pathogenesis of MS may, in part, be elucidated by animal studies. Recently, a CD46 transgenic mouse that can be infected by measles virus was described (Oldstone *et al.*, 1999). Measles virus infection in these transgenic mice was associated with immunosuppression and virus replication in the CNS. Measles virus infection was also associated with CNS disease in infected mice (Oldstone *et al.*, 1999). The generation of a CD46 transgenic mouse provides an excellent model for studying the role of measles virus infection in CNS disease. Additionally, the CD46 transgenic mouse may provide a model for studying the neuropathogenesis of HHV-6 infection in the CNS if in fact the virus, which has an extremely limited host range, may productively infect these mice that now express the HHV-6 receptor. Furthermore, a recent study has demonstrated that EAE may be inhibited by the use of a complement inhibitor, which indicates an important role for complement in EAE as well as in MS (Davoust *et al.*, 1999). Future studies investigating the interactions of measles virus and HHV-6 with CD46 may elucidate the role of both viruses and complement in the pathogenesis of MS.

IV. CONCLUSIONS

The pathogenesis and etiology of MS have yet to be well defined. Epidemiologic evidence suggests that it is a multifactorial disease that develops as a result of host genetics, immune response, and environment. Several lines of evidence, including the documentation of viruses that induce a variety of demyelinating diseases in both humans and animals, suggest that a virus may comprise the environmental component in the etiology of MS. While many viruses have been proposed as etiologic agents in MS, none of these viruses have been firmly associated with disease pathogenesis. Additionally, mechanisms by which virus–host interactions may lead to demyelination are not fully understood. Currently, HHV-6 and MSRV are receiving much attention as potential MS "triggers." However, the role of these viruses in the pathogenesis of MS is unclear. We suggest that multiple viral agents may induce a virus-specific and/or a cross-reactive autoimmune process resulting in clinical disease in a subset of genetically susceptible individuals. The involvement of multiple infectious agents in MS may explain the difficulty in identifying a single viral agent responsible for this highly variable and chronic disease. Moreover, we encourage extreme caution in attempts to readily associate viruses in a chronic, progressive neurologic disorder such as MS. As outlined in this review, it is difficult to determine cause from effect, particularly when a ubiquitous viral agent is suggested to play a role in disease pathogenesis. Uniformity in assay design, viral-isolation techniques, molecular probes, etc., must be employed by different research groups on a large number of MS cohorts to confirm these virus associations. Perhaps only through well-controlled, clinical, antiviral therapeutic trials, with defined clinical, virological, and radiographic outcome measures, can we ever determine the role that viruses, if any, may play in the pathogenesis of MS.

REFERENCES

Ablashi, D. V., Balachandran, N., Josephs, S. F., Hun, C. L., Krueger, G. R. F., Kramarsky, B., Salahuddin, S. Z., and Gallo, R. C. (1991). Genomic polymorphism, growth properties, and immunologic variations in human herpesvirus-6 isolates. *Virology* **184:**545–552.
Ablashi, D. V., Lapps, W., Kaplan, M., Whitman, J. E., Richert, J. E., and Pearson, G. R. (1998). Human herpesvirus-6 (HHV-6) infection in multiple sclerosis: a preliminary report. *Multiple Sclerosis* **4:**490–496.
Albright, A. V., Lavi, E., Black, J. B., Goldberg, S., O'Connor, M. J., and Gonzalez-Scarano, F. (1998) The effect of human herpesvirus-6 (HHV-6) on cultured human neural cells: oligodendrocytes and microglia. *J. Neurovirol.* **4:**486–494.
Allegretta, M., Nicklas, J. A., Siram, S., and Albertini, R. J. (1990). T cells responsive to myelin basic protein in patients with multiple sclerosis. *Science* **247:**718–721.
Alperovitch, A., Berr, C., Cambon-Thomsen, A., Puel, J., Dugoujon, J. M., Ruidavets, J. B., and Clanet, M. (1991). Viral antibody titers, immunogenetic markers, and their interrelations in multiple sclerosis patients and controls. *Hum. Immunol.* **31:**94–99.
Alter, M., Leibowitz, U., and Speer, J. (1966). Risk of multiple sclerosis related to age at immigration to Israel. *Arch. Neurol.* **15:**234–237.

Alter, M. and Okihiro, M. (1971). When is multiple sclerosis acquired? *Neurology* **21:**1030–1036.

Alter, M. Kahana, E. and Loewenson, R. (1978). Migration and risk of multiple sclerosis. *Neurology* **28:**1089–1093.

Appel, M. J. (1969). Pathogenesis of canine distemper. *Am. J. Vet Res.*, **30:**1167–1182.

Asano, Y., Yoshikawa, T., Kajita, Y., Ogura, R., Suga, S., Yazaki, T., Nakashima, T., Yamada, A., and Kurata, T. (1992). Fatal encephalitis/encephalopathy in primary human herpesvirus-6 infection. *Arch. Dis. Child* **67:**1484–1485.

Asano, Y, and Grose, C. (1994). Human herpesvirus type 6 infections. *In* "Herpes Virus Infections" (Glasser, R, Jones, J. F., eds.). Marcel Dekker, New York 227–244.

Bangham, C. R. M., Nightingale, S., Cruikshank, J. K., and Daenke S. (1989). PCR analysis of DNA from multiple slcerosis patients for the presence of HTLV-I. *Science* **246:**821–824.

Benoist C. and Mathis D. (1997). Retrovirus as trigger, precipitator or marker? *Nature* **388:**833–834.

Bergelson, J. M., Chan, M., Solomon, K. R., St. John, N. F., Lin, H., and Finberg, R. W. (1994). Decay-accelerating factor (CD5r), a glycosyl-phosphatidylinositol-anchored complement regulatory protein is a receptor for several echoviruses. *Proc. Natl. Acad. Sci. U.S.A.* **91:**6245–6249.

Bergelson, J. M., Mohanty, J. G., Crowell, R. L., St John, N. F., Lublin, D. M., and Fineberg, R. W. (1995). Coxsackie virus B3 adapted to growth in RD cells binds to decay-accelerating factor (CD55). *J. Virol.* **69:**1903–1906.

Berger, J. R., Kaszovitz, B., Post, M. J. D., and Dickinson, G. (1987). Progressive multifocal leukoencephalopathy associated with human immunodeficiency virus infection—a review of the literature with a report of sixteen cases. *Ann. Intern. Med.* **107:**78–87.

Bergstrom, T., Andersen, O., Vahlne, A. (1989). Isolation of herpes simplex virus type I during first attack of multiple sclerosis. *Ann. Neurol.* **26:**283–285.

Blond, J. L., Beseme, F., Duret, L., Bouton, O., Bedin, F., Perron, H., Mandrand, B., and Mallet, F. (1999). Molecular characterization and placental expression of HERV-W, a new human endogenous retrovirus family. *J. Virol.* **73:**1175–1185.

Bobowick, A. R., Kurtzke, J. F., and Brody, J. A. (1978). Twin study of multiple sclerosis: an epidemiological inquiry. *Neurology* **28:**978–987.

Borrow, P., Tonks, P., Welsh, C. J. R., and Nash, A. A. (1992). The role of CD8+ T cells in the acute and chronic phases of Theiler's virus-induced disease in mice. *J. Gen. Virol.* **73:**1861–1865.

Brahic M. and Bureau J. F. (1997). Multiple sclerosis and retroviruses [letter]. *Ann. Neurol.* **42:**984–985.

Braun, D. K., Dominguez, G. and Pellett, P. E. (1997). Human herpesvirus 6. *Clin. Microbiol. Rev.* **10:**521–567.

Brosnan, C. F., Selmaj, K., and Raine, C. S. (1988). Hypothesis: a role of tumor necrosis factor in immune-mediated demyelination and its relevance to multiple sclerosis. *J. Neuroimmunol.* **18:**87–94.

Brosnan, C. F., Racke, M. K., Selmaj, K. An investigational approach to disease therapy in multiple sclerosis. *In* "Multiple Sclerosis Clincal and Pathogenetic Basis" (Raine, C. S., McFarland, H. F., and Tourtellotte, W. W., eds.) 243–283. Chapman and Hall, London.

Burks, J. S., Devald, B. L., Jankovsky, L. D., and Gerdes, J. C. (1980). Two coronaviruses isolated from central nervous system tissue of two multiple sclerosis patients. *Science* **209:**933–934.

Burnham, J. A., Wright, R. R., Dreisbach, J. and Murray, R. S. (1991). The effect of high-dose steroids on MRI gadolinium enhancement in acute demyelinating lesions. *Neurology* **41:**1349–1354.

Burns, J. Rosenzweig, A., Zweiman, B., and Lisak, R. P. (1983). Isolation of myelin basic protein-reactive T-cell lines from normal human blood. *Cell. Immunol.* **81:**435–440.

Carrigan, D. R., Harrington, D., and Knox, K. K. (1996). Subacute leukoencephalitis caused by CNS infection with human herpesvirus 6 manifesting as acute multiple sclerosis. *Neurology* **47:**145–148.

Challoner, P. B., Smith, K. T., Parker, J. D., MacLeod, D. L., Coulter, S. N., Rose, T. M., Schultz, E. R., Bennett, J. L., Garber, R. L., and Chang, M. (1995). Plaque-associated expression of human herpesvirus 6 in multiple sclerosis. *Proc. Natl. Acad. Sci. U.S.A.* **92:**7440–7444.

Chen, I. S. Y., Haislip, A. M., and Myers L. W. (1990). Failure to detect human T cell leukemia virus-related sequences in multiple sclerosis blood. *Arch. Neurol.* **47:**1064–1065.

Colombo M. (1999). Natural history and pathogenesis of hepatitis C virus-related hepatocellular carcinoma. *J. Hepatol.* **31**Suppl:25–30.

Compston, A. (1994). The role of genetic factors in multiple sclerosis susceptibility. *Ann. Neurol.,* **Suppl**S211–217.

Conrad, B., Weissmahr, R. N., Boni, J., Acari, R., Schupbach, J., and Mach, B. (1997). A human endogenous retroviral superantigen as candidate autoimmune gene in type I diabetes. *Cell* **90:**303–313.

Cook, S. D. (1997). Multiple sclerosis and viruses. *Mult. Scler.* **3:**388–389.

Cuida, M., Legler, D. W., Eidsheim M., and Jonsson R. (1997). Complement regulatory proteins in the salivary glands and saliva of Sjogren's syndrome patients and healthy subjects. *Clin. Exp. Rheumatol.* **15:**615–623.

Davoust, N., Nataf S., Reiman R., Holers, M. V., Campbell, I. L., Barnum and S. R. (1999). Central nervous system-targeted expression of the complement inhibitor sCrry prevents experimental allergic encephalomyelitis. *J. Immunol.* **163:**6551–6556.

Dean, G., and Kurtzke, J. F. (1971). On the risk of multiple sclerosis according to age at immigration to South Africa. *Br. Med. J.,* **3:**725–729.

DeFreitas, E., Wroblewska, G., Sheremata, W., Ferrante, P., Lavi, E., Harper, M., Marzo-Veronese, F., and Koprowski, H. (1997). *AIDS Research and Human Retroviruses* **3:**19–32.

Detels, R., Visscher, B. R., and Malmgren, R. M., Coulson, A. H., Lucia, M. V., Dudley, J. P. (1977). Evidence for susceptibility to multiple sclerosis in Japanese Americans. *Am. J. Epidemiol.,* **105:**307–328.

Dore-Duffy, P., Newman, W., Balabanov, R., Lisak, R. P., Mainolfi, E., Rothlein, R., and Peterson, M. (1995). Circulating soluble adhesion proteins in cerebrospinal fluid and serum of patients with multiple sclerosis: correlation with clinical activity. *Ann. Neurol.* **37:**55–62.

Dorig, R., Marcel, A., Chopra, A., and Richardson, C. D. (1993). The human CD46 molecule is a receptor for measles virus (Edmonton strain). *Cell* **75:**295–305.

Dorries, K., and ter Meulen, V. (1983). Progressive multifocal leucoencephalopathy: detection of papovavirus JC in kidney tissue. *J. Med. Virol.* **11:**307–317.

Dostal, C, Newkirk, M. M., Duffy, K. N., Paleckova, A., Bosak, V., Cerna, M., Zd'arsky, E., and Zvarova, J. (1997). Herpesviruses in multicase families with rheumatoid arthritis and systemic lupus erythematosus. *Ann. N. Y. Acad. Sci.* **815:**334–337.

Drobyski, W. R., Knox, K. K., Majewski, D., and Carrigan, D. R. (1994). Brief report: fatal encephalitis due to variant B human herpesvirus-6 infection in a bone marrow-transplant recipient. *N. Engl. J. Med.,* **330:**1356–1360.

Ebers, G. C., and Sadovnick, A. D. (1994). The role of genetic factors in multiple sclerosis susceptibility. *J. Neuroimmunol* **54:**1–17.

Ebers, G. C., Sadovnick, A. D., and Risch, N. J. (1995). A genetic basis for familial aggregation in multiple sclerosis. *Nature,* **377:**50–51.

Enbom, M. (1999). Similar humoral and cellular immunological reactivities to human herpesvirus 6 in patients with multiple sclerosis and controls *Clin. Diag. Lab. Immunol.* **6:**545–549.

Field, E. J., Cowshall, S., Narang, H. K., and Bell, T. M. (1972). Viruses in multiple sclerosis? *Lancet,* **ii:**280–281.

Fillet, A. M., Lozeron, P., Agut, H., Lyon-Caen, O., and Liblau, R. (1998). HHV-6 and multiple sclerosis. *Nat. Med.* (letter) **4:**537.

Fogdell-Hahn, A., Ligers, A., Gronning, M., Hillert, J., and Ollerup, O. (2000). Multiple sclerosis: a modifying influence of HLA class I genes in an HLA class II associated autoimmune disease. *Tissue Antigens* (in press).

Forghani, B., Cremer, N. E., Johnson, K. P., Ginsberg, A. H., Likosky, W. H. (1978). Viral antibodies in cerebrospinal fluid of multiple sclerosis and control patients: comparison between radioimmunoassay and conventional techniques. *J. Clin. Microbiol.* **7:**63–69.

Friedman, J. E. (1999). The association of the human herpesvirus-6 and MS. *Mult. Scler.* **5:**355–362.

Fritz, R. B., and McFarlin, D. E. (1989). Encephalitogenic eptiopes of myelin basic protein. *In* "Antigenic Determinants and Immune Response" (E. E. Sercarz, ed.) *Chem. Immunol.* **46:**Karger, Basel, 101–125.

Fujinami, R. S., Oldstone, M. B., Wroblewska, A., Frenkel, M. E., and Koprowski, H. (1983). Molecular mimicry in virus infection. Cross-reaction of measles virus phosphoprotein herpes simplex virus protein with human intermediate filaments. *Proc. Natl. Acad. Sci. USA* **80:**2346–2350.

Fujinami, R. S., and Oldstone, M. B. A. (1985). Amino acid homology between the encephalitogenic site of myelin basic protein and virus: mechanism for autoimmunity. *Science,* **230:**1043–1045.

Fujinami, R. S., and Libbey, J. E. (1999). Endogenous retroviruses: are they the cause of multiple sclerosis? *Trends in Microbiology* **7:**263–264.

Garry, R. F. (1994). New evidence for involvement of retroviruses in Sjogren syndrome and other autoimmune diseases. *Arthis. Rheum.* **37:**465–469.

Garson, J. A., Tuke, P. W., Giraud, P., Paranhos-Baccala, G., and Peron, H. (1998). Detection of virion-associate MSRV-RNA in serum of patients with multiple sclerosis. *Lancet* **351:**33.

Gessain, A., Barin, F., Vernant, J. C., Gout, O., Maurs, L., Calender, A., and de The, G. (1985). Antibodies to human T-lymphotropic virus I in patients with tropical spastic paraparesis. *Lancet* **2:**407–410.

Gessain A. Epidemiology of HTLV-I and associated diseases. (1996). *In* "Human T-cell lymphotropic virus type I". (Höllsberg P, Hafler, D. A., eds.), 33–64. John Wiley & Sons London.

Ghali, M., and Schneider-Schaulies, J. (1998). Receptor (CD46) and replication-mediated interleukin-6 induction by measles virus in human astrocytoma cells. *J. Neurovirol.* **4:**521–530.

Greer, J. M., Sobel, R. A., Sette, A., Southwood, S, Lees, M. B., and Kuchroo, V. K. (1996). Immunogenic and encephalitogenic epitope clusters of myelin proteolipid protein. *J. Immunol.* **156:**371–379.

Grinnell, B. W., Padgett, B. L., and Walker D. L. (1983). Distribution of nonintegrated DNA from JC papovavirus in organs of patients with progressive multifocal leukoencephalopathy. *J. Infect. Dis.* **147:**669–675.

Gudnadottir, M., Helgadottir, H., Gjarnason, O., and Jonsdottir, K. (1964). Virus isolated from the brain of a patient with multiple sclerosis. *Exp. Neurol.,* **9:**85–95.

Haahr, S., Munch, M., Christensen, T., Meller-Larson, A., and Hvas, J. (1997). Cluster of multiple sclerosis patients from Danish community. *Lancet.* **349:**9056.

Haase, A. T., Ventura, P., Gibbs, C. J., and Tourtellotte, W. W. (1981). Measles virus nucleotides sequences—detection by hybridization in situ. *Science* **212:**672–675.

Hall, C. B., Long, C. E., Schnabel K., and Hall C. B. (1994). Human herpes virus 6 infection in children: prospective evaluation for complications and reactivation. *N. Engl. J. Med.* **331:**432–438.

Hall, C. B., Caserta, M. T., Schnabel, K. C., Long, C., Epstein, L. G., Insel, R. A., and Dewhurst, S. (1998). Persistence of human herpesvirus 6 according to site and variant: possible greater neurotropism of variant A. *Clin. Infect. Dis.* **26:**132–137.

Hall, W. W., and Choppin, P. W. (1982). Failure to detect measles virus proteins in brain tissue of patients with multiple sclerosis. *Lancet* **1:**957.

Hara, T., Kuriyama, S., Kiyohara, H., Nagase, Y., Matsumoto, M., and Seya. (1992). Soluble forms of membrane cofactor protein (CD46 MCP) are present in plasma tears, and seminal fluid in normal subjects. *Clin. Exp. Immunol.* **89:**490–494.

He, J., and McCarthy, M., Zhou, Y., Chandran, B., Wood, C. (1996). Infection of primary human fetal astrocytes by human herpesvirus 6. *J. Virol.* **70:**1296–1300.

Henson, T. E., Brody, J. A., Sever, J. L., Dyken, M. L., and Cannon, J. (1970). Measles antibody titers in multiple sclerosis patients, siblings, and controls. *JAMA* **211:**1985–1989.

Hinuma, Y., Nagata, K., Hanaoka, M., Nakai, M., Matsumoto, T., Kinoshita, K. I., Shirakawa, S., and Miyoshi, I. (1981). Adult T-cell leukemia: antigen in an ATL cell line and detection of antibodies to the antigen in human sera. *Proc. Natl. Acad. Sci. USA* **78:**6476–6480.

Hirsch, M. S., and Proffett, M. R. (1975) Autoimmunity in viral infection. *In* "Viral Immunology and Immunopathology" (A. L. Notkins, ed.), Academic Press, New York, 419.

Hoffmann, A., Kirn, E., Kuerten, A., Sander, C., Krueger, G. R., and Ablashi, D. V. (1991). Active human herpesvirus-6 (HHV-6) infection associated with Kikuchi-Fujimoto disease and systemic lupus erythematosus (SLE). *In Vivo* **5:**265–269.

Höllsberg, P., and Hafler, D. A. (1993). Pathogenesis of diseases induced by human lymphotropic virus type I infection. *N. Engl. J. Med.* **328:**1173–1182.

IFNβ Multiple Sclerosis Study Group (1993). Interferon beta-1b is effective in relapsing-remitting multiple sclerosis. I. Clinical results of a multicenter, randomized, double-blind, placebo-controlled trial. *Neurology* **43:**655–661.

Ijichi, S., Izumo, S., Eiraku, N. Machigashira, K., Kubota, R., Nagai. M., Ikegami, N., Kashio, N., Umehara, F., and Maruyama, I. (1993). An autoaggressive process against bystander tissues in HTLV-I infected individuals: a possible pathomechanism of HAM/TSP. *Med. Hypotheses* **41:**542–547.

Ishiguro, M. (1990). Meningo-encephalitis associated with HHV-6 related exanthem subitum. *Acta. Paediatr. Scand.* **79:**987–989.

Ito, M., Barron, A. L., Olszewski, W. A., and Milgrom, F. (1975). Antibody titers by mixed agglutination to varicella-zoster, herpes simplex and vaccinia viruses in patients with multiple sclerosis. *Proc. Soc. Exp. Biol. Med.* **149:**835–839.

Jacobson, S., Flerlage, M. L., and McFarland, H. F. (1985). Impaired measles virus-specific cytotoxic T cell response in multiple sclerosis. *J. Exp. Med.* **162:**839–850.

Jacobson, S., Gupta, A., Mattson, D., Mingioli, E., and McFarlin, D. E. (1990). Immunological studies in tropical spastic paraparesis. *Ann. Neurol.* **27:**149–156.

Jacobson, S. McFarlin, D. E., Robinson, S. Voshkuhl, R., Martin, R., Brewah, A., Newell, A. J., and Koenig, S. (1992). HTLV-I specific cytotoxic T-lymphocytes in the cerebrospinal fluid of patients with HTLV-I associated neurological disease. *Ann. Neurol.* **32**:651–657.

Jacobson, S. (1995). HTLV-I Myelopathy: An immunopathologically mediated chronic progressive disease of the central nervous system. *Current Opinion in Neurology* **8**:179–183.

Jacobson, S. (1998). Association of human herpesvirus-6 and multiple sclerosis: here we go again? *J. Neurovirol.* **4**:471–473.

Jacobson, S., Soldan, S. S., Berti, R. (1998b). HHV-6 and multiple sclerosis. *Nat. Med.* **4**:538.

Jahnke, U., Fischer, E. H., and Alvord, E. C. (1985). Sequence homology between certain viral proteins and proteins related to encephalomyelitis and neuritis. *Science* **229**:282–284.

Jeffery, K. J., Usuku, K., Hall, S. E., Matsumoto, W., Taylor, G. P., Procter, J., Bunce, M., Ogg, G. S., Welsh, K. I., Weber, J. N., Lloyd, A. L., Nowak, M. A., Nagai, M., Kodama, D., Izumo, S., Osame, M., and Bangham, C. R. (1999). HLA alleles determine human T-lymphotropic virus-I (HTLV-I) proviral load and the risk of HTLV-I associated myelopathy. *Proc. Natl. Acad. Sci. USA* **96**:3848–3453.

Johnson, K. P., Brooks, B. R., Cohen, J. A., Ford, C. C., Goldstein, J., Lisak, R. P., Myers, L. W., Panitch, H. S., Rose, J. W., Schiffer, R. B. (1995). Copolymer 1 reduces relapse rate and improves disability in relapsing–remitting multiple sclerosis results of a phase III multicenter, double-blind, placebo-controlled trial. *Neurology,* **45**:1268–1276.

Johnson, R. T., Narayan, O., Weiner, C. P., and Greenlee, J. E. (1977). Progressive multifocal leukoencephalopathy. *In* "Slow Virus Infections of the CNS" (V. ter Meulen and M. Katz). Springer Verlag, New York.

Johnson, R. T., Griffin, D. E., Hirsch, R. L. (1985). Measles encephalomyelitis: clinical and immunological studies. *N. Engl. J. Med.* **320**:1667–1672.

Johnson, R. T. (1994). The Virology of Demyelinating Diseases. *Ann. Neurol.* **36**:S54–D60.

Joshi, N., Usuku, K., and Hauser, S. L. (1993). The T-cell response to myelin basic protein in familial multiple sclerosis: diversity of fine-specificity restricting elements, and T-cell receptor usage. *Ann. Neurol.* **34**:385–393.

Kabat, E. A., Fredman, D. A., Murray, J. P., and Knaub, V. (1950). A study of the crystalline albumin, gamma globulin and total protein in the cerebrospinal fluid of one hundred cases of multiple sclerosis and in other diseases. *Am. J. Med. Sci.* **219**:55–64.

Kalman, B., Takacs, K., Gyodi, E. Kramer, J., Fust, G., Tauszik, T., Guseo, A., Kuntar, L., Komoly, S., and Nagy, C. (1991). Sclerosis multiplex in gypsies. *Acta. Neurol. Scand.* 181–185.

Kamei A., Ichinohe, S., Onuma, R., Hiraga, S., Fujiwara, T. (1997). Acute disseminated demyelination due to primary human herpesvirus-6 infection, *Eur. J. Pediatr.* **156**:709–712.

Karp, C. L., Wysocka, M., Wahl, L. M., Ahearn, H. M., Cuomo, P. J., Sherry, B., Trichieri, G., and Griffin, D. E. (1996). Mechanism of suppression of cell-mediated immunity by measles virus. *Science* **273**:228–231.

Kawano, M., Seya, T., Koni, I., and Mabuchi, H. (1999). Elevated serum levels of soluble membrane cofactor protein (CD46,MCP) in patients with systemic lupus erythematosus (SLE). *Clin. Exp. Immunol.* **116**:542–546.

Kim, J. -S., Park, J. -H., Lee, K. -H., and Lee, K. -S. (1999). Detection of human herpesvirus 6 variant A sequence in multiple sclerosis. AAN 51st Annual Meeting, Toronto, Canada, April 17–24. Abstract number P06.034.

Kitajima, I., Osame, M., Izumo, S., Igata, A. (1998). Immunological study of HTLV-I-associated myelopathy. *Autoimmunity* **1**:125–131.

Knox, K. K., and Carrigan, D. R. (1994). Disseminated active HHV-6 infections in patients with AIDS. *Lancet* **343:**577–578.

Knox, K. K., and Carrigan, D. R. (1995). Active human herpesvirus (HHV-6) infection of the central nervous system in patients with AIDS. *J. Acqui. Immune. Defic. Syndr. Hum. Retrovirol.* **9:**69–73.

Kondo, K., Nagafuji, H., Hata, A., Tomomori C, Yamanishi K. (1993). Association of human herpesvirus 6 infection of the central nervous system with recurrence of febrile convulsions. *J. Infect Dis.* **167:**1197–1200.

Kondo, K. (1998). Persistent/latent infection of human beta-herpesvirus. *Nippon Rinsho* **56:**83–89.

Koprowski, H. and DeFreitas, E. (1988). HTLV-I and chronic nervous diseases: present status and a look into the future. *Ann. Neurol.* **23(suppl):**S166–S170.

Kramer, A., Jacobson, S., Reuben, J. F., Murphy, E. L., Wiktor, S. Z., Cranston, B., Figueroa, J. P., Hanchard, B., McFarlin, D., and Blattner, W. A. (1989). Spontaneous lymphocyte proliferation is elevated in asymptomatic HTLV-I positive Jamaicans. *Lancet* **2:**923–924.

Krieg, A. M., Gourley, M. F., and Perl, A. (1992). Endogenous retroviruses: potential etiologic agents in autoimmunity. *FASEB. J.* **6:**2537–2544.

Kubota, R., Osame, M., and Jacobson, S. (2000). Retrovirus: Human T-cell lymphotropic virus type I-associated diseases and immune dysfunction. *In* "Effects of Microbes on the Immune System" (M. W. Cunningham and R. S. Fujinami, eds.), 349–371. Lippincott/Williams and Wilkins, Philadephia.

Kurtzke, J. F., Dean, G., and Botha, D. P. J. (1970). A method of estimating the age at immigration of white immigrants to South Africa, with an example of its importance. *S. Afr. Med. J.* **44:**663–669.

Kurtzke, J. F. (1995). MS epidemiology worldwide. One view of current status. *Acta. Neurol. Scand.* **Suppl 161:**23–33.

Kurtzke, J. F. (1997). The Epidemiology of Multiple Sclerosis. *In* "Mutliple Sclerosis: Clinical and Pathogenetic Basis" (Cedric D. Raine, ed.), 91–139. Chapman and Hall, London.

Kyuwa, S. and Stohlman, S. A. (1990). Pathogenesis of a neurotropic murine coronavirus, strain JHM in the central nervous system of mice. *Semin. Virol.,* **1:**273–280.

Leibert, U. G., Linington, C., and ter Meulen, V. (1988). Induction of autoimmune reactions to myelin basic protein in measles virus encephalitis in Lewis rats. *J. Neuroimmunol.* **17:**103–118.

Locatelli, G., Malnati, M. S., Franciotta, D., Furlan, R., Comi, G., Martino, G., and Lusso, P. (2000). Detection of human herpesvirus 6 by quantitative real-time PCR in serum and cerebrospinal fluid of patients with multiple sclerosis. *Mult. Scler.* (in press).

Lorentzen, J. C., Issazadeh, S., Storch, M., *et al.* (1995). Protracted relapsing and demyelinating experimental autoimmune encephalomyelitis in DA rats immunized with syngeneic spinal cord and incomplete Freund's adjuvant. *J. Neuroimmunol.,* **63:**193–205.

Lusso, P., di Marzo Veronese, F., Salahuddin, S. Z., Ablashi, D. V., Pahwa, S., Krohn, K., and Gallo, R. C. (1988). In vitro cellular tropism of human B-lymphotropic virus (human herpesvirus 6). *J. Exp. Med.* **167:**1659–1670.

Lynch, S. G., Rose, J. W., and Smoker, W. (1990). MRI in familial multiple sclerosis. *Neurology,* **40:**900–903.

Mackenzie, I. R., Carrigan, D. R., and Wiley, C. A. (1995). Chronic myelopathy associated with human herpesvirus-6. *Neurology* **45:**2015–2017.

Margulis, M. S., Soloviev, V. D., and Shubladze, A. K. (1946). Aetiology and pathogenesis of acute sporadic disseminated encephalomyelitis and multiple sclerosis. *J. Neurol. Neurosurg. Psychiatry.* **9:**63–74.

Martin, R, and McFarland, H. F. Immunology of multiple sclerosis and experimental allergic encephalomyelitis. *In* "Multiple Sclerosis: Clinical and Pathogenetic Basis" (C. S. Raine, H. F., McFarland, and W. W., Tourtellotte, eds.), 221–239. Chapman and Hall, London.

Massacesi, L., Joshi, N., Lee, P. D. Rombos, A., Letvin, N. L., Hauser, S. L. (1992). Experimental allergic encephalomyelitis in cynomolgus monkeys. Wquantitation of T cell responses in peripheral blood. *J. Clin. Invest.,* **90:**399–404.

McFarland, H. F., Martin, R., and McFarlin, D. E. (1997). Genetic influences in multiple sclerosis. *In* "Mutliple Sclerosis: Clinical and Pathogenetic Basis" (Cedric S. Raine, ed.), 205–219. Chapman and Hall, London.

McFarlin, D. E., and Blattner, W. A. (1991). Non-AIDS retroviral infections in humans. *Annu. Rev. Med.* **42:**97–105.

Miller, G. (1990). Epstein-Barr Virus Biology, Pathogenesis, and Medical Aspects. *In* "Virology," 2nd ed. (B. N. Fields, D. M. Knipe *et al.,* eds.), 1921–1958, Raven Press, New York.

Mitchell, D. N., Porterfield, J. S., Micheletti, R. (1978). Isolation of an infectious agent from bone marrows of patients with multiple sclerosis. *Lancet,* **II:**387–391.

Mochizuki, M., Wanatanabe, T., Yamaguchi, K., Takatsuki, K., Yoshimura, K., Shirao, M., Nakashima, S., Mori, S., Araki, S., and Miyata, N. (1992). HTLV-I uveitis: a distinct clinical entity caused by HTLV-I. *Jpn. J. Cancer. Res.,* 29–42.

Mokhtarian, F., Zhang, Z., Shi, Y., Gonzales, E., Sobel, R. A. (1999). Molecular mimicry between a viral peptide and a myelin oligodendrocyte glycoprotein peptide induces autoimmune demyelinating disease in mice. *J. Neuroimmunol.* **95:**43–54.

Morgan, O. S., Rodgers-Johnson, P., Mora, C., Char, G. (1989). HTLV-I and polymyositis in Jamaica. *Lancet* **2:**1184–1187.

Munch, M., Hvas, J., Christensen, T. Moller-Larsen, A., and Haahr, S. (1997). The implication of Epstein-Barr virus in multiple sclerosis—a review. *Acta. Neurol. Scand.* Suppl **1699:**59–64.

Nagai, M., Usuku, K., Matsumoto, W., Kodama, D., Takenouchi, N., Moritoyo, T., Hashiguchi, S., Ichinose, M., Bangham, C. R., Izumo, S., and Osame, M. (1998). HTLV-I proviral load in 202 HAM/SP patients and 243 asymptomatic HTLV-I carriers: High proviral load strongly predisposes to HAM/TSP. *J. Neuro. Virology.* **4:**586–593.

Nakagawa, K, and Harrison, L. C. (1997). The potential roles of endogenous retroviruses in autoimmunity. *Immunol. Rev.* **152:**193–236.

Nakagawa, M., Izumo, S., Ijichi, S. Kubota, H., Arimura, K., Kawabata, M., and Osame, M. (1995). HTLV-I-associated myelopathy: analysis of 213 patients based on clinical features and laboratory findings. *J. Neurovirol.* **1:**50–61.

Nishioka, K., Maruyama, I., Sato, K. Kitajima, I., Nakajima, Y., and Osame, M. (1989). Chronic inflammatory arthropathy associated with HTLV-I. *Lancet* **1:**441.

Norby E (1978). Viral antibodies in multiple sclerosis in progress. *In* "Medical Virology" (J. L. Melnich, ed.), Karger, Basel.

Novoa, L. J., Edwards-Lee, T., Tourtellotte, W. W., and Cornford, M. E. (1997). Fulminant demyelinating encephalomyelitis associated with productive HHV-6 infection in an immunocompetent adult. *J. Med. Virol.* **2:**301–308.

Oldstone, M. A., Lewicki, H., Thomas, D., Tishon, A., Dales, S., Patterson, J., Manchester, M., Homann, D., Naniche, D., and Holz, A. (1999). Measles virus infection in a transgenic model: Virus-induced immunosuppression and central nervous system disease. *Cell* **99:**629–640.

Olson, T., Wei Ahi, W., Hôjeburg, B. *et al.* (1990). Autoreactive T lymphocytes in multiple sclerosis determined by antigen-induced secretion of interferon-g. *J. Clin. Invest.,* **86:**981–985.

Ongradi, J., Rajda, C., Marodi, C. L., Csiszar, A., and Vecsei, L. (1999). A pilot study on the antibodies to HHV-6 variants and HHV-7 in CSF of MS patients. *J. Neurovirol.* **5:**529–532.

Osame, M., Usuku, K., Izumo, S., *et al.* (1986). HTLV-I associated myelopathy: a new clinical entity. *Lancet* **1:**1031–1032.

Osame, M., Matsumoto, M., Usuku, K. Izumo, S., Ijichi, N., Amitani, H., Tara, M., and Igata, A. (1987). Chronic progressive myelopathy associated with elevated antibodies to human T-lymphotropic virus type I and adult T-cell leukemia-like cells. *Ann. Neurol.* **21:**117–122.

Osame, M., Igata, A., Matsumoto, M., Kohka, M., Usuku, K., and Izumo, S. (1990a). HTLV-I-associated myelopathy (HAM), treatment trials, retrospective survey and clinical and laboratory findings. *Hematol. Rev.* **3:**271–284.

Osame, M., and McArthur, J. C. (1990b). Neurologic manifestations of infection with human T cell lymphotropic virus type I. *In:* "Disease of the Nervous System Clinical Neurobiology," 1921–1958 W. B. Saunders, Philadelphia.

Osame, M., Janssen, R., Kubota, H. Nishitani, H., Igata, A., Nagataki, S., Mori, M., Goto, I., Shimabukuro, H., and Khabbaz, R. (1990c). Nationwide survey of HTLV-I-associated myelopathy in Japan: association with blood transfusion. *Ann. Neurol.* **28:**51–56.

Ota, K., Matsui, M., Milford, E. L. (1990). T-cell recognition of an immunodominant myelin basic protein epitope in multiple sclerosis. *Nature* **346:**183–187.

Panitch, H. S., Hirsch, R. L., Schindler, J. and Johnson, K. P. (1987). Treatment of multiple sclerosis with gamma interferon: exacerbations associated with activation of the immune system. *Neurology.* **37:**1097–1102.

Paniton, H. (1994). Influence of infection on exacerbations of multiple sclerosis. *Ann. Neurol.* **36:**S25–S28.

Paty, D. W., Li, D. K. B., the UBC MS/MRI Study Group, and the IFNβ relapsing–remitting multiple sclerosis study group (1993). II. MRI analysis results of a multicenter, randomized, double-blind, placebo-controlled trial. *Neurology,* **43:**662–667.

Peron, H., Garson, J. A., Bedin, F., Beseme, F., Paranhos-Baccala, G., Komurian-and Pradel, F. (1997). Molecular identification of a novel retrovirus repeatedly isolated from patients with multiple sclerosis. *Proc. Natl. Acad. Sci. U.S.A.* **94:**7583–7588.

Pevear, D. C., Brokowski, J., Calenhogg, M., Oh, C. K., Ostrowski, B., and Lipton, H. L (1988). Insights into Theiler's virus neurovirulence based on a genomic comparison of the neurovirulent GDVII and less virulent Ben strains. *Virology,* **165:**253–259.

Poiesz, B. J., Ruscetti, F. W., Gazdar, A. F., and Gallo, R. C. (1980). Detection and isolation of type C retrovirus particles from fresh and cultured lymphocytes of a patient with cutaneous T cell lymphoma. *Proc. Natl. Acad. Sci. U.S.A.* **77:**7415–7419.

Raine, C. S. (1983) Multiple sclerosis and chronic–relapsing EAE: comparative ultrastructural neuropathology. *In* "Multiple Sclerosis" (J. F. Hallpike, C. W. Adams, and W. W. Tourtellotte, eds.). Williams and Wilkins, Baltimore.

Rand, K. H., Houck, H., Denslow, N. D., and Heilman, K. M. (2000). Epstein-Barr virus nuclear antigen-1 (EBNA-1)-associated oligoclonal bands in patients with multiple sclerosis. *J. Neurol. Sci.* **173:**32–39.

Rassmussen, H. B., Geny, C., Deforeges, L., Perron, H., Tourtelotte, W. W., Heberg, A., and Clausen, J. (1997). Expression of endogenous retroviruses in blood mononuclear cells and brain tissue from multiple sclerosis patients. *Multiple Sclerosis* **1:**82–89.

Reddy, E. P., Sandberg-Wollheim, M., Mettus, R. V., Ray, P. E., DeFreitas, E., and Koprowski, H. (1989). Amplification and molecular cloning of HTLV-I sequences from DNA of multiple sclerosis patients. *Science* **243(suppl):**S29–S33.

Richardson, J. H., Wucherpfennig, K. W., and Endo, N. (1989). PCR analysis of DNA from multiple sclerosis patients for the presence of HTLV-I. *Science* **246:**821–824.

Riekman, P., Albrecht, M., and Kitze, B. (1995). Tumor necrosis factor-α messenger RNA expression in patients with relapsing–remitting multiple sclerosis is associated with disease activity. *Ann. Neurol.,* **37:**82–88.

Rodriguez, M., Oleszak, E., and Liebowitz, J. (1987). Theiler's murine encephalomyelitis: a model of demyelination and persistence of virus. *Crit. Rev. Immunol.,* **7:**325–365.

Rosen, A., Gergley, P. Jondal, M., Klein, G, and Briton, S. (1977). Polyclonal Ig production after Epstein-Barr virus infection of human lymphocytes in vitro. *Nature* **267:**5254.

Rudge P, (1991). Does a retrovirally encoded superantigen cause multiple sclerosis? *J. Neurol. Neurosurg. Psychiatry.* **54:**853–855.

Sadovnick, A. D., Baird, P. A., and Ward, R. H. (1988). Multiple sclerosis: updated risk for relatives. *Am. J. Genet.,* **29:**533–541.

Sadovnick, A. D., and Ebers, G. C. (1993). Epidemiology of multiple sclerosis: a critical overview. *Can. J. Neurol. Sci.,* **20:**17–29.

Sadovnick, A. D. (1993). Familial recurrence risks and inheritance of multiple sclerosis. *Curr. Opin. Neurol. Neurosurg.* 189–194.

Sadovnick, A. D., Dircks, A., and Ebers G. C. (1999). Genetic counselling in multiple sclerosis: risks to sibs and children of affected individuals. *Clin. Genet.* **56:**118–122.

Salmi, A., Viljanen, M., and Reunanen, M. (1981). Intrathecal synthesis of antibodies to diphtheria and tetanus toxoids in multiple sclerosis patients. *J. Neuroimmunol.* **1:**333–41.

Salmi, A., Reunanen, M., Ilonen, J., and Panelis, M. (1983). Intrathecal antibody synthesis to virus antigens in multiple sclerosis. *Clin. Exp. Immunol.* **52:**241–249.

Santoro, F., Kennedy, P. E., Locatelli, G., Malnati, M. S., Berger, E. A., and Lusso, P. (1999). CD46 is a cellular receptor for human herpesvirus 6. *Cell* **99:**817–827.

Secchiero, P., Carigan, D. R., Asano, Y., Benedetti, L., Crowley, R. W., Komaroff, A. L., Gallo, R. C., Lusso, P. (1995). Detection of human herpesvirus 6 in plasma of children with primary infection and immunosuppressed patients by polymerase chain reaction. *J. Infect. Dis.* **171:**273–280.

Selmaj, K. and Raine, C. S. (1988). Tumor necrosis factor mediates myelin and oligodendrocyte damage in vitro. *Ann. Neurol.,* **23:**339–346.

Sibley, W. A., Bamford, C. R., and Clark, K. (1985). *Lancet* **184:**1313–1315.

Sindic, C. J., Monteyne, P., Laterre, E. C. (1994) The intrathecal synthesis of virus-specific oligoclonal IgG in multiple sclerosis. *J. Neuroimmunol.* **54:**75–80.

Söderberg-Nauclér, C. S., Larsson, S., Möller, E. (1996). A novel mechanism for virus-induced autoimmunity in humans. *Immunological Reviews* **152:**175–192.

Sola, P., Merelli, E., Marasca, R., Poggi, M., Luppi, M., Montorsi, M., Torelli, G. (1993). Human herpesvirus 6 and multiple sclerosis: survey of anti-HHV-6 antibodies by immunofluorescence analysis and of viral sequences by polymerase chain reaction. *J. Neurol. Neurosurg. Psychiatry.* **56:**917–919.

Soldan, S. S., Berti, R., Salem, N., Secchiero, P., Flamand, L., Calabresi, P. A., Brennan, M. B., Maloni, H. W., McFarland, H. F., Lin, H-C., Patnaik, M., and Jacobson, S. (1997). Association of human herpes virus 6 (HHV-6) with multiple sclerosis: Increased IgM response to HHV-6 early antigen and detection of serum HHV-6 DNA. *Nature Medicine* **3:**1394–1397.

Soldan, S. S., Leist, T. P., Juhng, K. N., McFarland, H. F., and Jacobson, S. (2000). Increased lymphoproliferative response to human herpesvirus type 6 A variant in multiple sclerosis patients. *Ann. Neurol.* (in press).

Sonoda, S., Yashiki, S., and Fujiyoshi, T. (1992). Immunogenetic factors involved in the pathogenesis of adult T-cell leukemia and HTLV-I associated myelopathy. *Gann.* **39:**81–93.

Sonoda, S., Fujiyoshi, T., Yashiki, S. (1996) Immunogenetics of HTLV-I/II and associated diseases. *J. Acquir. Immune. Defic. Syndr. Hum. Retrovirol.* **13**:S119–S123.

Sriram, S., Stratton, C. W., Yao, S., Tharp, A., Ding, L., Bannan, J. D., Mitchell, W. M. (1999). Chlamydia pneumoniae infection of the central nervous system in multiple sclerosis. *Ann. Neurol.* **46**:6–14.

Stepaniak, J. A., Gould, K. E., Sun, D. and Swanbourg, R. H. (1994). A comparative study of experimental autoimmune encephalomyelitis in Lewis and DA rats. *J. Immunol.,* **155**:2702–2709.

Stevens, J. G., Bastone, V. B., Ellison, G. W., and Myers, L. W. (1980). No measles virus genetic information detected in multiple sclerosis-derived brains. *Ann. Neruol.* **8**:625–637.

Stone, L. A., Frank, J. A., Albert, P. S. Bash, C., Smith, M. E., Maloni, H., and McFarland, H. F. (1995). The effect of interferon-b on blood-brain-barrier disruptions demonstrated by contrast-enhanced magnetic resonance imaging in relapsing–remitting multiple sclerosis. *Ann. Neurol.* **37**:611–619.

Striker, R. B., McHugh, T. M., and Moody D. J. (1987). An AIDS-related autoantibody reacts with a specific antigen on stimulated CD4+ T cells. *Nature,* **327**:710–713.

Sumaya, C. V., Myers, L. W., and Ellison, G. W. (1980). Epstein-Barr virus antibodies in multiple sclerosis. *Arch. Neurol.* **37**:94–96.

Szmuness, W., Stevens, C. E., Ikram, H., Much, M. I., Harley, E. J., Hollinger, and F. B. (1978). Prevalence of hepatitis B virus infection and hepatocellular carcinoma in Chinese-Americans. *J. Infect. Dis.* **137**:822–29.

ter Meulen, V., Koprowski, H., Iwasaki, T. (1972). Fusion of cultured multiple-sclerosis brain cells with indicator cells: presence of nucleocapsid and virion and isolation of parainfluenza -type virus *Lancet* **1**:1–5.

ter Meulen, V., Stephenson, J. R., and Kreth, H. W. (1983) Subacute sclerosing panencephalitis. *In* "Comprehensive Virology" (H. Frenkel-Conrat and R. R. Wagner, eds.), 105–59. Plenum Press, New York.

ter Meulen, V., Katz, M. The proposed viral etiology of multiple sclerosis and related demyelinating diseases. *In* "Multiple Sclerosis: Clinical and Pathogenetic Basis" (C. S. Raine, McFarland, H. F., and Tourtellotte, W. W., eds.), 287–301. Chapman and Hall, London.

Terada, K., Katamine, S., Eguchi, K., Moriuchi, R., Kita, M., Shimada, H., Yamashita, I., Iwata, K., Tsuji, Y., and Nagataki, S. (1994). Prevalence of serum and salivary antibodies to HTLV-I in Sjogren's syndrome. *Lancet* **344**:1116–1119.

Takahashi K, Sonoda, S., Higashi, K., Kondo, T., Takahashi, H., Takahashi, M., Yamanishi, K. (1989). Predominant CD4 T-lymphocyte tropism of human herpesvirus 6 related virus. *J. Virol.* **63**:3161–3163.

Tienari, P. J., Salonen, O., Wikstron, J., Valanne, L. and Pallo, J. (1992). Familial multiple sclerosis: MRI findings in clinically affected and unaffected siblings. *J. Neurol. Neurosurg. Psychiatry* **55**:883–886.

Tourtellotte, W. W. (1985). The cerebrospinal fluid in multiple sclerosis. *In* "Handbook of Clinical Neurology 3" (revised series), "Demyelinating Diseases" (P. J. Vinken, G. W. Bruyn, H. L. Klawans and J. C. Koetsier, eds.), 79–130, Elsevier, Amsterdam/New York.

Uchiyama, T., Yodoi, J., and Sagawa, K. (1977). Adult T-cell leukemia: Clinical and hematologic features of 16 cases. *Blood.* 481–92.

Umovitz, H. B., and Murphy, W. B. (1996). Human endogenous retroviruses: nature, occurrence, and clinical implications in human disease. *Clin. Microbiol. Rev.* **9**:72–99.

Usuku, K., Sonoda, S., Osame, M. Yashiki, S., Takahashi, K., Matsumoto, M., Sawada, T., Tsuji, K., Tara, M., Igata, A. (1981). HLA haplotype-linked high immune responsiveness against HTLV-I in HTLV-I associated myelopathy: compariosn with adult T cell leukemia/lymphoma. *Ann. Neurol.* **23**:S143–S150.

Vagabov, R. M., Skvortsova, T. M., Gofman, YuP, and Barinsky, I. F. Isolation of the tick-borne encephalitis virus from a patient with multiple sclerosis. *Acta. Virol.* **26**:403.

Vartdal, F., Vandvik, B., and Norby, E. (1980). Viral and bacterial antibody responses in multiple sclerosis. *Ann. Neurol.* **8**:248–255.

Wagner, M., Muller-Berghaus, J., Schroeder, R., Sollberg, S., Luka, J., Leyssens, N., Schneider, B., Krueger, G. R. (1997) Human herpesvirus-6 (HHV-6) associated necro-tizing encephalitis in Griscelli's syndrome. *J. Med. Virol.* **53**:306–312.

Wang, F.-Z., Dahl, H., Ljungman, P., Linde, A. (1999). Lymphoproliferative responses to human herpesvirus-6 in healthy adults. *J. Med. Virol.* **57**:134–139.

Weinshenker, B. G. Epidemiology of Multiple Sclerosis. *Neuro. Clin.* **14**(2):291–308.

Whitaker, J. N., Herrmann, K. L., Rogentine, G. N., Stein, S. F., and Kollins, L. L. (1976). Immunogenetic analysis and serum viral antibody titers in multiple sclerosis. *Arch. Neurol.* **33**:399–403.

Wilborn, F., Schmidt, C. A., Brinkmann, V. Jendroska, K., Oettle, H., and Siegert, W. (1994). A potential role for human herpesvirus type 6 in nervous system disease. *J. Neuroimmunol.* **49**:213–124.

Wisniewsky, H. M., Raine, C. S., and Kay, K. J. (1972). Observations on viral demyeli-nating encephalomyelitis: canine distemper. *Lab. Invest.,* **26**:589–599.

Wisniewsky, H. M. and Bloom, B. R. (1975). Primary demyelination as a nonspecific con-sequence of a cell-mediated immune reaction. *J. Exp. Med.* **141**:346–359.

Wrobleska, Z., Gilden, D., Devlin, M. (1979). Cytomegalovirus isolation from a chim-panzee with acute demyelinating disease after inoculation of multiple sclerosis brain cells. *Infect. Immun.* **25**:1008–1015.

Yakushijin Y, Yasukawa M, and Kobayashi Y (1991) T-cell immune response to human herpesvirus-6 in healthy adults. *Microbiol. Immunol.* **35**:655–660.

Yamada, M., Zubriggn, A., and Fujinami, R. S. (1990). Monoclonal antibody to Theiler's murine encephalomyelitis virus defines a determinant on myelin and oligodendro-cytes and augments demyelination in experimental allergic encephalomyelitis. *J. Exp. Med,* **171**:1893–1907.

Yaminishi, K., Okhuno, T., Shiraki, K., Takahashi, M., Kondo, T., Asano, Y., Kurata, T. (1988). Identification of human herpesvirus-6 as a causal agent for exanthem subi-tum. *Lancet* **1**:1065–1067.

Yamanishi K. (1992). Human herpesvirus. *Microbiol. Immunol.* **36**:551–561.

Yasukawa, M., Ohminami, H., Sada, E., Yakushijin, Y., Kaneko, M., Yanagisawa, K., Kohno, H., Bando, S., and Fujita, S. (1999). Latent infection and reactivation of human herpesvirus 6 in two novel myeloid cell lines. *Blood* **93**:991–999.

Yerfenol, E., Klein, G., Jondal, M., and Oldstone, M. B. (1976). Surface markers on human V and T-lymphocytes. IX. Two color immunofluorescence studies on the asso-ciation between EBV receptor and complement receptor on the surface of lymphoid cell lines. *Int. J. Cancer.* **17**:693–700.

Zink, M. C. (1992). The pathogenesis of lentiviral disease in sheep and goats. *Semin. Virol.,* **3**:147–155.

Zink, M. C., and Narayan, O. (1989). Lentivirus-induced interferon inhibits maturation and proliferation of monocytes and restricts the replication of caprine arthritis encephalitis virus. *J. Virol.,* **63**:2578–2584.

ADVANCES IN VIRUS RESEARCH, VOL 56

BORNAVIRUS TROPISM AND TARGETED PATHOGENESIS: VIRUS–HOST INTERACTIONS IN A NEURODEVELOPMENTAL MODEL

Mady Hornig, Thomas Briese, and W. Ian Lipkin

Emerging Diseases Laboratory
Gillespie Neuroscience Research Facility
University of California
Irvine, California 92697

I. Introduction

Outcomes following viral infections of the central nervous system (CNS) depend on a complex interplay of specific viral characteristics with host factors such as species, strain, age, and relative maturity and integrity of the immune and nervous systems at the time of infection. Among viral determinants, tropism of a virus for selected neuronal and/or glial populations plays a key role in influencing patterns of CNS injury or dysfunction. Borna disease virus (BDV), a newly classified RNA virus (Briese *et al.*, 1994; Cubitt *et al.*, 1994; Schneemann *et al.*, 1995), demonstrates tropism for, and induces damage in, monoaminergic circuits of Lewis rats following infection as juveniles or adults (Solbrig *et al.*, 1995, 1994, 1996a, 1996b, 1998); less is known about neurotransmitter selectivity of BDV following neonatal infections. It is associated with a spectrum of behavioral disturbances reminiscent of human neuropsychiatric disorders in both natural and experimental animal infections, depending on the timing of infection and the host species and strain (Lipkin *et al.*, 1995). Irrespective of lively controversy over the potential association of human neuropsychiatric disease with BDV infection (Hatalski *et al.*, 1997; Lipkin and Hornig, 1998), experimental infections of adult and neonatal rats with

BDV afford an intriguing system within which to examine the role of host maturity in determining patterns of selective virus-induced damage (Rubin et al., 1999a).

While BDV infection of adult immunocompetent rodents and ungulates results in marked CNS inflammation, loss of brain mass, and gliosis that is accompanied by dramatic disturbances in behavior, limbic circuitry, and monoamine neurotransmitter systems (Solbrig et al., 1995, 1994, 1996a, 1996b, 1998), in neonatally infected rats, more subtle disturbances of behavior, without robust inflammatory cell infiltration, are observed (Carbone et al., 1991; Hornig et al., 1999; Narayan et al., 1983a). Rats infected with BDV during the neonatal period have dysgenesis of hippocampal and cerebellar structures (Bautista et al., 1995; Bautista et al., 1994; Carbone et al., 1991; Eisenman et al., 1999; Hornig et al., 1999); disturbances of early postnatal locomotor development (Hornig et al., 1999); hyperactivity (Bautista et al., 1994; Hornig et al., 1999; Pletnikov et al., 1999a), abnormal play behavior (Pletnikov et al., 1999b); learning deficits (Dittrich et al., 1989; Rubin et al., 1999b); and sleep–wake cycles (Bautista et al., 1994); these parallel aspects of the neurodevelopmental abnormalities of autism (Rapin and Katzman, 1998), schizophrenia (Stevens, 1973; Weinberger, 1987), and bipolar mood disorder (Beckmann and Jakob, 1991; Gutierrez et al., 1998; van Os et al., 1997; Yolken and Torrey, 1995). The more modest immunopathology in the neonatal rat model suggests that damage is due to viral products that impact neuronal migration, connectivity, function, or viability. This chapter will first review mechanisms for neurotropism and viral persistence. We will then describe the immunopathologic and neuropharmacologic disturbances noted in the more widely studied, immune-mediated adult rodent model of Borna disease; this will be followed by an examination of recent findings relevant to neuropathogenesis following neonatal BDV infection.

II. Mechanisms for Neurotropism: phosphorylation of Borna Disease Virus Phosphoprotein by Protein Kinase Cε

BDV is neurotropic and disseminates from the site of infection by intraaxonal transport (Carbone et al., 1987; Rott and Becht, 1995). The integrity of the humoral immune response is critical to the restriction of virus to neural compartments (Stitz et al., 1998); however, this alone fails to explain limbic tropism as replication remains higher in limbic structures in animals with compromised humoral immunity. The extent to which tropism is influenced by the expression of viral receptors is not known.

Restricted distribution of the enzymatic machinery required for the virus' life cycle is postulated to contribute to the preferential replication of BDV in limbic circuits. Phosphorylation of the BDV phosphoprotein (P), for example, is thought to be important for its functional activation and ability to serve as a transcription factor, and is accomplished primarily by the epsilon isotype of protein kinase C (PKCε) (Schwemmle et al., 1997). PKCε is concentrated heavily in limbic circuitry, including intense staining in hippocampus and olfactory tubercle, and moderate staining in anterior olfactory nuclei, olfactory bulb, cerebellum, nucleus accumbens, lateral septal nuclei, and caudate putamen (Saito et al., 1993). Given the extensive overlap in the regional distributions of BDV and PKCε in rat brain, the possibility exists that the localization of PKCε and its associated phosphorylation effects may influence the tropism of BDV for limbic circuitry.

III. MECHANISMS FOR BDV PERSISTENCE

Persistent infection of the CNS typically requires evasion of host immune response mechanisms, attenuation and restriction of the virus' replicative capacity, and prolonged host survival. The persistence of BDV in adult-infected hosts is unusual in that the agent is capable of establishing persistent infection and maintaining an apparently normal repertoire of viral gene expression in the context of an intact immune response. Indeed, the cellular immune response is essential to expression of classical Borna disease: the appearance of CD8+ T cells in the brains of adult-infected rats occurs in concert with upregulation of major histocompatibility complex (MHC) Class I antigen expression in brain (Carbone et al., 1991) and parallels the onset of signs of neurologic dysfunction (Planz et al., 1995). Although the molecular basis for persistence is not known, induction of tolerance may play a role. Altered expression of viral genes or of host cellular immune responses are possible explanations. The first possibility is unlikely as CNS viral titers are not substantively changed over the course of disease (Carbone et al., 1987; Narayan et al., 1983a). Furthermore, brain levels of the two BDV proteins that can be readily quantitated, nucleoprotein (N) and phosphoprotein, and of mRNAs coding for other BDV proteins, do not shift from acute to chronic stages of infection (Hatalski, 1996). Modulation of the immune response over the course of BD, on the other hand, is supported by recent findings indicating the induction of BDV-specific Th1 tolerance in chronic infection. Whereas lymphocytes isolated from brains of acutely infected rats have potent cytolytic activity, lymphocytes from brains of chronically infected rats do not lyse BDV-infected target cells (Sobbe *et al.,* 1997). Induction of this BDV-specific tolerance in chronic infection may reflect the time course for presentation of viral antigens in the thymus (Rubin *et al.,* 1995). Alternatively, Th1 cells may become anergic or undergo apoptosis due to presentation of BDV antigens in brain without essential costimulatory signals (Karpas *et al.,* 1994; Khoury *et al.,* 1995; Schwartz, 1992). Support for this hypothesis is found in the observation that apoptosis of perivascular inflammatory cells is most apparent at 5–6 weeks postinfection, coincident with the onset of decline in encephalitis (Hatalski, 1996).

To identify the mechanisms by which inflammatory and immune responses are curbed in adult-infected Lewis rats during the late stages of disease, without changes in viral gene expression, mRNAs from the brains of acutely and chronically infected rats were compared for differences in host gene expression. Subtractive cloning approaches (representation-difference analysis) revealed a marked increase in mRNAs coding for immunoglobulin (Ig), suggesting that the apparent induction of Th1 tolerance in chronic infection might be based on a shift from Th1 to Th2-type immune responses. Confirming this hypothesis, assessment of whole brain RNAs by RNase protection assays (RPAs) identified a shift from Th1- to Th2-type cytokines, and analysis of immunoglobulin isotypes in peripheral blood revealed a switch from IgG to IgE (Hatalski *et al.,* 1998a; Hatalski *et al.,* 1998b), also consistent with Th2-type responses. Although this potentiation of the humoral immune response may play an important role in limiting viral gene expression in chronic disease, it does not induce viral clearance (Hatalski *et al.,* 1998b). Nonetheless, humoral immune capacity may be instrumental in restricting BDV replication to specific compartments: the intra-CNS production of IgG antibodies has been shown to parallel increases in peripheral blood of antibodies with neutralizing activity against BDV during the chronic phase of BD (Hatalski *et al.,* 1998a), and passive transfer of neutralizing antibodies results in limitation of viral replication within the CNS (Stitz *et al.,* 1998). Enhancement of the humoral immune response directed against N protein during acute infection may reduce viral load, but at the expense of increased morbidity; immunization of Lewis rats with a recombinant vaccinia virus expressing N protein, prior to challenge with infectious BDV, is associated with decreased viral burden but increased immune cell infiltration of CNS and increased disease severity (Lewis *et al.,* 1999). A role for humoral immunity in modulating viral tropism and persistence has been reported in other viral systems; for example, passive transfer of virus-specific antibodies limits viral replication in the CNS following infection with murine hepatitis virus type-4 (Buchmeier *et al.*, 1984) or measles virus (Liebert *et al.*, 1990), and induces viral clearance following infection with rabies (Dietzschold *et al.*, 1992) or Sindbis virus (Levine *et al.*, 1991).

IV. BORNA DISEASE RAT MODELS FOR HUMAN CENTRAL NERVOUS SYSTEM DISORDERS

The mechanisms by which viral infections alter the architecture and function of neural circuitry are not well understood. In rats infected as adults with BDV, the cellular arm of the immune response appears critical to the development of the dramatic neurologic manifestations (Planz *et al.*, 1995); in neonatally infected rats, on the other hand, a more subtle neurobehavioral syndrome ensues without invoking an extensive infiltration of inflammatory elements or immune-mediated destruction (Hornig *et al.*, 1999; Stitz *et al.*, 1995). Thus, differential timing of a viral insult relative to the specific stage(s) of neural and immune system maturation may yield a continuum of functional and structural outcomes. When considered in the context of the wide variability in clinical presentations of human neuropsychiatric disorders such as autism, schizophrenia, and affective illnesses, the heterogeneity in neurobehavioral consequences following BDV infection underscores the relevance and utility of neurodevelopmental animal models for understanding neuropsychiatric disease pathogenesis.

BDV infection is associated with two diseases in rodent models: (1) an immune-mediated syndrome characterized by dramatic disturbances in movement and behavior (infection of adult immunocompetent rats) (Ludwig *et al.*, 1988; Narayan *et al.*, 1983a; Solbrig *et al.*, 1995; Solbrig *et al.*, 1994; Solbrig *et al.*, 1996a; Solbrig *et al.*, 1996b; Solbrig *et al.*, 1996c; Solbrig *et al.*, 1998); and (2) a distinct syndrome characterized by cerebellar and hippocampal dysgenesis, hyperactivity, and social and learning disturbances (infection of neonatal rats) Bautista *et al.*, 1995; Bautista *et al.*, 1994; Carbone *et al.*, 1996; Dittrich *et al.*, 1989; Hornig *et al.*, 1999).

A. Adult Infection: a Viral Model for Movement Disorders

As in autism (Anderson, 1994; Ernst *et al.*, 1997), schizophrenia (Cooper *et al.*, 1991), and mood disorders (Hamner and Diamond, 1996; Kelsoe *et al.*, 1996; Partonen, 1996), disorders of movement and behavior in adult BD rats are linked to distinct changes in CNS dopamine (DA) systems (Solbrig *et al.*, 1999; Solbrig *et al.*, 1995; Solbrig *et al.*, 1994; Solbrig *et al.*, 1996a; Solbrig *et al.*, 1996b; Solbrig *et al.*, 1998) and may be further linked to serotonin (5HT) abnormalities (Solbrig *et al.*, 1995). The immune-mediated disorder in adult-infected rats presents clinically as hyperactivity and exaggerated startle responses 10–14 days after intracerebral infection (Narayan *et al.*, 1983a). This acute phase coincides with infiltration of monocytes into the brain, particularly in areas of highest viral burden, including the hippocampus, amygdala, and other limbic structures (Carbone *et al.*, 1987). Two to three weeks later, in parallel with the widespread distribution of virus in limbic and prefrontal circuits, rats demonstrate severe stereotyped motor behaviors (the continuous repetition of behavioral elements, including sniffing, chewing, scratching, grooming, and self-biting), dyskinesias, dystonia, and flexed seated postures (Solbrig *et al.*, 1994). Up to 10 percent of animals become obese, with body weights approaching 300% of normal (Ludwig *et al.*, 1988).

Functional disruptions and structural changes of several neuromodulator systems have been found following adult BDV infection. Disturbances of dopamine neurotransmission are most extensively characterized (Solbrig *et al.*, 1994; Solbrig *et al.*, 1996a; Solbrig *et al.*, 1996b), but alterations of serotonin (Solbrig *et al.*, 1995), opioid (Solbrig *et al.*, 1996c), acetylcholine (Gies *et al.*, 1998), somatostatin (Lipkin *et al.*, 1988b), cholecystokinin (Lipkin *et al.*, 1988b), and gamma-aminobutyric acid (Lipkin *et al.*, 1988b) neurocircuitry and/or metabolism are also reported. Dopaminergic abnormalities noted following adult BDV infection include enhanced sensitivity to DA agonists and antagonists. The mixed-acting DA agonist, dextroamphetamine, markedly increases locomotor and stereotypic behavior in infected animals (Solbrig *et al.*, 1994). Similarly, enhanced locomotion and stereotypes are seen in adult-infected rats in response to cocaine-induced DA reuptake inhibition, indicating dose-dependent potentiation of DA neurotransmission (Solbrig *et al.*, 1998). Biphasic responses are observed following administration of the direct DA agonist, apomorphine: at low, presynaptic, autoreceptor doses, activity levels are reduced, whereas at higher doses locomotion is increased. In keeping with the hypothesis that the adult neurobehavioral syndrome is related to selective disturbances of DA receptors, effects of treatment with DA antagonists vary depending on the DA receptor specificity of the agent. D2-selective antagonists (e.g., raclopride) do not affect locomotor responses in BD rats; high doses of selective D1 antagonists (e.g., SCH23390), and of atypical DA blocking agents with mixed D1 and D2 antagonist activity (e.g., clozapine) selectively reduce locomotor activity in BD rats but not in controls (Solbrig *et al.*, 1994).

The neuropathologic basis for these functional disturbances of DA neurocircuitry in the adult BD system has begun to be defined. DA neurochemistry is abnormal in high-performance liquid chromatography (HPLC) analyses. Levels of DA are decreased more substantially than levels of its major metabolite, dihydroxyphenylacetic acid (DOPAC) in caudate/putamen, nucleus accumbens, and olfactory tubercle (Solbrig et al., 1994), whereas in prefrontal cortex the reverse pattern is seen with a marked increase in DOPAC and no significant change in DA levels (Solbrig et al., 1996a). In addition, tyrosine hydroxylase-immunoreactive cells are depleted in the substantia nigra and ventral tegmental area (Solbrig et al., 1994). At the receptor level, damage to the DA transmitter system in caudate-putamen and nucleus accumbens appears to occur at both pre- and postsynaptic sites, consistent with the neurochemical and neuropharmacologic data described above. DA uptake sites, as measured by binding of mazindol, are reduced in nucleus accumbens (Solbrig et al., 1996b) and caudate-putamen (Solbrig et al., 1998). D2, but not D1, receptor binding is markedly reduced in caudate-putamen; D2 and D3 receptor binding are reduced in nucleus accumbens (Solbrig et al., 1994; Solbrig et al., 1996a; Solbrig et al., 1996b). In contrast, postsynaptic D1, D2, and D3 receptors remain intact in prefrontal cortex (Solbrig et al., 1996a). These findings suggest partial DA deafferentation with compensatory metabolic hyperactivity in mesolimbic DA systems.

Although the increased locomotor activity, stereotypic behaviors, and dyskinesias of the adult BD model are linked to distinct disturbances in DA pathways, additional neuromodulator abnormalities in adult BD have also been noted. The expression of genes for neuromodulatory substances and their associated synthesizing enzymes, including somatostatin, cholecystokinin, and glutamic acid decarboxylase, is greatly reduced during the acute phase but recovers toward normal in the chronic phase of adult BD (Lipkin et al., 1988a). The cholinergic system, a major participant in sensorimotor processing, learning, and memory, also appears to be affected in adult BDV infection. The number of choline acetyltransferase-positive fibers has been observed to decrease as early as day 6 postinfection (p.i.), progressing to nearly complete loss of cholinergic fibers in hippocampus and neocortex by day 15 p.i. (Gies et al., 1998). Preliminary work on dysregulation of serotonin (5HT) and norepinephrine (NE) systems suggests metabolic hyperactivity of 5HT in striatum (as evidenced by a modest increase in the metabolite 5-hydroxyindoleacetic acid [5HIAA]) and of NE in prefrontal and anterior cingulate cortex regions (as evidenced by a small increase in 3-methoxy-4-hydroxyphenethylene glycol [MHPG]) (Solbrig et al., 1995; personal communication, Solbrig and

Lipkin, 1997). These changes may reflect compensatory upregulation or heterotypic sprouting following partial loss of DA afferents to these brain regions. Selective effects of BDV on 5HT and NE pre- or postsynaptic receptors have not yet been investigated. Pharmacologic and neurotransmitter-specific molecular probes have also been used to characterize disturbances of endogenous opioid systems in the adult rat model. Infected animals respond abnormally to the opiate antagonist, naloxone, with hyperkinesis and seizures; increases in striatal preproenkephalin mRNA are also observed at 14 and 21 days (Fu et al., 1993), and 45 days after BDV infection (Solbrig and Lipkin, personal communication, 1997). However, the mechanisms by which these changes in endogenous opioid systems occur are unclear. The marked CNS inflammation in adult-infected rats makes it difficult to determine whether monoamine, cholinergic, and opiatergic dysfunction in BD results from direct effects of the virus, virus effects on resident cells of the CNS, or a cellular immune response to viral gene products. More recently, our efforts have turned toward the neonatal model in an effort to identify the functional and structural consequences of BDV in a system that is linked to more direct virus–CNS interactions.

B. Neonatal Infection: A Viral Model for Neurodevelopmental Disorders

Neonatal rat infection presents an intriguing model for neuropsychiatric illness; its immunopathologic correlates are more subtle and the cerebellar and hippocampal dysgenesis observed is consistent with the neurodevelopmental abnormalities reported in autism (Kemper and Bauman, 1993), schizophrenia (Altshuler et al., 1987; Fish et al., 1992), and affective disorders (Soares and Mann, 1997). Close parallels exist among the core and associated features of these psychiatric disorders and the wide range of physiologic and neurobehavioral disturbances described in neonatally infected animals; in particular, the overlap of signs of autism spectrum disorders with elements of the neonatal syndrome is especially striking.

Abnormal growth and physiologic profiles have been noted, although the mechanisms contributing to these disturbances are not known. Neonatally infected animals are stunted in growth as compared to uninfected littermates as early as day 4 postinfection (Bautista et al., 1994; Carbone et al., 1991; Hornig et al., 1999), without demonstrable alterations of glucose, growth hormone, insulin-like growth factor-1 (Bautista et al., 1994), or amount of food ingested (Bautista et al., 1995). They also display altered sleep–wake cycles and heightened taste preferences for salt-containing solutions (Bautista et al., 1994). Given reported alterations in other neuropeptide systems following neonatal infection of Lewis rats (Plata-Salamán et al., 1999), it is intriguing to speculate that disturbed salt preferences might result from dysregulation of the neuropeptide arginine vasopressin, either through direct infection of magnocellular neurons of hypothalamus or indirect effects on the magnocellular division through perturbations of mineralocorticoid responses (including potential effects of BDV on mineralocorticoid release by the adrenal glands, or on mineralocorticoid receptors in hypothalamus). Alternatively, BDV could influence salt taste preferences through effects on gustatory neurocircuitry, either at the level of primary gustatory transduction, sensory transmission via the chorda tympani branch of cranial nerve VII, or processing of taste signals in the rostral portions of the nucleus of the solitary tract, where primary gustatory nerves terminate for salt sensation (King et al., 1999). A role for a higher-order sensory processing defect may be a more likely explanation than salt depletion or aberrant taste discrimination capacity, given that neonatally infected animals do not differ in the amount of salt solution consumed in single-bottle taste acceptance experiments (Bautista et al., 1994); nonetheless, the alternative hypotheses have yet to be examined by neurohormonal and serum osmolality assessments and detailed neuropathologic analysis of neural systems regulating salt intake, including the nucleus of the solitary tract. Whether disturbances of circadian rhythms relate to direct or indirect effects of BDV infection is similarly unclear; however, a wide variety of inflammatory stressors is known to induce sleep-wake disturbances through shifts in cytokines, prostaglandins, and body temperature (Dunn, 1993; Dunn and Swiergiel, 1998; Wong et al., 1997). In studies of the retroviruses HIV (Opp et al., 1996) and feline immunodeficiency virus (Prospero-Garcia et al., 1994) in rats, sleep architecture changes are associated with the envelope glycoprotein (gp); in the case of HIV, intracerebroventricular administration of gp 120 also increases mRNA expression for the cytokines interleukin (IL)-1β and IL-10 in hypothalamus, a brain region crucial for sleep regulation (Opp et al., 1996). Intriguingly, increased brain levels of mRNA coding for IL-1β have been noted by several investigators following neonatal infection of Lewis rats (Hornig et al., 1999; Plata-Salamán et al., 1999; Sauder and de la Torre, 1999), suggesting the possibility that cytokines may contribute to sleep–wake cycle abnormalities in this model as well. The potential contribution of dysregulation of autocoids and temperature regulation to the pathogenesis of the neonatal syndrome remains to be studied.

Notably, many children with autism demonstrate abnormal taste preferences and sleep disturbances (Wing, 1997)

Persistent disturbances are also reported in the cognition, socioemotional behavior and communication, and motor development of neonatally BDV-infected animals. Cognitive deficits and unusual emotional reactivity were first described in Wistar rats, in which spatial and aversive learning impairments were found along with increased motor activity and decreased anxiety responses in open-field testing under bright light conditions from 15 weeks postinfection (Dittrich et al., 1989). Similar deficits in spatial learning and memory were more recently confirmed in neonatally infected Lewis rats 43 to 72 days postinfection, in association with the peak of dentate gyrus granule cell loss (Rubin et al., 1999b). Emotional responses of adult Lewis rats infected as neonates appear to depend both on testing conditions and age at testing. Dittrich et al., (1989) found locomotor hyperactivity and absence of freezing behavior in brightly lit open-field testing of 4-month-old Wistar rats infected with BDV as neonates, a finding which was interpreted as being indicative of abnormally low anxiety responses. However, based on a comparison of the aversiveness of bright and dim illumination conditions in 3- to 4-month-old, neonatally infected and sham-inoculated animals, Pletnikov et al. noted that hyperactivity of infected animals was greater in bright light conditions than in dim light, whereas the opposite pattern pertained to the control animals. Although freezing and thigmotaxic behaviors ("wall-seeking" behavior consistent with an attempt to escape from the testing apparatus) were more evident in neonatally infected animals in bright light, as opposed to "dim light" open-field testing, the mean time spent in freezing behavior during the testing period was consistently lower in infected than in control animals. The authors concluded that the increased motor activity in bright light represented hyperreactivity of neonatally infected animals to aversive stimulation (i.e., the bright light). Other indices of disturbed emotional reactivity in infected rats, including increased defecation in the brightly lit open field, were consistent with this hypothesis of increased fearfulness and drive to avoid aversive stimuli. Indeed, the more traditional measures of anxiety that were used in this study, freezing and thigmotaxis, actually showed decreases in infected animals compared to controls (Pletnikov et al., 1999a); however, these measures are less reliable as measures of anxiety at extremely high levels, where fearfulness and flight responses are induced (Davis and Shi, 1999; King, 1999). Additionally, during acoustic startle testing of neonatally infected animals, Pletnikov and colleagues found a dissociation between the amplitude of the startle response and autonomic reactivity: whereas infected animals had a significantly lower startle response amplitude as compared to sham-inoculated rats, and both habituation and foot-shock sensitization of their startle responses demonstrated a normal pattern, autonomic responses (defecation) were higher (Pletnikov *et al.*, 1999a). These findings point to a possible disturbance in integration of signals from neural circuits involved in higher-order sensory processing (glutamatergic afferents to, or efferents from, a critical site in primary startle response neurocircuitry, the caudal pontine reticular nucleus), anxiety/fear responses, and autonomic system outputs. Although the role of the amygdala in fear-potentiation of the classic motor aspects of startle responses of infected animals appears to be intact, exaggeration of responses of the central nucleus of the amygdala to noradrenergic, autonomic signals from locus coeruleus may be faulty (Koch, 1999), possibly accounting for the opposite effects of neonatal infection on the motor and autonomic components of the startle response. The relationship of these abnormalities of anxiety and fear responses to developmental maturity and the unfolding of CNS damage following neonatal infection was not reported. In this context, we have reported a transient, abnormal response to novelty (inhibition of locomotor activity responses upon introduction to novel environment) at 4 weeks postinfection, when brain levels of mRNAs for cytokines associated with decreased exploratory responses, such as IL-1, were at their peak (Dantzer *et al.*, 1998; Hornig *et al.*, 1999); this observation is also consistent with abnormal amygdaloid responses.

Disturbances of complex socioemotional behaviors and communication that are known to be altered in conditions such as autism have also been reported. Play behavior is decreased with respect to both infected animals' initiation of nondominance-related play interactions and response to initiation of play by noninfected, age- and gender-matched control animals or by infected littermates (Pletnikov *et al.*, 1999b). Given the selective effects of BDV on D2 receptors in the adult infection model (Solbrig *et al.*, 1994, 1996a, 1996b), it is interesting to note that this receptor subtype is thought to be the site of dopamine's effects on social play behavior (Vanderschuren *et al.*, 1997). Other neuromodulators thought to be disrupted in adult BDV infection of Lewis rats, including monoamines, acetylcholine, and opioids (Solbrig *et al.*, 1995), also play a role in regulating social play (Vanderschuren *et al.*, 1997). Studies of regional monoamine alterations following neonatal infections are under way to determine whether neurochemical aberrations similar to those in adult-infected animals may contribute to these social play abnormalities. Nucleus accumbens and striatum, two brain areas affected in both the adult and neonatal rat models, also impact initiation of play behaviors. Furthermore, these sites receive signals from the amygdaloid nuclei, areas involved in the selection of socially appropriate responses (Vanderschuren et al., 1997) and also observed to be heavily infected after adult and neonatal BDV infection. In another intriguing parallel with a core feature of autistic disorder, that of impaired social communication, infant–maternal communication appears altered following neonatal infections (Hornig et al., 1999). Ultrasonic vocalization distress calls induced by maternal separation are one of the earliest and most universal social communication responses that develop in mammals (Hofer, 1996; Winslow and Insel, 1990), and are normally reduced by social signals (e.g., presence of an anesthetized mother or littermate in the testing chamber) and by serotonergic agents, but are increased following administration of noradrenergic reuptake inhibitors (Hofer, 1996). In neonatally infected Lewis rats, ultrasonic vocalizations are remarkable for increased call frequency and abnormal waveforms (M. Hornig and W. I. Lipkin, unpublished data). The efficacy of these calls in eliciting appropriate maternal responses and their responsivity to social cues are currently being explored.

Even before neuronal subsets are observed to be depleted in cerebellum, the developmental progression of motor skills and activity levels is distorted. Neonatally infected rats show asymmetric protoambulatory ("pivoting") responses, with an increased frequency of falls into a supine position between days 4 and 9 postinfection—an unusual movement never observed in control animals. In addition, they exhibit significant delay in righting themselves on a flat surface, and explore the open-field chamber less widely through postnatal day 9 (Hornig et al., 1999). Such sensorimotor processing disturbances suggest early functional damage to motor circuitry, including cerebellum and striatum, and to acetylcholine systems; in addition to known infection of cerebellum and striatum in the neonatal system, adult infections have been associated with losses of choline acetyltransferase-positive fibers before the onset of encephalitis. Thus, it is conceivable that similar damage may occur to cholinergic neurons in the neonatal system, where immune cell infiltration also occurs later and is only fleeting. Given a recent report by Teitelbaum and colleagues indicating early, often transient, locomotor abnormalities in children with autism, including abnormalities of the righting response, crawling, and ambulation (Teitelbaum et al., 1998), further examination of the pathogenesis of neurodevelopmental disturbances in this infection-based animal model is warranted.

Because the cerebellum undergoes substantial postnatal development in many mammals, it is particularly vulnerable to injury from perinatal virus infection (Monjan *et al.*, 1973; Monjan *et al.*, 1971; Oster-Granite and Herdon, 1985). There is ample support for the role of the cerebellum in motor behavior and motor learning (for reviews see Llinas and Welsh, 1993; Roland, 1993; Thompson and Kim, 1996). However, in the neonatal BDV infection model, overt cerebellar dysfunction appears late and is mild to moderate in severity (Hatalski, 1996; Hornig *et al.*, 1999). Only 5% of neonatally infected Lewis rats had mild gait ataxia between 12 and 24 weeks postinfection, but all were impaired in skills requiring more complex coordination and balance, such as that required to maintain balance while walking across a dowel (Hornig *et al.*, 1999). These modest functional abnormalities stand in contrast to the reportedly dramatic losses in Purkinje cells by 6 weeks postinfection (Eisenman *et al.*, 1999; Hornig *et al.*, 1999; Weissenböck *et al.*, 2000), and imply the possibility of alternate, compensatory circuits for maintainence of gait and balance.

As seen in adult infected rats, locomotor activity is increased in neonatally infected rats beginning as early as 4 weeks of age. Interestingly, the degree of baseline motor activity in neonatally infected rats far exceeds the baseline levels seen in the adult system. As noted above, exploratory behavior patterns and adaptation to novel environment are abnormal at 4 weeks after neonatal infection, as evidenced by prolonged locomotor inhibition on introduction to a novel environment, and suggest dysfunction of the amygdaloid nuclei. This resistance to novel stimuli is not seen at any other time point through 12 weeks; indeed, this behavioral inhibition at 4 weeks is all the more striking, given that it is only apparent in the first 30 minutes of testing, after which infected animals progress to marked hyperactivity for the remainder of the 90-minute test session. Additionally, all infected animals fail to show the normal patterns of attenuation in exploratory activity at 60 and 90 minutes, consistent with spatial memory deficits and hippocampal dysfunction. Subtle increases are noted in mild stereotypic behaviors, such as sniffing and rearing, actions which can also serve to assist exploration of the environment (Ikemoto and Panksepp, 1994; Poltyrev and Weinstock, 1997), but not in more severe, self-mutilatory repetitive behaviors (Hornig *et al.*, 1999).

Our understanding of the mechanisms mediating functional damage to the Lewis rat host following neonatal infection is beginning to expand. Consistent with previous reports (Carbone *et al.*, 1991; Narayan *et al.*, 1983b), we found morphologic alterations in brains of rats infected as newborns, including loss of granule cells in dentate gyrus and thinning of cerebellar granule cell layers. Other cerebellar abnormalities included decreased overall size and foliation of cerebellum and loss of Purkinje cells. Hippocampal changes in neonatally and adult infected animals included distinct loss of granule cells in dentate gyrus. Previous investigators reported an absence of cellular inflammatory response to BDV following neonatal infection (Bautista et al., 1995; Bautista et al., 1994; Carbone et al., 1991; Gosztonyi and Ludwig, 1995; Herzog et al., 1985; Narayan et al., 1983b; Plata-Salamán et al., 1999; Rott and Becht, 1995; Rubin et al., 1999b; Sierra-Honigmann et al., 1993; Stitz et al., 1989), a phenomenon ascribed to the immaturity of the rat immune system in the postnatal period. Humoral immune responses to BDV in neonatally infected animals are also reported to be restricted, with anti-BDV antibody titers remaining below 1:10 through 133 days postinfection (Carbone et al., 1991). In contrast to the many published reports of persistent infection without an inflammatory response, close serial analysis has revealed transient, regionally restricted inflammation (Hornig et al., 1999; Sauder and de la Torre, 1999). These inflammatory infiltrates emerge at about 4 weeks postinfection in perivascular areas of motor and parietal cortex, with less marked infiltrates in nonhippocampal portions of temporal cortex, thalamus, and basal ganglia, and disappear completely by 6 weeks postinfection. Despite nearly complete loss of the dentate gyrus, marked loss of cerebellar Purkinje cells, and thinning of granular cell layers in cerebellum by this time point, no such infiltrates are present in these areas at 2, 4, or 6 weeks postinfection. Areas of inflammation are also distinct from those areas showing the greatest amount of apoptosis (periventricular germinal layer, dentate gyrus, granular layer of cerebellum). Immunohistochemical analysis of infiltrates indicates that they are comprised primarily of T cells with approximately equal numbers of CD4+ and CD8+ cells (Weissenböck et al., 2000). Curiously, brain lymphocytes from 4-week-old, neonatally infected rats do not demonstrate cytotoxic activity against BDV antigens (Hornig, 1999), suggesting that these T cells are not specific. Furthermore, neonatally thymectomized, infected animals do not differ from sham-thymectomized, infected animals in degree of dentate gyrus damage, Purkinje cell loss, or cortical atrophy, despite successful ablation of inflammation at the 4-week time point by the neonatal thymectomy procedure (Hornig et al., 1999). How these presumably activated, but nonspecific, T cells are attracted to, and transiently retained in, selected regions of CNS and not in others is an intriguing, but unanswered, question. Likely candidates for direct viral effects include cell adhesion molecules and chemokines. Levels and distribution of PECAM and ICAM-1, as measured by immunohistochemistry in neonatally infected and sham-inoculated animals, suggest they cannot be implicated. Staining for PECAM on endothelial cells does not differ between infected and uninfected groups, and although levels of ICAM-1 increase on endothelial and other vascular and perivascular cells of neonatal BD rats at about 4 weeks postinfection, the most prominent staining appears in hippocampus and cerebellum in addition to cerebral cortex. Thus, the regional distribution of ICAM-1 does not explain the apparent exclusion of inflammatory infiltrates from hippocampus and cerebellum at 4 weeks postinfection.

Our report of inflammatory infiltrates in cortex (Hornig et al., 1999) differs from another recent report (Sauder and de la Torre, 1999) of transient immune infiltration in that we do not find marked morbidity and mortality. Our neonatal BD rats have no mortality and do not manifest classical Borna disease. Instead, clinical assessments reveal only subtle signs of CNS dysfunction (minimally disheveled fur and dystonic postures; mild lethargy). Pilot studies indicate a more adult pattern of CNS disease if rats are inoculated after 12 hours of postnatal life or with lower titer stocks. Thus, developmental maturity of the immune and nervous systems at the time of exposure to the viral agent are critical to determining inflammatory responses and ultimate neurodevelopmental outcomes (Hornig et al., 1999; Rubin et al., 1999a).

In addition to gaps in our knowledge regarding the nature of the functional damage induced by neonatal BDV infection, the manner in which infection results in loss of select neuronal subsets is not yet known. Mechanisms may be both direct and indirect. Clearly, infection alone is an insufficient signal, as many neuronal populations remain persistently infected in neonatal BD without evident reduction of cell numbers. Cells that are lost due to neonatal infection, predominantly found in cerebral cortex, dentate gyrus, and the Purkinje cell layer of cerebellum, appear to be lost exclusively through apoptosis (Hornig et al., 1999; Weissenböck et al., 2000). Apoptotic cells, characterized by shrinkage, hypereosinophilic cytoplasm, nuclear pyknosis, and karyorrhexis, and signal on transferase dUTP-biotin nick end labeling (TUNEL) assay, peaked at four weeks postinfection, with most marked pathology in dentate gyrus and cortical layers 5 and 6 of retrosplenial and cingulate cortex. Although apoptosis is described in hippocampus of developing animals as late as days 7 to 10 of postnatal life, it is not found at later time points (Toth et al., 1998). It is intriguing to speculate that BDV may influence the expression of age-related programs associated with normal, development-associated apoptosis, through the direct or indirect elaboration of soluble factors.

We observe that reactivity and proliferation of microglial cells throughout the brain are even more striking than astrocytosis. Interestingly, microglia remain activated even at 76 weeks postinfection, despite the absence of any evidence of infection; in contrast, a small proportion of astrocytes do become infected, although astrocytic activation subsides between 4 and 24 weeks postinfection. Cerebellar size has been reported to be reduced, with evidence of reactive astrocytosis as demonstrated by glial fibrillary acidic protein (GFAP) reactivity as early as 3 days postinoculation, preceding the identification of BDV proteins in the cerebellum. Furthermore, reactivity of cerebellar astrocytes and loss of cerebellar granule cells occur without signs of BDV infection in those cell populations at all time points through 30 days postinfection. Curiously, Purkinje cells appear to be the predominant cerebellar cell population demonstrating BDV antigens, although these cells were previously reported to be retained through day 30 postinfection (Bautista *et al.*, 1995). However, our investigations indicate that by day 42 postinfection, Purkinje cell populations are selectively depleted. Similarly, at 7 months after neonatal infection, Eisenman and colleagues (1999) report 75% depletion of Purkinje cells.

Astrocytosis and microgliosis occur in all brain regions by 3 weeks postinfection. In cerebellum, reactive astrocytosis is present as early as 3 days postinoculation, preceding the identification of BDV proteins in the cerebellum. The onset and distribution of glial cell proliferation coincides with neuronal loss, yet is sustained through 24 weeks postinfection, after neuronal cell death wanes. Astrocyte and microglia morphology is consistent with activation (increased cytoplasm; short, thickened processes; additionally, in microglia, more intense staining with OX42). In hippocampus, gliosis is most evident in dentate gyrus. In cerebellum, gliosis is equally prominent in white and gray matter. Microgliosis persists throughout the brain as late as 76 weeks postinfection. Curiously, microglia express MHC Class I and II, CD4 and CD8 (OX8 and CD8b) molecules on their surface (Weissenböck *et al.*, 2000).

Astrocytes and microglia are activated in the absence of infection, be it directly by BDV or indirectly through elaboration of soluble factors by other cell types. Given the role of astrocytes in guiding migration of granule cells during cerebellar development, an assessment of the frequency of astrocyte reactivity without viral infection, in conjunction with studies of apoptosis in limbic structures, would aid our understanding of the relative contributions of migrational failure and programmed cell death in the evolution of BD neuropathogenesis. Even more intriguing may be the mechanism by which microglia are activated in the absence of infection. In neonatal human immunodeficiency virus type 1 (HIV-1) infection, for instance, diffuse microglial activation and reactive astrogliosis occur along with impaired brain growth and developmental delays (Epstein and Gelbard, 1999). However, in contrast to neonatal BD, microglia internalize virus in neonatal HIV-1 infection and replication can occur on chronic exposure to proinflammatory cytokines (Janabi et al., 1998). Two other findings regarding microglia-related injury in HIV-1 infection may be relevant to an understanding of mechanisms in BD pathogenesis: (1) production of the neurotoxin, quinolinic acid, is substantially greater in uninfected than in HIV-1 infected monocytes following lipopolysaccharide or interferon-γ stimulation (Nottet et al., 1996); and (2) the viral protein, gp41, may act indirectly to induce neurotoxicity by triggering IL-1β production in microglia, thereby stimulating inducible nitric oxide synthetase production and nitric oxide (NO) generation by astrocytes (Hu et al., 1999). General mechanisms of CNS injury following microglial activation also appear to relate to differential regulation of MHC and other molecules on microglia (Stoll and Jander, 1999). Whether the unusual pattern of upregulation of MHC, CD4, and CD8 molecules observed on uninfected, activated microglia in neonatal BDV infection might explain the pattern of damage observed is not known.

Higher levels of message for tissue factor (TF) have been found in infected hippocampus (Gonzalez-Dunia et al., 1996). TF is a member of the class II cytokine receptor family, primarily produced by astrocytes, that plays important roles in cellular signal transduction, brain function, and neural development through its effects on coagulation protease cascades. Although this may be one mechanism by which BDV may alter CNS development (Gonzalez-Dunia et al., 1996), cerebellar changes cannot be explained by this mechanism as TF upregulation is not observed in cerebellum despite prominent astrocytosis. Furthermore, BDV infection of astrocytes appears to be required for TF upregulation (Gonzalez-Dunia et al., 1996), and cerebellar astrocytes are rarely infected (Hornig et al., 1999).

One means by which a virus might disrupt neural function and development, in the absence of inflammation, is through the induction of neuronotrophic cytokines. Neuronotrophic cytokines comprise a burgeoning set of immunoregulatory molecules, including the hematolymphopoietic factors (e.g., interleukins, tumor necrosis factor family, interferons); the transforming growth factor (TGF)-β superfamily factors (including TGF-β1, 2, 3; GDNF); and the classic neurotrophic factors (NGF, BDNF, NT3, NT4/5). A large subset of the neuronotrophic, hematolymphopoietic cytokines may be roughly categorized according to their origin from one of two types of T-helper cells:

Th1 (cell-mediated immunity and stimulation of antigen-presenting cells) or Th2 (humoral or B-cell mediated immunity). The potential mechanisms of cytokine-mediated damage in the context of the developing brain include direct effects on neuronal elements; activation or suppression of second messenger/intracellular signaling pathways; induction of changes in excitotoxic elements such as quinolinic acid or acute phase proteins such as neopterin or β_2-microglobulin; direct alterations of neuronal function (e.g., inhibition of long-term potentiation in hippocampus); activation or suppression of glial cells; or alteration of glial cell proliferation or differentiation (including expression of adhesion molecules such as the integrins) (Benveniste, 1997; Mehler et al., 1996). Given that the postnatal expression of neuronotrophic cytokine and cytokine receptor mRNAs in brain differs for each cytokine (Benveniste, 1997), and that the sensitivity of neuronal populations to the trophic or apoptosis-inducing effects of cytokines changes during development, wide variation in the patterns of virus-induced, cytokine-related damage would be expected, depending on the relative maturity of the evolving nervous system at the time of infection. In addition, cell loss induced by either BDV or developmentally programmed changes may alter the capacity of resident CNS cells to both produce and respond to neuronotrophic cytokines.

One of the primary mechanisms of host defense following viral infection begins with the induction of interferon-γ (IFN-γ) and other cytokines, which in turn initiate a cascade of host responses in a wide variety of cell types. In the CNS, IFN-γ modulates oligodendrocyte, neuronal and glial cell functions, and is important in activating glial cells to produce mediators of cell damage or death, including toxic intermediates of nitrogen and oxygen, and complement components (St. Pierre et al., 1996). Viral damage to neurodevelopmental circuitry may thus parallel the production of these downstream mediators following IFN-γ induction, and provide a means by which BDV might disrupt brain cell differentiation and function without inflammatory cell infiltration. There are only limited published data concerning cytokine expression in adult infected animals. Dietzschold and coworkers found elevated levels of mRNA encoding proinflammatory cytokines in brains of adult infected rats with acute disease, including IFN-γ (Shankar et al., 1992). As noted above, we found changes in cytokine expression over the time course of adult infection consistent with a shift from a predominantly Th1-type pattern during the acute phase to a Th2 pattern in the chronic stages of disease (Hatalski et al., 1998a).

Recent studies concerning cytokine expression during neonatal infection provide a converging view of the potential importance of cytokines as mediators of BDV-related CNS injury in neonatally-infected rats (Hornig *et al.*, 1999; Plata-Salamán *et al.*, 1999; Sauder and de la Torre, 1999). Cytokine expression changes over time in different brain regions, reaching a maximum at 4 weeks. Higher levels of mRNAs for cytokine products of CNS macrophages/microglia (IL-1α, IL-1β, IL-6, TNF-α) are noted in hippocampus, amygdala, cerebellum, prefrontal cortex, and nucleus accumbens (Hornig *et al.*, 1999). Elevated levels of these proin-flammatory cytokines were first apparent at 2 weeks, peaked at 4 weeks, and then declined at 6 and 12 weeks. Alterations in other proin-flammatory cytokines, including IL-2, IL-3, TNF-β, and IFN-γ, were not observed. The fact that cell populations other than macrophages or microglia—T cells, B cells, mast cells, bone marrow stromal cells—are the primary sources for the proinflammatory cytokines that remained static following neonatal infection suggests a selective effect of BDV on cells of microglial or macrophage lineage.

Following neonatal infection, BDV influences the expression of apop-tosis-related products. Increased levels of mRNAs coding for FAS and ICE (caspase-1), two promotors of apoptosis, and decreased mRNA for bcl-x, a factor that inhibits apoptosis, were identified in hippocampus, amygdala, prefrontal cortex, nucleus accumbens, and cerebellum (Hornig *et al.*, 1999). These findings are consistent with promotion of apoptosis throughout the brains of rats neonatally infected with BDV by at least two strategies. A host of excitants or neurotoxins including arachidonic acid, platelet-activating factor, free radicals (NO, O_2^-), glu-tamate, quinolinate, cysteine, cytokines (TNF-α, IL-1β, IL-6), amines, and as yet unidentified factors arising from stimulated macrophages and possibly reactive astrocytes may also influence apoptosis by exces-sive activation of N-methyl-D-aspartate (NMDA) receptors (Lipton, 1996). Interestingly, Gosztonyi and Ludwig (1995) have proposed that the targeted pathology of BDV for two hippocampal cell layers, stratum oriens and stratum radiatum, may be due to their rich concentration of glutamate and aspartate receptors, and preliminary results from our laboratory indicate regional upregulation at 4 weeks postinfection of a non-NMDA glutamate receptor, the calcium-permeable AMPA receptor, GluR1 (M. Hornig and W. I. Lipkin, unpublished data). Consistent with our region-specific findings of apoptosis by TUNEL analysis, this upregulation is most evident on granule cells in dentate gyrus and at synapses of cells in molecular cell layer of cerebellum that terminate on Purkinje cells (presumably, basket cells or astrocytes).

Withdrawal of selected neurotrophic factors can also contribute to apoptotic losses (Takei *et al.*, 1999). In contrast to our findings of diffuse alterations in gene expression for cytokines described above, disturbances in neurotrophic factor mRNAs are restricted to hippocampus. Decreased mRNA coding for BDNF and NT3 is observed in hippocampus by 4 weeks, but is still evident by 12 weeks postinfection. Although decreased NT3 mRNA may reflect loss of the granule cell population in dentate gyrus, the role of BDNF in maintaining viability of cells suggests that its downregulation may be a more essential step in neonatal BDV pathogenesis. Nonetheless, if BDNF withdrawal is a potent influence on apoptosis, it is difficult to explain the abrupt dropoff in apoptotic losses in dentate gyrus after 5 to 6 weeks postinfection. Furthermore, this mechanism is unable to account for cell losses in cerebellum at any time point as BDNF is not expressed at substantive levels in normal cerebellum.

V. Summary

Animal models provide unique opportunities to explore interactions between host and environment. Two models have been established based on Bornavirus infection that provide new insights into mechanisms by which neurotropic agents and/or immune factors may impact developing or mature CNS circuitry to effect complex disturbances in movement and behavior. Distinct losses in DA pathways in the adult infection model, and the associated dramatic movement disorder that accompanies it, make it an intriguing model for tardive dyskinesia and dystonic syndromes. The neuropathologic, physiologic, and neurobehavioral features of BDV infection of neonates indicate that it not only provides a useful model for exploring the mechanisms by which viral and immune factors may damage developing neurocircuitry, but also has significant links to the range of biologic, neurostructural, locomotor, cognitive, and social deficits observed in serious neuropsychiatric illnesses such as autism.

References

Altshuler, L. L., Conrad, A., Kovelman, J. A., and Scheibel, A. (1987). Hippocampal pyramidal cell orientation in schizophrenia: a controlled neurohistologic study of the Yakovlev Collection. *Arch. Gen. Psychiatry.* **44:**1094–1098.

Anderson, G. M. (1994). Studies on the neurochemistry of autism. *In* "The Neurobiology of Autism" (M. L. Bauman and T. L. Kemper, Eds.), pp. 227–242. Johns Hopkins University Press, Baltimore.

Bautista, J. R., Rubin, S. A., Moran, T. H., Schwartz, G. J., and Carbone, K. M. (1995). Developmental injury to the cerebellum following perinatal Borna disease virus infection. *Brain Res. Dev. Brain. Res.* **90:**45–53.

Bautista, J. R., Schwartz, G. J., De La Torre, J. C., Moran, T. H., and Carbone, K. M. (1994). Early and persistent abnormalities in rats with neonatally acquired Borna disease virus infection. *Brain. Res. Bull.* **34:**31–40.

Beckmann, H., and Jakob, H. (1991). Prenatal disturbances of nerve cell migration in the entorhinal region: a common vulnerability factor in functional psychoses? *J. Neural. Transmission—Gen. Section.* **84:**155–164.

Benveniste, E. N. (1997). Cytokine expression in the nervous system. *In* "Immunology of the Nervous System" (R. W. Keane and W. F. Hickey, eds.), pp. 419–459. Oxford University Press, New York.

Briese, T., Schneemann, A., Lewis, A. J., Park, Y. S., Kim, S., Ludwig, H., and Lipkin, W. I. (1994). Genomic organization of Borna disease virus. *Proc. Natl. Acad. Sci. USA* **91:**4362–4366.

Buchmeier, M. J., Lewicki, H. A., Talbot, P. J., and Knobler, R. L. (1984). Murine hepatitis virus-4 (strain JHM)-induced neurologic disease is modulated in vivo by monoclonal antibody. *Virology* **132:**261–270.

Carbone, K. M., Duchala, C. S., Griffin, J. W., Kincaid, A. L., and Narayan, O. (1987). Pathogenesis of Borna disease in rats: evidence that intra-axonal spread is the major route for virus dissemination and the determinant for disease incubation. *J. Virol.* **61:**3431–3440.

Carbone, K. M., Park, S. W., Rubin, S. A., Waltrip, R. W., 2nd, and Vogelsang, G. B. (1991). Borna disease: association with a maturation defect in the cellular immune response. *J. Virol.* **65:**6154–6164.

Carbone, K. M., Silvas, P. M., Rubin, S. A., Vogel, M., Moran, T. H., and Schwartz, G. (1996). Quantitative correlation of viral induced damage to the hippocampus and spatial learning and memory deficits. *J. Neurovirol.* **2:**195.

Cooper, J. R., Bloom, F. E., and Roth, R. H. (1991). "The Biochemical Basis of Neuropharmacology." Oxford University Press, New York.

Cubitt, B., Oldstone, C., and de la Torre, J. C. (1994). Sequence and genome organization of Borna disease virus. *J. Virol.* **68:**1382–1396.

Dantzer, R., Bluthé, R. -M., Layé, S., Bret-Dibat, J. -L., Parnet, P., and Kelley, K. W. (1998). Cytokines and sickness behavior. *Ann. NY. Acad. Sci.* **840:**586–590.

Davis, M., and Shi, C. (1999). The extended amygdala: are the central nucleus of the amygdala and the bed nucleus of the stria terminalis differentially involved in fear versus anxiety? *Ann. NY. Acad. Sci.* **877:**281–291.

Dietzschold, B., Kao, M., Zheng, Y. M., Chen, Z. Y., Maul, G., Fu, Z. F., Rupprecht, C. E., and Koprowski, H. (1992). Delineation of putative mechanisms involved in antibody-mediated clearance of rabies virus from the central nervous system. *Proc. Natl. Acad. Sci.* **89:**7252–7256.

Dittrich, W., Bode, L., Ludwig, H., Kao, M., and Schneider, K. (1989). Learning deficiencies in Borna disease virus-infected but clinically healthy rats. *Biol. Psychiatry* **26:**818–828.

Dunn, A. J. (1993). Infection as a stressor: a cytokine-mediated activation of the hypothalamo–pituitary–adrenal axis? *Ciba. Found. Symp.* **172:**226–239.

Dunn, A. J., and Swiergiel, A. H. (1998). The role of cytokines in infection-related behavior. *Ann. NY. Acad. Sci.* **840:**51–58.

Eisenman, L. M., Brother, R., Tran, M. H., Kean, R. B., Dickson, G. M., Dietzschold, B., and Hooper, D. C. (1999). Neonatal Borna disease virus infection in the rat causes a loss of Purkinje cells in the cerebellum. *J. Neurovirol.* **5:**181–189.

Epstein, L. G., and Gelbard, H. A. (1999). HIV-1-induced neuronal injury in the developing brain. *J. Leukoc. Biol.* **65:**453–457.

Ernst, M., Zametkin, A. J., Matochik, J. A., Pascualvaca, D., and Cohen, R. M. (1997). Low medial prefrontal dopaminergic activity in autistic children. *Lancet* **350:**638.

Fish, B., Marcus, J., Hans, S. L., Auerbach, J. G., and Perdue, S. (1992). Infants at risk for schizophrenia: sequelae of a genetic neurointegrative defect: a review and replication analysis of pandysmaturation in the Jerusalem Infant Development Study. *Arch. Gen. Psychiatry* **49:**221–235.

Fu, Z. F., Weihe, E., Zheng, Y. M., Schafer, M. K., Sheng, H., Corisdeo, S., Rauscher, F. J. I., Koprowski, H., and Dietzschold, B. (1993). Differential effects of rabies and borna disease viruses on immediate-, early-, and late-response gene expression in brain tissues. *J. Virol.* **67:**6674–6681.

Gies, U., Bilzer, T., Stitz, L., and Staiger, J. F. (1998). Disturbance of the cortical cholinergic innervation in Borna disease prior to encephalitis. *Brain Pathol.* **8:**39–48.

Gonzalez-Dunia, D., Eddleston, M., Mackman, N., Carbone, K., and de la Torre, J. C. (1996). Expression of tissue factor is increased in astrocytes within the central nervous system during persistent infection with borna disease virus. *J. Virol.* **70:**5812–5820.

Gosztonyi, G., and Ludwig, H. (1995). Borna disease: neuropathology and pathogenesis. *In* "Current Topics in Microbiology and Immunology" (H. Koprowski and W. I. Lipkin, eds.), Vol. 190, pp. 39–73. Springer-Verlag, Berlin.

Gutierrez, B., Van, Os, J., Valles, V., Guillamat, R., M., C., and Fananas, L. (1998). Congenital dermatoglyphic malformations in severe bipolar disorder. *Psychiatry Res.* **78:**133–140.

Hamner, M. B., and Diamond, B. I. (1996). Plasma dopamine and norepinephrine correlations with psychomotor retardation, anxiety, and depression in non-psychotic depressed patients: a pilot study. *Psychiatry Res.* **64:**209–211.

Hatalski, C. G. (1996). Ph.D. thesis, University of California at Irvine, Irvine.

Hatalski, C. G., Hickey, W. F., and Lipkin, W. I. (1998a). Evolution of the immune response in the central nervous system following infection with Borna disease virus. *J. Neuroimmunol.* **90:**137–142.

Hatalski, C. G., Hickey, W. F., and Lipkin, W. I. (1998b). Humoral immunity in the central nervous system of Lewis rats infected with Borna disease virus. *J. Neuroimmunol* **90:**128–136.

Hatalski, C. G., Lewis, A. J., and Lipkin, W. I. (1997). Borna disease. *Emerg. Infect. Dis.* **3:**129–135.

Herzog, S., Wonigeit, K., Frese, K., Hedrich, H. J., and Rott, R. (1985). Effect of Borna disease virus infection on athymic rats. *J. Gen. Virol.* **66:**503–508.

Hofer, M. A. (1996). Multiple regulators of ultrasonic vocalization in the infant rat. *Psychoneuroendocrinol.* **21:**203–217.

Hornig, M., Weissenböck, H., Horscroft, N., and Lipkin, W. I. (1999). An infection-based model of neurodevelopmental damage. *Proc. Natl. Acad. Sci. USA* **96:**12102–12107.

Hu, S., Ali, H., Sheng, W. S., Ehrlich, L. C., Peterson, P. K., and Chao, C. C. (1999). gp-41-mediated astrocyte inducible nitric oxide synthase mRNA expression: involvement of interleukin-1 beta production by microglia. *J. Neurosci.* **19:**6468–6474.

Ikemoto, S., and Panksepp, J. (1994). The relationship between self-stimulation and sniffing in rats: does a common brain system mediate these behaviors? *Behav. Brain. Res.* **61:**143–162.

Janabi, N., DiStefano, M., Wallon, C., Hery, C., Chiodi, F., and Tardieu, M. (1998). Induction of human immunodeficiency virus type 1 replication in human glial cells after proinflammatory cytokines stimulation: effect of IFNgamma, IL 1beta, and TNFalpha on differentiation and chemokine production in glial cells. *Glia* **23:**304–315.

Karpas, W. J., Peterson, J. D., and Miller, S. D. (1994). Anergy in vivo: down regulation of antigen-specific CD4+ Th1 but not Th2 cytokine responses. *Int. Immunol.* **6:**721–730.

Kelsoe, J. R., Savodnick, A. D., Kristbjarnarson, H., Bergesch, P., Mroczkowski-Parker, Z., Drennan, M., Rapaport, M. H., Flodman, P., Spence, M. A., and Remick, R. A. (1996). Possible locus for bipolar disorder near the dopamine transporter on chromosome 5. *Am. J. Med. Genet.* **67:**533–540.

Kemper, T. L., and Bauman, M. L. (1993). The contribution of neuropathologic studies to the understanding of autism. *Neurol. Clin. North. Am.* **11:**175–187.

Khoury, S. J., Akalin, E., Chandraker, A., Turka, L. A., Linsley, P. S., Sayegh, M. H., and Hancock, W. W. (1995). CD28-B7 costimulatory blockade by CTLA4Ig prevents actively induced experimental autoimmune encephalomyelitis and inhibits Th1 but spares Th2 cytokines in the central nervous system. *J. Immunol.* **155:**4521–4524.

King, C. T., Travers, S. P., Rowland, N. E., Garcea, M., and Spector, A. C. (1999). Glossopharyngeal nerve transection eliminates quinine-stimulated fos-like immunoreactivity in the nucleus of the solitary tract: implications for a functional topography of gustatory nerve input in rats. *J. Neurosci.* **19:**3107–3121.

King, S. M. (1999). Escape-related behaviours in an unstable elevated and exposed environment: I. A new behavioural model of extreme anxiety. *Behav. Brain. Res.* **98:**113–126.

Koch, M. (1999). The neurobiology of startle. *Prog. Neurobiol.* **59:**107–128.

Levine, B., Hardwick, J. M., Trapp, B. D., Crawford, T. O., Bollinger, R. C., and Griffin, D. E. (1991). Antibody-mediated clearance of alphavirus infection from neurons. *Science* **254:**856–860.

Lewis, A. J., Whitton, J. L., Hatalski, C. G., Weissenböck, H., and Lipkin, W. I. (1999). Effect of immune priming on Borna Disease. *J. Virol.* **73:**2541–2546.

Liebert, U. G., Schneider-Schaulies, S., Baczko, K., and ter Meulen, V. (1990). Antibody-induced restriction of viral gene expression in measles encephalitis in rats. *J. Virol.* **64:**706–713.

Lipkin, W. I., Battenberg, E. L. F., Bloom, F. E., and Oldstone, M. B. A. (1988a). Viral infection of neurons can depress neurotransmitter mRNA levels without histologic injury. *Brain Res.* **451:**333–339.

Lipkin, W. I., Carbone, K. M., Wilson, M. C., Duchala, C. S., Narayan, O., and Oldstone, M. B. (1988b). Neurotransmitter abnormalities in Borna disease. *Brain Res.* **475:**366–370.

Lipkin, W. I., and Hornig, M. (1998). Microbes and the brain. *Lancet* **352:**SIV21.

Lipkin, W. I., Schneemann, A., and Solbrig, M. V. (1995). Borna disease virus: implications for human neuropsychiatric illness. *Trends Microbiol.* **3:**64–69.

Lipton, S. A. (1996). Similarity of neuronal cell injury and death in AIDS dementia and focal cerebral ischemia: potential treatment with NMDA open-channel blockers and nitric oxide-related species. *Brain Pathol.* **6:**507–517.

Llinas, R., and Welsh, J. P. (1993). On the cerebellum and motor learning. *Current Opin. Neurobiol.* **3:**958–965.

Ludwig, H., Bode, L., and Gosztonyi, G. (1988). Borna disease: a persistent virus infection of the central nervous system. *Prog. Med. Virol.* **35:**107–151.

Mehler, M. F., Goldstein, H., and Kessler, J. A. (1996). Effects of cytokines on CNS cells: neurons. *In* "Cytokines and the CNS" (R. M. Ransohoff and E. N. Benveniste, eds.), pp. 115–150. CRC Press, Boca Raton.

Monjan, A. A., Cole, G. A., Gilden, D. H., and Nathanson, N. (1973). Pathogenesis of cerebellar hypoplasia produced by lymphocytic choriomeningitis virus infection of neonatal rats: evolution of disease following infection at 4 days of age. *J. Neuropathol. Exp. Neurol.* **32:**110–124.

Monjan, A. A., Gilden, D. H., Cole, G. A., and Nathanson, N. (1971). Cerebellar hypoplasia in neonatal rats caused by lymphocytic choriomeningitis virus. *Science* **171:**194–196.

Narayan, O., Herzog, S., Frese, K., Scheefers, H., and Rott, R. (1983a). Behavioral disease in rats caused by immunopathological responses to persistent borna virus in the brain. *Science* **220:**1401–1403.

Narayan, O., Herzog, S., Frese, K., Scheefers, H., and Rott, R. (1983b). Pathogenesis of Borna disease in rats: immune-mediated viral ophthalmoencephalopathy causing blindness and behavioral abnormalities. *J. Infect. Dis.* **148:**305–315.

Nottet, H. S., Flanagan, E. M., Flanagan, C. R., Gelbard, H. A., Gendelman, H. E., and Reinhard, J. F. J. (1996). The regulation of quinolinic acid in human immunodeficiency virus-infected monocytes. *J. Neurovirol.* **2:**111–117.

Opp, M. R., Rady, P. L., Hughes, T. K. J., Cadet, P., Tyring, S. K., and Smith, E. M. (1996). Human immunodeficiency virus envelope glycoprotein 120 alters sleep and induces cytokine mRNA expression in rats. *Am. J. Physiol.* **270:**R963–R970.

Oster-Granite, M. L., and Herdon, R. M. (1985). The pathogenesis of parvovirus-induced cerebellar hypoplasia in the Syrian hamster, Mesociretus auratus: fluorescent antibody, foliation, cytoarchitectonic, golgi and electron microscopic studies. *J. Compar. Neurol.* **169:**481–522.

Partonen, T. (1996). Dopamine and circadian rhythms in seasonal affective disorder. *Med. Hyp.* **47:**191–192.

Planz, O., Bilzer, T., and Stitz, L. (1995). Immunopathogenic role of T-cell subsets in Borna disease virus-induced progressive encephalitis. *J. Virol.* **69:**896–903.

Plata-Salamán, C. R., Ilyin, S. E., Gayle, D., Romanovitch, A., and Carbone, K. M. (1999). Persistent Borna disease virus infection of neonatal rats causes brain regional changes of mRNAs for cytokines, cytokine receptor components and neuropeptides. *Brain. Res. Bull.* **49:**441–451.

Pletnikov, M. V., Rubin, S. A., Schwartz, G. J., Moran, T. H., Sobotka, T. J., and Carbone, K. M. (1999a). Persistent neonatal Borna disease virus (BDV) infection of the brain causes chronic emotional abnormalities in adult rats. *Physiol. Behav.* **66:**823–831.

Pletnikov, M. V., Rubin, S. A., Vasudevan, K., Moran, T. H., and Carbone, K. M. (1999b). Developmental brain injury associated with abnormal play behavior in neonatally Borna disease virus-infected Lewis rats: a model of autism. *Behav. Brain. Res.* **100:**43–50.

Poltyrev, T., and Weinstock, M. (1997). Effect of prenatal stress on opioid component of exploration in different experimental situations. *Pharmacol. Biochem. Behav.* **58:**387–393.

Prospero-Garcia, O., Herold, N., Waters, A. K., Phillips, T. R., Elder, J. H., and Henriksen, S. J. (1994). Intraventricular administration of an FIV-envelope protein induces sleep architecture changes in rats. *Brain Res.* **659:**254–258.

Rapin, I., and Katzman, R. (1998). Neurobiology of autism. *Ann. Neurol.* **43:**7–14.

Roland, P. E. (1993). Partition of the human cerebellum in sensory-motor activities, learning and cognition. *Canad. J. Neurol. Sci.* **20:**S75–S77.

Rott, R., and Becht, H. (1995). Natural and experimental Borna disease in animals. *Curr. Top. Microbiol. Immunol.* **190:**17–30.

Rubin, S. A., Bautista, J. R., Moran, T. H., Schwartz, G. J., and Carbone, K. M. (1999a). Viral teratogenesis: brain developmental damage associated with maturation state at time of infection. *Brain Res. Dev. Brain Res.* **112:**237–244.

Rubin, S. A., Sierra-Honigmann, A. M., Lederman, H. M., Waltrip, R. W., 2nd, Eiden, J. J., and Carbone, K. M. (1995). Hematologic consequences of Borna disease virus infection of rat bone marrow and thymus stromal cells. *Blood* **85:**2762–2769.

Rubin, S. A., Sylves, P., Vogel, M., Pletnikov, M., Moran, T. H., Schwartz, G. J., and Carbone, K. M. (1999b). Borna disease virus-induced hippocampal dentate gyrus damage is associated with spatial learning and memory deficits. *Brain. Res. Bull.* **48:**23–30.

Saito, N., Itouji, A., Totani, Y., Osawa, I., Koide, H., Fujisawa, N., Ogita, K., and Tanaka, C. (1993). Cellular and intracellular localization of ε-subspecies of protein kinase C in the rat brain; presynaptic localization of the ε-subspecies. *Brain Res.* **607:**241–248.

Sauder, C., and de la Torre, J. C. (1999). Cytokine expression in the rat central nervous system following perinatal Borna disease virus infection. *J. Neuroimmunol.* **96:**29–45.

Schneemann, A., Schneider, P. A., Lamb, R. A., and Lipkin, W. I. (1995). The remarkable coding strategy of borna disease virus: a new member of the nonsegmented negative strand RNA viruses. *Virology* **210:**1–8.

Schwartz, R. H. (1992). Costimulation of T lymphocytes: the role of CD28, CTLA-4 and B4/BB1 in interleukin-2 production and immunotherapy. *Cell* **71:**1065–1068.

Schwemmle, M., De, B., Shi, L., Banerjee, A., and Lipkin, W. I. (1997). Borna disease virus P-protein is phosphorylated by protein kinase Cepsilon and casein kinase II. *J. Biol. Chem.* **272:**21818–21823.

Shankar, V., Kao, M., Hamir, A. N., Sheng, H., Koprowski, H., and Dietzschold, B. (1992). Kinetics of virus spread and changes in levels of several cytokine mRNAs in the brain after intranasal infection of rats with Borna disease virus. *J. Virol.* **66:**992–998.

Sierra-Honigmann, A. M., Rubin, S. A., Estafanous, M. G., Yolken, R. H., and Carbone, K. M. (1993). Borna disease virus in peripheral blood mononuclear and bone marrow cells of neonatally and chronically infected rats. *J. Neuroimmunol.* **45:**31–6.

Soares, J. C., and Mann, J. J. (1997). The anatomy of mood disorders—review of structural neuroimaging studies. *Biol. Psychiatry* **41:**86–106.

Sobbe, M., Bilzer, T., Gommel, S., Noske, K., Planz, O., and Stitz, L. (1997). Induction of degenerative brain lesions after adoptive transfer of brain lymphocytes from Borna disease virus-infected rats: presence of CD8+ T cells and perforin mRNA. *J. Virol.* **71:**2400–2407.

Solbrig, M., Koob, G., and Lipkin, W. (1999). Orofacial dyskinesias and dystonia in rats infected with Borna disease virus; a model for tardive dyskinetic syndromes. *Mol. Psychiatry* **4:**310–312.

Solbrig, M. V., Fallon, J. H., and Lipkin, W. I. (1995). Behavioral disturbances and pharmacology of Borna disease. *Curr. Top. Microbiol. Immunol.* **190:**93–101.

Solbrig, M. V., Koob, G. F., Fallon, J. H., and Lipkin, W. I. (1994). Tardive dyskinetic syndrome in rats infected with Borna disease virus. *Neurobiol. Dis.* **1:**111–119.

Solbrig, M. V., Koob, G. F., Fallon, J. H., Reid, S., and Lipkin, W. I. (1996a). Prefrontal cortex dysfunction in Borna disease virus (BDV)-infected rats. *Biol. Psychiatry* **40:**629–636.

Solbrig, M. V., Koob, G. F., Joyce, J. N., and Lipkin, W. I. (1996b). A neural substrate of hyperactivity in borna disease: changes in brain dopamine receptors. *Virology* **222:**332–338.

Solbrig, M. V., Koob, G. F., and Lipkin, W. I. (1996c). Naloxone-induced seizures in rats infected with Borna disease virus. *Neurology* **46:**1170–1171.

Solbrig, M. V., Koob, G. F., and Lipkin, W. I. (1998). Cocaine sensitivity in Borna disease virus-infected rats. *Pharmacol. Biochem. Behav.* **59:**1047–1052.

St. Pierre, B. A., Merrill, J. E., and Dopp, J. M. (1996). Effects of cytokines on CNS cells: glia. *In* "Cytokines and the CNS" (R. M. Ransohoff and E. N. Benveniste, eds.), pp. 151–168. CRC Press, Boca Raton.

Stevens, J. R. (1973). An anatomy of schizophrenia? *Arch. Gen. Psychiatry* **29:**177–189.

Stitz, L., Dietzschold, B., and Carbone, K. M. (1995). Immunopathogenesis of Borna disease. *Curr. Top. Microbiol. Immunol.* **190:**75–92.

Stitz, L., Noske, K., Planz, O., Furrer, E., Lipkin, W. I., and Bilzer, T. (1998). A functional role for neutralizing antibodies in Borna disease: influence on virus tropism outside the central nervous system. *J. Virol.* **72:**8884–8892.

Stitz, L., Soeder, D., Deschl, U., Frese, K., and Rott, R. (1989). Inhibition of immune-mediated meningoencephalitis in persistently Borna disease virus-infected rats by cyclosporine A. *J. Immunol.* **143:**4250–4256.

Stoll, G., and Jander, S. (1999). The role of microglia and macrophages in the pathophysiology of the CNS. *Prog. Neurobiol.* **58:**233–247.

Takei, N., Tanaka, O., Endo, Y., Lindholm, D., and Hatanaka, H. (1999). BDNF and NT-3 but not CNTF counteract the Ca2+ ionophore-induced apoptosis of cultured cortical neurons: involvement of dual pathways. *Neuropharmacol.* **38:**283–288.

Teitelbaum, P., Teitelbaum, O., Nye, J., Fryman, J., and Maurer, R. G. (1998). Movement analysis in infancy may be useful for early diagnosis of autism. *Proc. Natl. Acad. Sci. USA* **95:**13982–13987.

Thompson, R. F., and Kim, J. J. (1996). Memory systems in the brain and localization of memory. *Proc. Natl. Acad. Sci. USA* **93:**13438–13444.

Toth, Z., Yan, X. X., Haftoglou, S., Ribak, C. E., and Baram, T. Z. (1998). Seizure-induced neuronal injury: vulnerability to febrile seizures in an immature rat model. *J. Neurosci.* **18:**4285–4294.

van Os, J., Jones, P., Lewis, G., Wadsworth, M., and Murray, R. (1997). Developmental precursors of affective illness in a general population birth cohort. *Arch. Gen. Psychiatry* **54:**625–631.

Vanderschuren, L. J., Niesink, R. J., and Van Ree, J. M. (1997). The neurobiology of social play behavior in rats. *Neurosci. Biobehav. Rev.* **21:**309–326.

Weinberger, D. R. (1987). Implications of normal brain development for the pathogenesis of schizophrenia. *Arch. Gen. Psychiatry* **44:**660–669.

Weissenböck, H., Hornig, M., Hickey, W. F., and Lipkin, W. I. (2000). Microglial activation and neuronal apoptosis in Bornavirus infected neonatal Lewis rats. *Brain Pathol.* **10:**260–271.

Wing, L. (1997). The autistic spectrum. *Lancet* **350:**761–766.

Winslow, J. T., and Insel, T. R. (1990). Serotonergic and catecholaminergic reuptake inhibitors have opposite effects on the ultrasonic isolation calls of rat pups. *Neuropsychopharmacol.* **3:**51–59.

Wong, M. L., Bongiorno, P. B., Rettori, V., McCann, S. M., and Licinio, J. (1997). Interleukin (IL) 1beta, IL-1 receptor antagonist, IL-10, and IL-13 gene expression in the central nervous system and anterior pituitary during systemic inflammation: pathophysiological implications. *Proc. Natl. Acad. Sci. USA* **94:**227–232.

Yolken, R. H., and Torrey, E. F. (1995). Viruses, schizophrenia, and bipolar disorder. *Clin. Microbiol. Rev.* **8:**131–145.

ADVANCES IN VIRUS RESEARCH, VOL 56

PARADIGMS FOR BEHAVIORAL ASSESSMENT OF VIRAL PATHOGENESIS

Michael R. Weed* and Lisa H. Gold†

*Department of Psychiatry
Johns Hopkins Medical School
Baltimore, Maryland 21224
† Department of Neuropharmacology
The Scripps Research Institute
La Jolla, California 92037

I. Introduction

Neurobehavioral dysfunction is a serious consequence of a number of viral infections in which the pathological and toxic changes induced by the virus affect nervous system structure and function (Mohammed, Norrby, and Kristensson, 1993). Viral infections have been associated with a wide range of functional and behavioral abnormalities, including those that affect basic physiology and homeostasis, motor performance, affective behavior, and complex cognition. A number of working hypotheses draw a specific link between viral infection and subsequent disturbance of neural function for a wide range of neurological diseases. In brief, various neurotropic viruses may trigger transient or irreversible disorders in the central nervous system (CNS) by causing neuronal death, altering the expression of neurotransmitters, neuropeptides, or receptors, or by triggering immune system mechanisms that interfere with neural function.

Infections in animals are associated with a variety of behavioral changes and substances that stimulate the immune system, such as

583

endotoxin, lipopolysaccharide (LPS) and cytokines, can mimic behavioral responses to sickness (Dantzer *et al.*, 1998). Thus, animal models of CNS viral disease are valuable for demonstrating that the neurologic disease evident in infected animals is sufficient to produce behavioral impairments like those associated with similar viral infection in humans. The term "behavioral virology," coined by Mohammed and colleagues (1990), describes the documentation of disturbances of behavior in virus–host systems. Animal models can then be experimentally manipulated in order to investigate potential mechanisms for the pathological states. Host genetics has been established as having an important role for viral pathogenesis, and the studies discussed below investigate the genetic determinants of behavioral effects of viruses. Similarly, critical differences in age of infection have been demonstrated for both physical and behavioral outcomes.

One primary objective of neurobehavioral studies of viral pathogenesis is to characterize specific behavioral phenotypes in infection models, in both rodents and nonhuman primates, that can then be correlated with the distinctive neuropathology and immunopathology observed in these same models. A second, equally important goal for behavioral virology is to use infection and consequent behavioral abnormalities to model diseases of unknown etiology, e.g., autism. The viral infection system may share behavioral pathologies with other disease/disorders and, conversely, the pathogenesis of a variety of diseases may possibly involve infectious components. In addition to investigating the host and viral factors that contribute to neuropathogenesis, behavioral paradigms are valuable for testing novel therapeutic treatments. This approach has been less often used in rodent models than in the nonhuman primate models.

A broad range of behavioral paradigms are used for investigating the neurobehavioral sequelae in animal models of viral pathogenesis. While some viruses produce relatively regionally specific pathology, for others the neuropathology is not restricted to a single brain region, but involves multiple brain regions and systems. In some cases, the host response to infection may initiate acute "sickness" behavior or more permanent pathologic changes. The extreme sensitivity of behavioral function to CNS insults is often illustrated in cases where animals appear normal on gross observation, yet behavioral performance is impaired. Analysis can involve comprehensive evaluation of behavioral performance in tasks that represent both spontaneous, unconditioned behaviors, and conditioned, learned behaviors. To supplement basal phenotypic assessment, pharmacologic and immunologic agents are used to perturb behavior and exaggerate potential differences between experimental groups. For example, drugs possessing known neurobehavioral actions can be used to probe selected neurotransmitter systems.

This chapter will use two approaches to describe the behavioral paradigms used in assessing viral pathology. Section II will approach the behavioral effects of viral infection in rodents by providing a broad overview of the types of studies conducted in rodents, and by using examples from a number of virus model systems. In contrast, Section III will discuss the behavioral effects of viral infection in nonhuman primates by focusing on the simian immunodeficiency virus (SIV) model of acquired immunodeficiency syndrome (AIDS) and elaborating on recent work involving a number of behavioral paradigms in this one model system. These two complementary approaches should provide both an outline of the current state of the field and illustrations of the variety of behavioral techniques that are in use in the newly emerging field of behavioral virology.

II. BEHAVIORAL EFFECTS OF VIRAL INFECTION IN RODENTS

A. Background

Behavioral paradigms in rodents have predominantly focused on three general behavioral domains and the following sections will be structured accordingly. Many involve basic physiology such as ingestive behavior and body weight, body temperature, sleep, and nociception (pain detection). Observable neurological signs will also be considered in this category. A second group of studies assesses motor activity (locomotion and coordination), exploration, and anxiety. These paradigms address the interaction of the organism with its environment and are sensitive to a variety of neural disturbances and sicknesses. Finally, many psychiatric and neurological diseases, as well as viral infections in humans, are associated with disturbances of learning and memory and social behavior. Such complex behavioral processes are also being evaluated in rodent models of viral pathogenesis. What follows does not exhaustively review the rodent literature, but provides illustrative examples of viral infection models and the paradigms used to quantify the consequent behavioral alterations (Table I). These studies demonstrate the value of this approach and support the further development and utilization of behavioral paradigms for the assessment of viral pathogenesis.

B. Basic Physiology

1. Rationale

Observational phenotypic assessment in adult or neonatal rodents provides a systematic method for comprehensively assessing and quan-

TABLE I

BEHAVIORAL ANALYSIS IN VIRAL INFECTION MODELS

Virus	Host	Behavioral Phenotype	
		Effect	Refs.
Borna disease (BDV)	Rat	salt preference, progressive learning impairment	Rubin et al. (1999, 1998)
	Rat	hyperactivity, decreased anxiety, learning impairments	Dittrich et al., 1989
	Rat	increased motor sensitivity to amphetamine and cocaine, effects of other pharmacologic agents	Solbrig et al., 1994, 1998, 1995, 1996, 1999
	Rat	disruption in play behavior	Pletnikov et al., 1999
	Mouse	neurological disease	Hallensleben et al., 1998
Canine distemper	Mouse	paralysis and abnormal turning and obesity	Bernard et al., 1999
Encephalomyocarditis (EMC)	Mouse	mild learning deficits	Yayou et al., 1993
Herpes simplex (HSV)	Mouse	strain-dependent activity differences, learning deficits	McFarland and Hotchin, 1983
	Mouse	increased activity during light period, impaired avoidance	Crnic and Pizer, 1988
	Mouse	social stress reactivated virus	Padgett et al., 1998
	Mouse	greater cytokine production in splenocytes with single housing	Karp et al., 1997
Human immuno-deficiency (HIV)	Mouse	learning impairment, reduced home cage activity	Avgeropoulos et al., 1998

Influenza	Mouse	daily food intake and body weight reductions, sweetened milk consumption unaffected	Swiergiel et al., 1997
	Mouse	decreases in food and water intake, body weight, temperature, and motor activity	Conn et al., 1995
	Mouse	sleep disruptions, temperature and motor activity reductions	Toth et al., 1995
	Mouse	restraint-reduced mortality in DBA	Hermann et al., 1993
Japanese encephalitis	Rat	motor impairment, bradykinesia	Ogata et al. 1997
Lymphocytic	Mouse	reduced ambulation and wheel running	Hotchin and Seegal, 1977
choriomeningitis (LCMV)	Mouse	hypoactivity during first motor test, impaired learning and increased sensitivity to scopolamine in both procedures	Gold, et al., 1994, Brot et al., 1997
	Rat	electrophysiological abnormalities	Pearce et al., 1996
LP-BM5 Murine	Mouse	open-field behavior and rotorod normal, learning impairments	Iida et al., 1999
leukemia retrovirus	Mouse	rotorod and tail suspension normal, learning impairment	Sei et al., 1992
mixture	Mouse	learning impairment, enhanced PCP discrimination	English et al., 1998, 1999
Moloney leukemia	Mouse	food restriction and lack of a social hierarchy in group housing, increased incidence of leukemia	Ebbesen et al., 1999
Tahyna (TAH)	Mouse	reduced motor activity	Hubalek and Rodl, 1994
Vesicular stomatitis	Rat	reductions in rearing, impaired learning	Mohammed et al., 1990
(VSV)	Rat	reductions in amphetamine-stimulated rearing	Andersson et al., 1993
Varicella zoster (VZV)	Rat	allodynia and hyperalgesia	Fleetwood-Walker et al., 1999
Yellow fever (YFV)	Mouse	reduced running speeds for food reward, transient impairments on rotating rod	Museteanu et al., 1979

tifying behavioral and physiologic status. Standard neurobehavioral protocols involve brief screening procedures reminiscent of a general neurological exam in human patients. They include monitoring of simple reflexes (inhibition and emergence) and each test provides information about the pattern of function of a particular system (Irwin, 1968). Observational screening techniques permit an initial assessment of gross gain or loss of function and these primary observations can point to further exploration of increasingly complex behaviors.

Measurement of physiologic parameters is an important aspect in charting the time course of disease progression. Food intake and body weight gain are critical aspects related to the general health of the animals. In general, mice are weighed before each behavioral manipulation and, at the very least, a minimum of once each week during experimentation. These data provide longitudinal information on body weight over time and during the course of infection. Differences in food consumption can also importantly interact with performance in food-motivated tasks. Preference for specific solutions (sucrose, salt) also informs about brain regions known to be important for regulating food intake and containing neurons that sense metabolic parameters. Fever and alterations in sleep patterns are additional concomitants of many viral infections (Darko, Mitler, and Henriksen, 1995; Kluger *et al.*, 1998). These particular physiologic dysregulations may have important consequences for behavioral performance across a wide range of domains.

Well-validated protocols that measure mechanical, electrical, and thermal sensitivity are also available. In addition to assessing hyper- and hypoanalgesia, they are used to establish similar levels of responsivity between groups, when these aversive stimuli are used to motivate behavior in other behavioral tasks.

2. Studies Reporting Observable Neurological Signs

The influence of genetic factors on disease susceptibility, following intracerebral (IC) inoculation of newborn mice from a variety of mouse strains with Borna disease virus (BDV), was recently illustrated by Hallensleben and colleagues (1998). Infection with BDV induced neurological disease characterized by a nonphysiological position of the hind limbs when lifted by the tail. Hunched posture, rough fur, and head tilt were often observed, and symptoms progressed rapidly and corresponded with pronounced loss of body weight and paraparesis at later stages. The incidence and severity of disease varied dramatically across mouse strains with neurological symptoms being greatest in the

MRL>C3H>BALB/c>CBA>C57 strains, yet they were not related to different levels of brain viral replication. Paralysis and abnormal turning activity were also observed in survivors of canine distemper virus infection that go on to develop late-stage disease (Bernard *et al.*, 1999).

3. Studies of Ingestive Behavior and Body Weight Regulation

The developmental stage of the brain at the time of virus infection also determines the selective vulnerability of specific neuroanatomic regions and behaviors. Areas of the brain maturing at the time of infection or afterward may be at higher risk for a disrupted normal development. For example, Lewis rats inoculated intracerebrally with Borna disease virus on postnatal day 15 develop a salt preference when given a 24-hour, two-bottle choice test involving salt solution and water at 90 days of age (Rubin *et al.*, 1999). These rats had been treated with cyclosporin A during the periinoculation period to prevent the development of intense meningoencephalitis. The expression of the salt preference was attributed to virally induced damage of the circuits mediating salt intake regulation, which continue to develop 3 and 4 weeks postnatally. Also, as commonly observed in infected animals, these rats demonstrated a reduction in body weight as compared to controls, when analyzed on postnatal day 45.

Ingestive behavior was quantified in CD-1 outbred mice inoculated intranasally with a sublethal dose of influenza virus (Swiergiel, Smagin, and Dunn, 1997). Daily food intake and body weight were reduced following infection, and returned to normal by days 9–10 postinoculation. Interestingly, 30 min consumption of sweetened milk was unaffected by infection, which may have been due to the enhanced palatability or ease of ingestion of the liquid. This study also demonstrated that immune modulators such as LPS and the cytokines interleukin (IL)-1α and β also altered these same behaviors. A decline in food and water intake and in body weight, as well as decreases in body temperature, were also measured in Swiss Webster mice that were given influenza inoculation (Conn *et al.*, 1995). Interestingly, Ebbesen and colleagues (1991) showed that severe food restriction enhanced the incidence of leukemia in DBA/2 mice inoculated intraperitoneally with Moloney virus.

While it is more common for viral infections to be associated with a reduced body weight gain, outbred Swiss mice that survive acute infection following intracerebral inoculation at 4 weeks of age with a neuroadaptive strain of canine distemper virus, exhibit a late-stage disease characterized by an obesity syndrome (Bernard *et al.*, 1999). Interestingly, obese mice were found to have persistence of virus in the hypothalamus and a downregulation of the leptin receptor in this brain area. Impaired response to leptin, thought to be an important satiety signal, may contribute to the obesity syndrome. The development of obesity was independent of the histocompatibility system since inbred BALB/c and C3H mice exhibited similar clinical disease.

4. Sleep Studies

Viral infection and elements of the immune system, such as cytokines, are known to affect sleep processes (Darko, Mitler, and Henriksen, 1995). Alterations in sleep produced by influenza virus inoculated intranasally were examined in adult C57BL/6 and BALB/c mice implanted with wire electrodes (Toth, Rehg, and Webster, 1995). C57 mice showed robust increases in slow-wave sleep and reductions in delta-wave activity, indicating lighter sleep and resulting in a loss of normal circadian rhythmicity of sleep. In contrast, infected BALB mice did not increase their slow-wave sleep. Concomitant reductions of core body temperature and locomotor activity were measured by implanted telemetric monitoring transmitters. Strain differences in immune responses to influenza virus may underlie the different sleep patterns.

5. Studies of Nociception

Because viruses may alter nociception (pain detection), and because electrical stimuli are sometimes used to motivate behavior, studies have examined sensitivity thresholds in infected mice. Female outbred Nylar mice inoculated intracerebrally with lymphocytic choriomeningitis virus neonatally were found to respond to lower levels of current than uninfected mice, when they were tested at 3.5–6 months of age (Hotchin and Seegal, 1977). However, no differences in sensitivity were reported in another study when male DBA mice were tested at 6–7.5 months of age (Gold et al., 1994).

A rodent viral model has also been developed to examine the longterm consequence of postherpetic neuralgia. Wistar rats inoculated in the footpad with varicella zoster virus developed a chronic infection and were examined for alterations in sensory thresholds (FleetwoodWalker et al., 1999). A paw withdrawal response was measured on exposure to graded innocuous and noxious mechanical stimuli applied by using a von Frey filament and noxious thermal stimuli. Thresholds for both mechanical and thermal stimuli showed significant and sustained declines for up to 33 days postinoculation. This measurable allodynia and hyperalgesia, respectively, support the further use of this rodent model to study mechanisms involved in the establishment of postherpetic neuralgia.

C. Motor Function, Exploration, Anxiety, and Stress

1. Rationale

Several procedures can be used to evaluate the function of neural systems mediating motor output. Spontaneous motor activity is thought to reflect exploratory drive, reactivity to novelty, and general level of arousal. Alterations in motor activity may indicate changes in these constructs that may then influence other behaviors. Motor ability and coordination can also be examined. For example, balancing on a rotating rod requires a variety of proprioceptive, vestibular, and fine-tuned motor abilities—the mouse must balance on a rotating rod that gradually accelerates during a short test, and this paradigm is often used to screen for motor deficits.

Anxiety, basal reactivity to stressors, and novelty/exploration can be characterized by using a variety of paradigms that manipulate the exploratory drive of rodents. In many cases the exploration of a novel or familiar environment is directly challenged by the fear of open, lighted places. As with motor activity, anxiety, and exploration, reactivity to external stressors involves the integration of multiple neural systems. Stressors often take the form of physical challenges, such as restraint.

2. Studies of Motor Function, Exploration, and Anxiety

A classic behavioral test for measuring motor activity and the tendency to explore a large arena is the open-field test. Wistar rats inoculated intracerebrally with Borna disease virus at one day of age, and tested at around 15 weeks of age, showed reduced resting behavior, reflective of hyperactivity, and increased exploration of the center of the field, which indicated decreased anxiety (Dittrich *et al.*, 1989). In these rats, decreased anxiety was also evident in a neophobia test measuring latency in exiting from a dark chamber into a lighted arena. However, rats inoculated at postnatal day 15, as described above, did not exhibit spontaneous hyperactivity when examined in a more continuous manner by using photobeam activity monitors over a 26-hour period starting on postnatal day 30 (Rubin *et al.*, 1999).

LP-BM5 infection produces profound immunosuppression that shares many immunological features with AIDS. C57BL/6 mice inoculated intraperitoneally and tested at 10 and 11 weeks postinoculation exhibited no differences in open-field activity or in ability to maintain balance on a rod rotating with a speed of 15 revolutions per minute compared to control mice (Iida *et al.*, 1999). Similarly, rotating rod performance and the duration of immobility recorded in a tail suspension test were found to be normal in LP-BM5, infected mice tested at 5–7 weeks postinoculation. (Sei *et al.*, 1992).

Open-field activity as well as exploration in a Y-maze have also been used to characterize herpes simplex virus (HSV) infection in Nya:NYLAR mice as compared with Nya:(SW) mice inoculated intracerebrally at 8 weeks of age (McFarland and Hotchin, 1983). Mice were tested 3 weeks postinoculation and infected NYLAR mice were significantly more active, while infected SW mice were less active, than their respective controls. Despite these activity differences, there was no effect of the virus on swimming speed measured in a straight-alley swim tank. A mutant HSV virus with limited virulence was used to infect (SC) BALB/c mice neonatally (Crnic and Pizer, 1988). Open-field activity was measured at 20 days of age and also at 50–66 days of age for a 24-hour period. Horizontal activity, or ambulation, was higher in infected animals than in controls and the increase was primarily during the light phase, when the mice are normally inactive.

Motor activity changes have also been reported in mice infected with lymphocytic choriomeningitis virus (LCMV) neonatally. Male outbred Nylar mice inoculated intracerebrally within 24 hours of birth exhibited reduced ambulation in an open field and a decreased amount of wheel running when tested at 3.5–6 months of age (Hotchin and Seegal, 1977). Inbred DBA mice inoculated neonatally also exhibited hypoactivity during the first exposure to locomotor activity test chambers, although infected mice did exhibit normal within-session and between-session habituation (Gold *et al.*, 1994). Both studies suggest a reduced tendency to explore a novel environment in LCMV-infected mice. Infected mice were also more sensitive to the stimulating effects of the cholinergic antagonist, scopolamine, pointing to a disruption of cholinergic systems in these mice (Gold *et al.*, 1994).

Open-field activity is also increased by intranasal inoculation of vesicular stomatitis virus (VSV) in 12-day-old rats evaluated as adults; however rearing was reduced, suggesting alterations in exploratory behavior (Mohammed *et al.*, 1990). The behavioral pharmacology of VSV hyperactivity was then addressed by examining the effects of amphetamine, a prototypical dopaminergic stimulant drug, on behavior (Andersson *et al.*, 1993). While amphetamine increased locomotor activity similarly in both inoculated rats and controls, the increase in duration and frequency of rearing behavior was diminished in infected rats. These results suggest that long-term changes may result from transient infection, as no viral antigens were detected at the time of testing.

Rats experimentally infected with Borna disease virus as immature adults develop a persistent CNS infection with hyperactive, dyskinetic, and stereotypic behavioral manifestations (Solbrig *et al.*, 1994). Because of the resemblance of these motor behaviors to syndromes of dopamine (DA) sensitivity or excess, and to animal models of psychostimulant abuse, a line of research using pharmacologic probes, and locomotor and observational data was undertaken to elucidate the BDV movement and behavior disorder. The effects of a viral encephalitis on dopaminergic locomotor dyskinetic and stereotypic behavior were examined, to test the hypothesis that BDV rats show differential sensitivity to the stimulant properties of DA agonists. BDV rats had increased sensitivity to the indirect DA agonists *d*-amphetamine (Solbrig *et al.*, 1994) and cocaine (Solbrig, Koob, and Lipkin, 1998), as manifested by increased psychostimulant-induced locomotion (measured by activity-chamber photocell counts) and stereotypical behaviors, such as sniffing, grooming, and oral–facial movements. The direct DA agonist, apomorphine, had an enhanced sedative effect at presynaptic (autoreceptor) doses, while the D1 antagonist SCH23390 and the atypical neuroleptic clozapine also selectively reduced activity in BD rats, but not in uninfected controls. The D2 antagonists haloperidol and raclopride had no behavioral effects at the doses tested. (Solbrig *et al.*, 1994). Additional pharmacologic and neurochemical studies have established the scope of noradrenergic, serotonergic, cholinergic, and opioidergic disturbances in infected animals (Solbrig, Koob, and Lipkin, 1999; Solbrig, Fallon, and Lipkin, 1995; Solbrig, Koob, and Lipkin, 1996).

In contrast to infections that cause hyperactivity, lethargy and reduced activity are often considered as nonspecific host responses to infection. Lethal intranasal inoculation of influenza virus in CD-1 outbred mice produces reductions in open-field rearings and ambulation prior to death (Swiergiel, Smagin, and Dunn, 1997). A minimally effective dose administered to Swiss Webster mice also produced reductions in motor activity over 24-hour periods measured by using telemetric monitoring devices (Conn *et al.*, 1995). Inoculation of C57BL/6 and BALB/c inbred mice produced similar decreases in locomotor activity in both strains, also measured over 24-hour periods by telemetric monitoring devices (Toth, Rehg, and Webster, 1995).

Fisher rats inoculated intracerebrally with Japanese encephalitis virus 13 days postnatally demonstrate clinicopathologic changes resembling those found in Parkinson's disease (Ogata *et al.*, 1997). Notably, the substantia nigra of infected rats was characterized by neuronal loss with gliosis and reduced tyrosine hydroxylase-positive neurons, reflecting damage to the dopaminergic system. Infected rats demonstrated bradykinesia and increased time to descend a pole (100 centimeters in height) to the floor. The motor deficit showed significant improvement following treatment for 7 days with L-DOPA, a dopamine precursor used therapeutically in Parkinson's disease. In addition to activity changes due to disrupted central nervous system function, direct effects on muscle can also lead to motor impairments. Such is the case following intraperitoneal Tahyna virus inoculation in 25-day-old C57BL/6 mice. Significantly lower activity in an open field was recorded 1–3 days postinoculation, consistent with a primary target of striated muscle following peripheral inoculation (Hubalek and Rodl, 1994).

Yellow fever virus inoculated intracerebrally in 21-day-old Swiss Webster mice resulted in longer latencies in obtaining food reward in a straight maze when tested 76 days postinoculation (Museteanu et al., 1979). Reductions in running speeds in the maze were proportional to the quantity of virus injected. Also, there was correspondence between extent of brain lesions and running time. In contrast, infected animals exhibited impairments in maintaining balance on a rotating rod when tested at 45–55 days postinoculation but had recovered to control levels when tested at 97–160 days postinoculation.

2. Studies of Stressors

The study of neuroimmunology supports an interaction between stress and the capacity of the immune system to respond to infection. Thus, stressors can be used to manipulate the outcome of viral infection in rodents. While social stress is commonly examined (as discussed below), physical stress, often in the form of periods of restraint, is also employed. Restraint in conical tubes for 12 hr/day starting 1 day prior to inoculation, and continuing for 10 days postinoculation was examined in C57BL/6, DBA/2, and C3H mice inoculated intranasally with influenza virus (Hermann et al., 1993). Because such a restraint protocol also involved restriction of food and water, similarly restricted groups were included as controls. There was no difference in the survival of food/water-restricted mice as compared to restraint-stressed C57 and C3H infected mice. However, DBA mice that were restrained exhibited reduced mortality, and restraint was associated with reduction of pulmonary inflammation in DBA mice. Interestingly, restraint did not increase levels of glucocorticoids above those associated with influenza infection in DBA mice (Hermann et al., 1994).

D. Social Interactions, Learning, and Memory

1. Rationale

Cognition, which is the culmination of the function of several underlying processes, can be examined in behavioral tasks that require integration of motivational, sensory, learning, memory, and/or motor processes. Multiple tasks of complex learning ability are often employed, in order to provide a convergence of information that supports dysfunction in a specific brain site or impairment in a functional construct. In many cases, debilitating consequences of viral infection and neuropsychiatric diseases are associated with effects on cognitive function. Thus, animal models of viral pathogenesis are frequently evaluated by employing paradigms designed to test learning and memory functions as well as other complex associative behaviors and neurological correlates, such as electrophysiology.

Nonspecific symptoms of infection include profound behavioral and psychological changes, including loss of interest in usual activities and in social contacts (Kent et al., 1992). These behavioral changes are now thought to be part of the natural homeostatic reaction that the body uses to fight infection. Group housing conditions are also potent manipulators of the host response to various insults.

2. Studies of Social Influences

Group housing situations allow for the development of social hierarchies, and disruptions of these social groups can have profound effects on behavior and the immune systems response to infection (see also, section III). In humans it is known that stress is often associated with reactivation of latent herpes virus infection. The effects of social stress versus restraint stress (16 hours per day) on latent herpes simplex virus infection (5 weeks postinoculation) in BALB/c mice inoculated intraocularly were examined at 4–6 weeks of age (Padgett et al., 1998). Mice were housed in groups of five, and dominant males in each cage were determined by observation of social interactions. Social stress was induced by repeated rearranging of the dominant males among cages. Irradiation-induced reactivation of HSV was found in around 45% of mice, and a similar result was obtained in the group subjected to social reorganization. However, restraint stress was associated with significantly less reactivation. Dominant animals are frequently involved in the greatest number of stressful social interactions and they were more likely than subordinates to exhibit reactivation. The absence of a social hierarchy in cages with large numbers of mice has also been shown to result in a higher incidence of virus-induced leukemia in DBA/2 mice inoculated with Moloney virus (Ebbesen et al., 1991). A subordinant position was also associated with a worse outcome. Mice were housed either singly, in groups of 3 where a dominance hierarchy can develop, or in groups of 9 or 15 that are too large for establishment of social rank. In contrast to the findings with HSV discussed above, dominant mice were protected against leukemia produced by Moloney virus. The effects of differential housing conditions on markers of cellular and humoral response to herpes simplex virus infection were investigated in BALB/c mice inoculated subcutaneously in the footpad at around 9 weeks of age (Karp, Moynihan, and Ader, 1997). Greater cytokine production in splenocytes was measured in singly housed mice as compared with group-housed mice, while there was no change in circulating immunoglobulin (Ig)M or IgG antibody titers.

Because play behavior is well characterized in rats, impoverished or atypical social behavior in Lewis rats infected with Borna disease virus as newborns has been proposed as an animal model for childhood autism (Pletnikov et al., 1999). Infected rats and controls were assigned to groups that were singly housed (residents) or group housed (intruders). Intruders were placed in residents' cages for 10 minutes and social behavior was recorded. There was no difference in nonsocial exploratory behavior (ambulation and rearing) in infected residents as compared with controls. Nonplay social behavior (approaching, following, sniffing, grooming) was increased in infected residents only on the first day, while play behavior was consistently reduced whether the resident or the intruder was infected. The selective disruption in play behavior, and no change in nonsocial behaviors or nonplay behaviors in infected rats, were interpreted as further support for this model of human autism.

3. Studies of Learning and Memory

Wistar rats inoculated intracerebrally with Borna disease virus at one day of age were tested, starting at 15 weeks postinoculation on a variety of learning tests (Dittrich et al., 1989). Infected mice made more errors in Y-maze discrimination and in hole board search tasks aimed at food reward. Infected mice also required more trials to learn a taste aversion when a lithium chloride injection was paired with saccharine consumption, and exhibited a more mild aversion. Similarly infected rats demonstrated a weaker suppression of operant responding under conditions of negative reinforcement (a shock condition) than did uninfected mice. This extensive characterization of learning deficits was conducted in Borna disease-infected rats that were clinically healthy, to highlight the utility of sensitive behavioral testing in revealing functional effects of viral pathogenesis. A progressive performance deficit was also measured by using a Morris water maze in Lewis rats inoculated as newborns with BDV and behaviorally tested at 43, 53, and 72 days of age (Rubin et al., 1998). On day 43, BDV rats took longer to get to the escape platform than did uninfected rats but demonstrated equivalent levels of learning over trials. At 53 days old, infected rats were again slower but at this time point they did not exhibit comparable learning over trials as controls. At 72 days old, infected rats were much slower than controls and BVD groups tested at earlier time points and no learning was evident across trials. At all ages, infected rats exhibited fewer target area crossings as compared with uninfected rats and progressively worsened from day 43 to 72 day. By 72 days of age, infected mice also exhibited reduced swimming speeds as compared to controls.

Severe combined immunodeficient mice inoculated intracerebrally with HIV-infected human monocytes exhibit HIV-positive macrophages, multinucleated giant cells, astrogliosis, microglial nodules, and neuronal dropout (Avgeropoulos et al., 1998). Mice were reinoculated every 4 weeks for 3 months in order to develop a model of chronic HIV encephalitis. Behavioral testing in a Morris water maze, in which mice have to swim to a hidden escape platform, is thought to be useful for evaluating both spatial reference and working memory functions, depending on the task parameters. HIV-infected mice tested 3.5 months after first inoculation showed a worse performance in acquisition and retention tests. There were no differences in swimming speed under conditions where the platform was visible, although motor activity measured in the home cage was reduced in infected mice.

An alternate model of neuroAIDS involves infection with LP-BM5 murine leukemia virus. C57BL/6 mice inoculated intraperitoneally were tested in various learning tasks 10–15 weeks postinoculation (Iida et al., 1999). For the water-finding task, which is thought to measure spatial attention and latent learning, mice are placed in an open field that contains an alcove that houses a water spout. After initial exposure to the apparatus, mice are water deprived and reintroduced into the open field and the latencies in approaching the alcove and in drinking are recorded. While there were no group differences in the latency in finding the water during the training trial, infected mice exhibited longer latencies during the test trial (24 hours later). Infected mice also exhibited a reduced rate of spontaneous alternation in a Y maze thought to index short-term memory and increased latencies in finding the platform in a Morris water maze. In an earlier study by another group of investigators, fewer infected mice met the criterion for acquisition in a Morris water maze when tested 5–7 weeks postinoculation and a reversal probe revealed an absence of spatial preference for the platform target area following training in the infected mice (Sei et al., 1992). Impairment in acquisition of a shuttle-box avoidance response was observed in C57BL/6 mice inoculated with LP-BM5 and tested 11 weeks postinoculation (English, Hemphill, and Paul, 1998). Interestingly, mice tested 7 weeks postinoculation were not impaired during acquisition or retention testing 4 weeks later, suggesting that performance was less easily disrupted than acquisition at 11 weeks postinoculation. LP-BM5 infection was also associated with increased sensitivity to phencyclidine (PCP) at 6 and 9 weeks postinoculation in a task in which C57BL/6 mice were trained to discriminate the subjective effects of PCP from vehicle injection (English, Bruce, and Paul, 1999). Alterations of the glutamatergic system following viral infection may contribute to this enhanced sensitivity to PCP.

Lymphocytic choriomeningitis virus infection of newborn mice produces disruptions of learning and memory performance when they are tested as adults. DBA/2 mice inoculated intracerebrally within 18 hours of birth were tested for acquisition and retention of a Y-maze-discriminated avoidance task (Gold et al., 1994). Mice had to learn to run into one of the two goal arms to avoid or escape a mild electrical stimulus. Infected mice exhibited a deficit in acquisition of the discrimination when tested at 2–4.5 months of age and were similarly impaired when tested for reacquisition 7.5 months later. As seen for motor activity, infected mice exhibited enhanced sensitivity to the effects of scopolamine, a cholinergic antagonist, in this task. LCMV-infected BALB/c mice also made more incorrect choices when tested at 10–12 weeks of age by use of a similar protocol (Brot et al., 1997). Retesting at about 5 months, after adoptive transfer of cytotoxic T cells to induce viral clearance, still revealed a performance deficient in previously infected mice, which suggested that the behavioral deficit does not depend on persistent viremia.

Studies with HSV and VSV also indicated that clearance of virus is not associated with restoration of behavioral functions. Following intranasal inoculation of vesicular stomatitis virus in 12-day-old rats, viral antigens are no longer detected 13–15 days postinoculation. However, on testing in a Morris water maze at 4 months postinoculation infected rats exhibit longer latencies in reaching the escape platform, while performance at 1, 2, and 9 months postinoculation was normal (Mohammed et al., 1990).

Herpes simplex virus infection of Nya:NYLAR and Nya:(SW) mice at 8 weeks of age did not result in Y-maze alternation deficits when they were tested at 3 weeks postinoculation (McFarland and Hotchin, 1983). However, acquisition and reversal of spatial learning in a Y-maze water escape task were impaired in SW mice tested 6 weeks postinoculation. A mutant of HSV with limited virulence in mice was employed to inoculate subcutaneously BALB/c mice neonatally. A passive-avoidance task in which mice had to inhibit their movement and remain on a small platform to avoid an electrical stimulus revealed deficits in infected mice

(Crnic and Pizer, 1988). In contrast, infection was not associated with altered learning motivated by water reward (in water-restricted mice) in a complex task involving visual spatial learning in an 8-arm radial maze.

A diabetogenic variant of encephalomyocarditis virus produces variable pathogenicity in different mouse strains with BALB/c mice manifesting more resistance (Yayou *et al.*, 1993). Performance in a Morris water maze was compared in DBA/2 and BALB/c mice inoculated intraperitoneally one day prior to testing. While infected mice learned the location of the platform over days, acquisition was mildly retarded as compared to controls. The hippocampus exhibited focal degeneration in both strains but the spinal cord was only affected in DBA/2 mice, suggesting that decreased mobility could not solely account for performance impairments.

E. Neurophysiological Correlates of Behavior

Brain electrophysiological activity is a correlate of behavior and, in particular, hippocampal activity is associated with learning and memory function. Lewis rats infected with lymphocytic choriomeningitis virus on postnatal day 4 were then evaluated 84–107 days postinoculation at a time when virus was cleared from the dentate gyrus of the hippocampus and the number of granule cells was dramatically reduced (Pearce *et al.*, 1996). The hippocampal electroencephalogram (EEG) was dominated by continuous theta activity and a dissociation of synaptic and cellular responses, and suppression of GABA-mediated recurrent inhibition in the dentate gyrus was recorded in infected rats. These results reveal abnormalities in synaptic function that persist after clearance of infectious virus from the central nervous system and suggest one mechanism for long-term behavioral changes consequent to viral pathogenesis.

III. BEHAVIORAL EFFECTS OF VIRAL INFECTION IN NONHUMAN PRIMATES

A. Background

It is widely appreciated that viral infections can lead to behavioral changes in humans and other animals. The behavioral effects of rabies infection in humans and dogs are well known for their dramatic nature in causing behavioral disturbances such as anxiety, irritability, and episodes of unpredictable, often violent behavior (Mohammed, Norrby, and Kristensson, 1993). Some viral infections can lead to encephalitis and have been reported to produce behavioral changes including psychosis (Mohammed, Norrby, and Kristensson, 1993). With the current pandemic of AIDS, the behavioral and motor effects of human immunodeficiency virus (HIV) infection are beginning to be more widely known. Approximately 40% of AIDS patients present some neurologic dysfunction, 15–33% develop AIDS minor cognitive/motor disorder and roughly 5–10% develop AIDS dementia complex (ADC), including frank dementia (Kelly et al., 1996; Price, 1994; Sacktor and McArthur, 1997). Additionally, a cognitive or motor disorder has been reported to be the first clinical symptom in approximately 7% of AIDS patients, and the only AIDS-defining illness in 3% of patients (Janssen et al., 1992). The behavioral changes in AIDS have been studied to a larger extent than those of other viral infections due to the seriousness of the pandemic and the development of excellent nonhuman primate models of the disease.

SIV infection of macaques produces an immune deficiency syndrome quite similar to that of HIV infection in humans (Desrosiers, 1990; Fox et al., 1997; Rausch, Murray, and Eiden, 1999; Zink et al., 1997). Decreases in T-cell $CD4^+$ numbers, development of lymphadenopathy, opportunistic infections, and a wasting disease occur in both humans with HIV infection and macaques with SIV infection. In the CNS, infiltrating macrophages and microglia, as well as brain capillary endothelial cells, are infected by both HIV and SIV (Flaherty et al., 1997; Mankowski et al., 1994; Moses and Nelson, 1994; Price et al., 1988; Simon, Chalifoux, and Ringler, 1992; Wiley et al., 1986). The similarity of the ensuing neuropathologies of HIV and SIV infection have been described in numerous studies (Chakrabarti et al., 1991; Lackner et al., 1991; Sharer et al., 1991; Simon, Chalifoux, and Ringler, 1992).

The precise mechanism by which HIV and SIV produce their neuropathologies is unknown. Since most studies do not find that HIV or SIV directly infects CNS neurons, current hypotheses are centered largely on indirect mechanisms related to the immunologic response of the host (e.g., release of neurotoxic cytokines), to degradation of the blood–brain barrier, or to increases in neurotoxic molecules such as quinolinic acid, and neurotoxic viral proteins (Fox et al., 1997; Lipton and Gendelman, 1995; Rausch, Murray, and Eiden, 1999). Further, the compressed time course of disease progression of SIV in macaques is measured in months and is much shorter than the progression of HIV in humans that is measured in years. This accelerated time course is being exploited by AIDS researchers in many disciplines, including the investigation of AIDS neuropathology, and in the functional effects of this neuropathology as manifested in behavioral changes.

Highly active antiretroviral therapies (HAARTs) have been credited with the recent decline in mortality from AIDS. These therapies have been reported to be beneficial in cases of AIDS dementia and in minor cognitive/motor disorders as well (Chang *et al.*, 1999; Gendelman *et al.*, 1998). However, as these therapies continue to be used, treatment failure has become a significant concern, whether failure is due to non-compliance, development of resistant strains, intolerable side effects, or other causes (d'Arminio Monforte *et al.*, 1998; Ledergerber *et al.*, 1999; Wit *et al.*, 1999). Additionally, the cost of these treatments reduces their availability for AIDS patients outside the Western world. Furthermore, the majority of the therapeutics do not penetrate the blood–brain barrier well, and the long-term effects of clearing peripheral viral loads, while sparing virus in the CNS, are not understood. Therefore, the investigation of the CNS/behavioral effects of AIDS remains of the utmost importance. The study of CNS/behavioral effects of AIDS can contribute to the development of AIDS therapies in at least two ways: (1) an understanding of the time course of CNS/behavioral effects will help in directing *in vitro* investigations of changes that occur in the CNS that are coincident with behavioral changes and may contribute to an understanding of the mechanism of the CNS insult that produces behavioral changes, and (2) behavioral assays can be used as expressions of CNS function in the study of AIDS therapeutics, indicating whether a given drug effectively reduces or delays CNS dysfunction.

Despite the similarity of AIDS in monkeys and in man, or perhaps because of this similarity, the characterization of the behavioral effects of SIV in monkeys has proven to be extremely challenging. A major difficulty is that frank AIDS dementia occurs in a relatively small percentage of AIDS cases; therefore, to study full-blown SIV dementia would require a very large number of subjects. Accordingly, most of the emphasis has been on developing behavioral and functional tests in nonhuman primates that are affected in a majority of SIV-infected animals. These tests are being developed in order to be sensitive to simian versions of both dementia and minor cognitive-motor disorder.

B. Simian Immunodeficiency Virus

1. Rationale

This section presents a brief review of the behavioral literature related to SIV. In addition, descriptions are provided for a variety of the behavioral methodologies used in behavioral studies of CNS viral infections or of other CNS insults such as discrete lesions or administration of neurotoxic drugs. A summary of the results for these studies is included in Table II.

2. Studies Reporting Results of Neurologic Examinations

Several studies have used relatively informal behavioral measures that constitute a neurologic exam as their behavioral assessments. Two such studies are described below. Details of the methods for examinations were not provided beyond references that the monkeys were "monitored clinically" or that they were "examined daily by trained

TABLE II

BEHAVIORAL MONKEY IMPAIRMENT DUE TO SIV DISEASE

Study	Procedure	Monkeys Impaired[a]
Heyes et al. (1992)	Neurologic exam	6 of 11 (55%)
Sopper et al. (1998)	Neurologic exam	
	in rapid progressors	6 of 12 (50%)
	in slow progressors	1 of 7 (14%)
	Overall	7 of 19 (37%)
Murray et al. (1992)	Motor skills (rotating turntable)	5 of 8 (63%)
	Visual discrimination	3 of 8 (38%)
	Recognition/recency memory (delayed matching to sample)	1 of 8 (13%)
	Home cage behavioral observation	0 of 8 (0%)
	Overall	8 of 8 (100%)
Marcario et al. (1999)	Simple and choice	8 of 9 (89%)
	Motor skills (rotating turntable)	6 of 9 (67%)
	Home cage behavioral observation	2 of 9 (22%)
	Overall	8 of 9 (89%)
Horn et al. (1998)	Gross motor activity	5 of 5 (100%)
Gold et al. (1996–1999)	Progressive-ratio (PR) task	7 of 8 (88%)
	Bimanual motor task	6 of 8 (75%)
	Choice reaction time (RT) task	3 of 4 (75%)
	Spatial working memory: self-ordered spatial search (SOSS)	5 of 8 (63%)
	Attentional set shifting, intra/extradimensional (ID/ED)	3 of 7 (43%)
	Recognition memory (delayed nonmatching to sample)	1 of 4 (25%)
	Overall	8 of 8 (100%)

[a] Percent impaired given in parentheses.

laboratory personnel and/or the authors." Neurologic examinations of nonhuman primates are typically less extensive than examinations of humans, primarily due to the verbal nature of the instructions given to the patient. Despite their disadvantages, these studies provide information as to what percentage of the animals displayed gross behavioral changes, and the timing of such changes.

Heyes *et al.* (1992) relate CNS changes such as brain atrophy (indicated by MRI), levels of CSF quinolinic acid (a potential neurotoxin), and the occurrence of neurologic symptoms in the progression of SIV disease. Ratings of neurologic signs included mild-to-severe lethargy, clonus, and loss of balance. Six of 11 SIV-infected monkeys presented neurologic symptoms. Four of 6 monkeys with neurologic signs had increased levels of quinolinic acid, increased brain atrophy, and greatly decreased survival times (within 12 weeks) relative to monkeys without neurologic signs (survival times greater than 37 weeks).

Sopper *et al.* (1998) reported such clinical neurological symptoms as ataxia, apathy, seizure, and opisthotonos (large spasm in back muscles) in 6 of 12 rhesus monkeys rapidly progressing to SIV disease (within 7 months of infection; mean survival time of 18 weeks) but in only 1 of 7 monkeys with slowly progressing AIDS (mean survival time of 58 weeks). Three of the 12 rapid progressors also displayed neurologic symptoms as the first and only clinical symptoms of AIDS (Sopper *et al.*, 1998), similar to previous reports in human AIDS patients (Janssen *et al.*, 1992). Sopper *et al.* (1998) also confirm previous reports that rapid progressors failed to produce virus-specific immune responses in either the periphery or CNS (Heyes *et al.*, 1992). Importantly, there was an apparent increase in neurologic/CNS effects of SIV in rapid progressors as compared to animals with slower disease progression.

3. Studies Using Instrumental Behavioral Techniques

Behavioral studies of SIV disease have drawn on human neuropsychological studies as guideposts for which tests may be sensitive to SIV-induced neuropathology. Neuropsychological assessments of AIDS-related cognitive and motor disruptions, including both AIDS dementia and minor cognitive/motor disorders associated with AIDS, have resulted in the working hypothesis that dysfunction in frontal cortical, and striatal systems play a large role in the cognitive/motor effects resulting from HIV infection (Berger and Nath, 1997; Kelly *et al.*, 1996). For example, neuropsychological tests sensitive to frontal cortical function have been disrupted in both AIDS patients and HIV-positive asymptomatic individuals (Heaton *et al.*, 1995; Sahakian *et al.*, 1995). There are numerous reports of slowing of reaction time (RT) in AIDS patients (Arendt *et al.*, 1990; Bornstein *et al.*, 1993; Dunlop *et al.*, 1992; Karlsen, Reinvang, and Froland, 1992; Martin *et al.*, 1992), and striatal dopaminergic systems have been shown to be important to normal RT functioning (Amalric *et al.*, 1993; Amalric and Koob, 1987; Brown and Robbins, 1991; Robbins *et al.*, 1990; Weed and Gold, 1998). Arendt and colleagues (Arendt *et al.*, 1992; Arendt *et al.*, 1989; Arendt *et al.*, 1994) explored the effects of HIV on motor control by measuring voluntary finger movements, postural tremor, and isometric force production; they demonstrated that finger movement and force production measures not only differed between HIV-positive and control groups, but were also predictors of speed of disease progression (Arendt *et al.*, 1994). Reports of dopamine deficits in the postmortem striatum of AIDS patients (Sardar, Czudek, and Reynolds, 1996), and reports of declines in cerebrospinal fluid dopamine levels in AIDS patients and SIV-infected monkeys (Berger *et al.*, 1994; Koutsilieri *et al.*, 1997) suggest that both HIV and SIV infection lead to impaired CNS dopaminergic function. Furthermore, the ability of levodopa treatment to ameliorate motor symptoms in AIDS patients also supports the involvement of the dopaminergic system in AIDS motor disruptions (Mintz *et al.*, 1996). The results of these studies strongly suggest that AIDS-related changes in CNS dopaminergic function underlie many of the motor effects of AIDS. The mechanism producing frontal cortical dysfunction is less well understood, but may also involve dopaminergic disruption or possibly selective neuropathology of specific cortical layers (Berger *et al.*, 1994; da Cunha, Eiden, and Rausch, 1994; da Cunha, Rausch, and Eiden, 1995; Weihe *et al.*, 1993).

Murray *et al.* (1992) published the first study that reported behavioral changes following infection of rhesus macaques with the Delta B670 strain of SIV. In this study, 15 monkeys were trained on a battery of food-reinforced behavioral tests and were also observed for changes in spontaneous behavior in their home cages. The behavioral battery employed tests to measure (1) visual memory with a delayed matching-to-sample paradigm using novel stimuli in each trial; (2) recency memory with a delayed matching-to-sample paradigm using the same two stimuli repeated in each trial; (3) visual discrimination learning and retention; (4) fine motor control with the removal of rewards from a rotating turntable; and (5) unconditioned behavior in the home cage. The advantages of this selection of tasks were twofold: (a) learning, memory, and fine motor control had been shown to be affected in some HIV-positive individuals and (b) the neural substrates of some of these tasks have been previously investigated in macaques. Therefore, changes in performance of these tasks may provide an indication of functional impairment of the CNS following SIV infection, as well as providing clues to the locations of the affected brain areas.

In the delayed matching-to-sample (DMS) task using novel stimuli, three sample stimuli were presented on a touch-sensitive computer monitor, one at a time, with 2.5 seconds between presentations, and it was required that each sample be touched for the trial to proceed. After an 8-second delay, the sample stimuli were repeated in reverse order and paired with novel stimuli. Touching the sample stimulus was rewarded with food and counted as a correct response. Touching the novel stimulus was not rewarded and was counted as incorrect. The next pair was presented after another 2.5-second delay, and so on, producing sample-test intervals of 8, 13, and 18 seconds. The recency memory task was similar, except that one sample stimulus was presented, followed by a 5-second delay, after which there were two choice stimuli. The recency test differed from the recognition in that the same two stimuli were used repeatedly. In trial after trial, each stimulus was randomly assigned as the sample stimulus and the incorrect stimulus. The recency memory task is thought to increase the difficulty of the test by introducing "interference" from previously performed trials (i.e., the same stimulus that was the "correct" stimulus in the previous trial may or may not be the "incorrect" stimulus in the current trial). Thus, this relatively subtle difference in task parameters engenders a different "cognitive load" on performance of the task.

The visual discrimination task used by Murray et al. (1992) consisted of 20 pairs of stimuli presented in the same order during each test day. As in the memory tasks cited above, left and right positions of correct stimuli varied randomly both within and across test sessions. One of the two stimuli was arbitrarily determined to be "correct" and touching this stimulus produced a food reward and blanked the screen, while touching the "incorrect" stimulus only blanked the screen. Performance on this part of the visual discrimination task was called "retention" as the same stimuli were maintained as correct or incorrect throughout testing. Once the animals were infected, this test was modified with new pairs of stimuli in addition to the pairs used previously, as a test for whether learning of the discriminations was changed after infection with SIV (discrimination learning).

The test of fine motor control consisted of a rotating turntable containing wells that were baited with food. The speed of rotation was titrated until a speed at which the monkey retrieved food on 50% of the trials was determined. The unconditioned home cage behavior was determined by sampling several 5-minute observation periods per month of behavioral testing for duration and frequency of locomotion,

interaction with an enrichment ring in the cage, environmental exploration, social interaction, self-grooming, and stereotyped behavior.
Eight monkeys became productively infected with SIV Delta B670.
The performance of all 8 infected monkeys displayed some impairment
on at least one test prior to either their death (2 monkeys) or their
scheduled sacrifice after 11 months (6 monkeys). A monkey's performance was defined as being impaired when it differed from the mean
of control performance by more than 2 standard deviations. For 5
infected monkeys, performance on the motor control task was the only
impairment demonstrated. The other 3 productively infected monkeys
had impaired performance on the visual-discrimination task, both
learning and retention, and one of these 3 monkeys was also impaired
on the DMS recency-memory test (two repeated stimuli). None of the
monkeys showed changes in the home cage behaviors observed.

The results of the study by Murray et al. (1992) provided the first
evidence of impairment of cognitive/motor behaviors following SIV
infection in macaques. Similar to the cognitive/motor impairment seen
with HIV infection, motor impairment occurred more frequently and
often preceded cognitive impairment.

The same battery of tests was used in a later study on the effects of
zidovudine (AZT) on survival and CNS effects of perinatal SIV infection (Rausch et al., 1995). Four groups of neonatal rhesus monkeys
were divided into uninfected, untreated controls (CON; $N=5$); SIV-
infected, untreated controls (SIV+AZT–; $N=5$); SIV infected treated
with AZT (SIV+AZT+; $N=8$). AZT was administered under 2 regimens
but as the dosing strategy did not affect the outcome, these groups
have been collapsed. SIV inoculations (or vehicle injections) and AZT
treatments were administered within 48 hours of birth. AZT administration continued for 6 months.

Beginning at approximately 10 weeks of age, the monkeys began
testing for fine motor control. Two of the 4 surviving SIV+AZT– monkeys had impaired motor control at this time; however, none of the
SIV+AZT+ monkeys displayed impaired motor control. At 6 months of
age, AZT administration was terminated in the treated groups, and
testing began for recognition memory and discrimination learning.
Methods were similar to those cited above; however, in the discrimination-learning task, new stimuli replaced the old after every 20 sessions. At the beginning of DMS and discrimination-learning testing,
only 3 of 5 SIV+AZT– monkeys survived while all 8 SIV+AZT+ monkeys survived to begin training. Behavioral training continued for all
surviving monkeys for an additional 6 months. During this period, 3 of
8 SIV+AZT+ and 1 of 3 surviving SIV+AZT– monkeys had transiently

impaired motor performance. Additionally, 2 of 8 SIV+AZT+ monkeys showed impaired performance of discrimination learning, but performance was not affected in any of the 3 SIV+AZT– monkeys for as long as they survived. The results of this study demonstrated that AZT treatment prolonged survival times and delayed the onset of motor impairment following infection with SIV.

These behavioral studies have been extended with investigations of possible mechanisms for SIV-induced neurotoxicity. One CNS change reported in both HIV and SIV infections is an increase in levels of the potentially neurotoxic quinolinic acid (QUIN) in cerebrospinal fluid (CSF). QUIN increases have been related to neuropsychological deficits in both adult and juvenile HIV patients (Brouwers et al., 1993; Heyes et al., 1991). Likewise, QUIN increases in the CSF of SIV-infected macaques have been related to severity of neurological symptoms and motor impairments (Heyes et al., 1992; Rausch et al., 1994; Rausch, Murray, and Eiden, 1999). The concentrations of QUIN following HIV and SIV infection are similar to, or higher than, concentrations shown to be neurotoxic experimentally (Heyes et al., 1992). Thus, QUIN levels appear to be a marker for CNS dysfunction with behavioral consequences. QUIN's neurotoxicicty may also implicate it as one of the mechanisms of AIDS neuropathology. The relation of QUIN levels and behavioral changes cited here is a good example of the contribution of behavioral studies to the understanding of AIDS neuropathology.

Marcario et al. (1999a) described an extensive analysis of reaction time (RT) performance of rhesus monkeys infected with SIVmac –R71/17E. Two RT paradigms were employed in chair-restrained monkeys: simple and choice. The simple RT paradigm was performed by both the left and right hands separately, with the alternate hand being restrained. The simple RT paradigm required a press on a switch plate positioned before a touch-sensitive computer monitor. A tone and stimulus on the monitor indicated that the monkey should press the switch plate until the auditory stimulus was terminated and the visual stimulus was changed to provide a target for the monkey to touch. Touching the target within 1.25 seconds was rewarded with food pellets. The choice RT paradigm involved the monkeys' pressing of two switch plates simultaneously, one with each hand. In the choice RT, different colored stimuli indicated whether the right or left hand was to be released from the switch plate and used to touch the target. The alternate hand was required to continue pressing the switch plate through the touch of the target by the other hand. The primary data presented for both RT tasks consisted of reaction time (time from target presentation to release of switch plate) and movement time (time from release of switch plate to touch of the target).

In 8 of the 9 monkeys, RT performance was impaired on at least one measure for the mean of the last 10 RT sessions prior to sacrifice, relative to 10 preinfection baseline sessions. Reaction times in the simple RT paradigm were slowed in more monkeys (5 of 9) than in the choice RT paradigm (3 of 9). In addition, movement times were slowed in more monkeys in the choice RT paradigm (6 of 9) than in the simple RT paradigm (4 of 9). Despite the similarity of the two RT paradigms, the overall motor requirements for responding to the screen did differ (i.e., in the choice RT paradigm the opposite hand had to continue to press the switch plate while the monkey touched the screen), and this slight difference may be related to the different sensitivities of the two paradigms. This last point is an example of how very subtle differences in behavioral paradigms (opposite hand being restrained in the simple RT paradigm versus opposite hand maintaining a press in the choice RT paradigm) may lead to differences in sensitivity of the tests, even within the same animals.

Although the literature on RTs in HIV patients is somewhat mixed, the results of the study by Marcario et al. (1999a) are in accord with the majority of HIV RT studies that show impaired RT performance in AIDS. Curiously, as these authors discuss, more HIV studies show performance declines in choice RT paradigms than in simple RT paradigms, but the monkeys were more often impaired in the simple RT paradigm. No other studies have been published that compare simple-to-choice RT in SIV-infected monkeys.

Recently, Marcario et al. (1999b) provided a supplement to the analysis of fine motor control using the rotating turntable test (Murray et al., 1992) in the same cohort of monkeys used in the previous study (Marcario et al., 1999). Overall, 6 of 9 SIV-infected monkeys showed impaired motor performance on this task, and for 5 of the 6 impaired monkeys the motor impairment preceded onset of clinical neurological signs. Interestingly, the 5 monkeys in which motor impairment came before clinical neurological impairment all had a rapid course of disease progression (i.e., within 4 months following infection).

An observational study of home cage behaviors was also used in this cohort. Eighteen behaviors were scored from 1-hour-per-week videotaping sessions. Behaviors scored included lying down, sitting, walking around the cage, and scratching. Two of the 9 SIV-infected monkeys showed statistically significant changes in home cage behaviors; however, these changes occurred after the changes in RT or motor control. The contribution of general malaise, lack of interest in food, or other "sickness behaviors" to behavioral changes seen following SIV infection is an important consideration in these studies. The observational analysis complemented the behavioral performance measures by providing evidence that the monkeys were not overtly sick during the time of RT or turntable performance changes.

Work at the Scripps Research Institute has taken the approach of Murray *et al.* (1992) one step farther by using additional behavioral tasks with highly studied neural substrates. A computerized behavioral battery, based on human neuropsychological tests, was developed to assess cognitive behaviors of nonhuman primates (Roberts, 1996; Weed *et al.*, 1999). Neuropsychological tests have traditionally relied on human-specific responses, usually employing paper-and-pencil responses. Recently, tests have been developed that use minimal verbal instruction and record responses with touch-sensitive computer monitors. The CANTAB (CAmbridge Neuropsychological Test Automated Batteries; CeNeS Ltd., Cambridge, UK) is a group of computerized tests developed to utilize advances, in the understanding of the neural substrates of cognitive functions, that have been gained from animal psychology, such as delayed matching-to-sample tests, and tests of spatial memory. Other battery tests are computerized analogues of standard human neuropsychological tests. Over the past decade, numerous human populations have been evaluated with this battery, including healthy controls, children, the elderly, neurosurgical patients, and populations with diseases such as Alzheimer's, Huntington's, Parkinson's, multiple-systems atrophy, and AIDS. Brain-imaging studies have contributed to the identification of brain areas involved in the performance of specific tasks, and pharmacological studies have investigated the neurotransmitter systems influencing performance. Pharmacotherapies for CNS diseases such as Alzheimer's and Parkinson's have been evaluated (see Owen *et al.*, 1995; Robbins, *et al.*, 1998; Weed *et al.*, 1999—these have reviews of the above). The nonverbal nature of the CANTAB has facilitated its use in nonhuman primates. Individual tests from the battery have been used to study the behavioral effects of brain lesions and toxin administration in marmosets (Collins *et al.*, 1998; Dias, Robbins, and Roberts, 1996; Pearce *et al.*, 1999; Roberts *et al.*, 1994), and of pharmacological manipulations in rhesus monkeys (Taffe *et al.*, 1999; Weed and Gold, 1998).

The goal in using the CANTAB tests in rhesus monkeys is the same goal sought by Murray *et al.* (1992) and the same goal as that in using CANTAB in human populations: to use performance profiles from the battery to infer functional impairment of brain regions. As a monkey's performance on a given task can decline for a number of reasons

(Ridley and Baker, 1993), the use of a number of tasks increases the validity of both the neuropsychological evaluation of a given subject and the comparative neuropsychology between species. A complementary approach focuses on understanding the qualitative and quantitative nature of performance variables that contribute to the behavioral outcome on a particular neuropsychological task. This approach is intrinsically represented within each CANTAB task as trials or stages are included that are designed to reveal deficiencies in such areas as motivation, perception, and rule comprehension.

The test battery used to study the effects of SIV on cognitive behaviors in rhesus monkeys includes tests for memory using (1) the self-ordered spatial search and (2) the delayed nonmatching-to-sample (DNMS) task; tests for attention and learning using (3) the intradimensional/extradimensional shift task; tests of motor performance using (4) the bimanual motor tasks and (5) reaction time tests; and a test for the reinforcing efficacy of food reward, using (6) a progressive-ratio (PR) schedule. Methods for each of these tasks are described in detail in Weed et al. (1999). Data summarized here (and in Table II) include previously published studies (Fox et al., 1997; Gold et al., 1998; Prospero-Garcia et al., 1996) and unpublished data.

The self-ordered spatial search (SOSS) task is an analogue of the radial-arm maze that is used to assess spatial working memory in rodents. Performance of other spatial working-memory tests has been impaired by lesions or dysfunction of the frontal cortex (Courtney et al., 1988; McCarthy et al., 1996; Miotto et al., 1996), as has performance on similar SOSS tasks (Roberts, 1996) by both marmosets (Collins et al., 1998) and man (Owen et al., 1996a; Owen et al., 1996b). Additionally, both asymptomatic and symptomatic HIV seropositive patients demonstrated increased errors in the SOSS task relative to HIV seronegative controls (Sahakian et al., 1995). In the SOSS task, a number of colored squares (2–5 squares or "boxes") were displayed in different positions on the monitor. The monkey was required to touch each box only once within a trial. Each correct touch was rewarded with a food pellet and was followed by a 2-second delay. Trials ended when all boxes were touched, or when one box was touched twice (an error). Variations on the SOSS task, "box probes" and "time probes," were periodically presented, and challenged the animal with additional boxes (up to 8 boxes per trial) and increased delays between touches (up to 5 seconds).

The intradimensional/extradimensional (ID/ED) shift task is a computerized analogue of the Wisconsin card-sort task used to test category abstraction in humans (Roberts, 1996; Roberts, Robbins, and Everitt, 1988). Lesions or dysfunction of the frontal cortex have been shown to impair performance on this attentional set-shifting task in both marmosets (Dias, Robbins, and Roberts, 1997) and man (Owen *et al.*, 1991). Additionally, both asymptomatic and symptomatic HIV seropositive patients have demonstrated increased errors on the ID/ED task relative to HIV seronegative controls (Sahakian *et al.*, 1995).

Tasks involving delayed matching (described above) or delayed nonmatching-to-sample have been used extensively in the study of memory, and studies of neuroanatomical lesions in monkeys strongly implicate temporal cortical involvement of this task (Murray, 1996). In human patients, temporal excisions, but not frontal excisions, produced impairments on delayed matching to sample (Owen *et al.*, 1995). However, lesions in some areas within the frontal cortex can impair performance of delayed matching or nonmatching tasks (Bachevalier and Mishkin, 1986; Bauer and Fuster, 1976; Fuster, Bauer, and Jervey, 1985; Passingham, 1975). When combined with other tasks more selectively sensitive to frontal cortical dysfunction, such as the SOSS or ID/ED tasks, DNMS performance can be used as an indicator of temporal cortical function. In the DNMS task the monkey performed an observing response, on the computer monitor, to one visually abstract, multicolored stimulus; following a delay, the initial stimulus was presented, and accompanied by a novel stimulus. Touching the novel stimulus resulted in a food reward. Trial types included simultaneous exposure (sample and choices appearing on the screen together and delays of 16, 32, and 64 seconds.

The bimanual motor task involved a modification of the tasks used by Brinkman (1981) and by Mark and Sperry (1968), in which the monkey must coordinate both hands to retrieve raisins from a hole board. Another test of motor control is the choice–reaction time task. In the CANTAB choice–RT task, 5 circles are drawn on the computer monitor. Depressing the lever produces a stimulus flash in one circle. Selecting the circle in which the flash appeared is reinforced and ends the trial. This RT paradigm has been shown to be sensitive to the effects of dopamine antagonists in rhesus monkeys (Weed and Gold, 1998).

A progressive-ratio schedule of reinforcement, wherein response requirements are progressively increased within a session, was included in the rhesus test battery. PR schedules have been used to measure the reinforcing effects of a variety of reinforcers including food, drugs, and electrical brain stimulation (Hodos, 1961; Hodos, 1965; Katz, 1990). Performance under the PR schedule provided a control for changes in motivation, such as changes in the reinforcer efficacy of the food reinforcer, and in the ability to perform the operant required for most battery tasks (i.e., touching the monitor).

A monkey's performance was determined to be impaired when it was outside the 95% confidence limits of the performance from the month prior to infection for that monkey for tasks including the SOSS, DNMS, RT, bimanual motor, and PR and when performance was outside confidence limits from all preinfection testing with the ID/ED. Data from 8 animals are included in this analysis. Three of the 8 monkeys were trained on the entire battery; 1 was trained on 5 of the 6 tasks, and 4 were trained on 4 tasks. Overall, each of the 8 monkeys was impaired on at least two of the tasks, following infection with microglia-passaged SIV_{mac} (Watry et al., 1995). PR performance declined in 7 of 8 SIV-infected monkeys. In 6 of 8 SIV-infected monkeys, bimanual performance worsened at some point in the SIV disease progression. Three of the 4 SIV-infected monkeys have been impaired on either reaction time or movement time in the RT procedure. In 5 of 8 SOSS-trained, SIV-infected monkeys, SOSS performance was disrupted at some stage. ID/ED performance declines (increased errors) were seen in 3 of 7 SIV-infected monkeys, and only 1 of 4 SIV-infected monkeys showed impairment on DNMS performance. These results suggest that subcortically mediated tasks are disrupted more often than cortically mediated tasks. Additionally, of the cortically mediated tasks, frontally mediated tasks appear more sensitive to disruption by SIV infection than do temporally mediated tasks.

The characteristics of long-term performance have been established for macaques on this battery (Weed et al., 1999). This characterization of long-term battery performance increases the validity of the longitudinal design of these studies because an individual's performance profile can be compared to the normative data. Additionally, because performance continues until a stable level is reached prior to infection with SIV, each monkey serves as its own control, controlling for individual differences in performance levels. Furthermore, functional or physiological measures, such as sensory-evoked potentials, can be obtained at multiple time points prior to and following SIV infection. Ultimately, performance profiles can be interpreted in the light of additional functional and immunological measures.

A performance profile for one monkey trained on the entire battery is presented in Fig. 1, which also includes tests of several immunological, viral, and physiological measures associated with SIV disease progression. The performance profile of this monkey fits, reasonably well, the model of frontostriatal dysfunction derived from human neuropsychological testing. Performance on striatally involved tasks (e.g., PR and RT) was impaired by week 5 and impairment continued through the disease progression until sacrifice in the 12th week. Starting in the 6th week, performance of the SOSS task began a progressive decline.

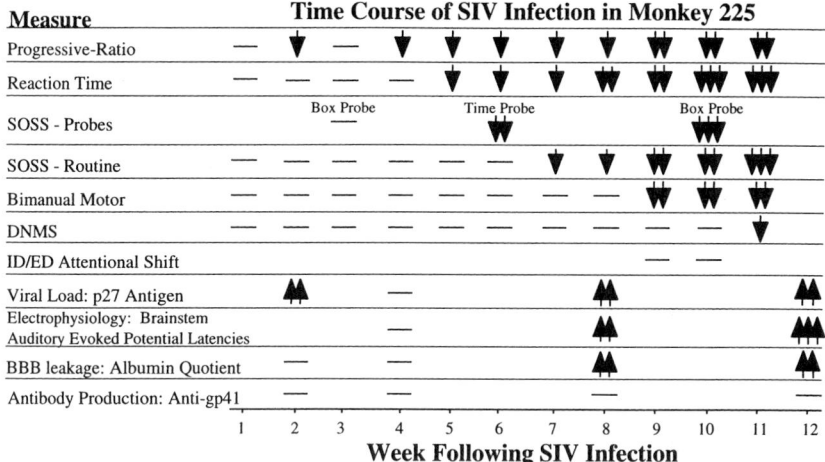

FIG 1. Summary of behavioral, neurophysiological, viral, and immunological effects in SIV-infected monkey 225 over the time course of disease. Arrows pointing downward indicate decreases; arrows pointing upward indicate increases; and dashes indicate no change in the various measures. The number of arrows reflects a qualitative estimate of progressive changes. SOSS, Self-ordered spatial search task: DNMS, delayed nonmatching-to-sample task; ID/ED, intradimensional/extradimensional shift task; BBB, blood–brain barrier. With permission from Gold et al. (1998).

Performance of the presumed frontostriatally mediated bimanual motor task deteriorated next, in week 9. In contrast, the temporally mediated DNMS task showed only minor disruption in the final weeks. The disease progression of monkey 225 was so rapid that the frontally mediated ID/ED test was administered only once, and at that time, ID/ED performance was not affected. Despite the lack of ID/ED impairment, the remainder of monkey 225's performance profile is consistent with early subcortical/striatal dysfunction, followed by impairment of frontal cortical function.

Consistent with the findings of other behavioral batteries, results from the Scripps Research Institute group suggest that subcortical/striatal dysfunction tends to occur earlier in the disease progression than impairment of "higher" cognitive functions, such as spatial working or recognition memory, thought to be cortically mediated. The performance profile in Fig. 1 illustrates the importance of multiple behavioral tests, especially when working with a disease such as AIDS, where individual differences in disease progression can be dramatic. Fig. 1 also highlights the interdisciplinary approach necessary for comprehensively studying the CNS effects of SIV infection in macaques.

The addition of behavioral markers to immunological, viral, and physiological markers of disease progression will facilitate the determination of which CNS changes actually lead to functional changes. For instance, brainstem auditory-evoked potentials become delayed early in SIV infection, indicating early changes in CNS neuronal function (Horn *et al.*, 1998; Prospero-Garcia *et al.*, 1996; Raymond *et al.*, 1998). Fig. 1 indicates that these evoked potentials were delayed in monkey 225 as well. Whether this change in neuronal function contributes to any behavioral changes is not yet known. Other markers of disease progression, such as the host's immune response to the virus, have been implicated in decreased survival times and in increased severity of neurological symptoms and encephalitis (Sopper *et al.*, 1998). As indicated in Fig. 1, monkey 225 failed to mount a significant antibody response to the SIV infection, as measured by anti-gp41 antibody levels. The rapidity of disease progression and the severity of behavioral deficits are all consistent with previous reports.

4. Studies Using Continuous Recording of Motor Activity

Horn *et al.* (1998) monitored gross motor activity in rhesus monkeys implanted with radiotelemetry equipment throughout the course of SIV disease progression. Postinoculation spontaneous motor activity decreased relative to the preinfection baseline in the second and third months, to approximately 80% and 40% of baseline, respectively (Horn *et al.*, 1998). This finding has also been replicated in pig-tailed macaques infected with SIV wearing collar activity monitors (Hienz, unpublished data). Interestingly, spontaneous activity can decrease to very low levels (e.g., 20–30% of baseline), while performance in behavioral tasks remains largely unchanged (Gold *et al.*, and Hienz *et al.*, unpublished data). The decrease in spontaneous motor activity following SIV infection appears to be an early and a reliable indicator of SIV disease progression. Although a reliable marker of disease progression, the neuroanatomic substrates of gross motor activity are not as well defined as those of the instrumental behaviors described above. Gross motor activity is considered to be modulated by various systems within the CNS and is thus a useful, albeit global, behavioral measure that complements the more discretely regulated behavioral measures in the behavioral batteries.

5. Sickness Behavior: A Factor That Potentially Confounds Viral Studies

One factor that potentially confounds the study of viral effects on behavior is the contribution of general malaise, or "sickness behavior," to observed behavioral effects. For example, 10–14 days following infection with HIV or SIV, there is an initial viremia that produces "flu-like" symptoms. Additionally, opportunistic infections in immunocompromised animals may also adversely affect behavioral performance through a non-SIV mechanism. Both of these conditions can be easily controlled for. In the former case, SIV viremia is transient, and data collected during the initial 2–3 weeks postinoculation are typically analyzed separately from data collected as SIV disease progresses. In the latter case, veterinary diagnosis of an animal's opportunistic infection may also exclude that animal's data from analysis. SIV studies normally include thorough veterinary monitoring of the animals that allows for the monitoring of potentially confounding variables such as non-SIV diseases and symptomologies that would interfere with performance in a non-SIV-mediated fashion. Additionally, Murray *et al.* (1992) and Marcario *et al.* (1999a,b) included videotaping of home cage behavior as a measure to control for sickness behavior. If home cage behavior changes in a manner indicating "sickness behavior," these observational data are considered along with performance changes. In fact, significant sickness behavior was seen mostly in very late stages of disease and usually after changes in task performance had already occurred.

6. Effects of Environmental Variables and Individual Differences on SIV Disease Progression

In an effort to account for some of the variation in SIV disease progression, Capitanio and colleagues investigated variables in the history of the host animal and related these to markers of AIDS progression in rhesus macaques. Two different classes of factors were investigated: environmental impacts on the host and individual differences in the host's behavioral histories. Capitanio and Lerche (1991, 1998) used archival methods, gathering data, from four Regional Primate Research Centers in the United States, on environmental variables including demographic, viral, clinical, and housing histories for a total of 260 rhesus macaques productively infected with several strains of SIV. Results of their statistical modeling indicated that housing changes between 90 days preinfection and 30 days postinoculation were associated with an increased death rate after controlling for SIV inoculum and medical variables (Capitanio and Lerche, 1998). Additionally, separations from familiar companions within 90 days prior to SIV inoculation was associated with reduced latencies to leukopenia and lymphopenia and with increased risk of weight loss greater than 10% of initial body weight (Capitanio and Lerche, 1991). These authors postulate that stress–response systems, such as increased corticosteroid secretion, may have affected either the host immune response or viral replication. Human AIDS patients often experience adverse environmental or psychosocial stressors, such as problems with social support, employment, health care, and housing (Capitanio and Lerche, 1998). Therefore, understanding the mechanisms by which environmental or psychosocial variables affect AIDS progression may significantly impact their quality of life.

Capitanio *et al.* (1999) investigated another variable in AIDS progression—the effects of an individual monkey's "personality dimensions" on markers of AIDS progression. Individual animals have unique behavioral repertoires, at least in terms of individual differences in the mix of the various behaviors characteristic of their species. In humans, these individual repertoires can be scored with adjective-based rating scales resulting in an assessment of the construct of one's personality. Scores on such scales can be given by the individual subject, or by other individuals familiar with the subject. Capitanio *et al.* (1999) used the latter approach to score rhesus macaques with a 25-adjective list, including such adjectives as "aggressive," "confident," "motherly," and "fearful." The 25 adjectives cluster into 4 "personality dimensions" including "equability," "sociability," "confidence," "excitability." Scores from this list have been shown to be predictive of social behavior in male rhesus monkeys over a 4.5-year time span (Capitanio, 1999). Eighteen male rhesus monkeys were rated with this list prior to inoculation with SIV-$_{mac}$251. The viral and physiological measures of plasma cortisol, anti-SIV immunoglobulin G (IgG), and antirhesus cytomegalovirus (RhCMV) IgG and SIV RNA (viral load) were related to the personality dimensions. Statistically, the 4 personality dimensions were treated as independent variables, and the dependent viral and physiological measures were regressed on them. Several personality dimensions were found to have physiological correlates. For instance, sociability, comprising scores of the adjectives "sociable," "playful," and "curious," was found to be substantially correlated with plasma cortisol (inversely), anti-RhCMV IgG (positively), and viral load (positively early on, then inversely later in infection). The effect of personality on physiology was in the "small to medium" range and statistically reliable. These results suggest that personality dimensions may interact with immune function to affect the course of SIV disease progression.

C. SIV/Macaque Model of AIDS

The SIV/macaque model. of AIDS is the most frequently used primate model to study the behavioral effects of viral infection. SIV infection in macaques produces a disease progression qualitatively similar to HIV infection in humans, indeed producing similar kinds and degrees of cognitive/motor impairments. SIV disease in macaques, like HIV disease in humans, can cause dysfunction in both cognitive and motor behaviors, with behaviors mediated by subcortical and frontal cortical areas at higher risk of impairment. In the light of the percentage of human AIDS patients that present minor cognitive/motor disorder or AIDS dementia complex (ADC), the behavioral tests employed to date in the SIV model have been relatively successful in identifying impaired cognitive or motor function in SIV-infected macaques.

An important goal of behavioral testing in AIDS models is to characterize the CNS injuries arising from AIDS in both a psychological and a pathophysiological fashion. Behavioral tests that provide markers for disease progression would allow *ex vivo* researchers to focus on how the CNS of impaired animals differs from those of the unimpaired. It is not necessary that any specific test of cognitive/motor function reveals impairments in 100% of animals tested in order to be useful, as groups of tests may be helpful in distinguishing impaired and unimpaired animals. Such tests, as described here, add to the model's potential to investigate the mechanism of neuropathology in AIDS. Additionally, the development of appropriate behavioral tests may help the SIV macaque model to further contribute to the investigation of AIDS therapeutics and AIDS vaccines.

The SIV/macaque model of AIDS has been discussed in detail because work with this model demonstrates the range of behavioral testing that can be applied to the study of viral infection in primates. Observational techniques, from clinical veterinary assessments to videotaped and scored behavioral indices, gross motor activity scores, and a wide range of instrumental behavior assessments of cognitive and motor function, are all part of the arsenal of behavioral assays available. In addition to work with the SIV model, these and other behavioral assays are useful for the investigation of other CNS insults, such as brain lesions (Collins *et al.*, 1998; Dias, Robbins, and Roberts, 1996; Murray, 1996; Passingham, 1975; Voytko *et al.*, 1994); neurotoxicity (Prendergast, Terry, and Buccafusco, 1998; Roberts *et al.*, 1994); aging (Baxter and Voytko, 1996; Voytko, 1998); and psychoactive drugs (Frederick *et al.*, 1997; Hienz *et al.*, 1989; Taffe *et al.*, 1999; Weed and Gold, 1998).

IV. Conclusion

The literature reviewed in this chapter confirms the sensitivity of behavioral paradigms to assays of functional changes in animal models of viral pathogenesis. While in some cases chronic or persistent infection may contribute to clinical pathology, in many cases virus has been cleared, and behavioral alterations may therefore be long-lasting or irreversible. These studies suggest that numerous viruses can cause significant behavioral abnormalities, in some cases in apparently clinically healthy animals. Repeated testing reliably documents the progression of behavioral changes in a longitudinal fashion. In addition, pharmacological and immunological challenges can be used to probe the function of selected neurotransmitter systems and to reveal immunopathological mechanisms. The further development and utilization of behavioral analyses in animal models of viral pathogenesis have the potential to provide new insights into the etiology of a variety of human behavioral disorders.

ACKNOWLEDGMENTS

Many thanks to Drs. R. J. Hienz and M. A. Taffe for their editorial comments. This work was supported in part by PHS Grants DA10190 and MH47680 (LHG), and DA05831 (MRW). This is publication number 12887-NP from The Scripps Research Institute.

REFERENCES

Amalric, M., Berhow, M., Polis, I., and Koob, G. F. (1993). Selective effects of low-dose D2 dopamine receptor antagonism in a reaction-time task in rats. *Neuropsychopharmacology* **8**(3):195–200.
Amalric, M., and Koob, G. F. (1987). Depletion of dopamine in the caudate nucleus but not in nucleus accumbens impairs reaction-time performance in rats. *J. Neurosci.* **7**(7):2129–34.
Andersson, T., Mohammed, A. K., Henriksson, B. G., Wickman, C., Norrby, E., Schultzberg, M., and Kristensson, K. (1993). Immunohistochemical and behaviour pharmacological analysis of rats inoculated intranasally with vesicular stomatitis virus. *J. Chem. Neuroanat.* **6**(1):7–18.
Arendt, G., Hefter, H., Buescher, L., Hilperath, F., Elsing, C., and Freund, H. J. (1992). Improvement of motor performance of HIV-positive patients under AZT therapy. *Neurology* **42**(4):891–896.
Arendt, G., Hefter, H., Elsing, C., Neuen-Jakob, E., Strohmeyer, G., and Freund, H. J. (1989). New electrophysiological findings on the incidence of brain involvement in clinically and neurologically asymptomatic HIV infections. *EEG EMG Z Elektroenzephalogr Elektromyogr Verwandte Geb.* **20**(4):280–287.
Arendt, G., Hefter, H., Elsing, C., Strohmeyer, G., and Freund, H. J. (1990). Motor dysfunction in HIV-infected patients without clinically detectable central-nervous deficit. *J. Neurol.* **237**(6):362–368.
Arendt, G., Hefter, H., Hilperath, F., von Giesen, H. J., Strohmeyer, G., and Freund, H. J. (1994). Motor analysis predicts progression in HIV-associated brain disease. *J. Neurol. Sci.* **123**(1–2):180–185.
Avgeropoulos, N., Kelley, B., Middaugh, L., Arrigo, S., Persidsky, Y., Gendelman, H. E., and Tyor, W. R. (1998). SCID mice with HIV encephalitis develop behavioral abnormalities. *J. Acquir. Immune. Defic. Syndr. Hum. Retrovirol.* **18**(1):13–20.

Bachevalier, J., and Mishkin, M. (1986). Visual recognition impairment follows ventro-medial but not dorsolateral prefrontal lesions in monkeys. *Behav. Brain Res.* **20**:249–261.

Bauer, R. H., and Fuster, J. M. (1976). Delayed-matching and delayed-response deficit from cooling dorsolateral prefrontal cortex in monkeys. *J. Comp. Physiol. Psychol.* **90**:293–302.

Baxter, M. G., and Voytko, M. L. (1996). Spatial orienting of attention in adult and aged rhesus monkeys. *Behav. Neurosci.* **110**(5):898–904.

Berger, J. R., Kumar, M., Kumar, A., Fernandez, J. B., and Levin, B. (1994). Cerebrospinal fluid dopamine in HIV-1 infection. *AIDS* **8**(1):67–71.

Berger, J. R., and Nath, A. (1997). HIV dementia and the basal ganglia. *Intervirology* **40**(2–3):122–131.

Bernard, A., Cohen, R., Khuth, S.-T., Vedrine, B., Verlaeten, O., Akaoka, H., Giraudon, P., and Belin, M.-F. (1999). Alteration of the leptin network in late morbid obesity induced in mice by brain infection with canine distemper virus. *Journal of Virology* **73**:7317–7327.

Bornstein, R. A., Nasrallah, H. A., Para, M. F., Whitacre, C. C., Rosenberger, P., and Fass, R. J. (1993). Neuropsychological performance in symptomatic and asymptomatic HIV infection. *AIDS* **7**(4):519–524.

Brinkman, C. (1981). Lesions in supplementary motor area interfere with a monkey's performance of a bimanual coordination task. *Neurosci. Lett.* **27**:267–270.

Brot, M. D., Rall, G. F., Oldstone, M. B., Koob, G. F., and Gold, L. H. (1997). Deficits in discriminated learning remain despite clearance of long-term persistent viral infection in mice. *J. Neurovirol.* **3**(4):265–273.

Brouwers, P., Heyes, M. P., Moss, H. A., Wolters, P. L., Poplack, D. G., Markey, S. P., and Pizzo, P. A. (1993). Quinolinic acid in the cerebrospinal fluid of children with symptomatic human immunodeficiency virus type 1 disease: relationships to clinical status and therapeutic response. *J. Infect. Dis.* **168**(6):1380–1386.

Brown, V. J., and Robbins, T. W. (1991). Simple and choice reaction time performance following unilateral striatal dopamine depletion in the rat. Impaired motor readiness but preserved response preparation. *Brain* **114**(Pt 1B):513–525.

Capitanio, J. P. (1999). Personality dimensions in adult male rhesus macaques: prediction of behaviors across time and situation. *Am. J. Primatol.* **47**(4):299–320.

Capitanio, J. P., and Lerche, N. W. (1991). Psychosocial factors and disease progression in simian AIDS: a preliminary report. *AIDS* **5**(9):1103–1106.

Capitanio, J. P., and Lerche, N. W. (1998). Social separation, housing relocation, and survival in simian AIDS: a retrospective analysis. *Psychosom. Med.* **60**(3):235–244.

Chakrabarti, L., Hurtrel, M., Maire, M. A., Vazeux, R., Dormont, D., Montagnier, L., and Hurtrel, B. (1991). Early viral replication in the brain of SIV-infected rhesus monkeys. *Am. J. Pathol.* **139**(6):1273–1280.

Chang, L., Ernst, T., Leonido-Yee, M., Witt, M., Speck, O., Walot, I., and Miller, E. N. (1999). Highly active antiretroviral therapy reverses brain metabolite abnormalities in mild HIV dementia. *Neurology* **53**(4):782–789.

Collins, P., Roberts, A. C., Dias, R., Everitt, B. J., and Robbins, T. W. (1998). Perseveration and Strategy in a Novel Spatial Self-Ordered Sequencing Task for Nonhuman Primates. Effects of Excitotoxic Lesions and Dopamine Depletions of the Prefrontal Cortex. *J. Cogn. Neurosci.* **10**:332–354.

Conn, C. A., McClellan, J. L., Maassab, H. F., Smitka, C. W., Majde, J. A., and Kluger, M. J. (1995). Cytokines and the acute phase response to influenza virus in mice. *Am. J. Physiol.* **268**(1 Pt 2):R78–R84.

Courtney, S. M., Petit, L., Maisog, J. M., Ungerleider, L. G., and Haxby, J. V. (1988). An area specialized for spatial working memory in human frontal cortex. *Science* **279**:1347–1351.

Crnic, L. S., and Pizer, L. I. (1988). Behavioral effects of neonatal herpes simplex type 1 infection of mice. *Neurotoxicol Teratol* **10**(4):381–386.

d'Arminio Monforte, A., Testa, L., Adorni, F., Chiesa, E., Bini, T., Moscatelli, G. C., Abeli, C., Rusconi, S., Sollima, S., Balotta, C., Musicco, M., Galli, M., and Moroni, M. (1998). Clinical outcome and predictive factors of failure of highly active antiretroviral therapy in antiretroviral-experienced patients in advanced stages of HIV-1 infection. *AIDS* **12**(13):1631–1637.

da Cunha, A., Eiden, L. E., and Rausch, D. M. (1994). Neuronal substrates for SIV encephalopathy. *Adv. Neuroimmunol* **4**(3):265–271.

da Cunha, A., Rausch, D. M., and Eiden, L. E. (1995). An early increase in somatostatin mRNA expression in the frontal cortex of rhesus monkeys infected with simian immunodeficiency virus. *Proc. Natl. Acad. Sci. U.S.A.* **92**(5):1371–1375.

Dantzer, R., Bluthe, R. M., Gheusi, G., Cremona, S., Laye, S., Parnet, P., and Kelley, K. W. (1998). Molecular basis of sickness behavior. *Annals of New York Academy of Sciences* **856**:132–138.

Darko, D. F., Mitler, M. M., and Henriksen, S. J. (1995). Lentiviral infection, immune response peptides and sleep. *Adv. Neuroimmunol* **5**(1):57–77.

Desrosiers, R. C. (1990). The simian immunodeficiency viruses. *Annu. Rev. Immunol.* **8**:557–578.

Dias, R., Robbins, T. W., and Roberts, A. C. (1996). Primate analogue of the Wisconsin Card Sorting Test: effects of excitotoxic lesions of the prefrontal cortex in the marmoset. *Behav. Neurosci.* **110**:872–886.

Dias, R., Robbins, T. W., and Roberts, A. C. (1997). Dissociable forms of inhibitory control within prefrontal cortex with an analog of the Wisconsin Card Sort Test: restriction to novel situations and independence from "on-line" processing. *J. Neurosci.* **17**:9285–9297.

Dittrich, W., Bode, L., Ludwig, H., Kao, M., and Schneider, K. (1989). Learning deficiencies in Borna disease virus-infected but clinically healthy rats. *Biol. Psychiatry.* **26**(8):818–828.

Dunlop, O., Bjorklund, R. A., Abdelnoor, M., and Myrvang, B. (1992). Five different tests of reaction time evaluated in HIV seropositive men. *Acta. Neurol. Scand.* **86**(3):260–266.

Ebbesen, P., Villadsen, J. A., Villadsen, H. D., and Heller, K. E. (1991). Effect of subordinance, lack of social hierarchy, and restricted feeding on murine survival and virus leukemia. *Exp. Gerontol.* **26**(5):479–486.

English, J. A., Bruce, K. H., and Paul, I. A. (1999). Increased discriminative stimulus potency of phencyclidine in C57B1/6 mice infected with the LP-BM5 retrovirus. *Eur. J. Pharmacol.* **367**(1):1–5.

English, J. A., Hemphill, K. M., and Paul, I. A. (1998). LP-BM5 infection impairs acquisition, but not performance, of active avoidance responding in C57B1/6 mice. *Faseb J* **12**(2):175–179.

Flaherty, M. T., Hauer, D. A., Mankowski, J. L., Zink, M. C., and Clements, J. E. (1997). Molecular and biological characterization of a neurovirulent molecular clone of simian immunodeficiency virus. *J. Virol.* **71**(8):5790–5798.

Fleetwood-Walker, S. M., Quinn, J. P., Wallace, C., Blackburn-Munro, G., Kelly, B. G., Fiskerstrand, C. E., Nash, A. A., and Dalziel, R. G. (1999). Behavioural changes in the rat following infection with varicella-zoster virus. *J. Gen. Virol.* **80**(Pt 9):2433–2436.

Fox, H. S., Gold, L. H., Henriksen, S. J., and Bloom, F. E. (1997). Simian immunodeficiency virus: a model for neuroAIDS. *Neurobiol. Dis.* **4**(3–4):265–274.

Frederick, D. L., Gillam, M. P., Lensing, S., and Paule, M. G. (1997). Acute effects of LSD on rhesus monkey operant test battery performance. *Pharmacol. Biochem. Behav.* **57**(4):633–641.

Fuster, J. M., Bauer, R. H., and Jervey, J. P. (1985). Functional interactions between inferotemporal and prefrontal cortex in a cognitive task. *Brain. Res.* **330**:299–307.

Gendelman, H. E., Zheng, J., Coulter, C. L., Ghorpade, A., Che, M., Thylin, M., Rubocki, R., Persidsky, Y., Hahn, F., Reinhard, J., Jr., and Swindells, S. (1998). Suppression of inflammatory neurotoxins by highly active antiretroviral therapy in human immunodeficiency virus-associated dementia. *J. Infect. Dis.* **178**(4):1000–1007.

Gold, L. H., Brot, M. D., Polis, I., Schroeder, R., Tishon, A., de la Torre, J. C., Oldstone, M. B., and Koob, G. F. (1994). Behavioral effects of persistent lymphocytic choriomeningitis virus infection in mice. *Behav. Neural. Biol.* **62**(2):100–109.

Gold, L. H., Fox, H. S., Henriksen, S. J., Buchmeier, M. J., Weed, M. R., Taffe, M. A., Huitron-Resendiz, S., Horn, T. F., and Bloom, F. E. (1998). Longitudinal analysis of behavioral, neurophysiological, viral and immunological effects of SIV infection in rhesus monkeys. *J. Med. Primatol.* **27**(2–3):104–112.

Hallensleben, W., Schwemmle, M., Hausmann, J., Stitz, L., Volk, B., Pagenstecher, A., and Staeheli, P. (1998). Borna disease virus-induced neurological disorder in mice: infection of neonates results in immunopathology. *J. Virol.* **72**(5):4379–4386.

Heaton, R. K., Grant, I., Butters, N., White, D. A., Kirson, D., Atkinson, J. H., McCutchan, J. A., Taylor, M. J., Kelly, M. D., Ellis, R. J., and et al. (1995). The HNRC 500—neuropsychology of HIV infection at different disease stages. HIV Neurobehavioral Research Center. *J. Int. Neuropsychol. Soc.* **1**(3):231–251.

Hermann, G., Tovar, C. A., Beck, F. M., Allen, C., and Sheridan, J. F. (1993). Restraint stress differentially affects the pathogenesis of an experimental influenza viral infection in three inbred strains of mice. *J. Neuroimmunol.* **47**(1):83–94.

Hermann, G., Tovar, C. A., Beck, F. M., and Sheridan, J. F. (1994). Kinetics of glucocorticoid response to restraint stress and/or experimental influenza viral infection in two inbred strains of mice. *J. Neuroimmunol.* **49**(1–2):25–33.

Heyes, M. P., Brew, B. J., Martin, A., Price, R. W., Salazar, A. M., Sidtis, J. J., Yergey, J. A., Mouradian, M. M., Sadler, A. E., Keilp, J., and et al. (1991). Quinolinic acid in cerebrospinal fluid and serum in HIV-1 infection: relationship to clinical and neurological status. *Ann. Neurol.* **29**(2):202–209.

Heyes, M. P., Jordan, E. K., Lee, K., Saito, K., Frank, J. A., Snoy, P. J., Markey, S. P., and Gravell, M. (1992). Relationship of neurologic status in macaques infected with the simian immunodeficiency virus to cerebrospinal fluid quinolinic acid and kynurenic acid. *Brain. Res.* **570**(1–2):237–250.

Hienz, R. D., Brady, J. V., Bowers, D. A., and Ator, N. A. (1989). Ethanol's effects on auditory thresholds and reaction times during the acquisition of chronic ethanol self-administration in baboons. *Drug Alcohol Depend* **24**(3):213–225.

Hodos, W. (1961). Progressive-ratio as a measure of reward strength. *Science* **134**:943–944.

Hodos, W. (1965). Motivational properties of long durations of rewarding brain stimulation. *J. Comp. Physiol. Psychol.* **59**:219–224.

Horn, T. F., Huitron-Resendiz, S., Weed, M. R., Henriksen, S. J., and Fox, H. S. (1998). Early physiological abnormalities after simian immunodeficiency virus infection. *Proc. Natl. Acad. Sci. U.S.A.* **95**(25):15072–15077.

Hotchin, J., and Seegal, R. (1977). Virus-induced behavioral alteration of mice. *Science* **196**(4290):671–674.

Hubalek, Z., and Rodl, P. (1994). Assessment of the mouse open-field activity after infection with Tahyna virus (California serogroup, Bunyaviridae). *Lab. Anim. Sci.* **44**(2):186–188.

Iida, R., Yamada, K., Mamiya, T., Saito, K., Seishima, M., and Nabeshima, T. (1999). Characterization of learning and memory deficits in C57BL/6 mice infected with LP-BM5, a murine model of AIDS. *J. Neuroimmunol.* **95**(1–2):65–72.

Irwin, S. (1968). Comprehensive observational assessment: Ia. A systematic, quantitative procedure for assessing the behavioral and physiologic state of the mouse. *Psychopharmacologia* **13**(3):222–257.

Janssen, R. S., Nwanyanwu, O. C., Selik, R. M., and Stehr-Green, J. K. (1992). Epidemiology of human immunodeficiency virus encephalopathy in the United States. *Neurology* **42**(8):1472–1476.

Karlsen, N. R., Reinvang, I., and Froland, S. S. (1992). Slowed reaction time in asymptomatic HIV-positive patients. *Acta. Neurol. Scand.* **86**(3):242–246.

Karp, J. D., Moynihan, J. A., and Ader, R. (1997). Psychosocial influences on immune responses to HSV-1 infection in BALB/c mice. *Brain Behav. Immun.* **11**(1):47–62.

Katz, J. L. (1990). Models of relative reinforcing efficacy of drugs and their predictive utility. *Beh. Pharmacol.* **1**:283–301.

Kelly, M. D., Grant, I., Heaton, R. K., and Marcotte, T. D. (1996). Neuropsychological findings in HIV infection and AIDS. *In* "Neuropsychological Assessment of Neuropsychiatric Disorders" (I. Grant and K. M. Adams, eds.), pp. 403–415. Oxford University Press, Oxford.

Kent, S., Bluthe, R. M., Kelley, K. W., and Dantzer, R. (1992). Sickness behavior as a new target for drug development. *Trends. Pharmacol. Sci.* **13**(1):24–28.

Kluger, M. J., Kozak, W., Conn, C. A., Leon, L. R., and Soszynski, D. (1998). Role of fever in disease. *Ann. N.Y. Acad. Sci.* **856**:224–233.

Koutsilieri, E., Gotz, M. E., Sopper, S., Stahl-Hennig, C., Czub, M., ter Meulen, V., and Riederer, P. (1997). Monoamine metabolite levels in CSF of SIV-infected rhesus monkeys (Macaca mulatta). *Neuroreport* **8**(17):3833–3836.

Lackner, A. A., Smith, M. O., Munn, R. J., Martfeld, D. J., Gardner, M. B., Marx, P. A., and Dandekar, S. (1991). Localization of simian immunodeficiency virus in the central nervous system of rhesus monkeys. *Am. J. Pathol.* **139**(3):609–621.

Ledergerber, B., Egger, M., Opravil, M., Telenti, A., Hirschel, B., Battegay, M., Vernazza, P., Sudre, P., Flepp, M., Furrer, H., Francioli, P., and Weber, R. (1999). Clinical progression and virological failure on highly active antiretroviral therapy in HIV-1 patients: a prospective cohort study. Swiss HIV Cohort Study. *Lancet* **353**(9156):863–868.

Lipton, S. A., and Gendelman, H. E. (1995). Seminars in medicine of the Beth Israel Hospital, Boston. Dementia associated with the acquired immunodeficiency syndrome [see comments]. *N. Engl. J. Med.* **332**(14):934–940.

Mankowski, J. L., Spelman, J. P., Ressetar, H. G., Strandberg, J. D., Laterra, J., Carter, D. L., Clements, J. E., and Zink, M. C. (1994). Neurovirulent simian immunodeficiency virus replicates productively in endothelial cells of the central nervous system in vivo and in vitro. *J. Virol.* **68**(12):8202–8208.

Marcario, J. K., Raymond, L. A., McKiernan, B. J., Foresman, L. L., Joag, S. V., Raghavan, R., Narayan, O., Hershberger, S., and Cheney, P. D. (1999a). Simple and choice reaction time performance in SIV-infected rhesus macaques. *AIDS Res. Hum. Retroviruses* **15**(6):571–583.

Marcario, J. K., Raymond, L. A., McKiernan, B. J., Foresman, L. L., Joag, S. V., Roghavan, R., Narayan, O., and Cheney, P. D. (1999a). Motor skill impairment in SIV-infected macaques with rapidly and slowly progressing disease. *J. Med. Primatol.* **28**(3):105–117.

Mark, R. F., and Sperry, R. W. (1968). Bimanual Coordination in Monkeys. *Exp. Neurol.* **21**:92–104.

Martin, A., Heyes, M. P., Salazar, A. M., Kampen, D. L., Williams, J., Law, W. A., Coats, M. E., and Markey, S. P. (1992). Progressive slowing of reaction time and increasing cerebrospinal fluid concentrations of quinolinic acid in HIV-infected individuals. *J. Neuropsychiatry Clin. Neurosci.* **4**(3):270–279.

McCarthy, G., Puce, A., Constable, R. T., Krystal, J. H., Gore, J. C., and Goldman-Rakic, P. (1996). Activation of human prefrontal cortex during spatial and nonspatial working memory tasks measured by functional MRI. *Cereb. Cortex* **6**:600–611.

McFarland, D. J., and Hotchin, J. (1983). Host genetics and the behavioral sequelae to herpes encephalitis in mice. *Physiol. Behav.* **30**(6):881–884.

Mintz, M., Tardieu, M., Hoyt, L., McSherry, G., Mendelson, J., and Oleske, J. (1996). Levodopa therapy improves motor function in HIV-infected children with extrapyramidal syndromes. *Neurology* **47**(6):1583–1585.

Miotto, E. C., Bullock, P., Polkey, C. E., and Morris, R. G. (1996). Spatial working memory and strategy formation in patients with frontal lobe excisions. *Cortex* **32**:613–630.

Mohammed, A. H., Norrby, E., and Kristensson, K. (1993). Viruses and behavioural changes: a review of clinical and experimental findings. *Rev. Neurosci.* **4**(3):267–286.

Mohammed, A. K., Magnusson, O., Maehlen, J., Fonnum, F., Norrby, E., Schultzberg, M., and Kristensson, K. (1990). Behavioural deficits and serotonin depletion in adult rats after transient infant nasal viral infection. *Neuroscience* **35**(2):355–363.

Moses, A. V., and Nelson, J. A. (1994). HIV infection of human brain capillary endothelial cells—implications for AIDS dementia. *Adv. Neuroimmunol.* **4**(3):239–247.

Murray, E. A. (1996). What have ablation studies told us about the neural substrates of stimulus memory? *Seminars in the Neurosciences* **8**:13–22.

Murray, E. A., Rausch, D. M., Lendvay, J., Sharer, L. R., and Eiden, L. E. (1992). Cognitive and motor impairments associated with SIV infection in rhesus monkeys. *Science* **255**(5049):1246–1249.

Museteanu, C., Welte, M., Henneberg, G., and Haase, J. (1979). Relation between decreased mental efficiency in mice and the presence of cerebral lesions after experimental encephalitis caused by yellow fever virus. *J. Infect Dis.* **139**(3):320–323.

Ogata, A., Tashiro, K., Nukuzuma, S., Nagashima, K., and Hall, W. W. (1997). A rat model of Parkinson's disease induced by Japanese encephalitis virus. *J. Neurovirol.* **3**(2):141–147.

Owen, A. M., Doyon, J., Petrides, M., and Evans, A. C. (1996a). Planning and spatial working memory: a positron emission tomography study in humans. *Eur. J. Neurosci.* **8**:353–364.

Owen, A. M., Morris, R. G., Sahakian, B. J., Polkey, C. E., and Robbins, T. W. (1996b). Double dissociations of memory and executive functions in working memory tasks following frontal lobe excisions, temporal lobe excisions or amygdalo-hippocampectomy in man. *Brain* **119**:1597–1615.

Owen, A. M., Roberts, A. C., Polkey, C. E., Sahakian, B. J., and Robbins, T. W. (1991). Extradimensional versus intra-dimensional set shifting performance following frontal lobe excisions, temporal lobe excisions or amygdalo-hippocampectomy in man. *Neuropsychologia* **29**:993–1006.

Owen, A. M., Sahakian, B. J., Semple, J., Polkey, C. E., and Robbins, T. W. (1995). Visuospatial short-term recognition memory and learning after temporal lobe excisions, frontal lobe excisions or amygdalo-hippocampectomy in man. *Neuropsychologia* **33**(1):1–24.

Padgett, D. A., Sheridan, J. F., Dorne, J., Berntson, G. G., Candelora, J., and Glaser, R. (1998). Social stress and the reactivation of latent herpes simplex virus type 1 [published erratum appears in *Proc. Natl. Acad. Sci. U.S.A.* 1998, Sep 29;95(20):12070]. *Proc. Natl. Acad. Sci. U.S.A.* **95**(12):7231–7235.

Passingham, R. (1975). Delayed matching after selective prefrontal lesions in monkeys (Macaca mulatta). *Brain Res.* **92**:89–102.

Pearce, B. D., Steffensen, S. C., Paoletti, A. D., Henriksen, S. J., and Buchmeier, M. J. (1996). Persistent dentate granule cell hyperexcitability after neonatal infection with lymphocytic choriomeningitis virus. *J. Neurosci.* **16**(1):220–228.

Pearce, P. C., Crofts, H. S., Muggleton, N. G., Ridout, D., and Scott, E. A. (1999). The effects of acutely administered low dose sarin on cognitive behaviour and the electroencephalogram in the common marmoset. *J. Psychopharmacol.* **13**(2):128–135.

Pletnikov, M. V., Rubin, S. A., Vasudevan, K., Moran, T. H., and Carbone, K. M. (1999). Developmental brain injury associated with abnormal play behavior in neonatally Borna disease virus-infected Lewis rats: a model of autism. *Behav. Brain. Res.* **100**(1–2):43–50.

Prendergast, M. A., Terry, A. V., Jr., and Buccafusco, J. J. (1998). Effects of chronic, low-level organophosphate exposure on delayed recall, discrimination, and spatial learning in monkeys and rats. *Neurotoxicol. Teratol.* **20**(2):115–122.

Price, R. W. (1994). Understanding the AIDS dementia complex (ADC). The challenge of HIV and its effects on the central nervous system. *Res. Publ. Assoc. Res. Nerv. Ment. Dis.* **72**:1–45.

Price, R. W., Brew, B., Sidtis, J., Rosenblum, M., Scheck, A. C., and Cleary, P. (1988). The brain in AIDS: central nervous system HIV-1 infection and AIDS dementia complex. *Science* **239**(4840):586–592.

Prospero-Garcia, O., Gold, L. H., Fox, H. S., Polis, I., Koob, G. F., Bloom, F. E., and Henriksen, S. J. (1996). Microglia-passaged simian immunodeficiency virus induces neurophysiological abnormalities in monkeys. *Proc. Natl. Acad. Sci. U.S.A.* **93**(24):14158–14163.

Rausch, D. M., Heyes, M. P., Murray, E. A., Lendvay, J., Sharer, L. R., Ward, J. M., Rehm, S., Nohr, D., Weihe, E., and Eiden, L. E. (1994). Cytopathologic and neurochemical correlates of progression to motor/cognitive impairment in SIV-infected rhesus monkeys. *J. Neuropathol. Exp. Neurol.* **53**(2):165–175.

Rausch, D. M., Murray, E. A., and Eiden, L. E. (1999). The SIV-infected rhesus monkey model for HIV-associated dementia and implications for neurological diseases. *J. Leukoc. Biol.* **65**(4):466–474.

Raymond, L. A., Wallace, D., Berman, N. E., Marcario, J., Foresman, L., Joag, S. V., Raghavan, R., Narayan, O., and Cheney, P. D. (1998). Auditory brainstem responses in a Rhesus Macaque model of neuro-AIDS. *J. Neurovirol.* **4**(5):512–520.

Ridley, R. M., and Baker, H. F. (1993). Assessing memory in monkeys. *In* "Behavioral Neuroscience: A Practical Approach" (A. Sahgal, ed.), vol. 1, chap. 12, pp. 149–163. Oxford University Press, Oxford.

Robbins, T. W., Giardini, V., Jones, G. H., Reading, P., and Sahakian, B. J. (1990). Effects of dopamine depletion from the caudate-putamen and nucleus accumbens septi on the acquisition and performance of a conditional discrimination task. *Behav. Brain Res.* **38**(3):243–261.

Roberts, A. C. (1996). Comparison of cognitive function in human and non-human primates. *Cognitive Brain Research* **3**:319–327.

Roberts, A. C., DeSalvia, M. A., Wilkinson, L. S., Collins, P., Muir, J. L., Everitt, B. J., and Robbins, T. W. (1994). 6-hydroxydopamine lesions of the prefrontal cortex in monkeys enhance performance on an analog of the Wisconsin Card Sort Test: Possible interactions with subcortical dopamine. *Journal of Neuroscience* **14**(5):2531–2544.

Roberts, A. C., Robbins, T. W., and Everitt, B. J. (1988). The effects of intradimensional and extradimensional shifts on visual discrimination learning in humans and nonhuman primates. *Quarterly Journal of Experimental Psychology* **40B**(4):321–341.

Rubin, S. A., Bautista, J. R., Moran, T. H., Schwartz, G. J., and Carbone, K. M. (1999). Viral teratogenesis: brain developmental damage associated with maturation state at time of infection. *Brain Res. Dev. Brain Res.* **112**(2):237–244.

Rubin, S. A., Sylves, P., Vogel, M., Pletnikov, M., Moran, T. H., Schwartz, G. J., and Carbone, K. M. (1998). Borna disease virus-induced hippocampal dentate gyrus damage is associated with spatial learning and memory deficits. *Brain Res. Bull.* **48**(1):23–30.

Sacktor, N., and McArthur, J. (1997). Prospects for therapy of HIV-associated neurologic diseases. *J. Neurovirol.* **3**(2):89–101.

Sahakian, B. J., Elliott, R., Low, N., Mehta, M., Clark, R. T., and Pozniak, A. L. (1995). Neuropsychological deficits in tests of executive function in asymptomatic and symptomatic HIV-1 seropositive men. *Psychological Medicine* **25**(6):1233–1246.

Sardar, A. M., Czudek, C., and Reynolds, G. P. (1996). Dopamine deficits in the brain: the neurochemical basis of parkinsonian symptoms in AIDS. *Neuroreport* **7**(4):910–912.

Sei, Y., Arora, P. K., Skolnick, P., and Paul, I. A. (1992). Spatial learning impairment in a murine model of AIDS. *Faseb. J.* **6**(11):3008–3013.

Sharer, L. R., Michaels, J., Murphey-Corb, M., Hu, F. S., Kuebler, D. J., Martin, L. N., and Baskin, G. B. (1991). Serial pathogenesis study of SIV brain infection. *J. Med. Primatol.* **20**(4):211–217.

Simon, M. A., Chalifoux, L. V., and Ringler, D. J. (1992). Pathologic features of SIV-induced disease and the association of macrophage infection with disease evolution. *AIDS. Res. Hum. Retroviruses.* **8**(3):327–337.

Solbrig, M., Koob, G., and Lipkin, W. (1999). Orofacial dyskinesias and dystonia in rats infected with Borna disease virus; a model for tardive dyskinetic syndromes. *Mol. Psychiatry* **4**(4):310–312.

Solbrig, M. V., Fallon, J. H., and Lipkin, W. I. (1995). Behavioral disturbances and pharmacology of Borna disease. *Curr. Top. Microbiol. Immunol.* **190**:93–101.

Solbrig, M. V., Koob, G. F., Fallon, J. H., and Lipkin, W. I. (1994). Tardive dyskinetic syndrome in rats infected with Borna disease virus. *Neurobiol. Dis.* **1**(3):111–119.

Solbrig, M. V., Koob, G. F., and Lipkin, W. I. (1996). Naloxone-induced seizures in rats infected with Borna disease virus. *Neurology* **46**(4):1170–1171.

Solbrig, M. V., Koob, G. F., and Lipkin, W. I. (1998). Cocaine sensitivity in Borna disease virus-infected rats. *Pharmacol. Biochem. Behav.* **59**(4):1047–1052.

Sopper, S., Sauer, U., Hemm, S., Demuth, M., Muller, J., Stahl-Hennig, C., Hunsmann, G., ter Meulen, V., and Dorries, R. (1998). Protective role of the virus-specific immune response for development of severe neurologic signs in simian immunodeficiency virus-infected macaques. *J. Virol.* **72**(12):9940–9947.

Swiergiel, A. H., Smagin, G. N., and Dunn, A. J. (1997). Influenza virus infection of mice induces anorexia: comparison with endotoxin and interleukin-1 and the effects of indomethacin. *Pharmacol. Biochem. Behav.* **57**(1–2):389–396.

Taffe, M. A., Weed, M. R., Polis, I., and Gold, L. H. (1999). The effects of scopolamine on cognitive performance in rhesus monkeys. *Cognitive Brain Research* **8**:203–212.

Toth, L. A., Rehg, J. E., and Webster, R. G. (1995). Strain differences in sleep and other pathophysiological sequelae of influenza virus infection in naive and immunized mice. *J. Neuroimmunol.* **58**(1):89–99.

Voytko, M. L. (1998). Nonhuman primates as models for aging and Alzheimer's disease. *Lab. Anim. Sci.* **48**(6):611–617.

Voytko, M. L., Olton, D. S., Richardson, R. T., Gorman, L. K., Tobin, J. R., and Price, D. L. (1994). Basal forebrain lesions in monkeys disrupt attention but not learning and memory [published erratum appears in *J. Neurosci.* 1995, Mar;15(3 Pt 2):following table of contents]. *J. Neurosci.* **14**(1):167–186.

Watry, D., Lane, T. E., Streb, M and Fox, H. S. (1995). Transfer of neuropathogenic simian immunodeficiency virus with naturally infected microglia. *Am. J. Pathol.* **146**:914–923.

Weed, M. R., and Gold, L. H. (1998). The effects of dopaminergic agents on reaction time in rhesus monkeys. *Psychopharmacology (Berl)* **137**(1):33–42.

Weed, M. R., Taffe, M. A., Polis, I., Roberts, A. C., Robbins, T. W., Koob, G. F., Bloom, F. E., and Gold, L. H. (1999). Performance norms for a rhesus monkey neuropsychological testing battery: acquisition and long-term performance. *Cognitive Brain Research* **8**:185–201.

Weihe, E., Nohr, D., Sharer, L., Murray, E., Rausch, D., and Eiden, L. (1993). Cortical astrocytosis in juvenile rhesus monkeys infected with simian immunodeficiency virus. *Neuroreport* **4**(3):263–266.

Wiley, C. A., Schrier, R. D., Nelson, J. A., Lampert, P. W., and Oldstone, M. B. (1986). Cellular localization of human immunodeficiency virus infection within the brains of acquired immune deficiency syndrome patients. *Proc. Natl. Acad. Sci. U.S.A.* **83**(18):7089–7093.

Wit, F. W., van Leeuwen, R., Weverling, G. J., Jurriaans, S., Nauta, K., Steingrover, R., Schuijtemaker, J., Eyssen, X., Fortuin, D., Weeda, M., de Wolf, F., Reiss, P., Danner, S. A., and Lange, J. M. (1999). Outcome and predictors of failure of highly active antiretroviral therapy: one-year follow-up of a cohort of human immunodeficiency virus type 1-infected persons. *J. Infect. Dis.* **179**(4):790–798.

Yayou, K., Takeda, M., Tsubone, H., Sugano, S., and Doi, K. (1993). The disturbance of water-maze task performance in mice with EMC-D virus infection. *J. Vet. Med. Sci.* **55**(2):341–342.

Zink, M. C., Amedee, A. M., Mankowski, J. L., Craig, L., Didier, P., Carter, D. L., Munoz, A., Murphey-Corb, M., and Clements, J. E. (1997). Pathogenesis of SIV encephalitis. Selection and replication of neurovirulent SIV. *Am. J. Pathol.* **151**(3):793–803.

INDEX

D